CHRISTOPHER A. NUNES

DEVELOPMENTS IN INDUSTRIAL MICROBIOLOGY

A Publication of the Society for Industrial Microbiology

DEVELOPMENTS IN INDUSTRIAL MICROBIOLOGY

Volume 26

Proceedings of the
Forty-First General Meeting
of the
Society for Industrial Microbiology
Held at Fort Collins, Colorado
August 11–17, 1984

P.O. BOX 12534
ARLINGTON, VIRGINIA
1985

LIBRARY OF CONGRESS CATALOG CARD NUMBER 60-13953

ISSN 0070-4563

Society for Industrial Microbiology
P.O. Box 12534, Arlington, Virginia 22209-8534
PRINTED IN THE UNITED STATES

Foreword

Developments in Industrial Microbiology, Volume 26

The Society for Industrial Microbiology presents in this volume the technical papers presented at its Forty-First Annual Meeting held at the Colorado State University, Fort Collins, Colorado, in August 1984.

The efforts of each and every one of the many individuals involved in producing this Volume 26 are most gratefully acknowledged. Particularly worthy of mention is the fine program arranged by the Program Chairman, Paula Myers-Keith. The symposia speakers and the authors of contributed papers, of course, provided the substance of this volume. Editorial Board members expedited the review of manuscripts in order to meet publication deadlines. The good work of our new Production Manager, Pat Gerkin, is also greatly appreciated.

It behooves us here to note and mourn the death of Margaret Chambers, who, for so many years, so ably guided the production work of our volumes.

LELAND A. UNDERKOFLER
Editor, Volume 26

SOCIETY FOR INDUSTRIAL MICROBIOLOGY

Editorial Board 1984–1985

Table of Contents

Introduction

C. H. Ward

President, SIM, 1984
Rice University
Houston, Texas

Publication of Volume 26 of *Developments in Industrial Microbiology* marks the end of a tradition for the Society for Industrial Microbiology and the beginning of a new initiative in publishing. *Developments* has been, for 26 years, the proceedings of the Annual Meeting. As the Society has grown, so has the volume. In the early days the technical program of the Annual Meeting was largely devoted to a series of distinguished symposia, all with invited speakers. Today, at least half of the program is devoted to contributed papers reporting the results of original research in the diverse aspects of industrial microbiology.

The Society's growth and the needs of its members for a more timely avenue for publication of original research lead the SIM Board, in 1984, to pursue the creation of a new peer-reviewed, primary research journal. I am pleased to announce that the first issue of *Industrial Microbiology* will be published in January, 1986, with Dr. George E. Pierce as Editor-in-Chief.

Publication of the new research journal should enhance both the recognition and cost effectiveness of *Developments* which, in future volumes, will contain only the proceedings of our distinguished symposia and invited lectures. The new journal will represent the fourth instrument the Society has created for publication in the field of industrial microbiology: *SIM News,* which contains feature articles and timely information for members; *SIM Special Publications,* a monographic series on selected topics; *Developments,* a distinguished symposium volume, and *Industrial Microbiology,* a primary journal for peer-reviewed research.

This volume of *Developments* records the proceedings of the 1984 Annual Meeting at Colorado State University in Fort Collins, Colorado, the largest to date and certainly one of the most successful. The Society is grateful to Dr. James Ogg and his committee for exceptional local arrangements and to Dr. Paula Keith for leadership in arranging an outstanding technical program, consisting of seven symposia, 58 contributed papers, and invited lectures by two internationally known industrial microbiologists. Dr. Richard P. Elander presented the Charles Thom Award lecture, "Present and Future Roles of Biotechnology in the Fermentation Industry," and the "Dean of Biotechnology in Australia," Professor Bernhard J. Ralph, University of New South Wales, presented the Office of Naval Research Lecture on "Geomicrobiology and the New Biotechnology." The meeting was preceded by a heavily subscribed workshop on "Plasmids in Biotechnology: Isolation and Applications," organized by President-Elect Dr. George A. Somkuti.

The SIM is indebted to many for the continued success of this volume and its Annual Meeting. Greatest recognition must go to the scientists that reported the results of their research, to DIM Editor, Dr. Leland A. Underkofler, and to the members of the Editorial Board for assembling this fine volume.

On behalf of the SIM I am proud to present this volume as a permanent record of the Society's commitment to the field of industrial microbiology. I feel privileged to have been associated with the events that led to its publication.

1984 Charles Thom Award Winner, Society for Industrial Microbiology

Richard P. Elander received the Charles Thom Award at the Annual Meeting of the Society for Industrial Microbiology, August 16, 1984. The award is given to scientists making outstanding contributions in the field of industrial microbiology.

Dr. Elander received his B.S. (honors) and M.S. degrees from the University of Detroit, and in 1960, he received his Ph.D. degree in Bacteriology and Botany from the University of Wisconsin-Madison. He did postdoctoral research in microbial genetics at the University of Minnesota-Minneapolis in 1966–67. He joined Eli Lilly and Co. as a Senior Microbiologist in fermentation research in 1960 and was promoted to a Research Scientist in the Lilly Research Laboratories in 1965.

Dr. Elander joined Wyeth Laboratories near Philadelphia in 1967 and, after serving in various functions, was appointed Associate Director of Antibiotic Development and Manufacturing in 1970. In 1972 he relocated to Smith, Kline and French Laboratories where he organized an antibiotic R&D program in Philadelphia and established a new antibiotics research program for the company in Bangalore, India. At SK&F he was also responsible for the scaling-up of cephalosporin fermentation manufacturing technology for companies in Japan and Spain. In 1975 Dr. Elander was appointed Director of Fermentation Development for the Bristol-Myers Company, Industrial Division, in Syracuse, New York, where he is currently responsible for directing fermentation R&D for Bristol's bulk fermentation products and antitumor antibiotic efforts. He also is an advisor to Bristol's antibiotic screening research program at the Bristol-Myers Research Institute in Tokyo. He was appointed Senior Director of Fermentation Research and Development in 1980 and was made responsible for the company's genetic engineering research programs. In 1983 Dr. Elander was appointed Vice President of Biotechnology, Industrial Division.

Dr. Elander has been actively involved in fermentation research for over 25 years, and his scientific interests are in the genetics and biochemical control mechanisms of antibiotic-producing microorganisms. He is currently directing research on the plasmid control of kanamycin synthesis, cell-free synthesis of

β-lactam antibiotics, immobilized enzyme technology, and in the application of R-DNA technology to fermentation microorganisms, including the microbial synthesis of enzymes, interferons, and interleukins. He has authored or coauthored over 90 technical papers and has several patents.

Dr. Elander is a member of the American Association for the Advancement of Science (AAAS), Society for Industrial Microbiology (SIM), American Society for Microbiology (ASM), American Chemical Society (ACS), and is a Fellow in the American Academy of Microbiology (AAM) and the Society for Industrial Microbiology. He has been Secretary (1968) and President (1974) of the Society for Industrial Microbiology and was Chairman of the Fermentation Microbiology Division of the ASM in 1978. He is also a member of Sigma Xi, Phi Sigma and Gamma Alpha. He has served on the editorial board of *Developments in Industrial Microbiology* and has been on the editorial board for *Applied & Environmental Microbiology* from 1973–1982.

Dr. Elander was a member of the International Advisory Committee for three of the International Symposia for the Genetics of Industrial Microorganisms-Kyoto and coconvened a symposium on secondary metabolic products with H. Umezawa at the XIII International Congress of Microbiology-Boston (1982). In 1983 he presented lectures at the Canadian Society for Microbiology (Ottawa), Lehigh University, State University of New York (Buffalo), Brazilian Congress of Microbiology (Sao Paulo), University of Sao Paulo, and at several companies in Japan (Tokyo and Osaka). He recently organized a symposium on "Implications of Genetic Engineering for Pharmaceutical Scientists" at the RXPO Congress held at the New York Coliseum.

Dr. Elander is currently a Research Professor at Syracuse University, and presents lectures in their biotechnology and biochemical engineering programs. He also lectures in a short course on "Advanced Biochemical Engineering" at the Rensselaer Institute of Technology (Troy, New York) each summer.

Thom Award Address

Present and Future Roles for Biotechnology in the Fermentation Industry

RICHARD P. ELANDER

Bristol-Myers Company, Industrial Division, Syracuse, New York 13221–4755

Biotechnology is the integration of a number of scientific disciplines that use living organisms (or systems of products from organisms) to make or modify useful products. The major disciplines involved are microbiology, genetics, biochemistry, and chemical engineering. Biotechnology has experienced a dramatic resurgence with the discovery and application of gene splicing technology and hybridoma/cell fusion technology. As a consequence, modern biotechnology now represents one of the most important emerging technologies of the 1980's. This paper highlights existing and new applications of biotechnology that will have a significant impact and assure the future growth of the fermentation industry.

INTRODUCTION

Biotechnology, the use of living cells or parts thereof, to produce commercial products, is not a new industry. Centuries ago in Asia and Africa, people from various cultures made bread, beverages, and other food products using microorganisms. These traditional microbial processes developed without any real scientific knowledge and without even the realization that microorganisms played a role in the fermentation process. With the development of the science of microbiology in the 19th century, specific roles were found for the essential microorganism(s) in the process, and small industries were founded with the commercialization of these microbial metabolites (Demain and Solomon 1981). During the first two decades of the 20th century, large-scale processes for the production of lactic acid, ethanol, acetone, and butanol were developed. Also, microbial processes for the production of riboflavin and certain enzymes, including proteases, amylases, and invertase, were discovered and commercialized.

Modern biotechnology, however, dates from the 1940's with the large-scale manufacture of penicillin and streptomycin. Intensive development in strain improvement and bioprocess methodology led to techniques designed to genetically manipulate and stabilize microbial strains, to aseptically handle microorganisms on a large-scale, to effect sterilization of large volumes of fermentation media, to provide for adequate oxygen supply and gas exchange, and to isolate efficiently pure crystalline products from microbial processes.

The large-scale manufacture of amino acids and nucleotides were also successfully accomplished along with the discovery, development, and commercialization of new, more effective antibiotics. Later, microorganisms were used for the biotransformation of chemically difficult steps required for the synthesis of medically important steroid hormones. Further exploitation of microbial enzymes led to the establishment of a vast array of processes for industrial, analytical, and medical purposes. An additional milestone in biotechnology occurred with the development of continuous fermentation processes for the production of food and feed from single-cell protein processes (Humphrey 1982).

Because of the impending scarcity and high costs of petroleum, ethanol fuels are being generated from fermentation processes using starch-containing raw materials. Today, starch-based fermentation processes generated 40 yr ago for the acetone/butanol fermentation are being revived for ethanol manufacture in Brazil and other developing nations. Indeed, cellulosic waste materials may have vast future potential for the microbial production of fuels and biodegradable plastics.

From a chemical point of view, microorganisms may be viewed as catalytic bioreactors. In the future, immobilized microbial cells or enzymes will be used more and more to carry out specific chemical reactions because of steric difficulties or substrate instability in existing chemical process technology. Procedures for immobilizing cells and enzymes have now been developed, which allow for the continuous or semicontinuous manufacture of antibiotic intermediates, with considerable savings in product costs (Harrison and Gibson 1984).

Of even greater promise for the future development of biotechnology is the use of recombinant DNA and cell fusion technology. Gene splicing has been described as the most important emerging technology of the 1980's. The techniques of gene splicing, together with hybridoma (cell-fusion) technology, form the basis of modern biotechnology (Elander 1985). These remarkable biological achievements represent a significant advance in man's ability to use microorganisms to improve life. The development of commercial and pilot-scale processes for the manufacture of human insulin, human growth hormone, and hepatitis vaccines has been accomplished, and the effective production-scale manufacture now depends on the careful application of traditional strain improvement and industrial microbiology practices.

Discussion

Biotechnology Processes

Bioprocesses are systems in which whole living cells or their components (organelles, enzymes, etc.) are used to effect desired physical or chemical changes. The basic steps in a biotechnological process are presented in Fig. 1. The substrate and nutrients are prepared in a sterile medium and introduced into the process with free or immobilized cells or enzymes. Under controlled conditions, the substrate is converted to product(s) and when the desired degree of conversion has been achieved, the byproducts and waste materials are separated (Fig. 2). The products are usually purified from dilute aqueous solutions.

Bioprocesses require a closely controlled environment as biocatalysts generally exhibit great sensitivity to changes in temperature, pH, and even concentrations of certain nutrients or metal ions. Oftentimes, the success of a biotechnology process depends on the extent to which these factors are controlled in the medium where interaction between the enzyme and substrate takes place.

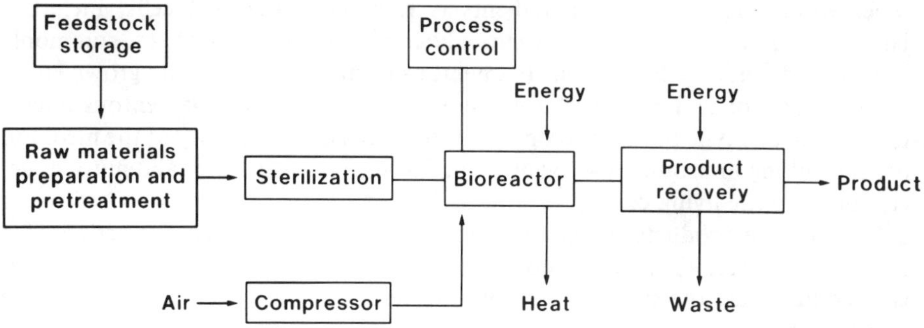

FIG. 1. Schematic overview of biotechnological process (from Cooney 1982).

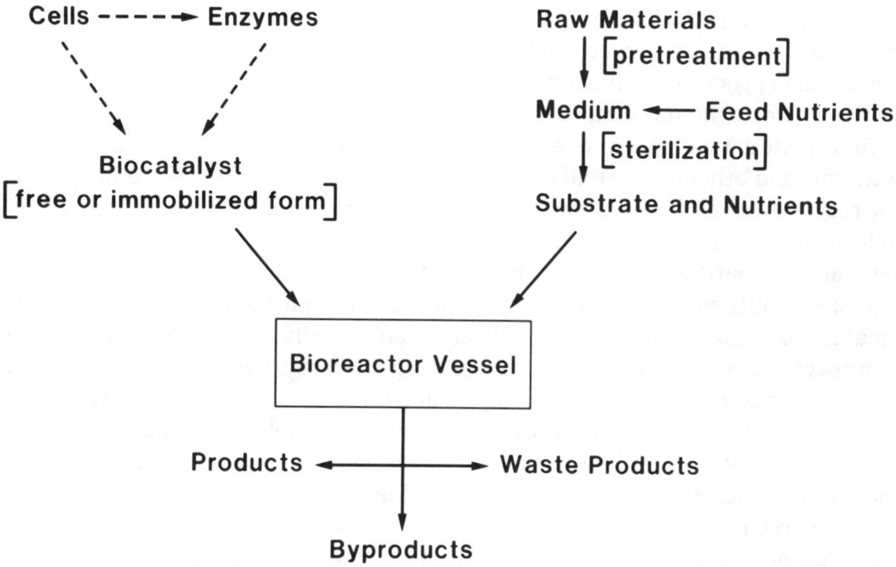

FIG. 2. Essential steps in bioprocess technology.

In addition to establishing a suitable environment, the medium must provide the essential nutrients for the living cell that contain the biocatalysts. A primary requirement of all living cells is carbon, which supplies energy for overall metabolism and protein synthesis. Carbon, usually in the form of sugars, starches, or glycerides, often contributes structural elements required for the synthesis of complex molecules. Other important nutrients required by most living cells are nitrogen, phosphorus, and certain metallic ions. Oxygen is essential for all aerobic bioprocesses.

In order to make the substrate and nutrient materials accessible to the biocatalyst, mixing of the ingredients is essential. Microbial cells, including bacteria and yeasts, organisms commonly employed in bioprocesses, commonly grow as individual cells, whereas filamentous bacteria and fungi are grown either as aggregates or as long mycelial strands. In the growth of filamentous microorganisms, this type of morphology tends to increase the viscosity of the medium, often resulting in problems associated with oxygen mass transfer and nutrient accessibility to the living cells.

Most of the products of biotechnology are formed through the action of a single biocatalyst, either a microorganism or an enzyme. If foreign microorganisms contaminate the system, they can interfere with the system and destroy the biocatalyst or the desirable byproduct. Foreign organisms may also generate undesirable metabolites, which make purification of the desired product difficult.

Most biotechnological processes use pure culture techniques to avoid biocatalyst contamination. The bioreactor vessel containing the necessary nutrients must be sterilized by heat, and a pure culture of the microorganism or sterile immobilized biocatalyst is introduced into the sterile system. All essential additions into the system must be appropriately sterilized.

Biotechnological processes generally use the operating modes of conventional chemical technology. These modes range from batch processing to continuous steady-state processing. In batch or fed-batch processing, the biocatalyst is added to the reactor containing the sterile nutrients and the conversion takes place over a time period ranging from several hours to many days. During this period, nutrients and other agents that effect necessary proper pH control are supplied to the reaction vessel. Volatile waste products are usually removed from the reactor. When the conversion is complete, the reaction is stopped, the bioreactor is emptied, and the purification process is initiated. In continuous steady-state processing, sterile nutrients are supplied to the bioreactor and the byproduct and spent medium are continuously withdrawn at volumetrically equal rates. Continuous bioprocessing offers advantages over batch processing. Costs are lower because of continuous reuse of the biocatalyst, greater ease of product recovery and, oftentimes, higher overall productivity rates. The simplest approach to the implementation of the continuous processing mode is to modify a batch reactor so that fresh substrate and nutrients can be added continuously while the product stream is removed. This arrangement has a serious drawback, however, in that the biocatalyst leaves the reactor continuously. Biocatalysts can now be fixed or immobilized in a suitable matrix, thereby avoiding loss of the catalyst and allowing for continued reuse of the biocatalyst. Thus, the development of immobilized enzyme technology has greatly expanded the possibilities for continuous bioprocesses.

Byproduct separation and purification techniques are most important aspects for the production of novel products such as antibiotics, proteins, enzymes, hormones, etc. Some of the newer purification strategies used for the recovery of bioproducts include the use of ultrafiltration, continuous chromatography, electrophoresis, and monoclonal antibodies. Although purification and separation processes have been developed for existing bioprocesses, new biotechnology is presenting new challenges. Genetic engineering and rDNA technology have already culminated in the commercialization of human insulin derived from microbial fermentations. Considerable technology was necessary to purify this hormone when isolated from the complex microbial fermentation broths.

Examples of commercial biotechnology processes currently in operation are (a) production of whole cell biomass (brewer's and baker's yeast, single-cell protein); (b) production of cell components (enzymes, nucleic acids); (c) production of chemical metabolites, including primary metabolites (ethanol, citric, and lactic acids) and secondary metabolites (antibiotics, plant hormones); (d) catalysis of specific, single-substrate conversions [glucose to fructose, penicillins to 6-aminopenicillanic acid (6-APA)]; (e) catalysis of multiple-substrate conversions (biological waste treatment).

Biotechnology processes offer a number of advantages over conventional process technology. Some advantages are milder reaction conditions (temperature, pH, pressure); use of abundant, cheap resources as raw materials; less hazardous operation and reduced environmental impact; greater specificity of catalytic reaction (stereospecific enzymes, etc.); less complex manufacturing facilities usually requiring smaller capital investment; catalytic reactions operative in aqueous environments at ambient temperatures; improved process efficiencies through biocatalyst reuse (higher process yield and reduced energy consumption); and the potential use of genetic engineering technology to tailor-make enzymes via site-directed mutagenesis, etc.

Some of the disadvantages of bioprocess technology compared with existing chemical technology include the generation of complex products, which require extensive separation and purification, especially when complex natural raw materials are used as substrates; problems arising from dilute aqueous environments in which most biocatalysts function; the susceptibility of most bioprocess systems to contamination by foreign microorganisms; the inherent variability of most biological processes because of the genetic instability of organisms; the inherent variability of most natural substrates; and the necessary costly containment required for recombinant microorganisms and the need to contain waste process streams from recombinant microorganisms.

The net advantages of biotechnology processes, however, will result in an ever-expanding biotechnology industry. Over the next several years, there will be an expanded research and development in the following areas: (a) continued development on the practical use and design of bioreactors for immobilized whole cell and enzyme systems; (b) development of a wide range of more sensitive, sterilizable sensor probes for process monitoring and controls; (c) development of improved product recovery techniques, especially for proteinaceous products; (d) improved bioreactor design providing for better mixing and improved mass transfer; (e) inhibition of intracellular and extracellular protein-degrading enzymes; (f) improved methods for heat dissipation during bioprocessing; (g) development of genetically engineered strains with greater genetic stability; and (h) development of microbial cultures with more efficient protein-secreting mechanisms.

Applications of Biotechnology to the Enzyme Industry

Microbial enzymes have been used commercially since the 1890's, when mold extracts were added to brewing vessels to facilitate the breakdown of starch to fermentable sugars. The size of the current world industrial enzyme market is 70,000 tons valued at $430 million (*Commercial Biotechnology—An International Analysis,* OTA-BA-218, U. S. Congress, Office of Technology Assessment, January, 1984).

A growth rate resulting in 75,000 tons valued at $600 million has been predicted for the end of 1985. Fewer than 20 enzymes comprise the large majority of this market. Economic sources of enzymes include a limited number of plants and animals and a few species of microorganisms (Godfrey and Reichelt 1983).

The enzyme industry is dominated by two European companies, Novo Industri (Denmark) and Gist-Brocades NV (Netherlands), which together have about 65% of the current world market (Eveleigh 1981). Other companies marketing or planning to market large-volume enzymes include CPC International (U.S.), ADM (a division of Clinton, U.S.), Miles (U.S.), Pfizer (U.S.), Dawi Kasi (Japan), Alko (Finland), Finnish Sugar (Finland), and Rohm (a division of Henkel, F.R.G.).

The leading enzymes on the world market in terms of volume are the proteases, amylases, and glucose isomerase (Eveleigh 1981). Alkaline protease is added to detergents as a cleaning aid and is widely used in western Europe. Trypsin, another type of protease, is important in the leather industry. Two amylases, alpha-amylase and glucoamylase, and glucose isomerase are corn-processing enzymes. The reactions catalyzed by these three enzymes represent the three steps by which starch is converted into high-fructose corn syrup (Fig. 3). Fructose is sweeter than glucose and can be used in place of table sugar (sucrose) in preparation of candy, bread, carbonated beverages, and canned goods. Historically the United States imported sugar, but with the commercial development of an economic process for converting glucose to fructose in the late 1960's, corn sweeteners have decreased the amount of sugar imported. About $1.3 billion in U. S. payments for sugar was saved in 1980 because of the domestic use of corn sweeteners (Corn Refiners Association, Inc., *The Amazing Maize,* Washington, DC, 1981).

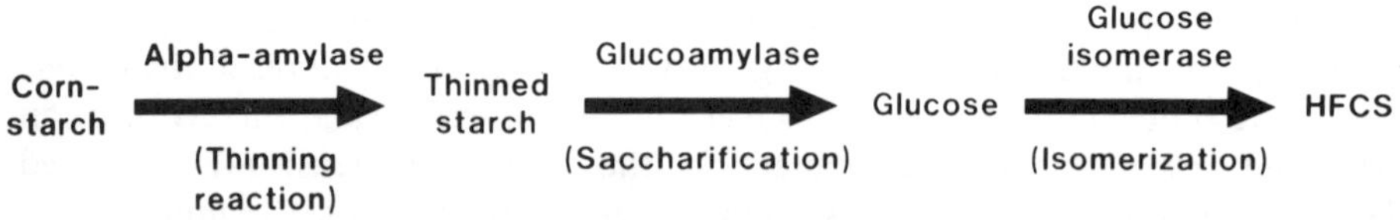

FIG. 3. Conversion of starch into high fructose corn syrup (HFCS).

The process for converting glucose to fructose is catalyzed by the enzyme glucose isomerase. Initially, the conversion was done using a batch reaction. In 1972 a continuous system using immobilized glucose isomerase was initiated (Hebeda 1983). The immobilized glucose isomerase process represents the largest immobilized enzyme process used in production in the world. A large processing plant can convert 2 million pounds of corn starch into high-fructose corn syrup per day (Danzig 1984).

The demand for enzymes will increase because of expanded sales in the detergent and high-fructose corn syrup markets. The application of rDNA techniques to microbial enzyme production is expected to facilitate the expansion of the enzyme industry. Additionally, enzymatic activities of higher organisms could be cloned into microorganisms, also expanding the enzyme industry. The fact that enzymes are direct gene products makes them good candidates for improved production through rDNA technology. For example, a 500-fold increase in the yield of a ligase, used for connecting DNA strands in rDNA research, was obtained by cloning the gene for that enzyme on an *E. coli* plasmid vector (Eveleigh 1981). Several research enzymes now on the market are produced by microorganisms modified using rDNA techniques. Some are restriction endonucleases used for cutting DNA, and others are DNA-modifying enzymes. Companies that market these enzymes include Bethesda Research Laboratories (U.S.), New England Biolabs (U.S.), P-L Biochemicals (U.S.), and Boehringer Mannheim (F.R.G.).

Recombinant DNA technology could potentially be used to increase glucose isomerase production in microorganisms and to improve the enzyme's properties. An improved glucose isomerase would have the following properties: (a) a lower pH optimum to decrease the browning reaction caused by the alkaline pH now required; (b) thermostability so that the reaction temperature can be raised, thus pushing the equilibrium of isomerization to a higher percentage fructose; and (c) improved reaction rates to decrease production time.

Improvements in glucose isomerase will first come from the cloning of its gene into vectors and microorganisms that have been developed for high production. It is also possible that screening a broad range of microorganisms will yield enzymes with some improved properties. Finally, it will be possible in the future to identify the regions of the enzyme that are responsible for its various properties, such as pH optimum, and to direct changes in the gene structure to modify these properties.

Rennet is an enzyme that is essential to the cheese industry because of its milk-clotting properties. The world market for rennet from various sources is valued at approximately $64 million, over half of which is the more valuable calf rennet (Eveleigh 1981). The increasing scarcity of calf rennet has made this enzyme a very attractive candidate for gene cloning and subsequent production in a microbial bioprocess. The first announcement of the cloning of the rennet gene came from Japan (Beppu et al. 1980). The first marketing of calf rennet produced by genetically manipulated bacteria is likely to occur in 1984 (from Genex Corp., *Impact of Biotechnology on the Specialty Chemicals Industry;* contract paper prepared for the Office of Technology Assessment, U. S. Congress, April, 1983).

Single-Cell Protein

The term "single-cell protein" (SCP) refers to cells, or protein extracts, of microorganisms grown in large quantities for use as human or animal protein supplements. Although SCP has a high protein content, it also contains fats, carbohydrates, nucleic acids, vitamins, and minerals. Interest in SCP production is not new, as evidenced by the fact that Dutch, German, and British patents for SCP production were issued as early as 1920 (Marstrand 1981). Interest in SCP has waxed and waned throughout the ensuing years, but SCP production has never achieved great significance, mostly because of economic considerations. With the advent of new biotechnology and the threat of potential world food shortages, interest in SCP may be rekindled (Litchfield 1983).

SCP can be used as a protein supplement for both humans and animals. In animal feed it is a replacement for more traditional supplements, such as soybean meal and fishmeal. For humans SCP is used either as a protein supplement or as a food additive to improve product functionality, for example, flavor, whipping action, or fat binding (Litchfield 1983). The use of SCP in human food presents a problem: humans have a limited capacity to degrade nucleic acids. Therefore, additional processing is necessary before SCP can be used in human food. The animal feed market is more attractive for SCP, not only because there is less processing of the product, but also because the regulatory approval process is less stringent.

Incentives for production of SCP are fourfold. First, in some parts of the world having high rainfall and tropical areas, agricultural feed and food products are high in carbohydrate content. In such places, there is a chronic shortage of protein, which results in deteriorated physical and mental health. The use of SCP would raise the protein content of food in these areas. Second the land in other regions, including the Middle East and Africa south of the Sahara, cannot produce sufficient food of any type to prevent hunger. Here also an SCP supplement would be an asset. Third, there is demand worldwide for very high protein ingredients for feeds in the aquaculture industry, i.e., in the production of shrimp, prawns, trout, salmon, and other finfish and shellfish. Finally, SCP does not rely on temperature, rainfall, or sun for survival. At least one of the variety of feedstocks is usually available in almost any country or region of the world. The security of having such an internal source of protein is attractive to many countries.

Economically feasible SCP production is dependent on the efficient use of an inexpensive feedstock by microorganisms. A large variety of feedstocks have been used for SCP production over the years, including carbon dioxide, methane, methanol, ethanol, sugars, petroleum hydrocarbons, and industrial and agricultural wastes. These feedstocks have been used industrially with different microorganisms, including algae, actinomycetes, bacteria, yeasts, molds, and higher fungi. The choice of a feedstock includes such considerations as cost, availability, efficient growth of the microorganisms, and requirements for pretreatment (Litchfield 1983).

SCP has yet to become an important source of protein, mainly because of high production costs. Some SCP-production processes that were economical at one time have not remained so because of changes in prices of competitive sources of protein, such as soybean meal or fishmeal. In comparison to SCP, these protein sources are quite inexpensive. In fact, the price of most SCP processes would have to be decreased by 20–50% for SCP to be competitive with soybean meal and fishmeal.

It is possible that the application of biotechnology will help to reduce the cost of production of SCP. Strains of microorganisms can be improved using rDNA techniques. Improvements include increasing the production of proteins with a better amino acid balance or improving the ability of the microorganism to use the feedstock more efficiently. Technological improvements in the process and recovery steps would also be important. The use of automated, continuous processes could improve the efficiency of production. Recovery steps could be aided by using microorganisms that have been genetically manipulated to excrete protein. Additionally, it is possible that an enzyme that degrades cell walls could be cloned and produced in large amounts. Its use would help in the production of a protein concentrate from cells. New technologies will probably improve the production of SCP, but widespread introduction of SCP will be governed by economic and regulatory factors.

The center of SCP technology is in England, especially at ICI (Waterworth 1981). The ICI process uses aerobic bacteria with methanol and ammonia as feedstocks. The bacteria are grown in the world's largest continuous bioprocess system with computerized control and monitoring of performance. The product, Pruteen contains 80% crude protein as well as a high content of essential micronutrients, especially B group vitamins. Pruteen is used in animal feed diets and as a milk replacement. In 1981, ICI had scaled up its process to produce 3,000 tons of SCP per month. It is beginning research using rDNA technology to facilitate protein harvesting (Litchfield 1983). So far, however, the production of Pruteen has not been economic even though it is twice as nutritious as soybean meal (in *New Scientist,* Protein process faces German competition, May 20, 1982, p. 495; Sherwood 1984).

The future of SCP depends largely on reduction in cost and improvement in quality. Means to meet these requirements involve lower cost feedstocks, improved engineering of the conversion and recovery processes, and upgrading the yield and quality of the product through conventional genetic and rDNA methods. The renewed interest in all of biotechnology, in part due to rDNA technology, is leading to increased effort in developing economically competitive SCP with improved qualities.

Microbial Production of Commodity Chemicals

The United States has abundant biomass resources. The largest potential amount of cellulosic biomass is from cropland residues such as corn stover and cereal straw, although the potential amount of cellulosic biomass from forest resources is also quite large. About 550 million dry tons of lignocellulose are easily collected and available for conversion to chemicals each year. In addition, some percentage of the 190 million dry tons of corn produced yearly could be converted to starch and used for chemical production (Bungay 1981).

Some commodity chemicals, including ethanol and acetic acid, are now produced in the United States with microbial bioprocesses (Ng et al. 1983), while other chemicals, such as ethylene and propylene, will probably continue to be made from petroleum feedstocks because of lower production costs. The commodity chemicals that are attractive targets for production from biomass include ethanol, acetone, isopropanol, acetic acid, citric acid, propanoic acid, fumaric acid, butanol, 2,3-butanediol. methylethylketone, glycerin, tetrahydrofuran, and adipic acid (Crawford et al. 1983). Additionally, some chemicals such as lactic and levulinic acids could be used as intermediates in the synthesis of polymers that might replace petrochemically derived polymers (Crawford et al. 1983). Some of the microorganisms that are presently used for the production of commodity chemicals are listed in Table 1.

TABLE 1. *Potentially important bioprocessing systems for the production of commodity chemicals (from Ng et al. 1983)*

Microorganism	Carbon source(s)	Major fermentation product(s)
Saccharomyces cerevisiae	Glucose	Ethanol
Saccharomyces cerevisiae	Glucose	Glycerol
Zymomonas mobilis	Glucose	Ethanol
Clostridium thermocellum	Glucose, lactic acid	Ethanol, acetic acid
Clostridium thermosaccharolyticum	Lactic acid	Glucose, xylose, ethanol, acetic acid
Clostridium thermohydrosulfuricum	Glucose, xylose	Ethanol, acetic acid, lactic acid
Schizosaccharomyces pombe	Xylulose	Ethanol
Kluyveromyces lactis	Xylulose	Ethanol
Pachysolen tannophilus	Glucose, xylose	Ethanol
Thermobacteroides saccharolyticum	Xylose, glucose	Ethanol
Thermoanaerobacter ethanolicus	Glucose, xylose	Ethanol, acetic acid, lactic acid
Clostridium acetobutylicum	Glucose, xylose, arabinose	Acetone, butanol
Clostridium aurianticum	Glucose	Isopropanol
Clostridium thermoaceticum	Glucose, fructose, xylose	Acetic acid
Clostridium propionicum	Alanine	Propionic acid, acetic acid, acrylic acid
Aeromonas hydrophilia	Xylose	Ethanol, 2,3-butanediol
Dunaliella sp.	Carbon dioxide	Glycerol
Aspergillus niger	Glucose	Citric acid
Aerobacter aerogenes	Glucose	2,3-butanediol
Bacillus polymyxa	Glucose	2,3-butanediol

Because the chemical composition of biomass differs from that of petroleum and because microorganisms are capable of a wide range of activities, it may be that the most important commodity chemicals produced from biomass will be, not chemicals that directly substitute for petrochemicals, but other chemicals that together define a new structure for the chemical industry. Microorganisms used to produce organic chemicals could be used with other microorganisms that fix nitrogen to produce nitrogeneous chemicals, either higher value-added compounds or ammonia, a high-volume commodity chemical. Other microorganisms, such as the methanogens or those that metabolize hydrogen sulfide, may be used to produce sulfur-containing chemicals (Chibata 1978).

Biotechnology will be a key factor in developing economic processes for the conversion of biomass to commodity chemicals. A number of priorities for research that will improve the efficiency of this conversion can be identified. (1) Bioprocess improvements, including the use of immobilized cell and enzyme systems and improved separation and recovery methods, an area especially important to the production of commodity chemicals because incremental improvements in bioprocess technology will be readily reflected in the price of these chemicals; (2) screening programs to identify microorganisms (and their biochemical pathways) useful to processes, such as commodity chemical synthesis, cellulose hydrolysis, lignin degradation, and catalysis of reactions that use byproducts that are currently unmarketable; developing host/vector systems that facilitate increased production of commodity chemicals by gene amplification and increased gene expression of desired products and that allow the transfer of genes into industrially important microorganisms; (3) understanding the structure and function of the cellulase and ligninolytic activities of microorganisms; (4) understanding the mechanism of survival of microorganisms in extreme environments, such as high temperature, high pressure, acid, or salt; (5) understanding the mechanism of cell tolerance to alcohols, organic acids, and other organic chemicals; (6) understanding the genetics and biosynthetic pathways for the production of commodity chemicals, especially for the strict anaerobic bacteria such as the methanogens and the clostridia; (7) understanding microbial interactions in mixed cultures; and (8) developing an efficient pretreatment system for lignocellulose.

Application of Biotechnology to the Pharmaceutical Industry

The domestic sales of prescription drugs by U. S. pharmaceutical companies exceeded $8.6 billion in 1982 (from *Commercial Biotechnology—An International Analysis,* Washington, DC, U. S. Congress Office of Technology Assessment, OTA-BA-218, January, 1984). Of these sales, approximately 20% were products for which fermentation biotechnology played a significant role. The fermentation-derived products included anti-infective agents, vitamins, and biologicals. The "new" biotechnology (biomolecular engineering) is expected to be particularly helpful in the production of pharmaceuticals and biologicals which, in the past, could only be obtained by extraction of animal and plant tissues, and which now can be obtained from microbial sources as well.

The pharmaceutical industry was probably the last industry to adopt traditional fermentation technologies. However, it was the first industry to make widespread use of the newer molecular biology techniques, including genetic engineering and protoplast fusion. There were two major factors that accelerated the use of molecular biology in the pharmaceutical industry. First, the biological sources of many pharmacologically active products were microbial in origin and, thereby, were more amenable to directed genetic manipulation. Second, the major advances in biomolecular engineering were made under an institutional structure that allocated funding to biomedical research. In fact, the Federal support system has tended to promote studies that have as their ostensible goal, the improvement of human health (National Institutes of Health, National Cancer Institute, etc.).

Historical Uses of Genetic Technology

The genetic manipulation of biological systems for the production of antibiotics and other pharmaceuticals has two major goals. These are (a) to increase the efficiency of pharmaceutical production with proven or potential value, and (b) to discover potentially new useful drugs not found in nature.

The first goal has had the greatest influence on the pharmaceutical industry. Genetic manipulation has been shown to increase, in an almost dramatic fashion, the productivity of pharmacologically active products found in nature. Three examples are outlined:

(1) The genetic improvement of penicillin production is an example of long-term efforts that lead to dramatic increases in strains of the penicillin fungus, *Penicillium chrysogenum*. The original Peoria isolate, NRRL-1951, was treated with toxic chemicals and ultraviolet radiation through successive stages until a very superior mutant strain was developed (Elander and Demain 1981). This commercially valuable mutant yielded a 100-fold improvement in fermentation productivity compared to the original Fleming strain (Fig. 4).

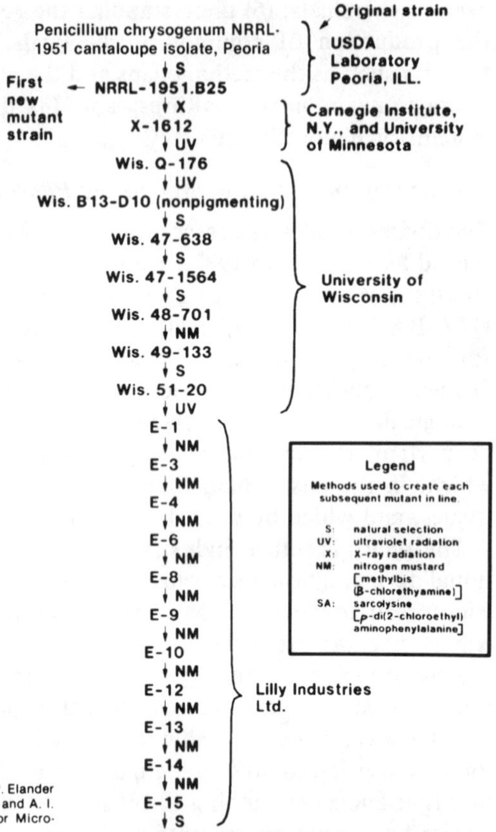

SOURCE: Adapted by Office of Technology Assessment from R. P. Elander in Genetics of Industrial Microorganisms. O. K. Sebek and A. I. Laskin (eds.) (Washington, D.C.: American Society for Microbioloogy, 1979), p. 23.

FIG. 4. The development of a high penicillin-producing strain via genetic manipulation.

(2) Chemically induced mutations improved a valuable strain of the bacterium, *Escherichia coli*, which produces an enzyme, L-asparaginase, an agent used to treat leukemia. Improved mutants were isolated, which produced 100-fold more enzyme, thereby resulting in easier isolation and increased purity of the enzyme preparation.

(3) Genetic manipulation of the gentamicin bacterium, *Micromonospora purpurea,* resulted in sufficient yield improvement that its manufacturer, Schering-Plough Corporation, did not have to build a scheduled manufacturing plant in Puerto Rico.

Major Areas for Biotechnology in the Pharmaceutical Industry

Antibiotics. Antibiotic agents for the treatment of infectious diseases have been the largest-selling prescription pharmaceuticals in the world for the past 30 yr. Most of these agents are antibiotic agents naturally produced by strains of microorganisms isolated from nature. Chemical synthesis and acylation of a variety of side-chain residues to naturally occurring antibiotic nuclei is the common method of commercial manufacture for many antibiotics, many of which belong to the semisynthetic b-lactam class. Chemical synthesis of a major naturally occurring antibiotic chloramphenicol is one exception to the more common microbial biosynthesis or semisynthesis manufacturing technology. Chemical synthesis will also probably be the more efficient technology for the manufacture of some of the future potent antibiotics of the carbapenem and monobactam classes of b-lactam antibiotics. These antibiotics are highly unstable in fermentation environments, and their yields are low via fermentation synthesis. U. S. pharmaceutical companies have been prominent in the development, production, and marketing of useful antibiotics. The present annual American market share for antimicrobial agents is about $2.7 billion of the $5.8 billion annual world market. The markets expand each year with the introduction of new, more effective antibiotic compounds.

For 30 yr, high-yielding, antibiotic-producing microorganisms have been identified by selection from among mutant strains (Fig. 4). Initially, organisms producing new antibiotics are isolated by soil sampling and other broad screening efforts. They are then cultured in the laboratory, and efforts are made to improve their productivity in pilot-scale fermentors.

Antibiotics are complex, usually nonprotein, substances that are generally the end products of a series of biological steps. While knowledge of molecular details in metabolism has made some difference, not a single antibiotic has had its biosynthetic pathway totally elucidated. This is partly because there is no single gene that can be isolated to produce an antibiotic. However, mutations can be induced within the original microorganism so that the level of production can be increased (Elander 1982).

Other methods can also increase production and possibly create new antibiotics. Microbial mating, for example, which leads to natural recombination, has been widely investigated as a way of developing vigorous, high-yielding antibiotic producers. However, its use has been limited by the mating incompatibility of many industrially important higher fungi, the presence of chromosomal aberrations in microorganisms improved by mutation, and a number of other problems. Furthermore, natural recombination is most advantageous when strains of extremely diverse origins are mated; the proprietary secrets protecting commercial strains usually prevent the sort of divergent "competitor" strains most likely to produce vigorous hybrids from being brought together.

The technique of protoplast or cell fusion provides a convenient method for establishing a recombinant system in strains, species, and genera that lack an efficient natural means for mating. For example, as many as four strains of the antibiotic-producing bacterium *Streptomyces* have been fused together in a single step to yield recombinants that inherit genes from four parents. The technique is applicable to nearly all antibiotic producers. It will help combine the benefits developed in divergent lines by mutation and selection.

In addition, Hamlyn and Ball (1979) compared the quality of an antibiotic-producing fungus, *Cephalosporium acremonium,* produced by mating to one produced by protoplast fusion. They concluded that protoplast fusion was far superior for that purpose. Moreover, protoplast fusion gave rise to hundreds of recombinants—including one isolate that consistently produced the antibiotic cephalosporin C in 40% greater yield than the best producer among its parents without loosing that parent strain's rare capacity to use inorganic sulfate as a source of sulphur rather than expensive methionine. It also acquired the rapid growth and sporulation characteristics of its less-productive parent. Thus, desirable attributes from different parents were combined in an important industrial organism that had proved resistant to conventional crossing.

Even more significant are the possibilities for cell fusion between different species, thereby creating novel hybrid strains that could have unique biosynthetic properties. One group is reported to have isolated a novel antibiotic, clearly not produced by either parent, in an organism created through fusion of actinomycete protoplasts.

The development of recombinant DNA technology and techniques for transformation and transfection of protoplasts as well as intact oprganisms has made possible the exploitation of these methods for gene cloning in actinomycetes and fungi. The rationale for attempting shotgun and self-cloning is that transformants may acquire new genes for enzymatic activities that can modify the chemical scructure of secondary metabolites normally produced by the organism to generate new antibiotics (Vournakis and Elander 1983).

The basic requirements for use of the recombinant DNA method to transfer and express segments of foreign DNA in a host organism include (1) an appropriate vector DNA molecule (plasmid or phage) compatible with the cell and carrying appropriately localized control elements such as promoters and ribosome binding sequences; (2) a convenient method for preparing the foreign DNA, cleaving it to a reasonable size range, and ligating it into the vector DNA molecule such that it will have a good probability of being expressed; (3) a method for introducing the recombinant DNA molecules into the host cell so that it is transformed to a new phenotype that includes the properties coded for by the recombinant; and (4) a method for assaying for the expression of the desired gene products.

There are many options available for altering the genetics of antibiotic-producing microorganisms to realize particular goals. Rationale selection methods based on an enhanced understanding of the biosynthetic pathways for antibiotic synthesis can lead to radical improvements in strain productivity. Protoplast fusion technology can be used to allow genetic recombinant to occur among different species of antibiotic producers to generate mixed synthetic pathways that may in turn give rise to hybrid antibiotics (Elander 1983). Recombinant DNA methods can be used to clone genes for strain improvement and novel antibiotic synthesis into existing microorganisms in either a semirandom or specific manner. The coming decade promises to be filled with excitement and rewards as many of the ideas discussed here and elsewhere are implemented in the laboratories, pilot plants, and ultimately, production systems of the pharmaceutical industry.

Regulatory Proteins

Human insulin. The first therapeutic agent produced by rDNA technology to achieve regulatory approval and market introduction was human insulin (hI). The methodology was generated by Genentech, Inc., and Eli Lilly and Company. The product now marketed in both the United States and the United Kingdom is known as Humulin® . The extent to which the new rDNA-derived products will be substituted in the marketplace for animal-derived insulin is still uncertain. Insulin derived from animals has long been the largest volume peptide hormone used in medicine. Chemically, human insulin differs only slightly from that of pigs and cows, and its incremental benefits have yet to be demonstrated. Human and porcine insulins differ in a single amino acid, while human and cattle insulins differ with respect to three. As far as is known, these slight variations do not impair the effectiveness of the insulin, but no meaningful comparative study has been undertaken because of the past insufficiency of human insulin.

In 1981, 0.75 tons of pure insulin for 1.5 million diabetics was sold in the United States. The number of American diabetics is expected to increase to 2.1 million by 1986 (Scrip, Oct. 4, 1982). Eli Lilly dominated the U. S. market in 1981 with $133 million in sales from a total U. S. market of $170 million. The 1985 estimated U. S. market is $345 million with Lilly's sales expected to be in excess of $200 million (*Commercial Biotechnology—An International Analysis,* Washington, DC, U. S. Congress, Office of Technology Assessment, OTA-BA-218, January 1984).

Interferons. The interferons (Ifns) represent a class of immune regulators or lymphokines that regulate the response of cells to viral infections and cancer proliferation. These extraordinarily potent proteinaceous substances are the subject of the most widely publicized, well-funded applications of rDNA technology to date, but the basic details of their functions remain largely unknown. Until recently, the study of Ifns was limited by the extremely small amounts that could be obtained from extracted human lymphocyte cells. However, the use of rDNA technology now allows for the large-scale production of nonglycosylated Ifns in bacteria and large clinical trials on a variety of them are currently in progess.

The gene cloning and microbial production of Ifns illustrate several important aspects of the current commercialization of biotechnology. These include the following: (1) the use of rDNA technology to produce a scarce product in quantities sufficient for research on the product's effects; (2) a massive, competitive scale-up campaign by pharmaceutical manufacturers in advance of demonstrated uses of the product; (3) the attempt to produce economically a functional glycoprotein (protein with attached sugar molecules) in an rDNA system; (4) a pattern of international R&D investment that reflects the differing needs and medical practices of various nations; and (5) the establishment of a U. S. national effort, via research grants and procurement contracts administered through the National Cancer Institute (NCI), the American Cancer Society (ACS), and other organizations, to support testing of Ifns toward a national goal (cure of cancer).

Ifns are being considered for various health-related applications but are not yet approved as pharmaceutical products. There is some evidence that Ifns are effective in certain viral infections, but more clinical trials are necessary. Ifns may prove useful in treatment of some viral diseases in combination with other drugs (Merigan 1982).

Many clinical trials are presently underway with a type of interferon, called gamma Ifn for the treatment of certain kinds of cancer. However, at present only limited conclusions can be drawn from the available data. In some cases, Ifns inhibit tumor cell growth and may stimulate immune cells to destroy cancer cells; their effect on inhibiting tumor metastasis is better established than their ability to effect actual regression of primary tumors. Also, most tumors that show some response to Ifns are also quite responsive to established chemotherapeutic agents. Several problems have also been noted in clinical trials using Ifns. The occurrence of fatigue and flu-like symptoms in patients following injection with Ifns were once thought to be reactions to impurities contained in the drug preparation, but highly purified preparations of Ifns show similar effects (Billeau 1981). Despite extensive research and vast ongoing clinical trials, numerous questions still remain concerning the anticancer potential for interferons.

Perhaps the most enlightening results stemming from Ifn research will concern cellular function during immune responses. Such results may prove extremely valuable in medicine. Better understanding of immune mechanisms, for example, may provide insight into the etiology of the recently problematic acquired immune deficiency syndrome (AIDS). Substantial supplies of Ifns to conduct such research can now be produced with rDNA technology.

Though most rDNA-made Ifns currently under evaluation are produced in the bacterium, *E. coli*, yeast is being increasingly employed as a production organism. Yeast requires less stringent culture conditions than do most bacteria, has long records of reliability and safety in large-scale bioprocessing, and is more adaptable to continuous culture production than are many bacteria. Furthermore, because yeast more closely resembles higher organisms than bacteria, yeast can add sugar molecules to protein when necessary. Thus, modified products made in yeast are more likely to be pharmaceutically useful than unmodified products made in bacteria.

Several biotechnology companies have reported significant progress using yeast for the manufacture of Ifns. Yeast has important advantages over bacterial synthesis in that glycosylated Ifns can be synthesized and secreted into the fermentation environment. Numerous genetic techniques are currently being employed in yeast strains to increase Ifn production. They include (1) amplification of the number of Ifn genes; (2) enhancement of gene expression by placing it under control of regulatory elements that can be varied without impeding cell growth; (3) limitation of production degradation by extracellular protease enzymes; (4) induction and enhancement of Ifn product secretion; and (5) the genetic stabilization of genetically engineered strains.

Human growth hormone. Genetic engineering technology is being used increasingly to produce large amounts of otherwise scarce biological compounds. Human growth hormone (hGH) is an excellent example of the future promise of biotechnology to produce in large amounts, hormones that are only naturally produced as several molecules per cell.

The development of hGH with rDNA methods is another model for biotechnology's use in the pharmaceutical industry. Human growth hormone is one of a family of at least three, closely related, large peptide hormones secreted by the pituitary gland. These peptide hormones are about four times larger than insulin (191 to 198 amino acids in length). All three hormones possess a wider variety of biological actions than do most other hormones. The primary function of hGH is apparently the control of postnatal growth in humans. Whereas insulin derived from slaughtered animals can be used for treating diabetics, only growth hormone derived from humans is satisfactory for reversing the deficiencies of hypopituitarism in children.

Although the established pharmaceutical market for hGH is small and current supplies for the treatment are sufficient, hGH is one of the first targets for the applications of rDNA technology. Human growth hormone is currently being evaluated for the following human maladies: (1) treatment of constitutionally delayed short stature; (2) improvement of the healing of burns, wounds, and bone fracture; (3) treatment of a disease condition known as cachexia or deficiency in nitrogen assimilation. It has been reported that 3% of all children have constitutionally delayed short stature and that as many as one-third of these may benefit from hGH administration (*Bioengineering News,* p. 7, Jan. 7, 1983).

Lymphokines. Lymphokines are proteins produced by lymphocytes that convey information among lymphocytes. With the exception of interferons, lymphokines are only beginning to be characterized, but these proteins appear to be crucial to immune reactions. Some lymphocytes produce lymphokines that engage other lymphocytes to boost the immune response to a foreign antigen and to repel foreign invasion. Other lymphocytes produce substances that act in conjunction with an antigen to stimulate the secretion of antibodies.

The increasing importance of lymphokines in preventing disease and understanding cellular function (including cancerous growth) is fostering widespread research on these important immune regulatory molecules. The recent establishment of discrete lymphocyte cell lines that produce various classes of lymphokines and the recent cloning of lymphokine-producing genes into rDNA systems for their large-scale production in microbial fermentation systems will increase the amount of research in this new important branch of immunology. rDNA technology now provides for the availability of pure lymphokine preparations that will allow immunologists to answer more fundamental questions concerning cell biology and the immune system. Eventually, these efforts may lead to the use of lymphokines in medicine to stimulate the patient's own immune system to combat disease.

Interleukin-2 (IL-2) or lymphocyte growth factor has been reported to be effective in eliciting an elevated immune response in patients having an impaired immune response (compromised patients undergoing cancer chemotherapy) or patients afflicted with acquired immune deficiency syndrome (AIDS). The genes responsible for IL-2 production have now been cloned successfully into microbial systems, and pure IL-2 protein is now readily available for research and clinical trial evaluation (*Biotechnology Newswatch,* p. 1, Feb. 21, 1983). IL-2 has been shown to stimulate the body's T-cells to mature to a point where they become natural killer cells (NK-cells), which are able to attack and diminish the size of discrete tumors. In test tube and animal experiments, IL-2 has greatly increased the number of T-cells available for the destruction of tumor cells.

A number of regulatory protein "growth factors" for a number of somatic cells have been isolated and are currently being characterized. Several of these have been sequenced as to their exact amino acid composition and soon will be candidates for production by rDNA technology. Two of these factors, epidermal growth factor (EGF) and platelet-derived growth factor (PDGF) appear to be stimulated by cancer-causing virus genomes known as oncogenes. Oncogenes are transformed normal cellular genes found in retroviruses and other cellular tissues that have been implicated in cancer (Slamon et al. 1984).

Oncogenes and Cancer

The notion that our own cells actually contain genes, now referred to as oncogenes or cancer-causing genes, having the potential of causing cancerous transformation of those cells is not a comfortable one. Nevertheless, some 20 oncogenes have now been identified that can, when appropriately activated, produce cancer in animals and cause the malignant transformation of special lines of human cultured tissue.

The recent advances in oncogene research and their implications in cancer come from studies on RNA-containing viruses that induce cancers. Such viruses are termed retroviruses. The oncogenes contained in certain retroviruses are not native to viruses nor are they even native to cancer cells. Rather, these genes are present and functioning in normal cells of vertebrates. In fact, they may be essential for the normal cellular function and development, as well as for the unrestrained growth of cancer tissue. The cellular counterparts of the viral oncogenes have been closely conserved throughout evolution. Indeed, they are found in species as diverse as fruit flies, fish, and mammals, including humans, and now have been recently discovered in common baker's yeast. Oncogenes have been defined simply as wayward copies of genes found in all metazoan organisms and probably evolved long before the advent of human kind.

Oncogenes code for proteins that appear to be capable of triggering a cascade of biochemical events that ultimately lead to a malignant transformation, eventually resulting in neoplasia. One oncogene product is known as p60 src protein, which consists of some 520 amino acid subunits. The protein is an enzyme (tyrosine kinase) that adds phosphate groups to the amino acid tyrosine and other protein subunits. The protein is bound to the plasma membrane of cells.

Cellular elements such as these oncogene proteins appear to be at the heart of the cancer process, but just how these proteins work remains unclear at this time. They appear to regulate the growth of cells, with the protein encoded by the normal cellular gene (proto-oncogene) controlling normal growth whereas the altered version of the gene (oncogene) forces cancerous growth (Hunter 1984).

The cancer-gene concept, supported by oncogene data presently suggests a unifying explanation for various forms of carcinogenesis. The common central element is a group of cellular genes required for normal growth and development. When the cellular gene is transplanted into a retrovirus genome, such a gene becomes an oncogene. Cancer can also result if the cellular gene is changed by any of a wide variety of mutagens and other carcinogens. A single point mutation in a single oncogene segment consisting of only 350 nucleotides can result in a gene transformation of a proto-oncogene which, when transformed, develops into an EJ-bladder carcinoma tissue (Weinberg 1982). Growth factors also appear to be involved in the transformation process and appear to stimulate the rapid cell division process typical of cancerous growth.

An understanding of the precise function of oncogene proteins and their exact roles in the phosphorylation cascade and combined growth factor involvement may make it possible to develop drug antagonists that inhibit key steps in these intricate biochemical interactions; in this way new therapy could be targeted to the few central defects of a transformed cancerous cell (Bishop 1982; Weinberg 1984).

Future Outlook

The past decade of biotechnology has resulted in a rapid development of a new biotechnology industry. A number of the major events important for the expansion of this technology are summarized in Table 2. Future expanded government training grants for molecular biologists and biotechnologists and increased funding for molecular biology programs at the NIH, NCI, and other government laboratories will accelerate biotechnology and its numerous applications for the improvement of man's well-being. The increased federal and industrial biotechnology funding, coupled with the greatly expanded pharmaceutical R&D

TABLE 2. *Major events in the commercialization of modern biotechnology (Adapted from OTA-BAS-218 report, January, 1984)*

1973	First gene cloned.
1974	First expression of a foreign gene cloned in bacteria.
	Recombinant DNA (rDNA) experiments first discussed in a public forum (Gordon Conference).
1975	U.S. guidelines for rDNA research outlined (Asilomar Conference).
	First hybridoma created.
1976	First firm to exploit rDNA technology found in U.S. (Genentech).
	Genetic Manipulation Advisory Group (U.K.) started in U.K.
1980	Diamond v. Chakrabarty—U.S. Supreme Court rules that microorganisms can be patented.
	Cohen/Boyer patent issued on construction of rDNA.
	U.K. targets biotechnology.
	Germany targets biotechnology.
	Initial public offering by Genentech sets Wall Street record ($35–$89 in 20 minutes).
1981	First monoclonal antibody diagnostic kits approved.
	First automated gene synthesizer marketed.
	Japan declares 1981 "The Year of Biotechnology".
	France targets biotechnology.
	Hoechst/Massachusetts General Hospital agreement.
	Initial offering by Cetus sets Wall Street record ($115 million).
	Industrial Biotechnology Association founded.
	DuPont commits $120 million R&D.
	Over 80 NBFs formed.
1982	First rDNA animal vaccine (for colibacillosis) approved for use in Europe.
	First rDNA product (human insulin) approved.
	First R&D partnership formed for clinical trials.
1983	First plant gene expressed in a different species.
	$500 million raised in U.S. public markets.
1984	Oncogene proteins characterized.
	T-cell receptor site sequenced.
	Factor VIII gene cloned and expressed in bacteria.

budgets for molecular biology and biotechnology will culminate in an increased understanding of many disease mechanisms. This understanding will result in the discovery and development of many new, useful pharmaceutical products.

It is clear that biotechnology developed initially from industrial microbiology and chemical engineering and thus, has been dominated by the development of large-scale fermentation processes. In the future, biotechnology will become more oriented toward highly specialized processes using immobilized cellular or enzymic bioreactors derived from microbial, plant, and animal cell cultures in which expensive substances will be made in smaller scale processes with sophisticated computerized process control. Gross sales of products derived from genetically engineered microorganisms are estimated in the billions of dollars for the late 1990's.

Other applications of genetically engineered microorganisms reside in the environmental and electronic industries. Detoxification and decontamination of pesticide residues, treatment of sewage waste and cleanup of oil spills are examples of how biotechnology will benefit mankind. The concept of the "biocomputer" is now close to reality as microprocessors will involve biochips that use proteins to provide a framework for molecules that act as semiconductors. Since molecules are the smallest known entities that can act as semiconductors, the protein biochip represents the ultimate in electronic miniaturization. Successful tests have already been conducted with polylysine, and it has been estimated that such chips would have 100,000 times the switching components as conventional microprocessor chips (Robinson 1983).

Although it is clear that new developments in gene splicing and molecular biotechnology will lead to important new industrial processes and possibly new fermentation-based industries, successful long-term commercialization may experience constraints depending upon changing world economics, dynamic university/industry technology transfer, application of antitrust law to biotechnology licensing agreements, trade barriers affecting biotechnology products and open federal, industrial, and academic policies, which foster intellectual freedom for scientists to explore the unknown.

Literature Cited

Beppu, T., K. Nishimori, Y. Kawaguchi, H. Uchiyama, and T. Uozomi. 1980. Cloning of cDNA of calf prorennin in *Escherichia coli*. *Abstr. VI Int. Ferment. Symp.* Canada, p. 179.

Billeau, A. 1981. Perspectives in cancer research: The clinical value of interferons as antitumor agents. *Eur. J. Cancer and Clin. Oncol.* 17:949–967.

Bishop, J. M. 1982. Oncogenes. *Sci. Am.* 246:80–92.

Bungay, H. R. 1981. *Energy, the Biomass Options.* John Wiley and Sons, New York.

Chibata, I. 1978 *In Immobilized Enzymes.* John Wiley and Sons, New York.

Cooney, C. L. 1983. Bioreactors: Design and operation. *Science* 219:728–733.

Crawford, D. L., A. L. Pometto, and R. L. Crawford. 1983. Lignin degradation by *Streptomyces viridosporus:* Isolation and characterization of a new polymeric lignin degradation intermediate. *Appl. Environ. Microbiol.* 45:898–904.

Danzig, M. 1984. CPC International, Argo, IL, *personal communication.*
Demain, A. L., and N. S. Solomon. 1981. Industrial microbiology. *Sci. Am.* 245:67–75.
Elander, R. P. 1982. Traditional versus current approaches for the genetic improvement of microbial strains. Pages 353–369 *in* V. Krumphanzl, B. Sikyta, and Z. Vanek, eds., *Overproduction of Microbial Metabolites,* Academic Press, New York.
______. 1983. Pages 97–146 *in* A. L. Demain and N. A. Solomon, eds. *Antibiotics containing the β-lactam structure, Handbook of Experimental Pharmacology* 67/1. Springer-Verlag, Berlin.
______. 1984. Biotechnology: Present and future roles in the pharmaceutical industry. *Drug Dev. Ind. Pharmacol.* 11: In press.
Elander, R. P., and A. L. Demain. 1981. Pages 237–277 *in* H. J. Rehm and G. Reed, eds., *Biotechnology,* Vol. 1, Academie-Verlag, Weinheim, F.G.R.
Eveleigh, D. E. 1981. The microbiological production of industrial chemicals. *Sci. Am.* 245:155–178.
Godfrey, T., and J. R. Reichelt. 1983. Introduction to industrial enzymology. *In* T. Godfrey and J. R. Reichelt, eds., *Industrial Enzymology,* Nature Press, New York.
Hamlyn, P. F., and C. F. Ball. 1979. Pages 185–191 *in* O. K. Sebek and A. I. Laskin, eds., *Genetics of Industrial Microorganisms,* American Society of Microbiology, Washington, DC.
Harrison, F. G., and E. D. Gibson. 1984. Approaches for reducing the manufacturing costs of 6-aminopenicillanic acid. *Process Biochemistry,* Feb. 1984.
Hebeda, R. E. 1983. High fructose syrups. *In* T. Kirk-Othmer, ed., *Encyclopedia of Chemical Technology,* 3rd ed., Vol. 22, John Wiley and Sons, New York.
Humphrey, A. E., 1982. Biotechnology. The way ahead. *J. Chem. Tech. Biotechnol.* 32:25–33.
Hunter, T. 1984. The proteins of oncogenes. *Sci. Am.* 249:70–79.
Litchfield, J. H. 1983. Single-cell proteins. *Science* 219:740–746.
Marstrand, P. K. 1981. Production of microbial protein: A study of the development and introduction of a new technology. *Res. Policy* 10:148–171.
Merigan, T. C. 1982. Present appraisal of and future hopes for the clinical utilization of human interferons. *In Interferon,* Vol. 3, Academic Press, New York.
Ng, T. K., R. M. Busche, C. C. McDonald, and R. W. F. Hardy. 1983. Production of feedstock chemicals. *Science* 219:733–740.
Robinson, A. L. 1983. Nanocomputers from organic molecules. *Science* 220:940–942.
Sherwood, M. 1984. The case of the money-hungry microbe *Biotechnology* 2:606–609.
Slamon, D. J., J. B. deKernion, I. M. Verma, and M. J. Kline. 1984. Expression of cellular oncogenes in human malignancies. *Science* 224:256–262.
Vournakis, J. N., and R. P. Elander. 1983. Genetic manipulation of antibiotic-producing microorganisms. *Science* 219:703–709.
Waterworth, D. G. 1981. Single-cell protein. *Outlook on Agriculture* 10:403–408.
Weinberg, R. A. 1982. Oncogenes of spontaneous and chemically induced tumors. *Adv. Cancer Res.* 36:149–163.
______. 1984. A molecular basis of cancer. *Sci. Am.* 248:126–142.

Invitational ONR Lecture

Geomicrobiology and the New Biotechnology

BERNHARD J. RALPH

School of Metallurgy, University of New South Wales,
Kensington, New South Wales, 2033, Australia

INTRODUCTION

The origins of geomicrobiology are somewhat obscure, but it is clear that the observation of phenomena attributable to the interaction of microorganisms with the geological fabric are found very early in the history of a number of civilized societies. The earliest record of these phenomena is in the historical volume named *Humainancius,* a book written by Liu-An (177–122 B.C.), who was king of Huainan Kingdom in the West Han Dynasty. It refers to gall-springs (that is, bitter-tasting blue waters), which yielded a blue material on evaporation. The observation is made that on "being attacked by iron, the blue vitriol turns to copper," which is probably the first recording of a cementation process for metal recovery. By the end of the first millennium, the exploitation of copper sulfate produced by natural biodegradation as a source of metallic copper had become well established in China (Yao Dun Pu 1982). Similar developments occurred in Europe and in Spain. Recovery of copper from the biodegradation of copper sulfide minerals was practiced at the Rio Tinto mines during the 17th century (Taylor and Whelan 1942). The degradations of sulfide minerals that gave rise to these early observations are now known to be brought about by a complex series of chemical and electrochemical reactions, some of which are spectacularly accelerated by microbial catalysis.

It has been observed by other authors that the main thrust of microbiological studies has been and continues to be concerned with the interactions of microorganisms with organic materials and that much lesser attention has been paid, until comparatively recently, to microbial interventions in the inorganic environment (Silverman and Ehrlich 1964). The ability of some microbial species to influence the course of inorganic chemical reactions was not suspected until the work of Winogradsky (1888) indicated that the growth of certain filamentous bacteria in ferruginous water was accompanied by the accumulation of oxidized and insoluble forms of iron. The scientific foundations of geomicrobiology and biogeochemistry were subsequently solidly laid by microbiologists such as Beijerinck, Kluyver, and Van Niel, and in areas of particular interest in respect of the practical applications of geomicrobiology, by the studies of Rudolfs and Helbronner (1922) on the microbial oxidation of zinc sulfide minerals and by the classical studies of Colmer and Hinkle (1947) on the central microbial role in the oxidation of ferrous to ferric iron.

The development of geomicrobiology and biogeochemistry has been stimulated over the past few decades by increasing awareness of the microbial contribution in environmental protection situations, by the need for greater understanding of the mechanisms of the global ecology, and by the potential for the development of useful applications and novel technologies for the recovery of metals and other inorganic materials from low grade sources.

The past few decades have seen the accumulation of a formidable store of fundamental information on the role of biological systems in the cycling of the elements and the mechanisms by which the ecological dynamics are affected (Trudinger and Swaine 1979). This corpus of knowledge provides the firm foundation for rationalization and improvement of a number of empirical technologies and the potential for development of novel processes of considerable economic significance. The application of basic geomicrobiological and biogeochemical information requires techniques similar to those involved for other biotechnological transformations of organic materials.

Biotechnology is the generic term describing processes of actual and potential industrial importance of which the most essential and characteristic feature is the involvement of biocatalytic systems (CSIRO 1981; Bull et al. 1982). Biotechnology is the intellectual and methodological pathway by which the demonstrated bioconversion abilities of biological systems can be translated into large-scale processes for the production of useful products and services. It is the nexus at which inputs from the relevant basic disciplines, such as chemistry and physics, biochemistry, microbiology and molecular biology, and engineering fuse to yield usable technology. The nature of biotechnology is well illustrated by the "Biotechnology Tree" (Fig. 1). The trunk of this tree is of paramount interest since it embodies the two most important facets of any biotechnological process, namely, the acquisition of the most appropriate biocatalytic system, and the provision of an optimized milieu in which the catalytic system can work. The rationalization of the working environment for the biocatalytic system has developed steadily with the emergence of biochemical engineering as a specialization within biotechnology, an area in which the concepts and techniques of chemical engineering (and other branches of engineering) have been applied to the particular requirements of biological and biochemical systems. The historical progression from the innovations of Chaim Weizmann in the acetone/butanol process during World War I and the deep tank methods for the production of penicillin evolved by the Wisconsin group during World War II, to the sophistication of modern fermentation processes, has been steady and continuous with respect to improvement and refinement. It is in the other facet of biotechnological processes, however, that the most spectacular and revolutionary changes have occurred over the past decade. The extension and consolidation of basic information on the hereditary mechanisms and the evolution of methodologies for the manipulation of the generic blueprint have placed tools of immense potency in the hands of the biotechnologist and enable, for the first time, the rational design, production, and control of biocatalytic systems appropriate to particular ends. The largely fortuitous and often labor-intensive procedures of selection from the natural environment and of random mutation and selection, while still most useful under appropriate circumstances, are in process of rapid displacement as means to the acquisition of biological systems for the

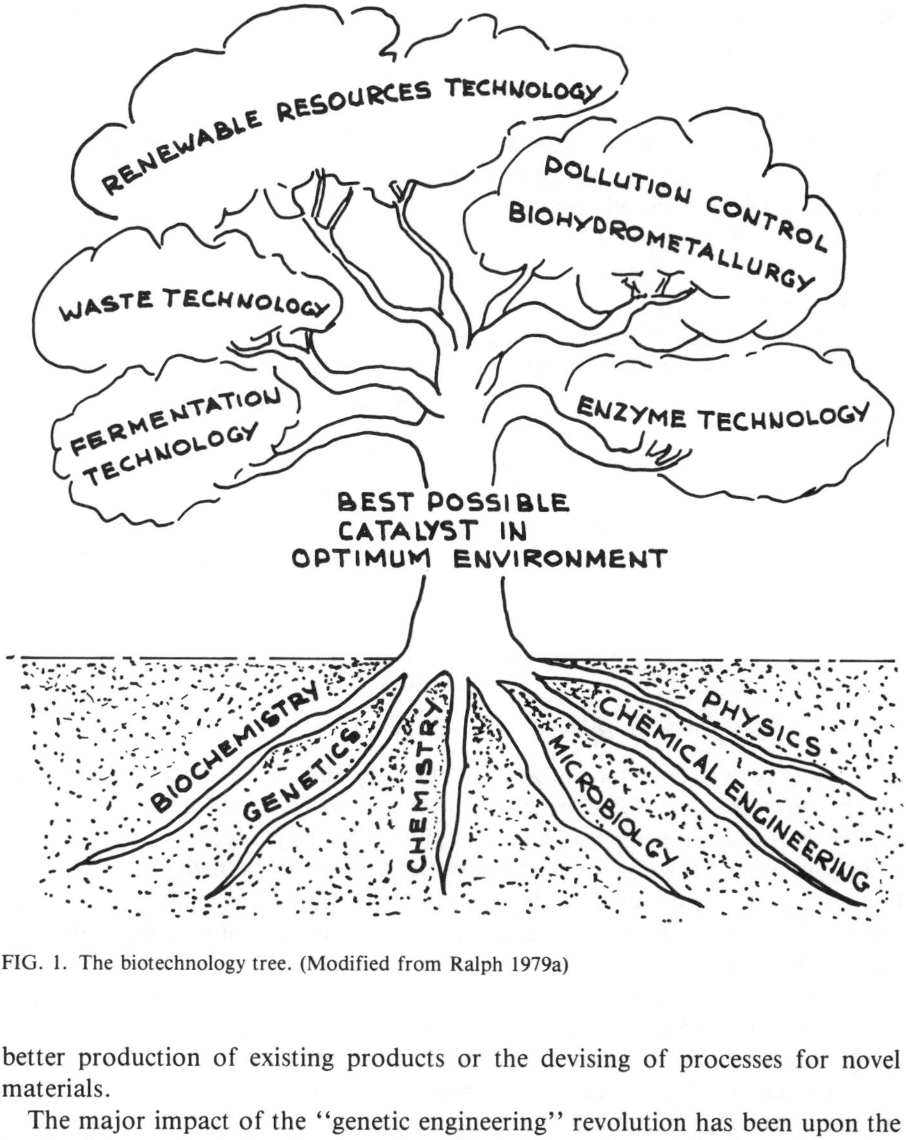

FIG. 1. The biotechnology tree. (Modified from Ralph 1979a)

better production of existing products or the devising of processes for novel materials.

The major impact of the "genetic engineering" revolution has been upon the improvement of existing fermentation processes and upon the development of novel routes for a wide range of new organic materials, ranging from vaccines and therapeutic agents to microbial polysaccharides and other chemical materials. The influence of the "new" biotechnology on processes involving the microbial modification of inorganic materials, and upon those systems with which geomicrobiology and biogeochemistry are concerned, has so far been slight but the potential impact is great.

The established practices of mining and metallurgy have been highly successful for the recovery of metals from a wide range of raw material sources, and have coped extremely well with generally declining ore grades and with increasingly difficult problems of access to ore bodies. There are strong indications, however, that public demand for the limitation of environmental damage during mining and metallurgical operations, the need for more complete recoveries from low-grade and recalcitrant ores, and long-standing, technical processing problems may be underlining some inherent limitations of conventional practices; these difficulties may provide the stimulus for the examination of innovative, alternative procedures for the recovery of metals and other inorganic materials. In spite of a long history as an empirical technology for the recovery of copper, biohydrometallurgy has emerged only recently as a rational and credible processing route. Biohydrometallurgy provides one of the alternate technologies to which the mining and metallurgical industries may be forced to turn in the future.

Biohydrometallurgy is the technological extrapolation of geomicrobiology and biogeochemistry and is essentially the application of biotechnology to mineral processing. The potential of biohydrometallurgy for the provision of highly efficient and environmentally acceptable methods of metal production is high and the need to press on with the relevant research and development increasingly urgent. Some areas of biohydrometallurgy are in vigorous ferment, and in the following sections an attempt is made to outline the current status, to indicate areas of potential development and to predict the likely outcome of the impact of the "new" biotechnology on the technologies derived from geomicrobiology and biogeochemistry.

Discussion

Areas of Development in Biohydrometallurgy

The best known applications of geomicrobiology and biogeochemistry are those related to the recovery of copper and uranium by bacterial leaching, in which copper and uranyl ions are brought into solution by the biodegradation of sulfide and oxide ores of these metals. The exercise is commonly effected by the irrigation of masses of broken ore, disposed in heaps or dumps (Fig. 2) of various geometrical configurations, with a recycled acidic aqueous solution that performs the dual function of transporting nutrient for microbial growth and chemical intermediates for other reactions to the mineral surfaces, and of carrying away the various soluble products of the chemical and biological degradation mechanisms. The scale of bacterial leaching for the recovery of copper is large; it is said to account for about 15% of total copper production in the United States at the present time (Bhappu 1982) and for perhaps as much as 20% of total world production (Solozhenkin 1980). An important variant of this type of leaching process is *in situ* leaching or solution mining in which the mineralized material is not extracted but made permeable—if necessary, by explosive shattering—to appropriate leaching reagents. The method is widely applicable (Schlitt and Hiskey 1982) and has the potential for metal recovery from ore deposits located at great depth (Davidson et al. 1981) (Fig. 3).

Similar practicable methods have been demonstrated, at laboratory and pilot scale, for a number of other metals, including bismuth, cobalt, nickel, and zinc (Torma and Bosecker 1982). These modes of biologically assisted leaching of metallic ores imply ready percolation by aqueous solutions, and the relatively large particle size (ca. 1 cm) in such situations imposes a limitation on reaction rates because the exposed areas of available mineral surfaces is small. The

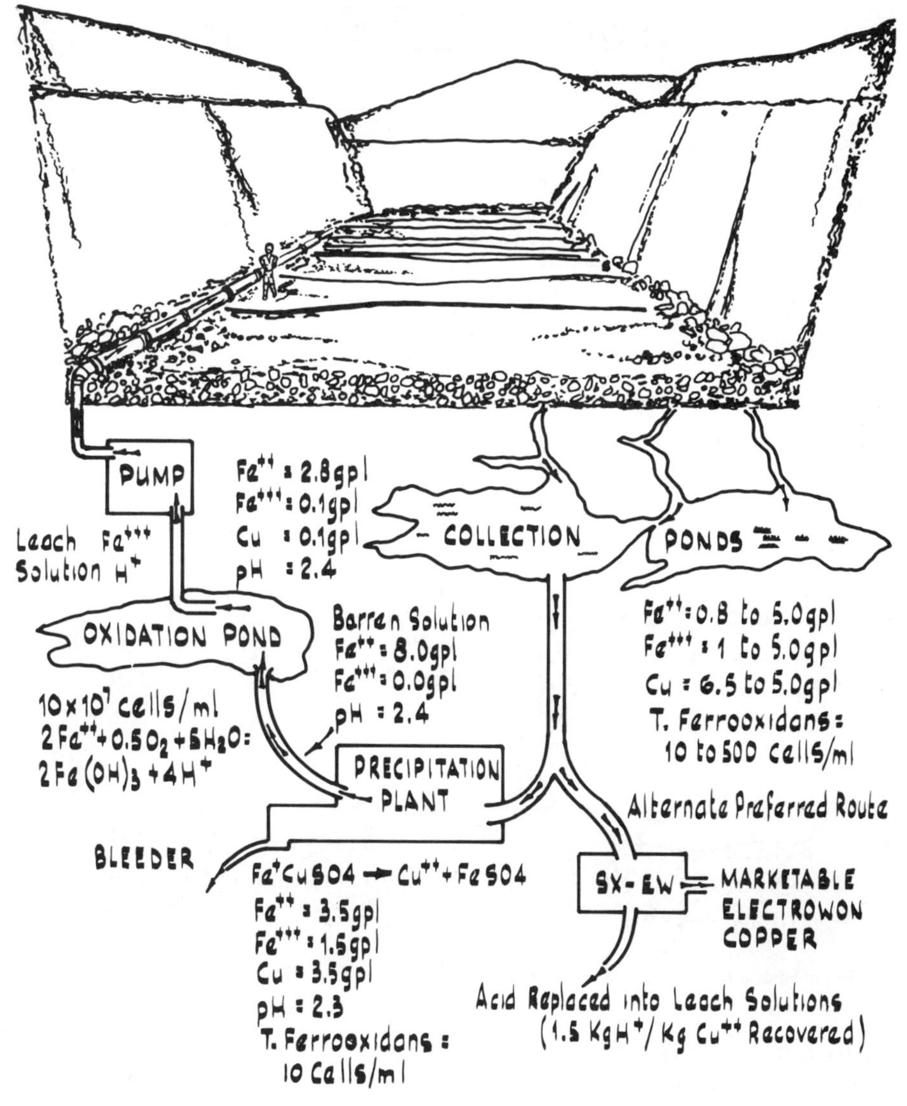

FIG. 2. Schematic representation of a heap leaching operation. (From Lundgren and Malouf 1983)

enhancement of the degradative reaction rates with diminishing particle size of the mineral has been clearly demonstrated by a number of workers (Torma 1977), and this circumstance, together with the desirability of usage of process tailings from orthodox processing as a raw material, has impelled the development of procedures in which finely divided mineral is treated in some form of agitated reactor. The use of reactor systems for biohydrometallurgical processes is advantageous, and the development of this mode is one of the growing points in the area.

The processes mentioned so far have in common the solubilization by biodegradative mechanisms of a valuable component of the raw material which is

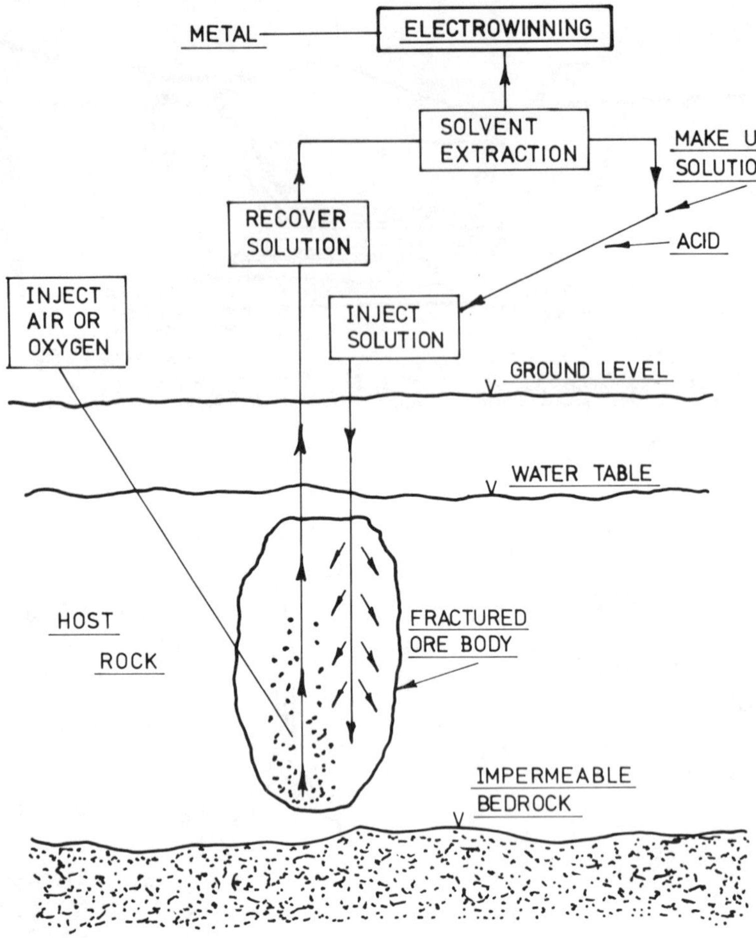

FIG. 3. Schematic representation of an *in situ* leaching operation. (From Davidson et al. 1981)

usually a minor constituent in the quantitative sense. Increasing attention has been directed to a different application of geomicrobiological phenomena in circumstances in which conventional modes for the extraction or separation of the minor valuable constituent are impeded or completely blocked by the associated major, low-value mineral components. If the latter are amenable to microbial degradation or modification, it is possible to carry out a reverse leaching procedure whereby the recalcitrant mineral assemblages are disrupted and orthodox recovery of the valuable component is facilitated. This procedure has been successfully applied to the recovery of metals such as gold and to minerals such as cassiterite (SnO_2), neither of which are readily solubilized by microbial processes but are frequently associated intimately with minerals that are (e.g., pyrite and pyrrhotite).

A potent and widely used leaching reagent in hydrometallurgy is acidic ferric sulfate solution.

$$2MeS + 2Fe_2(SO_4)_3 + 2H_2O + 3O_2 \rightarrow 2MeSO_4 + 4FeSO_4 + 2H_2SO_4$$

The reagent is reduced to ferrous sulfate and its reuse demands reoxidation to ferric sulfate. A favored method involves aerial oxidation in the presence of a bacterial catalyst. This regeneration process, in spite of wide usage, still presents some problems and is generally linked with other biological interventions in other parts of the leaching cycle. The process has attracted considerable attention not only for these reasons but because of its relevance to the usage of the reagent in *in situ* leaching or solution mining, particularly at greater depths (Davidson et al. 1981).

Other areas of wide significance in the development of biohydrometallurgy are concerned with the extension of the range of usable microbial systems, and attention has recently been focused on novel species, on the use of thermophilic and barophilic species and strains, on the utility of mixed populations, and on the potential of heterotrophic species. Some of the newer microbial systems are likely to be of considerable significance in the development of techniques for microbially enhanced oil recovery. Such systems will need to be able to effect changes in the relative viscosities of the oil/water system, to change the interfacial tension between oil and water, and to survive and reproduce under environmental conditions of extreme severity with respect to temperature and pressure.

Finally, a frequent roadblock to the economic feasibility of biohydrometallurgical processes has been the cost and the limited effectiveness of methods for the recovery of solubilized metals from dilute leach solutions, a problem that also arises in the treatment of metal-contaminated waste waters. While substantial progress continues to be made with methods such as solvent extraction, ion exchange, and membrane filtration, bioadsorption could provide a further alternative mode and particularly so if suitable bioadsorbents could be produced cheaply enough to become "throw-away" reagents. The possibilities for practicable processes based on microbial metal accumulation have been discussed recently by Lundgren and Malouf (1983).

Important though these areas of development may be, they are overshadowed and likely to be profoundly influenced by the progress toward a fuller understanding of the genetic patterns and mechanisms of the thiobacilli and other organisms prominent in the degradation and modification of mineral systems. Such understanding already offers the possibility of the securing of strains with significantly enhanced and extended capabilities.

These and closely related topics have been extensively reviewed by various authors during the past decade (C. L. Brierley 1978; Kelley et al. 1979; Murr 1980; Ralph 1979b, 1984; Torma and Bosecker 1982; Trudinger and Swaine 1979). The proceedings of international conferences in recent years provide evidence of the lively and increasing interest in both fundamental and applied aspects of mineral biodegradation phenomena (Schlitt 1977; Murr et al. 1978; Trudinger et al. 1980; Czegledi 1980; Rossi and Torma 1983).

Reactor Technology in Geomicrobiology and Biohydrometallurgy

Impellor-agitated or air-agitated reactor systems offer some advantages for geomicrobiological research by facilitating the continuous monitoring and adjustment of a considerable range of physicochemical parameters and the acquisition of basic kinetic data. A further advantage lies in the range of operational patterns that can be achieved with reactor systems; simple batch, semicontinuous flow, continuous flow, and continuous flow with recycle of organisms and/or substrate can be readily contrived. The variability of mineral substrates, the very real difficulties of valid sampling, and the frequent need for substrate charge sizes of sufficient dimension to allow extensive chemical and mineralogical examination of leached residues are requirements sometimes difficult to meet with simpler equipment such as shake flasks and percolation columns.

The design of satisfactory reactor systems for the study of mineral biodegradation presents some difficulties not commonly encountered in fermentors for organic substrates. The mineral charge is usually dense and not readily maintained as a homogeneous suspension; it is often highly abrasive to the reactor and to the impellor system, and it frequently has a facility for the tenacious coating of reactor walls and of the instrumental probes. Most of these difficulties are overcome in the numerous designs described in the literature (Torma et al. 1972; Gormely et al. 1975; McElroy and Bruynesteyn 1978; Babij and Ralph 1979). Babji et al. (1980) have described in detail an eight-unit reactor set (Fig. 4) for mineral biodegradation studies and have outlined operational details and monitoring procedures and the application of the equipment to various specific investigations. The modelling, scale-up, and design of reactors for the microbial desulfurization of coal (Fig. 5) have been discussed by Huber et al. (1983).

The particular advantages of continuous flow reactor systems for the study of mineral biodegradation were early recognized by some workers (Moss and Anderson 1968) and extended by other investigators (Gormely et al. 1975; McElroy and Bruynesteyn 1978; Sanmugasunderam 1981; Chang and Myerson 1982). Continuous reactor leaching processes have been little developed at full commercial scale; main preferences to date have been for stirred tank systems, but other types have been investigated. The mass transfer characteristics of an airlift fermentor with internal loop for the leaching of chalcopyrite concentrate has been studied by Yukawa (1975), and similar small reactors have been designed and evaluated by Ebner (1980) and Kiese et al. (1980). Comparison of the efficacy of various types of mineral bioreactors has been made by Atkins and Pooley (1983). The use of modified pachucas as airlift reactors has been advocated by Torma and Bosecker (1982).

The suitability of reactor leaching for large-scale mineral processing has recently been discussed in some detail by the present author (Ralph 1984). The raw material in dump and heap leaching is predominantly of large particle size and is extremely heterogeneous; the rate of leaching is severely limited by a number of factors, including gaseous and aqueous diffusion rates and the diluting effects of the large preponderance of gangue minerals. Materials for reactor leaching must of necessity be of small particle size and can arise from physical separation techniques such as flotation, which require reductions in particle size of primary raw material and the concentration of the valuable mineral components.

A number of factors have stimulated examination of the feasibility of reactor leaching of mineral concentrates and of process residues still containing significant amounts of valuable components. They include

- increases in smelting, refining, and freight costs;
- the pressures of environmental pollution control legislation;
- problems inherent in the application of orthodox technology to some complex ore types, such as those in which the mineralization is fine and intergrown;

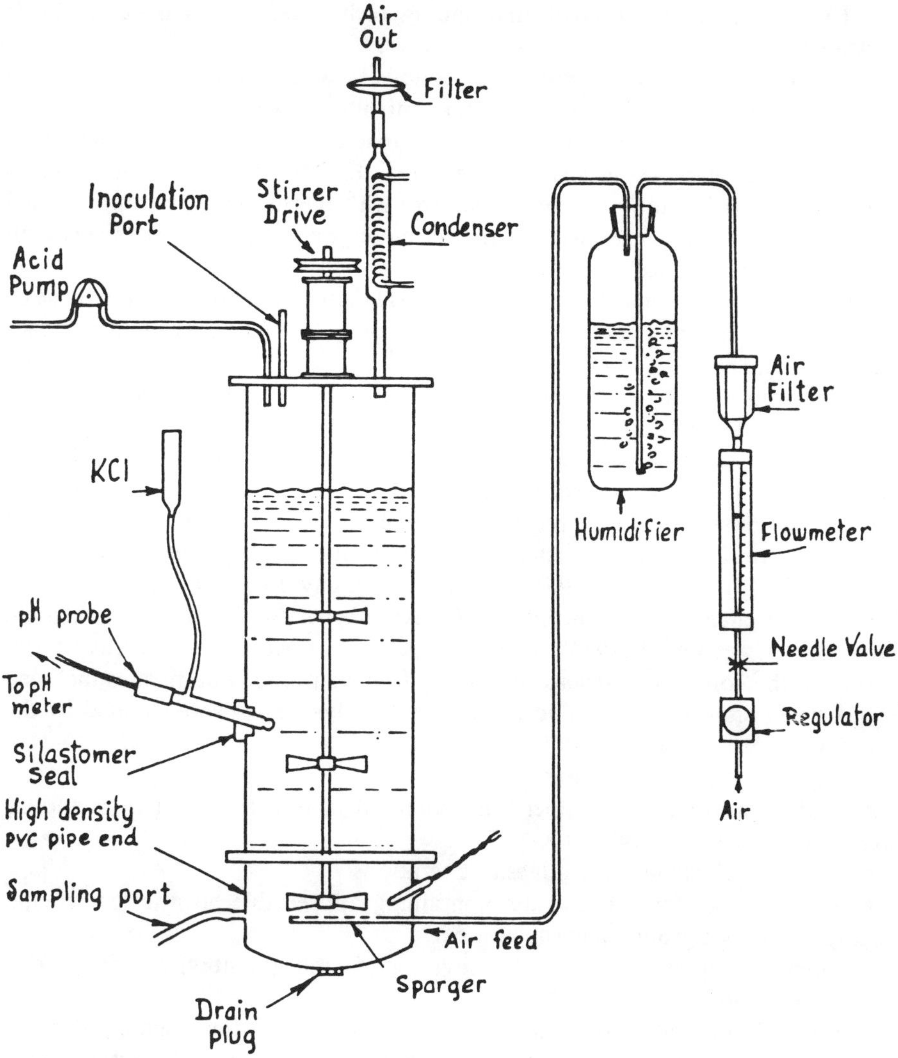

FIG. 4. Stirred reactor for mineral biodegradation studies. (From Babij et al. 1980)

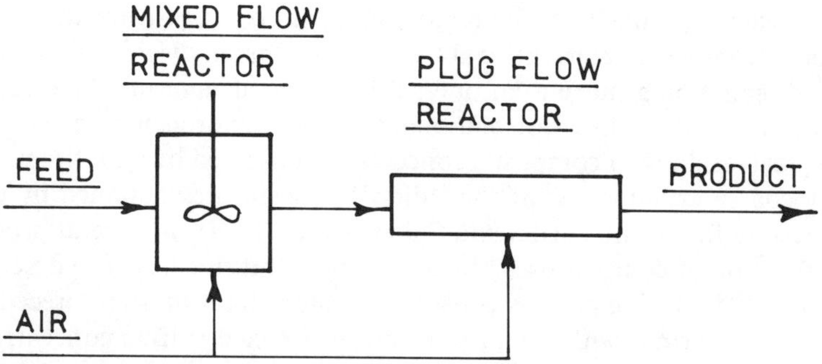

FIG. 5. Reactor design for microbial desulfurization. (From Huber et al. 1983)

• The development of attrition grinding techniques (Gerlach et al. 1973; Beckstead et al. 1976);

• The technical improvements in continuous flow, multistage, recycling reactor systems, with precise monitoring and control of process variables;

• the much higher level of control in a reactor system over the biological systems and their working environment, including the possibility of usage as process catalysts of monocultures, organisms with thermophilic, barophilic, and other tolerances, and in each case, the ultimate goal of use of organisms with genetically enhanced capabilities;

• the likelihood of achieving rates of mineral dissolution approaching those of conventional hydrometallurgical processes at lower levels of overall energy consumption; and

• improvements in the technologies for the recovery of mixtures of metallic ions from dilute solution.

These considerations have led a number of workers to examine the possibilities of reactor leaching for the processing of both primary ores and of concentrates and process residues. The search for alternate processing routes has stimulated the examination of reactor leaching for zinc sulfide concentrates (Torma et al. 1970, 1972; Gormely et al. 1975; Torma and Guay 1976; Sanmugasunderam 1981) and of chalcopyrite concentrates (Bruynesteyn 1978; Torma and Rozgonyi 1980; Bruynesteyn et al. 1983). The investigations of McElroy and Bruynesteyn (1978) on the continuous reactor leaching of chalcopyrite concentrates included economic feasibility studies, and the following advantages were claimed by the authors:

• negligible atmospheric emissions;

• production of refined copper (i.e., potentially lower transport costs and increased market flexibility);

• feasibility of small-scale minesite operations;

• reduced costs for concentrates containing appreciable bornite, chalcocite, etc., regardless of grade or pyrite content;

• potentially increased returns for silver, gold, and (possibly) molybdenite in concentrates; and

• production of dilute sulfuric acid as a byproduct for the leaching of oxide ores and/or mill tailings and/or acid-consuming waste dumps. The authors were of the opinion that where any of these factors were significant, biologically assisted reactor leaching should be considered as an option.

Reactor leaching is likely to find major application in the treatment of complex sulfide ores from which conventional treatments generally yield "dirty" concentrates (i.e., containing more than one sulfide mineral) and in which recoveries may be as low as 70%. Less conventional treatment approaches have advocated the biological leaching of composite concentrates obtained by bulk flotation after relatively coarse grinding, or of the "dirty" concentrates obtained by conventional selective flotation (Torma and Subramanian 1974). A general process for the beneficiation of complex ores of lead, copper, and zinc has been described by Carta et al. (1980a). The process consists of a bulk flotation step, preceded by a relatively coarse grind, with the production of a mixed sulfide concentrate at a recovery of more than 80%, followed by a regrinding step. The reground material is microbially leached in a reactor system with adapted strains of *T. ferro-oxidans* and yields a virtually zinc-free lead concentrate in which the lead content is upgraded 3-fold. Only minor removal of copper occurs. While this sort of process has the potential for the recovery of all metals present in the bulk concentrate, the results indicate that the relative abundances of different sulfides have an important influence on the rates of leaching and on the metal content of the leach liquor. Other interactions of this kind have been reported (between sphalerite and chalcopyrite, Carta et al. 1980b; between chalcopyrite and pyrrhotite, Rossi et al. 1983) and demand closer investigations.

Environmental control legislation has stimulated further investigation of the microbial desulfurization of coal, and various operational modes have been examined (Dugan and Apel 1978). The relative merits of dump leaching and processing plant techniques have been discussed by Pooley and Atkins (1983). The possibility of dump leaching after pelletizing of ground coal has not yet been closely examined, and the majority of investigations have favored reactor leaching techniques. Kargi (1982) has described a process for the removal of pyritic sulfur from pulverized coal by mixed enrichment cultures of acidophilic microorganisms, which were more effective than pure cultures of the component organisms. The superior performance of thermophilic organisms for coal desulfurization has been demonstrated by Murr and Mehta (1982) and Kargi and Robinson (1982). The removal of organic sulfur from coal by mixed cultures has been investigated by Mishra et al. (1983).

Reactor leaching has been a favored tool for investigation of "reverse" leaching processes in which microbial catalysis is used to modify a major, low-value component of the raw material in order to facilitate the recovery of minor, high-value components. Particular attention has been directed to the recovery of gold and silver from pyrite/arsenopyrite ores from which recovery of the precious metals is usually unacceptably low. The solubilization of the iron and arsenic components by bacterial pretreatment has been studied (Pinches 1975; Babij and Ralph 1979), and full processes for gold recovery have been described by Livesey-Goldblatt et al. (1983). A similar approach for the recovery of cassiterite from process tailings, using *T. ferro-oxidans* at low pH and a consortium of organisms at a higher pH, has been described by Harris et al. (1983). The method seems to have considerable applicability to recalcitrant mineral assemblages and to those in which a valuable component is in solid solution in a mineral lattice (e.g., silver in bismuth sulfide or nickel/cobalt in pyrite.

The use of the potent oxidant acidic ferric sulfate has been mentioned earlier as a much used leaching reagent in hydrometallurgy. In the context of biohydrometallurgy, its aerial regeneration with a bacterial catalyst has attracted a good deal of attention and considerable ingenuity has been expended on the devising of reactor systems for the most effective means of implementing this process. Multistage, continuous flow systems for this purpose have been described (Gorog et al. 1980) and fixed film reactors such as the Bacfox system (Livesey-Goldblatt et al. 1977) are in commercial scale usage. There seems to be considerable scope for further improvements in reactor systems for this purpose, but whichever mode is employed, it is important that the bacterial strain used has biochemical and physiological characteristics that are closely matched to the requirements of the particular leaching operation in which the reagent is used. In the case of the use of *T. ferro-oxidans* as the catalyst for reagent regeneration in the leaching of uranium ores, it is clear that improvements are needed in the organisms' capabilities in respect of (a) rate of ferrous iron oxidation; (b) tolerance to low pH (ca. pH 1-1.6) and to sudden changes in pH round the leaching circuit; (c) tolerance to high concentrations of uranium, thorium, and various common heavy metals; (d) ability to function under saline conditions; and (e) thermophilic and barophilic tolerance for deep mine operation.

It might be noted further that ferric sulfate has the capacity to act as an electron acceptor in lieu of molecular oxygen in some microbially mediated oxidations. The oxidation of sulfur (with concomitant reduction of ferric ions) (Brock and Gustafson 1976; Brock 1977), of chalcopyrite (Babij et al. 1981), and of marmatite (Goodman et al. 1983) (Table 1) under anaerobic conditions provide mechanisms by which metal sulfide degradation can proceed at a substantial rate even when oxygen is limiting. In natural and field situations, one can envisage cyclic mechanisms in which iron, under linked aerobic and anaerobic conditions and with microbial intervention in each circumstance, undergoes alternate reductions and oxidations and facilitates the degradation of sulfide minerals. Such mechanisms are probably significant in some pollutional situations (Goodman et al. 1981b) and offer prospects for the design of processes for metal recovery.

TABLE 1. *Anaerobic leaching of a zinc-iron sulfate mineral.* (From Goodman et al. 1983)

Comparison of amounts of iron and zinc solubilized from the zinc-iron sulfide under aerobic and anaerobic conditions at pH 2.5

Leaching Conditions	Maximum Concentration of Metal Solubilized as a % of its Concentration in the Ore	
	Iron	Zinc
Sterile aerobic[a] control	20	9
Sterile anaerobic[b] control	19	13
T. ferrooxidans aerobic[a]	48	100
T. ferrooxidans anaerobic[b]	86	63

Gas supplied: [a] air at flow rate of 600 ml/min
[b] 5% CO_2 in N_2 at flow rate of 35 ml/min

Geomicrobiological Systems of Actual or Potential Utility

Natural locales, undisturbed by man's activities, have been prolific sources of a wide range of microbial types, and have included lacustrine and marine sediments, salt lakes, hot springs, and other manifestations of geothermal activity. The study of such organisms has been crucial to the understanding of the great geochemical cycles.

Geomicrobiological investigations have identified a considerable diversity of microbial types concerned with primary or secondary roles in the processes of mineral biodegradation or modification. The most common source of such microbial systems has been field situations in which mineral breakdown is occurring; such milieu commonly arise as the result of the disturbance of ore bodies during mining operations, in the waste dumps or tailings dams for the disposal of the low-value byproducts of metallurgical processing, or in the drainage waters from mined ore bodies or residual waste materials. Well-established methodologies for the isolation, purification, identification, and characterization of such microorganisms have been described (Silverman and Lundgren 1959; Collins 1969; Tuovinen and Kelly 1973; Manning 1975; C. L. Brierley et al. 1980; Goodman et al. 1981a; Marsh and Norris 1983).

It is characteristic of both fractured ore bodies and the wastes from their extraction and processing that the degree of physical and compositional heterogeneity is likely to be very high and that a wide range of physicochemical conditions will occur in microenvironments. It is not surprising that ecological studies of the microbial populations occurring in such situations have revealed very complex associations of microorganisms, which show extensive variation in kind and in relative magnitude according to location and time (Goodman et al. 1981a, 1981b). However, circumstances can occur that favor the predominance of a limited number of species. Ecological successions commonly develop over time in locations such as sulfidic waste dumps, the earlier-developing populations modifying the microenvironments in respect of factors such as acidity, temperature and availability of biomass, thus paving the way for their own decline and the ascendancy of species better adapted to the new conditions. In the example quoted, the colonizing microbial populations are frequently species that can use the sulfide moiety of sulfide minerals (e.g., *Thiobacillus delicatus* and *T. rubellus*) (Mizoguchi et al. 1976; Mizoguchi and Okabe 1980) with concomitant generation of hydrogen ions but which themselves have low tolerance to acidity. Such species may substantially modify the composition of the sulfide minerals without release of soluble metallic ions but may increase the acidity of the microenvironments and generate biomass. There is little information available on the mechanisms by which mineral structure is modified without metal release (Silver and Torma 1974). The continuation of processes of this kind lead to a succession of increasingly acidophilic, sulfur-oxidizing species and eventually to conditions under which the stability of ferrous ions, derived from the ever-present iron minerals, is sufficiently high for them to be available as energy sources for iron-oxidizing species such as acid-tolerant *Gallionella* spp. and *T. ferro-oxidans.* The accumulation of dead biomass from autotrophic species occurring earlier in the succession may stimulate the development of heterotrophic species (e.g., some nitrogen-fixing types) and encourage mixotrophic modes of nutrition.

While the microbial communities associated with sulfide mineral degradation in field situations are dominated by organisms with sulfur- and iron-oxidizing abilities, these types are associated with a diverse array of metal- and acid-tolerant, heterotrophic bacteria, fungi, yeasts, algae, and protozoa. The taxonomic characteristics of the better known *thiobacilli* have been tabulated by Lundgren and Malouf (1983), and the growth characteristics of the principal sulfur-oxidizing and iron-oxidizing bacteria involved in sulfide mineral degradation have been extensively recorded (C. L. Brierley 1978; Ralph 1979a; Goodman et al. 1981a; Torma and Bosecker 1982). The paramount thiobacillus species involved in sulfide mineral degradation is the very well-known *Thiobacillus ferrooxidans,* and this remarkable organism deserves further comment. Its physiology and biochemistry have been intensively studied and extensively reviewed

(Tuovinen and Kelly 1972; Torma 1977). A number of workers have noted the population heterogeneity of this organism as it occurs in leaching dumps and other mineral degradation situations (Groudev et al. 1978) and as manifested by isolates in colony morphology on solid media (Manning 1975; Goodman et al. 1980) by the different capacities of adapted strains to oxidize particular substrates (Silver and Torma 1974), by the variation in DNA base compositions (Guay et al. 1976) and by the varying enzymic profiles of isolated and mutated strains (Groudeva and Groudev 1980; Groudeva et al. 1981). It is unlikely that the full range of sulfide mineral-degrading capabilities or the potential utility in other biogeochemical transformation of its numerous biochemical and physiological variants have yet been fully explored. A steady stream of new information continues to emerge in respect of this organism; for example, on the lower limits of its pH tolerance in respect of iron-oxidizing ability (Atkins 1978; Bruynesteyn et al. 1980a, 1980b), its tolerance to elevated hydrostatic pressures (Bosecker et al. 1979; Davidson et al. 1981), and its ability to fractionate sulfur isotopes during leaching processes (Karavaiko et al. 1980). Other recent studies have been concerned with novel chemical abilities, such as the volatilization of mercury compounds (Olsen et al. 1982) and the direct oxidation of uranyl ions (DiSpirito and Tuovinen 1982a, 1982b). The range of useful abilities of this organism is likely to be further extended when a more detailed understanding of its genetic pattern emerges from the considerable volume of recent and current studies.

Studies on the biochemistry and physiology of other members of the genus *Thiobacillus* continue and include the purification and characterization of two principal enzymes from the thiosulfate-oxidizing system of *Thiobacillus* A-2 (Lu and Kelly 1983) and on the use of carboxylic acids by the same species (Wood and Kelly 1983), purification of the sulfite oxidase from *T. novellus* (Toghrol and Southerland 1983a; Southerland and Toghrol 1983) and further investigations of the subunit structure and cofactors of the sulfite oxidase of *T. novellus* (Toghrol and Southerland 1983b). Comprehensive information on the occurrence, isolation, and identification of the genus *Thiobacillus* and *Thiomicrospira* has recently become available (Kuenen and Tuovinen 1981). Several new *Thiobacillus* species have been described and include *Thiosphaera pantotropha,* a facultatively anaerobic, facultatively autotrophic bacterium (Robertson and Kuenen 1983), *Thiobacillus albertis* (Bryant et al. 1983) and *Thiobacillus rapidicrescens* Katayama-Fujimura et al. 1983a).

The demonstrated capacity of some microorganisms to modify mineral structure without release of hydrogen ions or metallic ions into solution has important connotations for the devising of mineral processing routes with low pollutional hazards (Harris et al. 1983), and it is of interest to note a report of the catalysis of pyrite oxidation by *Thiobacillus thioparus* at near neutral pH (Shepard et al. 1983). Other candidate organisms, which might be similarly applicable, are *Gallionella* spp. (Hanert 1981a) and *Siderocapsa* and related species (Hanert 1981b). At the other extreme, in very low pH situations (ca. pH 1), the vibroid, iron-oxidizing bacterium *Leptospirillum ferro-oxidans* (Balashova et al. 1974; Norris and Kelly 1982; Norris 1983) offers some interesting possibilities for improved processes.

In addition to the high intrinsic interest of organisms able to survive and reproduce at higher temperatures, the implications of their possible technological application is considerable. Thermophilic strains of the thiobacilli have attracted attention over the past decade. A number of thermophilic, thiobacillus-like bacteria have been isolated and their applicability to the bacterial leaching of sulfide minerals demonstrated (Le Roux et al. 1977; J. A. Brierley 1978; J. A. Brierley and C. L. Brierley 1978; Le Roux and Wakerley 1980; J. A. Brierley 1980; Murr and Mehta 1982; Marsh and Norris 1983). Since the late 1960s there has been considerable interest in the extreme thermophiles of the genus *Thermus* (Brock and Freeze 1969; Brock and Edwards 1970; Darland et al. 1970; Brock 1981) and *Sulfolobus* (Brock et al. 1972; Fliermans and Brock 1972; C. L. Brierley and J. A. Brierley 1973; De Rosa et al. 1974; De Rosa et al. 1975; Bohlool 1975), whose habitats are hot springs and sulfur-rich, acidic soils. Studies on the biochemistry and physiology of these remarkable organisms have continued apace (Mosser et al. 1973, 1974; De Rosa et al. 1983; Collins and Langworthy 1983; Green et al. 1983; Deatherage et al. 1983). *Sulfolobus*-like organisms have been shown to be capable of oxidizing ferrous iron and metallic sulfides (C. L. Brierley and Murr 1973), to leach molybdenum from molybdenite (C. L. Brierley 1974). Some minerals rich in chalcopyrite have been more effectively leached by *Sulfolobus* than by *T. ferro-oxidans* at elevated temperatures (J. A. Brierley and C. L. Brierley 1978). Of particular interest is the ability of *Sulfolobus* to grow when the concentration of molybdenum is 7.82 m*M* (ca. 750 ppm), which greatly exceeds the moloybdenum tolerance of *T. ferro-oxidans* (5–8 ppm) (Tuovinen et al. 1971). Further growth and mineral oxidation studies (Fig. 6) with *Sulfolobus* strains (Marsh et al. 1983) have indicated considerable difference between strains with respect to mineral-oxidizing activity and the requirement for organic supplements. The removal of sulfur compounds from coal by *S. acidocaldarius* has been described by Kargi and Robinson (1982). More recently the anaerobic reduction of molybdenum by *Sulfolobus* species has been demonstrated (C. L. Brierley and J. A. Brierley 1982).

The tolerance of the important leaching organism *T. ferro-oxidans* to elevated temperature and pressure has been referred to earlier and leads to speculation as to whether microbial systems might be able to survive under more severe conditions. The availability of liquid water appears to be the prime nutritional requirement, and at temperatures above 100 C this presupposes sufficiently high hydrostatic or osmotic pressures. Natural habitats with such characteristics exist, and it is most exciting to note the recent isolations of bacteria from such hostile environments and the cultivation in the laboratory of complex bacterial communities at temperatures up to 250 C (Baross and Deming 1983). These findings open up possibilities for contrived biotechnological processes using biocatalysts at greatly elevated temperatures and pressures.

It is well known to workers on the biodegradation of sulfide minerals that higher rates of leaching are frequently achieved with mixed cultures from natural sources than with monocultures of the individual cultures from such mixed populations. There has been little systematic study of mixed culture phenomena in mineral biodegradation situations. Norris and Kelly (1978) have discussed the role of mixed populations of thiobacilli in pyrite oxidation, and the same authors (Norris and Kelly 1982) have examined the general question of the use of mixed cultures and have reviewed different types of interaction. Of interest also are potential mutualistic associations, involving the exchange of organic material and

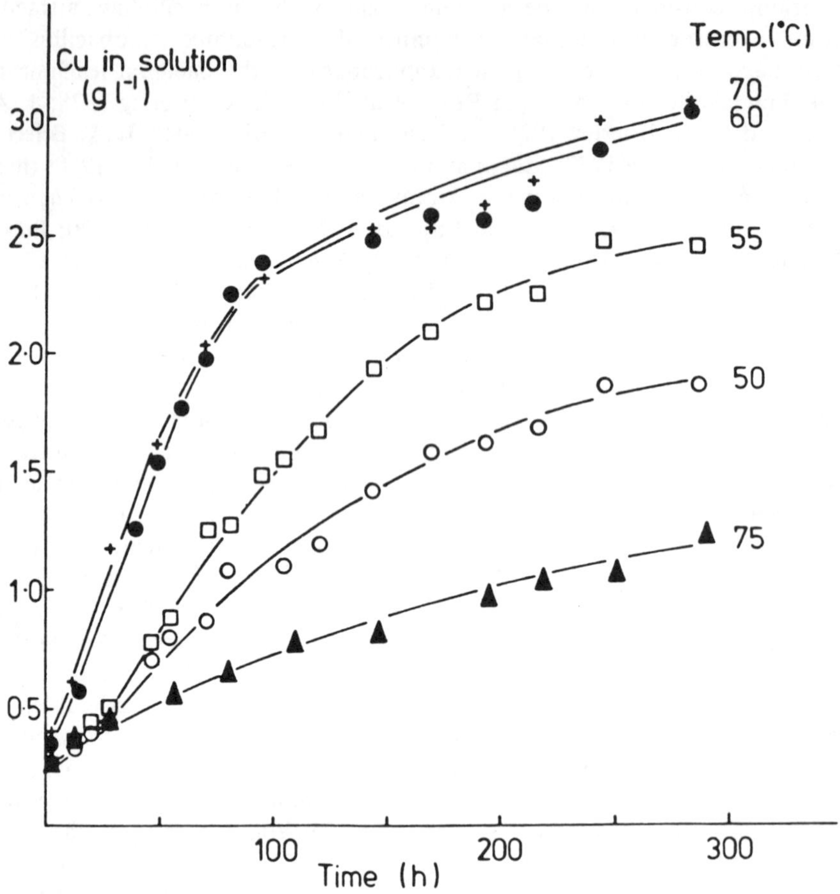

FIG. 6. Effect of temperature on copper solubilization from chalcopyrite during autotrophic growth of *Sulfolobus* (Lake Myvam). (From Marsh et al. 1983)

fixed nitrogen, between organisms such as *T. ferro-oxidans* and *Beijerinckia lacticogenes* (Tsuchiya et al. 1974; Tsuchiya 1977), between nitrogen-fixing and nonnitrogen-fixing strains of *T. ferro-oxidans* (Mackintosh 1978), and between nitrogen-fixing strains of *T. rubellus* and other thiobacilli (Goodman and Ralph 1978). Norris and Kelly (1982) have also pointed out that interactions can be envisaged between different bacterial types during sulfide mineral breakdown, which result in inorganic material recycling between the bacteria catalyzing the reactions. A number of mixed culture reaction cycles, involving chemical attack on sulfide minerals by ferric ions; mineral dissolution by iron-oxidizing organisms such as *Leptospirillum;* and sulfur removal by sulfur-oxidizing organisms such as *T. thio-oxidans, T. acidophilus* (Arkesteyn 1980), and *T. organoparus* (Markosyan 1973), under both aerobic and anaerobic conditions, can be envisaged.

The isolation and characterization of the autotrophic organism *Stibiobacter senarmontii,* which can use the Sb^{III}/Sb^{V} oxidation as an energy source has been reported by Lyalikova (1972, 1974), and this organism has some prospects of application to the processing of antimony ores and to nickel and gold antimonide minerals. The possibilities of various heterotrophic species for metal recovery have been examined; for example, the solubilization of gold by *Bacillus* spp., *Pseudomonas fluorescens, Serratia marcescens,* and *Agrobacterium tumefaciens* (Torma and Bosecker 1982). The involvement of other heterotrophic species in the degradation of carbonate and silicate ores of copper (Kiel 1977) and of copper-nickel sulfide concentrates (Le Roux et al. 1978) has been described. Heterotrophic microorganisms capable of leaching manganese ores have been reviewed by Marshall (1979) and Ehrlich (1980), and the possibility of microbially mediated processes for the refining of manganese dioxide slimes continues to attract attention (Mercz and Madgwick 1982; Holden and Madgwick 1983).

Of particular interest is the role of heterotrophic organisms in the biodegradation of silicate minerals (Agbim and Doxtader 1975) and for the beneficiation of high silica and high iron bauxites. The biodegradation of silicate minerals has been reviewed by Silverman (1979). Some systems have been applied by the leaching of lateritic ores and the beneficiation of low-grade bauxite (Groudev and Genchev 1978b; Silverman 1979; Torma and Bosecker 1982). A good deal of attention has been devoted to leaching of alumino-silicate minerals (Rossi 1978), but the mechanisms of aluminum concentration and silicon removal are still not clear.

The use of heterotrophic microorganisms for leaching processes implies the provision of reduced carbon energy sources, and this need has been a major impediment to the development of such processes. It should be noted, however, that paleontological evidence relating to the genesis of bauxite deposits suggests that, by the use of mixed culture systems, processes could, under some circumstances, be driven by solar energy (Ralph 1977). In some locations, agricultural and industrial wastes have been used as energy sources for such leaching processes (Kiel 1977).

Sulfate-reducing bacteria have found some application in mineral processing, and it has been demonstrated by Layalikova et al. (1977), Solozhenkin et al. (1979), and Solozhenkin and Lyubavina (1980) that improved extractions of oxide ores of antimony and bismuth can be achieved by prior treatment with such bacteria. Sulfidization occurs and significant improvement in the flotation of mixed concentrates is claimed. There are some effects on the flotation process per se in addition to sulfidization (Kupeyeva et al. 1977); this raises the possibility that excretory products from both these and other types of leaching organisms might yield clues as molecular structures, which could form the basis of improved flotation reagents. Effects on the flotation properties of sulfide minerals have been reported by Kim et al. (1981), and Tomizuka and Yagisawa (1978) have demonstrated the possibility of recovery of metals from leaching solutions and acidic mine drainage by the use of sulfate-reducing bacteria.

Microbiologically Enhanced Oil Recovery (MEOR)

The availability of economically extractable crude petroleum, like the other nonrenewable fossil fuels, is likely to progressively decline in the future. It is imperative, therefore, that the search for new oil deposits be intensified and that the yield from existing resources be considerably improved. It is estimated that, on average, only about 30% of the crude oil present in reservoirs is recoverable by the action of the "primary" or "endogenous" energy of the reservoir, and various methods are currently used to enhance the yield. Processes for enhanced oil recovery are based on the introduction into the well system of some additional energy source to free oil entrapped in the capillary spaces of the reservoir rocks. The methods used include the injection of gas, flooding with water, the introduction of surface-active agents and viscosifiers, the use of thermal and mining methods, and the employment of microbial systems.

The earlier field applications of microbially enhanced oil recovery methods in Hungary, Russia, and Poland have been described by Bubela (1978). The limited success of these earlier attempts at MEOR has stimulated systematic examination of the numerous problems and imaginative approaches to their solutions. The now high possibility of practicable biological procedures for the enhancement of oil recovery is one of the most exciting and important applications of biotechnology in the geomicrobiological area. A detailed overview of the numerous specific problems of MEOR, the rationale of the multidisciplinary approaches needed for practicable solutions, and the current status of development, have recently been authoritatively presented by Bubela (1984).

The basic requirements of MEOR are similar to those used in abiological approaches (Bubela 1983a). They are (a) reduction of the interfacial tension of the rock-water-oil system, (b) improvement of the mobility ratio of the fluids involved, and (c) increase in the permeability of the reservoir rocks.

A number of investigators have reported on the endogenous microbial populations of oil wells and have postulated various effects upon the petroleum substrates and other components, such as sulfate reduction and modification of hydrocarbons. Such activities are probably minimal at the reservoir rock sites and are limited by nutrient availability. Introduced microbial populations, if they are to have any significant effect on oil recovery, must be able to meet a number of requirements. They must be able to

- penetrate the rock formations and multiply and metabolize in such situations;
- survive in relatively severe environments in which the temperature may approach 100 C, the hydrostatic pressure may approximate 20,000 KpA, there may be significant concentrations of metallic ions and other solutes, and anaerobic or, at best, microaerophilic conditions may prevail;
- produce agents which reduce the interfacial tension of rock-water-oil systems and improve the mobility ratio of the fluids involved;
- modify the reservoir rocks so as to improve permeability; and
- not be of a morphology that produces pore blockage or not be generators of metabolites which cause the same effect. Diverse microbial species are likely to be needed to meet the unique requirements of individual oil reservoirs.

An important difference between chemical EOR and MEOR techniques is the establishment of active agents in the oil reservoir (and their proliferation); therefore, it is necessary to evaluate particular reservoir characteristics for compatibility with the microbial systems it is proposed to inject (Bubela and McKay 1983). Laboratory experimentation can yield essential information on the effects of the various physico-parameters on the behavior of selected microbial types. Simulation procedures are also essential and have provided an approach to the evaluation of reservoir characteristics (Fig. 7). The examples of these approaches,

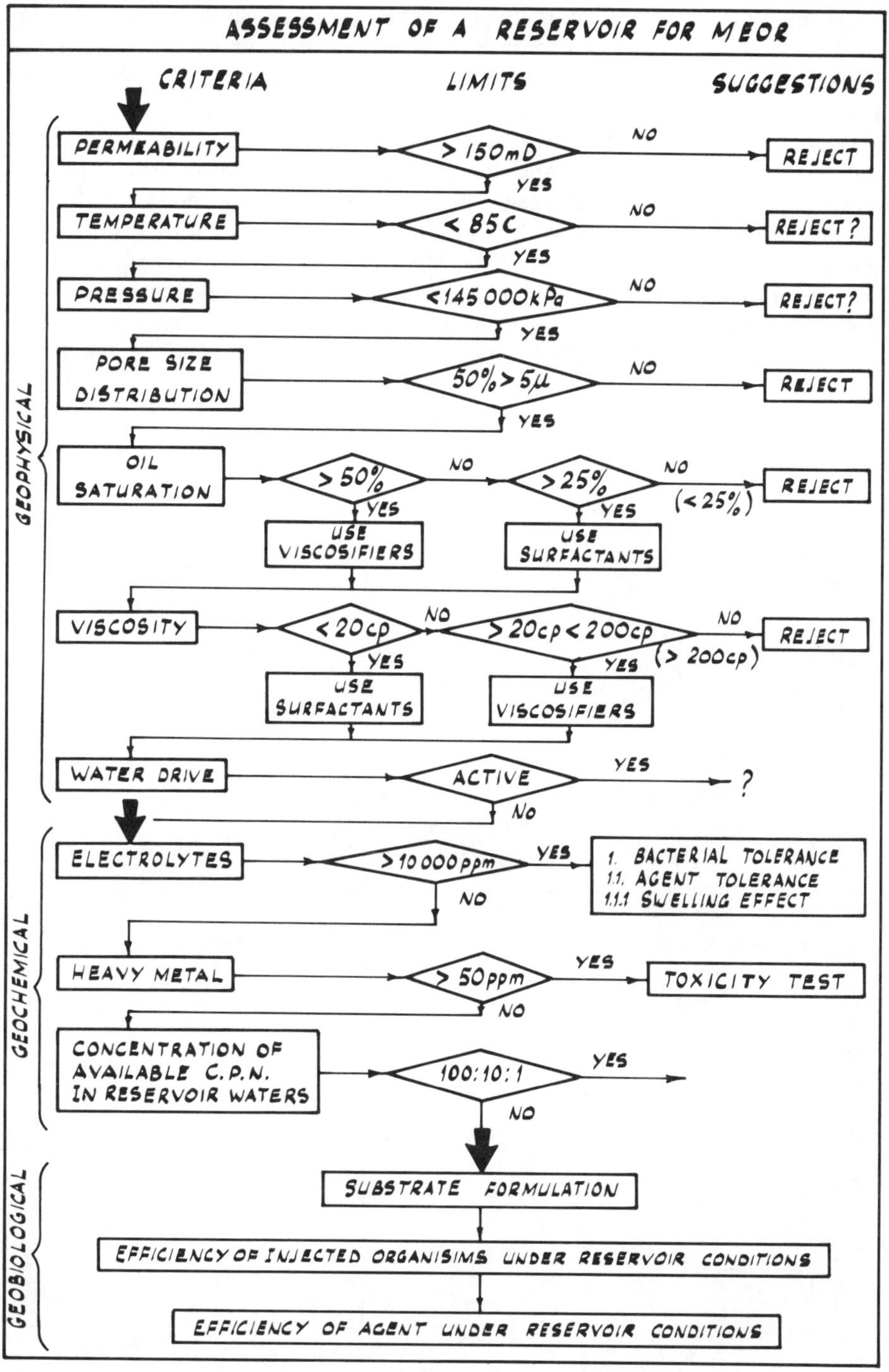

FIG. 7. Evaluation of oil reservoir for MEOR. (From Bubela and McKay 1983)

which follow, are taken from the work of the MEOR research group of the Baas Becking Geobiological Laboratory in Canberra, which, under the leadership of V. Bubela, has secured a great deal of relevant information over the past few years, using both laboratory and simulation techniques at various scale levels.

Using *Bacillus stearothermophilus* as a test organism, laboratory experimentation has clearly shown the manner in which a metallic ion (copper) can affect the morphology (Bubela 1970) and cell wall composition (Bubela and Powell 1973) of the organism and the way in which the temperature of cultivation can influence copper toxicity effects (Fig. 8). The combined effects of temperature and pressure have been studied with another organism, an anaerobic rod-shaped organism, 6–8 μm long and 3–4 μm wide, which produces a metabolite capable of reducing the interfacial tension of an oil-water system. This organism was isolated from anaerobic digesters in the Melbourne sewage treatment system. Under atmospheric pressure, the optimum temperature for growth was 50 C with minimum growth time (MGT) of 17 h. When grown in a continuous flow loop

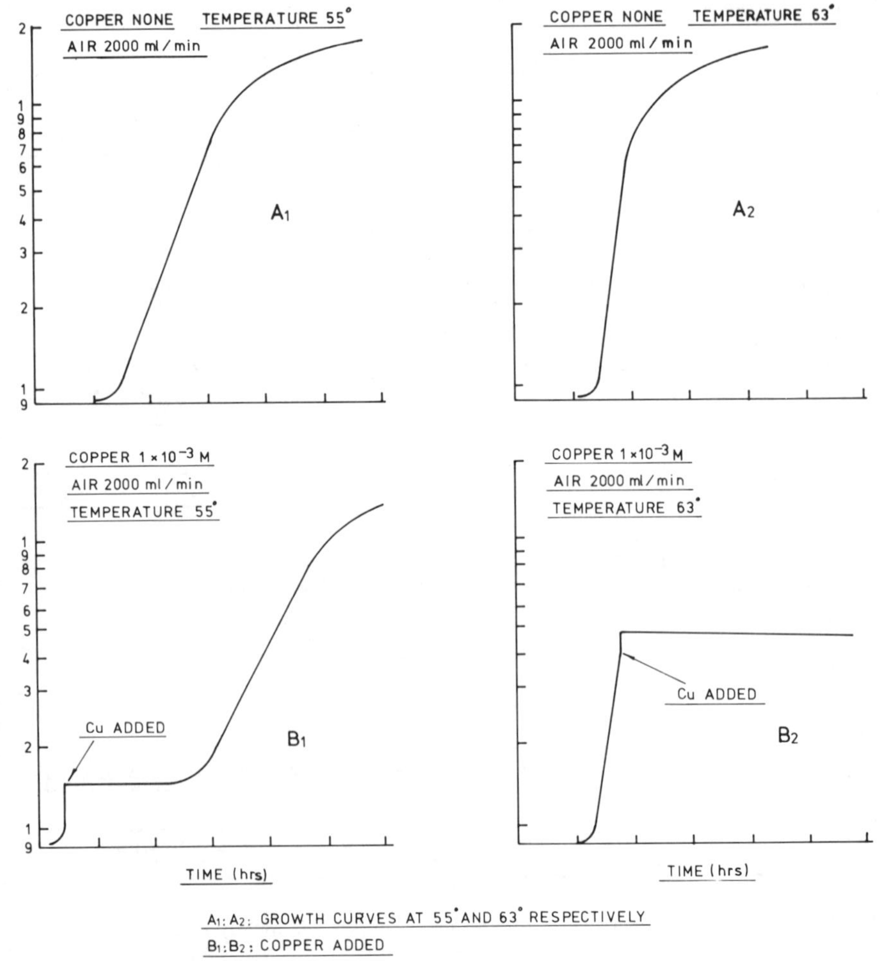

FIG. 8. Temperature: Copper toxicity effects in *Bacillus stearothermophilus*. (From Bubela 1982)

fermentor, capable of operation at pressures up to 25,000 KpA and temperatures up to 150 C, the maximum growth temperature was 65 C at a pressure of 20,000 KpA. The MGT under these conditions was 12 h, and the morphology of the organism changed to a coccoidal form about 5 μm in diameter. The organism grew normally in saline solutions of up to 3% NaCl concentration but at pressures above 15,000 KpA could not survive this salt concentration (Bubela 1982).

The effect of biological activity on the movement of fluids through porous rocks and sediments has been studied using a series of artificial materials of measured porosity, permeability, and particle size distribution (Bubela 1983b). These systems have enabled the evaluation of the effects of various microbial systems on the movement of oil through porous, permeable material that had been oil- or water-prewetted (Fig. 9). In other experiments the effects of

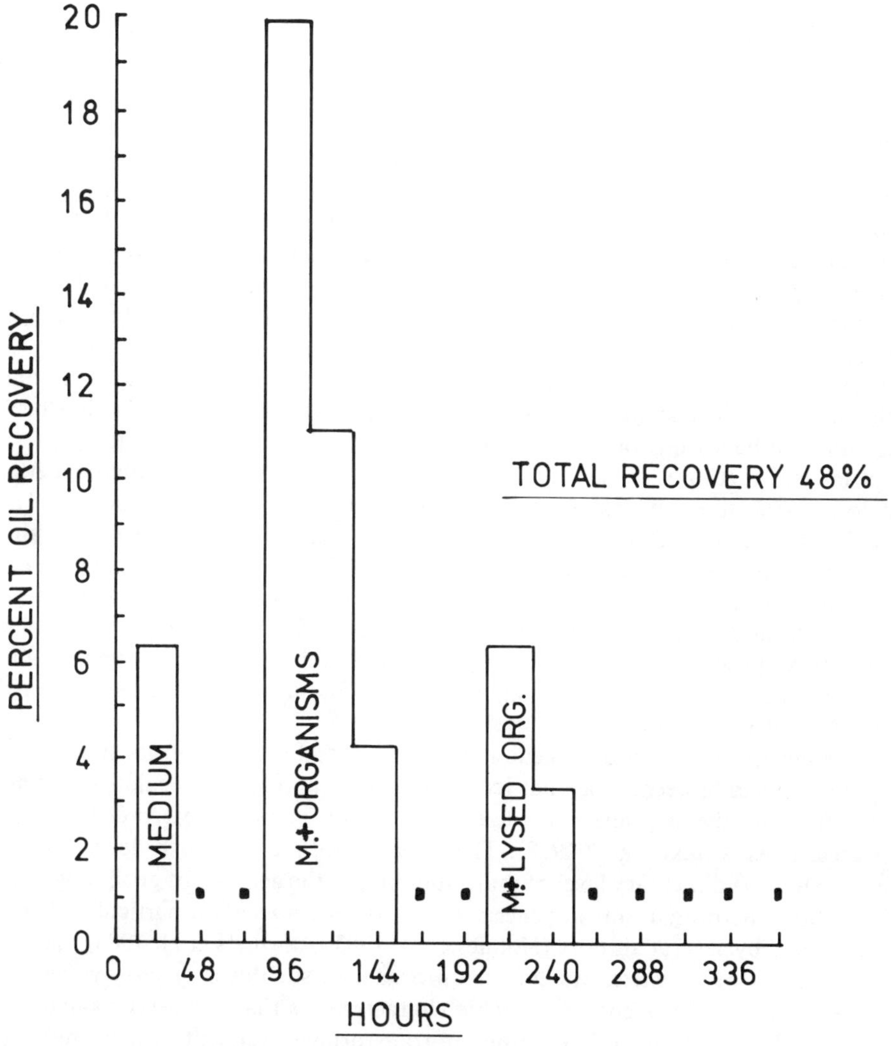

FIG. 9. Microbiologically enhanced oil recovery from simulated oil reservoir. (From Bubela 1983b)

biological activity on specific reservoir rock components, such as calcite and dolomite, have been examined, and increases in permeability and porosity correlated with the biological production of fatty acids and carbon dioxide. Useful information has been obtained with these simulation systems on the role of the morphology of the organisms in the plugging of pores and flow spaces. In general, rod-shaped organisms were more effective in decreasing the permeability of the system, and such a decrease was enhanced if a higher hydraulic pressure head was applied to the moving aqueous front. Coccoid organisms were less liable to produce blockages, and the flow-through could be maintained by an increase of the pressure head of the water front.

The developments in this area are promising. The techniques now being applied have considerable relevance to other geobiological problems, such as the genesis of oil deposits and the biogenesis of minerals.

The Impact of Genetic Engineering in Geomicrobiology and Biohydrometallurgy

In a number of applications of geomicrobiology, the overall effectiveness of the process is diminished by the inability of the microbial system(s) to function at an economically acceptable rate. This is probably due to constraints of a physicochemical kind, and to some extent these limitations may be modified by better engineering of the system or closer control of various environmental factors. The other principal impediment to improved performance lies within the inherent capabilities of the microbial system in characteristics such as its growth rate; its productive capacity with respect to key enzymatic systems and chemical metabolites; its reaction to particular levels of temperature, pressure, pH, Eh, and water activity; and its tolerance to various cationic and anionic species. All these characteristics are a reflection of the genetic make-up of the organism, and the observable spectrum of abilities within strains of particular species has been assumed to be a function of variations in the structure of the genome.

Classical methods of strain improvement have been used to obtain better cultures of *T. ferro-oxidans* for the recovery of copper by bacterial leaching. The methods used have included selection from the natural environment of strains with high copper tolerance (Groudeva et al. 1981) and the possession of enzymatic profiles likely to favor high rates of degradation of sulfide minerals (e.g., with high rhodanese levels) (Groudev and Genchev 1978a). Stable mutant strains have been produced by UV irradiation and chemical mutagenic agents, and strains with leaching capabilities superior to the parent strains have been selected out (Groudev et al. 1978).

The emergence of modern gene manipulation technology and its spectacular applications in biotechnology has led a number of authors to speculate on its applicability to the organisms involved in biohydrometallurgical processes and related areas (Chakrabarty 1978). The main impediment to such developments has been the rudimentary level of understanding of the genetics of geobiologically important microorganisms in general and of the *Thiobacilli* in particular. It has been noted by several authors (Holmes et al. 1983; Rawlings et al. 1983) that the development of a genetic system for an organism such as *T. ferro-oxidans* requires several major steps: (a) suitable characteristics that can serve as markers for the identification and selection of transformed recipient strains must be found; (b) vector DNA molecules that can replicate in *T. ferro-oxidans* and are able to accommodate inserted DNA must be identified and characterized; (c) DNA coding for the desired phenotypic characteristics must be isolated and purified and vector DNA molecules constructed; and (d) a means of transforming host cells with the cloned DNA must be developed.

Heavy metal tolerance is a well-marked characteristic of the thiobacilli, and since it is known to be encoded by plasmid DNA in a number of heterotrophic organisms, attention has been directed to the demonstration of plasmid occurrence in thiobacilli and to the nature of the information that the plasmids carry. Plasmids were first isolated from *T. ferro-oxidans* and *T. acidophilus* by Mao et al. (1980) and were shown to be distinctly different in the two strains. Plasmid patterns were determined in 15 strains of iron-oxidizing *T. ferro-oxidans* (Martin et al. 1981). In four of these strains plasmid DNA was not detected, and in each of the strains in which plasmids occurred, a different plasmid composition was manifest. A change of growth substrate from ferrous iron to tetrathionate did not affect the plasmid pattern in the *T. ferro-oxidans* strains or in a *T. acidophilus* strain, and from the circumstance and the absence of plasmid DNA in some strains, it was deduced that plasmids are unlikely to encode the iron-oxidizing function.

Biochemical and physiological differences between individual strains of *T. ferro-oxidans* and other thiobacilli have not been studied in as fine detail as some other industrially important organisms. There is still considerable uncertainty as to the purity of a number of the described strains of the thiobacilli, and there is accumulating evidence of contamination in a number of cases with heterotrophic species and with closely related strains of the same species. The mixotrophic behavior of some *T. ferro-oxidans* strains, for example, has been attributed to contamination with the obligatory heterotroph *Acidophilium cryptum* (Harrison 1981); recently, however, a true mixotrophic strain of *T. ferro-oxidans* has been described (Barros et al. 1984). It is clear that great efforts will be needed to ensure the purity of strains if the conclusions from researches on the genetic system are to be unequivocal. The vexed question of the purity of thiobacilli strains has been investigated by the systematic examination of DNA homologies by Harrison (1982) and Katayama-Fujimura et al. (1983b). DNA sequence homology studies between plasmids isolated from several acidophilic thiobacilli have recently been reported (Lobos et al. 1984).

Unequivocal evidence is still sought that metal tolerance in the thiobacilli is encoded on plasmid DNA. Strong circumstantial evidence is provided by the observation of Martin et al. (1981) that in one strain of *T. ferro-oxidans*, the disappearance of a ca. 18 kilobase plasmid was positively correlated with the disappearance of uranium resistance. Holmes et al. (1984) have succeeded in cloning a 6.7 Kb plasmid from a *T. ferro-oxidans* strain into *E. coli* plasmid pBR322 and in obtaining large quantities of the plasmid for structural and genetic analysis and for the construction of a cloning vehicle that will replicate and be expressed in *E. coli* and in *T. ferro-oxidans*. The same investigation demonstrated for the first time, by use of restriction analysis, that a purified strain of *T. ferro-oxidans* contains multiple plasmids. Other recent work has demonstrated the transfer of plasmid RP1, a broad-host-range, incompatibility group P1 plasmid specifying multiple drug resistance into the chemolithotrophic bacterium

T. neapolitanus and shows that this autotroph can accept, replicate, and express heterologous plasmid DNA from a heterotrophic bacterium (Kulpa et al. 1983). Other wide-host-range plasmids have been transferred from *E. coli* to *T. novellus* (Davidson and Summers 1983) but were not transmissable to other thiobacillus species. Of some interest is the development of a cosmid cloning system, which could be useful for the construction of genomic libraries and the introduction of clones into a broad range of bacterial species, including the thiobacilli (Frey et al. 1983). Further progress in the construction and characterization of a genomic library for *T. ferro-oxidans* has been described by Roskey and Kulpa (1984).

It is evident that considerable progress has been made in the last few years toward a greater understanding of the genetics of geobiologically important organisms, but there is a long way to go before improved strains for practical purposes are likely to be available.

It has been noted by C. L. Brierley (1982) that "from a biological viewpoint, dump leaching remains an essentially uncontrolled process." In this type of geobiological process, in which the heterogeneity of the substrate provides a diversity of microenvironments, the prospect of spectacularly improved performance by the use of genetically improved organisms seems rather limited. The introduction of selected, high-performing strains of *T. ferro-oxidans* as "seed" in commercial leaching heaps proved unsuccessful in improving performance, and the wild flora rapidly overwhelmed the introduced strains (Groudev and Genchev 1978b).

However, as noted earlier, one of the advantages of reactor leaching techniques is the possibility of usage of a considerable range of microbial systems, both natural and contrived, and the assurance of tight control and substantial exclusion of competing species. When further progress is made with the understanding of the genetics of geobiological organisms, it should be possible to construct organisms with optimal characteristics in respect of process requirements and possibly to achieve rates of mineral degradation or modification comparable to those obtained in high pressure/high temperature hydrometallurgical processes. To a limited extent, using the classical methods of selection and adaptation, superior organisms have already been used in this way in some processes.

Similar comments apply to the microbial catalysts for the regeneration of ferric sulfate leaching reagent, and in this case the needs for improved organisms have been mentioned earlier. When the mechanisms of metal accumulation by microorganisms are more fully understood, it should be possible to design organisms with enhanced capacity as bioadsorbers, and in this area also, genetic engineering should have considerable impact.

As in the case of dump leaching, considerably greater sophistication in the manipulation of the genes controlling specific chemical activities and those characteristics related to survival and competitiveness will need to be achieved before genetic engineering can have a significant impact on *in situ* leaching processes and upon microbially enhanced oil recovery. However, since the ore substrate in *in situ* leaching is frequently more concentrated and homogeneous than that in dump leaching, the establishment of populations of engineered organisms may be somewhat easier, and similar considerations could well apply to their use in oil reservoirs.

Conclusions

The development of the fundamental aspects of geomicrobiology and of associated technologies such as biohydrometallurgy has been extensive over the past four decades. Further progress promises to be stimulated by the emergence of genetic engineering and by the advances of biotechnology on a wide front. The applications of geomicrobiology hold promise for more complete and less energy-intensive modes of mining and metal recovery and for minimization of environmental pollution from these and other industrial activities.

Literature Cited

Agbim, N. N., and K. G. Doxtader. 1975. Microbial degradation of zinc silicates. *Soil Biol. Biochem.* 7:275–280; 15:259–264.

Arkesteyn, G. J. M. W. 1980. Contribution of microorganisms to the oxidation of pyrite. Doctoral thesis, University of Wageningen.

Atkins, A. S. 1978. Studies on the oxidation of sulphide minerals (pyrite) in the presence of bacteria. Pages 403–426 *in* L. E. Murr, A. E. Torma, and J. A. Brierley, eds., *Metallurgical Applications of Bacterial Leaching and Related Microbiological Phenomena.* Academic Press, New York.

Atkins, A. S., and F. D. Pooley. 1983. Comparison of bacterial reactors employed in the oxidation of sulphide concentration. Pages 111–125 *in* G. Rossi and A. E. Torma, eds., *Recent Progress in Biohydrometallurgy.* Associazione Mineraria Sarda-09016 Iglesias-Italy.

Babij, T., and B. J. Ralph. 1979. Assessment of metal recovery and metal pollution potentiality of sulphidic mine wastes. *Proc. GIAM V 1979,* 476–478.

Babij, T., R. B. Doble, and B. J. Ralph. 1980. A reactor system for mineral leaching investigations. Pages 563–572 *in* P. A. Trudinger, M. R. Walter, and B. J. Ralph, eds., *Biogeochemistry of Ancient and Modern Environments.* Australian Academy of Science, Canberra.

Babij, T., A. Goodman, A. M. Khalid, and B. J. Ralph. 1981. Microbial Ecology of Rum Jungle III. Leaching behaviour of sulphidic waste material under controlled conditions. *AAEC/E520.* Australian Atomic Energy Commission.

Balashova, V. V., I. Y. Vedenina, G. E. Markosyan, and G. A. Zavarzin. 1974. The autotrophic growth of *Leptospirillum ferro-oxidans. Mikrobiologiya* 43:581–585 (English translation pp. 491–494).

Baross, J. A., and J. W. Deming. 1983. Growth of "black smoker" bacteria at temperatures of at least 250° C.

Barros, M. E. C., D. E. Rawlings, and D. R. Woods. 1984. Mixotrophic growth of a *Thiobacillus ferro-oxidans* strain. *Appl. Environ. Microbiol.* 47:593–595.

Beckstead, L. W., P. B. Munoz, J. L. Sepulvada, J. A. Herbst, J. D. Miller, F. A. Olsen, and M. E. Wadsworth. 1976. Acid ferric sulfate leaching of attrition-ground chalcopyrite concentrates. Pages 611–632 *in* J. C. Yannapoulos and J. C. Agarval, eds., *Extractive Metallurgy and Electrowinning.* Port City Press, Baltimore.

Bhappu, R. B. 1982. Past, present and future of solution mining. Pages 1–9 *in* W. J. Schlitt, ed., *Interfacing Technologies in Solution Mining.* Proc. Second SME-SPE Internat. Solut. Min. Symp., Nov. 18–20, 1981, Denver, CO.

Bohlool, B. B. 1975. Occurrence of *Sulfolobus acidocaldarius,* an extremely thermophilic bacterium in New Zealand hot springs. *Arch. Mikrobiol.* 106:171–174.

Bosecker, K., A. E. Torma, and J. A. Brierley. 1979. Microbiological leaching of a chalcopyrite concentrate and the influence of hydrostatic pressure on the activity of *Thiobacillus ferro-oxidans. Eur. J. Appl. Microbiol. Biotechnol.* 7:85–90.

Brierley, C. L. 1974. Molybdenite leaching: Use of a high temperature microbe. *J. of Less-common Metals* 36:237–247.

______. 1977. Thermophilic microorganisms in extraction of metals from ores. *Dev. Ind. Microbiol.* 18:273–284.

______. 1978. Bacterial leaching. *CRC Critical Reviews in Microbiology* 207–262.

______. 1982. Microbiological mining. *Sci. Am.* 247:42–51.

Brierley, C. L., and J. A. Brierley. 1973. A chemoautotrophic and thermophilic microorganism isolated from an acid hot spring. *Can. J. Microbiol.* 19:183–188.

______. 1982. Anaerobic reduction of molybdenum by *Sulfolobus* species. *Zentralbl. Bakteriol. Hyg. C. Orig.* 3:289–294.

Brierley, C. L., and L. E. Murr. 1973. Leaching: Use of a thermophilic and chemoautotrophic microbe. *Science* 179:488–490.

Brierley, C. L., J. A. Brierley, P. R. Norris, and D. P. Kelly. 1980. Metal-tolerant microorganisms of hot, acid environments. Pages 39–51 *in* G. W. Gould and J. E. Corry, eds., *Microbial Growth and Survival in Extremes of Environment.* Soc. Appl. Bacteriol. Tech. Ser. No. 15, Academic Press, London.

Brierley, J. A. 1978. Thermophilic iron-oxidizing bacteria found in copper leaching dumps. *Appl. Environ. Microbiol.* 36:523–525.

______, 1980. Facultative thermophilic *Thiobacillus*-like bacteria in metal leaching. Pages 445–450 *in* P. A. Trudinger, M. R. Walter, and B. J. Ralph, eds., *Biogeochemistry of Ancient and Modern Environments,* Australian Academy of Science, Canberra.

Brierley, J. A., and C. L. Brierley. 1978. Microbial leaching of copper at ambient and elevated temperatures. Pages 477–490 *in* L. E. Murr, A. E. Torma, and J. A. Brierley, eds., *Metallurgical Applications of Bacterial Leaching and Related Microbiological Phenomena.* Academic Press, New York.

Brock, T. D. 1977. Ferric iron reduction by sulphur- and iron-oxidizing bacteria. Page 47 *in* W. Schwartz, ed., *GBF Monograph Series,* No. 4 (August, 1977), *Conference Bacterial Leaching 1977,* Verlag Chemie, Weinheim.

______. 1981. Extreme thermophiles of the genera *Thermus* and *Sulfolobus.* Pages 978–984 *in* M. P. Starr, H. Stolp, H. G. Truper, A. Balows, and H. G. Schlegler, eds., *The Prokaryotes: A Handbook of Habits, Isolation and Identification of Bacteria.* Springer-Verlag, Berlin.

Brock, T. D., and M. R. Edwards. 1970. Fine structure of *Thermus aquatious,* an extreme thermophile. *J. Bacteriol.* 104:509–517.

Brock, T. D., and H. Freeze. 1969. *Thermus aquatious* gen. n. and sp. n., a non-sporulating extreme thermophile. *J. Bacteriol.* 98:289–297.

Brock, T. D., and J. Gustafson. 1976. Ferric iron reduction by sulphur- and iron-oxidizing bacteria. *Appl. Environ. Microbiol.* 32:567–571.

Brock, T. D., K. M. Brock, R. T. Belly, and L. R. Weiss. 1972. *Sulfolobus:* A new genus of sulphur-oxidizing bacteria living at low pH and high temperature. *Arch. Microbiol.* 84:54–68.

Bruynesteyn, A., and D. W. Duncan. 1971. Microbiological leaching of sulphide concentrates *Can. Metall. Quart.* 10:57–63.

Bruynesteyn, A., A. Vizsolyi, and R. Vos. 1980a. Effect of low pH on the rate of ferrous iron oxidation by *Thiobacillus ferro-oxidans.* Pages 155–160 *in* B. Czegledi, ed., *Proc. Int. Conf. on Use of Micro-organisms in Hydrometallurgy.* Pecs, 4–6 Dec., 1980, Hungarian Academy of Sciences.

______. 1980b. The ability of *Thiobacillus ferro-oxidans* to withstand changes in pH. *Ibid.,* pp. 151–154.

Bruynesteyn, A., R. W. Lawrence, A. Vizsolyi, and R. Hackl. 1983. An elemental sulphur producing hydrometallurgical process for treating sulphide concentrates. Pages 151–168 *in* G. Rossi and A. E. Torma, eds., *Recent Progress in Biohydrometallurgy.* Associazione Mineraria Sarda-09016 Iglesias-Italy.

Bryant, R. D., K. M. McGroarty, J. W. Costerton, and E. J. Laishly. 1983. Isolation and characterization of a new acidophilic *Thiobacillus* species *Thiobacillus albertis,* new species. *Can. J. Microbiol.* 29:1159–1170.

Bubela, B. 1970. Chemical and microbiological changes in *Bacillus stearothermophilus* induced by copper. *Chem.-Biol. Interactions* 2:107–116.

______. 1978. Role of geomicrobiology in enhanced recovery of oil: Status quo. *The APEA J.* 1978:161–166.

______. 1982. Combined effects of temperature and other environmental stresses on microbiologically enhanced oil recovery. *Proc. Int. Conf. Microbiol. Enhanced Oil Recov.,* pp. 118–123, 16–21 May, 1982, Univ. Oklahoma and U.S. Dept. Energy.

______. 1983a. Microbial enhanced oil recovery—the problems underground. *Int. Conf. on Applic. and Implic. Biotechnol.,* pp 423–435, Biotech 83, May 1983. London, Online Publications, Norwood, U.K.

______. 1983b. Physical simulation of microbiologically enhanced oil recovery. Pages 1–7 *in* J. E. Zajic et al., eds., *Microbial Enhanced Oil Recovery.* Penn Well Publishing Co., Tulsa, OK.

______.1984. Geobiology and microbiologically enhanced oil recovery. Chapter in *Biotechnology and the Oil Industry.* In press.

Bubela, B., and B. A. McKay. 1983. Assessment of oil reservoirs for microbiological enhanced oil recovery. Report, Bass-Becking Geobiological Laboratory and Bureau of Mineral Resources, Canberra.

Bubela, B., and T. C. Powell. 1973. Effect of copper on the growth characteristics of *Bacillus stearothermophilus, Zentralbl. Bakteriol.* 128:457–466.

Bull, A. T., G. Holt, and M. D. Lilly. 1982. Biotechnology—International trends and perspectives. *OECD Report.*

Carta, M., M. Ghiani, and G. Rossi. 1980a. Beneficiation of a complex sulphide ore by an integrated process of flotation and bioleaching. *Proc. Complex Sulphide Ores Conf.* pp. 178–185, Rome, 5–8 Oct., 1980. The Institution of Mining and Metallurgy, London.

______. 1980b. Complex sulphides concentrates bioleaching performance as related to feed composition. Pages 203–209 *in* B. Czegledi, ed., *Proc. Int. Conf.*

on Use of Micro-organisms in Hydrometallurgy. Pecs, 4–6 Dec., 1980, Hungarian Academy of Sciences.

Chakrabarty, A. M. 1978. Genetic mechanism in metal-microbe interactions. Pages 137–149 *in* L. E. Murr, A. E. Torma, and J. A. Brierley, eds., *Metallurgical Applications of Bacterial Leaching and Related Microbiological Phenomena*. Academic Press, New York.

Chang, Y. C., and A. S. Myerson. 1982. Growth models of the continuous bacterial leaching of iron pyrite by *Thiobacillus ferro-oxidans*. *Biotechnol. Bioeng*. 24:889–902.

Collins, V. G. 1969. Pages 1–52 *in* J. R. Norris and D. W. Ribbons, eds., *Methods in Microbiology* Vol. 38. Academic Press, London.

Collins, M. D., and T. A. Langworthy. 1983. Respiratory quinone composition of some acidophilic bacteria. *Syst. Appl. Microbiol*. 4:295–304.

Colmer, A. R., and M. E. Hinkle. 1947. The role of microorganisms in acid mine drainage. *Science* 106:253–256.

C.S.I.R.O. 1981. Biotechnology Research and Development. Commonwealth Scientific and Research Organization, Canberra, Australia.

Czegledi, B. ed. 1980. *Proceedings of International Conference on Use of Micro-organisms in Hydrometallurgy*. Pecs, 4–6 Dec., 1980, Hungarian Academy of Sciences.

Darland, G., T. D. Brock, W. Samsonoff, and S. F. Conti. 1970. A thermophilic, acidophilic mycoplasma isolated from a coal refuse pile. *Science* 170:1416–1418.

Davidson, M. S., and A. D. Summers. 1983. Wide-host-range plasmids function in the genus *Thiobacillus*. *Appl. Environ. Microbiol*. 46:565–572.

Davidson, M. S., A. E. Torma, J. A. Brierley, and C. L. Brierley. 1981. Effects of elevated pressures on iron- and sulphur-oxidizing bacteria. *Biotechnol. Bioeng. Symp. No. 11* 603–618.

Deatherage, J. F., K. A. Taylor, and L. A. Amos. 1983. Three dimensional arrangement of the cell-wall protein of *Sulfolobus acidocaldarius J. Mol. Biol*. 167:823–848.

De Rosa, M., A. Gambacorta, and J. D. Bu'Lock. 1975. Extremely thermophilic acidophilic bacteria convergent with *Sulfolobus acidocaldarius*. *J. Gen. Microbiol*. 86:156–164.

De Rosa, M., A. Gambacorta, G. Millonig, and J. D. Bu'Lock. 1974. Convergent characters of extremely thermophilic acidophilic bacteria. *Experientia* 30:866–868.

De Rosa, M., A. Gambacorta, B. Nicolaus, B. Chappe, and P. Albrecht. 1983. Isoprenoid ethers backbone of complex lipids of the archaebacterium *Sulfolobus solfatarious*. *Biochim. Biophys. Acta* 753:249–256.

DiSpirito, A. A., and O. H. Tuovinen. 1982a. Uranous ion oxidation and carbon dioxide fixation by *Thiobacillus ferro-oxidans*. *Arch. Mikrobiol*. 133:28–32.

______. 1982b. Kinetics of uranous ion and ferrous iron oxidation by *Thiobacillus ferro-oxidans*. *Arch. Mikrobiol*. 133:33–37.

Dugan, P. R., and W. A. Apel. 1978. Microbiological desulphurization of coal. Pages 223–250 *in* L. E. Murr, A. E. Torma, and J. A. Brierley, eds., *Metallurgical Applications of Bacterial Leaching and Related Microbiological Phenomena*. Academic Press, New York.

Ebner, H. G. 1980. Bacterial leaching in an airlift reactor. Pages 211–217 *in* B. Czegledi, ed., *Proc. Int. Conf. on Use of Micro-organisms in Hydrometallurgy*. Pecs. 4–6 Dec., 1980, Hungarian Academy of Sciences.

Ehrlich, H. L. 1980. Bacterial leaching of manganese ores. Pages 609–614 *in* P. A. Trudinger, M. R. Walter, and B. J. Ralph, eds., *Biochemistry of Ancient and Modern Environments*. Australian Academy of Science, Canberra.

Fliermans, C. G., and T. D. Brock. 1972. Ecology of sulfur-oxidizing bacteria in hot acid soils. *J. Bacteriol.* 111:343–350.

Frey, J., M. Bagdasarian, D. Feiss, F. Christopher, H. Franklin, and J. Deshusses. 1983. Stable cosmid vectors that enable the introduction of cloned fragments into a wide range of gram-negative bacteria. *Gene* 24:299–308.

Gerlach, J. K., E. D. Gock, and S. K. Ghosh. 1973. Activation and leaching of chalcopyrite concentrates with dilute sulphuric acid. Pages 403–416 *in* D. J. I. Evans and R. S. Shoemaker, eds., *Extractive Metallurgy of Copper*. Hydrometallurgy and Electrowinning. American Institute of Mining, Metallurgical and Petroleum Engineers, Inc., New York.

Goodman, A., and B. J. Ralph. 1978. Suppression and control of microbial leaching, Progress Report No. 6. The isolation and characterization of organisms representative of some ecologically important bacterial species in the Rum Jungle area. *Australian Atomic Energy Commission Research Contract* 74/F/40.

Goodman, A. E., T. Babij, and A. I. M. Ritchie. 1983. Leaching of a sulphide ore by *Thiobacillus ferro-oxidans* under anaerobic conditions. Pages 361–376 *in* G. Rossi and A. E. Torma, eds., *Recent Progress in Biohydrometallurgy*. Associazione Mineraria Sarda-09016 Iglesias-Italy.

Goodman, A. E., A. M. Khalid, and B. J. Ralph. 1980. A scanning electron microscopy study of the colony morphology of *Thiobacillus ferro-oxidans*. Pages 459–467 *in* P. A. Trudinger, M. R. Walter, and B. J. Ralph, eds., *Biogeochemistry of Ancient and Modern Environments*. Australian Academy of Science, Canberra.

______. 1981a. Microbial ecology of Rum Jungle. Part I. Environmental study of sulphidic overburden dumps, experimental heap-leach piles and tailings dam area. Australian Atomic Energy Commission. *AAEC/E531*.

______. 1981b. Microbial ecology of Rum Jungle. Part II. Environmental study of two flooded opencuts and smaller associated water bodies. Australian Atomic Energy Commission. *AAEC/E527*.

Gormely, L. S., D. W. Duncan, R. M. R. Branion, and K. L. Pinder. 1975. Continuous culture of *Thiobacillus ferro-oxidans* on a zinc sulphide concentrate. *Biotechnol. Bioeng.* 17:31–49.

Gorog, J., G. Pap, J. Hollo, and T. Lakatos. 1980. An efficient method for the production of leaching medium. Pages 601–607 *in* P. A. Trudinger, M. R. Walter, and B. J. Ralph, eds., *Biogeochemistry of Ancient and Modern Environments*. Australian Academy of Science, Canberra.

Green, G. R., D. G. Searcy, and R. J. Delange. 1983. Histone-like protein in the archaebacterium *Sulfolobus acidocaldarius*. *Biochim. Biophys. Acta* 741:251–257.

Groudev, S. N., and F. Genchev. 1978a. Mechanisms of bacterial oxidation of chalcopyrite. *Mikrobiologija* 139–152.

______. 1978b. Bioleaching of bauxites by wild and laboratory-bred microbial strains. *Proc. 4th Int. Congr. for the Study of Bauxites, Alumina and Aluminum.* Athens, October 1978, Vol. 1, pp. 271–278.

Groudev, S. N., F. N. Genchev, and S. S. Gaidarjiev. 1978. Observations on the microflora in an industrial copper dump leaching operation. Pages 253–274 *in* L. E. Murr, A. E. Torma, and J. E. Brierley, eds., *Metallurgical Applications of Bacterial Leaching and Related Microbiological Phenomena.* Academic Press, New York.

Groudeva, V., and S. Groudev. 1980. Strain improvement of *Thiobacillus ferro-oxidans* for the purpose of leaching of copper sulphide minerals. *Proc. 12th Meet. of Miners and Metallurgists.* Technical Faculty, Bor (Yugoslavia) and Institute for Copper, Bor; Bor, October 1980, Vol. II, pp. 354–364.

Groudeva, V. I., S. N. Groudev, and K. I. Markov. 1981. Selection of *Thiobacillus ferro-oxidans* mutants tolerant to high concentrations of copper ions. *C. R. Acad. Bulg. Sci.* 34 (No. 3):375–378.

Guay, R., M. Silver, and A. E. Torma. 1976. Base composition of DNA isolated from *Thiobacillus ferro-oxidans* grown on different substrates. *Rev. Can. Biol.* 35:61–67.

Hanert, H. H. 1981a. The genus *Gallionella.* Pages 509-515 *in* M. P. Starr, H. Stolp, H. G. Truper, A. Balows, and H. G. Schlegel, eds., *The Prokaryotes: A Handbook on Habitats, Isolation and Identification of Bacteria.* Springer-Verlag, Berlin.

______. 1981b. The genus *Siderocapsa* and other iron-oxidizing or manganese-oxidizing eubacteria. Pages 1949–2060 *in* M. P. Starr et al., eds., *The Prokaryotes: A Handbook on Habitats, Isolation and Identification of Bacteria.* Springer-Verlag, Berlin.

Harris, B., A. M. Khalid, B. J. Ralph, and R. Winby. 1983. Biohydrometallurgical beneficiation of process tailings. Pages 595–616 *in* G. Rossi and A. E. Torma, eds., *Recent Progress in Biohydrometallurgy.* Associazione Mineraria Sarda-09016 Iglesias-Italy.

Harrison, A. P., Jr. 1981. *Acidophilium cryptum,* gen. nov., spec. nov., heterotrophic bacterium from acidic mineral environments. *Int. J. Syst. Bacteriol.* 31:327–332.

______. 1982. Genomic and physiological diversity amongst strains of *Thiobacillus ferro-oxidans* and genomic comparison with *Thiobacillus thio-oxidans. Arch. Mikrobiol.* 131:68–76.

Holden, P. J., and J. C. Madgwick. 1983. Mixed culture bacterial leaching of manganese dioxide. *Proc. Australia Inst. Min. Metall.* 286:61–63.

Holmes, D. S., J. H. Lobos, and L. H. Bopp. 1983. Cloning of *Thiobacillus ferro-oxidans* plasmids in *Escherichia coli.* Pages 541–554 *in* G. Rossi and A. E. Torma, eds., *Recent Progress in Biohydrometallurgy.* Associazione Mineraria Sarda-09016 Iglesias-Italy.

Holmes, D. S., J. H. Lobos, L. H. Bopp, and G. C. Welch. 1984. Cloning of a *Thiobacillus ferro-oxidans* plasmid in *Escherichia coli. J. Bacteriol.* 157:324–326.

Huber, T. F., C. H. Kos, P. Bos., N. W. F. Kossen, and J. G. Kuenen. 1983. Modelling, scale-up and design of bioreactor for microbial desulphurisation of coal. Pages 279–289 *in* G. Rossi and A. E. Torma, eds., *Recent Progress in Biohydrometallurgy.* Associazione Mineraria Sarda-09016 Iglesias-Italy.

Karavaiko, G. I., Y. U. M. Miller, O. A. Kapustin, and T. A. Pivovarova. 1980. Fractionation of stable isotopes of sulphur during its oxidation by *Thiobacillus ferro-oxidans. Microbiology* (Engl. Transl. Mikrobiologiya 1981) 46:667–671.

Kargi, F. 1982. Microbiological coal desulphurization. *Enzyme Microb. Technol.* 4:13–19.

Kargi, F., and J. M. Robinson. 1982. Removal of sulfur compounds from coal by the thermophilic organism *Sulfolobus acidocaldarius. Appl. Environ. Microbiol.* 44:878–883.

Katayama-Fujimura, Y., I. Kawashima, N. Tsuzaki, and H. Kuraishi. 1983a. Re identification of *Thiobacillus perometabolis* and *Thiobacillus* sp. strain A-2 with reference to a new species *Thiobacillus rapidicrescens. Int. J. Syst. Bacteriol.* 33:532–538.

Katayama-Fujimura, Y., Y. Enokizono, T. Kaneko, and H. Kuraishi. 1983b. DNA homologies among species of the genus *Thiobacillus. J. Gen. Appl. Microbiol.* 29:287–296.

Kelly, D. P., P. R. Norris, and C. L. Brierley. 1979. Microbiological methods for the extraction and recovery of metals. Pages 263–308 *in* A. T. Bull, D. C. Ellwood, and C. Ratledge, eds., *Microbial Technology*. Soc. Gen. Microbiol. Symp. 29, Society for General Microbiology Ltd., Great Britain.

Kiel, H. 1977. Laugung von kupferkarbonat-und kupfersilikatersen mit heterotrophen mikroorganismen. Pages 261–270 *in* W. Schwartz, ed., *GBF Monograph Series, No. 4* (August 1977), Verlag Chemie, Weinheim.

Kiese, S., H. G. Ebner, and U. Onken. 1980. A simple laboratory airlift fermentor. *Biotechnol. Lett.* 2:345–351.

Kim, D. K. H., N. G. Klimenko, G. I. Karavaiko, and N. D. Klyueva. 1981. Effect of *Thiobacillus ferro-oxidans* on flotation properties of sulphide minerals. *Appl. Biochem. Microbiol.* (Engl. trans. Prikl. Biokhim. Mikrobiol.) 17:210–213.

Kuenen, J. G., and O. H. Tuovinen. 1981. The general *Thiobacillus* and *Thiomicrospira.* Pages 1023–1036 *in* M. P. Starr, H. Stolp, H. G. Truper, A. Balows, and H. G. Schlegler, eds., *The Prokaryotes: A Handbook on Habitats, Isolation and Identification of Bacteria.* Springer-Verlag, Berlin.

Kulpa, C. F., M. T. Roskey, and M. T. Travis. 1983. Transfer of plasmid RPl into chemolithotrophic *Thiobacillus neapolitanus. J. Bacteriol.* 156:434–436.

Kupeyeva, R. D., P. M. Solozhenkin, and G. A. Khan. 1977. Use of bacteria for xanthogenate desorption of collective concentrate from the mineral surface. *Ore Flotation.* Irkutsk, N5, pp. 158–164.

Le Roux, N. W., and D. S. Wakerley. 1980. The leaching of sulphide ores by a thermophilic bacterium. Pages 451–457 *in* P. A. Trudinger, M. R. Walter, and B. J. Ralph, eds., *Biogeochemistry of Ancient and Modern Environments.* Australian Academy of Science, Canberra.

Le Roux, N. W., D. S. Wakerley, and S. D. Hunt. 1977. Thermophilic *Thiobacillus*-type bacteria from Icelandic thermal areas. *J. Gen. Microbiol.* 100:197–201.

Le Roux, N. W., D. S. Wakerley, and V. S. Perry. 1978. Leaching of minerals using bacteria other than *Thiobacilli.* Pages 167–191 *in* L. E. Murr, A. E. Torma, and J. A. Brierley, eds., *Metallurgical Applications of Bacterial Leaching and Related Microbiological Phenomena.* Academic Press, New York.

Livesey-Goldblatt, E., T. H. Tunley, and I. F. Nagy. 1977. Pilot-plant bacterial film oxidation (Bacfox Process) of recycled acidified uranium plant ferrous sulphate leach solution. Pages 175–190 *in* W. Schwartz, ed., *GBF Monograph Series, No. 4* (August 1977), *Conference Bacterial Leaching 1977.* Verlag Chemie, Weinheim.

Livesey-Goldblatt, E., P. Norman, and D. R. Livesey-Goldblatt. 1983. Gold recovery from arsenopyrite/pyrite ore by bacterial leaching and cyanidation. Pages 627–641 *in* G. Rossi and A. E. Torma, eds., *Recent Progress in Biohydrometallurgy.* Associazione Mineraria Sarda-09016 Iglesias-Italy.

Lobos, J. H., H. Grimmond, G. C. Welch, M. Newell, and D. S. Holmes. 1984. DNA sequence homology studies between plasmids isolated from several acidophilic thiobacilli. *Abstract 024, 84th Annu. Meet. Am. Soc. Microbiol.* 4–9 March 1984, St. Louis, MO.

Lu, W. P., and D. P. Kelly. 1983. Purification and some properties of two principal enzymes of the thiosulphate-oxidising multi-enzyme system from *Thiobacillus* A-2. *J. Gen. Microbiol.* 129:3549–3564.

Lundgren, D. G., and E. E. Malouf. 1983. Microbial extraction and concentration of metals. *Adv. Biotechnol. Processes* 1:223–249.

Lyalikova, N. N. 1972. Oxidation of trivalent antimony to higher oxides as an energy source for the development of a new autotrophic organism *Stibiobacter* gen. nov. *Doklady Akademii Nauk SSSR* 205.

______. 1974. *Stibiobacter senarmontii* a new antimony oxidising microorganism. *Mikrobiologiya* 43:941–943.

Lyalikova, N. N., L. L. Lyubavina, and P. M. Solozhenkin. 1977. Application of sulphate-reducing bacteria for the enrichment of ores. Pages 93–100 *in* W. Schwartz, ed., *GBF Monograph Series, No. 4* (August 1977), Conference Bacterial Leaching 1977. Verlag Chemie, Weinheim.

McElroy, R. O., and A. Bruynesteyn. 1978. Continuous biological leaching of chalcopyrite concentrates: Demonstration and economic analysis. Pages 441–462 *in* L. E. Murr, A. E. Torma, and J. A. Brierley, eds., *Metallurgical Applications of Bacterial Leaching and Related Microbiological Phenomena.* Academic Press, New York.

Mackintosh, M. E. 1978. Nitrogen fixation by *Thiobacillus ferro-oxidans. J. Gen. Microbiol.* 105:215–218.

Manning, H. L. 1975. New medium for isolating iron-oxidising and heterotrophic acidophilic bacteria from acid mine drainage. *Appl. Microbiol.* 30:1010–1016.

Mao, M. W. H., P. R. Dugan, P. A. W. Martin, and O. H. Tuovinen. 1980. Plasmid DNA in chemo-organotrophic *Thiobacillus ferro-oxidans* and *T. acidophilus. FEMS Microbiol. Lett.* 8:121–125.

Markosyan, G. E. 1973. A new mixotrophic sulphur bacterium developing in acid media, *Thiobacillus organoparus,* sp.n. *Doklady Akademii Nauk SSSR.* 211:1205–1208 (Engl. transl. pp., 318–320).

Marsh, R. M., and P. R. Norris. 1983. The isolation of some thermophilic, autotrophic, iron- and sulphur-oxidizing bacteria. *FEMS Microbiol. Lett.* 17:311–315.

Marsh, R. M., P. R. Norris, and N. W. Le Roux. 1983. Growth and mineral oxidation studies with *Sulfolobus.* Pages 71–81 *in* G. Rossi and A. E. Torma, eds., *Recent Progress in Biohydrometallurgy.* Associazione Mineraria Sarda-09016 Iglesias-Italy.

Marshall, K. C. 1979. Biogeochemistry of manganese minerals. Pages 253–292 *in* P. A. Trudinger and D. J. Swaine, eds., *Biogeochemical Cycling of Mineral-forming Elements.* Elsevier, Amsterdam.

Martin, P. A., P. Dugan, and O. H. Tuovinen. 1981. Plasmid DNA in acidophilic, chemolithotrophic *Thiobacilli. Can. J. Microbiol.* 27:850–853.

Mercz, T. I., and J. C. Madgwick. 1982. Enhancement of bacterial manganese leaching by microbial growth products. *Proc. Australas. Inst. Min. Metall.* 283:43–46.

Mishra, A. K., P. Roy, S. S. Mahapatra, and D. Chandra. 1983. The role of *Thiobacillus ferro-oxidans* under microbes in biohydrometallurgy with special reference to desulphurization of coal. Pages 491–510 *in* G. Rossi and A. E. Torma, eds., *Recent Progress in Biohydrometallurgy.* Associazione Mineraria Sarda-09016 Iglesias-Italy.

Mizoguchi, T., and T. Okabe. 1980. Isolation of iron- and sulphur-oxidizing bacteria from mine water in Japan and some investigations on the leaching of sulphide ores. Pages 505–513 *in* P. A. Trudinger, M. R. Walter, and B. J. Ralph, eds., *Biogeochemistry of Ancient and Modern Environments.* Australian Academy of Science, Canberra.

Mizoguchi, T., T. Sato, and T. Okabe. 1976. New sulphur-oxidizing bacteria capable of growing heterotrophically, *Thiobacillus rubellus* nov. sp. and *Thiobacillus delicatus* nov. sp. *J. Ferment. Technol.* 54:181.

Moss, F. J., and J. E. Andersen. 1968. The effects of environment on bacterial leaching rates. *Proc. Australas. Inst. Min. Metall.* 225:15–25.

Mosser, J. L., B. B. Bohlool, and T. D. Brock. 1974. Growth rates of *Sulfolobus acidocaldarius* in nature. *J. Bacteriol.* 118:1075–1082.

Mosser, J. L., A. G. Mosser, and T. D. Brock. 1973. Bacterial origin of sulphuric acid in geothermal habitats. *Science* 179:1323–1324.

Murr, L. E. 1980. Theory and practice of copper sulphide leaching in dumps and *in situ. Minerals Sci. Eng.* 12:121–189.

Murr, L. E., A. E. Torma, and J. A. Brierley, eds. 1978. *Metallurgical Applications of Bacterial Leaching and Related Microbiological Phenomena.* Academic Press, New York.

Murr, L. E., and A. P. Mehta. 1982. Coal desulphurization by leaching involving acidophilic and thermophilic microorganisms. *Biotechnol. Bioeng.* 24:743–748.

Norris, P. R. 1983. Iron and mineral oxidation with *Leptospirillum*-like bacteria. Pages 83–96 *in* G. Rossi and A. E. Torma, eds., *Recent Progress in Biohydrometallurgy.* Associazione Mineraria Sarda-09016 Iglesias-Italy.

Norris, P. R., and D. P. Kelly. 1978. Dissolution of pyrite (FeS_2) by pure and mixed cultures of some acidophilic bacteria. *FEMS Microbiol. Lett.* 4:143–146.

______. 1982. The use of mixed microbial cultures in metal recovery. Pages 443–474 *in* A. T. Bull and J. H. Stater, eds., *Microbial Interactions and Communities.* Academic Press, London.

Olsen, G. J., F. D. Porter, J. Rubinstein, and S. Silver. 1982. Mercuric reductase enzyme from a mercury volatilising strain of *Thiobacillus ferro-oxidans*. *J. Bacteriol.* 15:1230–1236.

Pinches, A. 1975. Bacterial leaching of an arsenic-bearing sulphide concentrate. Pages 28–35 *in* A. R. Burkin, ed., *Leaching and Reduction in Hydrometallurgy*. The Institution of Mining and Metallurgy, London.

Pooley, F. D., and A. S. Atkins. 1983. Desulphurization of coal using bacteria by both dump and process plant techniques. Pages 511–526 *in* G. Rossi and A. E. Torma, eds., *Recent Progress in Biohydrometallurgy*. Associazione Mineraria Sarda-09016 Iglesias-Italy.

Ralph, B. J. 1977. Photosynthetic approaches to ore leaching. *Proc. Symp. on Practical Applic. of Photosynthesis*. 26 August 1977, University of New South Wales and International Solar Energy Society (ANZ Section).

______. 1979a. Biotechnology—The state of the art. *Chem. Aust.* 46:45–49.

______. 1979b. Oxidative reactions in the sulfur cycle. Pages 369–400 *in* P. A. Trudinger and D. J. Swaine, eds., *Biogeochemical Cycling of Mineral-forming Elements*. Elsevier, Amsterdam.

______. 1984. Biotechnology applied to raw mineral processing. In press *in* M. Moo-Young, ed., *Comprehensive Biotechnology and Bioengineering*. Pergamon Press, Oxford.

Rawlings, D. E., C. Gawith, A. Petersen, and D. R. Woods. 1983. Characteristics of plasmids and genetic markers in *Thiobacillus ferro-oxidans*. Pages 555–570 *in* G. Rossi and A. E. Torma, eds., *Recent Progress in Biohydrometallurgy*. Associazione Mineraria Sarda-09016 Iglesias-Italy.

Robertson, L. A., and J. G. Kuenen. 1983. *Thiosphaera pantotropha,* new genus, new species. A facultatively anaerobic facultatively autotrophic sulphur bacterium. *J. Gen Microbiol.* 129:2847–2856.

Roskey, M. T., and C. F. Kulpa. 1984. Construction and characterization of a genomic library of *Thiobacillus ferro-oxidans. Abstract H59, 84th Annu. Meet. Am. Soc. Microbiol.,* 4–9 March 1984, St. Louis, MO.

Rossi, G. 1978. Potassium recovery through leucite bioleaching: Possibilities and limitations. Pages 297–419 *in* L. E. Murr, A. E. Torma, and J. A. Brierley, eds., *Metallurgical Applications of Bacterial Leaching and Related Microbiological Phenomena*. Academic Press, New York.

Rossi, G., and A. E. Torma, eds. 1983. *Recent Progress in Biohydrometallurgy*. Associazione Mineraria Sarda-09016 Iglesias-Italy.

Rossi, G., A. E. Torma, and P. Trois. 1983. Bacteria-mediated copper recovery from a cupriferous pyrrhotite ore: Chalcopyrite/pyrrhotite interactions. Pages 185–200 *in* G. Rossi and A. E. Torma, eds., *Recent Progress in Biohydrometallurgy*. Associazione Mineraria Sarda-09016 Iglesias-Italy.

Rudolfs, W., and A. Helbronner. 1922. Oxidation of zinc sulphide by micro-organisms. *Soil Sci.* 14:459–464.

Sanmugasunderam, V. 1981. *The continuous microbiological leaching of zinc sulphide concentrate with recycle*. Ph.D. Thesis, University of British Columbia.

Schlitt, W., and J. D. Hiskay, eds. 1982. *Interfacing Technologies in Solution Mining*. Proc. 2nd SME-SPE Int. Solut. Min. Symp., Denver, CO, Nov. 18–20, 1981.

Schlitt, W., ed. 1977. *Bacterial Leaching Conference 1977. GBF Monograph Series, No. 4 (August 1977).* Verlag Chemie, Weinheim.

Shepard, C., M. B. Goldhaber, and D. M. Updegraf. 1983. Catalysis of pyrite oxidation by *Thiobacillus thioparus* at near neutral pH. *Abstract 038, 83rd Annu. Meet. Am. Soc. Microbiol.,* 6–11 March 1983, New Orleans.

Silver, M., and A. E. Torma. 1974. Oxidation of metal sulphides by *Thiobacillus ferro-oxidans* grown on different substrates. *Can. J. Microbiol.* 20:141–147.

Silverman, M. P. 1979. Biological and organic chemical decomposition of silicates. Pages 445–465 *in* P. A. Trudinger and D. J. Swaine, eds., *Biogeochemical Cycling of Mineral-forming Elements.* Elsevier, Amsterdam.

Silverman, M. P., and H. L. Ehrlich. 1964. Microbial formation and degradation of minerals. *Adv. Appl. Microbiol.* 6:153–206.

Silverman, M. P., and D. G. Lundgren. 1959. Studies of the chemoautotrophic bacterium *Ferrobacillus ferro-oxidans.* I. An improved medium and a harvesting procedure for securing high cell yields. *J. Bacteriol.* 77: 642–647.

Solozhenkin, P. M. 1980. Brief report on bacterial leaching of nonferrous metals. Pages 11–25 *in* B. Czegledi, ed., *Proc. Int. Conf. on Use of Micro-organisms in Hydrometallurgy.* Pecs, 4–6 December 1980, Hungarian Academy of Sciences.

Solozhenkin, P. M., and L. L. Lyubavina. 1980. The bacterial leaching of antimony- and bismuth-bearing ores and the utilization of sewage waters. Pages 615–621 *in* P. A. Trudinger, M. R. Walter, and B. J. Ralph, eds., *Biogeochemistry of Ancient and Modern Environments.* Australian Academy of Sciences, Canberra.

Solozhenkin, P. M., L. L. Lyubavina, L. F. Samopkhvalova, and V. S. Pupkov. 1979. Studies of flotation properties of sulphate-reducing bacteria at ore treatment. *Izv. Vuzov, Tsventnaya Metall.* 3:13–21.

Southerland, W. M., and F. Toghrol. 1983. Sulfite oxidase activity in *Thiobacillus novellus. J. Bacteriol.* 156:941–944.

Starr, M. P., H. Stolp, H. G. Truper, A. Balows, and H. G. Schlegler, eds. 1981. *The Prokaryotes: A Handbook on Habitats, Isolation and Identification of Bacteria.* Springer-Verlag, Berlin.

Taylor, J. H., and P. F. Whelan. 1942. The leaching of cupreous pyrites and the precipitation of copper at Rio Tinto, Spain. *Trans. Inst. Min. Metall.* 52:35–71,

Toghrol, F., and W. M. Southerland. 1983a. Purification of *Thiobacillus novellus* sulphite oxidase—evidence for the presence of heme and molybdenum. *J. Biol. Chem.* 258:6762–6765.

______. 1983b. *Thiobacillus novellus* sulfite oxidase sub-unit structure and enzymic cofactors. *Abstract 811, 74th Annu. Meet. Am. Soc. Biol. Chem.,* San Francisco, 5–9 June 1983.

Tomizuka, N., and M. Yagisawa. 1978. Optimum conditions for leaching of uranium and oxidation of lead sulphide with *Thiobacillus ferro-oxidans* and recovery of metals from bacterial leaching solution with sulphate-reducing bacteria. Pages 321–344 *in* L. E. Murr, A. E. Torma, and J. A. Brierley, eds., *Metallurgical Applications of Bacterial Leaching and Related Microbiological Phenomena.* Academic Press, New York.

Torma, A. E. 1977. The role of *Thiobacillus ferro-oxidans* in hydrometallurgical processes. Pages 1-37 *in* T. K. Ghose, A. Fiechter, and N. Blakebrough, eds., *Advances in Biochemical Engineering*. Springer Verlag, Berlin.

Torma, A. E., and K. Bosecker. 1982. Bacterial Leaching. Pages 77–118 *in* M. J. Bull, ed., *Progress in Industrial Microbiology, Vol. 16*. Elsevier, Amsterdam.

Torma, A. E., and R. Guay. 1976. Effect of particle size on the biodegradation of a sphalerite concentrate. *Nat. Can.* 103:133–138.

Torma, A. E., and T. Rozgonyi. 1980. Influence of attrition grinding on the recovery of copper from a high-grade chalcopyrite concentrate by the BSE-process. Pages 583–588 *in* P. A. Trudinger, M. R. Walter, and B. J. Ralph, eds., *Biogeochemistry of Ancient and Modern Environments*. Australian Academy of Science, Canberra.

Torma, A. E., and K. N. Subramanian. 1974. Selective bacterial leaching of a lead sulphide concentrate. *Int. J. Min. Proc.* 1:125–134.

Torma, A. E., P. R. Ashman, T. M. Olsen, and K. Bosecker. 1979. Microbiological leaching of chalcopyrite concentrate and recovery of copper by solvent extraction and electrowinning. *Metallurgy* 33:479–484.

Torma, A. E., C. C. Walden, and R. M. R. Branion. 1970. Microbiological leaching of a zinc sulphide concentrate. *Biotechnol. Bioeng.* 12:501–517.

Torma, A. E., C. C. Walden, D. W. Duncan, and R. M. R. Branion. 1972. The effect of carbon dioxide and particle surface area on the microbiological leaching of a zinc sulphide concentrate. *Biotechnol. Bioeng.* 14:777–786.

Trudinger, P. A., and D. J. Swaine, eds. 1979. *Biogeochemical Cycling of the Mineral-forming Elements*. Elsevier, Amsterdam.

Trudinger, P. A., M. R. Walter, and B. J. Ralph, eds. 1980. *Biogeochemistry of Ancient and Modern Environments*. Australian Academy of Science, Canberra.

Tsuchiya, H. M. 1977. Leaching of Cu-Ni sulphide concentrate from the Dulith Gabbro. Pages 101–106 *in* W. Schwartz, ed., *GBF Monograph Series No. 4* (August 1977), *Conference Bacterial Leaching 1977*. Verlag Chemie, Weinheim.

Tsuchiya, H. M., N. C. Trivedi, and M. L. Schuler. 1974. Microbiol mutualism in ore leaching. *Biotechnol. Bioeng.* 16:991–995.

Tuovinen, O. H., and D. P. Kelly. 1972. Biology of *Thiobacillus ferro-oxidans* in relation to the microbiological leaching of sulphide ores. *Z. Allg. Mikrobiol.* 12:311–346.

______. 1973. Studies on the growth of *Thiobacillus ferro-oxidans*. I. Use of membrane filters and ferrous iron agar to determine viable numbers, and comparison with $^{14}CO_2$ fixation and iron oxidation as measures of growth. *Arch. Mikrobiol.* 88:285–298.

Tuovinen, O. H., S. I. Niemela, and H. G. Gyllenberg. 1971. Tolerance of *Thiobacillus ferro-oxidans* to some metals. *Antonie van Leeuwenhoek J. Microbiol. Serol.* 37:489–497.

Winogradsky, S. N. 1888. Uber eisenbacterin. *Bot. Ztg.* 46:262–270.

Wood, A. P., and D. P. Kelly, 1983. Use of carboxylic acids by *Thiobacillus* A-2. *Microbios* 38:15–26.

Yao Dun Pu. 1982. The history and present status of practice and research work on solution mining in China. Pages 13–20 *in* W. J. Schlitt, ed., *Interfacing*

Technologies in Solution Mining. Proc. 2nd SME-SPE Int. Solut. Min. Symp., Denver, CO, Nov. 18–20, 1981.

Yukawa, T. 1975. *Mass Transfer Studies in Microbial Systems*. Ph.D. Thesis, University of New South Wales.

I
SYMPOSIUM: PLASMID VECTORS, CONSTRUCTION AND EXPRESSION

G. E. Pierce and G. A. Somkuti, *Conveners*
Battelle Memorial Institute, Columbus, Ohio, and Eastern Regional Research Center, USDA, Philadelphia, Pennsylvania

PANELISTS

M. J. Duncan
R. P. Evans
D. I. Johnson
K. R. Jones
T. Kohno
F. L. Macrina
J. Mao
D. T. Moir
R. A. Smith
R. L. Somerville
J. A. Tobian
R. B. Winter

CHAPTER 1

Genetic Analysis of Streptococci: Useful Recombinant Plasmids

R. PAUL EVANS**, ROBERT B. WINTER*, JANET A. TOBIAN***, KEVIN R. JONES, AND FRANCIS L. MACRINA

*Department of Microbiology and Immunology, Virginia Commonwealth University, Richmond, Virginia 23298 and *Department of Molecular, Cellular, and Developmental Biology, University of Colorado, Boulder, Colorado 80309*

The isolation and characterization of streptococcal plasmid and chromosomal genes important in the industrial (e.g., food fermentation) and clinical (e.g., caries production and other diseases) settings have been accelerated by the development of molecular cloning systems that use a streptococcal host. Conjugation and transformation based streptococcal gene transfer systems have been essential in such studies. Novel plasmid shuttle vectors, pVA838 and pVA856, consisting of the entire pACYC184 (*Escherichia coli;* $Cm^r Tc^r$)[1] and pVA749 (*Streptococcus sanguis*; Em^r) molecules joined in vitro were shown to replicate in both *S. sanguis* and *E. coli*. These plasmids afford several different cleavage-ligation strategies for cloning streptococcal genes in *E. coli* and for introducing chimeras into transformation-competent streptococcal hosts. The conjugative R plasmid pIP501 ($Cm^r Em^r$) is capable of self-mediated transfer to a broad range of streptococcal and other gram-positive hosts. Plasmid pVA797 was constructed in vitro from pIP501. pVA797 ($Cm^r Em^s$) was shown to mobilize Em^r streptococcal cloning vectors. pVA797 was cloned as a single *Eco*RI fragment into the *E. coli* vector pOP203(A_2^+) to obtain pVA904. This shuttle plasmid retained conjugal transfer ability in streptococci. Proteins encoded by the pVA797 portion of pVA904 were detected in *E. coli* minicell lysates. Using pVA904 as a model replicon, we should be able to define the genetic and biochemical basis of streptococcal conjugation by coupling mutagenesis protocols and minicell analyses in *E. coli* with evaluation of transfer function in streptococci.

INTRODUCTION

The genus *Streptococcus* encompasses organisms whose properties can be decidedly beneficial or detrimental to man and animals. The study of the genetic basis of virulence and the industrially important processes mediated by members

**Present address: Agronomy Department, Purdue University, West Lafayette, IN 47907.

***Present address: Human Genetics Branch, National Institute of Child Health and Human Development, NIH, Bethesda, MD 20205.

[1]Abbreviations: Cm^r, (*E. coli*) growth on 20 μg of chloramphenicol per ml; Cm^r (*S. sanguis*) growth of 5 μg of chloramphenicol per ml; Tc^r (*E. coli*) growth on 10 μg of tetracycline per ml; Em^r (*E. coli, S. sanguis*) growth on 100 μg of erythromycin per ml; Tra^- does not exhibit genetic transfer by a conjugation-like mechanism when cells are cocultivated on filters.

of this genus has been facilitated by the development of streptococcal genetic transfer systems. As in other genera, these developments hinged on the discovery of resident plasmids in the streptococci and the elaboration of methods for the introduction and subsequent construction of chimeric plasmids in streptococcal hosts.

Since the first report of plasmid DNA in streptococci by Courvalin et al. (1972), plasmids have been identified in all streptococcal species examined to date. As reviewed by Clewell (1981), some functions ascribed to plasmids in streptococci include drug resistance, conjugal transfer, production of bacteriocins, hemolysins, and proteases as well as the ability to metabolize citrate, lactose, and other carbon sources.

This abundance of potential vectors coupled with the natural genetic competence of *S. sanguis* led to the description of the first streptococcal host (*S. sanguis*)-vector molecular cloning systems in 1980 (Behnke and Gilmore 1981; Macrina et al. 1980, 1982a; Malke et al. 1981). These systems have been valuable in the study of streptococcal plasmid sequences. However, transformation of genetically competent *S. sanguis* cells appears to require two-hit kinetics with plasmid monomers or the entrance of oligomeric plasmid forms (Macrina et al. 1981; Saunders and Guild 1981). As significant numbers of identical molecules or oligomeric forms are low in chromosomal shotgun cloning, such experiments have been limited in value as a number of recombinant molecules suffer deletions.

As in gram-negative bacteria, streptococcal self-transmissible (conjugal transfer) plasmids offer an additional avenue for the development of gene transfer systems in streptococci. Conjugative plasmids are capable of mediating their physical transfer to new hosts by a process that requires cell-cell contact. Some conjugative plasmids also mediate the transfer of chromosomal DNA and/or other resident nontransferable plasmids (mobilization). Conjugative plasmids that can mobilize resident plasmids into a broad range of hosts have been used for the development of gene transfer systems. These self-transferable elements can be used to manipulate genetic material in hosts for which natural or induced transformational competency is unknown.

Horodniceanu et al. (1976) isolated an Em^rCm^r conferring conjugative plasmid pIP501 (30 kb) from *S. agalactiae* (Lancefield group B). Based on host range and restriction endonuclease fragment similarities, Hershfield (1979) suggested that pIP501 was a member of an apparent family of streptococcal conjugative plasmids. These related plasmids are generally 23 to 37 kb in size, share common Em^r determinants, and transfer to similar hosts. Inter- and intrageneric transfer of pIP501 to Lancefield groups A (Malke 1979), C (Bougueleret et al. 1981), D (Malke 1979), F (Hershfield 1979) H (Malke 1979), N (Gonzalez and Kunka 1983), *S. pneumoniae* (Smith et al. 1980), and to *Pediococcus sp.* (Gonzalez and Kunka 1983) has been reported. In addition to self-transmissibility, pIP501 is capable of mobilizing nonself-transmissible plasmids (Hershfield 1979). pIP501 and other streptococcal broad-host-range resistance plasmids have been studied (Evans and Macrina 1983; Lee and LeBlanc 1983), not only because of their clinical importance but also for their use as genetic tools.

In this communication we report development of *E. coli-S. sanguis* transgeneric shuttle vectors. By constructing recombinant plasmids that can be isolated as homogeneous entities from *E. coli* transformants, deletion events that occur upon transformation of *S. sanguis* are minimized. We have previously characterized the conjugal plasmid pIP501 with regard to the organization of its antibiotic resistance determinants, replication region, and its restriction endonuclease site map (Evans and Macrina 1983). In addition, a conjugative derivative of pIP501, pVA797 (30.7 kb), was constructed such that it bears only the Cm^r determinant with a single *Eco*RI site in a nonessential region of the plasmid. Because of its Em^sCm^r phenotype and conjugative ability, pVA797 has been useful as a mobilizing plasmid for Em^r native or recombinant plasmids. We also report the cloning of the entire pVA797 molecule as a 30.7 kb *Eco*RI fragment into the *E. coli* positive selection vector pOP203(A_2^+). This shuttle plasmid, designated pVA904, retains conjugal transfer ability in streptococci. pVA904 serves as a model replicon for studying the genetic basis of conjugative donor ability conferred by broad-host-range streptococcal plasmids.

Materials and Methods

Bacterial strains and media. The strains used in this work are listed in the text. Conditions, drug concentrations, and media for growth of *E. coli* and streptococcal strains were as previously published (Macrina et al. 1982a). Selective media for transconjugants contained 50 μg/ml of rifampin (Sigma, St. Louis, MO). Isopropyl-thio-β-galactoside (IPTG; BRL, Rockville, MD) was used at a final concentration of 10 m*M* in solid (1.5% agar) media.

Genetic techniques. Plasmid isolation, efficiency of plating, restriction endonuclease site mapping, cloning methods, and transformation protocols were as previously published (Macrina et al. 1982b). Conjugal matings on filters (6 h duration) were performed as previously described (Evans and Macrina 1983).

Minicell purification and minicell lysate analysis. E. coli minicells were purified and labeled according to Kennedy et al. (1977) except that cycloserine (40 μg/ml; Sigma) treatment was used to reduce vegetative cell contamination prior to radioisotopic labeling. Following labeling (20 μci/ml ^{35}S-methionine; ICN, Cleveland, OH), minicell lysates were electrophoresed through SDS-polyacrylamide (10% acrylamide) by standard methods (Laemmli and Favre 1973). Autoradiography was for 48 h at −70 C using Kodak X-OMAT R film. Molecular weight standards (high and low) were purchased from Pharmacia Fine Chemicals.

Results and Discussion

Escherichia-Streptococcus *Plasmid Shuttle Vectors*

The molecular requirements for plasmid transformation in streptococci (identical monomers or oligomeric molecules) have led to the development of alternative strategies for the cloning of streptococcal chromosome fragments. One such system, termed "helper-plasmid cloning" (Tobian and Macrina 1982; Behnke

1982), allows for recombinational rescue of incoming chimeric plasmids by a resident homologous plasmid. The most widely used alternative strategy employs shuttle vectors capable of replication in both *Escherichia* and *Streptococcus* hosts.

We have previously described the construction of one such shuttle vector designated pVA838 (Macrina et al. 1982b). Two separate plasmids were used in the construction of this vector, namely pVA749 (streptococcal plasmid; Em^r; Macrina et al. 1981) and pACYC184 (*E. coli* plasmid; Cm^r, Tc^r; Chang and Cohen 1978) (Fig. 1). pVA749 and pACYC184 were joined at their unique *Hin*d III sites and the resulting chimeric plasmid selected in *E. coli* strain DB11 (an erythromycin-sensitive mutant of *E. coli* kindly supplied by Julian Davies) following transformation. pVA838 is 9.2 kb in size and confers resistance to chloramphenicol (25 μg/ml) and erythromycin (10 μg/ml) in *E. coli*. Only the erythromycin resistance determinant is expressed in streptococci. As anticipated, insertion of pVA749 in the *Hin*d III site of pACYC184 resulted in the inactivation of the tetracycline resistance determinant. The *Bam*HI, *Sph*I, *Sal*I, *Nru*I, *Pvu*II, and *Eco*RI sites of pVA838 can be used for the insertion of passenger DNA. However, inserts at the *Eco*RI and *Pvu*II (fragment replacement) sites can easily be monitored owing to inactivation of Cm^r.

Ligation of *Ava*I linearized pVA749 and pACYC184 resulted in a new shuttle plasmid designated pVA856, which was identical in size to pVA838 (Fig. 1). In contrast to pVA838, pVA856 retains tetracycline resistance in *E. coli* yet is unstable under nonselective conditions. Thus, passenger DNA insertion at the *Sph*I, *Bam*HI, *Sal*I, and *Nru*I sites can be monitored by Tc^r inactivation. The uninterrupted Cm^r or Tc^r determinants of pVA856 offer two modes of insertional inactivation allowing for ready selection of chimeric plasmids in a wide range of *E. coli* strains. The observed instability of the pVA856 molecules may be due to the proximity of the *Ava*I site to some of the pACYC184 replicative machinery (Meacock and Cohen 1980). With continuous selective pressure, however, the shuttle plasmid pVA856 is retained in all cells.

Both pVA838 and pVA856 have been successfully used by us and others to clone in *E. coli* both plasmid and chromosomal fragments of streptococcal origin. The purified monomer and/or oligomeric chimeras manufactured in *E. coli* transformationally enter *S. sanguis* cells with minimal risk for deletion formation. Upon establishment in *S. sanguis*, the recombinant plasmid can be mobilized by a conjugative plasmid into streptococcal species and other gram-positive bacteria for which transformational gene transfer systems are unavailable.

Construction of pVA797

The location of restriction endonuclease cleavage sites and functional regions (e.g., Cm^r, Em^r, replication region; Fig. 2) on plasmid pIP501 followed from standard mapping procedures and from the use of recombinant DNA methodology (Evans and Macrina 1983). The effective use of the pIP501 replicon as a genetic tool (e.g., mobilization of nonself-transmissible plasmids) required that the Em^r determinant be eliminated or inactivated since most streptococcal cloning vectors carry an Em^r determinant. To this end, pIP501 plasmid DNA was sub-

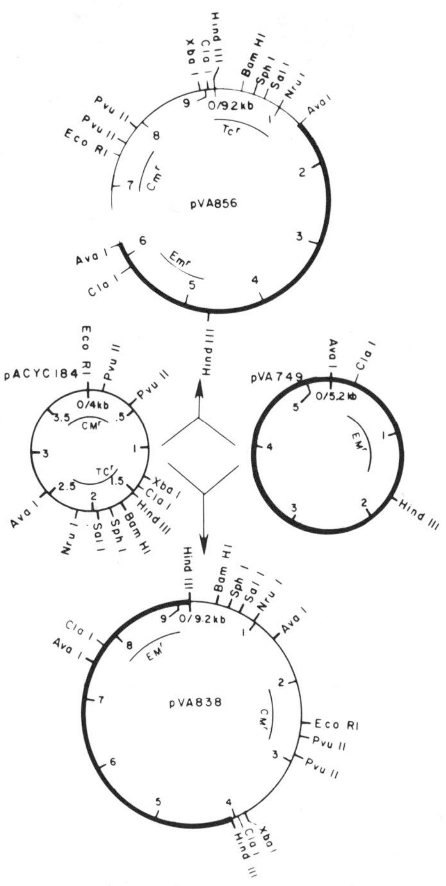

FIG. 1. Shuttle plasmid restriction maps. Coordinates (kb) are on the inside of each circular map. pVA838 (ligation of *Hind* III fragments) and pVA856 (ligation of *Ava*I fragments) may be propagated in *E. coli* or streptococcal hosts. pVA749 (thick lines) and pACYC184 (thin lines) replication is restricted to streptococcal or *E. coli* hosts, respectively. The approximate location of the antibiotic resistance determinants (Em^r, Cm^r, Tc^r) is indicated.

jected to *Hpa*II-*Ava*I cleavage and ligated with similarly treated pVA380-1 (a cryptic, multicopy streptococcal plasmid; Macrina et al. 1980). A resultant Em^sCm^r plasmid, designated pVA797 (Evans and Macrina 1983), was found to consist of the 3.4 kb pVA380-1 *Hpa*II-*Ava*I fragment inserted at the unique *Hpa*II and *Ava*I sites of pIP501 (Fig. 2). Plasmid pVA797 was able to transfer Cm^r during filter matings to *S. faecalis* at frequencies comparable to pIP501 (4.3×10^{-3} and 5.5×10^{-2} transconjugants/donor cell, respectively). pVA797 has been used to mobilize the Em^r *E. coli-S. sanguis* shuttle plasmid, pVA838, to a *S. faecalis* recipient (Smith and Clewell 1984). The observed transfer of the pIP501 molecule to other bacteria, such as *S. lactis* and *Pediococcus* spp. (Gonzalez and Kunka 1983), has obvious implications for the use of plasmid pVA797 in the development of genetic exchange systems for these and other gram-positive bacteria.

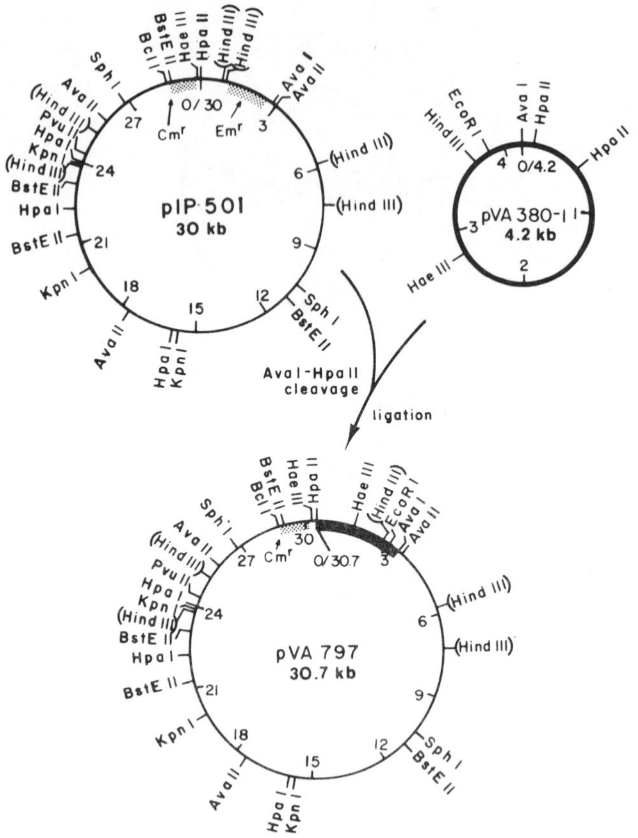

FIG. 2. Construction of pVA797. pVA797 was formed by the ligation of *Hpa*II-*Ava*I cleaved molecules of pIP501 (thin line) and pVA380-1 (thick line). pVA797 containing *S. sanguis* cells are Cm^r but Em^s because of the deletion of the pIP501 Em^r determinant. The *Hin*d III sites (in parentheses) are shown; additional sites exist between kb coordinates 7 and 23. A more complete map of pIP501 and pVA797 is found in Evans and Macrina (1983).

Construction and Characterization of pVA904

Earlier attempts to clone specific plasmid fragments that encompass the genes governing conjugation in plasmid pIP501 or the related plasmid pAMβ1 had not met with success (Lee and LeBlanc 1983; Evans and Macrina 1983). In vivo induced transfer deficient deletion derivatives of pIP501 or pAMβ1 also were not instructive in terms of locating specific regions that encode conjugal transfer ability. We decided to clone the entire pVA797 molecule using the unique *Eco*RI site (of pVA380-1 origin) that was known to reside in a region nonessential for plasmid replication or conjugal transfer ability (Evans and Macrina 1983). pOP203(A_2^+) (7.0 kb, Tc^r), a novel positive selection vehicle constructed by Winter and Gold (1983), was chosen as the carrier of the 30.7 kb pVA797 insert. This plasmid contains the A_2 (maturation protein) gene of phage Qβ inserted into the *lac*UV5 promoter vector pOP203 (Tc^r, Fuller 1982) such that expression of the A_2 gene product that mediates cellular lysis (Winter and Gold 1983) is under the control of the lactose regulatory signals. pOP203(A_2^+) may be maintained in *E. coli* i^q (hyper-production of lactose repressor) mutants, which ensures that transcription of the A_2 gene does not occur. Expression of the A_2 gene may also be prevented by the in vitro insertion of passenger DNA at any of several internal restriction endonuclease sites (e.g., *Eco*RI *Sst*I, *Xho*I, or *Bgl*II). In the presence of IPTG, only transformants with chimeric plasmids carrying inserts (in the A_2 gene) will survive. pOP203(A_2^+) and pVA797 were linearized by *Eco*RI cleavage, ligated, and transformed into *E. coli* with selection for Tc^r colonies. A resultant 37.7 kb plasmid, designated pVA904, was shown to consist of the pOP203(A_2^+) and pVA797 molecules joined at their *Eco*RI sites (Fig. 3). pVA904 plasmid DNA readily transformed *S. sanguis* V288 cells to Cm^r. As expected, pVA904 conjugal donor proficiency was unaltered as compared to pVA797 in streptococci (4.3×10^{-3} and 1.9×10^{-3} transconjugants/donor cell, respectively). Transfer of

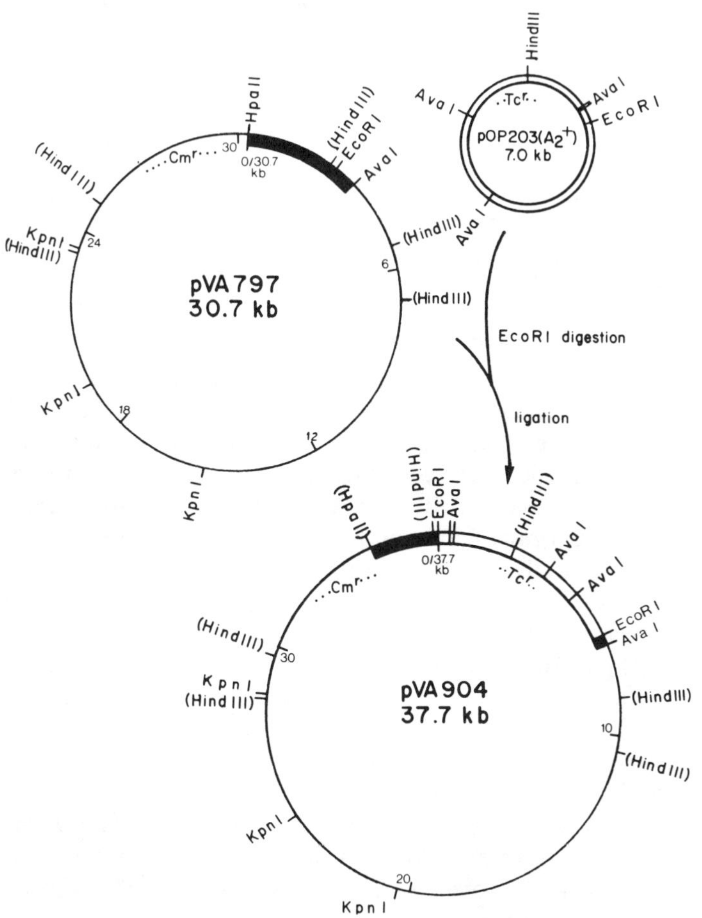

FIG. 3. Construction of pVA904. *Eco*RI-linearized pOP203(A_2^+) (double line) was ligated with similarly treated pVA797 (cleavage site within the pVA380-1 [thick line] portion). The single thin line of pVA797 refers to regions of pIP501 origin. The restriction enzyme sites shown in parentheses indicate that other such unmapped sites exist on the plasmids. A more complete restriction endonuclease site map of pOP203(A_2^+) and pVA797 may be found in Winter and Gold (1983) and Evans and Macrina (1983), respectively.

pVA904 in *Escherichia-Escherichia* or in intergeneric *Escherichia-Streptococcus* matings was not detected. Using a colony forming unit assay, resistance to tetracycline and chloramphenicol antibiotics in pVA904 containing and isogenic plasmidless strains was compared. *S. sanguis* or *S. faecalis* strains harboring pVA904 were no more resistant to tetracycline than their plasmidless counterparts (data not shown). However, pVA904-containing derivatives of *E. coli* displayed reproducible elevation of Cm^r (Table 1). These latter data indicated that streptococcal genes and their products are functional in *E. coli*.

TABLE 1. *Efficiency of plating on chloramphenicol*

Chloramphenicol Concentration (μg/ml)	V900[a]	V992[b]
0	100	100
0.1	80	89
0.2	73	82
0.5	53	77
1	3	71
2	0	68
5	0	1
10	0	0

[a] Minicell producing *E. coli* strain M2141 from F. Neidhardt.
[b] V900 strain that harbors pVA904.

Expression of pVA904 in E. coli *Minicells*

pVA904 and a transfer defective (Tra^-) deletion derivative (designated pVA1035) of pVA904 missing the region between kb coordinates 11 and 26 (Fig. 3) were introduced by transformation into an *E. coli* minicell-producing mutant. Using standard methods, plasmid directed ^{35}S-labeled polypeptides synthesized in purified minicells were detected by autoradiography following SDS-polyacrylamide electrophoresis. Lane 1 in Fig. 4 displays the proteins encoded by pVA1035 in *E. coli* minicells. The nonconjugative pVA1035 plasmid was unable to direct the synthesis of the majority of proteins specified by pVA904 (lane 2). At least 13 differently sized polypeptides were specified by pVA904. Because of the Tra^- phenotype of pVA1035, it is reasonable to conclude that one or more of the missing proteins may be conjugal transfer-associated gene products. At least one of pVA904-specific proteins synthesized in *E. coli* minicells is immunoprecipitated by immune sera raised against *S. sanguis* cells carrying pVA904 (Hartley and Macrina, in preparation).

The plasmid pVA904 advances the means by which streptococcal plasmid function, including conjugal transfer, may be explored. pVA904 plasmids mutagenized (e.g., controlled deletion formation, transposon insertion) in *E. coli* can be transformed into *S. sanguis* where plasmid function (e.g., conjugal transfer) can be determined. In conjunction with assessment of specific protein synthesis in *E. coli* minicells and immunoprecipitation patterns, it will be possible to correlate map region with phenotype and gene product. Using these multiple approaches, we hope to be able to delineate the genetic and biochemical basis of streptococcal broad-host-range conjugal transfer as represented by the plasmid pIP501.

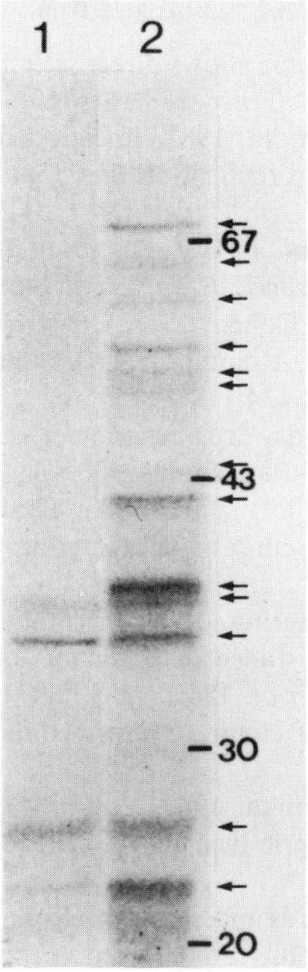

FIG. 4. Plasmid pVA904 encoded protein profile in *E. coli* minicells. The locations of molecular weight reference molecules are denoted on the figure (kdal). Lane 1. Analysis of pVA1035-containing minicells. Lane 2. Analysis of pVA904-containing minicells. Presumed plasmid pVA797 specific proteins (as compared to a pOP203(A_2^+) derivative plasmid (pOP203(A_2^+) *Eco*Rl fill-in; Winter and Gold 1983) with an altered A_2 protein because of a reading frame shift at the unique *Eco*Rl site (data not shown) are denoted by arrowheads.

Acknowledgments

We thank Donna Hartley for helpful discussions and Crescentia Motzi for expert typing of the manuscript. This work was supported by a grant #DE04224 from the National Institute of Dental Research, National Institutes of Health, U.S.A. F.L.M. was the recipient of a Research Career Development Award from the National Institute of Dental Research (Grant #DE00081). R.P.E. was the recipient of a Virginia Commonwealth University School of Graduate Studies Fellowship Award. R.B.W. was a postdoctoral fellow of the Damon Runyon–Walter Winchell Cancer Fund (Grant #DRG-340-F). Work performed in Boulder was supported by National Institutes of Health (U.S.A.) grant #GM28685 to Larry Gold.

Literature Cited

Behnke, D. 1982. Marker rescue allows direct selection for recombinant plasmids in streptococci. *Mol. Gen. Genet.* 188:161–163.

Behnke, D., and M. S. Gilmore. 1981. Location of antibiotic resistance determinants, copy control, and replication functions on the double-selective cloning vector pGB301. *Mol. Gen. Genet.* 184:115–120.

Bougueleret, L., G. Beith, and T. Horodniceanu. 1981. Conjugative R plasmids in group C and G streptococci. *J. Bacteriol.* 145:1102–1105.

Chang, A. C. Y., and S. N. Cohen. 1978. Construction and characterization of amplifiable multicopy DNA cloning vehicles derived from the P15A cryptic miniplasmid. *J. Bacteriol.* 134:1141–1156.

Clewell, D. B. 1981. Plasmids, drug resistance, and gene transfer in the genus *Streptococcus. Microbiol. Rev.* 45:409–436.

Courvalin, P. M., C. Carlier, and Y. A. Chabbert. 1972. Plasmid-linked tetracycline and erythromycin resistance in group D. *Streptococcus. Ann. Inst. Pasteur* 123:755–759.

Evans, R. P., and F. L. Macrina. 1983. Streptococcal R plasmid pIP501: Endonuclease site map, resistance determinant location, and construction of novel derivatives. *J. Bacteriol.* 154:1347–1355.

Fuller, F. 1982. A family of cloning vectors containing the *lac*UV5 promoter. *Gene* 19:43–54.

Gonzalez, C., and B. S. Kunka. 1983. Plasmid transfer in *Pediococcus* spp.: Intergeneric and intrageneric transfer of pIP501. *Appl. Environ. Microbiol.* 46:81–89.

Hershfield, V. 1979. Plasmids mediating multiple drug resistance in group B Streptococcus: Transferability and molecular properties. *Plasmid* 2:137–149.

Horodniceanu, T., D. Bouanchaud, G. Biet, and Y. Chabbert. 1976. R plasmids in *Streptococcus agalactiae* (group B). *Antimicrob. Agents Chemother.* 10:795–801.

Kennedy, N., L. Beutin, M. Achtman, R. Skurray, U. Rahmsdorf, and P. Herrlich. 1977. Conjugation proteins encoded by the F sex factor. *Nature* 270:580–585.

Laemmli, U. K., and M. Favre. 1973. Maturation of the head of bacteriophage T4. I. DNA packaging events. *J. Mol. Biol.* 80:575–599.

Lee L., and D. LeBlanc. 1983. Physical and genetic analyses of streptococcal plasmid pAMβ1 and cloning of its replication region. *J. Bacteriol.* 157:445–453.

Macrina, F. L., K. R. Jones, and R. A. Welch. 1981 Transformation of *Streptococcus sanguis* with monomeric pVA736 plasmid deoxyribonucleic acid. *J. Bacteriol.* 146:826–830.

Macrina, F. L., K. R. Jones, and P. H. Wood. 1980. Chimeric streptococcal plasmids and their use as molecular cloning vehicles in *Streptococcus sanguis* (Challis). *J. Bacteriol.* 143:1425–1435.

Macrina, F. L., K. R. Jones, J. A. Tobian, and R. P. Evans. 1982a. Molecular cloning in the streptococci. Pages 195–210 *in* A. Hollaender, R. DeMoss, S. Kaplan, J. Konisky, D. Savage, and R. Wolfe, eds., *Genetic Engineering of Micro-organisms for Chemicals.* Plenum Publishing Corp., New York.

Macrina, F. L., J. A. Tobian, K. R. Jones, R. P. Evans, and D. B. Clewell. 1982b. A cloning vector able to replicate in *Escherichia coli* and *Streptococcus sanguis*. *Gene* 19:345–353.

Malke, H. 1979. Conjugal transfer of plasmids determining resistance to macrolides, lincosamides, and streptogramin B-type antibiotics among group A, B, D and H streptococci. *FEMS Microbiol. Lett.* 5:335–338.

Malke, H., L. G. Burman, and S. E. Holm. 1981. Molecular cloning in streptococci: Physical mapping of the vehicle plasmid pSM10 and demonstration of intergroup transfer. *Mol. Gen. Genet.* 181:254–267.

Meacock, P. A., and S. N. Cohen. 1980. Partitioning of bacterial plasmids during cell division: A cis-acting locus that accomplishes stable plasmid inheritance. *Cell* 20:529–542.

Saunders, C. W., and W. Guild. 1981. Pathway of plasmid transformation in *Pneumococcus*: Open circular and linear molecules are active. *J. Bacteriol.* 146:517–526.

Smith, M. D., and D. B. Clewell. 1984. Return of *Streptococcus faecalis* DNA cloned in *Escherichia coli* to its original host via transformation of *Streptococcus sanguis* followed by conjugative mobilization. *J. Bacteriol.* 160:1109–1114.

Smith, M. D., N. D. Shoemaker, V. Burdett, and W. R. Guild. 1980. Transfer of plasmids by conjugation in *Streptococcus pneumoniae*. *Plasmid* 3:70–79.

Tobian, J. A., and F. L. Macrina. 1982. Helper plasmid cloning in *Streptococcus sanguis*: Cloning of a tetracycline resistance determinant from the *Streptococcus mutans* chromosome. *J. Bacteriol.* 152:215–222.

Winter, R. B., and L. Gold. 1983. Overproduction of bacteriophage Qβ maturation (A_2) protein leads to cell lysis. *Cell* 33:877–885.

CHAPTER 2

Production of Calf Chymosin by the Yeast *S. cerevisiae*

Donald T. Moir, Jen-i Mao, Margaret J. Duncan, Robert A. Smith, and Tadahiko Kohno

Department of Molecular Genetics, Collaborative Research, Inc., Lexington, Massachusetts 02173

Calf chymosin is the preferred milk-coagulating agent for cheese production; however, because of the scarcity of its source, the fourth stomach of an unweaned calf, various substitute enzymes are currently in use. By means of recombinant DNA technology, we have developed strains of the common baker's yeast *S. cerevisiae*, which synthesize authentic calf prochymosin, providing a virtually unlimited source for the chymosin precursor. Factors demonstrated to influence the efficiency of production include the nature of the plasmid vector, the promoter of transcription and its junction to the prochymosin gene, and the presence of a secretion signal. Unlike prochymosin synthesized and residing in the yeast cytoplasm, all secreted prochymosin is fully active in catalysis of milk-coagulation.

Introduction

The enzyme chymosin, also known as rennin, is an acid protease used in the milk-clotting step of cheese production. A mixture of chymosin and its zymogen prochymosin, which may be converted to chymosin by a simple low pH incubation, are currently obtained from the abomasum of an unweaned calf; however, because of the scarcity of that source, several substitute enzymes are presently in use. Calf chymosin is still the preferred milk-clotting agent because of its high specificity for a particular peptide bond in kappa-casein, and because of its heat-sensitivity. A high level of nonspecific proteolysis is undesirable because it affects the texture and flavor of the final cheese product. A heat-sensitive enzyme is important because of the need to stop the proteolytic activity after clotting, so that the whey byproduct is not destroyed. Therefore, authentic calf chymosin is superior to the known substitutes; however, its availability is limited because of its source. We have undertaken to generate new sources of authentic calf chymosin by means of recombinant DNA technology. The work described here concerns the use of the common baker's yeast *Saccharomyces cerevisiae* as a host cell for the production of calf prochymosin, which is easily converted to active chymosin.

A copy of the calf prochymosin gene was obtained by our group using reverse transcription of mRNA isolated from the fourth stomach. Since that time, several publications have described the isolation and characterization of the gene (Moir et al. 1982; Harris et al. 1982; Nishimori et al. 1982). From the point of view of industrial production, two important points emerge from analysis of the cDNA gene. First, the protein is apparently synthesized as a preprochymosin molecule containing a typical hydrophobic secretion signal sequence at the amino terminus. Second, there appears to be only one copy of the preprochymosin gene per haploid genome of the cow (Moir et al. 1982), but there are at least two alleles of that gene, resulting in the A and B forms of prochymosin previously described (Foltmann et al. 1979). Because the prochymosins and chymosins derived from the two known alleles A and B are quite similar in all measured properties and because prochymosin is stable in yeast cells and easily activatable to chymosin, we have chosen to concentrate on the production of prochymosin and have used the A allele.

As a host cell, we have chosen *Saccharomyces cerevisiae*. Yeast is a logical choice because it is a common constituent of man's diet, and chymosin is used in food processing. Furthermore, recent advances in recombinant DNA techniques as applied to *S. cerevisiae* make it a convenient organism for genetic manipulation. Also important is the fact that the technology for yeast fermentation is highly developed, allowing for growth to densities beyond 50 dry g/liter at controlled physiological conditions. In this report, we describe the production of calf prochymosin inside yeast cells and also the secretion of calf prochymosin from yeast cells.

Materials and Methods

Strains, media, and plasmid constructions. Yeast strains CGY238 and CGY461 containing the *GAL*1-promoted prochymosin gene on a beta vector and a delta vector followed by the *SUC*2 transcription terminator have been described previously (Goff et al. 1984). Galactose-utilizing host strains are CGY150 (*MATα ura*3-52 *leu*2-3), CGY566 (*MATα ura*3-52 *trp*1 *prb*1 *prc*1 *pep*4-3) and CGY1112 (*MATα ura*3-52 *leu*2-3 *leu*2-112 *his*4 *pep*4-3). Yeast strains were grown in standard minimal medium (0.67% yeast nitrogen base) or rich medium (YEP) containing 2% glucose or 2% galactose depending on the regulatory requirements of the promoter used for prochymosin expression.

Yeast strain CGY457 is host strain CGY150 containing plasmid pCGS240, the *GAL*1-promoted preprochymosin gene on a beta vector. Yeast strain CGY942 is host strain CGY566 containing plasmid pCGS466 with the prochymosin gene fused in translational reading frame to the yeast PGK structural gene at the *Bgl*II site. Strain CGY828 is host strain CGY150 containing plasmid pCGS419 with the prochymosin gene fused in translational reading frame to the yeast TPI structural gene at a site created by nuclease *BAL*31 treatment. Strain CGY701 is host strain CGY150 containing plasmid pCGS368 with the prochymosin gene fused in translational reading frame to the yeast *GAL*1 structural gene at the *Ava*I site.

Construction of the beta and delta vectors has been described in detail (Goff et al. 1984). The phi vector was constructed in an analogous manner by using the entire two micron plasmid DNA plus the inserted *LEU*2 gene from plasmid pJDB219 (Beggs 1978). Transcription of *LEU*2 and *URA*3 is clockwise in all three vectors. Yeast promoters and the prochymosin gene have been inserted at the *Eco*RI site of pBR322 in a clockwise orientation. Efficient termination of all prochymosin transcripts is achieved by terminators present in the *URA*3 DNA downstream; more proximal termination by an added *SUC*2 terminator has no effect on prochymosin production levels (Goff et al. 1984).

Yeast transformation and vector stability measurements. Yeast transformation was carried out by the spheroplasting method of Hinnen et al. (1978) or the LiCl method of Ito et al. (1983). For the stability experiment the vectors contained the *GAL*1 promoted prochymosin gene at the *Eco*RI site and were in the following host strains: beta and delta were transformed with uracil selection into CGY150; delta was transformed with leucine selection into CGY1112; and phi was transformed with uracil selection into an isogenic cir° derivative of CGY1112, CGY1153. Stability was measured after growth in the appropriate medium for the specified number of generations by plating on solid rich medium and replica-plating on selective medium.

Assays of prochymosin production. Milk-clotting activity assays and SDS-polyacrylamide gel-transfer immunoblot analysis of prochymosin antigen were as previously described (Goff et al. 1984). Prochymosin and galactokinase-prochymosin fusion protein stabilities were measured by following both activity and antigen levels, and the two measurements agreed. Secreted prochymosin levels were also determined by both methods after concentration of the medium through an Amicon ultrafiltration membrane.

RNA isolation and quantitation. Yeast RNA isolation and quantitation of prochymosin-specific mRNA was carried out as previously described by Goff et al. (1984). Briefly, purified hemoglobin mRNA (BRL) was used as a standard, and total yeast RNA amounts were estimated by A_{260} reading. The ethidium bromide-stained lanes 2–5 of Fig. 2 indicate that roughly equivalent amounts of yeast RNA were applied to the gel. Poly-A-containing mRNA was assumed to represent about 1–2% of the total yeast RNA so that the 5 μg of yeast RNA used in Fig. 2 corresponds to about 50 ng mRNA. Thus, the hybridization signal seen with 5 ng of globin mRNA corresponds to 10% of the yeast mRNA present in each yeast RNA sample in Fig. 2.

Results and Discussion

Vectors for Maintenance of a Gene in Yeast

The basic requirements for a vector are a yeast origin of replication and a selectable marker to allow both identification of transformants and selective growth of cells containing the plasmid. In general, an *E. coli* origin of replication and selectable marker are also added to allow convenient preparation of vector DNA in *E. coli*. Desirable features of a vector for purposes of use and production are stability, high copy number, and availability of unique restriction sites for addition of genes and controlling elements. Evidence from several laboratories indicates that copy number plays a role in stability since high copy number vectors are more stable than low copy number vectors (Erhart and Hollenberg 1983; Futcher and Cox 1984). Factors that appear to influence the copy number and stability are the

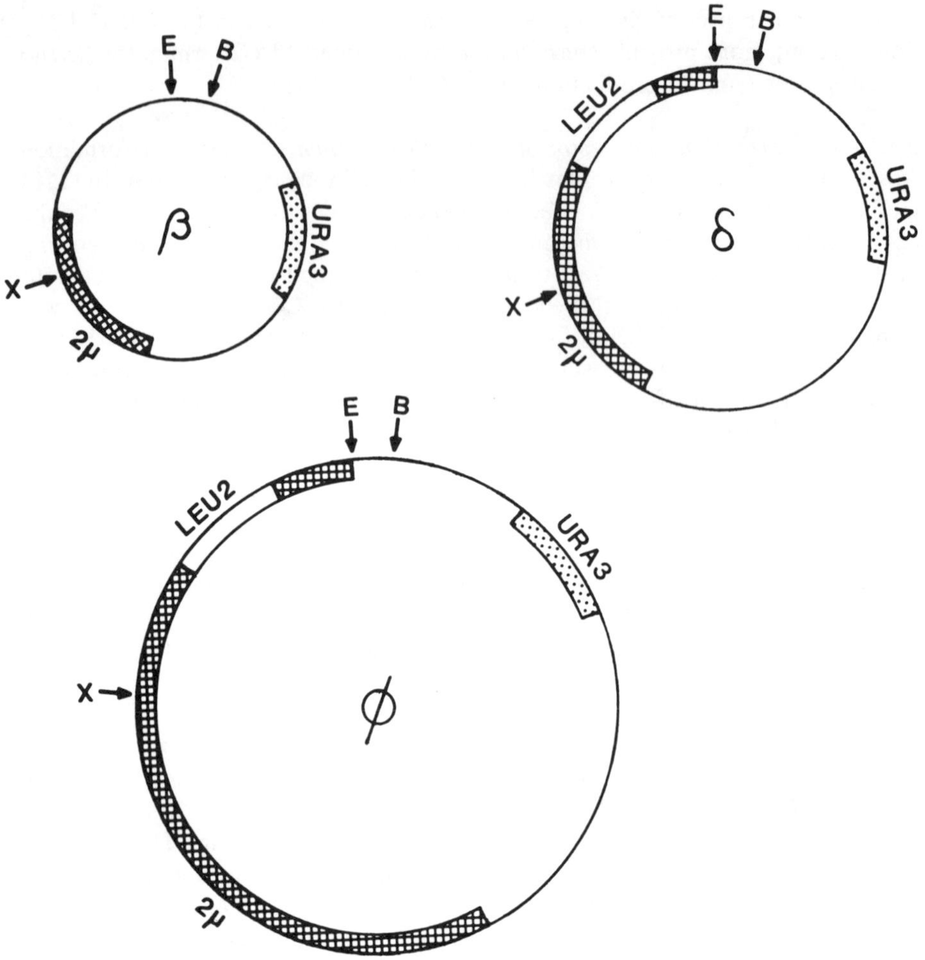

FIG. 1. Three vectors for introduction and maintenance of genes in yeast. The vectors are drawn approximately to scale: beta = 7.0 kb; delta = 9.1 kb; phi = 13.0 kb. The single line represents pBR322, all of which is present in the three vectors; double lines represent yeast chromosomal or two micron plasmid DNA. *URA*3 is inserted in the *Ava* II site of pBR322; the two micron plasmid DNA is in the *Pvu*I site of pBR322 in the beta vector and in the *Eco*RI site of pBR322 in the delta and phi vectors. Restriction endonuclease sites are shown for orientation purposes and are not necessarily the only such sites in each vector: E = *Eco*RI; B = *Bam*HI; X = *Xba*I.

source of the origin of replication and the presence of other sequences important for replication, the amount of gene product synthesized from the selectable marker, and the size of the plasmid. We have constructed three types of yeast vectors, and results from their use reveal some interesting facts about the effects on plasmid stability caused by the production level from the selectable marker and caused by sequences, other than the replication origin, which are important for efficient replication.

In all cases we have used the yeast replication origin from the two micron circle rather than an autonomously replicating segment (ARS) cloned from the yeast chromosomes. In general, yeast plasmids having replication origins from the two

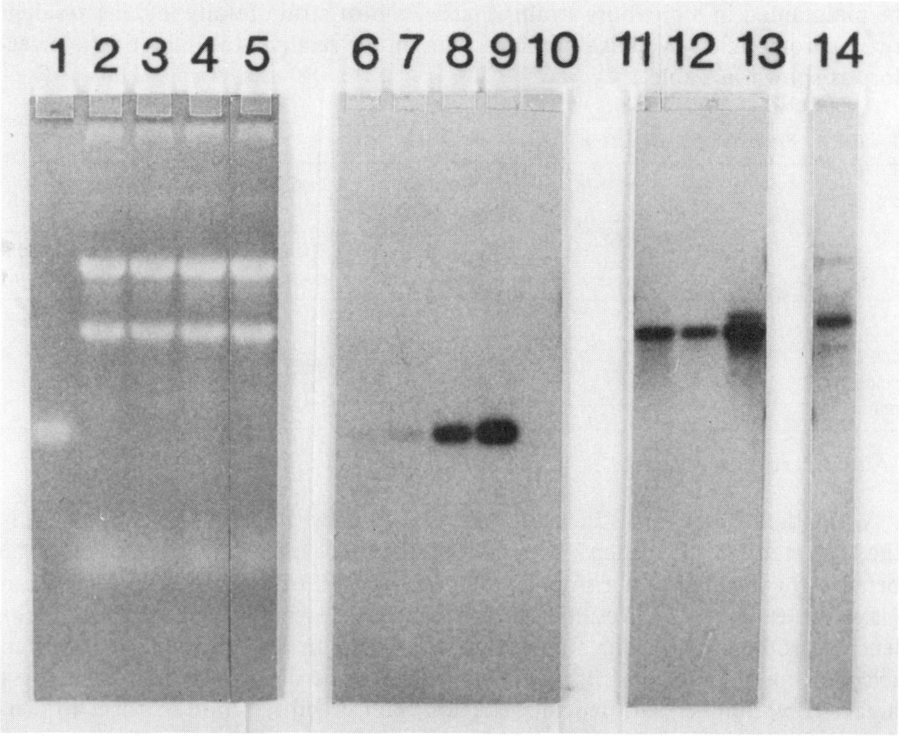

FIG. 2. Analysis of prochymosin-specific mRNA in yeast. Denaturing agarose gel electrophoresis of purified hemoglobin mRNA and yeast total RNA. Lanes 1–5 show the ethidium bromide stained gel of hemoglobin mRNA (0.5 μg) and yeast total RNA (5 μg each) from strains CGY150, CGY238, CGY457, and CGY701, respectively. Lanes 6–14 are an autoradiograph from a gel-transfer hybridization probed with a mixed globin and prochymosin DNA probe labeled at the same specific activity. Purified globin mRNA (0.5, 1, 5, and 10 ng) was applied in lanes 6–9, respectively, and total yeast RNA (5 μg each) from strains CGY150, CGY238, CGY457, CGY461, and CGY701 was applied in lanes 10–14, respectively.

micron circle are more stable than those having ARS replication origins (Derynck et al. 1983). We have used two different yeast genes as selectable markers: *LEU2* and *URA3*. The vectors are designated beta, delta, and phi, and they differ as follows (see Fig. 1). First, the beta vector is the smallest and the most convenient because it contains a short region (*Hind*III to *Hpa*I; 1.6 kb) from the two micron plasmid replication origin so that there are a number of unique restriction sites remaining for later manipulations (see pCGS40 in Goff et al. 1984). The delta vector differs in that it contains a larger two micron segment (*Eco*RI to *Eco*RI; 2.2 kb), and in addition, it contains another selectable marker, the *LEU2* gene (1.3 kb) within the two micron sequences, in addition to the *URA3* gene that is also present on the beta vector (see pCGS242 in Goff et al. 1984). This particular *LEU2* fragment is missing most of its promoter so it may rely on fortuitous transcription from a nearby two micron promoter for its expression (Erhart and Hollenberg 1983). Finally, the phi vector is identical to the delta vector except that it contains the entire two micron plasmid instead of only the 2.2 kb *Eco*RI fragment containing the replication origin region. As a result the phi vector can be maintained in a cir° host strain: that is, a host strain totally lacking resident two micron circles. We have obtained data on the relative stabilities of these vectors as shown in Table 1.

TABLE 1. *Stability of yeast vectors*

Vector	Selection in transformation	Host	Stability after 18 generations (% cells with vector)	
			Rich media (no selection)	Minimal media (with selection)
beta	URA	CGY150(cir⁺)	18%[a]	>95%[a]
delta	URA	CGY150(cir⁺)	28%	>95%
delta	LEU	CGY1112(cir⁺)	50%	>95%
phi	URA	CGY1153(cir°)	82–95%	99%

[a] After only eight generations.

While the beta vector is ideal for manipulations with restriction enzymes, it is the least stable of the group. The small piece of two micron DNA present in the beta vector is missing a cis-acting sequence (REP3) important for replication (Jayaram et al. 1983). Results with the delta vector indicate that selection for leucine prototrophy during transformation results in a more stable vector than selection for uracil prototrophy. This result may be due to a requirement for a higher copy number with leucine selection than uracil selection in order to produce sufficient *LEU2* gene product from the promoter-less *LEU2* fragment on the plasmid. The phi vector, on the other hand, is quite stable even when it is introduced into cells by selection for uracil prototrophy. The fact that 90% of such cells still had the plasmid even after 18 generations of growth in rich media with no selection permits the use of a wide variety of different media that may be required for different industrial applications. Furthermore, it is still possible that the stability of the phi vector can be improved even more by selection for leucine prototrophy during transformation.

Preliminary results on the effect of the various vector types on prochymosin production levels in yeast indicate that more prochymosin protein (compare CGY238 with CGY461 in Table 2) and more prochymosin-specific mRNA (compare lanes 10 and 13 in Fig. 2) are produced from the same *GAL1*-promoted construction on a delta vector than on a beta vector. These results are consistent with the view that increased stability of a vector is correlated with a higher copy number and a correspondingly higher level of production of mRNA and protein from genes encoded by the vector. Current work is aimed at extending this analysis to other vectors and promoters.

Effect of Promoters on Prochymosin Production

Several yeast genes, which are transcribed at high levels, yield protein products that constitute in excess of 1% of the total soluble protein. For example, the yeast glycolytic enzymes phosphoglycerate kinase (*PGK*) and triosephosphate isomerase (*TPI*) as well as the carbohydrate utilization enzyme galactokinase (*GAL1*) are present at high levels (1% to 2% of the soluble protein) in wild type yeast strains growing under certain physiological conditions. The transcriptional

TABLE 2. *Effect of the yeast promoter on prochymosin production*

Strain	Vector	Promoter	Fusion joint[a]	Prochymosin (% of soluble protein)	Activatable Fraction (% of Prochymosin)
CGY942	beta	PGK	387aa	5%	0.1%
CGY828	delta	TPI	57aa	2%	0.1%
CGY701	beta	GAL1	29aa	0.5%	0.5%
CGY238	beta	GAL1	0	0.1%	0.5%
CGY461	delta	GAL1	0	0.3%	0.5%

[a] Refers to the number of amino acids present in the fusion protein from the yeast gene whose promoter was used in the experiment.

promoter regions from those genes may be borrowed for production of other mRNAs such as prochymosin, and when placed on high copy number stable plasmids, should allow production of prochymosin at high levels.

The results from such an experiment are shown in Table 2. The prochymosin gene was fused either to a particular promoter or to both the promoter and a portion of its corresponding structural gene. The two types of constructions result in the production of either authentic prochymosin or prochymosin fused to part of a yeast protein. Three points are apparent from analysis of these data. First, attachment of high level transcriptional promoters to the prochymosin gene does, in fact, result in production of significant levels of prochymosin. There were differences seen with the three promoters. The *GAL*1 promoter resulted in less prochymosin being produced than did either the *PGK* or *TPI* promoter. However, the *GAL*1 promoter, besides being a strong promoter, is also tightly regulated (St. John and Davis 1981). In the absence of galactose or in the presence of glucose, prochymosin production was at least 20-fold lower. This property should make the *GAL*1 promoter extremely useful for production of proteins that are toxic to yeast cells.

Second, with the *GAL*1 promoter, the presence of a portion of the galactokinase structural gene resulted in more prochymosin being produced than in its absence. Qualitatively similar results have also been obtained with the *TPI* and *PGK* promoters (unpublished observations). Third, the majority of the prochymosin produced, either as a fusion or nonfusion protein, is not activatable to yield biologically active chymosin. These last two findings are unexpected and are dealt with in more detail below.

In order to determine if the improved production efficiency seen when prochymosin is fused to a yeast protein is due to an effect on transcription, translation or protein stability, we examined the prochymosin mRNA levels and protein half-lives for comparison with the steady-state prochymosin level shown in Table 2. Prochymosin from yeast strain CGY461 (authentic nonfused prochymosin) and galactokinase-prochymosin fusion protein from strain CGY701 exhibit very similar stabilities (half-lives of 1.5 and 2.5 h) inside the yeast cell following turn-off of synthesis by addition of glucose to the medium (R. G. Knowlton, unpublished observations). In addition, when prochymosin-specific mRNA levels were examined from strains CGY238 and CGY701 (Fig. 2, lanes 2 and 11 compared to lanes 5 and 14), which carry either the nonfusion or the fusion construction on the same vector type (beta), virtually the same mRNA level (about 2–5% of the cell's mRNA) was found (see Fig. 2). The nonfusion construction on a different vector (delta) did yield a higher mRNA level (Fig. 2, lane 13); however, this level is consistent with the increased stability and presumed higher average copy number of the delta vector as opposed to the beta vector. From these results we conclude that prochymosin mRNA fused to the yeast *GAL*1 mRNA is more efficiently translated in yeast than a nonfused prochymosin mRNA. Work is currently in progress to determine if this is also the explanation for the enhanced prochymosin production observed for fusions of prochymosin to other yeast promoters and structural genes.

In spite of the high production level of prochymosin in the yeast strains shown in Table 2, it is striking that very little of the prochymosin (whether fused to a yeast protein or not) is activatable by the standard pH 2 activation protocol effective for activation of prochymosin derived from the calf. This fact, coupled with the observation that most of the prochymosin produced in yeast and *E. coli* is not soluble in buffer but requires strong denaturing conditions such as TCA/SDS, 8 *M* urea, or 6 *M* guanidine•HCl to solubilize it (Goff et al. 1984; Emtage et al. 1983), suggests an aberrant structure for the prochymosin produced in microorganisms. For this reason, we sought an alternative to production of prochymosin in the yeast cytoplasm. Since calf prochymosin is normally secreted from calf stomach cells, it seemed logical that secretion might be important for proper folding of the protein and formation of disulfide bonds. To test this hypothesis, we developed a yeast secretion system for prochymosin.

Secretion of Prochymosin from Yeast

It is clear from numerous examples of secreted proteins from *E. coli*, yeast, and animal cells that a relatively hydrophobic sequence of amino acids usually located at the amino terminus of the protein plays a key role in directing the secretion of the molecule. In order to direct yeast cells to secrete prochymosin, we examined the effect of using the natural calf secretion signal present on preprochymosin as well as the secretion signals from the secreted yeast proteins, invertase (*SUC*2), and alpha factor (*MF*α). The prochymosin coding sequence was joined to the genes for invertase and alpha factor at convenient restriction endonuclease access sites (*Ava*II and *Hind*III, respectively) to yield fusions of prochymosin to the secretion signal plus a few coding nucleotides of the mature yeast protein (Carlson et al. 1983; Kurjan and Herskowitz 1982). A diagram of the invertase construction and the nucleotide sequence at the fusion joint are shown in Fig. 3. In both cases, glycosylation acceptor sites of the form ASN-X-SER/THR, derived from invertase (denoted by a "+" in Fig. 3) or pro-alpha factor, are present on the resulting fusion protein. Prochymosin itself contains two such acceptor sites, but apparently they are not glycosylated in the calf because purified calf prochymosin contains no detectable carbohydrate (unpublished observations).

Results indicate that both the invertase and the alpha-factor secretion signals direct the glycosylation and secretion of prochymosin-fusion proteins from yeast cells into the medium. The invertase sequence results in secretion of 10% of the prochymosin made, while the alpha factor sequence is slightly less efficient. Although secretion signals are apparently quite conserved through evolution, the natural calf secretion signal for chymosin does not direct secretion of detectable levels of the protein from yeast.

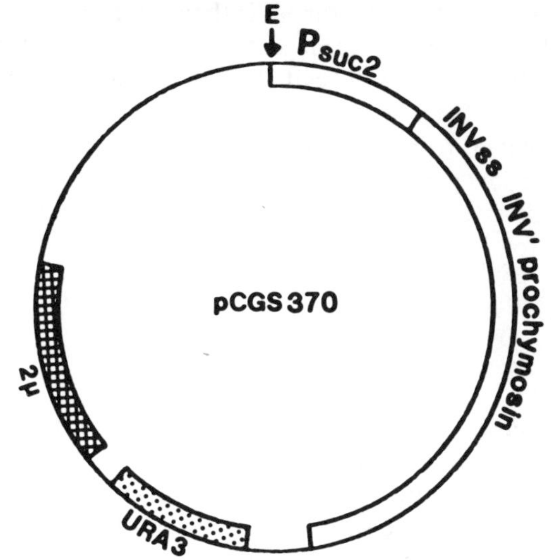


```
      *                                                 met leu
CTCAGAGAAACAAGCAAAACAAAAAGCTTTTCTTTTCACTAACGTATATG ATG CTT
     -40        -30        -20        -10        -1 1

leu gln ala phe leu phe leu leu ala gly phe ala ala lys ile
TTG CAA GCT TTC CTT TTC CTT TTG GCT GGT TTT GCA GCC AAA ATA
 10          20          30           40             50

        ↓           +
ser ala ser met thr asn glu thr ser asp arg pro leu|met ala
TCT GCA TCA ATG ACA AAC GAA ACT AGC GAT AGA CCT TTG ATG GCT
         60          70          80          90

glu ile
GAG ATC ... PROCHYMOSIN
 100
```

FIG. 3. A diagram of the yeast invertase-prochymosin fusion protein expression vector, and the nucleotide and amino acid sequence at the fusion junction. The symbols are as follows: * the site of transcription initiation of the *SUC2* gene; ↓ the site of proteolytic cleavage that removes the secretion signal from yeast invertase; + the first glycosylation recognition sequence (ASN-GLU-THR) of yeast invertase; the fusion joint between invertase and prochymosin.

Glycosylation of the invertase- and alpha factor-prochymosin fusion proteins is consistent with their passage through the secretion system, where accessible ASN-X-SER/THR sequences are normally glycosylated in yeast. Acid activation of the fusion proteins yields chymosin that is not glycosylated. Therefore, the carbohydrate must have been added to the appropriate sites on the invertase and alpha-factor portions of the fusion proteins and not to chymosin itself. These data are consistent with the fact that no carbohydrate can be detected on chymosin secreted from calf stomach cells.

The key finding from this experiment is that all of the secreted prochymosin is activatable to chymosin by the same activation protocol effective for activation of prochymosin derived from the calf. Prochymosin secreted from yeast, unlike prochymosin residing in the yeast cytoplasm, exhibits the same specific activity as prochymosin isolated from the calf. Secreted proteins may require passage through the secretion pathway for correct folding and full biological activity. This would be consistent with evidence from *E. coli*, which indicates that secretion is required for formation of disulfide bonds in beta-lactamase (Pollitt and Zalkin 1983).

Conclusions

A number of structural elements are required for efficient production of prochymosin from yeast. These include a stable, high copy number vector and an efficient promoter of transcription. The experiments described here also indicate that while these elements are sufficient for high level production of prochymosin, little of the material so produced is biologically active. Addition of a yeast secretion signal results in secretion of some of the prochymosin produced, and all of the secreted prochymosin is activatable by conventional methods. Therefore, secretion itself appears to be critical for the production of activatable prochymosin in yeast. This is substantiated by the fact that recent modifications of the secretion system described here have allowed recovery of activatable prochymosin at levels up to 40% of the total prochymosin produced, or almost 80 times the level obtainable from cytoplasmically localized prochymosin (Table 2).

Acknowledgments

We thank Dayle Holmes for preparation of the manuscript. We thank the other members of the yeast expression group whose efforts have been critical to the success of this project: Tina Gill, Iris Klein, Jay Lillquist, Doug Lovern, Susan Porteous, Mary Ellen Rhinehart, Kristin Stashenko, Cathy Stillman, and Edith Yamasaki. We thank Rob Knowlton and Bob Stearman for unpublished data and Gerald Vovis for critically reading the manuscript. We thank Dow Chemical Company for supporting this research and development.

Literature Cited

Beggs, J. D. 1978. Transformation of yeast by a replicating hybrid plasmid. *Nature* 275:104–109.

Carlson, M., R. Taussig, S. Kustu, and D. Botstein. 1983. The secreted form of invertase in *Saccharomyces cerevisiae* is synthesized from mRNA encoding a signal sequence. *Mol. Cell. Biol.* 3:439–447.

Derynck, R., A. Singh, and D. V. Goeddel. 1983. Expression of the human interferon-γ cDNA in yeast. *Nucleic Acids Res.* 11:1819–1837.

Emtage, J. S., S. Angal, M. T. Doel, T. J. R. Harris, B. Jenkins, G. Lilley, and

P. A. Lowe. 1983. Synthesis of calf prochymosin (prorennin) in *Escherichia coli*. *Proc. Natl. Acad. Sci. USA* 80:3671–3675.

Erhart, E., and C. P. Hollenberg. 1983. The presence of a defective *LEU2* gene on 2 micron DNA recombinant plasmids of *Saccharomyces cerevisiae* is responsible for curing and high copy number. *J. Bacteriol.* 156:625–635.

Foltmann, B., V. B. Pedersen, D. Kauffman, and G. Wybrandt. 1979. The primary structure of calf chymosin. *J. Biol. Chem.* 254:8447–8456.

Futcher, A. B., and B. S. Cox. 1984. Copy number and the stability of 2 μm circle-based artificial plasmids of *Saccharomyces cerevisiae*. *J. Bacteriol.* 157:283–290.

Goff, C. G., D. T. Moir, T. Kohno, T. C. Gravius, R. A. Smith, E. Yamasaki, and A. Taunton-Rigby. 1984. Expression of calf prochymosin in *Saccharomyces cerevisiae*. *Gene* 27:35–46.

Harris, T. J. R., P. A. Lowe, A. Lyons, P. G. Thomas, M. A. W. Eaton, T. A. Millican, T. P. Patel, C. C. Bose, N. H. Carey, and M. T. Doel. 1982. Molecular cloning and nucleotide sequence of cDNA coding for calf preprochymosin. *Nucleic Acids Res.* 10:2177–2187.

Hinnen, A., J. B. Hicks, and G. R. Fink. 1978. Transformation of yeast. *Proc. Natl. Acad. Sci. USA* 75:1929–1933.

Ito, H., Y. Fukuda, K. Murata, and A. Kimura. 1983. Transformation of intact yeast cells treated with alkali cations. *J. Bacteriol.* 153:163–168.

Jayaram, M., Y. Y. Li, and J. R. Broach. 1983. The yeast plasmid 2 micron circle encodes components required for its high copy propagation. *Cell* 34:95–104.

Kurjan, J., and I. Herskowitz. 1982. Structure of a yeast pheromone gene (MFα): A putative α-factor precursor contains four tandem copies of mature α-factor. *Cell* 30:933–943.

Moir, D., J. Mao, J. W. Schumm, G. F. Vovis, B. L. Alford, and A. Taunton-Rigby. 1982. Molecular cloning and characterization of double-stranded cDNA coding for bovine chymosin. *Gene* 19:127–138.

Nishimori, K., Y. Kawaguchi, M. Hidaka, T. Uozumi, and T. Beppu. 1982. Nucleotide sequence of calf prorennin cDNA cloned in *Escherichia coli*. *J. Biochem. (Tokyo)* 91:1035–1038.

Pollitt, S., and H. Zalkin. 1983. Role of primary structure and disulfide bond formation in β-lactamase secretion. *J. Bacteriol.* 153:27–32.

St. John, T. P., and R. W. Davis. 1981. The organization and transcription of the galactose gene cluster of *Saccharomyces*. *J. Mol. Biol.* 152:285–315.

CHAPTER 3

Controlled High-Level Expression of Genes Directed by the *trp* Promoter-Operator of *Escherichia coli*

DOUGLAS I. JOHNSON AND RONALD L. SOMERVILLE

Department of Biochemistry, Purdue University, West Lafayette, Indiana 47907

Because of its strength and the ease with which it can be regulated, the *trp* promoter of *E. coli*, and related enterics, is widely used for the hyperexpression of genes that encode proteins of potential economic value. Features that contribute to the strength of the *trp* promoter include the statistically favored "perfect" consensus sequence TTGACA, situated 35 base pairs upstream from the startpoint of transcription. In contrast, the sequence at −10 (TTAACT) differs markedly from the predominant consensus sequence (TATAAT) for this region. Astride the −10 region lies a 22 base pair operator able to form stable complexes with Trp repressor, provided that the effector molecule, L-tryptophan, is present. Such binding excludes RNA polymerase, thereby preventing the initiation of transcription. Specific examples wherein the *trp* promoter has been optimally positioned near coding sequences, then controlled by genetic and physiological means during production runs, are presented. Within other non-*trp* promoters lie secondary targets for *trp* holorepressor. Related regulatory networks that include newly identified repressor genes are currently being studied.

INTRODUCTION

The *trp* regulon of *Escherichia coli* consists of several genes or gene clusters whose promoters contain some version of a DNA sequence that can interact with Trp holorepressor. These genes include the *trp* operon (Morse and Yanofsky 1969), the *aroH* gene (Brown 1968), and the *trpR* gene (Bogosian et al. 1981). The *trp* operon is a stretch of about 7kb of DNA that contains five structural genes *(trpE, D, C, B, A)* whose protein products catalyze the final steps in tryptophan biosynthesis in *E. coli*.

The *trp* operon is under dual control by repression, mediated by Trp repressor-operator interactions, and attenuation. Repression acts as an on-off switch for transcription initiation allowing 70–80-fold range of expression. Attenuation is a fine-tuning mechanism for transcription elongation, which regulates over an 8–10-fold range in response to severe tryptophan starvation (Yanofsky et al. 1984). Together these two mechanisms provide a 500–600-fold range of expression.

The sites of action of both repression and attenuation lie upstream (5′ region) of the *trp* structural genes (Fig. 1). Trp repressor protein, when complexed with

*This is Journal Paper Number 9942 from the Purdue University Agricultural Experiment Station.

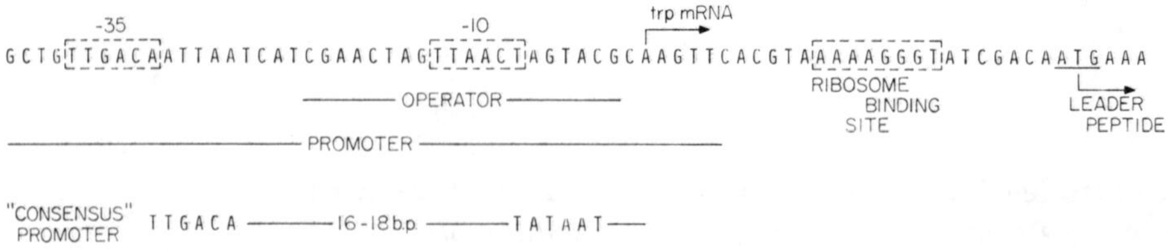

FIG. 1. Structure of the *trp* promoter-operator region of *Escherichia coli*. Numbering is relative to the start of transcription (+1).

tryptophan, binds to a 20–22 base pair target in a DNA termed the operator. The *trp* operator lies within the RNA polymerase recognition site, or promoter (Bennett et al. 1978). The binding of Trp repressor to the *trp* operator excludes the binding of RNA polymerase, thereby preventing transcription into the structural genes (Squires et al. 1975). Attenuation involves the coupling of transcription of the total operon to the translation of a short peptide encoded by DNA upstream of the first structural gene. Premature termination of transcription results when this short tryptophan-rich polypeptide can be readily made by the cell (Yanofsky 1981).

The initiation of transcription of the *trp* structural genes occurs at the *trp* promoter. The signal for RNA polymerase recognition of this region of DNA resides within the nucleotides that comprise the promoter region. There are two blocks of nucleotides that are believed to be critical for promoter recognition by RNA polymerase and the efficient initiation of transcription. These segments of DNA are the "−10" region or Pribnow box and the "−35" region. This numbering system designates the start of transcription as +1. How efficiently RNA polymerase interacts with various promoter sequences is correlated with the degree of homology between a given sequence and a perfect or "consensus" promoter sequence. In addition, the spacing between the "−10" and "−35" region plays a significant role.

The greater the degree of homology between a promoter and the consensus sequence, the stronger the promoter seems to be (Mulligan et al. 1984). The *trp* promoter exhibits 100% homology with the consensus sequence in the "−35" region but only 50% homology in the "−10" region (Fig. 1). Mulligan et al. (1984) have surveyed the known promoters and developed an algorithm to correlate promoter structure and promoter strength. According to their analysis, the *trp* promoter should be approximately 100 times weaker than the strongest characterized promoter, the *strA* promoter, and over 1,000 times stronger than one of the weakest promoters, the *malK* promoter (Table 1). The *trp* promoter is predicted to be about 10 times stronger than the *lac* promoter; this has been shown to be the case experimentally (Windass et al. 1982). Although the *thr* promoter is predicted to be stronger than the *trp* promoter and the *trp* promoter stronger than the λP_L promoter (Table 1), this has not been conclusively proven to date. In fact, *in vivo* studies indicate the λP_L promoter is stronger than the *trp* promoter (Ward and Murray 1979). Nevertheless, it is clear that the *trp* promoter is a strong prokaryotic promoter subject to control by regulatory elements that can be readily manipulated in production systems of current interest in the field of biotechnology. The *trp* promoter-operator system is currently the subject of patent literature and review.

TABLE 1. *Hierarchy of promoter strengths*[a]

Promoter	Homology Score[b]
str	79.9
thr	64.5
trp	61.5
λPl	58.0
lac	49.7
malK	32.0

[a] Promoter strength is defined as K_Bk_2 = rate constant for "open" complex formation (Mulligan et al. 1984).

[b] With each increase of 10 points in homology score, promoter strength increases by a factor of 10.

Results and Discussion

Production Considerations

Recombinant DNA technologies have led to efficient schemes to express mammalian and viral genes in *E. coli* and to produce proteins with potential economic value. In essence, the approach is to fuse the mammalian or viral gene to a strong, regulated *E. coli* promoter together with an appropriate ribosome binding site, thereby bringing the expression of the foreign gene under the control of prokaryotic transcription and translational signals.

The *trp* promoter of *E. coli* has been widely used in such systems. Not only is the *trp* promoter one of high signal strength, but it is also tightly regulated. The production of mammalian gene products fused to the *trp* promoter can thus be turned on or off either genetically or physiologically. Genetically, the approach is to utilize *trpR* mutant strains. Such cells cannot produce active Trp repressor. This leads to high-level constitutive expression from the *trp* promoter. Most *trp* promoter-driven production systems include deletions of the attenuator region, thereby allowing transcription to proceed unimpeded from the *trp* promoter into the fused structural genes. Physiologically, one can increase expression of the *trp* promoter in a *trpR*$^+$ cell by either imposing starvation for tryptophan or through the use of the gratuitous inducer indole acrylic acid (IAA). Indole acrylic acid combines with Trp repressor to form a complex known as a pseudorepressor. This species cannot bind to the *trp* operator resulting in relief of repression. Physiological control of the *trp* promoter is well worth considering in the design of a new system because of the sometimes deleterious effects of overproduction of mammalian gene products in *E. coli* (Rose and Shafferman 1981). Moreover, severe tryptophan starvation has been shown to adversely affect the expression of tryptophan-containing proteins; therefore, it is important to know the tryptophan content of the gene products to be produced.

Gene Products under trp *Control: Some Examples*

A wide variety of mammalian and viral genes have been placed under the control of the *trp* promoter and expressed in *E. coli* (Tables 2–4) (Boss et al. 1984; Boss and Emtage 1984; Emtage et al. 1980; Kleid et al. 1981; Tacon et al. 1983; McGrath and Levinson 1982; Yelverton et al. 1983; Lawn et al. 1981; Rose and Shafferman 1981; Rosenberg et al. 1984; Emtage et al. 1983; Martial et al. 1979;

Edman et al. 1981; Burnett 1983; Gray et al. 1984b; Leung et al. 1984; Seeburg et al. 1983; Goeddel et al. 1980a; Goeddel et al. 1980b; Pennica et al. 1983; Weck et al. 1981; Devos et al. 1983; Kenten et al. 1984). In addition to the *trp* promoter-operator control region, most of these constructions use the ribosome binding site and start codon of the *trp* leader peptide. The absence of the *trp* attenuator in these constructions greatly increases the amount of transcription into the cloned gene and usually leads to enhanced production of the desired gene product (Tables 2–4). Besides the products listed, a synthetic interferon gene has been fused to a chemically synthesized *trp* promoter (Windass et al. 1982), the bacterial exotoxin A gene has been fused to the *E. coli trp* promoter (Gray et al. 1984a), and a human leukocyte-type interferon has been fused to the *S. marcescens trp* promoter and an artificial ribosome binding site (Dworkin-Rastl et al. 1983).

TABLE 2. *Human gene products*

Human Gene Product	*trp* Attenuator	Amount Produced
Interleukin-2[a,b]	–	5–10% Total protein
β-Urogastrone[c]	–	10–14% Total protein
β-Interferon[c]	–	15% Total protein
Growth Hormone[d]	+	3% Total protein
Leukocyte-type Interferon[e,f]	–	12,000 Molecules/cell
Fibroblast Interferon[g]	–	4,500 Molecules/cell
Plasminogen Activator[h]	–	2,000 Molecules/cell
Serum Albumin[i]	–	"Modest"
Insulin[j]	+	?

[a] Rosenberg et al. (1984); [b] Devos et al. (1983); [c] Tacon et al. (1983); [d] Martial et al. (1979); [e] Goeddel et al. (1980b); [f] Weck et al. (1981); [g] Goeddel et al. (1980a); [h] Pennica et al (1983); [i] Lawn et al. (1981); [j] Burnett (1983).

TABLE 3. *Mammalian Gene Products*

Mammalian Gene Product	*trp* Attenuator	Amount Produced
Murine λ light chain[a]	–	13% Total protein
Murine μ heavy chain[a]	–	1–2% Total protein
Porcine growth hormone[b]	–	1.5 g/Liter cells
Bovine growth hormone[b]	–	1.5 g/Liter cells
Bovine prorennin[c]	–	1–5% Total protein
Bovine interferon[d]	–	200,000 Molecules/cell

[a] Boss et al. (1984); [b] Seeburg et al. (1983); [c] Emtage et al. (1983); [d] Leung et al. (1984).

Upon induction, one can produce up to 15% of the total cellular protein as the cloned gene product (Tables 2–4). The amount of production varies over a wide range and is probably attributable to either poor translation of the eukaryotic gene or instability of the gene product *in vivo*. The problem of protein instability has been overcome through the use of in-frame fusions to prokaryotic proteins such as the *trp* leader polypeptide and the *trpE* or *trpD* gene products. The purification of one such polypeptide, the viral protein vaccine for Foot and Mouth Disease (Kleid et al. 1981), is greatly enhanced by the inherent insolubility of this fusion polypeptide in cell extracts.

TABLE 4. *Viral gene products*

Viral Gene Products	*trp* Attenuator	Amount Produced
VSV Glycoprotein[a]	+	<1% Total protein
FPV Haemagglutinin[b]	+	2–3% Total protein
Hepatitis B antigens[c]	−	10% Total protein
Foot and mouth disease vaccine[d]	−	12% Total protein
Rabies virus glycoprotein[e]	−	2–3% Total protein
RSV *src* gene[f]	−	5% Total protein

[a] Rose and Shafferman (1981); [b] Emtage et al. (1980); [c] Edman et al. (1981); [d] Kleid et al. (1981); [e] Yelverton et al. (1983); [f] McGrath and Levinson (1982).

Other Repression Systems and Potential Complications

While studying the regulation of the *trpR* gene, we uncovered a complex regulatory network involving several other amino acid biosynthetic operons (Bogosian and Somerville 1983; Johnson and Somerville 1983; Johnson and Somerville 1984). Strains harboring plasmids that lead to the hyperproduction of Trp repressor exhibited a new compound nutritional requirement upon growth in the presence of tryptophan. This requirement (the "TROP" syndrome) was satisfied by the addition of threonine, isoleucine, leucine, valine, phenylalanine, tyrosine, and serine. Low-affinity operators similar to *trpO* were found within the promoter regions of these amino acid biosynthetic operons (Bogosian and Somerville 1983) suggesting that Trp repressor, when hyperproduced, could act as a surrogate repressor for the previously mentioned operons.

A mutation defining a repressor gene involved in the regulation of the *thr* and *ilv* operons has been described (Johnson and Somerville 1983; Johnson and Somerville 1984). This gene, designated *ileR*, encodes a repressor protein of MW 15,000 (Yandle et al., in preparation), which acts independently of attenuation, to modulate transcription of the *thr* and *ilv* operons (Johnson and Somerville 1984). These two operons, therefore, more closely resemble the *trp* operon than previously assumed, with respect to transcriptional regulation. During this work, it was also found that the *ilv* and *thr* operons were under positive regulation at the level of transcription initiation, mediated by diffusible factors (Johnson and Somerville 1984). This raises the possibility that other biosynthetic operons, such as the *trp* operon, could also be positively regulated and, therefore, expression from the *trp* promoter, like that from the *thr* and *ilv* promoters, could be increased significantly above the levels that have so far been observed. Studies to determine if this is the case are presently underway.

CONCLUSIONS

The *trp* promoter of *E. coli*, and other enterics, is a strong, regulated promoter that can be easily manipulated to drive the expression and production of foreign genes and gene products in *E. coli*. The *trp* promoter has advantages over other promoters such as the *lac* promoter in that it is about 10-fold stronger than the *lac* promoter and it is less expensive to induce expression from the *trp* promoter (tryptophan starvation vs. addition of IPTG). Recent work has shown that the *thr* and *ilv* operons are regulated in a way that is similar to the *trp* operon.

According to Mulligan et al. (1984), the *thr* promoter is predicted to be 3–4 times stronger than the *trp* promoter (Table 1). Whether this is the case *in vivo* and whether the *thr* promoter might in time supplant the *trp* promoter in production systems are questions that are being addressed. Previous work with the *trp* system provides us with useful guideposts.

Acknowledgments

Work described herein was supported by a grant (GM 22131) from the U.S. Public Health Service. D. I. Johnson was supported by an N.I.H. training grant (GM 07211-08).

Literature Cited

Bennett, G., M. Schweingruber, K. Brown, C. Squires, and C. Yanofsky. 1978. Nucleotide sequence of the promoter-operator region of the tryptophan operon of *Escherichia coli*. *J. Mol. Biol.* 121:113–137.

Bogosian, G., and R. L. Somerville. 1983. Trp repressor is capable of intruding into other amino acid biosynthetic systems. *Mol. Gen. Genet.* 191:51–58.

Bogosian, G., K. Bertrand, and R. L. Somerville. 1981. Trp repressor protein controls its own structural gene. *J. Mol. Biol.* 149:821–825.

Boss, M. A., and J. S. Emtage. 1984. Expression of an immunoglobulin light chain gene in *Escherichia coli*. Pages 513–522 *in* D. H. Hamer and M. J. Rosenberg, eds., *Gene Expression*. Alan R. Liss, Inc. New York.

Boss, M. A., J. H. Kenton, C. R. Wood and J. S. Emtage. 1984. Assembly of functional antibodies from immunoglobulin heavy and light chains synthesized in *Escherichia coli*. *Nucleic Acids Res.* 12:3791–3806.

Brown, K. 1968. Regulation of aromatic amino acid biosynthesis in *Escherichia coli* K-12. *Genetics* 60:31–48.

Burnett, J. P. 1983. Commercial production of recombinant DNA-derived products. Pages 259–277 *in* M. Inouye, ed., *Experimental Manipulation of Gene Expression*. Academic Press, New York.

Devos, R., G. Plaetinck, H. Cheroutre, G. Simons, W. Degrave, J. Tavernier, E. Remaut, and W. Fiers. 1983. Molecular cloning of human interleukin-2 cDNA and its expression in *E. coli*. *Nucleic Acids Res.* 11:4307–4323.

Dworkin-Rastl, E., P. Swetly, and M. B. Dworkin. 1983. Construction of expression plasmids producing high levels of human leukocyte-type interferon in *Escherichia coli*. *Gene* 21:237–248.

Edman, J. C., R. A. Hallewell, R. Valenzuela, H. M. Goodman, and W. J. Rutter. 1981. Synthesis of hepatitis B surface and core antigens in *Escherichia coli*. *Nature* 291:503–506.

Emtage, J. S., S. Angal, M. T. Doel, T. J. R. Harris, B. Jenkins, G. Lilley, and P. A. Lowe. 1983. Synthesis of calf prochymosin in *Escherichia coli*. *Proc. Natl. Acad. Sci. USA*. 80:3671–3675.

Emtage, J. S., W. C. A. Tacon, G. H. Catlin, B. Jenkins, A. G. Porter, and N. H. Carey. 1980. Influenza antigenic determinants are expressed from haemagglutinin genes cloned in *E. coli*. *Nature* 283:171–174.

Goeddel, D. V., H. M. Shepard, E. Yelverton, D. Leung, and R. Crea. 1980. Synthesis of human fibroblast interferon by *E. coli. Nucleic Acids Res.* 80:4057–4074.

Goeddel, D. V., E. Yelverton, A. Ullrich, H. L. Heyneker, G. Miozzari, W. Holmes, P. H. Seeburg, T. Dull, L. May, N. Stebbing, R. Crea, S. Maeda, R. McCandliss, A. Sloma, J. M. Tabor, M. Gross, P. C. Familletti, and S. Pestka. 1980. Human leukocyte interferon produced by *E. coli* is biologically active. *Nature* 287:411–416.

Gray, G. L., K. A. McKeown, A. J. S. Jones, P. H. Seeburg, and H. L. Heyneker. 1984. *Pseudomonas aeruginosa* secretes and correctly processes human growth hormone. *Bio/Technology* 2:161–165.

Gray, G. L., D. H. Smith, J. S. Baldridge, R. N. Harkins, M. L. Vasil, E. Y. Chen, and H. L. Heyneker. 1984. Cloning, nucleotide sequence, and expression in *E. coli* of the exotoxin A structural gene of *Pseudomonas aeruginosa. Proc. Natl. Acad. Sci. USA*. 81:2645–2649.

Johnson, D. I., and R. L. Somerville. 1983. Evidence that repression mechanisms can exert control over the *thr, leu,* and *ilv* operons of *Escherichia coli* K-12. *J. Bacteriol.* 155:49–55.

Johnson, D. I., and R. L. Somerville. 1984. New regulatory genes involved in the control of transcription initiation at the *thr* and *ilv* promoters of *Escherichia coli* K-12. *Mol. Gen. Genet.* 195:70–76.

Kenten, J., B. Helm, T. Ishizaka, P. Cattins, and H. Gould. 1984. Properties of a human immunoglobulin ϵ-chain fragment synthesized in *Escherichia coli. Proc. Natl. Acad. Sci. USA*. 81:2955–2959.

Kleid, D. G., D. Yansura, B. Small, D. Borobenko, D. M. Moore, M. J. Grubman, P. D. McKercher, D. O. Morgan, B. H. Robertson, and H. L. Bachrach. 1981. Cloned viral protein vaccine for foot and mouth disease: Response in cattle and swine. *Science* 214:1125–1128.

Lawn, R. M., J. Adelman, S. C. Bock, A. E. Franke, C. M. Houck, R. C. Najarian, P. H. Seeburg, and K. W. Wion. 1981. The sequence of human serum albumin cDNA and its expression in *Escherichia coli. Nucleic Acids Res.* 9:6103–6114.

Leung, D. W., D. J. Capon, and D. V. Goeddel. 1984. The structure and bacterial expression of three distinct bovine interferon-β genes. *Bio/Technology* 2:458–464.

J. A. Martial, R. A. Hallewall, J. D. Baxter, and H. M. Goodman. 1979. Human growth hormone: Complementary DNA cloning and expression in bacteria. *Science* 205:602–607.

McGrath, J. P., and A. D. Levinson. 1982. Bacterial expression of an enzymatically active protein encoded by RSV *src* gene. *Nature* 295:423–425.

Morse, D. E., and C. Yanofsky. 1969. Amber mutants of the *trpR* regulatory gene. *J. Mol. Biol.* 44:185–193.

Mulligan, M. E., D. K. Hawley, R. Entriken, and W. R. McClure. 1984. *Escherichia coli* promoter sequences predict *in vitro* RNA polymerase selectivity. *Nucleic Acids Res.* 12:789–800.

Pennica, D., W. E. Holmes, W. J. Kohr, R. N. Harkins, G. N. Vehar, C. A. Ward, N. F. Bennett, E. Yelverton, P. H. Seeburg, H. L. Heynecker, D. V.

Goeddel, and D. Collen. 1983. Cloning and expression of human tissue-type plasminogen activator cDNA in *E. coli*. *Nature* 301:214–221.

Platt, T. 1978. Regulation of gene expression in the tryptophan operan *Escherichia coli*. Pages 263–302 *in* J. Miller and W. Reznikoff, eds., *The Operon*. Cold Spring Harbor Laboratory, Cold Spring Harbor, NY.

Rose, J. K., and A. Shafferman. 1981. Conditional expression of the vesicular stomatitis virus glycoprotein gene in *Escherichia coli*. *Proc. Natl. Acad. Sci. USA*. 78:6670–6674.

Rosenberg, S. A., E. A. Grimm, M. McGrogan, M. Doyle, E. Kawasaki, K. Koths, and D. F. Mark. 1984. Biological activity of recombinant human interleukin-2 produced in *Escherichia coli*. *Science* 223:1412–1415.

Seeburg, P. H., S. Sias, J. Adelman, H. A. deBoer, J. Hayflick, P. Jhurani, D. V. Goeddel, and H. L. Heyneker. 1983. Efficient bacterial expression of bovine and porcine growth hormones. *DNA* 2:37–45.

Squires, C., F. Lee, and C. Yanofsky. 1975. Interaction of the *trp* repressor and RNA polymerase with the *trp* operon. *J. Mol. Biol.* 92:93–111.

Tacon, W. C. A., W. A. Bonass, B. Jenkins, and J. S. Emtage. 1983. Expression plasmid vectors containing *Escherichia coli* tryptophan promoter transcriptional units lacking the attenuator. *Gene* 23:255–265.

Ward, D. F., and N. E. Murray. 1979. Convergent transcription in bacteriophage λ: Interference with gene expression. *J. Mol. Biol.* 133:249–266.

Weck, P. K., S. Apperson, N. Steebing, P. W. Gray, D. Leung, H. M. Shepard, and D. V. Goeddel. 1981. Antiviral activities of hybrids of two major human leukocyte interferons. *Nucleic Acids Res.* 9:6153–6166.

Windass, J. D., C. R. Newton, J. DeMeyer-Guignard, V. E. Moore, A. F. Markham, and M. D. Edge. 1982. The construction of a synthetic *E. coli trp* promoter and its use in the expression of a synthetic interferon gene. *Nucleic Acids Res.* 10:6639–6657.

Yanofsky, C. 1981. Attenuation in the control of expression of bacterial operons. *Nature* 289:751–758.

Yanofsky, C., R. L. Kelley, and V. Horn. 1984. Repression is relieved before attenuation in the *trp* operon of *Escherichia coli* as tryptophan starvation becomes increasingly severe. *J. Bacteriol.* 158:1018–1024.

Yelverton, E., S. Norton, J. F. Obijeski, and D. V. Goeddel. 1983. Rabies virus glycoprotein analogs: Biosynthesis in *Escherichia coli*. *Science* 219:614–620.

II
SYMPOSIUM: MICROBIAL TRANSFORMATIONS OF ANTIBIOTICS AND ANTI-TUMOR AGENTS

V. P. Marshall, *Convener*
The Upjohn Company, Kalamazoo, Michigan

PANELISTS

M. W. Duffel
F. M. Eckenrode
J. R. Fang
F. Filippelli
W. Z. Jin
D. A. Lowe
V. P. Marshall
C. J. Pearce
K. L. Rinehart
J. P. Rosassa
F. S. Sariaslani
K. I. Tadano
T. Toyokuni
P. F. Wiley

CHAPTER 4

Microbial Conversion of Macrolides

PAUL F. WILEY

Research Laboratories, The Upjohn Company,
Kalamazoo, Michigan 49001

Microbial conversion of macrolide antibiotics has been quite extensive with the 16-membered lactone ring type being investigated most frequently. Various species of *Streptomyces* have been used most often, but representatives of some dozens of other genera have been employed. The most common conversions have been acylation, deacylation, oxidation, and reduction. Macrolide aglycones have been glycosylated to give compounds, particularly the (8S)-8-fluoroerythromycins, which may be superior to their parent antibiotics. In many cases, selective reactions occur that would be very difficult to do chemically.

DISCUSSION

The original definition of macrolide antibiotics was meant to include those antibiotics having 12-, 14-, or 16-membered lactone rings with an attached amino sugar. Subsequently the term was expanded to include similar antibiotics with neutral sugars but containing no amino sugar. Still later the term was applied to any antibiotic having a large number of atoms in a lactone ring but otherwise quite dissimilar chemically. In this discussion the emphasis will be on antibiotics fitting the original macrolide definition with only two others, lankamycin (*1*) and lankacidins (*2*), being included. In addition to antibiotic conversions a number of aglycone conversions to antibiotics and antibiotic-related compound conversions will be discussed. The literature is covered through 1983.

Structures of macrolide antibiotics are shown here.

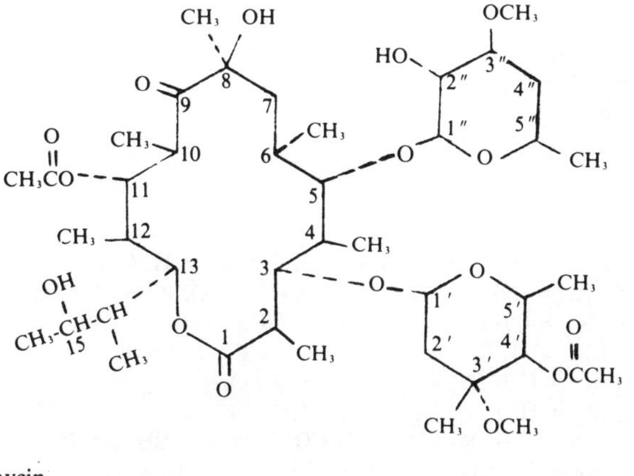

FIG. 1. Lankamycin

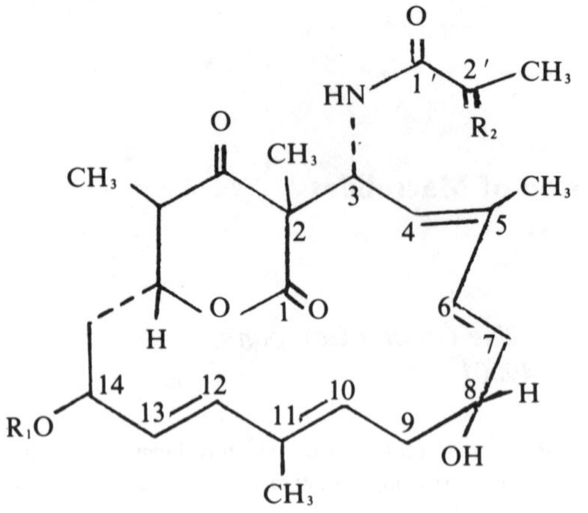

FIG. 2. Lankacidin C: R_1 = H; R_2 =O
Lankacidin A: R_1 = CH_3CO; R_2 = O
Lankacidinol A: R_1 = CH_3CO: R_2 = H, OH
Lankacidinol: R_1 = H; R_2 = H, OH

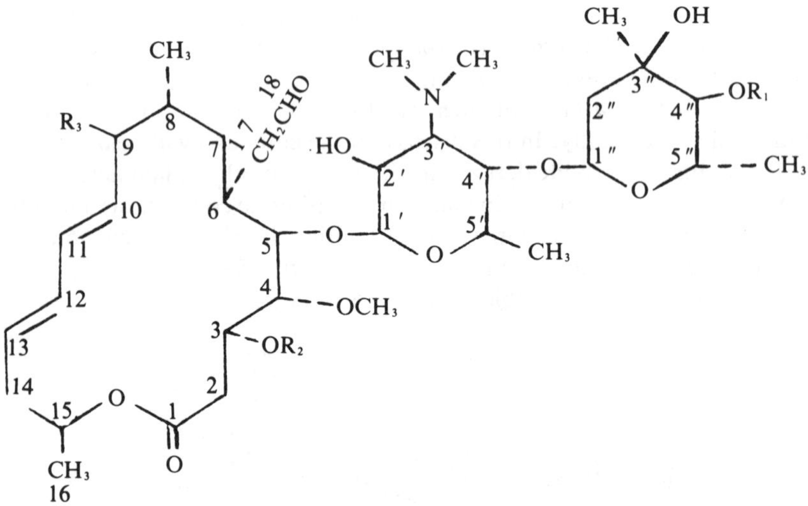

FIG. 3. Leucomycin A_1: R_1 = H, R_2 = $(CH_3)_2CHCH_2CO$, R_3 = H, OH.
Leucomycin A_3: R_1 = CH_3CO, R_2 = $(CH_3)_2CHCH_2CO$, R_3 = H, OH.
Leucomycin A_4: R_1 = CH_3CO, R_2 = $CH_3CH_2CH_2CO$, R_3 = H, OH.
Leucomycin A_5: R_1 = H, R_2 = $CH_3CH_2CH_2CO$, R_3 = H, OH.
Leucomycin U: R_1 = CH_3CO, R_2 = H, R_3 = H, OH.
Leucomycin V: R_1 = H, R_2 = H, R_3 = H, OH.
Midecamycin A_1, SF-837A: R_1 = CH_3CH_2CO, R_2 = CH_3CH_2CO, R_3 = H, OH.
Carbomycin B: R_1 = CH_3CO, R_2 = $(CH_3)_2CHCH_2CO$, R_3 = O.
Midecamycin A_3, SF-837A_3: R_1 = CH_3CH_2CO, R_2 = CH_3CH_2CO, R_3 = O.
Niddamycin: R_1 = H, R_2 = $(CH_3)_2CHCH_2CO$, R_3 = O.

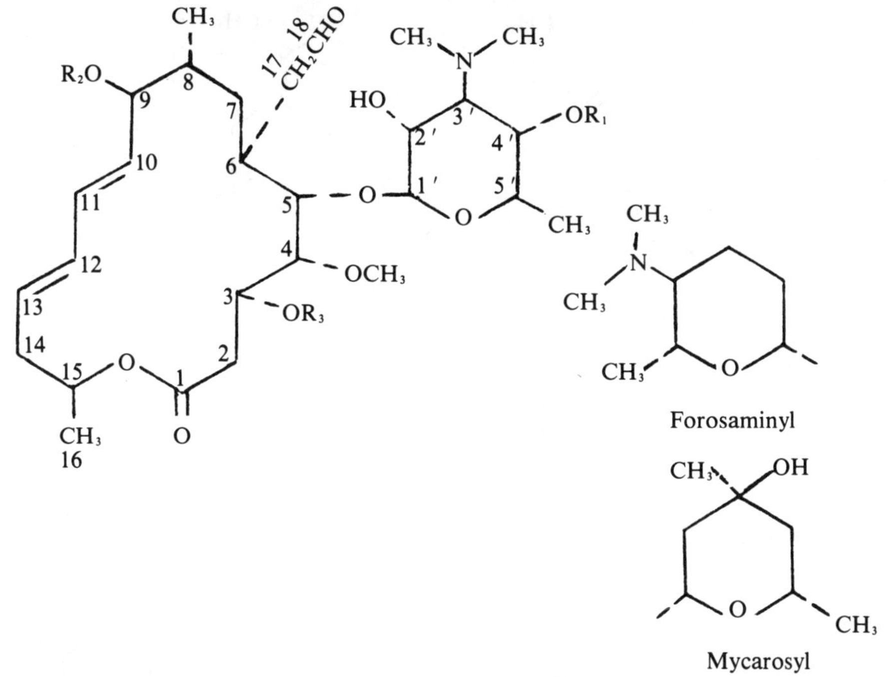

FIG. 4. Forocidins: $R_2 = R_3 = H$; I, $R_1 = H$; II, $R = CH_3CO$; III, $R_1 = CH_3CH_2CO$
Neospiramycin: R_2 = Forosaminyl; $R_3 = H$; I, $R_1 = H$; II, $R_1 = CH_3CO$; III, $R = CH_3CH_2CO$
Spiramycins: R_2 = Forosaminyl; R_3 = mycarosyl; I, $R_1 = H$; II, $R_1 = CH_3CO$; III, $R_1 = CH_3CH_2CO$

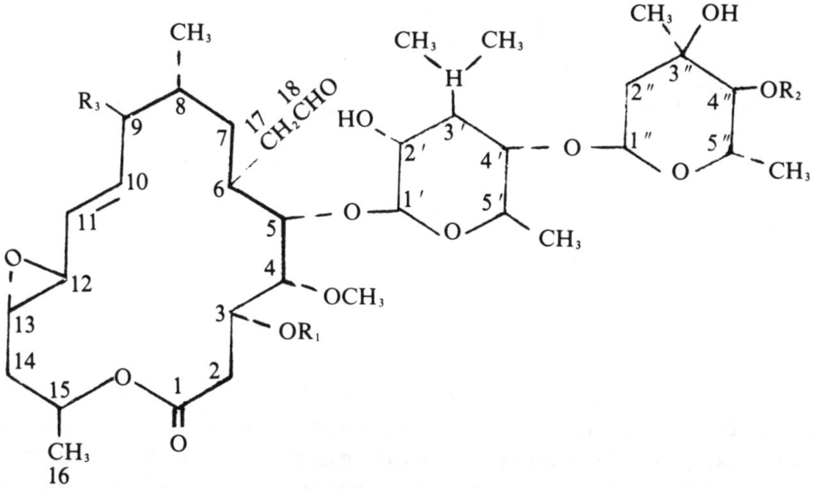

FIG. 5. Carbomycin A: $R_1 = CH_3CO$, $R_2 = (CH_3)_2CHCH_2CO$, $R_3 = O$
Maridomycin I: $R_1 = CH_3CH_2CO$, $R_2 = (CH_3)_2CHCH_2CO$, R_3 = H, OH
Maridomycin II: $R_1 = CH_3CO$, $R_2 = (CH_3)_2CHCH_2CO$, R_3 = H, OH
Maridomycin III: $R_1 = CH_3CH_2CO$, $R_2 = CH_3CH_2CO$, R_3 = H, OH
Maridomycin V: $R_1 = CH_3CH_2CO$, $R_2 = CH_3CO$, R_3 = H, OH

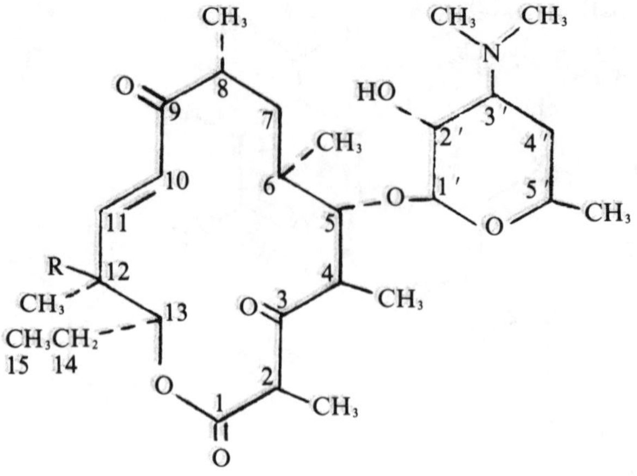

FIG. 6. Picromycin: R = OH
Narbomycin: R = H

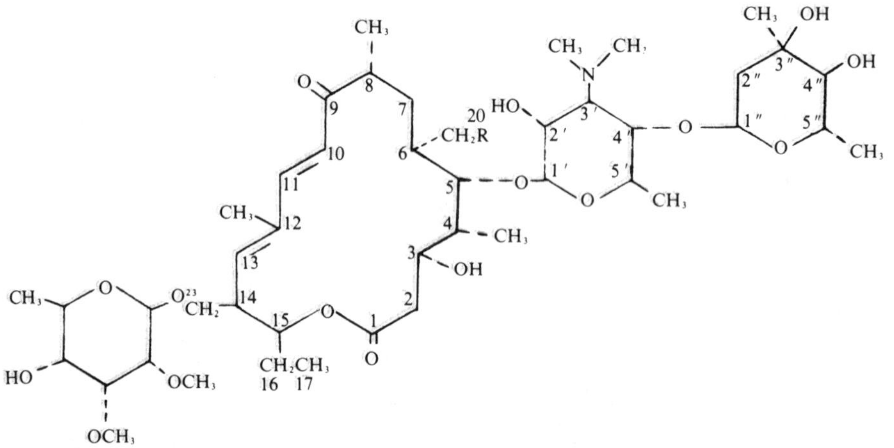

FIG. 7. Tylosin: R = CHO
Relomycin: R = CH_2OH

Many, if not most, macrolide conversions have been done in studies of macrolide biosynthesis (Kitao et al. 1979b; Omura et al. 1982) rather than in attempting to improve the properties of a known antibiotic. For the most part, conversions have been relatively simple ones such as acylation, deacylation, oxidation, and reduction. Although these reactions are rather routine, some interesting specificities have been observed. Substrates for conversion have been overwhelmingly 16-membered ring macrolides. For the most part mutant *Streptomyces* or enzymes derived from them have been used, but a large number of other genera have been utilized.

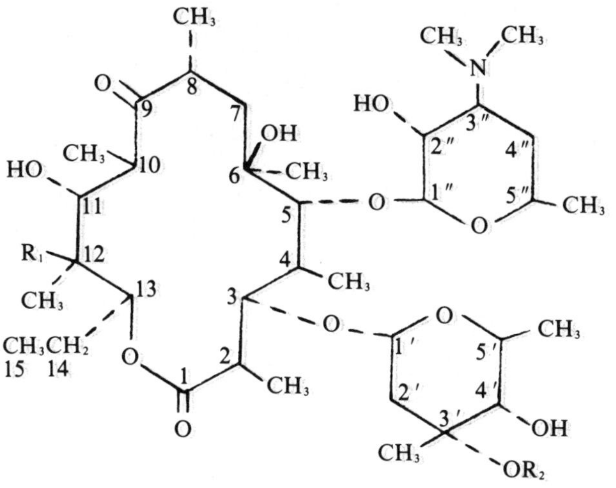

FIG. 8. Erythromycin A: R_1 = OH, R_2 = CH_3
Erythromycin B: R_1 = H, R_2 = CH_3
Erythromycin C: R_1 = OH, R_2 = H
Erythromycin D: R_1 = H, R_2 = H

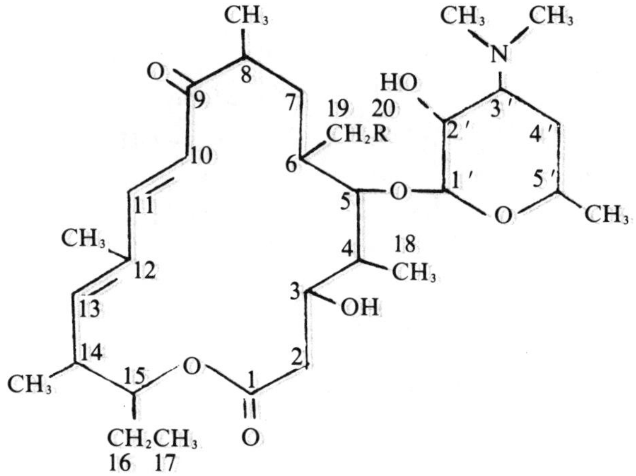

FIG. 9. Juvenimicin B_1: R = CH_2OH
M-4365G_2: R = CHO

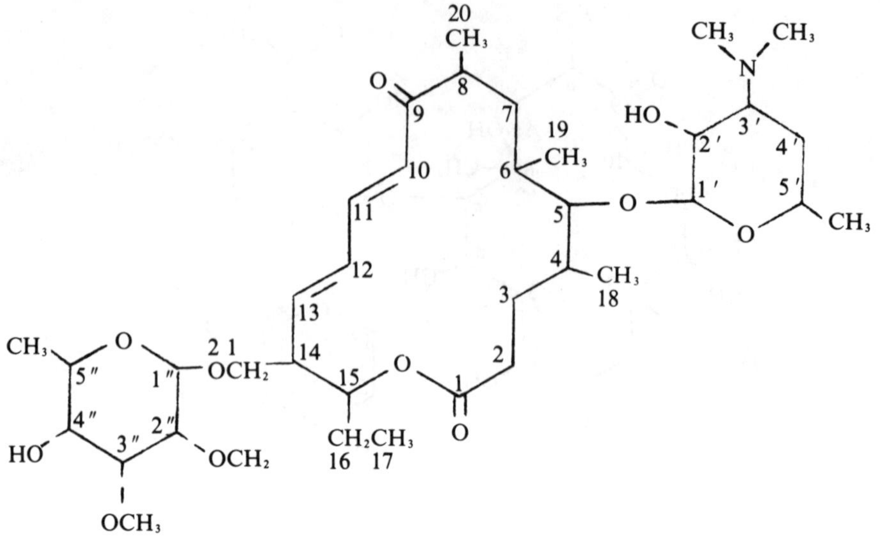

FIG. 10. Mycinamicin IV

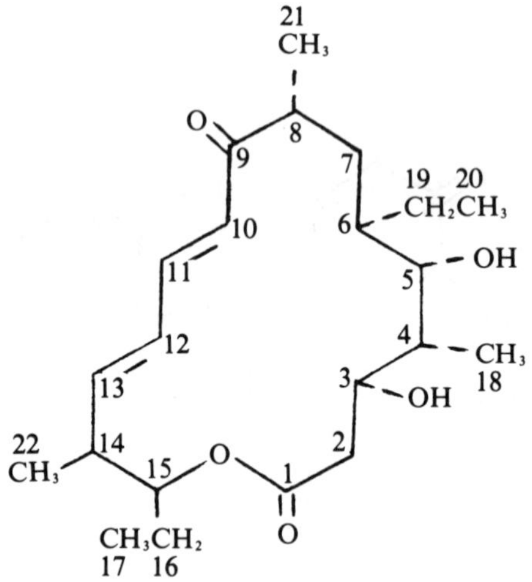

FIG. 11. Protylonolide

As many macrolide subgroups contain members that differ from each other in degree and kind of acylation, conversions involving acylation and deacylation (Table 1) alone and in combination with other changes (Table 2) have been the ones most commonly reported. Acylation of lankacidin (*2*) at C-8 by several acyl groups by adding a compound containing the acyl group to a fermentation of *Bacillus megatherium* IFO 12108 or using cell-free extracts with lankacidin as the substrate has been reported by Nakahama et al. (1975). Harada et al. (1973), using *Streptomyces rochei* var. *volubilis* cell-free extracts, report similar results as do Fugono et al. (1970), using *Aspergillus sojae, Aspergillus niger,* and *Trametes sanguinea.* These same organisms have been used (Harada et al. 1973; Fugono et al. 1970, 1971) to deacylate similar compounds. With lankacidins acylated at C-8 and C-14, deacylation occurred at C-14 but not at C-8. As the leucomycins (*3*) contain one acyl group at C-4″ and usually a second at C-3, they afford a considerable field for adding and removing acyl groups. Acylation (Kitao et al. 1979a) has usually been done with the organism *Streptoverticillium kitasatoensis* 66–14–3. This same organism deacylates leucomycin A_3 to leucomycin A_1. Theriault (1974) has reported deacylation using eight different genera. Acylation and deacylation of the spiramycin group (*4*), the carbomycins, and maridomycins (*3,5*), and a few other minor macrolides have been reported.

Only a few oxidations have been achieved, but these represent a number of different types and considerable specificity. For example Suzuki et al. (1977) have reported oxidation of 12,13-olefinic bonds to the corresponding epoxides with no attack on the neighboring double bond. Oxidation of isovaleryl groups in maridomycin I (*5*) and leucomycin A_3 (*3*) resulted in introduction of a hydroxyl at C-3‴ (Nakahama et al. 1974a). Hydroxylation of C-12 in narbomycin (*6*) to give picromycin using *Streptomyces zaomyceticus* MCRL-0405 also occurs (Maezawa et al. 1973). However, the most common oxidation is conversion of a C-9 hydroxyl group to a ketone carbonyl or conversion of a C-18 or C-20 hydroxyl to an aldehyde.

Only a few reports of reduction have appeared. These have been, almost entirely, reduction of a carbonyl group to a hydroxyl, usually involving the C-9 carbonyl. Omura et al. (1980b) and Feldman et al. (1963) have reduced the aldehyde in tylosin (*7*) using *Streptomyces ambofaciens* KA-1028, *Streptomyces hygroscopicus, Streptomyces espiralis,* and *Nocardia corallina.* A reduction of tylosin at both the C-9 and C-20 carbonyls has been achieved by Omura et al. (1980b).

A few cases of glycosylation have been reported in the erythromycin group (Toscano et al. 1983b), the spiramycin group (Omura et al. 1982), and of the tylosin type (Omura et al. 1982). For the most part, the organisms used were ones that produce macrolide antibiotics, such as mutants of *Streptomyces erythreus* and *Streptomyces fradiae.*

A few cases of O-methylation and O-demethylation are known usually involving, in the case of the O-methylation, a final step in a macrolide synthesis (Corcoran 1975).

Some conversions have been reported in which more than one chemical change has occurred. In all examples the changes taking place were ones already discussed above. The organisms used were normally mutants of macrolide-producing organisms.

TABLE 1. *Single biotransformations*

Substrate	Transformation Product	Biological Agent	Reference
1. Acylation			
4″-O-Depropionyl-maridomycin III	Maridomycin V	*Streptomyces* sp. Strain No. K-432[a]	Uyeda et al. (1977)
5-O-(4′,6′-Dideoxy-3′-C-acetyl-β-D-xylohexopyranosyl)-platenolide II	3-O-Acetyl-5-D-(4′,6′-dideoxy-3′-C-acetyl-β-D-xylohexo-pyranosyl)-platenolide II	*S. hygroscopicus* IMETJA 6599-NG-33–354r[a]	Gräfe et al. (1980a)
	3-O-Propionyl-5-O-(4′,6′-di-deoxy-3′-C-acetyl-β-D-xylo-hexopyranosyl)-platenolide II		
9-Dihydrojuvenimicin B	3-O-Acetyl-9-dihydro-juvenimicin B_1	*S. ambofaciens* KA-1028[a]	Sadakane et al. (1982)
Forocidin I	Forocidin III	*S. ambofaciens*[a]	Omura et al. (1979b)
Lankacidin C + Formate	14-O-Formyllankacidin C	*A. niger*[a], *A. sojae*[a], *T. sanguinea* [a]	Fugono et al. (1970)
Lankacidin C + Acetate	Lankacidin A	*A. niger*[a], *A. sojae* [a], *T. sanguinea*[a]	Fugono et al. (1970)
Lankacidin C + pro-pionate	14-O-Propionyllankacidin C	*A. niger*[a], *A. sojae*[a], *T. sanguinea*[a]	Fugono et al. (1970)
Lankacidin C	8-O-Acetyllankacidin C	*B. megatherium* IFO 12108[a,b]	Nakahama et al. (1975)
Lankacidin C + Butyrate	8-O-Butyryllankacidin C	*B. megatherium* IFO 12108[a,b]	Nakahama et al. (1975)
Lankacidin C + Isobutyrate	8-O-Isobutyryllankacidin C	*B. megatherium* IFO 12108[a,b]	Nakahama et al. (1975)
Lankacidin C + Valerate	8-O-Valeryllankacidin C	*B. megatherium* IFO 12108[a,b]	Nakahama et al. (1975)
Lankacidin C + Isovalerate	8-O-Isovaleryllankacidin C	*B. megatherium* IFO 12108[a,b]	Nakahama et el. (1975)
Lankacidinol + Acetate	Lankacidinol A 2′-O-Acetyllankacidinol A	*S. rochei* var. *volubilis*[b]	Harada et al. (1973)
Lankacidinol + Propionate	14-O-Propionyllankacidinol	*S. rochei* var. *volubilis*[b]	Harada et al. (1973)

Substrate	Transformation Product	Biological Agent	Reference
Leucomycin A_1	Leucomycin A_3	Macrolide-3-acyltransferase[b]	Okamoto et al. (1977)
		S. kitasatoensis 66–14–3[a]	Kitao et al. (1979b, 1979a), Omura et al. (1976, 1977)
Leucomycin U	Leucomycin A_3	*S. kitasatoensis* 66–14–3[a]	Kitao et al. (1979b)
Leucomycin V	Leucomycin A_3	*S. kitasatoensis* 66–14–3[a]	Kitao et al. (1979b)
Leucomycin V	Leucomycin U	*S. kitasatoensis* 66–14–3[a]	Kitao et al. (1979b)
M-4365G_2	3-O-Acetyl M-4365G_2	*S. kitasatoensis* 66–14–3[a]	Sadakane et al. (1982)
Neospiramycin I	Neospiramycin III	*S. ambofaciens* KA-1028[a]	Kitao et al. (1979b) Omura et al. (1979b)
Spiramycin I	Spiramycin II	*S. ambofaciens*[a]	Omura et al. (1979a)
Spiramycin I	Spiramycin III	*S. ambofaciens*[a]	Omura et al. (1979b)
2. Deacylation 8-O-Acetyllankacidin A	8-O-Acetyllankacidin C	*S. rochei* var. *volubilis*[b]	Harada et al. (1973)
3″-O-Acetylleucomycin A_5	Leucomycin V	*Nocardia* sp. A-2440[a]	Toyo Jozo Co., Ltd. (1981)
8-O-Benzoyllankacidin A	8-O-Benzoyllankacidin C	*S. rochei* var. *volubilis*[b]	Harada et al. (1973)
8-O-Butyryllankacidin A	8-O-Butryllankacidin C	*S. rochei* var. *volubilis*[b]	Harada et al. (1973)
Carbomycin A	4″-O-Deisovalerylcarbomycin A	8 Genera	Theriault (1974)
Carbomycin B	4″-O-Deisovalerylcarbomycin B	8 Genera	Theriault (1974)
8-O-Crotonyllankacidin A	8-O-Crotonyllankacidin C	*S. rochei* var. *volubilis*[b]	Harada et al. (1973)
8,14-Di-O-acetyl-lankacidin C	8-O-Acetyllankacidin C	*A. niger*[b], *A. sojae*[b], *T. sanguinea*[b]	Fugono et al. (1970)
2′,8-Di-O-butyryl-	2′,8-Di-O-acetyllankacidinol	*S. rochei* var. *volubilis*[b]	Harada et al. (1973)
lankacidin A	8-O-Acetyllankacidinyl		

TABLE 1. *Single biotransformations (Continued)*

Substrate	Transformation Product	Biological Agent	Reference
8,14-Di-O-butyryl-lankacidin C	8-O-Butyryllankacidin C	*B. megatherium* IFO 12108[a,b]	Nakahama et al. (1975)
8,14-Di-O-propionyl-lankacidin C	8-O-Propionyllankacidin C	*A. niger*[b], *A. sojae*[b], *T. sanguinea*[b]	Fugono et al. (1970)
Lankacidin A	Lankacidinol	*A. niger*[a], *A. sojae*[a], *T. sanguinea*[a]	Fugono et al. (1970)
Lankacidin A	Lankacidin C	*S. rochei* var. *volubilis* var. Nov[b]	Higashide et al. (1971)
Lankacidinol A	Lankacidinol	*S. rochei* var. *volubilis*[a,b] *T. sanguinea*[a,b]	Fugono et al. (1971)
Lankacidinol A	Lankacidinol	*A. niger*[b], *A. sojae*[b], *T. sanguinea*[b]	Fugono et al. (1974)
Leucomycin A_1	Leucomycin V	8 Genera	Theriault (1974)
Leucomycin A_1, A_3-A_9	a,b	*M. spinesciens* IAM 6071[a]	Shomura et al. (1973)
Leucomycin A_3	Leucomycin U	*B. megatherium* 91277[b]	Nakahama et al. (1974c)
Leucomycin A_3	Leucomycin U	8 Genera[a]	Theriault (1974)
Leucomycin A_3	Leucomycin A_1	*S. kitasatoensis*[a]	Kitao et al. (1979a)
Leucomycin A_4	Leucomycin A_5	*B. subtilis* ATCC 14593[a]	Okamoto et al. (1979)
Leucomycin A_5	Leucomycin V	*A. missouriensis*[a]	Singh and Rakhit (1979)
Maridomycin III	4″-O-Depropionyl-maridomycin III	*B. megatherium* 91277[b] *S. pristinaespiralis* IFO 13074[a]	Nakahama et al. (1974c) Nakahama et al. (1974b)
Maridomycin III	4″-O-Depropionyl-maridomycin III	*S. lavendulae* strain no. K-122a *S. lavendulae* strain no. K-245a	Uyeda et al. (1980) Shibata et al. (1975) Shibata et al. (1976)
8-O-Nicotinyllankacidin A	8-O-Nicotinyllankacidin C	*S. rochei* var. *volubilis*[b]	Harada et al. (1973)
Niddamycin	4″-O-Deisovalerylniddamycin	8 Genera	Theriault (1974)
9-O-Propionyl-4″-O-depropionylmaridomycin III	4″-O-Depropionyl-maridomycin III	*S. pristinaespiralis* IFO 13074[a]	Nakahama et al. (1974b)

Substrate	Transformation Product	Biological Agent	Reference
8-O-Propionyl-lankacidin A	8-O-Acetyllankacidin C	*S. rochei* var. *volubilis*[b]	Harada et al. (1973)
9-O-Propionyl-maridomycin III	9-O-Propionyl-4″-O-depropionyl-maridomycin III	*B. megatherium* 91277[b] *S. pristinaespiralis* IFO 13074[a] *S. olivaceus* 219[a]	Nakahama et al. (1974c) Nakahama et al. (1974b)
SF-837	SF-837M_1	*M. spinescens*[b]	Narimatsu et al. (1977)
SF-387 and acylated analogs	Deacylated compounds	11 Genera[a]	Niida et al. (1973)
8-O-Valeryllankacidin A	8-O-Valeryllankacidin C	*S. rochei* var. *volubilis*[b]	Harada et al. (1973)
3. Oxidation			
Carbomycin B	Carbomycin A	*S. hygroscopicus* B-5050-HA-M-1382[a]	Suzuki et al. (1977)
Lankacidinol A	Lankacidin A	*A. niger*[b], *A. sojae*[b], *T. sanguinea*[b]	Fugono et al. (1970)
Lankamycin	15-Deoxy-15-oxolankamycin	*S. erythreus* Abbott 2NU153[a]	Goldstein et al. (1978)
Leucomycin A_3	3″-Hydroxyleucomycin A_3	*S. olivaceus* 219[a]	Nakahama et al. (1974a)
Leucomycin A_3	Carbomycin B	*S. hygroscopicus* B-5050-HA[a]	Suzuki et al. (1977)
Leucomycin A_3	Maridomycin II	*S. hygroscopicus* B-5050-HA[a]	Suzuki et al. (1977)
M-4365G_2	Juvenimicin B_1	*S. ambofaciens* KA-1028	Sadakane et al. (1982)
Maridomycin I	3″-Hydroxymaridomycin I	*S. olivaceus* 219[a]	Nakahama et al. (1974a)
Maridomycin II	Carbomycin A	*S. hygroscopicus* B-5050-HA[a]	Suzuki et al. (1977)
Midecamycin A_1	Midecamycin A_3	*S. mycarofaciens* No. 070-19[b]	Matsuhashi et al. 1979)
Narbomycin	Picromycin	*S. zaomyceticus* MCRL-0405[a]	Maezawa et al. (1973)
4. Reduction			
Carbomycin A	Maridomycin II	*S. hygroscopicus* B-5050-HA[a]	Suzuki et al. (1977)
Carbomycin B	Leucomycin A_3	*S. hygroscopicus* B-5050-HA[a]	Suzuki et al. (1977)

TABLE 1. *Single biotransformations (Continued)*

Substrate	Transformation Product	Biological Agent	Reference
Maridomycin III	18-Dihydromaridomycin III	*Streptomyces* sp. strain no. K-245[a] *S. lavendulae* strain no. K-122[a]	Shibata et al. (1976) Uyeda et al. (1980) Shibata et al. (1975)
Midecamycin A_3	Midecamycin A_1	*S. mycarofaciens* no. 070-19[b]	Matsuhashi et al. (1979)
Tylosin	Relomycin	*S. ambofaciens* KA-1028[a] *S. hygroscopicus*[a] *S. griseospiralis*[a] *N. corrallina*[a]	Feldman et al. (1963)
Tylosin	9,20-Tetrahydrotylosin	*S. ambofaciens* KA-1028[a]	Omura et al. (1980b)
5. Glycosylation			
5-O-Dedesosaminyl-(8S)-8-fluoroerythromycin B	(8S)-8-Fluoroerythromycin B	*S. erythreus* ATCC 31772[a]	Toscano et al. (1983b)
5-O-Dedesosaminyl-(8S)-8-fluoroerythromycin D	(8S)-8-Fluoroerythromycin D	*S. erythreus* ATCC 31772[a]	Toscano et al. (1983b)
4′-O-Demycarosyl-20-deoxyrelomycin	20-Deoxyrelomycin	*S. fradiae* KA-427[a]	Omura et al. (1982)
Forocidin I	Neospiramycin III	*S. ambofaciens*[a]	Omura et al. (1979b)
5-O-Mycamino-syltylonolide	Tylosin	*S. fradiae* KA-427[a]	Omura et alo. (1980d, 1982)
23-O-Mycinosyl-20-deoxyrelonolide	20-Deoxyrelomycin	*S. fradiae* KA-427[a]	Omura et al. (1982)
6. Methylation			
Erythromycin C	Erythromycin A	*S. erythreus*[b]	Corcoran (1975)
Erythromycin D	Erythromycin B	*S. aureus* CA-340 Abbott[b]	Majer et al. (1977)
7. Demethylation			
Lankamycin	3′-O-Demethyllankamycin	*S. erythreus* Abbott 2NU153[a]	Goldstein et al. (1978)

Substrate	Transformation Product	Biological Agent	Reference
8. Other			
Erythromycin A	Erythronolide Erythrolosamine 2 unknown products	*Pseudomonas* 56[a,b]	Flickinger and Perlman (1975)
Maridomycin III	A_1, A_2, A_3, A_4	*Streptomyces* sp. strain No. K-241-41[a]	Shibata et al. (1976)
Maridomycin III	Maridomycins-S_1 and S_2	*S. marcescens*[a]	Uyeda et al. (1983)

[a] whole cells
[b] cell extracts and purified enzymes

TABLE 2. *Multiple biotransformations*

Substrate	Transformation Product	Type of Transformation	Biological Agent	Reference
3″-Acetylleucomycin A_5	Leucomycin A_1	Acylation, Deacylation	*Nocardia* sp. A-2440[a]	Toyo Jozo Co., Ltd. (1981)
5-O-Desosaminyl-mycinamicin IV	5-O-Dedesosaminyl-5-O-mycosaminyl-10,11-dihydromycinamicin IV	Glycosylation, Reduction	*S. fradiae* X5-174[a]	Lotvin et al. (1982)
Deepoxycirramycin B_1	Tylosin	Glycosylation, O-Methylation	*S. fradiae* KA-427[a]	Omura et al. (1982)
20-Deoxy-5-O-mycaminosylrelonide	Tylosin	Oxidation, Glycosylation	*S. fradiae* KA-427[a]	Omura et al. (1982)
2′,8-Di-O-acetyl-lankacidin A	8-O-Acetyllankacidin A	Deacylation, Oxidation	*S. rochei* var. *volubilis*[b]	Harada et al. (1973)

TABLE 2. *Multiple biotransformations*

Substrate	Transformation Product	Type of Transformation	Biological Agent	Reference
5-O-(4′,6′-Dideoxy-3′-C-acetyl-β-D-xylohexopyranosyl)-platenolide II	3-O-acetyl-14-hydroxy-5-O-(4′,6′-dideoxy-3′-C-acetyl-β-D-xylohexo-pyranosyl)-platenolide II 3-O-Propionyl-14-hydroxy-5-O-(4′,6′-dideoxy-3′-C-acetyl-β-D-xylohexo-pyranosyl-platenolide II	Acylation, Oxidation	*S. hygroscopicus* IMETJA 6599-14G-33-354r[a]	Gräfe et al. (1980a)
Forocidin I	Neospiramycin III Spiramycin III	Acylation, Glycosylation	*S. ambofaciens*[a]	Omura et al. (1979b)
Lankacidinol	Lankacidin A	Acylation, Oxidation	*S. rochei* var. *volubilis*[b]	Harada et al. (1973)
Lankamycin	4″-O-Deacetyl-15-deoxy-15-oxolankamycin	Deacylation, Oxidation	*S. erythreus* Abbott 2 NU155[a]	Goldstein et al. (1978)
Maridomycin III	4″-O-Depropionyl-18-dihydromaridomycin III	Deacylation, Reduction	*Streptomyces* strain No. K245[a] *S. lavendulae* K-122[a]	Shibata et al. (1975, 1976) Uyeda et al. (1980)
5-O-Mycaminosyl-protylonolide	Tylosin	Glycosylation, O-Methylation, Oxidation	*S. fradiae* KA-427[a]	Omura et al. (1982)
Neospiramycin I	Spiramycin III	Acylation, Glycosylation	*S. ambofaciens* KA-1028[a]	Kitao et al. (1979c)

[a] whole cells
[b] cell extracts and purified enzymes

TABLE 3. *Directed biosynthesis*

Substrate	Transformation Product	Biological Agent	Reference
Erythronolide A	Erythromycin A	*S. erythreus* ATCC 31722	Spagnoli and Toscano (1983)
Erythronolide A oxime	3-O-Oleandrosyl-5-O-desosaminylerythronolide oxime	*S. antibioticus* ATCC 18191	Le Mahieu et al. (1976)
Erythronolide B	3-O-Oleandrosyl-5-O-desosaminyl-15-hydroxy-erythromycin B 3-O-Oleandrosyl-5-O-desosaminyl-erythronolide B 3-O-Oleandrosyl-5-O-desosaminyl-(8S)-8-hydroxyerythronolide B 3-O-Oleandrosyl-5-O-desosaminyl-(8S)-8,19-epoxyerythronolide B	*S. antibioticus* ATCC 31771	Spagnoli et al. (1983)
(8S)-8-Fluoro-erythronolide A	(8S)-8-Fluoroerythromycin A (8S)-8-Fluoroerythromycin C	*S. erythreus* ATCC 31772	Toscano et al. (1983b)
(8S)-8-Fluoro-erythronolide A	3-O-Oleandrosyl-3-O-decladinosyl-(8S)-8-fluoroerythromycin A	*S. antibioticus* ATCC 31771	Toscano et al. (1983a)
(8S)-8-Fluoro-erythronolide B	(8S)-8-Fluoroerythromycin B (8S)-8-Fluoroerythromycin D	*S. erythreus* ATCC 31772	Toscano et al. (1983b)
(8S)-8-Fluoro-erythronolide B	3-O-Oleandrosyl-3-O-decladinosyl-(8S)-8-fluorerythromycin B	*S. antibioticus*ATCC 31771	Toscano et al. (1983a)
23-Hydroxy-protylonolide	20-Deoxyrelomycin	*S. fradiae* KA-427	Omura et al. (1982)
Narbonolide	Narbomycin, Picromycin	*S. narboensis* ISP-501[b]	Maezawa et al. (1973)
Narbonolide	5-O-Mycaminosylnarbonolide 5-O-Mycaminosyl-9-dihydronarbonolide	*S. platensis* MCRL-0388	Maezawa et al. (1976)
Platenolide I	Turimycin	*S. hygroscopicus* IMET JA 6599, strains Re9, UV53, NG11	Gräfe et al. (1980b)
Platenolide II	Turimycin	*S. hygroscopicus* IMET JA 6599, strains Re9, UV53, NG11	Gräfe et al. (1980b)

TABLE 3. *Directed biosynthesis (continued)*

Substrate	Transformation Product	Biological Agent	Reference
Protylonolide	M-4365G	*Streptomyces* sp. No. 4900	Omura et al. (1980c)
Protylonolide	Tylosin	*S. fradiae* and mutant strains *S. fradiae* KA-427	Omura et al. (1980a) Omura et al. (1982)
Protylonolide	4′-O-Propionyl-5-O-mycarosylprotylonolide	*S. kitasatoensis* KA-249	Sadakane et al. (1982)

Perhaps the most interesting microbial conversions have been carried out using aglycones of macrolide antibiotics (Table 3). Most of this work has involved studies of the biosynthesis of already known antibiotics, but in recent cases modified antibiotics have been prepared that seem to be quite promising. The earliest work in this field was reported by Maezawa et al. (1973, 1976), who converted narbonolide [the aglycone of narbomycin (*6*)] to narbomycin, picromycin, and analogs of these compounds. Omura et al. (1980c, 1982) have used protylonolide (*11*) as a substrate in fermentations containing cerulenin to prevent aglycone formation and strains of *Streptomyces fradiae* and *S. kitasatoensis* as the microorganisms. The products were M-4365G_2 (*9*), tylosin, and various analogs of these. Toscano et al. (1983b) prepared (8S)-8-fluoroerythronolides A and B chemically and by the action of *S. erythreus* ATCC 31772 on the aglycones converted them to (8S)-8-fluoroerythromycins A, B, C, and D. The resulting fluorinated erythromycins have *in vitro* antimicrobial activity comparable to the parent compounds. Somewhat surprisingly these fluorinated analogs were substantially more stable to acids than were the parent compounds. Their half-life at pH 4 is greater than 100 h while that of erythromycin A is 2 h.

Up to now comparatively little effort has been applied to preparing modified macrolide antibiotics by bioconversion relative to the chemical effort that has been expended. It appears that this could become a very fruitful field of research.

Literature Cited

Corcoran, J. W. 1975. S-Adenosylmethionine: Erythromycin C O-methyltransferase. Pages 487–498 *in* J. H. Hash, ed., *Methods in Enzymology.* Academic Press, Inc., New York.

Feldman, L. J., J. K. Dill, C. E. Holmlund, H. A. Whaley, E. L. Patterson, and N. Bohonos. 1963. Microbiological transformations of macrolide antibiotics. *Antimicrob. Agents Chemother.* 54–57.

Flickinger, M. C., and D. Perlman. 1975. Microbial degradation of erythromycins A and B. *J. Antibiot.* 28:307–311.

Fugono, T., S. Harada, E. Higashide, and T. Kishi. 1971. Studies on T-2636 antibiotics. III. A new component T-2636F. *J. Antibiot.* 24:23–38.

Fugono, T. E. Higashide, T. Suzuke, H. Yamamoto, S. Harada, and T. Kishi. 1970. Interconversion of T-2636 antibiotics produced by *Streptomyces rochei* var. *volubilis. Experientia* 26:26–27.

Goldstein, A. W., R. S. Egan, S. L. Mueller, J. R. Martin, and W. Keller-Schierlein. 1978. Biotransformation of lankamycin, darcanolide, and 11-acetyllankolide by a blocked mutant of the erythromycin producing organism *Streptomyces erythreus. J. Antibiot.* 31:63–69.

Grafe, U., G. Reinhardt, W. Schade, P. Muhlig, and H. Thrum. 1980a. Bioconversion of 5-O-(4′-6′-dideoxy-3′-C-acetyl-β-D-xylohexopyranosyl)-platenolide II by *Streptomyces hygroscopicus* IMET JA 6599-NG-33-354r. *J. Antibiot.* 33:1083–1085.

Grafe, U., W. F. Fleck, W. Schade, G. Reinhardt, D. Tresselt, and H. Thrum. 1980b. The platenolides I and II as precursors of turimycin. *J. Antibiot.* 33:663–664.

Harada, S., F. Yamazaki, K. Hatano, K. Tsuhiya, and T. Kishi. 1973. Studies on lankacidin-group (T-2636) antibiotics. VII. Structure-activity relationships of lankacidin-group antibiotics. *J. Antibiot.* 26:647–657.

Higashide, E., T. Fugono, K. Hatanoe, and M. Shibata. 1971. Studies of T-2636 antibiotics. I. Taxonomy of *Streptomyces rochei* var. *volubilis. J. Antibiot.* 24:1–12.

Kitao, C., J. Miyazawa, and S. Omura. 1979a. Induction of the bioconversion of leucomycin by glucose and its regulation by butyrate. *Agric. Biol. Chem.* 43:833–839.

Kitao, C., H. Hamada, H. Ikeda, and S. Omura. 1979b. Bioconversion and biosynthesis of 16-membered macrolide antibiotics. XV. Final steps in the biosynthesis of leucomycins. *J. Antibiot.* 32:1055–1057.

Kitao, C., H. Ikeda, H. Hamada, and S. Omura. 1979c. Bioconversion and biosynthesis of 16-membered macrolide antibiotics. XIII. Regulation of spiramycin I 3-hydroxylacylase formation by glucose, butyrate, and cerulenin. *J. Antibiot.* 32:593–599.

LeMahieu, R. A., H. A. Ax, J. F. Blount, M. Carson, C. W. Despraux, D. L. Prues, J. D. Scannell, T. Weish, and R. W. Kierstad. 1976. A new semisynthetic macrolide antibiotic 3-O-oleandrosyl-5-O-desosaminylerythronolide A oxime. *J. Antibiot.* 29:728–734.

Lotvin, J., M. S. Puar, M. Patel, B. K. Lee, D. Schumacher, and J. G. Waitz. 1982. Preparation and characterization of dedesosaminyl-5-O-mycaminosyl-10,11-dihydromycinamicin IV. *J. Antibiot.* 35:1407–1408.

Maezawa, I., A. Kinumaki, and M. Suzuki. 1976. Biological glycosidation of macrolide aglycones. I. Isolation and characterization of 5-O-mycaminosylnarbonolide and 9-dihydro-5-O-mycaminosylnarbonolide. *J. Antibiot.* 29:1203–1208.

Maezawa, I., T. Hori, A. Kinumaki, and M. Suzuki. 1973. Biological conversion of narbonolide to picromycin. *J. Antibiot.* 26:771–775.

Majer, J., J. R. Martin, R. S. Egan, and J. W. Corcoran. 1977. Antibiotic glycosides. 8. Erythromycin D, a new macrolide antibiotic. *J. Am. Chem. Soc.* 99:1620–1622.

Matsuhashi, Y., H. Ogawa, and K. Nagaoka. 1979. The enzymatic interconversion between midecamycin A_1 and A_3. *J. Antibiot.* 32:777–779.

Nakahama, K., T. Kishi, and S. Igaresi. 1974a. Microbial conversion of antibiotics. III. Hydroxylation of maridomycin I and josamycin. *J. Antibiot.* 27:443–441.

________. 1974b. Microbial conversion of antibiotics. II. Deacylation of maridomycin by *Actinomycetes. J. Antibiot.* 17:487–489.

Nakahama, K., M. Isawa, M. Muroi, T. Kishi, M. Uchida, and S. Igaresi. 1974c. Microbial conversion of antibiotics. I. Deacylation of maridomycin by bacteria. *J. Antibiot.* 27:425–432.

Nakahama, K., S. Harada, and S. Igaresi. 1975. Studies on lankacidin-group (T-2636) antibiotics. X. Microbial conversion of lankacidin-group antibiotics. *J. Antibiot.* 28:390–394.

Narimatsu, Y., M. Uyeda, and K. Watanabe. 1977. Production of deacylated derivatives of macrolide antibiotics by immobilized deacylase. *Jap. Kokai* 77 108, 085.

Niida, F., S. Inoue, T. Shomura, S. Omoto, H. Watanabe, B. Nomiya, and T. Koeda. 1973. Biological production of 9-acyl-4″-depropionyl-SF-837 compounds. *Jap. Kokai* 73 72, 389.

Okamoto, R., T. Fukumoto, T. Inui, and T. Takeuchi. 1979. 16 M-Macrolide-3-hydrolase. *Jap. Kokai* 79 28, 892.

Okamoto, R., T. Fukumoto, A. Takamatsu, and T. Takeuchi. 1977. Biochemical production of josamycin. *Jap. Kokai* 77 41, 294.

Omura, S., H. Ikeda, and C. Kitao. 1979a. Isolation and properties of spiramycin I 3-hydroxyl acylase from *Streptomyces ambofaciens. J. Biochem.* 86:1753–1758.

Omura, S., C. Kitao, H. Hamada, and H. Ikeda. 1979b. Bioconversion and biosynthesis of 16-membered macrolide antibiotics. X. Final steps in the biosynthesis of spiramycin using enzyme inhibitor: Cerulenin. *Chem. Pharm. Bull.* 17:176–182.

Omura, S., C. Kitao, and H. Matsubara. 1980a. Isolation and characterization of a new 16-membered lactone, protylonolide from a mutant of tylosin-producing strain, *Streptomyces fradiae* KA-427. *Chem Pharmacol. Bull.* 28:1963-1965.

Omura, S., C. Kitao, and N. Sadakane. 1980b. The microbial transformation of tyosin by the spiramycin-producing strain, *Streptomyces ambofaciens* KA-1028. *J. Antibiot.* 33:911–912.

Omura, S., H. Ikeda, K. Matsubara, and N. Sadakane. 1980c. Hybrid biosynthesis and absolute configuration of macrolide antibiotic M-4365G. *J. Antibiot.* 33:1570–1572.

Omura, S., J. Miyazawa, H. Takeshima, C. Kitao, and M. Aizawa. 1977. Induction of the bioconversion of leucomycins by glucose in a producing strain. *J. Antibiot.* 30:192–193.

Omura, S., J. Miyazawa, H. Takeshima, C. Kitao, K. Atsumi, and M. Aizawa. 1976. Bioconversion of leucomycins and its regulation by butyrate in a producing strain. *J. Antibiot.* 29:1131–1133.

Omura, S., N. Sadakane, C. Kitao, H. Matsubara, and A. Nakagawa. 1980d. Production of mycarosyl protylonolide by a mycaminose idiotroph from the tylosin-producing strain *Streptomyces fradiae* KA-427. *J. Antibiot.* 33:913–914.

Omura, S., N. Sadakane, and H. Matsubara. 1982. Bioconversion and biosynthesis of 16-membered macrolide antibiotics. XXII. Biosynthesis of tylosin after protylonolide formation. *Chem. Pharm. Bull.* 30:223–229.

Sadakane, N., Y. Tanake, and S. Omura. 1982. Hybrid biosynthesis of derivatives of protylonolide and M-4365 by macrolide-producing microorganisms. *J. Antibiot.* 35:680–687.

Shibata, M., M. Uyeda, and S. Mori. 1975. Microbial transformation of antibiotics. I. Isolation and characterization of the transformation products of maridomycin IV. *J. Antibiot.* 28:434–441.

________. 1976. Microbial transformation of antibiotics. II. Additional transformation products of maridomycin III. *J. Antibiot.* 29:824–828.

Shomura, T., S. Inoue, N. Ezaki, T. Tsurioka, T. Amano, H. Watanabe, and T. Niida. 1973. Deacylation of leucomycin and antibiotic SF-837. *Jap. Kokai* 73 29, 148.

Singh, K., and S. Rakhit. 1979. Microbial transformation of leucomycin A_5. *J. Antibiot.* 32:78–80.

Spagnoli, R., L. Cappelletti, and L. Toscano. 1983. Biological conversion of erythronolide B, an intermediate of erythromycin biogenesis, into new "hybrid" macrolide antibiotics. *J. Antibiot.* 36:365–375.

Spagnoli, R., and L. Toscano. 1983. Erythronolide A glycosidation to erythromycin A by a blocked mutant of *Streptomyces erythreus*. *J. Antibiot.* 36:435–437.

Suzuki, M., T. Takamaki, K. Miyagawa, H. Ono, E. Higashide, and M. Ochida. 1977. Interconversion among leucomycin A_3, carbomycin A, carbomycin B, and maridomycin II. *Agric. Biol. Chem.* 42:419–442.

Theriault, R. J. 1974. Mycarosyl macrolide antibiotics by fermentation. U.S. Pat. 3,784,447.

Toscano, L., G. Fioriello, R. Spagnoli, and L. Cappelletti. 1983a. New semi-synthetic fluorinated "hybrid" macrolides. *J. Antibiot.* 36:1585–1588.

Toscano, L., G. Fioriello, R. Spagnoli, L. Cappelletti, and G. Zanuso. 1983b. New fluorinated erythromycin obtained by mutasynthesis. *J. Antibiot.* 36:1439–1450.

Toyo Jozo Co., Ltd. 1981. Macrolide antibiotics by microbial conversion. *Jap. Kokai* 81 154, 998.

Uyeda, M., K. Kirbata, S. Miyamura, and M. Shibata. 1983. Microbial transformations of maridomycin III by *Serratia marcescens*. *J. Antibiot.* 36:1772–1728.

Uyeda, M., K. Nakamichi, K. Shigemi, and M. Shibata. 1980. Transformation of maridomycin III by maridomycin III insensitive *Streptomycetes*. *Agr. Biol. Chem.* 44:1399–1403.

Uyeda, M., S. Mori, M. Morita, T. Ogata, M. Mori, and M. Shibata. 1977. Microbial transformation of antibiotics. III. Reacylation of 4″-depropionyl maridomycin IV into maridomycin V (maridomycin K) by *Streptomyces* sp. strain no. 342. *J. Antibiot.* 30:1130–1131.

CHAPTER **5**

Biotransformations and Biosynthesis of Aminocyclitol Antibiotics

KENNETH L. RINEHART, JR., JIN-RUI FANG, WEN-ZAO JIN, CEDRIC J. PEARCE, KIN-ICHI TADANO, AND TATSUSHI TOYOKUNI

University of Illinois at Urbana-Champaign, Urbana, Illinois 61801

Biotransformation of clinically useful aminocyclitol antibiotics has not provided improved antibiotics. However, chemical transformations as well as studies of bioinactivation and of biosynthesis have led to superior aminocyclitols.

INTRODUCTION

Aminocyclitol antibiotics, compounds containing a cycloalkane ring substituted by amino and hydroxyl groups, are some of the longest known antibiotics, yet they remain among the most clinically useful (Rinehart and Suami 1980). Some, such as streptomycin and spectinomycin, contain aminocyclitols other than deoxystreptamine (all-*trans*-1,3-diamino-4,5,6-trihydroxycyclohexane), but deoxystreptamine is present in the majority of the clinically useful aminocyclitol antibiotics, including gentamicin, sisomicin, netilmicin, kanamycin, tobramycin, amikacin, neomycin, paromomycin, lividomycin, and ribostamycin. Because of the great importance of the deoxystreptamine-containing antibiotics, the following discussion will be restricted for the most part to them, with special emphasis on neomycin in view of our long personal involvement (Rinehart 1964).

DISCUSSION

The present symposium is titled "Microbial Transformations of Antibiotics and Antitumor Agents." For antibiotics, obviously, the most important microbial transformations would be those leading to a superior product. Unfortunately, we must note at the outset that no useful microbial transformation of a clinically important aminocyclitol antibiotic has ever been reported. This is not to say that useful modifications of aminocyclitol antibiotics are impossible or, indeed, unknown, but that they have all been carried out chemically rather than biologically.

By far the greatest effort has been expended on that very successful antibiotic gentamicin and the related sisomicin, with most of the studies being carried out at the Schering Corporation. This work has been reviewed extensively elsewhere (Daniels et al. 1980) and can be summarized for sisomicin in Fig. 1, where it is seen that (1) introduction of a hydroxyl group at C-2 of deoxystreptamine by mutasynthesis, as described later, greatly reduces the toxicity of the antibiotic; (2)

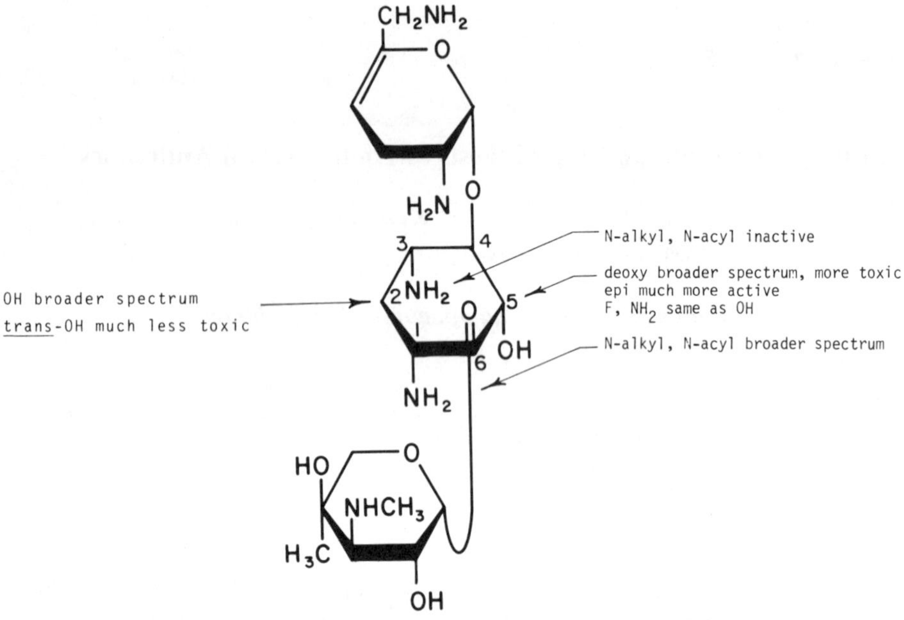

FIG. 1. Structure-activity relationships of modified sisomicins (Rinehart 1981).

chemical attachment of an N-alkyl group (N-ethyl in netilmicin) or an N-acyl group (3-amino-2-hydroxypropionyl) to the 1-amino nitrogen gives a broader spectrum antibiotic (against resistant strains, *cf.* later) of reduced toxicity; (3) epimerization at C-5 gives a more active antibiotic; and (4) replacement of the hydroxyl group at C-5 by a hydrogen or fluorine atom or an amino or azido group retains or enhances the bioactivity.

To return to the theme of the symposium, microbial transformations have in fact been much studied in the aminocyclitol antibiotics. However, this has been mainly as a result of the development of resistance by target microorganisms, which are able to transform or derivatize aminocyclitol antibiotics so as to render them inactive. Several reviews in this area are available, and Fig. 2 illustrates the variety of ways in which resistant organisms can inactivate kanamycin B (Haas and Dowding 1975; Davies 1980; Daniels et al. 1980; Davies and Yagisawa 1983). Although studies of the inactivation of aminocyclitols have focused mainly on

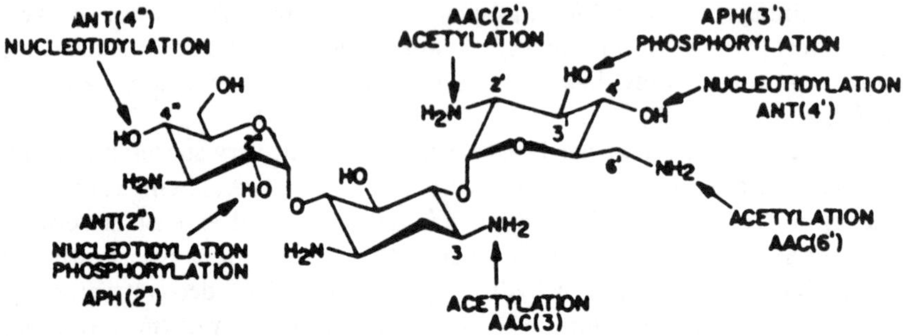

FIG. 2. Enzymatic inactivation of kanamycin B (Daniels et al. 1980).

target microorganisms, the producing organisms can apparently protect themselves in much the same way. Indeed, the first report of a biotransformed aminocyclitol was that in 1962 of LP-neomycin (low-potency neomycin), which was N-acetylated on the 3-amino group (see Fig. 3 for numbering system) (Rinehart 1964; Chilton 1963). Although inactivating microbial transformations

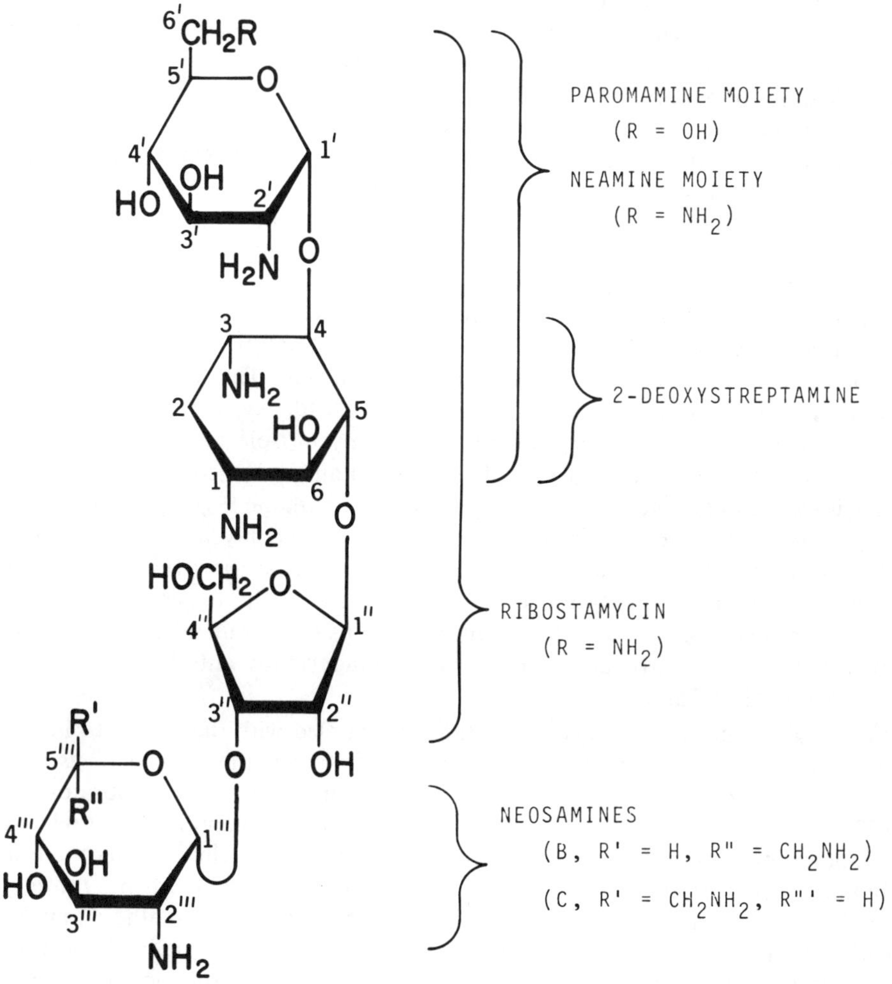

	R	R'	R"
NEOMYCIN B	NH_2	H	CH_2NH_2
NEOMYCIN C	NH_2	CH_2NH_2	H
PAROMOMYCIN I	OH	H	CH_2NH_2
PAROMOMYCIN II	OH	CH_2NH_2	H

FIG. 3. Neomycins B and C, paromomycins I and II, and their subunits (Fang et al. 1984).

do not provide improved antibiotics per se, they have been useful in designing modified antibiotics effective against resistant strains. In this connection, a great deal of work has been carried out on kanamycin. For example, 3′,4′-dideoxykanamycin is effective against resistant strains carrying the APH(3′) or ANT(4′) enzymes, as is 1-(γ-amino-α-hydroxybutyryl)kanamycin A (Amikacin), which is resistant against a variety of inactivating enzymes: ANT (2″), APH (3′ and 2″), and AAC (3, 2′, and 6′) (Davies 1980).

Interestingly enough, study of the biosynthesis of aminocyclitol antibiotics (in principle, the opposite of their biotransformation) has also led to useful modifications of aminocyclitol antibiotics, as well as to fundamental information regarding the mode of microbial synthesis of the compounds (Rinehart and Stroshane 1976; Rinehart 1979, 1980; Pearce and Rinehart 1981). If we examine the antibiotic neomycin B (Fig. 3) we can see that there are only two general problems regarding these antibiotics, one dealing with the mechanisms of the biosynthesis of the individual subunits, the other with the order and mechanisms of linking those subunits via glycosidic bonds.

Among the subunits of neomycin, deoxystreptamine itself is certainly the most important since it is the aminocyclitol. The first question with regard to deoxystreptamine is whether it is incorporated into neomycin. That it is incorporated has been demonstrated in two ways. First, [1-^{14}C]deoxystreptamine was administered to *Streptomyces fradiae* and the neomycin isolated was degraded to show that the label resided in C-1 of deoxystreptamine in neomycin, the same place it was found in the precursor (Rinehart and Stroshane 1976). Second, a mutant strain of *S. fradiae*, which normally produces neomycin, was prepared, which showed the same morphological and growth characteristics as the wild strain but was unable to produce neomycin except in the presence of deoxystreptamine (Shier et al. 1969). Such mutant strains (D^- strains) have been identified for all or nearly all of the other aminocyclitol antibiotic-producing microbial strains (Table 1).

The inability to produce deoxystreptamine, coupled with the ability to incorporate it into an antibiotic, led to the concept of mutasynthesis, which was first demonstrated for neomycin, paromomycin, and kanamycin in our laboratory in 1967 (Shier et al. 1969). By replacing the added deoxystreptamine with a related aminocyclitol (a mutasynthon), we produced a new aminocyclitol antibiotic. For example, when streptamine itself is added to D^- mutant of *S. fradiae*, 2-hydroxyneomycin is obtained. Since the initial report on mutasynthesis, mutasynthetic antibiotics related to essentially all of the clinically important aminocyclitol antibiotics have been produced (Table 1). For the most part these mutasynthetic compounds are less active than the parent antibiotic, but in some cases they are either considerably more active or considerably less toxic. The most important of them is hydroxygentamicin (Daum et al. 1977a, 1977b), which is only one-half as active as gentamicin but only one-fourth to one-sixth as toxic and has been brought to clinical trials.

An interesting modification of the mutasynthetic method involves administering a mutasynthon related to neamine (the pseudodisaccharide-containing deoxystreptamine, Fig. 3) to produce mutasynthetic compounds such as 3′,4′-dideoxykanamycin B from 3′,4′-dideoxyneamine (Kojima and Satoh 1973). The latter mutasynthetic conversions are, in fact, microbial transformations of actual antibiotics, since neamine and related mutasynthons are antibiotics themselves, albeit weaker than the pseudo-tri- or -tetrasaccharide antibiotics like ribostamycin, kanamycin, and neomycin.

We can turn now to the biosynthesis of deoxystreptamine itself. It was proposed 10 yrs ago that a deoxyinosose was the precursor of deoxystreptamine (Rinehart et al. 1974), and this has been demonstrated more recently by two groups, employing D^- mutants of the organisms producing butirosin and gentamicin. Administering deoxyinosose (or deoxystreptamine) to a D^- mutant of *Bacillus circulans,* which normally gives butirosin, indeed gives butirosin (Furumai et al. 1979), as does administering the same compound to the D^- mutant of *Micromonospora purpurea,* which normally produces gentamicin (Daum et al. 1977a, 1977b). Thus, deoxyinosose can be assumed to be a precursor of deoxystreptamine, as can 3-aminodideoxyinositol, which also gave butirosin (Furumai et al. 1979).

The demonstration of inosose's conversion to deoxystreptamine provided an important fringe benefit—the ability to use a nonaminated aminocyclitol in the preparation of hydroxygentamicin (Daum et al. 1977a). In particular, inosose has an advantage over streptamine in the mutasynthetic procedure in that inosose can be prepared cheaply from inositol, thus lowering the cost of hydroxygentamicin.

The primary metabolite that is converted to deoxystreptamine was shown some time ago to be D-glucose (Rinehart et al. 1974). D-Glucosamine had previously been proposed as the precursor but was not as well incorporated as glucose into deoxystreptamine, and label from [^{15}N]glucosamine was not incorporated at all.

Intermediates between glucose and deoxyinosose have not been isolated, but some indication of their nature is shown in the recent work of Kakinuma (Kakinuma et al. 1981; Kakinuma 1982). Both [6,6-$^2H_2$]- and 6*S*-[6-^{2}H]glucose were administered to *S. ribosidificus,* demonstrating that in the former case both deuterium atoms are retained, while in the latter case the 6*S*-deuterium atom becomes the equatorial deuterium in the deoxystreptamine unit of ribostamycin. Similar results were obtained for neomycin (Ewad et al. 1983) with [6-$^3H_2$,U-$^{14}C_6$]glucose. In view of the retention of both deuterium atoms from dideuterated glucose, the enol intermediate between glucose and deoxy-*scyllo*-inosose in Fig. 4a was proposed by Kakinuma et al. (1981).

It is not known at present whether the transformation takes place on nucleotide-bound glucose or on the free sugar. However, some time earlier [1-^{14}C]glucose-dTDP was administered by Byrne in our laboratory to *S. fradiae* and three ^{14}C-labeled biotransformation products were obtained (Byrne, K.M., personal communication; Rinehart 1979). Two of these were identified as 6-deoxy-D-glucose-dTDP and 4,6-dideoxy-4-keto-D-glucose-dTDP, but the third product could not be identified. From the stereospecificity in the conversion of glucose to deoxystreptamine it appears that neither 6-deoxyglucose nor 4,6-dideoxy-4-keto-glucose can be considered as a precursor.

Among the other subunits of neomycin B, [1-^{14}C]ribose was shown to be incorporated directly into neomycin, since the ribose recovered from neomycin had the label in the same position (Schimbor 1966; Rinehart and Schimbor 1967). In the normal biosynthesis of neomycin, glucose is converted to ribose. Apparently this occurs by two pathways—the hexose monophosphate, and, perhaps, the glucuronate pathways—since both 6- and 1-labeled glucose are incorporated into the

TABLE 1. *Summary of mutant organisms, mutasynthons, and mutasynthetic antibiotics*[a]

Mutant Ref.	Normal Antibiotic	Mutasynthon	Mutasynthetic Antibiotic
S. fradiae (D^-)	Neomycin	Streptamine	2-Hydroxyneomycins B, C (hybrimycins A1, A2)
Shier et al. (1969);		2-Epistreptamine	2-Epihydroxyneomycin (hybrimycins B1, B2)
Cleophax et al. (1976);		2,6-Dideoxystreptamine	6-Deoxyneomycins B, C
Rubenstein (1978).		2,5-Dideoxystreptamine	Not isolated
		6-O-Methyldeoxystreptamine	Not isolated
		3-N-Methyldeoxystreptamine	Not isolated
		2-Bromo-2-deoxystreptamine	Not isolated
		6-Bromo-6-deoxystreptamine	Not isolated
S. rimosus (D^-)	Paromomycin	Streptamine	2-Hydroxyparomomycins I, II (hybrimycins C1, C2)
Cleophax et al. (1976);		2,6-Dideoxystreptamine	6-Deoxyparomomycins I, II
Shier et al. (1974).			
S. ribosidificus (D^-)	Ribostamycin	Streptamine	2-Hydroxyribostamycin
Kojima and Satoh (1973).		Epistreptamine	2-Epihydroxyribostamycin
		1-N-Methyldeoxystreptamine	1-N-Methylribostamycin
		Gentamine C_{1a}	3′,4′-Dideoxyribostamycin
B. circulans (D^-, N^-)	Butirosin	Streptamine	2-Hydroxybutirosins A, B
Taylor and Schmitz (1976);		2,5-Dideoxystreptamine	5-Deoxybutirosamine
Takeda et al. (1978b, 1978c, 1978d).		Gentamine C_{1a}	3′, 4′-Dideoxybutirosins A, B
		6′-N-Methylneamine	6′-N-Methylbutirosins A, B
		6′-N-Methylgentamine C_{1a}	3′,4′-Dideoxy-6′-N-methylbutirosins A, B
		Gentamine C_2	3′,4′-Dideoxy-6′-C-methylbutirosins A, B
S. kanamyceticus (D^-)	Kanamycin	2-Epistreptamine	6′-Hydroxy-6′-deamino-2-epihydroxykanamycin A
Kojima and Satoh (1973).		1-N-Methylstreptamine	6′-Hydroxy-6′-deamino-1-N-methylkanamycin A

M. purpurea (D^-)	Gentamicin	Streptamine	2-Hydroxygentamicins C_1, C_2, C_{2a}
Rosi et al. (1977);		*Scyllo*-inosose	2-Hydroxygentamicin
Daum et al. (1977a, 1977b).		*Acyllo*-inosose pentaacetate	2-Hydroxygentamicin
		2,5-Dideoxystreptamine	5-Deoxygentamicins C_1, C_2, C_{2a}
		4,6-Hydrazino-1,3-cyclohexanediol	Not isolated
		1,3-Di-N-benzylidene-2,5-dideoxystreptamine	Not isolated
		Epistreptamine	Not isolated
M. inyoensis (D^-)	Sisomicin	Streptamine	2-Hydroxysisomicin (Mu 1)
Daniels (1978);			3″-N-Demethyl-3″-N-acetyl-2-hydroxysisomicin (Mu 1a)
Daniels and Rane (1979);			3″-N-Demethyl-2-hydroxysisomicin (Mu 1b)
Daniels et al. (1980);		2,5-Dideoxystreptamine	5-Deoxysisomicin (Mu 2)
Testa et al. (1974);			5-Deoxygentamicin A (Mu 2a)
Waitz et al. (1978).		2-Epistreptamine	2-Epihydroxysisomicin (Mu 4)
		5-Amino-2,5-dideoxystreptamine	5-Amino-5-deoxysisomicin (Mu 5)
		5-Epi-2-deoxystreptamine	5-Episisomicin (Mu 6)
		3-N-Methyl-2-deoxystreptamine	3-N-Methylsisomicin (Mu 7)
		1-N-Methyl-2,5-dideoxystreptamine	1-N-Methyl-5-deoxysisomicin (Mu 8)
		5-Epifluoro-2-deoxystreptamine	5-Epifluorosisomicin (Mu X)
S. griseus (S^-)	Streptomycin	2-Deoxystreptidine	Not isolated (streptomutin A)
Nagaoka and Demain (1975).			
S. spectabilis (A^-)	Spectinomycin	Streptamine	Not isolated (bioinactive)
Slechta and Coats (1974).		N,N′-Dimethylstreptamine	Not isolated (bioinactive)

[a] Reprinted with permission from Rinehart and Suami (1980), p. 363.

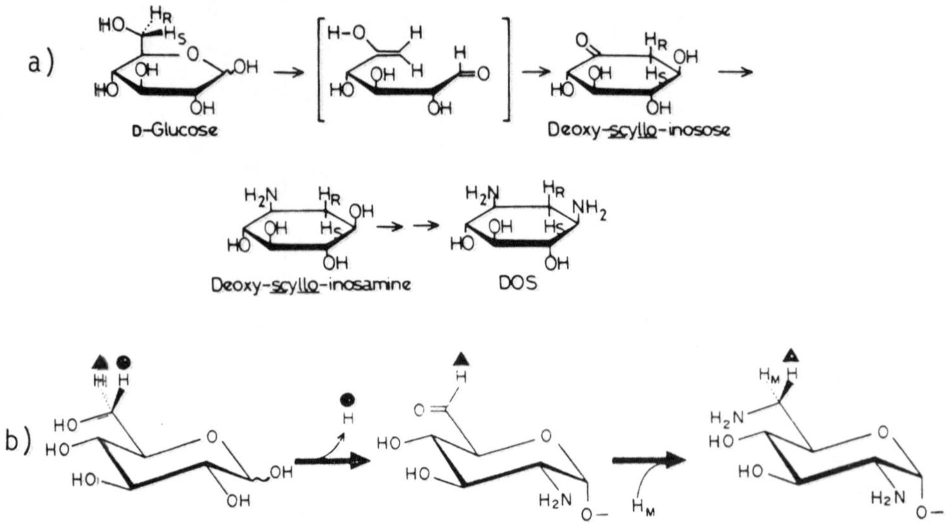

FIG. 4. Proposed biosynthetic pathways from D-glucose: (a) to deoxystreptamine (Kakinuma et al. 1981), and (b) to neosamine C (Kakinuma and Yamaya 1983).

antibiotic (Rinehart and Stroshane 1976). The results of Kakinuma et al. (1981), Kakinuma (1982), and Ewad et al. (1983) are also in agreement with a hexose monophosphate pathway in that the two deuterium atoms of [6,6-$^{2}H_2$]glucose or the two tritium atoms of [6-$^{3}H_2$,U-$^{14}C_6$]glucose are retained in the ribose of ribostamycin and neomycin. Nothing is known about the incorporation of ribose into ribostamycin or neomycin, specifically whether it is converted first to a nucleotide diphosphate. An interesting observation has been made that D-ribose (in ribostamycin) can apparently be converted to D-xylose (in the analogous antibiotic xylostasin) while attached to the deoxystreptamine unit (Takeda et al. 1978a).

In contrast to the demonstrated incorporation of both deoxystreptamine and ribose into neomycin, neosamine C is apparently not incorporated as such (Rinehart and Stroshane 1976). This has been interpreted as being the result of neosamine C's formation as a nucleotide-bound derivative, which then transfers neosamine C directly to deoxystreptamine. However, it could also be the result of conversion of paromamine (glucosaminyldeoxystreptamine, Fig. 3) to neamine (neosaminyldeoxystreptamine, Fig. 3), as discussed below. In any event, glucosamine has been demonstrated to be the precursor of both neosamine C and neosamine B, since [^{15}N]glucosamine is incorporated into [^{15}N]neosamines B and C (Rinehart et al. 1974). Glucose is also a precursor, but is less well incorporated into neosamines than glucosamine. The conversion of glucose to neosamine C in ribostamycin has also been studied with [6,6-$^{2}H_2$]-, 6R-[6-^{2}H]-, and 6S-[6-^{2}H]glucose (Kakinuma et al. 1981; Kakinuma 1982; Kakinuma and Yamaya 1983). In this case, the 6S-deuterium atom is lost in the conversion of glucose to neosamine C and the overall reaction proceeds with inversion (Fig. 4b). Similar results for neosamine C in neomycin have been obtained (Al-Feel et al. 1983) from [6-$^{3}H_2$,U-$^{14}C_6$]glucosamine. This has been interpreted as involving a

6-aldehydoglucosamine, which would then be transaminated. Again, whether this occurs on a nucleotide-bound glucosamine or on glucosamine attached to deoxystreptamine is uncertain.

Finally, the question of the formation of the glycosidic links in neomycin has been studied, at least for the formation of the bond between neosamine C and deoxystreptamine, employing three separate techniques. The question here is whether a nucleoside diphosphate-bound glucosamine is converted first to paromamine, which in turn is converted to neamine, or whether the nucleoside diphosphate-bound glucosamine is converted first to nucleoside diphosphate-bound neosamine, which is then attached to deoxystreptamine. The most fundamental approach to this question would involve synthesis of radioactively labeled nucleotides, glucosamine-NDP and neosamine-NDP, to see whether they are incorporated into neomycin. Five of the eight possible glucosamine nucleotides (Rinehart 1979) and three of the eight possible neosamine nucleotides (Tadano et al. 1982) have been prepared by Tadano. Unfortunately, initial biosynthetic studies were unrewarding, perhaps because of failure of the nucleotides to pass the cell wall barrier. Cell-free studies or protoplasts may be required.

Other evidence is, however, available, even though conflicting. In one study, the formation of butirosin was investigated, and it appeared that paromamine (with a 6′-hydroxyl group) was converted to butirosin (with a 6′-amino group) (Takeda et al. 1978a). Similar results have been obtained for gentamicin (Testa and Tilley 1976) and sisomicin (Testa and Tilley 1975) biosyntheses. A more direct investigation involved preparing highly radioactive neamine and administering it to *S. fradiae*, which demonstrated that neomycin was highly labeled in this process (again, the microbial transformation of a weak antibiotic to a strong antibiotic) (Fang et al. 1984). Since in theory the neamine could have been degraded and reassembled before incorporation into neomycin, it was first demonstrated that only the neamine portion was labeled in neomycin, and the relative activities of the neosamine C and deoxystreptamine portions of the neamine moiety were the same before and after the biotransformation.

In view of the very practical outcome of the mutasynthetic studies involving D^- mutants, one can wonder whether a similar approach could be used with mono- and disaccharides, such as neosamine C-NDP or neobiosamine-NDP, to incorporate relatives of those nucleoside-bound saccharides to novel neomycins. Future efforts in this direction may be rewarding.

Beyond neamine, little is known. It has been argued from indirect evidence that ribostamycin is an intermediate in neomycin biosynthesis (Baud et al. 1977), but ribostamycin was not converted to neomycin by the D^- mutant of *S. fradiae* (Shier et al. 1969). Clearly much remains to be done in this area.

To summarize the foregoing material, we can note that it is indeed possible to improve the aminocyclitol antibiotics but that this has not yet been accomplished by microbial transformation, except for the neamine-neomycin conversion. An alternative type of biotransformation, inactivation of the antibiotics, has proved useful in the past in guiding synthetic efforts designed to overcome this inactivation. Finally, studies arising from the biosynthesis of the aminocyclitol antibiotics have led to the concept of mutasynthesis, a technique used to prepare a number of improved antibiotics.

ACKNOWLEDGMENT

This work was supported in part by a grant from the National Institute of Allergy and Infectious Diseases (AI 01278).

LITERATURE CITED

Al-Feel, W., M. J. S. Ewad, C. J. Herbert, and M. Akhtar. 1983. Mechanism and stereochemistry of the elaboration of the neosamine C ring in the biosynthesis of neomycins. *J. Chem. Soc., Chem. Commun.*, pp. 18–20.

Baud, H., A. Betencourt, M. Peyre, and L. Penasse. 1977. Ribostamycin, as an intermediate in the biosynthesis of neomycin. *J. Antibiot.* 30:720–723.

Chilton, W. S. 1963. The structures of neomycins LP_B and LP_C. Ph.D. Thesis, University of Illinois at Urbana-Champaign, Urbana, IL.

Cleophax, J., S. D. Gero, J. Leboul, M. Akhtar, J. E. G. Barnett, and C. J. Pearce. 1976. A chiral synthesis of D-(+)-2,6-dideoxystreptamine and its microbial incorporation into novel antibiotics. *J. Am. Chem. Soc.* 98:7110–7112.

Daniels, P. J. L. 1978. Synthetic and mutasynthetic antibiotics related to sisomicin. *Symp. on Mutasynth. of Antibiot., 18th Intersci. Conf. on Antimicrob. Agents and Chemother.* American Society for Microbiology, Atlanta, GA, October 1–4.

Daniels, P. J. L., and D. F. Rane. 1979. Synthetic and mutasynthetic antibiotics related to sisomicin. Pages 314–317 *in* D. Schlessinger, ed., *Microbiology—1979.* American Society for Microbiology, Washington, DC.

Daniels, P. J. L., D. F. Rane, S. W. McCombie, R. T. Testa, J. J. Wright, and T. L. Nagabhushan. 1980. Chemical and biological modification of antibiotics of the gentamicin group. Pages 371–392 *in* K. L. Rinehart, Jr., and T. Suami, eds., *Aminocyclitol Antibiotics.* ACS Symp. Ser. 125. American Chemical Society, Washington, DC.

Daum, S. J., D. Rosi, and W. A. Goss. 1977a. Mutational biosynthesis by idiotrophs of *Micromonospora purpurea.* II. Conversion of non-amino containing cyclitols to aminoglycoside antibiotics. *J. Antibiot.* 30:98–105.

______. 1977b. Production of antibiotics by biotransformation of 2,4,6/3,5-pentahydroxycyclohexanone and 2,4/3,5-tetrahydroxycyclohexanone by a deoxystreptamine-negative mutant of *Micromonospora purpurea. J. Am. Chem. Soc.* 99:283–284.

Davies, J. 1980. Enzymes modifying aminocyclitol antibiotics and their roles in resistance determination and biosynthesis. Pages 323–334 *in* K. L. Rinehart, Jr. and T. Suami, eds., *Aminocyclitol Antibiotics.* ACS Symp. Ser. 125. American Chemical Society, Washington, DC.

Davies, J. E., and M. Yagisawa. 1983. The aminocyclitol glycosides (aminoglycosides). Pages 329–354 *in* L. C. Vining, ed., *Biochemistry and Genetic Regulation of Commercially Important Antibiotics.* Addison-Wesley, Reading, MA.

Ewad, M. J. S., W. Al-Feel, and M. Akhtar. 1983. Mechanistic studies on the biosynthesis of the 2-deoxystreptamine ring of neomycins. *J. Chem. Soc., Chem. Commun.* pp. 20–22.

Fang, J. -R., C. J. Pearce, and K. L. Rinehart, Jr. 1984. Neomycin biosynthesis: The involvement of neamine and paromamine as intermediates. *J. Antibiot.* 37:77–79.

Furumai, T., K. Takeda, A. Kinumaki, Y. Ito, and T. Okuda. 1979. Biosynthesis of butirosins. II. Biosynthetic pathway of butirosins elucidated from cosynthesis and feeding experiments. *J. Antibiot.* 32:891–899.

Haas, M. J., and J. E. Dowding. 1975. Aminoglycoside-modifying enzymes. *Methods Enzymol.* 43:611–628.

Kakinuma, K. 1982. Biosynthesis of ribostamycin. Application of the deuterium labeling. Pages 185–194 *in* H. Umezawa, A. L. Demain, T. Hata, and C. R. Hutchinson, eds., *Trends in Antibiotic Research. Genetics, Biosyntheses, Actions & New Substances.* Japan Antibiotics Research Association, Tokyo.

Kakinuma, K., and S. Yamaya. 1983. Transamination stereochemistry in the formation of neosamine C of ribostamycin. *J. Antibiot.* 36:749–750.

Kakinuma, K., Y. Ogawa, T. Sasaki, H. Seto, and N. Otake. 1981. Stereochemistry of ribostamycin biosynthesis. An application of ^{2}H NMR spectroscopy. *J. Am. Chem. Soc.* 103:5614–5616.

Kojima, M., and A. Satoh. 1973. Microbial semi-synthesis of aminoglycosidic antibiotics by mutants of *S. ribosidificus* and *S. kanamyceticus. J. Antibiot.* 26:784–786.

Nagaoka, K., and A. L. Demain. 1975. Mutational biosynthesis of a new antibiotic, streptomutin A, by an idiotroph of *Streptomyces griseus. J. Antibiot.* 28:627–635.

Pearce, C. J., and K. L. Rinehart, Jr. 1981. Biosynthesis of aminocyclitol antibiotics. Pages 74–100 *in* J. W. Corcoran, ed., *Antibiotics IV Biosynthesis.* Springer-Verlag, Berlin.

Rinehart, K. L., Jr. 1964. *The Neomycins and Related Antibiotics.* Wiley & Sons, Inc., New York.

______. 1979. Biosynthesis and mutasynthesis of aminocyclitol antibiotics. *Jap. J. Antibiot.* S-32:S-32—S-46.

______. 1980. Biosynthesis and mutasynthesis of aminocyclitol antibiotics. Pages 335–370 *in* K. L. Rinehart, Jr. and T. Suami, eds., *Aminocyclitol Antibiotics.* ACS Symp. Ser. 125. American Chemical Society, Washington, DC.

______. 1981. Mutasynthesis of Antibiotics. Page 87 *in* T. P. Singer and R. N. Ondarza, eds., *Molecular Basis of Drug Action.* Elsevier North Holland, New York.

Rinehart, K. L., Jr., and R. F. Schimbor. 1967. Neomycins. Pages 359–372 *in* D. Gottlieb and P. D. Shaw, eds., *Antibiotics II Biosynthesis.* Springer-Verlag, Berlin.

Rinehart, K. L., Jr., and R. M. Stroshane. 1976. Biosynthesis of aminocyclitol antibiotics. *J. Antibiot.* 29:319–353.

Rinehart, K. L., Jr., and T. Suami, eds. 1980. *Aminocyclitol Antibiotics,* ACS Symp. Ser. 125. American Chemical Society, Washington, DC.

Rinehart, K. L., Jr., J. M. Malik, R. F. Nystrom, R. M. Stroshane, S. T. Truitt, M. Taniguchi, J. P. Rolls, W. J. Haak, and B. A. Ruff. 1974. Biosynthetic incorporation of [1-^{13}C]glucosamine and [6-^{13}C]glucose into neomycin. *J. Am. Chem. Soc.* 96:2263–2265.

Rosi, D., W. A. Goss, and S. J. Daum. 1977. Mutational biosynthesis by idiotrophs of *Micromonospora purpurea*. I. Conversion of aminocyclitols to new aminoglycoside antibiotics. *J. Antibiot.* 30:88–97.

Rubenstein, H. M. 1978. Biosynthetic and mutasynthetic studies of neomycin. Ph.D. Thesis, University of Illinois at Urbana-Champaign, Urbana, IL.

Schimbor, R. F. 1966. I. The microbiological incorporation of labeled intermediates into the neomycin antibiotics. II. Mass spectral studies of N-acetyl derivatives of amino sugars and related compounds. Ph.D. Thesis, University of Illinois at Urbana-Champaign, Urbana, IL.

Shier, W. T., K. L. Rinehart, Jr., and D. Gottlieb. 1969. Preparation of four new antibiotics from a mutant of *Streptomyces fradiae*. *Proc. Natl. Acad. Sci. USA*. 63:198–204.

Shier, W. T., P. C. Schaefer, D. Gottlieb, and K. L. Rinehart, Jr. 1974. Use of mutants in the study of aminocyclitol antibiotic biosynthesis and the preparation of the hybrimycin C complex. *Biochemistry* 13:5073–5078.

Slechta, L., and J. H. Coats. 1974. Studies of the biosynthesis of spectinomycin. *Abstr. 14th Intersci. Conf. on Antimicrob. Agents and Chemother.*, San Francisco, CA; p. 294.

Tadano, K. -I., T. Tsuchiya, T. Suami, and K. L. Rinehart, Jr. 1982. Synthesis of cytidine, uridine, and adenosine 5′-(2,6-diamino-2,6-dideoxy-α-D-glucopyranosyl diphosphates) (Neosamine C-CDP, -UDP, and -ADP). *Bull. Chem. Soc. Jap.* 55:3840–3846.

Takeda, K., K. Aihara, T. Furumai, and Y. Ito. 1978a. An approach to the biosynthetic pathway of butirosins and the related antibiotics. *J. Antibiot.* 31:250–253.

Takeda, K., S. Okuno, Y. Ohashi, and T. Furumai. 1978b. Mutational biosynthesis of butirosin analogs. I. Conversion of neamine analogs into butirosin analogs by mutants of *Bacillus circulans*. *J. Antibiot.* 31:1023–1030.

Takeda, K., A. Kinumaki, H. Hayasaka, T. Yamaguchi, and Y. Ito. 1978c. Mutational biosynthesis of butirosin analogs. II. 3′,4′-Dideoxy-6′-N-methylbutirosins, new semisynthetic aminoglycosides. *J. Antibiot.* 31:1031–1038.

Takeda, K., A. Kinumaki, S. Okuno, T. Matsushita, and Y. Ito. 1978d. Mutational biosynthesis of butirosin analogs. III. 6′-N-methylbutirosins and 3′,4′-dideoxy-6′-C-methylbutirosins, new semisynthetic aminoglycosides. *J. Antibiot.* 31:1039–1045.

Taylor, H. D., and H. Schmitz. 1976. Antibiotics derived from a mutant of *Bacillus circulans*. *J. Antibiot.* 29:532–535.

Testa, R. T., and B. C. Tilley. 1975. Biotransformation, a new approach to aminoglycoside biosynthesis. I. Sisomicin. *J. Antibiot.* 28:573–579.

______. 1976. Biotransformation, a new approach to aminoglycoside biosynthesis. II. Gentamicin. *J. Antibiot.* 29:140–146.

Testa, R. T., G. H. Wagman, P. J. L. Daniels, and M. J. Weinstein. 1974. Mutamicins: Biosynthetically created new sisomicin analogues. *J. Antibiot.* 27:917–921.

Waitz, J. A., G. H. Miller, E. Moss, Jr., and P. J. S. Chiu. 1978. Chemotherapeutic evaluation of 5-episisomicin (Sch 22591), a new semisynthetic aminoglycoside. *Antimicrob. Agents Chemother.* 13:41–48.

CHAPTER 6

Microbial Transformations of Anthracycline Antibiotics and Analogs

VINCENT P. MARSHALL

Research Laboratories, The Upjohn Company, Kalamazoo, Michigan 49001

Microbial transformation provides a means involving rather mild reaction conditions to bring about highly specific chemical changes. As such, microbial transformation has been a useful technique in the modification of antibiotics and antitumor agents. However, the high degree of specificity can be disadvantageous when considered in terms of narrow substrate specificities. Antibiotic transformations in the context of this review are considered as metabolic alterations of completely synthesized anthracyclines and anthracyclinones (direct transformations) and as manipulations involving their biosyntheses (indirect transformations). The direct transformations are oxidation, reduction, acylation, and alkylation. The indirect transformations involve precursor directed biosynthesis, interrupted biosynthesis, and genetic recombination.

INTRODUCTION

Microorganisms and their enzymes have been effective agents in synthesis of analogs of members of various classes of antibiotics (Perlman 1971; Perlman and Sebek 1971; Sebek and Perlman 1971; Sebek 1974, 1975, 1980; Shibata and Uyeda 1978; Oki 1982; Marshall and Wiley 1982; and Marshall 1982). Antibiotic transformations in the context of this review are considered both as metabolic alterations of completely synthesized anthracyclines and anthracyclinones (direct transformations) and as other manipulations involving their biosyntheses (indirect transformations). The direct transformations are considered on the basis of reaction type. These include oxidation, reduction, acylation, and alkylation. The indirect transformations considered are precursor directed biosynthesis, interrupted biosynthesis, and genetic recombination.

Some compounds discussed as substrates are not anthracyclines but are closely related to them and were included. In several instances which are included, it was not shown by isolation that a new compound had been formed, but only that the substrate had been changed. These cases were considered on an individual basis with some similar ones being excluded. The reference list is not exhaustive, although representative coverage of anthracycline biotransformation was attempted. Some transformations were discussed in more than one publication and some of these were not included in the references.

The structures presented in Figs. 1 and 2 and in Table 1 are representative of the types of compounds modified in the reported studies. Fig. 2 furnishes the usual ring numbering system employed.

TABLE 1. *Substitutions in rings*[a]

	Substitution at Ring Position								
Anthracycline	A	B	C	D	E	W	X	Y	R
Daunomycin	H	H	CH_3O	HO	HO	H,H	CH_3CO	H	daunosamine
Adriamycin	H	H	CH_3O	HO	HO	H,H	CH_2OHCO	H	AS ABOVE
Nogalamycin	$C_6H_{15}NO_4$		HO	H	HO	H,$COOCH_3$	CH_3	H	nogalose
Steffimycin	H	CH_3O	HO	H	HO	=O	CH_3	CH_3O	o-methylrhamnose
Steffimycin B	H	CH_3O	HO	H	HO	=O	CH_3	CH_3O	di-o-methylrhamnose
Steffimycinone	H	CH_3O	HO	H	HO	=O	CH_3	CH_3O	H
Cinerubin	HO	H	HO	H	HO	H,$COOCH_3$	C_2H_5	H	trisaccharide

[a] See Fig. 1.

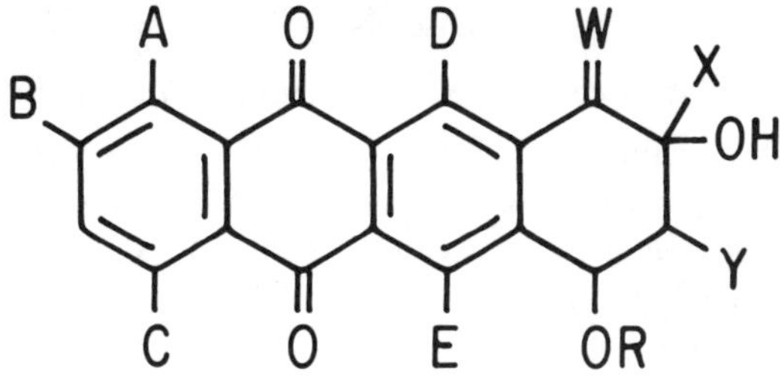

FIG. 1. General structure of anthracyclines and anthracyclinones (see Table 1).

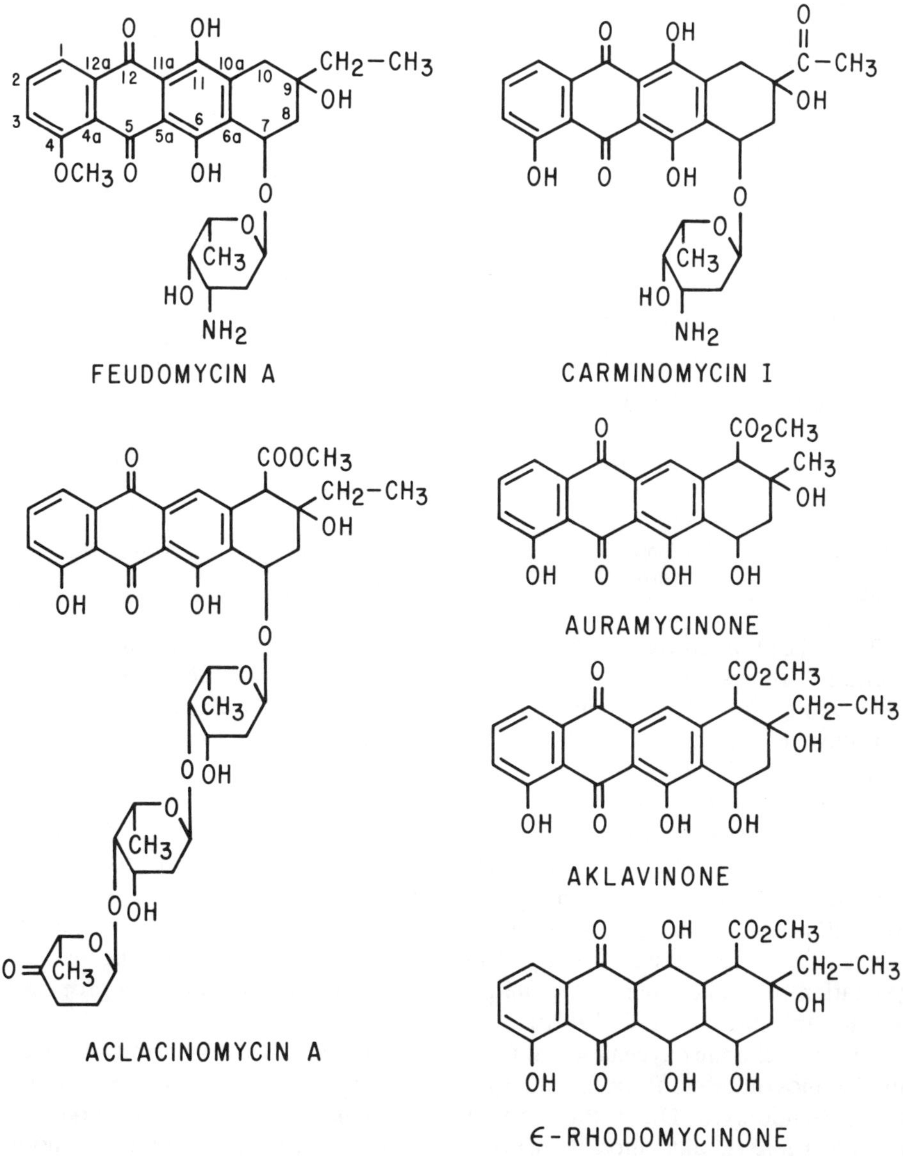

FIG. 2. Structures of specific anthracyclines and anthracyclinones.

Discussion

Biotransformation of Anthracyclines and Anthracyclinones by Oxidation

Oxidations of anthracyclines and anthracyclinones as reported to occur in microorganisms are of two general types (Table 2). The first is represented by the conversion of aclacinomycin A to Y as catalyzed by an oxidoreductase from *Streptomyces galilaeus* (Yoshimoto et al. 1979). In this conversion, the terminal sugar of aclacinomycin A known as L-cinerulose is oxidized to L-aculose with the removal of two hydrogen atoms. Studies revealed that the enzyme is able to modify several anthracyclic trisaccharides through oxidation of their terminal sugars. The isolated enzyme was found to be weakly acidic with a molecular wt of 72,000. Interestingly, aclacinomycin Y is reported to be approximately 10-fold more potent than aclacinomycin A.

TABLE 2. *Biotransformation of anthracyclines and anthracyclinones by oxidation*

Substrate	Transformation Product	Biological Agent	Reference
Aclacinomycin A	Aclacinomycin Y	*Streptomyces galilaeus*	Yoshimoto et al. 1979
Aklavinone	ϵ-Rhodomycinone	*Streptomyces coeruleorubidus*	Yoshimoto et al. 1980c; Blumauerova et al. 1979a
Auramycinone	11-Hydroxy-auramycinone	*S. coeruleorubidus*	Hoshino and Fujiwara 1983
Daunomycin	Adriamycin	*Streptomyces peucetins* var. *caesius*	Oki et al. 1981b
ϵ-Pyrromycinone	ϵ-Isorhodo-mycinone	*S. coeruleorubidus*	Yoshimoto et al. 1980a

The other type of oxidation reported to occur through microbial biotransformation involves oxygenation. Specific examples include the conversion of aklavinone to ϵ-rhodomycinone (Yoshimoto et al. 1980c), daunomycin to adriamycin (Oki et al. 1981b), and auramycinone to 11-hydroxyauramycinone (Hoshino and Fujiwara 1983).

In the conversion of daunomycin to adriamycin by a mutant of *Streptomyces peucetius* var. *caesius*, a saturated carbon alpha to a ketone is oxidized to yield an α-hydroxyketone. These data were reported in conjunction with a proposal for the biosynthesis of adriamycin where daunomycin is produced by 10-decarbomethoxylation, 4-O-methylation, and oxidation at C-13 of the intermediate glycoside *via* ϵ-rhodomycinone from aklavinone. The reported oxygenation of daunomycin is the final step in the biosynthetic pathway proposed for adriamycin (Oki et al. 1981b).

Anthracyclinone oxygenation at C-11 has been reported as an intermediate step in the biosyntheses of both daunomycinone and 9-methyl-10-hydroxydaunomycin (feudomycin D). In the biosynthesis of daunomycinone by *Streptomyces coeruleorubidus*, aklavinone is oxygenated at C-11 to yield ϵ-rhodomycinone (Yoshimoto et al. 1980c). In a similar study, *S. coeruleorubidus* was reported to oxygenate auramycinone to form 11-hydroxyauramycinone in the synthesis of feudomycin D. Auramycinone and aklavinone differ only at C-9 where the former is methylated and the latter contains an ethyl group.

Biotransformation of Anthracyclines and Anthracyclinones by Reduction

The reported biological reductions of anthracyclines and anthracyclinones are also of two general types. The first is carbonyl reduction (Table 3) and the second is reductive cleavage of carbon-oxygen bonds, usually at C-7 (Table 4).

Reduction of the C-13 ketonic carbonyl of daunomycin and its analogs has been the most frequently reported of the transformations by carbonyl reduction. These conversions have been performed by diphtheroids, streptomycetes, and

TABLE 3. *Biotransformation of anthracyclines and anthracyclinones through carbonyl reduction*

Substrate	Transformation Product	Biological Agent	Reference
Daunomycin	Daunomycinol (13-dihydro-daunomycin)	*Corynebacterium simplex*	Ninet et al. 1976
Daunomycin	Daunomycinol	*Corynebacterium equi*	Aszalos et al. 1977
Daunomycin	Daunomycinol	*Mucor spinosus*	Marshall et al. 1978
N-Acetyl-daunomycin	N-Acetyl-daunomycinol	*C. equi*	Aszalos et al. 1977
Daunomycinone	Daunomycinol aglycone (13-dihydrodauno-mycinone)	*Streptomyces aureofaciens*	Karnetova et al. 1976
Daunomycinone	Daunomycinol aglycone	*S. coeruleorubidus* and *S. galilaeus*	Blumauerova et al. 1979a
7-Deoxyadria-mycinone	7-Deoxyadria-mycinol aglycone	*Streptomyces steffisburgensis*	Marshall et al. 1978
7-Deoxy-daunomycinone	7-Deoxydauno-mycinol aglycone	*S. steffisburgensis*	Marshall et al. 1976b
ϵ-Isorhodo-mycinone	1-Hydroxy-13-dihydro-daunomycin	*S. coeruleorubidus*	Yoshimoto et al. 1980a
Rubeomycin A	Rubeomycin B	*Rhodotorula glutinis*	Ogawa et al. 1983
Steffimycin	10-Dihydro-steffimycin	*Actinoplanes utahensis*	Wiley et al. 1980
Steffimycin B	10-Dihydro-steffimycin B	*A. utahensis* and *Chaetomium* sp.	Wiley et al. 1980
Steffimycinone	Steffimycinol (10-dihydro-steffimycinone)	*Streptomyces peucetius* var. *caesius* and *Streptomyces nogalater*	Wiley et al. 1977

TABLE 4. *Biotransformation of anthracyclines and anthracyclinones through reductive carbon-oxygen bond cleavage*

Substrate	Transformation Product	Biological Agent	Reference
Aclacinomycin B	Aclacinomycin A	*Streptomyces galilaeus*	Hoshino and Fujiwara 1983
Adriamycin	7-Deoxyadria-mycinone	*Streptomyces steffisburgensis*	Marshall et al. 1978
Aklavinone	7-Deoxyaklavinone	*Streptomyces coeruleorubidus*	Blumauerova et al. 1979a
Cinerubin A	zeta-Pyrromycinone	*Aeromonas hydrophila*	Marshall et al. 1976a
Daunomycin	7-Deoxy-daunomycinone	*A. hydrophila*	Wiley and Marshall 1975; Marshall et al. 1976a
Daunomycin	7-Deoxy-daunomycinone	*S. steffisburgensis*	Marshall et al. 1976b
Daunomycinol	7-Deoxydauno-mycinol aglycone	*S. steffisburgensis*	Marshall et al. 1978
Daunomycinol ablycone	7-Deoxydauno-mycinol aglycone	*S. coeruleorubidus* and *S. galilaeus*	Blumauerova et al. 1979a
Nogalamycin	7-Deoxynogalarol	*A. hydrophila*	Wiley and Marshall 1975; Marshall et al. 1976a
Nogalamycin	7-Deoxynogalarol	*Streptomyces nogalater*	Reuckert et al. 1979
Steffimycin	7-Deoxy-steffimycinone	*A. hydrophila*, *Escherichia coli* and *Citrobacter freundii*	Wiley and Marshall 1975; Marshall et al. 1976a
Steffimycin	7-Deoxy-steffimycinone	*A. hydrophila*	McCarville and Marshall 1977

fungi, as well as by mammals. One commonly reported conversion is that of daunomycin to daunomycinol (13-dihydrodaunomycin). In addition to its role in microbial biotransformations, this C-13 carbonyl reduction is the initial reaction in the primary mammalian route of daunomycin metabolism (Takanashi and Bachur 1975) and is linked to NADPH (Felsted et al. 1974). Conversion of daunomycin to daunomycinol is also known to be linked to NADPH in *Mucor spinosus* (Marshall et al. 1978).

Another frequently reported microbial transformation of this type is the reduction of the C-10 ketonic carbonyl of steffimycin and related compounds. Table 3 lists examples of this conversion as performed by several actinomycetes and a species of *Chaetomium*. When investigated biochemically, carbonyl reduction was shown again to be linked to NADPH (Wiley et al. 1977, 1980).

Table 4 lists examples of reductive cleavage of carbon-oxygen bonds as reported among anthracyclines and anthracyclinones. When this conversion was investigated using crude enzyme preparations of *Aeromonas hydrophila* (Marshall et al. 1976a) and of several streptomycetes (Marshall et al. 1976b; Reuckert et al. 1979; Wiley and Marshall 1975), it was found to require either NADH or NADPH. When catalyzed by mammalian microsomal preparations, the same reaction requires NADPH (Asbell et al. 1972; Bachur and Gee 1976). The enzyme from *A. hydrophila* was purified *ca.* 100-fold and was found to have a molecular wt of *ca.* 35,000 (McCarville and Marshall 1977). The purified enzyme displayed an absolute requirement for NADH, indicating the presence of transhydrogenase activity in the crude enzyme preparations. Both the purified *A. hydrophila* enzyme and the microsomal enzyme required anaerobic reaction conditions.

Reductive cleavage is known to be the initial reaction of the minor mammalian route of daunomycin metabolism (Takanashi and Bachur 1975). The mammalian enzyme was purified and was found to be a flavoprotein. In one instance it was designated as cytochrome c reductase (Oki et al. 1977) and in another cytochrome P450 reductase (Komiyama et al. 1979). Fig. 3 shows a mechanism proposed for reductive cleavage of such carbon-oxygen bonds (Komiyama et al. 1979).

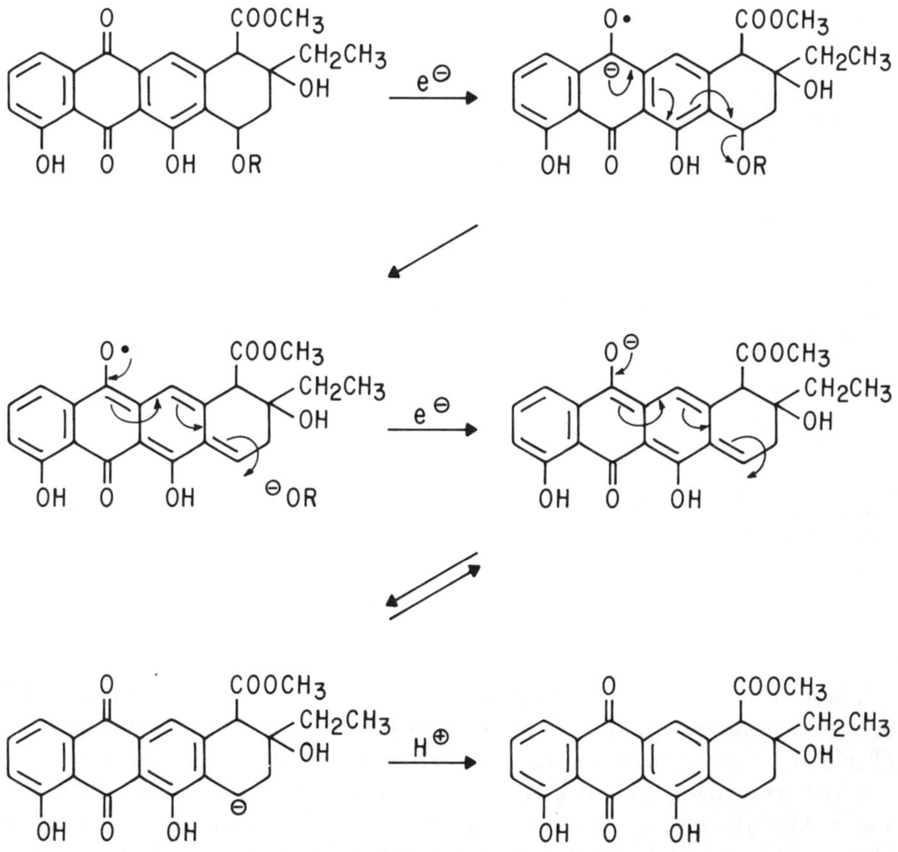

FIG. 3. The proposed reaction mechanism for reductive glycosidic cleavage of aclacinomycin A (Komiyama et al. 1979).

Biotransformation of Anthracyclines and Anthracyclinones by Acylation and Alkylation

The majority of microbial acylations of antibiotics have involved aminocyclitols, β-lactams, and macrolides. However, there have been several reported instances of such transformations of anthracyclines and anthracyclinones (Table 5). In one case, daunomycin and daunomycinol were shown to be converted to their N-acetyl derivatives (Hamilton et al. 1977). These conversions were brought about by *Bacillus subtilis* var. *mycoides* and involved N-acetylation of the daunosamine moieties of both antibiotics. Another case involved the N-formylation of internally synthesized 1-hydroxy-13-dihydrodaunomycin (Yoshimoto et al. 1980a). In these experiments, a mutant of *S. coeruleorubidus* was shown to synthesize 1-hydroxy-13-dihydrodaunomycin from ϵ-pyrromycinone or from ϵ-isorhodomycinone. The synthesized anthracycline was then converted to its N-formyl derivative by the mutant streptomycete.

TABLE 5. *Biotransformation of anthracyclines and anthracyclinones through acylation and alkylation*

Substrate	Transformation Product	Biological Agent	Reference
A. *Acylation*			
Daunomycin	N-Acetyl-daunomycin	*Bacillus cereus* var. *mycoides*	Hamilton et al. 1977
Daunomycinol	N-Acetyl-daunomycinol	*B. cereus* var. *mycoides*	Hamilton et al. 1977
1-Hydroxy-13-dihydro-daunomycin	N-Formyl-1-hydroxy-13-dihydrodaunomycin	*Streptomyces coeruleorubidus*	Yoshimoto et al. 1980a
B. *Alkylation*			
Carminomycin I	Daunomycin	*Streptomyces peucetius* var. *caesius*	Oki et al. 1981b
Daunosaminyl-aklavinone	Aklavin	*S. coeruleorubidus*	Oki et al. 1980
11-Hydroxy-auramycinone	Feudomycin D	*S. coeruleorubidus*	Hoshino and Fujiwara 1983

The microbial alkylations reported to occur among the anthracyclines and their aglycones have been performed by streptomycetes and are biosynthetic in nature (Table 5). In experiments on the bioconversion of aklavinone, aklavin was synthesized from daunosaminylaklavin *via* N-methylation (Oki et al. 1980; Oki 1982). Methylation has also been reported in the biosynthesis of adriamycin. Here, carminomycin I was converted to adriamycin *via* O-methylation at C-4 by *S. peucetius* var. *caesius* (Oki et al. 1981b). In another report, 11-hydroxy-auramycinone was shown to be converted to feudomycin D *via* several steps, one of which involved O-methylation at C-4 (Hoshino and Fujiwara 1983).

Indirect Biotransformation of Anthracyclines and Anthracyclinones through Interrupted Biosynthesis

One means of indirect transformation of these compounds involves the interruption of their biosyntheses (Table 6). Biosynthetic interruption can be achieved through genetic manipulation or through chemically mediated metabolic inhibition.

TABLE 6. *Indirect biotransformation of anthracyclines and anthracyclinones through interrupted biosynthesis*

Means of Interruption	Antibiotic Produced	Biological Agent	Reference
Mutation: defective oxidoreductase	Aclacinomycin analog MA 144-Nl	*Streptomyces galilaeus* mutant	Yoshimoto et al. 1981a; Matsuzawa et al. 1981
Mutation: defective glycosidation	Aclacinomycin analogs MA 144-U1, U2, U5-U9, and S1	*S. galilaeus* mutant	Yoshimoto et al. 1981a; Matsuzawa et al. 1981
Mutation: defective glycosidation	Aklavinone	*S. galilaeus* mutant	Yoshimoto et al. 1981a; Matsuzawa et al. 1982
Mutation: multiple defects	2-Hydroxyaklavinone and 2-hydroxy-7-deoxyaklavinone	*S. galilaeus* mutant	Yoshimoto et al. 1981a; Matsuzawa et al. 1981; Tobe et al. 1982
Mutation: oxygenation	Adriamycin	*Streptomyces peucetius* var. *caesius*	DiMarco et al. 1969
Mutation: multiple defects	Feudomycins A and B	*Streptomyces coeruleorubidus* mutant	Oki et al. 1981a
Mutation: multiple defects	Feudomycinones C and D	*S. coeruleorubidus* mutant	Oki et al. 1981a
Mutation: multiple defects	Maggiemycin and anhydromaggiemycin	Mutant streptomycete	Pandy et al. 1981
Ethionine	Carminomycinone	*S. coeruleorubidus*	Blumauerova et al. 1979b

As a result of a genetic study directed toward the isolation of aclacinomycin over-producing mutants, a series of mutants of *S. galilaeus* were isolated (Yoshimoto et al. 1981a; Matsuzawa et al. 1981). The mutant organisms included those that produced only specific components of the parental glycosides or that produced biochemically altered anthracyclinones. Among the compounds produced were 2-deoxyfucosyl-2-deoxyfucosyl-rhodosaminylaklavinone (MA 144 U1), 2-deoxyfucosyl-rhodosaminylaklavinone (MA 144 U2), and five aklavinone glycosides devoid of an amino sugar (MA 144 U5-U9). In addition, certain mutants produced the anthracyclinones, 2-hydroxyaklavinone and 2-hydroxy-7-deoxyaklavinone. The latter compounds were investigated in similar studies (Tobe et al. 1982).

Two novel anthracyclines termed the feudomycins A and B were produced by mutants of *S. coeruleorubidus* (Oki et al. 1981a). The feudomycins are glycosides of daunomycinone related anthracyclinones. Feudomycinone A contains 9-acetonyldaunomycinone. In addition to the feudomycins A and B, two novel anthracyclinones were produced, the feudomycinones C and D. These anthracyclinones were also variations of daunomycinone altered at C-9 and C-10.

Chemically mediated inhibition of antibiotic biosynthesis also has been a means for the production of anthracycline analogs (Blumauerova et al. 1979b). DL-ethionine was reported to inhibit 4-O-methylation and 10-carboxyl esterification in *S. coeruleorubidus*. The blockage of these reactions subsequently caused this organism to produce substances belonging to the carminomycinone series with concomitant reduced levels of daunomycinone and ϵ-rhodomycinone.

Indirect Biotransformation of Anthracyclinones through Precursor-directed Biosynthesis

The direction of antibiotic biosynthesis through the introduction of analogs of natural precursors has been an effective procedure in biological semisynthesis. Because of tolerances in substrate specificities of biosynthetic enzymes, it is often possible for antibiotic-producing microorganisms to incorporate these precursor analogs into the structures of various antibiotics.

Among the anthracyclines, glycosides have been formed from anthracyclinones supplied to microorganisms capable of causing their glycosidation (Table 7). An early example involved the feeding of ϵ-pyrromycinone to *S. galilaeus* (Vanek et al. 1973). As a result, the galirubins A and B were synthesized. In an extension of these studies, several anthracyclines were formed through the feeding of anthracyclinones to *S. galilaeus* and *S. coeruleorubidus* (Blumauerova et al. 1979a). Specifically, the anthracyclinones employed as precursors were daunomycinone, aklavinone, ϵ-isorhodomycinone, ϵ-rhodomycinone and ϵ-pyrromycinone.

The novel anthracycline 1-hydroxy-13-dihydrodaunomycin was produced through the addition of ϵ-pyrromycinone or ϵ-isorhodomycinone to fermentations of a mutant strain of *S. coeruleorubidus* (Yoshimoto et al. 1980a). In similar experiments, baumycin negative mutants of *S. coeruleorubidus* were shown not to convert daunomycinone to daunomycin (Yoshimoto et al. 1980b). However, these mutants convert aklavinone and ϵ-rhodomycinone to daunomycin. Further, daunomycinone was shown to be synthesized from aklavinone and ϵ-rhodomycinone (Yoshimoto et al. 1980c).

An aclacinomycin negative strain of *S. galilaeus* was also employed in the glycosidation of various anthracyclinones (Oki et al. 1980). ϵ-, α-, and β-Rhodomycinones, ϵ-isorhodomycinone, as well as ϵ- and β-pyrromycinones were found to be transformed into biologically active anthracyclines. However, daunomycinone, 13-deoxydaunomycinone, and steffimycinone did not function as substrates for glycosidation. These experiments also revealed that aclacinomycin is synthesized by sequential glycosidation from aklavinone *via* aklavin.

Trisarubicinol was synthesized by strain KE 303 of *S. galilaeus* through the addition of carminomycinone or 13-dihydrocarminomycinone to its fermentations (Yoshimoto et al. 1981b). Trisarubicinol, a novel anthracyclic trisaccharide, contains 13-dihydrocarminomycinone linked at C-7 to cinerubosyl-2-deoxyfucosyl-rhodosamine. Trisarubicinol is active vs. L1210 leukemia in CDF_1 mice and was

TABLE 7. *Indirect biotransformation of anthracyclinones through precursor-directed biosynthesis*

Precursor	Antibiotic Produced	Biological Agent	Reference
Aklavinone	Glycosides of aklavinone	*Streptomyces coeruleorubidus*	Blumauerova et al. 1979a
Aklavinone	Aclacinomycin A	*Streptomyces galilaeus*	Oki et al. 1980
Aklavinone and ϵ-rhodomycinone	Daunomycin	*S. coeruleorubidus*	Yoshimoto et al. 1980b
4-O-Methyl-aklavinone	4-O-Methyl-aclacinomycin A	*S. galilaeus*	Oki et al. 1980
10-Decarbomethoxy-aklavinone	10-Decarbomethoxy-aclacinomycin A	*S. galilaeus*	Oki et al. 1980
2-Hydroxy-aklavinone	2-Hydroxy-aclacinomycin A	*S. galilaeus*	Yoshimoto et al. 1981a
Carminomycinone or 13-dihydrocar-minomycinone	Trisarubicinol	*S. galilaeus*	Oki et al. 1980; Matsuzawa et al. 1980
Daunomycinone	Glycosides of daunomycinone	*S. coeruleorubidus*	Blumauerova et al. 1979a
Daunomycinol aglycone	Glycosides of daunomycinol aglycone	*S. coeruleorubidus*	Blumauerova et al. 1979a
ϵ-Pyrromycinone	Glycosides of ϵ-pyrromycinone	*S. coeruleorubidus*	Blumauerova et al. 1979a
ϵ-Pyrromycinone	Galirubins A and B	*S. galilaeus*	Vanek et al. 1973
ϵ-Pyrromycinone or ϵ-isorhodo-mycinone	1-Hydroxy-13-di-hydrodaunomycin and N-formyl-1-hydroxy-13-dihydro-daunomycin	*S. coeruleorubidus*	Yoshimoto et al. 1980a
ϵ-Isorhodomycinone	Glycosides of ϵ-Isorhodo-mycinone	*S. coeruleorubidus*	Blumauerova et al. 1979a
ϵ-Isorhodomycinone and ϵ-pyrro-mycinone	Anthracycline glycosides	*S. galilaeus*	Oki et al. 1980; Matsuzawa et al. 1980
Rhodomycinones	Cinerulosyl-deoxyfucosyl-rhodosaminyl-rhodomycinone	*S. galilaeus*	Oki et al. 1980; Matsuzawa et al. 1980
ϵ-Rhodomycinone	Glycosides of rhodomycin	*S. coeruleorubidus*	Blumauerova et al. 1979a

reported to be a more potent inhibitor of DNA synthesis than carminomycin I. Other examples of such transformations are presented in Table 7.

As many of the examples included here involve precursor-directed biosynthesis using mutationally blocked microorganisms, they could have been discussed in terms of mutational biosynthesis (mutasynthesis). However, in the interest of reducing the categories of indirect transformations presented, these were considered as precursor-directed biosynthesis.

Indirect Biotransformation of Anthracycline Antibiotics through Genetic Recombination

Recombination is a novel approach in the discovery of anthracyclines through genetic manipulation. In one of these studies, *Streptomyces violaceus* subsp. *iremyceticus* was prepared through hybridizations between different streptomycetes blocked in anthracycline biosynthesis (Ihn et al. 1980). This organism produced an anthracycline termed iremycin. Although the application of genetic recombination to the synthesis of antibiotics is in its early stages of development, this means appears to be extremely promising.

Literature Cited

Asbell, M. A., E. Schwartzbach, F. J. Bullock, and D. W. Yesair. 1972. Daunomycin and adriamycin metabolism *via* reductive glycosidic cleavage. *J. Pharmacol. Exp. Ther.* 182:63–69.

Aszalos, A. A., N. R. Bachur, B. K. Hamilton, A. Langlykke, P. P. Roller, M. Y. Sheikh, M. S. Sutphin, M. C. Thomas, D. A. Wareheim, and L. H. Wright. 1977. Microbial reduction of the side-chain carbonyl of daunorubicin and N-acetyldaunorubicin. *J. Antibiot.* 30:50–58.

Bachur, N. R., and M. Gee. 1976. Microsomal reductive glycosidase. *J. Pharmacol. Exp. Ther.* 197:681–686.

Blumauerova, M., E. Kralovocova, J. Mateju, J. Jizba, and Z. Vanek. 1979a. Biotransformation of anthracyclinones in *Streptomyces coeruleorubidus* and *Streptomyces galilaeus*. *Folia Microbiol.* 18:117–127.

Blumauerova, M., J. Jizba, K. Stajner, and Z. Vanek. 1979b. Effect of DL-ethionine on the biosynthesis of anthracyclines. *Biotechnol. Lett.* 471–476.

DiMarco, A., M. Gaetani, and B. Scarpinato. 1969. Adriamycin (NSC-123, 127): A new antibiotic with antitumor activity. *Cancer Chemother. Rep.* 53:33–37.

Felsted, R. L., M. Gee, and N. R. Bachur. 1974. Rat liver daunorubicin reductase. *J. Biol. Chem.* 249:3672–3679.

Hamilton, B. K., M. S. Sutphin, M. C. Thomas, D. A. Wareheim, and A. A. Aszalos. 1977. Microbial N-acetylation of daunorubicin and daunorubicinol. *J. Antibiot.* 30:425–426.

Hoshino, T., and A. Fujiwara. 1983. Microbial conversion of anthracycline antibiotics. *J. Antibiot.* 36:1463–1467.

Ihn, W., B. Schlegel, W. F. Fleck, and P. Sedmera. 1980. New anthracycline antibiotics produced by interspecific recombinants of streptomycetes. *J. Antibiot.* 33:1457–1461.

Karnetova, J., J. Mateju, P. Sedmera, J. Vokoun, and Z. Vanek. 1976. Microbial transformation of daunomycinone by *Streptomyces aureofaciens* B96. *J. Antibiot.* 29:1199–1202.

Komiyama, T., T. Oki, and T. A. Inui. 1979. A proposed reaction mechanism for the enzymatic reductive cleavage of glycosidic bond in anthracycline antibiotics. *J. Antibiot.* 32:1219–1222.

Marshall, V. P., E. A. Reisender, L. M. Reineke, J. H. Johnson, and P. F.

Wiley. 1976a. Reductive microbial conversion of anthracycline antibiotics. *Biochemistry* 15:4139–4145.

Marshall, V. P., E. A. Reisender, and P. F. Wiley. 1976b. Bacterial metabolism of daunomycin. *J. Antibiot.* 29:966–968.

Marshall, V. P., J. P. McGovren, F. A. Richard, R. E. Richard, and P. F. Wiley. 1978. Microbial metabolism of anthracycline antibiotics daunomycin and adriamycin. *J. Antibiot.* 31:336–342.

Marshall, V. P. 1982. Microbial metabolism of anthracycline antibiotics. Pages 75–83 *in* L. N. Ornston and S. G. Sligar, eds., *Experiences in Biochemical Perception.* Academic Press, New York and London.

Marshall, V. P., and P. F. Wiley. 1982. Microbial transformations of antibiotics. Pages 45–80 *in* J. R. Rosazza, ed., *Microbial Transformations of Bioactive Compounds.* CRC Press, Boca Raton, FL.

McCarville, M., and V. P. Marshall. 1977. Partial purification and characterization of a bacterial enzyme catalyzing reductive cleavage of anthracycline glycosides. *Biochem. Biophys. Res. Commun.* 74:331–335.

Matsuzawa, Y., A. Yoshimoto, T. Oki, H. Naganawa, T. Takeuchi, and H. Umezawa. 1980. Biosynthesis of anthracycline antibiotics by *Streptomyces galilaeus. J. Antibiot.* 33:1341–1347.

Matsuzawa, Y., A. Yoshimoto, N. Shibamoto, H. Naganawa, T. Takeuchi, T. Oki, and H. Umezawa. 1981. New anthracycline metabolites from mutant strains of *Streptomyces galilaeus* MA144-MI. *J. Antibiot.* 34:959–964.

Ninet, L., J. Florent, J. Lunel, J. Renaut, A. Abraham, B. Lombardi, and R. Tissier. 1976. Daunorubicinol preparation by biochemical reduction of daunorubicin. *Proc. 5th Intl. Ferment. Symp.* 17.06.

Ogawa, Y., S. Mizukoshi, and H. Mori. 1983. Microbial transformation of rubeomycin A to rubeomycin B. *J. Antibiot.* 36:1561–1563.

Oki, T., T. Komiyama, H. Tone, T. Inui, T. Takeuchi, and H. Umezawa. 1977. Reductive cleavage of anthracycline glycosides by microsomal NADPH-cytochrome C reductase. *J. Antibiot.* 30:613–615.

Oki, T., A. Yoshimoto, Y. Matsuzawa, T. Takeuchi, and H. Umezawa. 1980. Biosynthesis of anthracycline antibiotics by *Streptomyces galilaeus. J. Antibiot.* 33:1331–1340.

Oki, T., Y. Matsuzawa, K. Kiyoshima, A. Yoshimoto, H. Naganawa, T. Takeuchi, and H. Umezawa. 1981a. New anthracyclines, feudomycins, produced by the mutant from *Streptomyces coeruleorubidus* ME130-A4. *J. Antibiot.* 34:783–790.

Oki, T., Y. Takatsuki, H. Tobe, A. Yoshimoto, T. Takeuchi, and H. Umezawa. 1981b. Microbial conversion of daunomycin, carminomycin I and feudomycin A to adriamycin. *J. Antibiot.* 34:1229–1231.

Oki, T. 1982. Microbial transformation of anthracycline antibiotics and development of new anthracyclines. Pages 75–96 *in* H. S. El Khadem, ed., *Anthracycline Antibiotics.* Academic Press, New York and London.

Pandy, R. C., M. W. Toussaint, J. C. McGuire, and M. C. Thomas. 1981. Maggiemycin and anhydromaggiemycin: Two new anthracyclinone antitumor antibiotics. *Proc. 21st ICAAC:*182.

Perlman, D. 1971. Microbial transformations of antibiotics. *Process Biochem.* 13:13.

Perlman, D., and O. K. Sebek. 1971. Microbial transformations of antibiotics. *Pure Appl. Chem.* 28:637–648.

Reuckert, P. W., P. F. Wiley, J. P. McGovren, and V. P. Marshall. 1979. Mammalian and microbial conversions of anthracycline antibiotics and analogs. *J. Antibiot.* 32:141–147.

Sebek, O. K., and D. Perlman. 1971. Microbial transformations of antibiotics. *Adv. Appl. Microbiol.* 14:123–150.

Sebek, O. K. 1974. Microbial conversion of antibiotics. *Lloydia* 37:115–133.

Sebek, O. K. 1975. The scope and potential of antibiotic conversion by microorganisms. *Acta Microbiol. Acad. Sci. Hung.* 22:381–388.

Sebek, O. K. 1980. Microbial transformation of antibiotics. *Econ. Microbiol.* 5:575–612.

Shibata, M., and M. Uyeda. 1978. Microbial transformations of antibiotics. *Annu. Rep. Ferment. Processes* 2:267–303.

Takanashi, S., and N. R. Bachur. 1975. Daunorubicin metabolites in human urine. *J. Pharmacol. Exp. Ther.* 195:41–49.

Tobe, H., A. Yoshimoto, T. Ishikura, H. Naganawa, T. Takeuchi, and H. Umezawa. 1982. New anthracycline metabolites from two blocked mutants of *Streptomyces galilaeus* MA144-MI. *J. Antibiot.* 35:1641–1645.

Vanek, Z., J. Tax, I. Komersova, and K. Eckardt. 1973. Glycosidation of ϵ-pyrromycinone using the strain *Streptomyces galilaeus* JA3043. *Folia Microbiol.* 18:524–526.

Wiley, P. F., and V. P. Marshall. 1975. Microbial conversion of anthracycline antibiotics. *J. Antibiot.* 28:838–840.

Wiley, P. F., J. M. Koert, D. W. Elrod, E. A. Reisender, and V. P. Marshall. 1977. Bacterial metabolism of anthracycline antibiotics. Steffimycinone and steffimycinol conversions. *J. Antibiot.* 30:649–654.

Wiley, P. F., D. W. Elrod, J. M. Slavicek, and V. P. Marshall. 1980. Microbial conversion of steffimycin and steffimycin B to 10-dihydrosteffimycin and 10-dihydrosteffimycin B. *J. Antibiot.* 33:819–823.

Yoshimoto, A., T. Ogasawara, I. Kitamura, T. Oki, and T. Inui. 1979. Enzymatic conversion of aclacinomycin A to Y by a specific oxidoreductase in *Streptomyces. J. Antibiot.* 32:472–481.

Yoshimoto, A., Y. Matsuzawa, T. Oki, H. Naganawa, T. Takeuchi, and H. Umezawa. 1980a. Microbial conversion of ϵ-pyrromycinone and ϵ-isorhodomycinone to 1-hydroxy-13-dihydrodaunomycin and N-formyl-1-hydroxy-13-dihydrodaunomycin and their derivatives. *J. Antibiot.* 33:1150–1157, 1980a.

Yoshimoto, A., T. Oki, T. Takeuchi, and H. Umezawa. 1980b. Microbial conversion of anthracyclinones to daunomycin by blocked mutants of *Streptomyces coeruleorubidus. J. Antibiot.* 33:1158–1166.

Yoshimoto, A., T. Oki, and H. Umezawa. 1980c. Biosynthesis of daunomycinone from aklavinone and ϵ-rhodomycinone. *J. Antibiot.* 33:1199–1201.

Yoshimoto, A., Y. Matsuzawa, T. Oki, T. Takeuchi, and H. Umezawa. 1981a. New anthracycline metabolites from mutant strains of *Streptomyces galilaeus* MA144-MI. *J. Antibiot.* 34:951–958.

Yoshimoto, A., Y. Matsuzawa, Y. Matsushita, T. Oki, T. Takeuchi, and H. Umezawa. 1981b. Trisarubicinol, new anthracycline antibiotic. *J. Antibiot.* 34:1492–1494.

CHAPTER 7

Industrial Importance of Biotransformations of β-Lactam Antibiotics

D. A. Lowe

Bristol-Myers Co. Industrial Division
Syracuse, New York 13221-4755

Immobilized enzymes have now been established as important manufacturing tools in the large-scale production of β-lactam antibiotics. Immobilized penicillin acylase enzymes currently account for more than half the 6-aminopenicillanic acid (6-APA) produced and can yield up to 250 kg 6-APA per kg enzyme activity used. Similar enzyme systems can be used to yield 7-amino desacetoxycephalosporanic acid (7-ADCA) from ring expanded penicillins. According to published reports penicillin acylases and associated enzymes can function in the reverse direction to synthesize semisynthetic penicillins and cephalosporins. 7-Aminocephalosporanic acid (7-ACA) appears to be more difficult to obtain by enzymic means. Industrially useful enzymes capable of hydrolyzing the fermented cephalosporin C to 7-ACA have not been developed. However, enzyme activities have been reported that yield 7-ACA from N-acylated cephalosporin C and from a deaminated product, glutaryl-7-ACA. The ester group on cephalosporin C can be hydrolyzed by enzymic means to yield the desacetyl derivatives, useful intermediates in the synthesis of certain injectable cephalosporins. The synthesis and resolution of the side chains structures D(-)-phenylglycine and D(-)*p*-hydroxy phenylglycine can be conveniently carried out by enzymic systems that compete economically with alternate chemical procedures.

Introduction

The past 15 years have seen a dramatic increase in the application of enzymes, particularly in their immobilized form, as biological catalysts to replace convention chemical reactions. Immobilized enzymes have become one of the cornerstones of the recent biotechnology revolution.

The first commercial application of immobilized enzyme technology was realized in 1969 in Japan with the use of *Aspergillus oryzae* amino-acylase for the industrial production of L-amino acids (Chibata et al. 1972). During this same period, pilot plant processes were introduced for 6-aminopenicillanic acid (6-APA) production from penicillin G and for glucose to fructose conversion by immobilized glucose isomerase. Currently the annual production of 6-APA is thought to exceed 8,000 tons of which over half is by enzymic hydrolysis of the parent-fermented penicillin (Poulsen 1984). The economics of the enzymatic route include lower energy and raw material (i.e., solvents and chemicals) costs, and low environmental impact because of the absence of toxic intermediates and waste products. The use of immobilized penicillin acylases is particularly relevant to the pharmaceutical industry because they have had positive economic, environmental, and quality impacts on the manufacture of the important semisynthetic penicillin intermediate, 6-APA. The production has stimulated both the search for additional enzyme systems in other areas of β-lactam antibiotic manufacture and the continued development of immobilized enzyme and cell technologies. 6-APA manufacturing processes have recently been described (Harrison and Gibson 1984). The breakdown of costs for a 275,000 kg 6-APA per annum plant indicated a variable raw material cost of $30.27/kg out of a total of $52.27/kg 6-APA. The enzyme cost was $3.54/kg 6-APA.

Discussion

Hydrolysis of penicillin to 6-APA. Penicillin is fermented either as benzyl penicillin (penicillin G) or as phenoxymethyl penicillin (penicillin V) depending on the precursor used, i.e., phenylacetate or phenoxyacetate, respectively. Both penicillin types can be hydrolyzed chemically to 6-APA by similar methodology (Fosker et al. 1971); however, no single enzyme source is capable of the practical hydrolysis of these two penicillins (Fig. 1). Penicillin acylase activities have been demonstrated in a wide variety of microorganisms, and three types are clearly recognized. Penicillin G acylase activity is typically found in bacteria and yeasts and penicillin V acylase activity occurs in fungi and streptomycetes (Vandamme 1980). A third type, designated ampicillin acylase, occurs in bacteria and actinomycetes (Nara et al. 1971; Okachi et al. 1973). Many notable exceptions to these relationships occur, for example penicillin V acylase activity in *Erwinia aroideae* (Vandamme 1980), *Pseudomonas acidovorans* (Lowe et al. 1981), *Bacillus sphaericus* (Carlsen and Emborg 1981), or gram-negative bacillus (Gestrelius 1980).

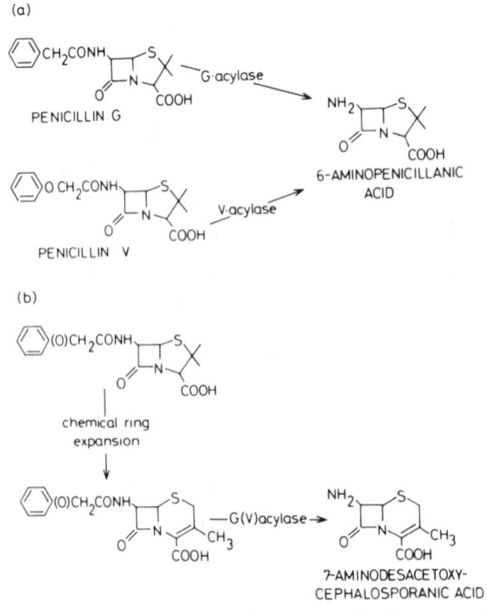

FIG. 1. Enzymic formation of 6-aminopenicillanic acid (a), and 7-aminodesacetoxycephalosporanic acid (b).

Enzymes suitable for the industrial production of 6-APA must have certain prerequisite properties, foremost of which must be long-term stability, especially under working conditions; tolerance of high substrate concentrations; and low product inhibition. These and other important properties are summarized in Table 1.

TABLE 1. *Enzyme properties most suitable for industrial use*

Stable at working temperature (long half life, >100 days)
Stable to prolonged storage in hydrated state
Stable to desiccation
Stable to organic solvents, detergents, surface active agents
Stable over broad pH range
Active over broad pH range
No cofactor or metal ion requirements
No dependence upon reducing groups
Not sensitive to oxidation
Not inhibited by high substrate or product(s) concentrations
Not reversible (in practice)
High affinity for substrate (low Km)
Available from suitable microorganism (nontoxic, nonpathogenic)
Easy to extract and purify to the required activity in good yield (>50%)
Small molecular weight—suitable for genetic engineering studies
Adaptable to various immobilization techniques, i.e., tolerance of glutaraldehyde, polymer solvents
High activity as a whole cell with no relevant contaminating enzymes (for whole cell immobilization)

Immobilized isolated enzymes. Penicillin acylase enzyme activities from *Escherichia coli, Bacillus megaterium, Proteus rettgeri,* and *Bovista plumbea* have been studied and documented in detail in published and patent literature. These activities have properties suitable for commercial use. For industrial use, the enzyme activity is best used in an immobilized form in order that it can be easily removed from the reaction mixture and reused. In this form the enzyme protein does not contaminate the products. Enzymatic splitting of penicillin to 6-APA represents one of the first successful applications of immobilized technology, and this hydrolysis continues to serve as an important system for the design of new immobilization techniques and reactor design (Vandamme 1983; Savidge 1984). The penicillin acylases have been immobilized by adsorption, cross-linking with absorption, covalent attachment, and physical entrapment, and many examples of these techniques have been published (Vandamme 1983; Savidge 1984).

The extracellular enzyme from *B. megaterium* has been adsorbed onto bentonite or diatomaceous earth and used in a continuous flow stirred reactor with an ultrafiltration unit (Ryu et al. 1971). *E. coli* acylase has been adsorbed onto various ion-exchange resins and cross-linked by the use of the bifunctional cross-linking reagent glutaraldehyde (Savidge and Powell 1977).

Several methods of covalent attachment have been described. The *E. coli* enzyme has been studied in detail. Cyanogen bromide activation of carriers such as cellulose, Sepharose 4B, has been described (Lagerlof et al. 1976), and the Sephadex-acylase has been used by Astra Lakemiddel in Sweden for large-scale 6-APA production in stirred-tank reactors (Vandamme 1983). The acylase has been successfully entrapped in cellulose triacetate fibers at Snamprogetti, Italy, and used in a recirculated column reactor (Marconi et al. 1973). This system was capable of yielding 85% hydrolysis of a 12% penicillin G solution in 190 min with only a 20% loss in its initial activity after 4 months.

The purified penicillin G acylase from *Proteus rettgeri* has been copolymerized with acrylamide and found to be suitable for the hydrolysis of penicillins and cephalosporins and related substrates (Robak and Szewczuk 1981). Enzyme extracts from *Bovista plumbea* have been immobilized by treatment with polyacrylic resins (Brandl et al. 1973) or by cellulose triacetate entrapment (Brandl and Knauseder 1975) to yield granulated preparations capable of hydrolyzing 8% penicillin V solutions in a recirculated column in 3 h to 6-APA at a 91.5% yield.

Immobilized whole cells. There is current interest in the use of intact whole cells as sources of enzyme activity. Such use obviates the costly process of enzyme extraction and purification. However, immobilized whole cell systems typically have low specific activities, and the complexity of the biological product can lead to the production of unwanted side reactions and reactants that can contribute to poor product quality. *E. coli* whole cells exhibited 80% of their acylase activity following entrapment in cellulose triacetate (Dinelli 1972). The system was active over 4 months. *E. coli* cells have also been entrapped in polyacrylamide gels (Sato et al. 1976). Klein and Eng (1979) concluded that epoxy bead entrapment of *E. coli* cells was superior to alginate, polymethylacrylate, or polyurethane treatments because of its high cell loading and good mechanical strength.

Glutaraldehyde-treated *Proteus rettgeri* cells have been entrapped in glycidyl methacrylate polymers and used in combination with a filter in a filter press reactor. In a recirculation mode, this system hydrolyzed 2.3% penicillin G solutions to 6-APA at 94% product yield (Hamsher and Lozanov 1978). A gram-negative coccus that exhibits penicillin V acylase activity has been immobilized as whole cells by polyethyleneimine and glutaraldehye treatment. The preparation was used in both stirred tank and in recirculated column reactors (Gestrelius 1980).

Hydrolysis of cephems from penicillin ring expansion. The oral cephalosporins, cephalexin and cefadroxil are manufactured from 7-aminodesacetoxycephalosporanic acid (7-ADCA). This cephem nucleus originates from the hydrolysis of the parent cephems made by ring expanding penicillin G or V, which yield the corresponding 7-phenylacetamide or 7-phenoxyacetamide desacetoxycephalosporanic acid (Fig. 2). These latter substrates can be hydrolyzed to yield 7-ADCA enzymically by either penicillin G or penicillin V acylase (Fig. 1). For example, 7-phenylacetamide-ADCA can be hydrolyzed by *B. megaterium* (Fujii et al. 1976), *P. rettgeri* (Toyo Jozo Co. 1973), *E. coli* (Nys et al. 1980), and 7-phenoxyacetamide-ACA can be hydrolyzed by the penicillin V acylase from *Erwinia aroideae* (Flemming et al. 1977). The reactions are carried out in a similar manner to those for penicillin splitting. The 7-ADCA can be recovered in high yield because of its relatively high stability and low solubility.

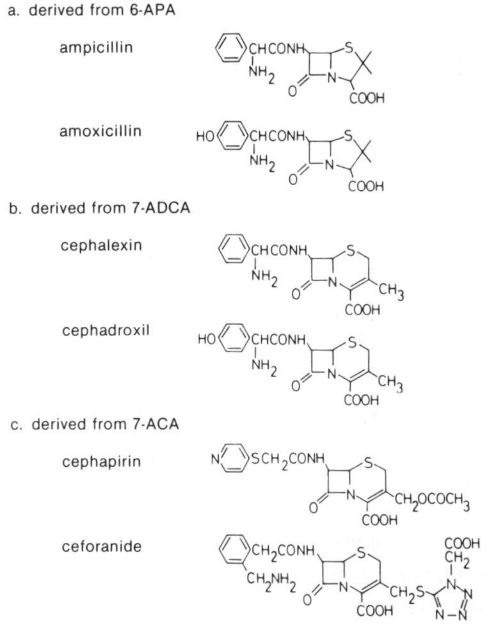

FIG. 2. Semisynthetic β-lactam antibiotics.

Synthesis of semisynthetic penicillins and cephalosporins. Penicillin G acylases were found to be reversible and, at lower pH values of 4–5, could be used in the synthetic direction, especially when using an energy-rich derivative of the substituted carboxylic acid such as the methyl ester (Rolinson et al. 1960). Cole (1969) has reported conversions up to 60% of ampicillin in the presence of a four molar excess of the side chain methyl ester. Such low conversions are not economical compared to chemical synthesis, especially in the presence of excess unreacted substrates.

Screening programs for ampicillin acylases have provided more suitable enzymes, e.g., *Kluyvera citrophila,* which yielded 62% conversions to ampicillin with a 2.5-fold excess of methyl ester (Nara et al. 1971). A more specific enzyme was isolated by the same laboratory from *Pseudomonas melanogenum* (Okachi et al. 1973).

Screening methodologies using α-amino acid derivatives such as phenylglycine methyl ester have yielded new types of acylases termed D-α-amino acid ester hydrolases, the most active of which occur in *Xanthomonas citri* (Kato et al. 1980) and *Acetobacter turbidans* (Takahashi et al. 1974). The enzyme from *X. citri* has been studied extensively using penicillinase deficient mutants (Kato et al. 1980) and can be used to synthesize ampicillin and amoxicillin from 6-APA and cephalexin for 7-ADCA in the presence of the appropriate side chain. Conversion efficiencies of 96% for amoxicillin with overall recovery to amoxicillin trihydrate of 65% were reported, and Amberlite XAD-2 coupled enzyme systems have given overall ampicillin trihydrate yields of 71% (Kato et al. 1980). These systems could possibly become economical with further process optimization.

Hydrolysis of cephalosporins to 7-ACA. 7-Aminocephalosporanic acid (7-ACA) is the cephem nucleus for the synthesis of injectable cephalosporins, for example cephapirin and ceforanide (Fig. 2). Unlike 6-APA, 7-ACA has never been detected in fermentations of the major cephalosporin-producing culture *Acremonium chrysogenum (Cephalosporium acremonium)*. The presence of the D-α-aminoadipyl side chain appears to make the amide bond recalcitrant to enzymic hydrolysis. This in part explains the inability to produce precursored solvent extractable cephalosporins. Currently cephalosporin C is hydrolyzed chemically to 7-ACA (Weissenburger and Gijsbertus Vander Hoeven 1971).

The first reports of cephalosporin C acylase activity in *Achromobacter, Brevibacterium* and *Flavobacterium* have not been confirmed (Walton 1964). The cultures have high esterase activity, which hydrolyzes the acetoxy group to yield the desacetyl derivative. Researchers at Meiji Seika Kaisha Co. Ltd. have described a *Pseudomonas* sp. related to *Ps. putida* capable of converting a 0.5% cephalosporin C solution at 37 C in 7 h to 7-ACA in 37% yield (Goi et al. 1978). Other Japanese scientists (Niwa et al. 1979) have isolated *Aspergillus* and *Alternaria* spp., which have cephalosporin C acylase activity. Such cultures also possess high esterase activity as the products of the reactions are 7-ACA and des 7-ACA. The same cultures can also hydrolyze N-acylated cephalosporins such as N-formyl, N-(2-chloroethoxycarbonyl), and N-phthaloyl cephalosporin C. A strain of *Bacillus megaterium* has been described that is able to hydrolyze N-(N′-1-phenylthiocarbamyl)-cephalosporin C to 7-ACA in 42% yield (Nakagawa et al. 1977). Current low product yields and the presence of esterase and cephalosporinase activities are not conducive to commercial use.

Indirect routes for 7-ACA enzymatic production have been described that involve the formation of the intermediate 7-β-(4-carboxybutaneamido)-cephalosporanic acid (glutaryl 7-ACA) by oxidative deamination of the D-α-aminoadipyl group followed by specific hydrolysis by glutaryl 7-ACA acylases of *Pseudomonas* or *Comamonas* origin (Fig. 3).

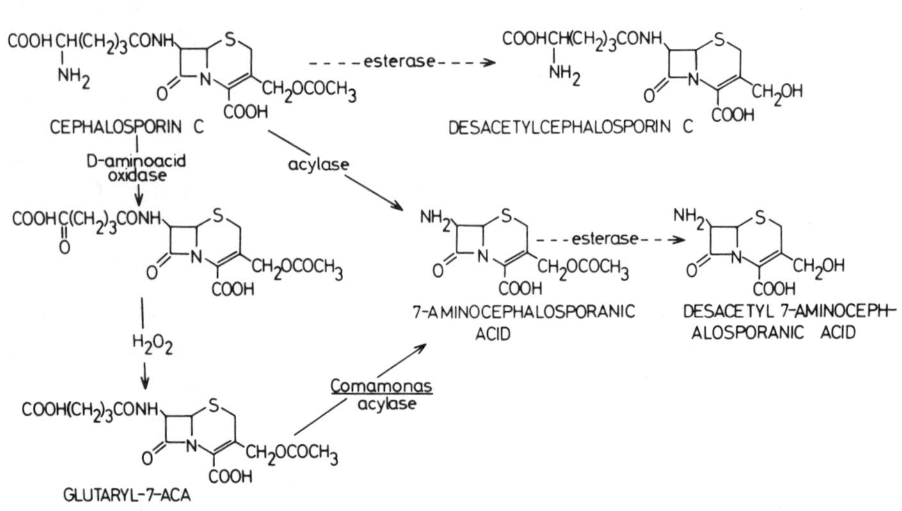

FIG. 3. Enzymic formation of 8-aminocephalosporanic acid.

D-α-Amino acid oxidase enzymes have been isolated from several fungal sources. Activities have been detected in *Aspergillus, Penicillium,* and *Neurospora* spp. (Arnold et al. 1972) and in *Fusarium* and *Cephalosporium* spp. (Matsumoto et al. 1977). *Trigonopsis variabilis* has been studied in detail (Matsumoto et al. 1977; Fildes et al. 1974) by Glaxo scientists for the deamination of cephalosporin C. They concluded that catalase inhibitors such as sodium azide were required to permit the peroxide-induced oxidative decarboxylation of the immediate product of the enzyme reaction 7-β-(5-carboxy-5-oxopentane-amido)-cephalosporanic acid to glutaryl 7-ACA. Yields of 72% were claimed. D-Amino acid oxidase from *T. variabilis* has also been reported to convert 7-α-methoxycephalosporins produced from *Streptomyces* to the corresponding glutaryl derivative (Naito et al. 1977). *Gliocladium deliquescens* has been described as a source of D-amino acid oxidase by Banyu workers (Takeda et al. 1977).

Chemical procedures have been described for the deamination of cephalosporin C using glyoxyloic acid, copper acetate, and pyridine (Suzuki et al. 1978).

Glutaryl 7-ACA acylase from *Pseudomonas* SY-77-1 *(Comamonas)* has been studied in detail (Matsuda et al. 1976; Shibuya et al. 1981). Constitutive and lactamase negative mutants have been isolated (Ichikawa et al. 1981). Cellulose triacetate entrapped microcapsules of *Pseudomonas* cells have been reported to convert glutaryl 7-ACA to 7-ACA in 85% yield (Matsuda et al. 1976).

Cephalosporin C esterase. Semisynthetic cephalosporins can be made by nucleophilic displacement of the acetate at C-3. However, for certain substitutions a free alcohol (desacetyl) group is required. Chemical hydrolysis to remove the acetate is difficult because at alkali pH double bond migration occurs and at low pH lactone formation occurs. Enzyme hydrolysis at neutral pH is more convenient. Acetylesterases have been detected and isolated from a wide variety of biological sources including citrus peel (Jeffrey et al. 1961), mammalian tissue (O'Callaghan and Muggleton 1963), *Fusarium oxysporum* (Singh et al. 1980), *Bacillus subtilis* (Konecny and Sieber 1980; Abbott and Fukuda 1975), *Rhodotorula rubra* (Goulden and Smith 1977), *Rhodosporidium toruloides* (Smith and Larner 1976), and *Aureobasidium pullulans* (Imanaka et al. 1983). Typically the esterases have broad substrate tolerance and can hydrolyze cephalosporin C, cephamycin C, and 7-ACA to the corresponding desacetyl derivatives. The esterase from *B. subtilis* has been used immobilized on brick powder and controlled pore glass (Konecny and Sieber 1980) and bentonite (Abbott and Fukuda 1975) and used in packed bed and stirred tank reactors.

Other esterase enzymes have been applied to cephalosporin bioconversions, namely the removal of carboxylic acid blocking groups. Chemical deblocking of such protective groups often requires acid hydrolysis or even hydrogenolysis; the use of enzymic deblocking, therefore, avoids such adverse conditions. The *p*-nitrobenzyl ester is a popular blocking group and it can be conveniently removed by a specific esterase from *B. subtilis* (Brannon et al. 1976). The enzyme was found to be useful for the preparation of cephalexin and 7-ADCA from the corresponding *p*-nitrobenzyl esters.

Methyl esters are also commonly used to protect carboxylic acids. They have the advantage of being less expensive and do not contribute to waste disposal problems. However, such esters also required harsh conditions for their removal.

From the screening of 700 microorganisms, a new streptomycete designated *Streptomyces capillispira* was isolated due to its ability to hydrolyze the methyl ester from 7-ADCA methyl ester (Berry et al. 1982). However, the enzyme had poor thermal stability.

Resolution of side chain amino acids. The D(-)-isomers of both phenylglycine and *para*-hydroxyphenylglycine are necessary for the synthesis of ampicillin and cephalexin, and amoxicillin and cefadroxil, respectively. Several enzymic procedures have been developed to resolve DL-amino acids.

The stereospecificity of the *E. coli* penicillin G acylase for acylated L-amino acids has enabled an Amberlite XAD-immobilized enzyme to be used to hydrolyze the L-isomer of N-phenylacetyl-DL-phenylglycine to the free amino acid. The unhydrolyzed D-isomer can be recovered by solvent extraction and chemically hydrolyzed (Savidge et al. 1974). In a similar manner the amides of L-amino acids can be hydrolyzed by *Ps. putida* (Neilson 1980), by *Ps. reptilivova* and *Ps. arrilla* (Novo Industri 1979), and by pig kidney L-leucine aminopeptidase (Boesten 1976), and the D-amino acid amide extracted and hydrolyzed chemically. Bacterial proteases from *B subtilis* and *B. licheniformis* (subtilisin) have been used to specifically hydrolyze the methyl ester from N-acetyl-L-(+)-phenylglycine methyl ester in a similar manner (Schutt 1981). The above methods have the common drawback of being indirect. The L-isomer is the target for the enzyme and the free D-isomer of the amino acid is only made available by further chemical treatment.

Specific D-amino acid acylases from *Streptomyces* spp. have been described that can yield D(-)-phenylglycine directly from N-acetyl-DL-phenylglycine (Sugie and Suzuki 1980). The unreacted L-isomer can easily be recycled by chemical racemization.

Enzymic hydrolysis of hydantoin derivatives currently shows the greatest potential for large-scale production of these side chain amino acids (Fig. 4).

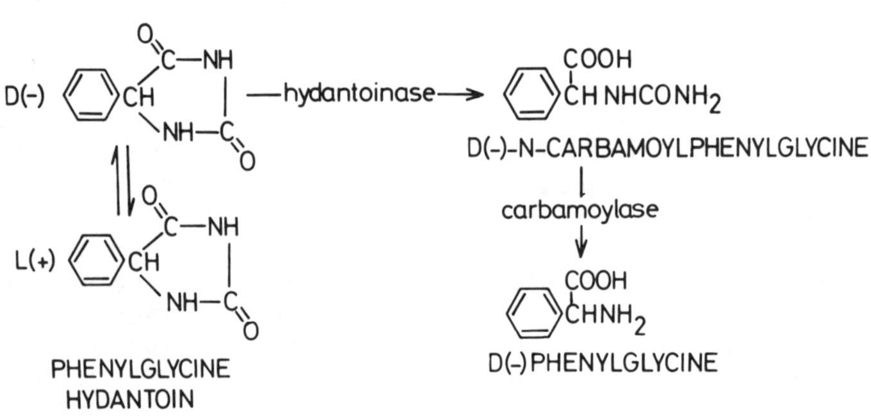

FIG. 4. Enzymic resolution of phenylglycine.

Racemic hydantoins can be conveniently synthesized chemically. The D-isomer of the substituted hydantoins can be hydrolyzed by hydantoinase activity. Such enzymes have been detected in a wide variety of microbial sources. Yamada et al.

(1978) reported hydantoinase activity in representatives of 31 different genera of bacteria, yeasts, and fungi, with highest activities occuring in *Ps. striata, Nocardia corallina, Corynebacterium sepedonicum,* and *Aerobacter cloacae.* Scientists at Snamprogetti isolated several *Pseudomonas* spp. with high hydantoinase activity (Degen et al. 1976). In all cases conversions from the racemic mixture to the D-carbamoyl amino acid were almost complete due to the spontaneous racemization of the L-hydantoin isomer under the conditions of the enzymic hydrolysis. D-Carbamoylase activities have been isolated from microorganisms with the high activity in a *Bacillus macroides,* which can complete the conversion to the free D-amino acid (Takahashi et al. 1980). More direct approaches, using a single microbial enzyme source, have been developed by Ajinomoto Company (Nakamori et al. 1980) with the isolation of bacteria from the *Pseudomonas, Achromobacter, Alcaligenes, Moraxella, Paracoccus,* and *Arthrobacter* spp. that possess both D-hydantoinase and D-carbamoylase activities. *Agrobacterium radiobacter* has been reported to possess both enzymes for the conversion of racemic hydantoins to the D-amino acid (Olivieri et al. 1979).

Conclusions

Although the actual industrial use of enzymes in the pharmaceutical industry cannot be assessed accurately because of the proprietary nature of such competitive processes, various authors (Vandamme 1983; Savidge 1984; Poulsen 1984) have concluded that the majority of companies making 6-APA and 7-ADCA do so enzymically, and economics alone will continue the trend. The industrial role of enzymes for the production of 7-ACA is uncertain and awaits the discovery of cephalosporin C acylases suitable for industrial use. The enzymic resolution of hydantoins to D-amino acid side chains appears to be a commercially viable process (Poulsen 1984). In the syunthesis direction, the reported good yields to semisynthetic penicillins using α-amino acid ester hydrolases suggest that with process optimization, such immobilized enzymes could become a practical alternative to present chemical methods for the synthesis of cephalexin and related structures.

Recently many new natural β-lactam structures have been reported (Elander and Aoki 1982), and several associated hydrolytic activities have been described, e.g., *Ps. schuylkillensis* for the hydrolysis of nocardicin C to 3-aminocardicinic acid (Komori et al. 1978) and *Ps. sp.* 1158 for the deacylation of the olivanic acid elated PS-5 antibiotic (Fukugawa et al. 1980). Dependent upon the commercial success of these new β-lactams or their derivatives, we can be certain that enzyme biotransformations will always be considered as possible integral stages in enzyme manufacture.

Literature Cited

Abbott, B. J., and D. S. Fukuda. 1975. Preparation and properties of a cephalosporin acetylesterase adsorbed onto bentonite. *Antimicrob. Agents Chemother.* 8:282–288.

Arnold, B. H., R. A. Fildes, and D. A. Gilbert. 1972. Cephalosporin derivatives. U.S. Pat. 3,658,649 April.

Berry, D. R., D. S. Fukuda, and B. J. Abbott. 1982. Enzymatic removal of a cephalosporin methyl ester blocking group. *Enzyme Microb. Technol.* 4:80–84.

Boesten, W. H. J. 1976. Process for the enzymatic resolution of DL-phenylglycineamide into its optically active antipodes. U.S. Pat. 3,971,700 July.

Brandl, E., and F. Knauseder. 1975. Process for the production of 6-aminopenicillanic acid. German Pat. 25 03 584.

Brandl, E., W. Kleiber, and F. Knauseder. 1973. Process for the production of 6-aminopenicillanic acid. U.S. Pat. 3,737,375 June.

Brannon, D. R., J. A. Mabe, and D. S. Fukuda. 1976. De-esterification of cephalosporin *para*-nitrobenzyl esters by microbial enzymes. *J. Antibiot.* 22:121–124.

Carlsen, F., and C. Emborg. 1981. *Bacillus sphaericus* V-penicillin acylase. I. *Ferment. Biotechnol. Lett.* 3:375–378.

Chibata, I., T. Tosa, T. Mori, and Y. Matuo. 1972. Preparation and industrial application of immobilized aminoacylases. Pages 383–389 *in* G. Terui, ed., *Fermentation Technology Today.* Soc. Ferm. Technol., Osaka, Japan.

Cole, M. 1969. Penicillins and other acylamino compounds synthesized by the cell-bound penicillin acylase of *Escherichia coli. Biochem. J.* 115:747–756.

Degen, L., A. Viglia, E. Fascetti, and E. Perricone. 1976. Enzymic complexes capable of converting racemic hydantoins into optically active aminoacids, and their applications. South Africa Pat. 76 3892.

Dinelli, D. 1972. Fiber-entrapped enzymes. *Proc. Biochem.* 7:9–12.

Elander, R. P., and H. Aoki. 1982. β-Lactam-producing microorganisms: Their biology and fermentation behavior. Pages 83–153 *in* R. B. Morin and M. Gorman, eds., *The Chemistry and Biology of β-Lactam Antibiotics.* Academic Press, New York.

Fildes, R. A., J. R. Potts, and J. E. Farthing. 1974. Process for preparing cephalosporin derivatives. U.S. Pat. 3,801,458 April.

Fleming, I. D., M. K. Turner, and E. J. Napier. 1977. Method of deacylating cephalosporins. British Pat. 1,473,100 May.

Fosker, G. R., K. D. Hardy, J. H. C. Nayler, P. Seggery, and E. R. Stover. 1971. Derivatives of 6-aminopenicillanic acid, Part X. A non-enzymic conversion of benzyl penicillin into semi-synthetic penicillins. *J. Chem. Soc. Sec. C, Org. Chem.* 1917–1919.

Fujii, T., K. Matsumoto, and T. Watanabe. 1976. Enzymatic synthesis of cephalexin. *Proc. Biochem.* 11:21–24.

Fukugawa, Y., K. Kubo, T. Ishikura, and K. J. Kouno. 1980. Deacetylation of PS-5, a new β-Lactam compound. I. Microbial deacetylation of PS-5. *J. Antibiot.* 33:543–549.

Gestrelius, S. 1980. Immobilized penicillin acylase for production of 6-APA from penicillin-V. Pages 439–442 *in* H. H. Weetall and G. P. Royer, eds., *Enzyme Engineering,* Volume 5. Plenum, New York.

Goi, H., T. Niwa, C. Nojiri, S. Miyado, M. Seki, and Y. Yamada. 1978. 7-Aminocephalosporanic acid and its derivatives. Japan Pat. 78 094093.

Goulden, S. A., and A. Smith. 1977. Enzymic hydrolysis of 3-acyloxymethyl cephalosporins. British Pat. 1,474,519 May.

Hamsher, J. J., and M. Lozanov. 1978. Process for preparing 6-aminopenicillanic acid. U.S. Pat. 4,113,566 Sept.

Harrison, F. G., and E. D. Gibson. 1984. Approaches for reducing the manufacturing costs of 6-aminopenicillanic acid. *Proc. Biochem.* Feb. 33–36.

Ichikawa, S., Y. Shibuya, K. Matsumoto, T. Fujii, K. Komatsu, and R. Kodaira. 1981. Purification and properties of 7β-(4-carboxybutanamido)-cephalosporanic acid acylase produced by mutants derived from *Pseudomonas. Agric. Biol. Chem.* 45:2231–2236.

Imanaka, H., T. Miyoshi, T. Konomi, Y. Kubochi, S. Hattori, and T. Kawakita. 1983. Process for the preparation of deacetylcephalosporin C. U.S. Pat. 4,414,328 Nov.

Jeffrey, J. D'A., E. P. Abraham, and G. G. F. Newton. 1961 Deacetylcephalosporin C. *Biochem. J.* 81:591–596.

Kato, K., K. Kawahara, T. Takahashi, and A. Kakinuma. 1980. Substrate specificity of α-amino acid ester hydrolase from *Xanthomonas citri. Agric. Biol. Chem.* 44:1075–1081.

Klein, J., and H. Eng. 1979. Immobilization of microbial cells in epoxy carrier systems. *Biotechnol. Lett.* 1:171–176.

Komori, T., K. Kunugita, K. Nakahara, H. Aoki, and H. Imanaka. 1978. Production of 3-amino-nocardicinic acid from nocardicin C by microbial enzymes. *Agric. Biol. Chem.* 42:1439–1440.

Konecny, J., and M. Sieber. 1980. Continuous deacetylation of cephalosporins. *Biotechnol. Bioeng.* 22:2013–2029.

Lagerlof, E., L. Nathorst-Westfeld, B. Ekstrom, and B. Sjoberg. 1976. Pages 759–768 *in* K. Mosbach, ed., *Methods in Enzymology,* Vol. 44. Academic Press, New York.

Lowe, D. A., G. Romancik, and R. P. Elander. 1981. Penicillin acylases. A review of existing enzymes and the isolation of a new bacterial penicillin V acylase. *Dev. Ind. Microbiol.* 22:163–180.

Marconi, W., F. Cecere, G. Della Penna, and B. Rappuoli. 1973. The hydrolysis of penicillin G to 6-aminopenicillanic acid by entrapped penicillin acylase. *J. Antibiot.* 26:228–232.

Matsuda, T., T. Yamaguchi, T. Fujii, K. Matsumoto, M. Morishita, M. Fukushima, and Y. Shibuya. 1976. Process for the production of 7-amino-cephem compounds. U.S. Pat. 3,960,662 June.

Matsumoto, K., S. Yamamoto, and T. Fujii. 1977. 7-(4-Carboxybutane amido)-3-cephem4-carboxylic acid. Japan Pat. 77 125 696.

Naito, A., N. Serizawa, and I. Seki. 1977. New 7 α-Methoxycephalosporins and their preparation. British Pat. 1,488,103 Oct.

Nakagawa, N., T. Yamaguchi, and T. Watanabe. 1977. 7-Amino compounds. Japan Pat. 77 82 791.

Nakamori, S., K. Yokozeki, K. Mitsugi, C. Eguchi, and H. Iwagami. 1980. Method for producing D-α-amino acid. U.S. Pat. 4,221,840 July.

Nara, T., M. Misawa, R. Okachi, and M. Yamamoto. 1971. Enzymatic synthesis of D(-)-α-aminobenzylpenicillin. Part I. Selection of penicillin acylase-producing bacteria. *Agric. Biol. Chem.* 35:1676–1682.

Neilson, M. H. 1980. Enzyme technology and enzyme production. Pages 41–58 *in* A. Verbraeck, ed., *13th International TNO Conf.,* Netherlands Central

Organization for Applied Scientific Research, The Hague.
Niwa, T., C. Nojiri, H. Goi, S. Miyado, F. Kai, S. Seki, Y. Yamada, and T. Niida. 1979. Process for the preparation of 7-amino-cephem compounds using mold fungi. U.S. Pat. 4,141,790 Feb.
Novo Industri. 1979. Enzyme preparation having L-α-aminoacylamidase activity. British Pat. 1,552,542 Sept.
Nys, P. S., D. E. Satarova, L. Podshibyakina, V. B. Korchagin, and E.M. Savitskaya. 1980. Penicillin amidase from *Escherichia coli. Antibiotiki* 25:914–921.
O'Callaghan, C. H., and P. W. Muggleton. 1963. The formation of metabolites from cephalosporin compounds. *Biochem. J.* 89:304–308.
Okachi, R., F. Kato, Y. Miyamura, and T. Nara. 1973. Selection of *Pseudomonas melanogenum* KY 3987 as a new ampicillin-producing bacteria. *Agric. Biol. Chem.* 31:1953–1957.
Olivieri, R., E. Fascetti, L. Angelini, and L. Degen. 1979. Enzymatic conversion of N-carbamoyl-D-amino acids to D-amino acids. *Enzyme Microb. Technol.* 1:201–204.
Poulsen, P. B. 1984. Current applications of immobilized enzymes for manufacturing purposes. *Biotechnol. Gen. Eng. Revs.* 1:121–140.
Robak, M., and A. Szewczuk. 1981. Penicillin amidase from *Proteus rettgeri. Acta Biochem. Pol.* 28:275–284.
Rolinson, G. N., F. R. Batchelor, D. Butlerworth, J. Cameron-Wood, M. Cole, C. G. Eustace, M. V. Hart, M. Richards, and E. B. Chain. 1960. Formation of 6-aminopenicillanic acid from penicillin by enzymic hydrolysis. *Nature* 197:236–237.
Ryu, D. Y., C. F. Bruno, B. K. Lee, and K. Venkatasubramanian. 1972. Microbial penicillin amidohydrolase and the performance of a continuous enzyme reactor system. Pages 307–314 *in* G. Terui, ed., *Fermentation Technology Today.* Soc. Ferm. Technol., Osaka, Japan.
Sato, T., T. Tosa, and I. Chibata. 1976. Continuous production of 6-aminopenicillanic acid from penicillin by immobilized microbial cells. *Eur. J. Appl. Microbiol.* 2:153–160.
Savidge, T. A. 1984. Enzymatic conversions used in the production of penicillins and cephalosporins. Pages 171–224 *in* E. J. Vandamme, ed., *Biotechnology of Industrial Antibiotics.* Marcel Dekker, New York.
Savidge, T. A., and L. W. Powell. 1977. Enzyme complexes and their use. U.S. Pat. 4,001,264 Jan.
Savidge, T. A., L. W. Powell, and M. D. Lilly. 1974. Enzymes. British Pat. 1,357,317.
Schutt, H. 1981. Stereoselective resolution of phenylglycine derivatives and 4-hydroxyphenylglycine derivatives with enzyme resins. U.S. Pat. 4,260,684 April.
Shibuya, Y., K. Matsumoto, and T. Fujii. 1981. Isolation and properties of 7-(4-carboxybutanamido) cephalosporanic acid acylase-producing bacteria. *Agric. Biol. Chem.* 45:1561–1567.
Singh, K., S. Sun, and S. Rakhit. 1980. Cephalosporin acetylesterase activity in *Fusaria. Eur. J. Appl. Microbiol. Biotechnol.* 9:15–18.
Smith, A., and R. W. Larner. 1976. Cephalosporins. U.S. Pat. 3,976,546 Aug.

Sugi, M., and H. Suzuki. 1980. Optical resolution of DL-amino acids with D-aminoacylase of Streptomyces. *Agric. Biol. Chem.* 44:1089–1095.

Suzuki, N., T. Sowa, and M. Murakami. 1978. Process for preparing 7-aminocephalosporanic acid derivatives. U.S. Pat. 4,079,180 Mar.

Takahashi, T., Y. Yamasaki, and K. Kato. 1974. Substrate specificity of an α-amino acid ester hydrolyase produced by *Acetobacter turbidans* ATCC 9325. *Biochem. J.* 137:497–503.

Takahashi, H., S. Takahashi, T. Ohashi, K. Yoneda, and K. Watanabe. 1980. Process for preparing D-α-amino acids. British Pat. 2,042,531 Sept.

Takeda, H., I. Matsumoto, and K. Matsuda. 1977. 7-Aminocephalosporanic acids. Japan Pat. 77 38 092.

Toyo Jozo Company. 1973. 7-Aminodesacetoxycephalosporanic acid by enzymic deacylation of desacetoxycephalosporins. Belgium Pat. 801 044.

Vandamme, E. J. 1980. Penicillin acylases and *beta*-lactamases. Pages 467–522 *in* A. H. Rose, ed., *Economic Microbiology,* Volume 5. Academic Press, New York.

Vandamme, E. J. 1983. Immobilized enzyme and cell technology to produce peptide antibiotics. Pages 237–270 *in* R. M. Lafferty, ed., *Enzyme Technology*. Springer-Verlag, New York.

Walton, R. B. 1964. Search for microorganisms producing cephalosporin C amidase. *Dev. Ind. Microbiol.* 5:349–353.

Weissenburger, H. W. O., and M. Gijsbertus Vander Hoeven. 1971. Process for preparation of 7-aminocephalosporanic acid compounds. U.S. Pat. 3,575,970 Apr.

Yamada, H., S. Takahashi, Y. Kii, and H. Kumagai. 1978. Distribution of hydantoin hydrolyzing activity in microorganisms. *J. Ferment. Technol.* 56:484–491.

CHAPTER 8

Biotransformations of Nonantibiotic Antineoplastic Agents

J. P. Rosazza,* M. W. Duffel, F. S. Sariaslani, F. M. Eckenrode, and F. Filippelli

Division of Medicinal Chemistry and Natural Products, College of Pharmacy, The University of Iowa, Iowa City, Iowa 52242

Microorganisms contain enzymes capable of achieving highly selective chemical changes in the structures of foreign organic compounds. Metabolic systems from microorganisms have two major and equally important uses in the development of new antitumor compounds. These are the preparation of difficult-to-synthesize derivatives for biological evaluation and structure-activity-relationship study; and the elaboration of pathways of metabolism likely to occur in mammals treated with antitumor drugs. Biotransformations of Vinca alkaloids reveal new metabolic pathways of potential pharmacological and toxicological significance. The Vinca alkaloids undergo enzymatic conversion to free-radical intermediates by copper oxidases and peroxidase and are converted into the same compounds by bacterial systems using unknown enzymes. The same pathways involve the formation of carbinolamine, iminium, and enamine intermediates. The implications of these findings are discussed.

Introduction

The microbial transformation field experienced its greatest burst of research activity during the 1950's through the early 1970's when it was widely exploited in the synthesis of steroid hormones (Charney and Herzog 1967; Iizuka and Naito 1982; Laskin and Lechevalier 1974; Peterson et al. 1952; Wallen et al. 1959). In many ways, applications of microbial technology in the steroid field laid the foundation for later developments in biotechnology such as the application of microbial cells and enzymes in the transformations of antibiotics (Wiley and Marshall 1982), alkaloids (Holland 1979), and numerous other groups of compounds (Sariaslani et al. 1984a).

Microbial transformations have also been applied as a useful tool in antitumor drug development. This approach has been particularly fruitful in a vigorous program sponsored by the Natural Products Branch of the Developmental Therapeutics Program of the National Cancer Institute. One major objective of this program was the identification of new and novel structural types of antitumor compounds from higher plants, the microbial world, and from the sea. As with numerous other classes of drugs, natural product prototype compounds provide

*To whom correspondence should be directed.

vide an essential starting point for studies including total or partial syntheses; drug metabolism; and structure-activity-relationships.

Useful Features of Microbial Transformations

Microbial transformations may serve in numerous ways in the development of antitumor drugs. Enzymes or microbial cells achieve highly selective chemical transformations of prototype compounds to give products which are obtainable only in very low yields by traditional chemical synthetic approaches or by isolation from natural sources (Sariaslani et al. 1984a). New analogs may be prepared for purposes of studying structure-activity-relationships (SAR) without having to resort to new and cumbersome total synthetic chemical methods. Microbial transformations may also serve as useful models of mammalian, plant and microbial metabolic systems by affording metabolites of natural products similar to those produced by these other types of complex living organisms. In addition, the structural complexities of naturally occurring antitumor agents provide a rich playground in which to observe novel enzymatic reactions.

Microorganisms and the enzymes they contain possess many attractive features in drug development work. These biocatalysts display high regio- and/or enantioselectivities or specificities when they react with substrates, and reactions occur under very mild reaction conditions of neutral pH, and room temperature. Biocatalysts result in the formation of few waste products like strong acids, bases or heavy metals. An advantage realized in biocatalysis with intact cells is that multienzyme conversions are possible. The utility of enzymes and cells has been greatly extended by their application in nonaqueous media. Best of all, it is possible to use microorganisms and their enzymes to introduce new functional groups into positions of substrate molecules which are not chemically activated (Sariaslani et al. 1984a).

Microbial Transformations and Natural Antitumor Agents

The rationale for using microbial transformations in antitumor drug development is supported by the idea that subtle changes in the structures of complex antitumor compounds may result in great changes in the degree of toxicity, the mechanism of action, and the scope of use of the antitumor agent. As examples (Rosazza 1978), it is useful to consider the subtle structural differences between different Catharanthus alkaloid dimers such as vinblastine and vincristine; and between the anthracycline antibiotics adriamycin and daunomycin. The Vinca alkaloids differ only in the state of oxidation of the carbon atom attached to the indole nitrogen group of the lower (Aspidosperma) alkaloid portion of the dimer. Similarly, the anthracyclines differ in the state of oxidation of the 14-carbon atom of an ethyl side chain. Despite these subtle structural differences, it is emphasized that the compounds differ significantly in their potency, their scope of antitumor activity and in their toxicity. Microbial transformations are capable of introducing subtle structural changes, which may cause significant changes in antitumor drug activity.

Natural Product Models

Once significant antitumor activity is discovered, new drugs are subjected to different types of studies including partial or total synthesis; structure modification;

and drug metabolism. These efforts are extended in order to optimize the activities of antitumor compounds while diminishing unwanted side effects. Drug metabolism studies play a central role in the development of new therapeutic entities, but several difficulties are encountered by those interested in determining the metabolic fate of structurally complex and highly potent antitumor compounds from nature. When a drug enters a living system, it may become metabolically activated, metabolically inactivated, converted to a more toxic derivative, or simply undergo conjugation and excretion. The importance of drug metabolism work is that it may yield important information as to how the antitumor compound exerts its antitumor and/or toxic side effects. Knowledge of the chemical changes experienced by a drug once it passes through the body may also provide useful suggestions for chemical modifications.

Problems encountered by those conducting traditional drug metabolism studies have focused on analytical methodology; the species variation observed in the different pathways of metabolism of drugs by animals such as rats, mice, guinea pigs and humans; and by the difficulty encountered in the preparation of drug metabolites for complete structure elucidation and biological evaluation. Rosazza and Smith (1979) and Smith and Rosazza (1983) forwarded the concept of microbial models for mammalian drug metabolism (Rosazza and Smith 1979; Smith and Rosazza 1983), and have demonstrated the value of this approach in the preparation of mammalian drug metabolites. The basis for the models approach is found in comparative biochemistry which demonstrates near homology in the types of metabolic reactions demonstrated by these different living systems. Among the advantages obtained by using the Microbial Models of Mammalian Drug Metabolism approach is that sufficient quantities of metabolites may be obtained by routine fermentation scale-up techniques. Furthermore, by use of microbial metabolic systems, it is possible to obtain compounds that are formed only fleetingly by mammalian metabolic systems.

Discussion

Applications of Microbial Transformations

Our efforts in the microbial transformation of antitumor alkaloids and related compounds have been very rewarding, and illustrate the wide range of applications possible with microbial transformations. Fermentation type-reactions observed with the Catharanthus (Vinca) alkaloids have included nearly every conceivable metabolic transformation ascribable to microorganisms or mammals. In this paper, attention is focused on transformations observed within this interesting group of compounds. They should be considered not as alkaloids, or as any single class of organic compounds, but rather they should be regarded as members of a much larger class of nitrogen heterocyclic compounds which may enter into and elicit an effect on other living systems. Reactions observed with vindoline are indicative of the metabolic richness of microbial transformations, and they illustrate how microbial metabolic experiments may be used to shed light on entirely new mechanisms by which the drugs might exert their antitumor effects.

Studies with Vindoline: O-Desmethylvindoline

Structures of various Vinca alkaloids are shown in Fig. 1. The structure of vindoline *(1a)* is found as one-half of the clinically active dimeric compounds vinblastine *(2a)* and vincristine *(2b)*. Vindoline is an Aspidosperma alkaloid, while the upper half of the dimeric structure is a representative Iboga alkaloid. Until we began our investigations, few metabolism studies of monomeric Aspidosperma or Iboga alkaloids, or dimeric alkaloids had been reported in the

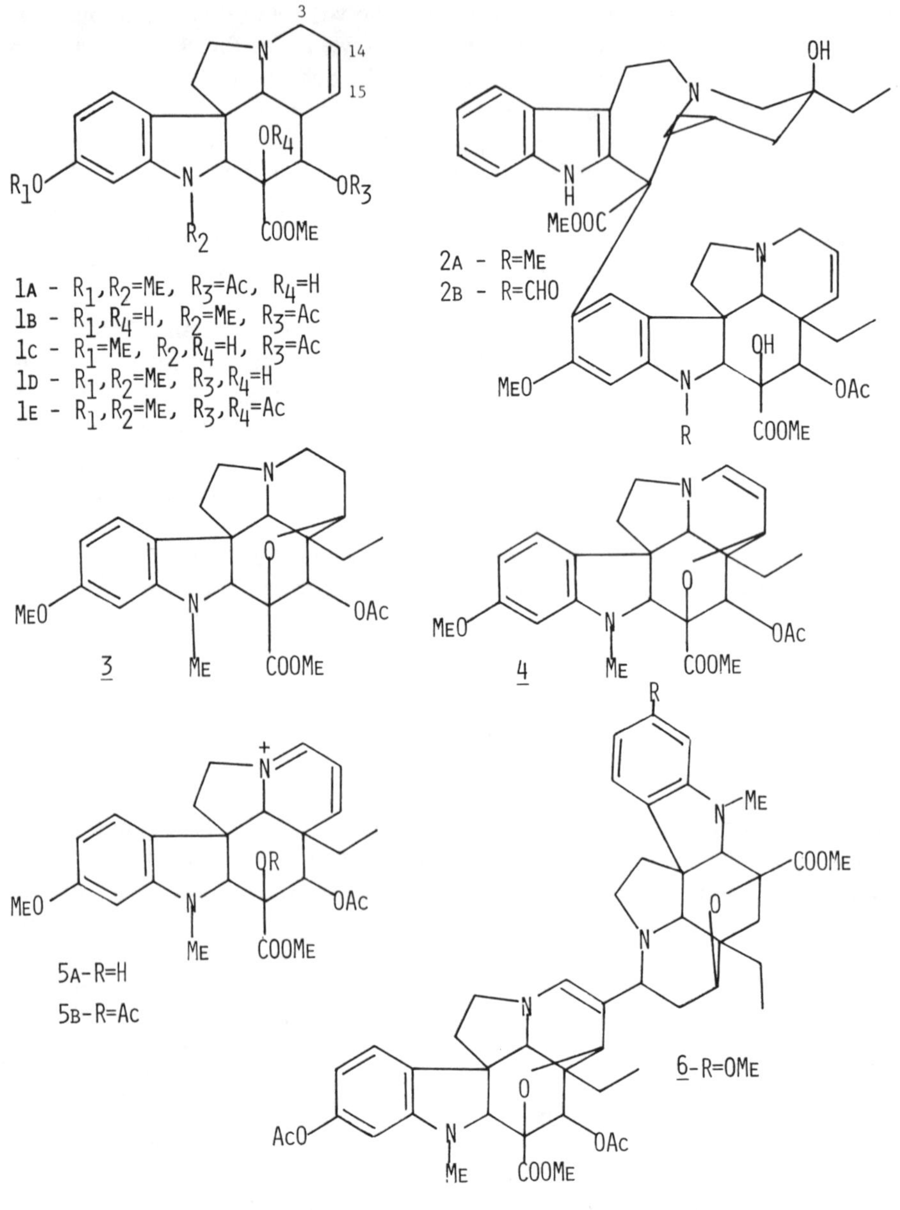

FIG. 1 The structures of various Vinca alkaloids used in biotransformation studies.

literature. Since vindoline was abundantly available, we began working with this compound as a model substrate for this class of antitumor compounds. One of the first microbial transformation reactions observed was the O-demethylation of vindoline by cultures of *Sepedonium chrysopermum* (Wu et al. 1978) which achieved the biotransformation reaction in good yield without optimization. A total of 650 mg of O-desmethylvindoline was obtained from a 2 g incubation of vindoline. Compounds like O-desmethylvindoline are not exciting in themselves, but they do become interesting intermediates or synthons for application in the semi-synthetic preparation of new types of Catharanthus alkaloids. O-Desmethylvindoline *(1b)* would be of value in the biomimetic synthesis of Catharanthus alkaloid dimers.

Dihydrovindoline ether and a dihydrovindoline ether dimer. Some of the earliest work in microbial transformations of vindoline was conducted by people at Eli Lilly (Mallett et al. 1964; Neuss et al. 1973, 1974). Reactions first reported included the conversion of *1a* to N-demethylvindoline *(1c)*, desacetylvindoline *(1d)*, and the structurally novel compound dihydrovindoline ether *(3)*. The pathway by which vindoline was converted to *(3)* was of interest in that several possible mechanisms for the bioconversion reaction could be postulated (Gustafson and Rosazza 1979; Rosazza et al. 1983).

We began investigations into this biotransformation reaction using a strain of *Streptomyces griseus* (Gustafson and Rosazza 1979; Nabih et al. 1978). The major metabolic compound was the dimeric dihydrovindoline ether dimer *(6)*. This compound was identified by Nabih et al. (1978) by a combination of proton and carbon-13 NMR spectroscopy and high resolution mass spectrometry, and we postulated the involvement of an enamine intermediate *(4)* in the formation of *(6)*. Gustafson and Rosazza (1979) were able to isolate and identify the unstable *(4)* by use of direct NMR spectral measurements of the olefinic proton signals, and by conversion of *(4)* to a deuterated derivative followed by mass spectrometry (Gustafson and Rosazza 1979).

The immediate precursor to *(4)* was identified as an iminium derivative *(5a)* in studies using *S. griseus*, and additional metabolic systems including the copper oxidases (Eckenrode et al. 1982; Sariaslani et al. 1984b). The iminium derivative *(5)* was too unstable to be isolated by itself. However, when the compound 16-O-acetyl-vindoline was subjected to metabolic transformations with the same systems, the iminium *(5b)* could be isolated and characterized. The iminium *(5b)* was reduced with sodium borodeuteride to give the mono-deuterated derivative as the sole product. This compound was identified by mass spectrometry and by high field proton and deuterium NMR. Hydrolysis of the 16-O-acetyl functional group from the iminium derivative resulted in the formation of *(5a)* which goes on to give the dimer *(6)* by intramolecular etherification through the reactive enamine *(4)*. These findings are significant in providing the first direct evidence for the existence of iminium intermediates in the metabolism of nitrogen heterocycles like vindoline. The iminium *(5b)* is remarkably stable in aqueous media, and can be isolated by preparative layer chromatography. The facile reduction of the iminium derivative with hydride reagents underlines the potential for such compounds to react with other nucleophiles of biological significance.

Unfortunately, the mechanism by which *S. griseus* achieves oxidation of vindoline remains a mystery, largely because it has not yet been possible to isolate and identify the enzymes responsible for catalysing these reactions.

Other metabolic systems—copper oxidases and peroxidases. Eckenrode et al. (1982) discovered that copper oxidases were capable of oxidizing vindoline to form the dihydrovindoline ether dimer *(6)*. The laccase from *Polyporus anceps* was successfully used to transform vindoline through *(5a)* and *(4)* to the dimer *(6)*. This reaction was catalyzed by plant and fungal laccases, and by the mammalian equivalent of these enzymes, human serum ceruloplasmin. Vindoline is a direct substrate for the copper oxidases, but oxidations were enhanced when reactions were conducted in the presence of chlorpromazine as a "cofactor." The involvement of chlorpromazine in copper oxidase reactions is illustrated in Fig. 2.

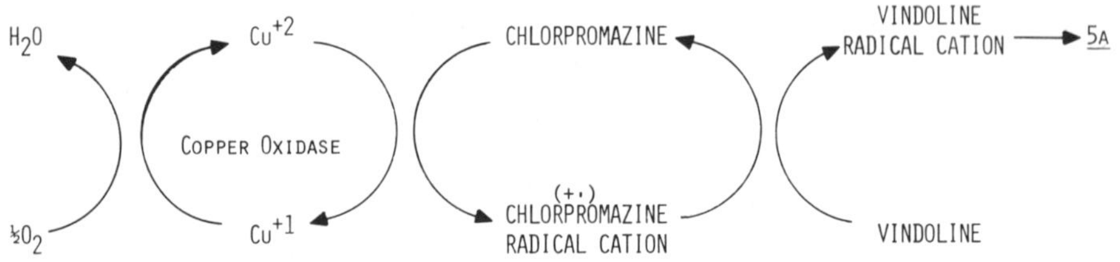

FIG. 2. The cascade of electrons from vindoline to the copper oxidases as the alkaloid is transformed into an enamine dimer *(6)*.

The importance of this finding is that the reaction is actually a free radical oxidation process facilitated by the copper component of the copper oxidases. When the oxidation reaction is in progress, it is possible to observe the formation of a red pigment with a measured absorption maximum at 529 nm. This red color indicates the existence of a radical cation species formed when chlorpromazine, and later vindoline, are oxidized by these enzymes. It was possible to mimic the copper oxidase reaction with 2,3-dichloro-5,6-dicyano-1,4-benzoquinone (DDQ) using 16-O-acetylvindoline as substrate, but neither this reaction, nor that with *S. griseus* enabled the confirmation of the involvement of a free radical oxidation mechanism in the vindoline oxidation.

Two systems were exploited in order to clearly demonstrate the involvement of free radical oxidation mechanisms with vindoline as substrate. These were the enzyme peroxidase, and photochemistry. Both vindoline and 16-O-acetylvindoline were substrates for horseradish peroxidase. Vindoline was converted into the enamine dimer *(6)*. During the reaction, an Fe(+5) radical cation form of the enzyme known as HRP-I is reduced to an Fe(+4) form of the enzyme known as HRP-II by abstraction of an electron from vindoline. The vindoline cation radical thus formed eliminates a second electron and a proton to produce the iminium product *(5a)* which undergoes intramolecular etherification and dimerization. Oxidation of *(1e)* by HRP-I results in the production of the iminium derivative *(5b)* which was isolated and characterized. The stoichiometry of the oxidation reaction was determined by spectral measurement of the reaction in which

HRP-I is reduced to HRP-II by vindoline. A summary of the oxidation reaction of vindoline by HRP-I is given in Fig. 3 (Sariaslani, Duffel and Rosazza, in press, *J. Med. Chem.)*

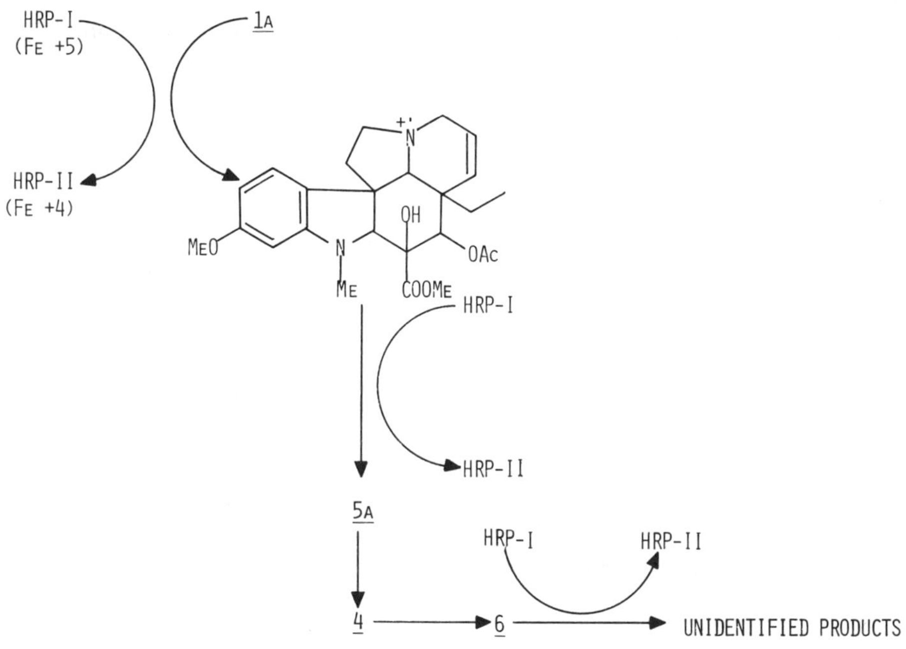

FIG. 3. The redox cycle involved as horseradish peroxidase transforms vindoline and acetoxyvindoline into an enamine dimer *(6)*.

As a pure chemical model of the enzymes accomplishing the free radical oxidation of vindoline, we next exploited photochemical oxidations of vindoline and 16-O-acetylvindoline (Filippelli, Duffel and Rosazza, unpublished results). In photochemistry, solutions of compounds are exposed to high-energy light to generate radical intermediates. Free radical intermediates may undergo single path reactions, or multiple path reactions. When vindoline was dissolved in methanol (1–2 mg/ml) in pyrex containers and exposed to high-energy light from a Hanovia UV lamp, no discernible products could be measured. The same results were obtained when vindoline was dissolved in solutions containing photosensitizing agents which are used to catalyze photochemical reactions much like chlorpromazine does with the copper oxidases. Under either type of condition, vindoline is simply utilized, and no products accumulate. However, when 16-O-acetylvindoline is photolyzed in methanol, a single product, *(5b)*, is isolated and completely characterized by NMR and by chemical reduction to a deuterated derivative. Together with the results obtained with horseradish peroxidase, the photochemical oxidation confirms the involvement of free radical processes in Vinca alkaloid transformations.

Oxidations of dihydrovindoline by Streptomyces griseus. 14,15-Dihydrovindoline *(7a)* (Fig. 4) was prepared by chemical reduction of vindoline and used as a substrate with *S. griseus* (Eckenrode and Rosazza 1983) for the purpose of elaborating pathways by which this organism metabolizes nitrogen heterocyclic compounds. Dihydrovindoline was incubated with growing cultures and resting cells of the microorganism, and metabolites were isolated from preparative scale incubations. A major metabolite is 16-O-desmethyl-14,15-dihydrovindoline *(7b)*.

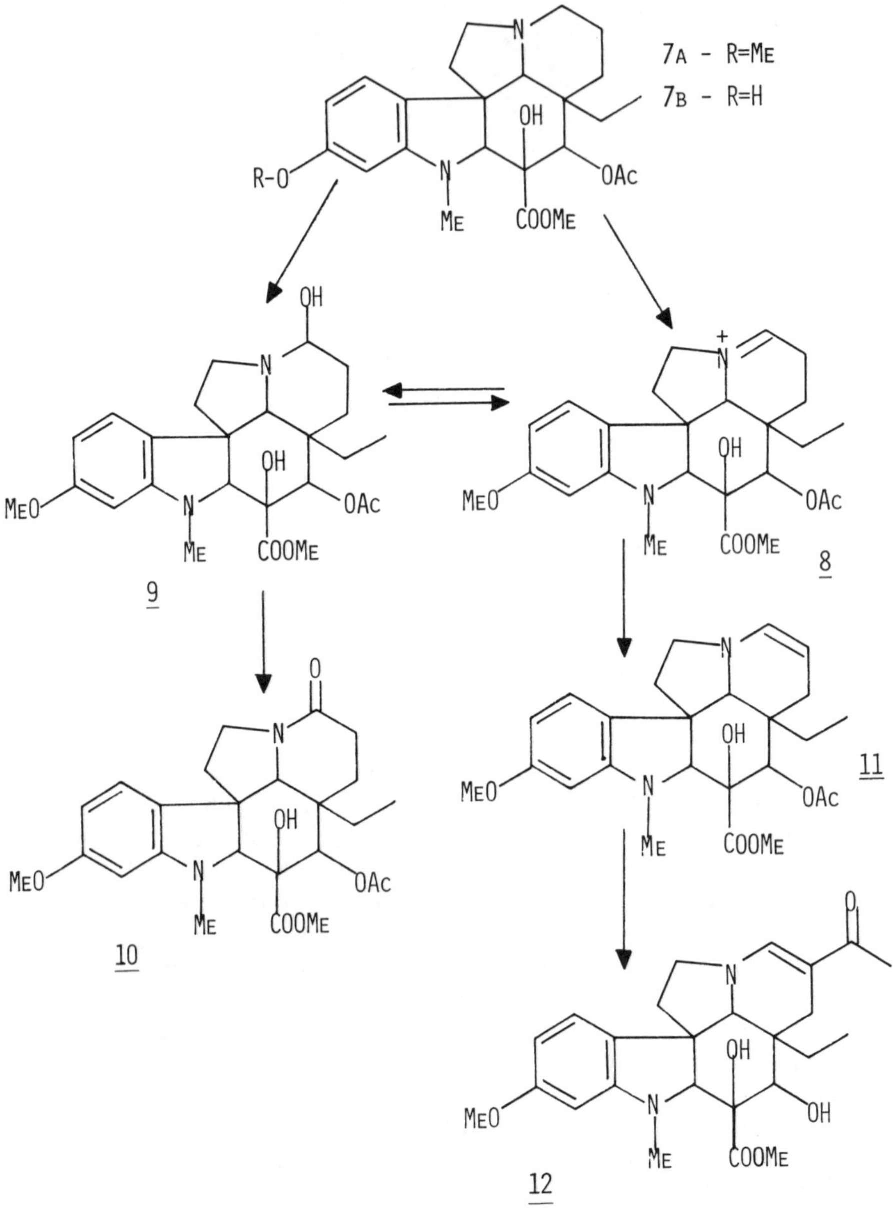

FIG. 4. Oxidations of 14,15-dihydrovindoline by *Streptomyces griseus*.

This reaction is interesting because *S. griseus* does not perform the O-demethylation reaction with vindoline itself. This result is a good example of the subtle control on reaction pathways exerted by a simple functional group such as a double bond. Other products isolated included 3-oxo-14,15-dihydrovindoline *(10)*; delta (3,14)-14-acetyl-14,15-dihydro-17-desacetylvindoline *(12)*; and the carbinolamine metabolite 3-hydroxy-14,15-dihydrovindoline *(9)*. The nature of these metabolites suggests the involvement of enamine *(11)* and iminium *(8)* intermediates, and a pathway for the oxidation of dihydrovindoline was suggested as shown in Fig. 4.

These results with dihydrovindoline are interesting from several perspectives. The observation of a carbinolamine and an iminium product indicate that oxidations of nitrogen heterocycles by *S. griseus* follow pathways similar to those observed in mammals. Carbinolamines and iminium species are essentially interconvertible in aqueous media, and it is impossible to tell which compound was the first metabolite formed by the enzymes of the microorganism. The enamine derivative *(12)* most probably forms by the pathway shown, but it is not known whether the acetyl group derives from an intramolecular or an intermolecular acylation process.

Oxidations of other Vinca alkaloids. Cleavamine and leurosine are also oxidized by the copper oxidases (Rosazza et al. 1983). Cleavamine undergoes complex rearrangements to form a completely new structural class of alkaloids. We are still investigating pathways involved in this interesting bioconversion. Leurosine undergoes oxidation at the Iboga nitrogen atom to form a carbinolamine product with the copper oxidases. The Iboga portion of the Vinca dimers appears to be more susceptible to oxidation by the copper oxidases as measured by oxygen uptake experiments (Rosazza et al. 1983). This was confirmed by measurements of the rates of oxidation of compounds including dimers like leurosine, and vinblastine; and monomers like cleavamine and 16-carbomethoxycleavamine; and vindoline. Leurosine is also hydroxylated at position 10 by *Streptomyces punipalus* (Schaumberg and Rosazza 1981). This result is analogous to the earlier report from Eli Lilly laboratories that vinblastine is also hydroxylated on the aromatic ring of the Iboga alkaloid portion by a Streptomycete.

Conclusions

This work illustrates the broad metabolic potential of microorganisms and their enzymes in the development of antitumor compounds within the Vinca alkaloid class. Work beginning with the bacterium *Streptomyces griseus* revealed a novel metabolic pathway involving enamine and iminium intermediates. More importantly, the use of other microbial enzymes from fungi, the laccases, enabled the identification of free radical pathways of oxidation for this group of compounds. The involvement of free radical oxidative processes in Vinca alkaloid transformations was confirmed by using the well-studied horseradish peroxidase redox cycle, and by using a photochemical model of enzymatic transformations of the alkaloids. With microorganisms as tools, we have been able to conclusively demonstrate the existence of chemically reactive intermediates that form enzymatically. Imines, carbinolamines, and free radical species all hold the potential for interacting with molecules critical to life processes within the cell. The existence of such compounds raises new questions as to mechanisms by which Vinca alkaloids exert their antitumor effects. Do the Vinca alkaloids function through heretofore unknown mechanisms in addition to the well-accepted process of inhibiting tubulin polymerization during the mitotic process? Is it possible that enzymatic transformations such as those observed in this work lead to undesirable toxic side-effects which limit the application of the Vinca alkaloids? The results of this work shed new light on these questions and suggest further metabolic and chemical studies in the development of the Vinca alkaloids.

Acknowledgments

We are thankful for financial support for this work through the National Cancer Institute grant CA-13786-10, and we are grateful to Eli Lilly and Company for samples of vindoline, Catharanthine, and leurosine, which were used in this work.

Literature Cited

Charney, W. and H. L. Herzog. 1967. *Microbial Transformation of Steroids.* Academic Press, New York.

Eckenrode, F., W. Peczynska-Czoch, and J. P. Rosazza. 1982. Microbial transformations of natural antitumor agents. XVIII. Conversions of vindoline with copper oxidases. *J. Pharm. Sci.* 71:1246–1250.

Eckenrode, F. M., and J. P. Rosazza. 1983. Oxidative transformations of 14,15-dihydrovindoline by *Streptomyces griseus. J. Nat. Prod.* 46:884–893.

Gustafson, M. E., and J. P. Rosazza. 1979. Microbial transformations of vindoline by *Streptomyces griseus.* Characterization of a reactive enamine intermediate. *J. Chem. Res.* (S.) 1979:166–167.

Holland, H. L. 1979. Cellular and enzymatic transformations of the alkaloids. Pages 324–400 *in* R. G. A. Rodgrigo, ed., *The Alkaloids,* Academic Press, New York.

Iizuka, H., and A. Naito. 1982. *Microbial Transformation of Steroids and Alkaloids.* Springer-Verlag Publishers, New York.

Laskin, A. I., and F. Lechevalier, eds. 1974. *CRC Handbook of Microbiology,* CRC Press, Inc., Boca Raton, FL.

Mallett, G. E., D. S. Fukuda, and M. Gorman. 1964. Microbial conversion of Catharanthus alkaloids. *Lloydia* 27:334–339.

Nabih, T., L. Youel, and J. P. Rosazza. 1978. Microbial transformations of natural antitumor agents. 5. Structure of a novel vindoline-dimer produced by *Streptomyces griseus. J. Chem. Soc., Perkin Trans I* 1978:757–762.

Neuss, N., D. S. Fukuda, G. E. Mallett, D. R. Brannon, and L. L. Huckstep. 1973. Vinca alkaloids. XXXII. Microbiological conversions of vindoline, a major alkaloid from *Vinca rosea* L. *Helv. Chim. Acta.* 56:2418–2426.

Neuss, N., D. S. Fukuda, D. R. Brannon, L. L. Huckstep. 1974. Vinca alkaloids. XXXIV. Preparation of des-N-methylvindoline by microbiological conversion of vindoline, a major alkaloid of *Vinca rosea* L. *Helv. Chim. Acta.* 57:1891–1893.

Peterson, D. H., H. C. Murray, S. H. Eppstein, L. M. Reineke, A. Weintraub, P. D. Meister, and H. M. Leigh. 1952. Microbiological transformations of steroids. 1. Introduction of oxygen at C-11 of progesterone. *J. Amer. Chem. Soc.* 74:5933–5939.

Rosazza, J. P., F. Eckenrode, M. E. Gustafson, T. Nabih, W. Peczynska-Czoch, and J. P. Schaumberg. 1983. Biotransformations of natural antitumor agents. Studies with the Catharanthus alkaloids. *Cancer Treatment Symp.* 1:51–58.

Rosazza, J. P., and R. V. Smith. 1979. Microbial systems in the preparation of drug metabolites. *Adv. Appl. Microbiol.* 25:169–207.

Rosazza, J. P. 1978. Microbial transformations of natural antitumor agents. *Lloydia* 41:297–313.

Sariaslani, F. S., and J. P. Rosazza. 1984a. Biocatalysis in natural products chemistry. *Enzyme Microb. Technol.* 6:242–253.

Sariaslani, F. S., F. M. Eckenrode, J. M. Beale, Jr., and J. P. Rosazza. 1984b. Formation of a reactive iminium derivative by enzymatic and chemical oxidations of 16-O-acetylvindoline. *J. Med. Chem.* 27:749–754.

Schaumberg, J. P., and J. P. Rosazza. 1981. Microbial transformations of natural antitumor agents. 16. Aromatic hydroxylation of leurosine by *S. punipalus. J. Nat. Prod.* 44:478–481.

Smith, R. V., and J. P. Rosazza. 1983. Microbial models of mammalian metabolism. *J. Nat. Prod.* 46:79–91.

Wallen, L. L., F. Stodola, and R. W. Jackson. 1959. Type reactions in fermentation chemistry, ARS Bulletin ARS-71-13, Peoria, IL.

Wiley, P. A., and V. Marshall. 1982. Microbial transformation of antibiotics. *In* J. P. Rosazza, ed., *Microbial Transformations of Bioactive Compounds,* CRC Press, Inc., Boca Raton, FL.

Wu, G. S., T. Nabih, L. Youel, W. Peczynska-Czoch, and J. P. Rosazza. 1978. Microbial transformations of natural antitumor agents. O-Demethylation of vindoline by *Sepedonium chrysospermum. Antimicrob. Agents Chemother.* 14:601–604.

III
SYMPOSIUM: EUKARYOTIC GENETIC ENGINEERING

J. W. Bennett and L. Lasure, *Conveners*
Tulane University, New Orleans, Louisiana, and
Miles Laboratories, Elkhart, Indiana

PANELISTS

S. Y. Chan
M. C. Gorman
J. Polazzi
J. C. Yarger
B. E. Whitted

CHAPTER 9

Construction of cDNA Libraries

S. Y. CHAN AND B. E. WHITTED

Recombinant DNA-Biology, Miles Laboratories, Inc., P.O. Box 932, Elkhart, Indiana 46515

Using the RNase H-DNA polymerase I-mediated second-strand CDNA synthesis approach, a simple and efficient cDNA cloning method has been developed. This procedure circumvents the use of S_1 nuclease and promotes full-length cDNA synthesis. Cloning efficiency as high as 1×10^7 clones per μg cDNA has been obtained. We have been able to construct cDNA libraries (30,000–50,000 clones) on a routine basis using 100–200 ng of mRNA. The size distribution of cDNA inserts ranges from 1.5–4.0 kb for mRNA isolated from mammalian cell culture, and 0.5–3.0 kb for mRNA from solid tissues. This method is useful in cloning of full-length cDNA of both rare and abundant mRNAs.

INTRODUCTION

The recent advances in the understanding of gene structure and function have been due largely to the discovery of DNA modifying enzymes such as restriction endonucleases, terminal transferases, ligases, and reverse transcriptase. Since the first reports on the successful cloning of complementary DNA copies of mRNA (Rougeon and Mach 1976; Efstratiadis et al. 1976), cDNA cloning has evolved into a powerful tool for analyzing the structure, organization, and expression of eukaryotic genes. This technique has been refined and improved in recent years and has had a tremendous impact on the rapid development of genetic engineering.

In many instances a cDNA clone bank is preferable over a genomic clone bank. It has been estimated that 100,000 to 1,000,000 genomic clones are needed to represent the whole genome while a bank of 10,000 to 30,000 cDNA clones is sufficient to represent all mRNA molecules. The number of false positives in a genomic clone bank is much higher than in a cDNA clone bank. Genomic clone banks are probed with either mRNA or cDNA, both containing sequences from the ribosomal genes. These multiple-copy ribosomal genes contribute significantly to the number of false positives. Other sources of false positives include polydeoxythymidylate tracts and sequence homologies between conserved genes (Williams 1981). Moreover, the use of genomic clones with intervening sequences requires the use of eukaryotic host cells for the processing of mRNA before expression can occur.

The number of genes studied through the use of cDNA clones has increased in the last few years (Schneider et al. 1983; Mayo et al. 1983; Heidecker and Messing 1983; Imam et al. 1983). The study of the structure, replication, and expression of

RNA viruses such as reoviruses (Cashdollar et al. 1984; Sleigh et al. 1979) has been furthered by the availability of cDNA clones. Initiation, coding, and termination sequences have been defined in mRNA molecules encoding β-globin, ovalbumin, and rat growth hormone (Catterall et al. 1979; Efstratiadis et al. 1977; Harpold et al. 1978; Jeffreys and Flavell 1979). This analysis is simplified by the lack of intervening sequences in cDNA clones. Regulation of transcription has also been studied through the use of cDNA clones (Magnuson and Nikodem 1983; Muthukrishnan et al. 1983; Orkin et al. 1983; Wiginton et al. 1983). The recent isolation of cDNA clones that code for dihidrofolate reductase (Costanzo et al. 1984) and apoferritin (Melera et al. 1984) will allow the study of the structure and expression of these genes.

The use of cDNA clones has furthered the understanding of gene structure for genes that produce very little mRNA or protein. As attempts are made to clone these rare mRNAs it becomes increasingly important to optimize the conditions for efficient cDNA cloning.

The hair-pin loop method of cDNA cloning (Efstratiadis and Villa-Komaroff 1979) has been the method of choice for years. Because of the use of S_1 nuclease to destroy the single-stranded DNA loop, this method rarely yields full-length cDNA clones and has a rather poor efficiency of cloning. Okayama and Berg (1982) have described an elegant method, which permits high recovery of plasmid recombinants with full-length cDNA inserts. The procedure uses a specifically designed plasmid DNA vector, which itself serves as the primer for first- and ultimately second-strand cDNA synthesis. Although this method has a high cloning efficiency, it is elaborate and involves a number of enzymatic manipulations involving two plasmids. Heidecker and Messing (1983) also reported on a similar approach using the plasmid pUC-9. A simplified procedure has been described (Gubler and Hoffman 1983; Chan and Whitted, unpublished data) that yields full-length cDNA clones at high efficiency. We have been using this procedure to construct routinely all of the cDNA libraries in our laboratory.

Materials and Methods

Isolation of polyadenylated mRNA. Total RNA was isolated from tissues or cultured cells using the guanidinium thiocyanate procedure (Chirgwin et al. 1979). Polyadenylated mRNA was isolated by a 2-cycle selection on oligo-(dT)-cellulose as described by Aviv and Leder (1972).

Resolution analysis of mRNA preparations was done using acid 6 *M* Urea gel (Cirgwin et al. 1979) and size exclusion HPLC using a Bio-Sil TSK 400 column. The mRNA preparations were also tested for biological activity by translation in a cell-free system as described by Pelham and Jackson (1976) using the rabbit reticulocyte lysate system (New England Nuclear). In addition, a first-strand cDNA was prepared from these samples using AMV reverse transcriptase and analyses were performed as described by Maniatis et al. (1982).

cDNA synthesis. Synthesis of first-strand cDNA was carried out essentially as described by Maniatis et al. (1982). Second-strand cDNA synthesis was performed using *E. coli* RNase H, polymerase I, and ligase. The conditions of the reaction were as described by Gubler and Hoffman (1983). Incubations were performed sequentially at 12 C for 60 min and at 22 C for 60 min.

Tailing and transformation. An outline of the cloning scheme is depicted in Fig. 1. The double-stranded cDNAs were tailed with dGTP and the *EcoRV*-linearized pBR322 was tailed with dCTP. The tailing reaction was done according

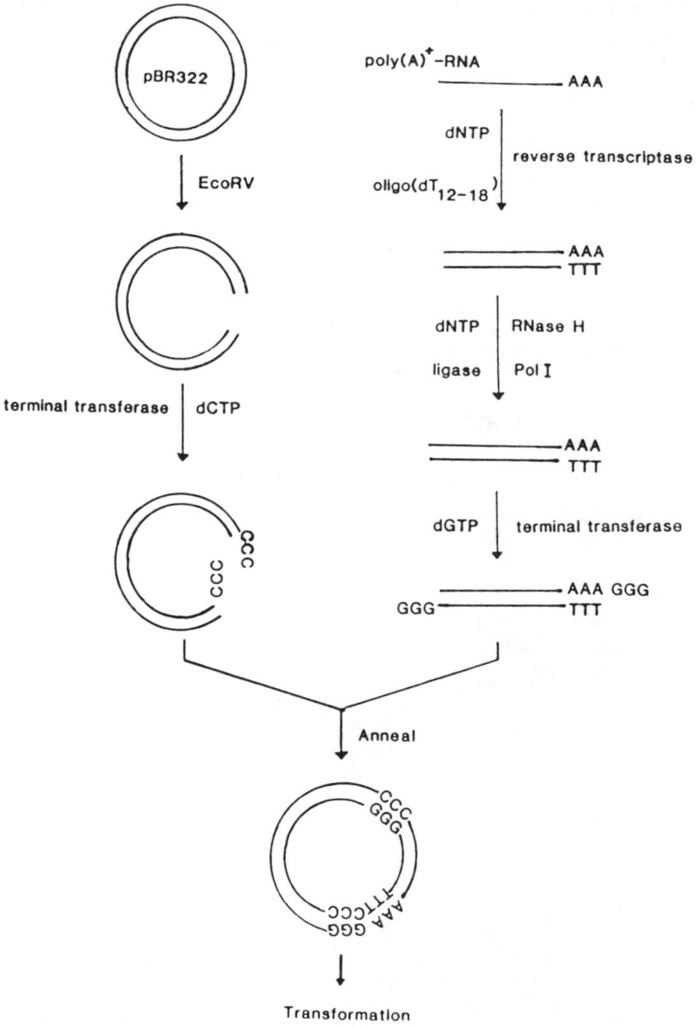

FIG. 1. Schematic presentation of the cDNA cloning procedure. The second strand cDNA synthesis is mediated by *E. coli* RNase H, ligase and PolI.

to Peacock et al. (1981) with minor modifications. Competent *E. coli* RR1 cells were prepared using the $CaCl_2$ procedure (Maniatis et al. 1982). Annealing and transformation was done as described by Peacock et al. (1981).

Miscellaneous. Enzymes and chemicals were purchased from P. L. Biochemicals or New England Biolabs. Oligo-dT-cellulose was purchased from Collaborative Research. Mini-lysates were prepared according to Birnboim and Doly (1979).

RESULTS

Isolation of mRNA and Synthesis of First Strand cDNA

Poly(A^+)-RNA was isolated from both solid mammalian tissue and mammalian tissue culture cells. High yields of poly(A^+)-RNA were obtained with no detectable degradation. 200–400 μg of poly(A^+)-RNA was isolated from approximately 50 g of solid tissue and yields of 1.5 mg were obtained for tissue culture cells ($\sim 10^9$ cells). The poly(A^+)-RNA size distribution as seen on 6 *M* urea gels was between 10–30S with no degradation detected (Fig. 2). A single sharp peak was obtained when the poly(A^+)-RNA was run through a TSK-400 column. Distribution of the poly(A^+)-RNA was usually between the 16S and 23S RNA markers.

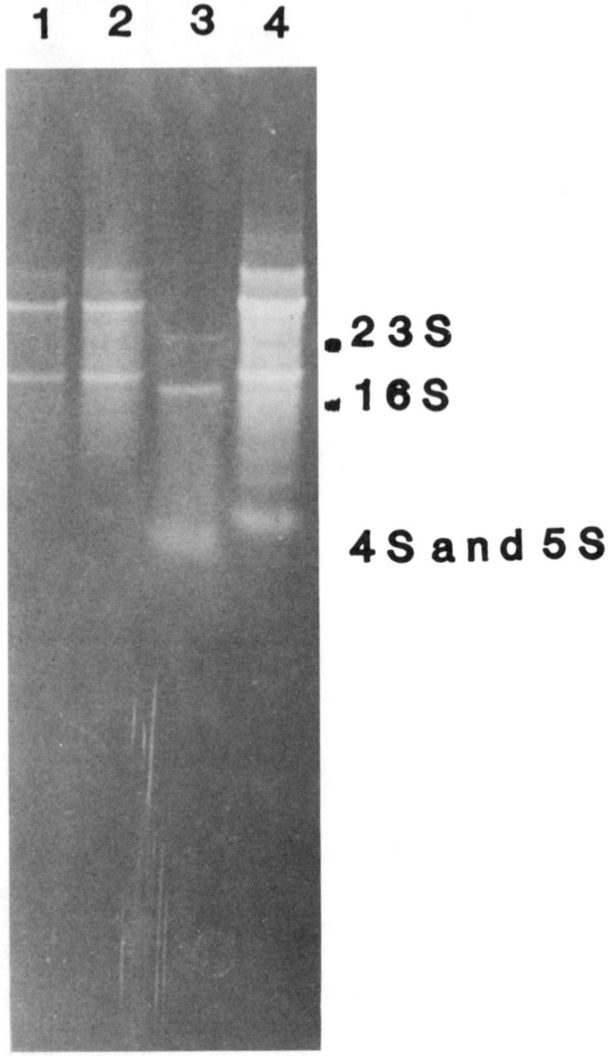

FIG. 2. 6 *M* urea gel electrophoresis of poly-(A^+)-mRNA from B-16_f melanoma cells. Lane 3, *E. coli* RNA markers (23s, 16s, 4s, and 5s); Lanes 1 and 2, poly-(A^+)-mRNA; Lane 4, total RNA from B-16_f cells.

The quality of the poly(A^+)-RNA was further confirmed by both *in vitro* translation and first-strand cDNA synthesis. A wide distribution of translation products was usually observed with molecular weight ranging from 10,000 to 150,000 daltons. The first-strand cDNA synthesized was between 80 bases and 4,000 bases long.

Addition of Homopolymer Tails

The number of dC-residues added to the 3′-OH end of *EcoR V*-cut pBR322 was estimated by digesting the tailed plasmid with *SalI*. The length of the *EcoR V-SalI* fragment (~600 bp) was determined by electrophoresis on a 2.5% agarose gel to analyze the tailing reaction. The addition of dG-tails is self-limiting to approximately 20 dG-residues. Since the greatest number of transformants is obtained by annealing dG- and dC-tails of equal lengths, the dC tailing reaction must be carefully controlled to limit the addition of dC-residues to 20. Also, the distribution of tail lengths must be limited, and the number of untailed plasmid molecules must be minimized (Fig. 3).

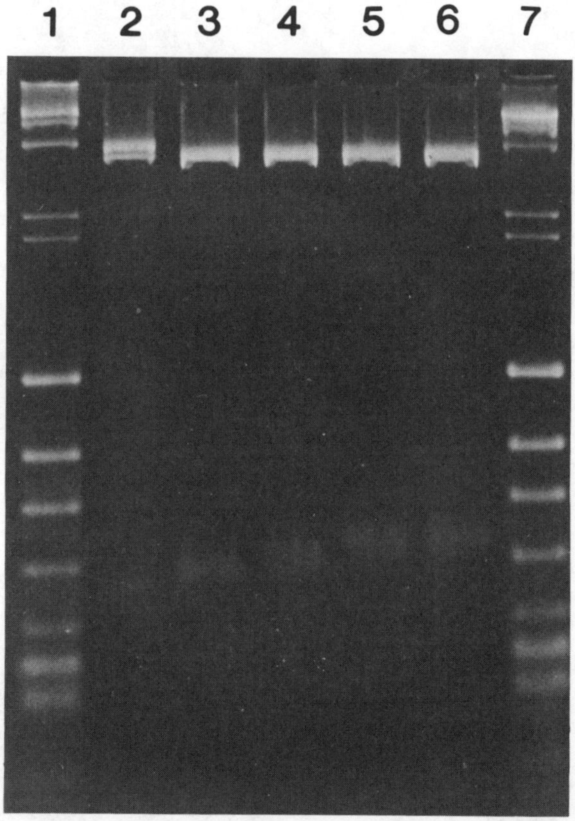

FIG. 3. Agarose gel electrophoresis of dC-tailed pBR322 DNA. Plasmid DNA was cut with *EcoR V* and tailed for various periods of time with dC using terminal transferase. The tailed plasmid DNA was then cut with *SalI*. The increased length of *EcoR V/SalI* was determined by electrophoresis on a 2.5% agarose gel. Lanes 1 and 7, *HindIII*-λ/*HincII*-ϕX-174 DNA markers; Lanes 2–6 dC-tailed *EcoRv/SalI* Fragment.

Analysis of cDNA Clone Banks

The inserts were removed from pBR322 by *BamHl* digestion. The two sites that were used were the unique *BamHl* site in pBR322 and a *BamHl* site newly constructed at the *EcoR V* site by the addition of dC-residues. The size of the inserts ranged between 1.5 and 4.0 kb for mRNA isolated from tissue culture cells and between 0.5 and 3.0 kb for mRNA isolated from solid tissues (Fig. 4). The heteroduplex method for cDNA cloning has been quite efficient, and the size of the inserts is dependent upon the quality of the mRNA.

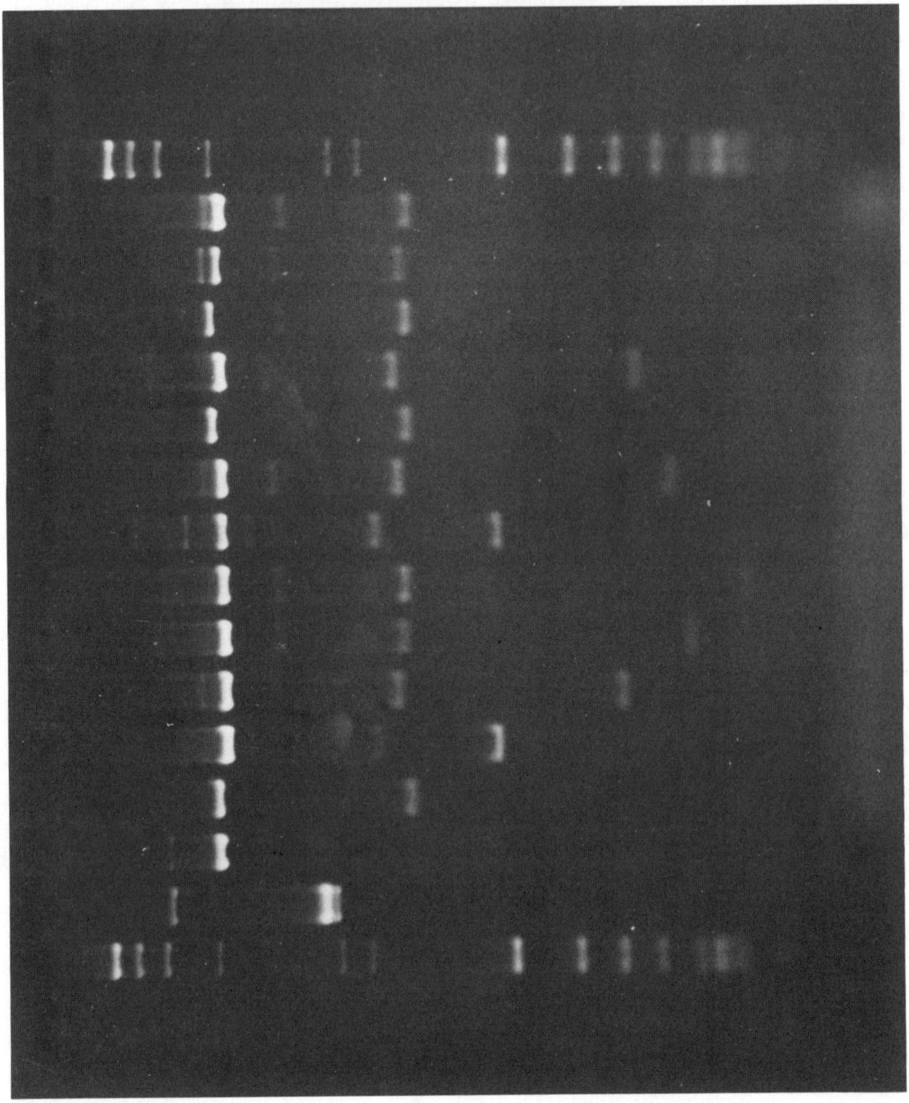

FIG. 4. *BamHI* digest of mini-lysate of recombinant clones from the B-16_f library. Lanes 1 and 16, *HindIII-X/HincII*-ϕX-174 DNA markers; Lane 2, supercoiled pBR322. Lane 3, linearized pBR322, Lanes 4–15, cDNA inserts of various clones.

The percentage of transformants without inserts (Table 1) is dependent mostly upon the number of molecules of pBR322 that are not linearized by *EcoR V*. Even with purification of the linearized pBR322 from the undigested plasmid a background of one colony/ng is to be expected (Grubler and Hoffman 1983). However, we have obtained the same background without further purification.

TABLE 1. *cDNA libraries*

mRNA Origin	Size of Library	Approximate % of Transformants with Inserts	Insert size Distribution (kb)	Transformation Efficiency (Transformants/μg cDNA)
B-16 $_f$ Mouse (Melanoma)	2,200	94%	1.5–4.0	1.25×10^5
Linings of Calf Stomach	33,000	90%	0.5–2.5	1.56×10^6
Human Liver	25,000	95%	0.5–3.0	1.0×10^7

Table 1 summarizes several cDNA clone banks that have been constructed. During optimization of the cloning procedure, the transformation efficiency has increased and 1×10^7 transformants/μg cDNA is routinely obtained.

Discussion

The successful construction of cDNA libraries depends mainly upon three factors: (1) the availability of high quality mRNA, (2) full-length cDNA synthesis, and (3) properly controlled tailing reactions. It has been estimated that in order to clone low-abundance sequences, a cDNA library should consist of at least 40,000 clones for a 99% probability of obtaining a given clone in an mRNA population. It is of utmost importance that high cloning efficiency such as 1×10^6 clones per μg cDNA be obtained for the cloning of rare mRNAs.

Using the guanidinium thiocyanate and CsCl procedure, we have been able to isolate polyadenylated mRNA from various tissue and cell sources in mg quantities with minimum degradation. The size distribution and biological function of mRNA is usually analyzed by standard techniques such as (1) size-exclusion HPLC, (2) acid 6 *M* urea gel electrophoresis, (3) *in vitro* translation, and (4) first-strand cDNA synthesis. The size distribution of most mRNA preparations ranges from 10–30S.

By combining the classical oligo-(dT)-primed first-strand cDNA synthesis and the RNase H-DNA polymerase I-mediated second-strand cDNA synthesis, a simple and efficient cDNA cloning method has been reported (Gubler and Hoffman 1983; Chan and Whitted, unpublished data). This procedure circumvents the use of S_1 nuclease and promotes synthesis of full-length cDNA. Instead of alkaline hydrolysis, *E. coli* RNase-H was used to introduce nicks and gaps in the mRNA of the heteroduplex. (Leis et al. 1973). *E. coli* polymerase I was then used to replace the RNA segment by nick translation (Kornberg 1980), and *E. coli* ligase was used to join the newly synthesized DNA fragments into a continuous second strand. Cloning efficiencies of up to 1×10^7 clones per μg cDNA were obtained.

Using this method we have been able to generate cDNA libraries (30,000–50,000 clones) from different mRNA sources on a routine basis. For unfractionated mRNA, the size of cDNA inserts ranges from 1.5–4.0 kb for mRNA isolated from tissue culture cells, and 0.4–3.0 kb for mRNA from solid tissues. This represents a considerable improvement in transcript lengths using nonenriched mRNA when compared to published data (Gubler and Hoffman 1983).

The potential of such an efficient cDNA cloning method in biotechnology is tremendous. The coding sequences of several clinically important proteins such as tissue plasminogen activator (Pennica et al. 1983), insulin (Cordel et al. 1979; Villa-Komaroff et al. 1978), α- and γ-interferon (Goeddel et al. 1980; Gray et al. 1982), and α_1-antitrypsin (Bollen et al. 1983) have been cloned and expressed in *E. coli* and other host systems. These coding sequences were cloned using the hair-pin loop method, which utilizes the initial reverse transcript as both the primer and template for second-strand cDNA synthesis. This method rarely yielded recombinant clones containing full-length cDNA inserts. The use of S_1 nuclease to destroy the single-stranded DNA loop further complicates this procedure by removing a portion of the 5′-ends of the cDNA sequence. Most of these sequences have to be reconstructed from several clones because of the inherent problems of the cloning method used. The procedure described in this communication will facilitate the cloning of full-length cDNA of both rare and abundant mRNAs.

Acknowledgments

We thank C. Paliganoff, J. Jazdewski, and C. Terry for excellent technical assistance, and Drs. J. C. Grosch and G. A. Wilson for support and critical comments of the manuscript. We also thank T. Nowicki for typing the manuscript.

Literature Cited

Aviv, H., and P. Leder. 1972. Purification of biologically active globin messenger RNA by chromatography on oligothymidylic acid-cellulose. *Proc. Natl. Acad. Sci. U.S.A.* 69:1408–1412.

Birnboim, H. C.., and J. Doly. 1979. A rapid alkaline extraction procedure for screening recombinant plasmid DNA. *Nucleic Acids Res.* 7:1573–1523.

Bollen, A., A. Herzog, A. Cravador, P. Herion, P. Chuchana, A. V. Straten, R. Lorian, P. Jacobs, and A. V. Elsen. 1983. Cloning and expression in *Escherichia coli* of full-length complementary DNA coding for human α_1-antitrypsin. *DNA* 2:255–264.

Cashdollar, L. W., R. Chmelo, J. Esparza, G. R. Hudson, and W. K. Joklik. 1984. Molecular cloning of the complete genome of reovirus serotype 3. *Virology* 133:191–196.

Catterall, J. F., J. P. Stein, E. C. Lai, S. L. C. Woo, M. L. Mace, A. R. Means, and B. W. O'Malley. 1979. The ovomucoid gene contains at least six intervening sequences. *Nature* (London) 278:323–327.

Chirgwin, J. M., A. E. Przybyla, R. J. MacDonald, and W. J. Rutter. 1979.

Isolation of biologically active ribonucleic acid from sources enriched in ribonuclease. *Biochemistry* 18:5294–5299.

Cordell, B., G. Bell, E. Tischer, F. DeNoto, A. Ullrich, R. Dictet, W. J. Rutter, and H. M. Goodman. 1979. Isolation and characterization of a cloned rat insulin gene. *Cell* 18:533–543.

Costanzo, F., C. Santoro, V. Colantuoni, G. Bensi, G. Raugei, V. Romano, and R. Cortese. 1984. Cloning and sequencing of a full length cDNA coding for a human apoferritin AH chain: Evidence for a multigene family. *EMBO J.* 3:23–27.

Efstratiadis, A., F. C. Kafatos, and T. Maniatis. 1977. The primary structure of rabbit β-globin mRNA as determined from cloned DNA. *Cell* 10:571–585.

Efstratiadis, A., F. C. Kafatos, A. M. Maxam, and T. Maniatis. 1976. Enzymatic in vitro synthesis of globin genes. *Cell* 7:279–288.

Efstratiadis, A., and L. Villa-Komaroff. 1979. Cloning of double-stranded cDNA. Pages 15–36 *in* J. K. Setlow and A. Hollaender, eds., *Genetic Engineering,* Vol. 1. Plenum Press, New York.

Goeddel, D. V., E. Yelverton, A. Ullrich, H. L. Heynecker, G. Miozzari, W. Holmes, P. H. Seeburg, T. Dull, L. May, N. Stebbing, R. Crea, S. Maeda, R. McCandliss, A. Sloma, J. M. Tabor, M. Gross, P. C. Familetti, and S. Petska. 1980. Human leukocyte interferon produced by *E. coli* is biologically active. *Nature* 287:411–416.

Gray, P. W., D. W. Leung, D. Pennica, E. Yelverton, R. Najarian, C. C. Simonsen, R. Derynck, P. J. Sherwood, D. M. Wallace, S. L. Berger, A. D. Levinson, D. V. Goeddel. 1982. Expression of human immune interferon cDNA in *E. coli* and monkey calls. *Nature* 295:503–508.

Gubler, U., and B. J. Hoffman. 1983. A simple and very efficient method for generating cDNA libraries. *Gene* 25:263–269.

Harpold, M. M., P. R. Dobner, R. M. Evans, and E. C. Bancroft. 1978. Construction and identification by positive hybridization-translation of a bacterial plasmid containing rat growth hormone structural gene sequence. *Nucleic Acids Res.* 5:2039–2053.

Heidecker, G., and J. Messing. 1983. Sequence and analysis of zein cDNAs obtained by an efficient mRNA cloning method. *Nucleic Acids Res.* 11:4891–4906.

Imam, A. M. A., M. A. W. Eaton, R. Williamson, and S. Humphries. 1983. Isolation and characterization of cDNA clones for the Aα- and γ-chains of human fibrinogen. *Nucleic Acids Res.* 11:7427–7434.

Jeffreys, A. J., and R. A. Flavell. 1979. The rabbit β-globin gene contains a large insert in the coding sequence. *Cell* 12:1097–1108.

Kornberg, A. 1980. DNA polymerase I of *E. coli*. Pages 101–166 *in DNA replication.* H. Freeman and Co., San Francisco.

Leis, J. P., I. Berkower, and J. Hurwitz. 1973. Mechanism of action of ribonuclease H isolated from avian myeloblastosis virus and *Escherichia coli. Proc. Natl. Acad. Sci. U.S.A.* 70:466–470.

Magnuson, M. A., and V. M. Nikodem. 1983. Molecular cloning of a cDNA sequence for rat malic enzyme. *J. Biol. Chem.* 258:12712–12717.

Maniatis, T., E. F. Fritsch, and J. Sambrook. 1982. *Molecular Cloning.* Cold Spring Harbor Laboratory, Cold Spring Harbor, New York.

Mayo, K. E., W. Vale, J. Rivier, M. G. Rosenfeld, and R. M. Evans. 1983. Expression—cloning and sequencing of a cDNA encoding human growth hormone releasing factor. *Nature* 306:86–88.

Melera, P. W., J. P. Davide, C. A. Hession, and K. W. Scotto. 1984. Phenotypic expression in *Escherichia coli* and nucleotide sequence of two Chinese Hamster Long cell cDNAs encoding different dihydrofolate reductases. *Mol. and Cell. Biol.* 4:38–48.

Muthukrishnan, S., G. R. Chandras, and E. S. Maxwell. 1983. Hormonal control of α-amylase gene expression in barley. *J. Biol. Chem.* 258:2370–2375.

Okayama, H., and P. Berg. 1982. High-efficiency cloning of full-length cDNA. *Mol. and Cell Biol.* 2:161–170.

Orkin, S. H., P. E. Daddona, D. S. Shewack, A. F. Markham, G. A. Bruns, S. C. Goff, and W. N. Kelley. 1983. Molecular cloning of human adenosine deaminase gene sequences. *J. Biol. Chem.* 258:12753–12756.

Peacock, S. L., C. M. McIver, and J. J. Monahan. 1981. Transformation of *E. coli* using homopolymer-linked plasmid chimeras. *Biochim. Biophys. Acta* 655:243–250.

Pelham, H. R. B., and R. J. Jackson. 1987. An efficient mRNA dependent translation system from reticulocyte lysates. *Eur. J. Biochem.* 67:247–256.

Pennica, D., W. E. Holmes, W. J. Kohr, R. N. Harkins, G. A. Vehar, C. A. Ward, W. F. Bennett, E. Yelverton, P. H. Seeburg, H. L. Heynecker, D. V. Goeddel, and D. Collen. 1983. Cloning and expression of human tissue-type plasminogen activator cDNA in *E. coli. Nature* 301:214–221.

Rougeon, F., and B. Mach. 1976. Stepwise biosynthesis in vitro of globin genes from globin mRNA by DNA polymerase of avian myeloblastosis virus. *Proc. Natl. Acad. Sci. U.S.A.* 73:3418–3422.

Schneider, C., M. Kurkinen, and M. Greaves. 1983. Isolation of cDNA clones for the human transferring receptor. *EMBO J.* 12:2259–2263.

Sleigh, M. J., G. W. Both, and G. G. Brownlee. 1979. The influenza virus haemagglutinin gene: Cloning and characterization of a double-stranded DNA copy. *Nucleic Acids Res.* 7:879–893.

Villa-Komaroff, L., A. Efstratiadis, S. Broome, P. Lomedico, R. Tizard, S. P. Naber, W. L. Chick, and W. Gilbert. 1978. A bacterial clone synthesizing proinsulin. *Proc. Natl. Acad. Sci. U.S.A.* 75:3727–3731.

Wiginton, D. A., G. S. Adrian, R. L. Friedman, D. P. Suttle, and J. J. Hutton. 1983. Cloning of cDNA sequences of human adenosine deaminase. *Proc. Natl. Acad. Sci. U.S.A.* 80:7481–7485.

Williams, J. G. 1981. The preparation and screening of a cDNA clone bank. Pages 2–61 *in Genetic Engineering,* Vol. 1. Academic Press, New York.

CHAPTER **10**

Regulation of *GAL*7 Gene Expression in the Yeast *Saccharomyces cerevisiae*

J. G. YARGER*[,1,2], M. C. GORMAN[2], AND J. POLAZZI[2]

[1]*Department of Biochemistry and Molecular Biology, 7 Divinity Ave., Harvard University, Cambridge, Massachusetts 02138*
[2]*Miles Laboratories, Inc., Elkhart, Indiana 46515*

Regulation of the *GAL*7 gene from *Saccharomyces cerevisiae* was examined by modifying a sequence of the *GAL*7 coding region and the *GAL*7 promoter that had been fused to *lac*Z of *Escherichia coli*. We obtained a fusion of the *GAL*7 coding region and *GAL*7 promoter to *lac*Z. A set of deletions was constructed in vitro that removed varying portions of the *GAL*7 regulatory sequence from the *GAL*7-*lac*Z fusion. These deletions defined an A-T rich upstream activation site (UAS) region of approximately 150 bp that contained sequences essential for full induction from the *GAL*7 promoter. Sequence data are presented for the DNA fragment consisting of 315 base pairs (bp) of the *GAL*7 promoter region and 40 bp of the *GAL*7 coding sequence. The *GAL*7 UAS is composed of multiple elements, several of which are required for full induction capacity. We have compared the *GAL*7 UAS sequence to a UAS region determined for *GAL*1 and *GAL*10 and present information concerning the structure and function of the *GAL* UAS sequences and *GAL* promoters.

INTRODUCTION

The mechanisms of gene regulation in some *prokaryotic* systems are now understood at the molecular level (Hochschild et al. 1983; Johnston et al. 1981; Ptashne et al. 1980). In contrast, relatively little is known about eukaryotic gene regulation. Prokaryotic promoters are regions of DNA that direct RNA polymerase binding and initiation of transcription. Mutations in these promoters map almost exclusively to two domains. The first domain is the Pribnow box with a consensus sequence of TATAAT centered −10 bp from the site of transcription initiation (Pribnow 1975; Rosenberg and Court 1979; Siebenlist et al. 1980). The second domain constitutes a region required for RNA polymerase binding and transcription initiation (Wasylyk et al. 1980) and shows a consensus sequence TTGACA. The spacing between the −10 region and −35 region is important in prokaryotic cells. The most frequently found spacing is 17 bp (Rosenbert and Court 1979; Siebenlist et al. 1980), and only minor changes are tolerated without abolishing promoter activity (Stefano and Gralla 1982). The sequences that mediate the effect of positive and negative regulatory proteins either overlap or are immediately adjacent to the promoter (Hochschild et al. 1983; Johnston et al. 1981).

*To whom reprint requests should be addressed.

What then is the molecular nature of a eukaryotic promoter? Is it simply an RNA polymerase binding site analogous to a prokaryotic promoter? Recent experiments have suggested that many of the principles of gene regulation and promoter function derived from the study of prokaryotes must be quite extensively modified when they are applied to eukaryotes. We do not know at the molecular level how a eukaryotic promoter works, but we are able to describe in moderate detail some of the essential components of a eukaryotic promoter.

Although two regulatory domains have been identified in eukaryotic promoters, it has not been determined whether both domains are essential in all cases. For example, Grosschedl and Birnstiel (1980a,b, 1982) have identified a TATAA box within the cloned sea urchin histone H2A promoter. Deletion of the TATAA box results in the synthesis of H2A mRNAs with heterogeneous 5′-ends instead of the unique 5′-end normally found for histone H2A mRNA. However, the significance of a TATAA box is unclear since the yeast *HIS*3 and *HIS*4 genes contain an abundance of TATAA boxes yet retain unique 5′-mRNA ends (Hinnebusch and Fink 1983).

The second regulatory domain has been identified as a region far upstream of the TATAA box. When a 30 bp region of DNA centered 125 bp upstream of the histone H2A TATAA box is deleted, histone H2A gene transcription is diminished 100-fold (Grosschedl and Birnstiel 1980b, 1982).

Similar promoter arrangements have been found for other eukaryotic genes. For example, from two to four upstream elements have been found upstream of the yeast *HIS*1, *HIS*3, *HIS*4, and *TRP*5 genes (Donahue et al. 1983; Hinnebusch and Fink 1983; Struhl 1981, 1982a,b). In all these cases, deletion analysis showed that at least two contiguous repeats were needed for proper gene regulation (Hinnebusch and Fink 1983). Essential upstream sequences located over 100 base pairs upstream from the transcription start point have also been found for early genes of SV40 (Darnell 1982), yeast *CYC*1 (Guarente and Ptashne 1981; Guarante et al. 1982), yeast *GAL*10 (Guarante et al. 1982; Yocum et al. 1984), yeast *GAL*1 (Yocum et al. 1984), and yeast *LEU*2 genes (Martinez-Arias et al. 1984) to name only a few. It is unknown how the upstream elements function.

Prompted by our desire to define and explain in general terms the working of a eukaryotic promoter, we have initiated an analysis of the inducible yeast *GAL*7 promoter. This analysis, in conjunction with the recent studies on the upstream domains for *GAL*10 (Guarante et al. 1982; Yocum et al. 1984; West et al. 1984) and *GAL*1 (Yocum et al. 1984; West et al. 1984), should allow us to arrive at a general statement of the molecular mechanisms regulating promoter function for one eukaryotic system.

Galactose System

S. cerevisiae is capable of growth on galactose as a sole carbon source. External galactose is transported into the cell via a specific galactose permease (*GAL*2), where it is converted to galactose 1-phosphate by the enzyme galactokinase (*GAL*1). α-D-Galactose-1-phosphate uridyltransferase (*GAL*7) converts galactose-1-phosphate to glucose-1-phosphate, which enters the glycolytic pathway. The third enzyme, uridine diphosphoglucose-4-epimerase (*GAL*10), recycles UDP-galactose, a byproduct in the formation of glucose-1-phosphate, to UDP-glucose. When cells grown on glucose are switched to galactose, the transcription of *GAL*1, *GAL*7, and *GAL*10 is coordinately induced at least 1,000-fold (Hopper et al. 1978; St. John and Davis 1981; Yarger et al. 1984). The galactose system is regulated through both positive (*GAL*4) and negative (*GAL*80) regulatory components (Douglas and Hawthorne 1972) and responds to carbon catabolite repression (Adams 1972; Matsumoto et al. 1981).

The current regulatory model suggests that in the absence of galactose, the transcription of *GAL*1, *GAL*7, and *GAL*10 is efficiently repressed by the negative regulatory protein *GAL*80 (Douglas and Hawthorne 1972). Both recessive *gal*80 and dominant *GAL*80^s mutations have been found. The *gal*80 mutations allow constitutive expression of *GAL*1, *GAL*7, and *GAL*10, whereas *GAL*80^s mutations confer an uninducible phenotype (Douglas and Hawthorne 1966, 1972; Douglas and Pelroy 1963; Nogi et al. 1977). Even in the absence of functional *GAL*80 protein, *GAL*4 gene product is required for positive regulation of *GAL*1, *GAL*7, and *GAL*10 gene expression (Douglas and Hawthorne 1966, 1972; Douglas and Pelroy 1963; Nogi et al. 1977). Dominant *GAL*4^c mutations confer constitutive expression, whereas recessive *gal*4 mutations confer an uninducible phenotype (Douglas and Hawthorne 1966; Nogi et al. 1977). Both *GAL*4 and *GAL*80 gene products are constitutively produced (Hashimoto et al. 1983; Johnston and Hopper 1982; Laughon and Gestland 1982; Perlman and Hopper 1979) and appear to interact at the protein level (Nogi et al. 1977; Perlman and Hopper 1979). According to the current model, when the *GAL*4/*GAL*80 protein complex binds the inducer (probably galactose), the *GAL*4 protein is released and acts to turn on transcription of *GAL*1, *GAL*10, and *GAL*7 by binding to sites near the 5′ ends of the genes (Guarante et al. 1982; Hopper et al. 1978; Matsumoto et al. 1980; Perlman and Hopper 1979; Yocum et al. 1984). Another form of regulation involves catabolite repression. Growth on galactose plus glucose results in substantial repression of *GAL*1, *GAL*7, and *GAL*10 activities (Adams 1972; Matsumoto et al. 1981). In a third form of regulation, the accumulation of *GAL*1 enzyme (and perhaps of *GAL*7 and *GAL*10 enzymes as well) is regulated through cell cycle controls acting at a post-translational level (Halvorson et al. 1984; Yarger et al. 1984).

*GAL*1, *GAL*7, and *GAL*10 genes have been cloned, sequenced, and characterized by transcriptional analysis (Schell and Wilson 1979; St. John and Davis 1979, 1981; St. John et al. 1981). The three genes are tightly linked and contained on a 6.5 kb DNA fragment. *GAL*1 and *GAL*10 are on separate strands and are transcribed from divergent promoters (St. John and Davis 1981; St. John et al. 1981).

A 365 bp fragment of DNA from the middle of the *GAL*1/*GAL*10 divergent promoter region contains all the information necessary for galactose-induced transcription (Guarente et al. 1982). For example, when this DNA fragment is placed upstream of the yeast *CYC*1 gene, the expression of *CYC*1 now comes under galactose-induced control (Guarente et al. 1982).

*GAL*7 is independently transcribed from a region downstream of the *GAL*10 gene. We have analyzed the *GAL*7 promoter region in an effort to learn more about the regulation of *GAL*7 at the molecular level. To do this, we used a fusion between the *GAL*7 promoter and the *E. coli lac*Z gene to allow a simple assay for *GAL*7 promoter function. We constructed a set of deletions in vitro to delimit the *GAL*7 UAS region involved in regulation of *GAL*7 transcription. We then compared our data to those recently obtained for the divergent *GAL*1/*GAL*10 promoters (Yocum et al. 1984; West et al. 1984) to help to clarify the structure and function of the *GAL* UAS regions.

Materials and Methods

Strains and media. S. cerevisiae strain DBY745 (*ade*1 *ura*3, *leu*2–100, *leu*2–112) was obtained from David Botstein. Minimal or selective medium contained, per liter: 6.7 g Difco Yeast Nitrogen Base w/o amino acids, 0.7 g complete amino acid mix minus uracil (Sherman et al. 1981) and 2% glucose (MIN GLU) or 2% glucose plus 2% galactose (MIN GLU GAL) or 2% galactose (MIN GAL). Solid medium contained the above plus 2% agar. Indicator plates for scoring β-galactosidase activity contained the above plus 40 mg/l of 5-bromo-4-chloro-3-indolyl-β-D-galactopyranoside (X-gal) indicator (Boehringer/Mannheim) and 70 m*M* potassium phosphate buffer, pH 7.0, both added after the medium was autoclaved and cooled to 60 C. The fusion plasmid pBD6 was constructed by Barbara Dunn.

DNA manipulations and DNA sequencing. All restriction enzymes, T4 DNA ligase, DNA polymerase Klenow fragment, synthetic linkers, and *Bal*31 nuclease, were purchased from New England Biolabs and were used according to the supplier's specifications. DNA sequences were determined by the chemical cleavage method (Maxam and Gilbert 1977). Manipulation of DNA sequences and homology searches were performed with the aid of the Intelligenetics <SEQ> computer program. Plasmid constructions and yeast transformations were performed by standard methods (Maniatis et al. 1982; Sherman et al. 1981).

Construction of deletions. Deletions were constructed by cleaving the unique *Sal*1 site at the *GAL*10 gene in pBD6, digesting the DNA ends for various times with *Bal*31 exonuclease, and then ligating in the presence of *Xho*1 linkers (Maniatis et al. 1982). Unidirectional deletions were constructed in the following manner. The unique pBD6 *Sal*1 site was changed to an *Xho*1 site using *Xho*1 linkers (Maniatis et al. 1982). The plasmid pBD6-Xho was digested with *Xho*1 and *Bam*H1 and the pBD6-Xho plasmid (backbone) isolated by gel electrophoresis. The original pBD6 deletion plasmids were then digested with *Xho*1 and *Bam*H1, and the short promoter fragments containing the deletions were isolated by gel electrophoresis. The pBD6-Xho (backbone) was then ligated to the short *Xho*1 to *Bam*H1 deletion fragments previously generated. This resulted in unidirectional deletions beginning at the *GAL*10 *Xhol*1 site (previously *Sal*1) and extending toward the*GAL*7 gene.

β-Galactosidase assays. β-Galactosidase assays were performed as follows. Single colonies of yeast transformants from MIN GLU plates were grown to saturation in 5 ml of MIN GLU. Cells were then diluted into MIN GLU (1:50), MIN GLU GAL (1:50), or MIN GAL (1:25), and grown at 30 C until an OD_{600} of about 1.0 was reached. Between 0.01 and 0.2 ml of culture was added to Z buffer (Miller 1972) to give a total volume of 1.0 ml. Each sample was then vortexed with

0.05 ml of 0.1% sodium dodecyl sulfate and three drops of chloroform. The remainder of the assay and the calculation of units of activity normalized to OD_{600} of the culture was exactly as described by Miller (1972). All assays were done in triplicate with reproducibility of ±15%.

Results and Discussion

In Vitro Deletions

A partial restriction map of the *GAL7/lac*Z fusion plasmid pBD6 is shown in Fig. 1. This fusion plasmid contains the first 555 bp of *GAL*7 coding sequence fused in frame to the *lac*Z gene, 730 bp of noncoding DNA between the carboxyl end of *GAL*10 and the coding region of *GAL*7, and 407 bp from carboxyl terminal of the *GAL*10 gene. The upstream region of *GAL*7 fused to pBD6 contains all of the information required for normal *GAL*7 regulation by galactose since pBD6 shows no β-galactosidase activity in the absence of galactose, good expression in the presence of galactose alone, and repressed expression in the presence of glucose and galactose (Table 1).

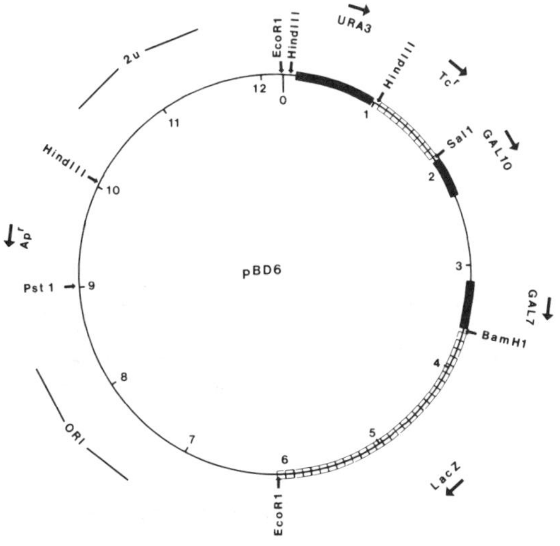

FIG. 1. Partial restriction map of the fusion plasmid pBD6. The *GAL7-lacZ* fusion is described in the text. The numbers indicate kilobases of DNA.

In order to define the upstream region of the *GAL*7 gene responsible for galactose-induced expression, we linearized pBD6 with *Sal*1 and then constructed a set of deletions in vitro using *Bal*31 as described in Materials and Methods. The deletion plasmids were used to transform the yeast strain DBY745, and the transformants were then analyzed for β-galactosidase activity (Fig. 2 and Table 1). The endpoint of each deletion was determined by sizing the *Bam*H1/*Xho*1 restriction fragments on agarose gels and/or sequencing the DNA. As shown in Fig. 2, there was no significant loss of β-galactosidase activity in plasmids carrying deletions ranging from −1,100 bp to −400 bp. However, deletions that removed upstream DNA to −325 bp (Δ137 and Δ136) showed a 25%

TABLE 1. *β-Galactosidase activity of GAL7-lacZ deletion plasmids in yeast DBY745 grown in minimal selective medium containing various carbon sources*

Deletion	Units of β-galactosidase activity		
	Carbon Source		
	gal	glu + gal	glu
pBD6	340	24	0
pJYΔ90	330	24	0
pJYΔ156	325	—	0
pJYΔ137	249	5	0
pJYΔ118	143	4	0
pJYΔ192	0	—	0
pJYΔ202	2	—	—

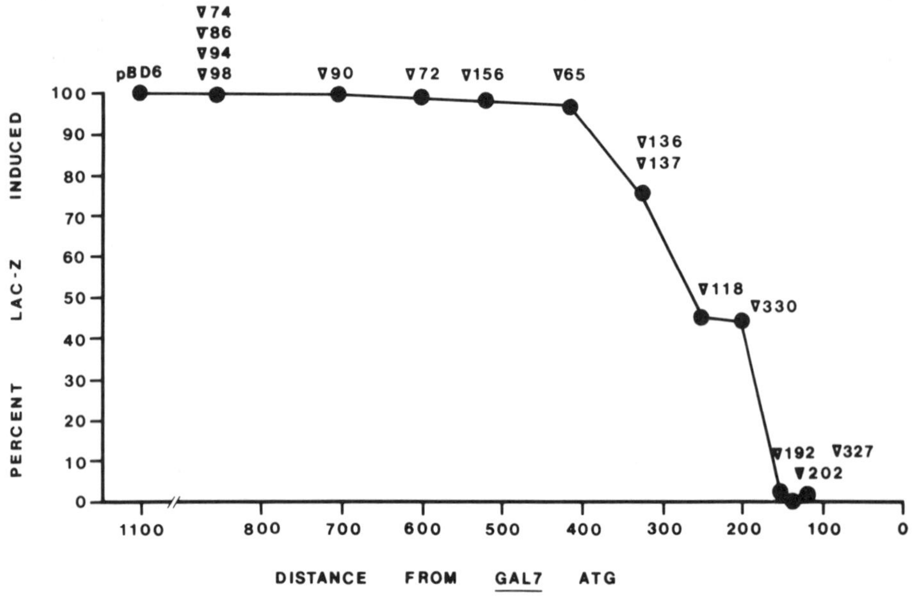

FIG. 2. β-Galactosidase enzyme activity of *GAL*7 promoter deletion mutants. Enzyme activity is expressed as a percent of the original pBD6 enzyme activity in each case (see Table 1).

reduction in β-galactosidase activity. Additional deletions extending up to −250 bp (Δ118) and −200 bp (Δ330) showed a 55% reduction in β-galactosidase activity and deletions beyond base −150 (Δ192 and Δ202) showed essentially no inducible enzyme activity (Fig. 2).

These deletions have defined a UAS region that contains sequences required for full induction of the *GAL*7 gene and that is located approximately −150–350 bp upstream of the *GAL*7 translational start site. Deletions across this region result in a gradually decreasing β-galactosidase activity, including one plateau, rather than an abrupt transition from inducibility to lack of inducibility. Such a gradient of decreasing activity suggests that the *GAL*7 UAS is composed of multiple elements, several of which are required for full induction.

DNA Sequence

We have sequenced 44 bp of *GAL7* coding region and 315 bp of *GAL7* upstream DNA (Fig. 3). A TATAA box is located 84 bp upstream of the *GAL7* translational start. When our sequence is compared to the *S. cerevisiae* sequence of Nogi and Fukasawa (1983) and to the equivalent region from *S. carlsbergensis* (Citron and Donelson 1984), we find 97% homology to the former and 91% homology to the latter (Fig. 3). Most of the differences are restricted to the noncoding region with the exception of two coding region base changes for *S. carlsbergensis*. In the first case, the C→T transition would not change the amino acid usage. The second would result in a tyrosine substitution for histidine in *S. carlsbergensis*.

The sequence differences in the noncoding region are of interest since the cloned *GAL7* gene from *S. carlsbergensis* is expressed in *S. cerevisiae* (Citron and Donelson 1984). Since the *S. cerevisiae GAL4* protein is capable of recognizing the regulatory regions upstream of the *S. carlsbergensis GAL* genes, the bases that constitute the regulatory regions in the two species could be very similar. The

```
Yarger  3'ATACCCTTAC+30CGATCTTTTT+20AGTTTAAGAA+10GTCGTCAGTA  +0AAAACTCCCT-10
Carl               T     T
Nogi

Yarger    TATAAGTTGA-20CAAAAAAAAA-30TAGTACAACT-40ACGAGACGTA -50TTATTCACGG-60
Carl                                                                d
Nogi                            d                                   d

Yarger    GTATTTATAA-70AGGCTGGACG-80AAAATATAGA-90AACGATCGGT-100TGATGACTTG-110
                                    .......
Carl                      T                               T TT    T
                                                            **    *
Nogi                                                            T T
                                                                  *

Yarger    TACGATTTCG-120AATGTGTAAT-130AAAAGTCGAA-140CCGATAAAAC-150ACTTGTGACA-160
Carl        T  ddddd    ddd
            *
Nogi        T    C      d
            *    *

Yarger    TATCGGTCAG-170GAAGCCTAGT-180GCCAGTTGTC-190AACAGGCTCG-200CGAAAAACCT-210
Carl                                    T   C
                                        *
Nogi

Yarger    GGGAGAGGGA-220ACTAAAAACC-230CAATTCCTTT-240TACTGCCTTT-250TATATAGATT-260
Carl          d          d                               T
Nogi                     d

Yarger    ACCCGGAAGC-270GAGTTGTCAC-280GAGGCTTCAT-290ATCGAAAGGT-300TTTCCTCCCC-310
Carl        T    GAG     G AG                                            T
Nogi
```

FIG. 3. Partial DNA sequence of the *Saccharomyces cerevisiae GAL7* reading 3′ to 5′ as compared to previous sequences of the same region: YARGER refers to the DNA sequence of the *GAL7* clone used in this study; CARL refers to the *S. carlsbergensis* sequence of Citron and Donelson (1984); NOGI refers to the *S. cerevisiae* sequence obtained by Nogi and Fukasawa (1983). Sequences are identical except where noted by a letter. The letter "d" refers to a base deletion in the corresponding sequence whereas a letter plus an asterisk (*) refers to a base insertion in front of the corresponding base under which it appears.

regions of major sequence divergence thus constitute an initial mutational analysis. In this regard, within the *GAL*7 UAS region of DNA (150 to 350 bp upstream of the translational start), the region from −265 to −275 contains bases that are not required for *full* UAS activity (see Fig. 3). The proximal boundary of the equivalent UAS region in *S. carlsbergensis* is located 5 bp closer to the translational start site (see Fig. 1). Thus, small variations in spacing must be tolerable for the *GAL*7 UAS activity. This would be similar to what has been found for the Herpes virus thymidine kinase UAS (McKnight et al. 1981; McKnight and Kingsbury 1982).

A region of DNA present between the *GAL*7 TATAA box and the start site of transcription bears a striking resemblance to a reverse transcription terminator. The sequence $3'\text{-}TAAGN_{13}$ $TAGTAN_{12}$ *TATTATTA* CGGG *TATTTTAT-5′* starting at base pair −12 and extending to base pair −70 (Fig. 3) is similar to Sherman's consensus sequence for transcription termination $5'TAGN_{5}$ $TAGTAN_{7}$ *TTATTTATAT-3′* (Zaret and Sherman 1982) (Fig. 4). It is interesting to note that in the *S. carlsbergensis* sequences (Citron and Donelson 1984), a Sherman terminator sequence can be found on the correct strand within the promoter for *GAL*1, i.e., bp 3048 to 3085, while a reverse terminator sequence is found on the correct strand within the promoter for *GAL*10, i.e., bp 2710 to 2751. In each case, the terminator-like sequence appears slightly upstream of the transcription start site.

It is unclear yet whether the terminator-like sequences found in the *GAL* TATAA box regions have any function. It is possible, however, that the presence of a functional transcription terminator upstream of the *GAL*7 transcriptional start site prevents unregulated *GAL*7 gene expression arising from read-through transcriptional events. On the other hand, these terminator-like sequences may not be functional since several terminator-like sequences can also be found within the *GAL*7, *GAL*10, and *GAL*1 coding regions (data shown for *GAL*1 only; see Fig. 4). Thus, an interesting paradox arises. If the terminator-like sequences are functional, then attenuation could be part of the model of regulation of *GAL*

GENE	SEQUENCE
cyc1-:	TAGTTATGTTAGTATTAGAACGTTATTTATATTT
(P)GAL7:	TATAAGTTGACAAAAAAAAATAGTACAACTACGAGACGTATTATTACGGGTATTTAT
(P)GAL10:	TATGTATAGGTATAGATTAGAATGAATATACAACACCTTTACATTTCTC
(T)GAL10:	TAGTTCTTAATTGCAACACATAGATTTGCTGTATAACGAATTTTATGCTATTTTT
(P)GAL1:	TAGGATGATAATGCGATTAGTTTTTTAGCCTTATTTCTGGGGTAA
(C)GAL1:	GATCGTTGTTTTAGGCCAAATCGTAGTATTCGCGAATATTTAAAGAATTAATAC

FIG. 4. Comparison of transcription termination sequences. The sequence for *CYC*1 is from Zaret and Sherman (1982). The DNA sequence of *GAL*10 and *GAL*1 are from Citron and Donelson (1984) whereas the *GAL*7 sequence is from Fig. 1. The letter (T) preceding the gene name refers to a carboxyl terminus location, whereas the letters (P) and (C) refer to promoter and to coding regions, respectively.

gene expression. However, if the terminator-like sequences are not functional, we could reconsider what sequences are involved in yeast transcription termination.

Comparison to GAL1 and GAL10 Promoters

We have compared the sequence of our *GAL7* UAS to the *GAL1/GAL10* UAS sequence as described by Yocum et al. (1984). The *GAL7* UAS is 44% G+C whereas the *GAL10/GAL1* UAS is 64% G+C. Thus, in a general sense, *GAL* UAS regions need not necessarily be G+C rich for full induction activity as has been alluded to previously (Yocum et al. 1984).

The region between Δ118 and Δ330 (Fig. 2) results in no further loss of β-gal enzyme activity when it is deleted. Thus the drop in β-gal activity between Δ330 and Δ192 (Fig. 2) could represent the location of one *GAL7* UAS element (UAS1) sufficient for partial GAL7 gene expression. Within this 40 bp region we find the 21 bp sequence 5′-179 CACGGTCAACAGTTGTCCGAG-199. Except for 2 thymidines and 2 adenines (underlined) in that sequence, there is perfect dyad symmetry around a central adenine (asterisked). In a similar fashion, the drop in activity between Δ118 and Δ136 could define the location of a second *GAL7* UAS element (UAS2) that acts in concert with UAS1 to increase GAL7 gene expression. Within this 60 bp region is the 21 bp UAS2 sequence 5′-266 TCGCTCAACAGTGCTCCGAAG-288. The UAS2 sequence shows exact homology at 16 out of 25 bases with UAS1. UAS1 and UAS2 are separated by approximately 60 bp (Fig. 2). The drop in activity between Δ65 and Δ136 could represent the location of a UAS3 although we are unable to detect a sequence within this region showing homology to UAS1 and UAS2. UAS1 alone is sufficient for 50% of the wild type level of *GAL7* gene expression. However, UAS1 and UAS2 are both required for full induction of the GAL7 promoter. It is unlikely that there are any cooperative effects between UAS1 and UAS2. The deletion data suggest that a purely arithmetic relationship exists. Within the *GAL1/GAL10* divergent promoter region (West et al. 1984) are several sequences that could be related to the *GAL7* UAS1 and UAS2 sequences. The sequence we have found in common between the GAL7 UAS1 and UAS2 elements is very similar to the sequence 5′-TCGGAGCACTGTTGAGCG found by Fukasawa (personal communication). It is interesting to note that UAS1 is very similar between *S. cerevisiae* and *S. carlsbergensis* (Fig. 3). In contrast, UAS2 shows a large degree of sequence divergence in one half-site between *S. cerevisiae* and *S. carlsbergensis*.

It is not clear whether the *GAL* UAS structures constitute a part of the eukaryotic promoter as defined by the RNA polymerase binding site. RNA polymerase II probably does not bind simultaneously to both promoter elements. Yeast RNA polymerase II/DNA interactions could involve only about 40–50 base pairs of DNA (Struhl 1981). Thus, the *GAL* UAS sequences are probably too far upstream of the TATAA box to allow simultaneous binding by RNA polymerase II. How could the UAS structures work? Possibly the *GAL* UAS regions in the presence of *GAL4* protein could facilitate the entry of RNA polymerase II following which the polymerase would scan the DNA for a proper initiation sequence. Alternatively, the *GAL* UAS in the presence of *GAL4* may alter chromatin structure locally to render the transcription initiation region more accessible to RNA polymerase II.

Acknowledgments

The authors would like to thank Mark Ptashne in whose laboratory this work was started, Barbara Dunn who constructed the fusion plasmid pBD6 as a rotation project in Mark Ptashne's laboratory, George Boguslawski for helpful conversations, and Theresa Nowicki for diligently typing this manuscript.

Literature Cited

Adams, B. 1972. Induction of galactokinase in *Saccharomyces cerevisiae:* Kinetics of induction and glucose effects. *J. Bacteriol.* 111:308–313.

Citron, B., and J. Donelson. 1984. The sequence of the yeast GAL region and its transcription *in vivo. J. Bacteriol.* 158:269–278.

Darnell, J. E. 1982. Variety in the level of gene control in eukaryotic cells. *Nature* 297:365–371.

Donahue, T. F., R. S. Daves, G. Lucchini, and G. R. Fink. 1983. A short nucleotide sequence required for regulation of HIS4 by the general control system of yeast. *Cell* 32:89–98.

Douglas, H. C., and D. C. Hawthorne. 1966. Regulation of genes controlling synthesis of the galactose pathway enzymes in yeast. *Genetics* 54:911–918.

______. 1972. Uninducible mutants in the gal i locus of *Saccharomyces cerevisiae. J. Bacteriol.* 109:1139–1143.

Douglas, H. C., and G. Pelroy. 1963. A gene controlling inducibility of the galactose pathway enzymes in *Saccharomyces. Biochim. Biophys. Acta* 68: 155–156.

Grosschedl, R., and M. L. Birnstiel. 1980. Identification of regulatory sequences in the prelude sequences of an H2A histone gene by the study of specific deletion mutants *in vivo. Proc. Natl. Acad. Sci. USA* 77:1432–1436.

______. 1980. Spacer DNA sequences upstream of the TATAATA sequence are essential for promotion of H2A histone gene transcription *in vivo. Proc. Natl. Acad. Sci. USA* 77:7102–7106.

______. 1982. Delimitation of far upstream sequences required for maximal *in vitro* transcription of an H2A histone gene. *Proc. Natl. Acad. Sci. USA* 79:297–301.

Guarente, L., and M. Ptashne. 1981. Fusion of *Escherichia coli LacZ* to the cytochrome C gene of *Saccharomyces cerevisiae. Proc. Natl. Acad. Sci. USA* 78:2199–2203.

Guarente, L., R. R. Yocum, and P. Gifford. 1982. A *GAL*10-*CYC*1 hybrid yeast promoter identifies the *GAL*4 regulatory region as an upstream site. *Proc. Natl. Acad. Sci. USA* 79:7410–7414.

Guarente, L., B. Lalonde, P. Gifford, and E. Alani. 1984. Distinctly regulated tandem upstream activation sites mediate catabolite repression of the *CYC*1 gene of *S. cerevisiae. Cell* 36:503–511.

Halvorson, H., K. Bostian, J. G. Yarger, and J. E. Hopper. 1984. Enzyme expression during growth and cell division in *Saccharomyces cerevisiae*: A study of galactose and phosphorus metabolism. Pages 49–96 *in* G. Stein and J. Stein, eds., *Recombinant DNA Approaches to Studying Control of Cell Proliferation.* Academic Press, New York.

Hashimoto, H., Y. Kikuchi, Y. Nogi, and T. Fukasawa. 1983. Regulation of expression of the galactose gene cluster in *Saccharomyces cerevisiae*: Isolation and characterization of the regulatory gene *GAL4*. *Mol. Gen. Genet.* 191:31–38.

Hinnebusch, A. G., and G. R. Fink. 1983. Repeated DNA sequences upstream from *HIS*1 also occur at several other co-regulated genes in *Saccharomyces cerevisiae. J. Biol. Chem.* 258:5238–5247.

Hochschild, A., C. Irwin, and M. Ptashne. 1983. Repressor structure and the mechanism of positive control. *Cell* 32:319–325.

Hopper, J. E., J. Broach, and L. Rowe. 1978. Regulation of the galactose pathway in *Saccharomyces cerevisiae*: Induction of uridyl transferase mRNA and dependency on *GAL*4 gene function. *Proc. Natl. Acad. Sci. USA* 75:2878–2882.

Johnston, A. D., A. R. Poteete, G. Lauer, R. T. Sauer, G. K. Ackers, and M. Ptashne. 1981. λ repressor and cro—components of an efficient molecular switch. *Nature* 294:217–223.

Johnston, S., and J. Hopper. 1982. Isolation of the yeast regulatory gene *GAL*4 and analysis of its dosage effects on the galactose/melibiose regulon. *Proc. Natl. Acad. Sci. USA* 79:6971–6975.

Laughon, A., and R. Gestland. 1982. Isolation and preliminary characterization of the *GAL*4 gene, a positive regulator of transcription in yeast. *Proc. Natl. Acad. Sci. USA* 79:6827–6831.

Maniatis, T., E. Fritsch, and J. Sambrook. 1982. *Molecular cloning: A Laboratory Manual.* Cold Spring Harbor Press, Cold Spring Harbor, New York.

Matsumoto, K., A. Toh-e, and Y. Oshima. 1981. Isolation and characterization of dominant mutations resistant to carbon catabolite repression of galactokinase synthesis in *Saccharomyces cerevisiae. Mol. Cell. Biol.* 1:83–93.

Matsumoto, K., Y. Adachi, A. Toh-e, and Y. Oshima. 1980. Function of positive regulatory gene *GAL*4 in the synthesis of galactose pathway enzymes in *Saccharomyces cerevisiae*: Evidence that the *GAL*81 region codes for part of the *GAL*4 protein. *J. Bacteriol.* 141:508–527.

Maxam, A., and W. Gilbert. 1977. A new method for sequencing DNA. *Proc. Natl. Acad. Sci. USA* 74:560–564.

Martinez-Arias, A., H. J. Yost, and M. J. Casadaban. 1984. Role of an upstream regulatory element in leucine repression of the *Saccharomyces cerevisiae* leu2 gene. *Nature* 307:740–742.

McKnight, S. L., and R. Kingsbury. 1982. Transcriptional control signals of eukaryotic protein-coding gene. *Science* 217:316–324.

McKnight, S. L., E. R. Gavis, R. Kingsbury, and R. Axel. 1981. Analysis of transcriptional regulatory signals of the HSV thymidine kinase gene: Identification of an upstream control region. *Cell* 25:385–398.

Miller, J. 1972. *Experiments in Molecular Genetics.* Cold Spring Harbor Press, Cold Spring Harbor, New York.

Nogi, Y., and T. Fukasawa. 1983. Nucleotide sequence of the transcriptional initiation region of the yeast *GAL*7 gene. *Nucl. Acid Res.* 11:8555–8568.

Nogi, Y., K. Matsumoto, A. Toh-e, and Y. Oshima. 1977. Interaction of superrepressible and dominant constitutive mutations for the synthesis of galactose pathway enzymes in *Saccharomyces cerevisiae*. *Mol. Gen. Genet.* 152: 137–144.

Perlman, D., and J. Hopper. 1979. Constitutive synthesis of *GAL*4 protein, a galactose pathway regulator. *Cell* 16:89–95.

Pribnow, D. 1975. Nucleotide sequence of an RNA polymerase binding site at an early promoter T7. *Proc. Natl. Acad. Sci. USA* 75:784–788.

Ptashne, M., A. Jeffrey, A. D. Johnson, B. Maurer, B. J. Meyer, C. O. Pabo, T. M. Roberts, and R. T. Sauer. 1980. How the λ repressor and cro work. *Cell* 19:1–11.

Rosenberg, M., and D. Court. 1979. Regulatory sequences involved in the promotion and termination of RNA transcription. *Annu. Rev. Genet.* 13:319–353.

Schell, M., and D. Wilson. 1979. Cloning and expression of the *Saccharomyces cerevisiae* galactokinase gene in an *Escherichia coli* plasmid. *Gene* 5:291–303.

Sherman, F., G. Fink, and J. Hicks. 1981. *Methods in Yeast Genetics,* Cold Spring Harbor Press, Cold Spring Harbor, New York.

Siebenlist, U., R. B. Simpson, and W. Gilbert. 1980. *Cell* 20:269–281.

St. John, T., and R. Davis. 1979. Isolation of galactose inducible sequences from *Saccharomyces cerevisiae* by differential plaque filter hybridization. *Cell* 16:443–452.

______. 1981. The organization and transcription of the galactose gene cluster of *Saccharomyces. J. Mol. Biol.* 152:285–315.

St. John, T., S. Scherer, M. McDonnell, and R. Davis. 1981. Deletion analysis of the GAL gene cluster: Transcription from three promoters. *J. Mol. Biol.* 152:317–337.

Stefano, J. E., and J. D. Gralla. 1982. Spacer mutations in the lac p^s promoter. *Proc. Natl. Acad. Sci. USA* 79:1069–1072.

Struhl, K. 1981. Deletion mapping a eukaryotic promoter. *Proc. Natl. Acad. Sci. USA* 78:4461–4465.

______. 1982a. Regulatory sites for HIS 3 gene expression in yeast. *Nature* 300:284–287.

______. 1982b. The yeast HIS3 promoter contains at least two distinct elements. *Proc. Natl. Acad. Sci. USA* 79:7385–7389.

Wasylyk, B., C. Kedinger, J. Corden, O. Brison, and P. Chambon. 1980. Specific *in vitro* initiation of transcription on con albumin and ovalbumin genes and comparison with adenovirus-2 and late genes. *Nature* 285:367–373.

West, R. W., R. R. Yocum, and M. Ptashne. 1984. *Saccharomyces cerevisiae* GAL1-GAL10 divergent promoter region: Location and function of the upstream activating sequence UAS_G *Mol. Cell Biol.* 4:2467–2478.

Yarger, J. G., H. O. Halvorson, and J. E. Hopper. 1984. Regulation of galactokinase (*GAL*1) enzyme accumulation in *Saccharomyces cerevisiae. Mol. Cell. Biochem.* 61:173–182.

Yocum, R. R., S. Hanley, and R. West. 1984. Use of *lac*Z fusions to delimit regulatory elements of the inducible divergent *GAL*1-*GAL*10 promoter in *Saccharomyces cerevisiae. Mol. Cell. Biol.,* in press.

Zaret, K. S., and F. Sherman. 1982. DNA sequence required for efficient transcription termination in yeast. *Cell* 28:563–573.

IV
SYMPOSIUM: BIOCONVERSION OF WASTE MATERIALS TO USEFUL INDUSTRIAL PRODUCTS

W. E. Gledhill, *Convener*
Monsanto Company, St. Louis, Missouri

PANELISTS

D. P. Chynoweth
B. E. Dale
L. L. Henk
H. Ikemoto
D. E. Jerger
A. P. Leuschner
P. F. Levy
T. Matsunaga
A. Mitsui
B. R. Renuka
M. Shiang
D. L. Wise

Introduction to Symposium IV

Bioconversion of Waste Materials to Useful Industrial Products

WILLIAM E. GLEDHILL, PH.D.

Monsanto Company, St. Louis, Missouri

Since the advent of the industrial revolution, mass production of practically all commodity items has accelerated almost exponentially. Large-scale industrial production inevitably leads to generation and release of major quantities of unwanted by-products (wastes) that present significant disposal problems. Various physical, chemical, and biological methods are used to treat most of these wastes, but the fact that we must resort to use of landfills, deepwells, and incineration indicates that ideal solutions to all treatment problems still do not exist. Moreover, increasing regulatory pressure and/or cost will make the latter three practices less desirable in the future. Consequently, alternative means are being sought for treating wastes now disposed of via these three methods.

Most current treatment processes are designed simply to get rid of the nuisance waste materials. Quite often, wastes contain valuable raw materials that potentially could be converted by microorganisms to useful products. The emerging field of biotechnology offers promise for developing new microbial processes for handling wastes, which currently are not amenable to conventional biological treatment. Various novel processes have been reported in which microorganisms not only metabolize undesirable waste materials but also convert them to beneficial products. Several industries are currently examining the feasibility of such processes for their particular waste systems.

This symposium presents four papers discussing such processes that provide attractive alternatives to conventional waste treatment practices. We hope that examples such as these will stimulate further development of more efficient, economical, and environmentally acceptable processes for solving complex waste treatment problems.

CHAPTER **11**

Suppressed Methane Fermentation of Selected Industrial Wastes: A Biologically Mediated Process for Conversion of Whey to Liquid Fuel

D. L. WISE, A. P. LEUSCHNER, AND P. F. LEVY

Dynatech Research and Development Company, 99 Erie Street, Cambridge, Massachusetts 02139

The process described produces a liquid fuel, composed primarily of *n*-octane and *n*-decane, from aqueous organic waste streams. The process consists of three steps. First, the waste stream undergoes anaerobic digestion under conditions that inhibit methane formation and encourage formation of higher molecular weight organic acids (C–4 to C–6). Second, these organic acids are both selectively removed and concentrated by liquid-liquid extraction, first into a hydrocarbon phase and then back into aqueous base. Third, the concentrated organic acid stream is fed to an electrolysis cell in which Kolbe electrolysis is used to form the desired liquid alkane product. Both the scientific and technological aspects of this overall process have been demonstrated in the laboratory. A preliminary engineering analysis was carried out on a whey stream from a cheese processing plant with a daily flow of 100,000 gal containing 5% of lactose by weight. A project payback period of 4.0 yr was calculated based on liquid fuel production, without credits for disposal costs.

INTRODUCTION

The process described is applicable to liquid industrial waste streams with dissolved or finely suspended biodegradable solids (BOD > 25,000 mg/liter). The products are a liquid fuel composed primarily of *n*-octane and *n*-decane, a gaseous fuel containing olefins (propylene, butenes, and pentenes) and hydrogen, carbon dioxide, and an aqueous liquid effluent with the BOD (biological oxygen demand) reduced by approximately 80%.

The process consists of three steps. First, the waste stream containing dissolved and/or suspended organic material is fed to an anaerobic digester. The digester is operated under conditions that inhibit methane formation, encouraging formation of "higher weight organic acids" (C–4 to C–6). These higher weight acids may be selectively removed and concentrated in the second processing step by liquid-liquid extraction, first into a hydrocarbon phase and then back into aqueous base. In the final step the concentrated organic acid stream is fed to an electrolysis cell where, through "Kolbe Electrolysis," a liquid alkane product is formed. Olefin, hydrogen, and carbon dioxide are the side-products.

Estimates have been made of industrial wastewater production and quality in relevant food processing and paper industries (Mueller Associates, Inc. 1979; EPA 1979; National Food Processors' Association 1979; Levy et al. 1983; EPA

1974; American Paper Institute 1980). Table 1 lists the flow, BOD, and suspended solids (TSS) of wastewater production in relevant industries. Since it is suggested that the process require a minimum BOD of 25,000 mg/l (0.209 lb/gal), the applicable waste streams, based on average BOD, are those from the poultry industry and whey streams from the dairy industry. Isolated streams from other industries may meet the minimum concentration requirement as well, but these are excluded for purposes of this estimation. Since the quantity of whey far outweighs poultry wastes, this discussion will focus on use of whey streams. Based on data in Table 1 (1978 levels), whey represents 57% of the disposed BOD in the food processing industries listed and 30% of BOD including the paper industry.

TABLE 1. *Industrial wastewater production (1978 data)*

	Wastewater		
Industry	Flow (10^6 gal)	BOD (10^3 tons)	TSS (10^3 tons)
Potato	24,600	315	425
Tomato	9,690	17.3	23.8
Poultry	566	133	3.5
Dairy: without whey	53,000	195	100
with whey	4,422	1,440	1,290
Malt beverage	45,700	260	103
Beet sugar	37,650	31	210
Wet corn milling	10,600	51.9	26.5
Paper	1,138,000	2,310	1,950

The potential impact of energy production from whey streams is estimated assuming that 50% of the whey is available for processing (the other 50% being used for other products), and for every pound of whey in the waste stream (1.123 lb BOD/lb whey) the net gain of energy in the process products (after accounting for process energy requirements and yields) is 3,500 Btu/lb whey. The available whey is 6.41×10^5 tons per year, which contains 9.1×10^{12} Btu. The net energy yield of converting this whey to fuels by the proposed process is 4.5×10^{12} Btu per year (6.6×10^{12} Btu/year contained in product fuels).

Application of the proposed process to other waste streams may also be possible. The process must be adapted to use much more dilute wastes and, in the case of paper wastes, to metabolize a different type of substrate. While these types of process modifications are possible, they have not been demonstrated and are not included in the projections made in this presentation.

Discussion

Process Background and Rationale

Fermentation processes are well-suited for conversion of organic materials associated with water. The process described produces a liquid fuel from waste material. The product is recovered simply by phase separation. Electricity is necessary for the final step of the process, but the required process energy (assuming 30% conversion efficiency for electricity) is only 32% of the energy present in process products. Each of the three process steps has been shown to be scientifically and technically feasible for dissolved sugar substrates at the bench scale (Levy et al. 1983; Levy et al. 1981; Sanderson et al. 1979b.)

Suppressed Methane Anaerobic Fermentation

The fermentation employed in the process may be described as a mixed-culture, methane-suppressed, anaerobic fermentation whose products are the organic acids acetic through caproic (hexanoic), carbon dioxide, and hydrogen (Sanderson et al. 1979b). It is the higher acids produced (butyric, valeric, and caproic) that may be selectively extracted from the fermentor and used to form the final product. The smaller organic acids (acetic and propionic) are more difficult to separate from the fermentor broth.

The fermentation is similar to methane fermentation used in sewage treatment plants. Inocula are obtained from sewage digester liquor and adapted to digestion of the substrates being tested. Methane inhibition was accomplished by using the specific inhibitor 2-bromoethane sulfonic acid (BES) (Sanderson et al. 1979a; Balch and Wolfe 1976). Tables 2 and 3 show organic acid production from glucose in the presence of BES in a fixed film type anaerobic fermentor. Hydrogen was also produced at 0.12 liter (STP)/g glucose feed.

On a commercial scale, use of BES would be prohibitively expensive. Other techniques for methane suppression could be employed. These include operating the digester at low pH (4–5), moderately low redox potential, and relatively short retention times All these conditions favor higher organic acid formation in preference to methane production.

Product Removal by Liquid-Liquid Extraction

As an alternative to energy intensive product recovery processes, such as distillation, liquid-liquid extraction is employed to recover the desired organic acids in a concentrated aqueous salt solution suitable for electrolytic oxidation to the alkane product. The protonated form of the organic acids is transferred from the fermentor liquid to the organic solvent (kerosene) in the extractor. Longer chain organic acids (butyric and higher) are selectively removed from the fermentor. Selectivity is based on the relative partition coefficients of the organic acids, as given in Table 4. In practice, negligible quantities of acetic and propionic acids are transferred by the kerosene phase. In a second extractor, the organic acids are transferred from the kerosene into an aqueous base, where the ionized form of the acids predominates and prevents transfer of the product back into the organic solvent. The organic acid salts may then be electrolytically oxidized to the alkane product, regenerating base in the process. Concentrations of organic acid salts exceeding 2.0 molar in the aqueous base have been achieved in our bench scale fermentor/extractor system (Levy et al. 1981). This statistic represents a greater than 10-fold increase in concentration of the higher acids than in the fermentor liquid. The pH difference between the fermentor (4–5) and aqueous base in the second extractor (8 or higher) provides the driving force for this separation. Base is regenerated in the electrolysis step.

TABLE 2. *Daily organic acid levels in methane-suppressed packed bed column digester*

Feed Solution: 20 g/liter glucose = 5×10^{-4} *M* BES
Retention Time: 1.8 days

Day	Concentration in MEQ/liter						
	Acetic	Propionic	Butyric	Valeric	Caproic	R[a]	%C[b]
4/19	8.7	0.8	7.0	1.3	20.9	117.5	35.3
4/20	11.7	1.1	7.4	1.1	29.1	152.1	45.6
4/21	12.6	1.1	7.4	1.1	29.1	153.0	45.9
4/22	12.1	0.8	10.9	1.5	34.0	181.6	54.5
4/23	16.7	0.0	14.0	0	51.3	256.8	77.0
4/24	16.7	t	9.6	0.8	52.8	254.5	76.4
4/25	15.8	0.8	8.0	1.0	43.3	213.5	64.1
4/26	16.2	0.8	9.7	0.8	43.3	217.7	65.3
4/27	15.8	0.5	10.7	0.7	46.1	230.2	69.1
4/28	15.5	t	10.7	0.6	44.2	221.21	66.4
4/29	14.7	t	11.7	0.6	46.5	231.9	69.6
4/30	18.3	0.5	11.6	0.8	41.6	217.4	65.2
5/1	16.3	0.5	12.8	1.0	42.5	222.4	66.7
5/2	17.9	0.8	16.2	1.0	47.6	253.6	76.0
5/3	17.5	0.8	18.2	1.4	48.7	263.8	79.1
5/4	18.8	0.8	14.2	0.8	38.2	211.0	63.3
5/5	13.7	t	13.3	0	35.1	187.3	56.2
5/6	17.6	t	20.9	0	32.2	198.7	59.6
5/7	11.3	1.3	23.4	2.1	35.6	221.4	66.4
5/8	— NO DATA; NOT FED —						
5/9	17.7	1.8	27.9	2.5	37.9	250.3	75.1
5/10	9.3	0.8	16.1	1.8	28.8	171.8	51.5
5/11	8.7	0.0	12.2	1.1	25.8	145.9	43.8
5/12	5.7	0.0	16.1	0.7	27.5	158.1	47.4
5/13	7.7	0.0	29.0	0.5	30.7	204.7	61.4
5/14	8.7	0.0	33.3	0.5	34.3	230.9	69.3

[a] R represents "acetic acid equivalents per liter" and is calculated from measured concentrations (meq/liter) of acetic (A), propionic (P), butyric (B), valeric (V), and caproic (C) acids:

$$R = A + 1.75P + 2.5B + 3.25V + 4.0C$$

[b] Percent conversion, %C, is calculated from R:

$$\%C = \frac{R \times .06}{S} \times 100$$

Where S is substrate concentration in g/liter.

Electrolytic Oxidation

The decarboxylation and dimerization of aliphatic organic acids to produce linear alkanes was first described by Kolbe (1849). The stoichiometry for the Kolbe reaction is

$$RCOOH + R'COOH \rightarrow R\text{-}R' + 2CO_2 + H_2$$

The reaction is predominant at smooth platinum anodes with acetic and the longer chain fatty acids (C_6 and up) (Weedon 1952). The main side-product is olefin:

$$RCOOH \rightarrow R'HC = CHR'' + CO_2 + H_2$$

Where R′ and R″ are either hydrogen or organic radicals.

Recent work at Dynatech has demonstrated conditions that result in high yields of alkane dimer (Sanderson et al. 1983; Levy et al. 1984). Operating the electrolysis cell at elevated reactor pressure or in the presence of 0.35 *M* or greater caproic acid results in high yields of alkane from all reacting organic species. The electrolysis run results shown in Table 5 give the product distribution from a mixture of butyric and caproic acids and is the basis for the yield projections made for the material balance and economic analysis, discussed later.

TABLE 3. *Suppressed methane fermentation of glucose in PBC reactor*

Retention Time:	1.8 d
Feed Concentration:	20 g/liter
Effluent pH:	5

Average Organic Acid Concentrations:

	MEQ/liter	Number of Points
Acetic	14.1 ± 4.1	19
Propionic	0.5 ± 0.5	19
Butyric	16.6 ± 7.1	19
Valeric	0.9 ± 0.6	19
Caproic	38.4 ± 7.2	19

Wt yield of 22.3% Caproic Acid
7.3% Butyric Acid
Maximum Yield Caproic Acid:

$$C_6H_{12}O_6 \rightarrow 0.75\ CH_3(CH_2)_4COOH + 1.5\ H_2O + 1.5\ CO_2$$

$$\frac{116 \times 0.75}{180} \times 100 = 48.3\%$$

TABLE 4. *Measured partition coefficients of organic acids between kerosene and water*

Organic Acid	Partition Coefficient $\frac{\text{(Acid) kerosene}}{\text{(Acid) water}}$
Acetic	~0
Propionic	<0.1
n-Butyric	0.2
n-Valeric	0.7
n-Caproic	2.9

Areas of Process Development

The suppressed-methane fermentation has been carried out in packed bed fermentators using 2% glucose feed solutions in defined media. Results of these experiments have shown the technical feasibility of the process and have been used to project yields in the engineering analysis. Based on these results we believe these are the areas of laboratory development still required before commercial implementation of the process will be considered:

- operate digesters on whey streams to obtain rate and yield data and define nutrient requirements;
- construct and operate complete laboratory scale system including fermentor,

TABLE 5. *Product distribution from electrolytic oxidation of 1.4* M *butyric/0.7* M *caproic acid mixture on platinum anode*

Product	mMoles Recovered	% of Acid Consumed	
		Butyric	Caproic
Propane	1.9	2.5	—
Propylene	8.0	10.7	—
Pentane	6.2	—	4.9
Pentene	17.2	—	13.7
Hexane	11.8	31.5	—
Octane	33.9	45.3	26.9
Decane	31.4	—	49.8
Propanol	2.2	2.9	—
Pentanol	0.4	—	0.3
Ester	5.4	7.0	4.4

liquid-liquid extractors, and electrolysis cell to obtain actual yield data and simulate commercial operating conditions; and

- define electrolytic cell operating conditions including pressure, temperature, current density, applied voltage, and product yields.

Characterization of Applicable Waste Streams

As noted earlier, a minimum concentration of biodegradable material is required in the waste stream to obtain sufficient higher acid yields in the process. The minimum concentration appears to be 2–2.5% by weight, limiting the resource to whey streams (~5% by weight lactose), poultry waste streams, and a limited number of streams in the sugar processing and corn processing industries. The estimate of available resources, discussed earlier, is based solely on whey streams; application of this technology to other streams requires a technological breakthrough that allows high yields of high weight acids from much more dilute waste streams.

Conventional Alternatives for Processing Whey Streams

Disposal of whey streams without recovery of energy products is accomplished most economically by land application, if this option is available. Fertilizer value (nitrogen and phosphorus content) may be realized through this type of disposal. Other options include vaporization of the liquid stream and recovery of whey for sale as a food product or aeration to accomplish BOD reduction. The food market for whey is saturated (based on recovery of half the whey produced), so vaporization becomes a very costly and energy intensive option. Aeration of concentrated waste streams is also costly and generally not practiced.

Methane fermentation of the whey can be considered as another option. The drawbacks to this approach are

- technical difficulties of methane fermentation of whey caused by presence of lactic acid-forming bacteria and subsequent souring of digester;
- production of relatively low-value product (biogas containing 60% methane); and
- high cost of upgrading biogas to pipeline quality methane.

The present process extracts an energy product from the whey stream that can be used onsite as a liquid fuel, transported to other locations for use, or sold. The presence of lactic acid-forming bacteria and potential digester souring is not a problem in suppressed-methane fermentations. If fertilizer value is realized from the stream through land application, this value is not diminished by anaerobic digestion. Sufficient BOD reduction is achieved in the process so that aeration of the stream prior to disposal becomes a viable option.

Process Flowsheet: Mass-Energy-Carbon Balances

It will be assumed that the suppressed-methane fermentation is carried out on a 5% (by weight) lactose stream and that conversion of 90% of lactose is accomplished and the product spectrum is extrapolated from experimental results as presented earlier in Tables 2 and 3. The basis for the flowsheet is a facility that processes 100,000 gal/d containing 41,650 lb/d of whey. With these assumptions the process mass flows are shown in Fig. 1. The weight yield of liquid products is 17.6% and gaseous products (excluding CO_2) is 6.2% of whey in the feed stream. The carbon and energy balances are indicated in Table 6. The energy content of liquid products is 47.3% and gaseous products, 25.1% of the energy contained in the whey in the feed stream. Table 7 shows a net gain (after accounting for process energy requirements) of 3,500 Btu/lb whey in the feed stream.

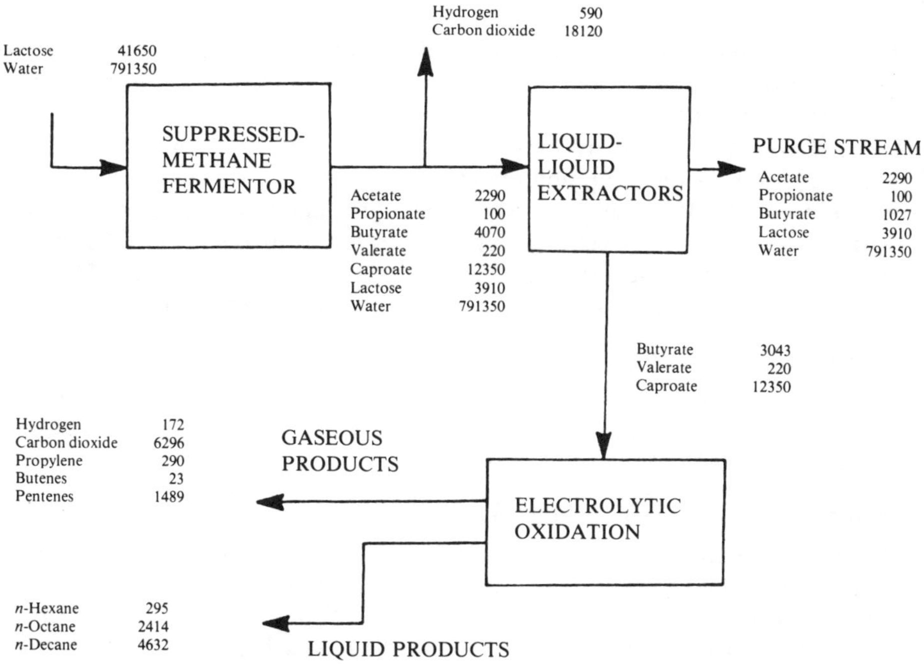

FIG. 1. Process flowsheet (lb/d basis).

Product Recovery and Quality

The liquid product is assumed to be recovered from the liquid-liquid extractor so

TABLE 6. *Carbon and energy balance*

Stream	Carbon Content (lb/d)	Energy Content (Btu/d)
Input	17,537	2.96×10^8
Liquid purge	3,171	5.41×10^7
Gaseous products	1,545	7.42×10^7
Carbon dioxide	6,659	0
Liquid products	6,194	1.40×10^8
Energy lost as heat of reaction		2.78×10^7

TABLE 7. *Process energy balance*

Process Energy Requirements	
Electrolysis	6,130 kwh/d
Pumping	370
	6,500
Fuel required at 10,500 Btu/Kwh	6.825×10^7 Btu/d
Energy content of recoverable products	2.14×10^8 Btu/d
Net energy gain	1.458×10^8 Btu/d
Net energy gain/lb whey feed	3,500 Btu

that residual organic acids are removed. This product, containing normal alkanes (decane, octane, hexane, and traces of heptane and nonane) is easily recovered after phase separation.

The gaseous products are recovered as mixtures. Hydrogen from the fermentor is recovered mixed with CO_2 and olefins, hydrogen and CO_2 are recovered as a mixture from the electrolysis cell. This product is suitable as a heating fuel. The heating value of gaseous electrolysis products as a mixture (including CO_2) is 470 Btu/scf. The hydrogen-containing digester gas has a heating value of 150 Btu/scf. No credit is taken for the digester product gas in the economic analysis.

Environmental Impacts

No adverse environmental impacts are anticipated from implementing this process or using the fuels produced by this process. Further, it must be pointed out that the liquid fuel produced will not contain sulfur.

Preliminary Design and Economic Assessment

The process design and capital cost estimates are based on extrapolation of experimental data from suppressed-methane fermentation of glucose and electrolytic oxidation of mixed organic acids and the process flowsheet presented earlier in Fig. 1. The whey stream considered in this analysis has a daily flow of 100,000 gal and contains 5% lactose by weight. Major capital items required are listed in Table 8. The total plant investment (TPI) including installation of equipment and supporting facilities is $685,700. The total capital requirement, which includes interest during construction (12% of TPI), start-up costs (8% of TPI), and working capital (one month's operating costs) is $847,200. Annual operating costs are listed in Table 9. In Table 10 is shown the cash flow on an annual basis. A 0.9

service factor is assumed. No credit is taken for the H_2 in the digester gas since its heating value is too low.

From these assumptions and calculations it was found that the project payback period, based on gross profit, is 4.0 yr; the payback period, based on after tax net profit, is 4.4 yr. If reduction in disposal costs associated with the whey stream or credit for the digester gas are appropriate, the payback period may be shortened.

The fuel products may be used in the dairy industry to provide process fuel or liquid fuel suitable for storage or as a diesel motor fuel. There are no apparent formal institutional barriers to implementing this process, though companies that are not in the business of marketing fuels may be reluctant to enter that market on such a small scale. It appears that the best use for the fuel produced, therefore, is to displace fuels presently used by the dairy industry. The potential net energy gain, as shown in Table 7, is 3,500 Btu/lb of whey in the waste stream.

TABLE 8. *Capital Items; 100,000 gal/d, 5% whey*

Item	Unit Size	Number Units	Unit Cost	Total
Fermentor	80,000 ft^3	1	280,000	$280,000
HC/AQ mixers	1,150 gal	4	6,100	24,400
HC/AQ settler	6,900 gal	4	8,800	35,200
HC/base mixer	1,110 gal	1	6,000	6,000
HC/base settler	5,580 gal	1	7,800	7,800
Electrolysis cell	370 ft^2	1	37,000	37,000
DC Transformer/ rectifier	260 kw	1	36,400	36,400
Liquid product storage	1,500 gal	2	4,500	9,000
Gas product storage	1,000 gal	1	10,000	10,000
Pumps	16.5 BHP	—	—	7,500
Total Capital Items				$453,300

TABLE 9. *Annual operating costs*

Materials	
Buffer	$ 10,000
Miscellaneous chemical supplies	5,000
Utilities	
Electric (5¢/kwh)	106,800
Labor	
Operating ($7/h)	61,000
Maintenance (1.5% TPI)	10,300
Supervision (15% op. & maint.)	10,700
Administration and overhead (60% labor)	49,200
Supplies	
Operating (15% operating labor)	9,200
Maintenance (1.5% TPI)	10,300
Local taxes and insurance (2.7% TPI)	18,500
Gross annual operating cost	$291,000

TABLE 10. *Balance sheet—annual basis*

Total Capital Requirement	\$847,200
Pounds liquid alkane/yr[a]	2.38×10^6 lb
Market value of liquid fuel	\$0.19/lb (\$1.25/gal)
MMBtu olefin and H_2 recovered	14,380
Market value/MMBtu	\$3.50
Annual gross income	\$502,530
Annual costs	\$291,000
Gross profit	\$211,530
Net profit[b]	\$192,170

[a] Assume 1% of liquid hydrocarbon product used as makeup solvent in liquid-liquid extractors.
[b] After tax (46%) with depreciation credit for 20% of Total Capital Requirement.

Literature Cited

American Paper Institute. 1980. *Raw materials and energy division data.* New York.

Balch, W. E., and R. S. Wolfe. 1976. New approach to the cultivation of methanogenic bacteria: 2-mercaptoethanesulfonic acid (HS-CoM)-dependent growth of *Methanobacterium ruminantium* in a pressurized atmosphere. *Appl. Environ. Microbiol.* 32:781–791.

EPA. 1974. Development document for effluent guidelines and new source performance standards for the unbleached kraft and semichemical pulp segment of the pulp, paper, and paperboard mills point source category. Washington, DC.

______. 1979. Overview of the environmental control measures and problems in the food processing industry, EPA-600/2/9-009. Cincinnati, OH.

Kolbe, H. 1849. Untersuchungen ueber die Elektrolyse organischer Verbindungen. *Ann. Chem.* 69:257–294.

Levy, P. F., J. E. Sanderson, and L. K. Cheng. 1984. Kolbe electrolysis of mixtures of aliphatic organic acids. *J. Electrochem. Soc.* 131:773–777.

Levy, P. F., A. P. Leuschner, and J. H. Stoddart, Jr. 1983. Biorefining of selected industrial wastes to liquid fuels and organic chemicals. *Dynatech Report No. 2238,* for DOE Contract No. DE–AC02 82CE40556. Cambridge, MA.

Levy, P. F., J. E. Sanderson, E. Ashare, S. R. DeRiel, and D. L. Wise. 1981. Liquid fuels production from biomass. *Dynatech Report No. 2147,* prepared for SERI Contract No. XB–0–9291. Cambridge, MA.

Mueller Associates, Inc. 1979. Whey and dry milk products as feedstocks for ethanol production. *Report DOE/CS/2098-03,* for U.S. Department of Energy, Office of Transportation Programs. Washington, DC.

National Food Processors' Association. 1979. Canned food pack statistics, 1978–1979. Washington, DC.

Sanderson, J. E., P. F. Levy, L. K. Cheng, and G. W. Barnard. 1983. The effect of pressure on the product distribution in Kolbe electrolysis. *J. Electrochem. Soc.* 130:1844–1848.

Sanderson, J. E., D. V. Garcia-Martinez, J. J. Dillon, G. S. George, and D. L. Wise. 1979a. Liquid fuel production from biomass. *Intl. Solar Energy Soc. Meet.* Atlanta, GA.

Sanderson, J. E., D. V. Garcia-Martinez, G. S. George, J. J. Dillon, M. S. Molyneux, G. W. Barnard, and D. L. Wise. 1979b. Liquid fuels production from biomass. *Dynatech Report No. 1931,* for DOE Contract No. EG-77-C-02-4388. Cambridge, MA.

Weedon, B. C. L. 1952. Anodic syntheses with carboxylic acids. *Q. Rev. Biophys.* (London) 6:380–398.

CHAPTER 12

Organic and Inorganic Waste Treatment and Simultaneous Photoproduction of Hydrogen by Immobilized Photosynthetic Bacteria

A. MITSUI, T. MATSUNAGA, H. IKEMOTO, AND B. R. RENUKA

School of Marine and Atmospheric Science, University of Miami, 4600 Rickenbacker Causeway, Miami, Florida 33149

Photosynthetic bacterial strains isolated from tropical and subtropical marine environments that have high growth and hydrogen production were selected and immobilized in agar. Immobilization significantly stabilized H_2 production by preventing the inhibitory effect from O_2. Immobilized *Chromatium* sp. removed sulfide from medium effectively and simultaneously produced H_2. Immobilized *Rhodopseudomonas* sp., which had wide organic substrate specificity, were also used to treat waste water from an orange-processing plant. Removal of total organic carbon (TOC), near-complete reduction of BOD, and simultaneous H_2 photoproduction were also demonstrated. Continuous H_2 photoproduction was carried out for several weeks with periodic addition of waste in indoor artificial light. These experiments were successfully applied in outdoor natural sunlight conditions. These data clearly demonstrated that immobilized selected strains of photosynthetic bacteria can be effectively used simultaneously for waste treatment and clean energy (H_2) production.

INTRODUCTION

Photosynthetic bacteria are capable of using light to grow, fix nitrogen, and produce hydrogen gas (Gest and Kamen 1949). The process of photosynthesis in this group is a one-photosystem reaction requiring an electron donor more reduced than water. There are two major types of photosynthetic bacteria, nonsulfur and sulfur strains. Photosynthetic nonsulfur bacteria (Rhodospirillaceae) have the capability to use a wide range of organic substances, such as carbohydrates, organic acids, and fatty acids for growth and hydrogen production, although the substrate specificity depends on the strain (Mitsui et al. 1980; Kumazawa and Mitsui 1982). Many food processing, brewery, chemical, plastic, and pulp industries produce these waste substances, and they should be removed before the waste is released back into the environment.

Photosynthetic sulfur bacteria (Chromatiaceae and Chlorobiaceae) have the unique ability to use sulfide as an electron donor for growth and hydrogen production. Sulfide is one of the major pollutants in the environment and can be found in sewage, industrial and agricultural wastes, polluted rivers, marine water, and sediments. Coal gasification process plants produce large amounts of sulfide also. Therefore, hydrogen production systems using photosynthetic bacteria with removal of organic waste and sulfide (Fig. 1) could have significant benefits (Mitsui 1980; Mitsui et al. 1980; Mitsui et al. 1983; Mitsui et al. 1984).

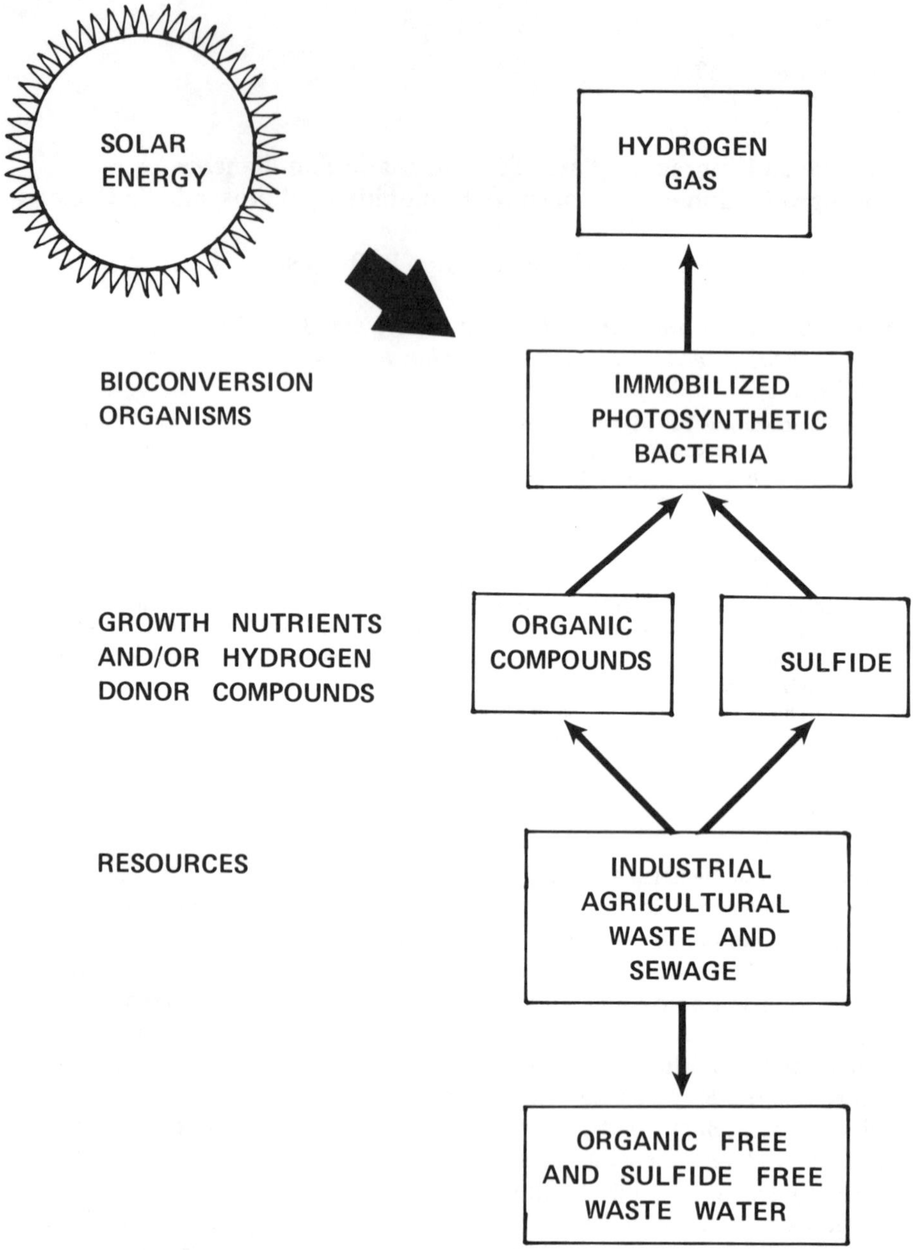

FIG. 1. General scheme of hydrogen photoproduction and waste treatment by immobilized photosynthetic bacteria.

In this laboratory, numerous strains of photosynthetic bacteria were isolated from tropical and subtropical environments including marine water, marine sediments, salt marsh, salt production fields, living and decaying sea grasses, mangrove leaves, seaweeds, etc. (Mitsui 1975, 1976a,b, 1979; Mitsui et al. 1980; Mitsui et al. 1984). Many of these strains exhibited high growth and hydrogen photoproduction rates from these substances (Mitsui 1976a,b; Mitsui 1981; Mitsui et al. 1980; Mitsui et al. 1983; Mitsui et al. 1984; Matsunaga and Mitsui 1982; Ohta and Mitsui 1981; Ohta et al. 1981). Aqueous suspensions of photosynthetic bacteria could remove these substances and produce hydrogen; however, it is difficult to separate the photosynthetic bacteria after use. Immobilized cells can be easily separated from waste water. Therefore, immobilized cells are more suitable for waste water treatment. Immobilization also has the advantages of obtaining a dense photosynthetic bacterial culture inside a gel matrix and stabilizing the hydrogen production (Matsunaga and Mitsui 1982; Mitsui et al. 1983; Ikemoto and Mitsui 1984; Mitsui et al. 1984.)

In this paper selection of sulfur and nonsulfur photosynthetic bacterial strains for hydrogen photoproduction is described, and, using two of these selected marine photosynthetic bacterial strains from the laboratory's collection, prototype systems for removal of organic substances and sulfide and simultaneous clean energy (hydrogen) production are described.

Materials and Methods

Marine photosynthetic bacteria were obtained from the culture collection of this laboratory (Mitsui et al. 1980). Marine photosynthetic bacteria were cultured under conditions previously described for nonsulfur strains (Matsunaga and Mitsui 1982) and for sulfur strains (Ohta and Mitsui 1981). Immobilization in agar was performed for nonsulfur photosynthetic bacteria (Matsunaga and Mitsui 1982) and for sulfur photosynthetic bacteria (Ikemoto and Mitsui 1984).

Hydrogen was measured by gas chromatography by the method described previously (Kumazawa and Mitsui 1981). Cellular bacterial chlorophyll (BChl) was measured after immobilized cells were crushed and ground in a glass mortar and extracted with acetone-methanol (7:2, v/v). Sulfide in the medium was estimated by Ag^{+}/S^{2-} solid state electrode (Orion model 94–16, Orion Res., Inc.) using a digital pH meter and/or the colorimetric method (ALPHA 1975). Uptake hydrogenase activity was measured (Kumazawa et al. 1983), and light intensity was measured (Phlips and Mitsui 1983). The orange-processing waste water was obtained from Plymouth Citrus Products (Plymouth, FL).

Results

Hydrogen Photoproduction by Various Marine Nonsulfur Photosynthetic Bacteria

The marine photosynthetic bacteria were isolated from samples collected at Atlantic tropical and subtropical marine environments (Mitsui et al. 1980). More than 400 strains have been isolated. Among these strains of marine photosynthetic bacteria, 20 strains were chosen for comparative studies of hydrogen photoproduction because of high growth rate and biomass yield. Table 1 shows the hydrogen production rate of these photosynthetic bacteria from malate. Miami PBE 2271 (identified as *Rhodopseudomonas* sp.) showed the highest hydrogen production rate (8 μmol/mg protein/h).

TABLE 1. *Hydrogen production from malate by various marine nonsulfur photosynthetic bacteria*

	Strain	Hydrogen Production μmol/mg protein/h
Miami	PBE 115	1.6
	PBE 136	2.5
	PBE 217	6.4
	PBE 2271	8.0
	PBE 228	0.0
	PBE 230	0.0
	PBE 242	1.4
	PBE 245	0.0
	PBE 258	0.0
	PBE 2591	0.0
	PBE 2592	0.1
	PBE 266	6.9
	PBE 285	1.4
	PBE 301	0.3
	PBE 320	5.4
	PBE 336	1.1
	PBE 338	0.7
	PBE 361	4.0
	PBE 370	0.0
	PBE 380	5.4

The optimum temperature for hydrogen production by PBE 2271 was found to be 35 C. The optimum pH was 7.6, which is similar to that of seawater. Hydrogen production was observed over a wide salinity range, with an optimum salinity of 25–30 °/oo, which again is similar to seawater. The hydrogen production rate increased with increasing light intensity and it was saturated at 100 μEinsteins/m²/s. There was no inhibition at 300 μEinsteins/m²/s, the highest light intensity tested. These results showed that the marine photosynthetic bacterium *Rhodopseudomonas* sp. Miami PBE 2271 is well suited for hydrogen production based on seawater and solar energy.

Hydrogen Photoproduction by Immobilized Nonsulfur Photosynthetic Bacteria

Marine photosynthetic bacterium Miami PBE 2271 was immobilized on agarose coated polyester film with 2% agar. The nitrogenase of the immobilized photosynthetic bacteria was protected from the inhibitory effect of oxygen observed with native cells. Immobilized cells retained 95% of initial activity even when they were exposed to air for 5 h. Hydrogen production was carried out in the seawater-based reactor using immobilized marine photosynthetic bacteria. The immobilized marine photosynthetic bacteria continuously produced hydrogen at rates of 3.75 μmol/g gel/h (180 μmol/mg BChl/h, 3.96 μmol/mg protein/h, or 1.98 μmol/mg dry wt/h) for more than one month. Fig. 2 indicates the results over a 20 d period.

Substrate Specificity for Hydrogen Production by Immobilized Nonsulfur Photosynthetic Bacteria

A wide variety of substrates can be utilized for hydrogen production by immobilized nonsulfur photosynthetic bacterium *Rhodopseudomonas* sp. Miami

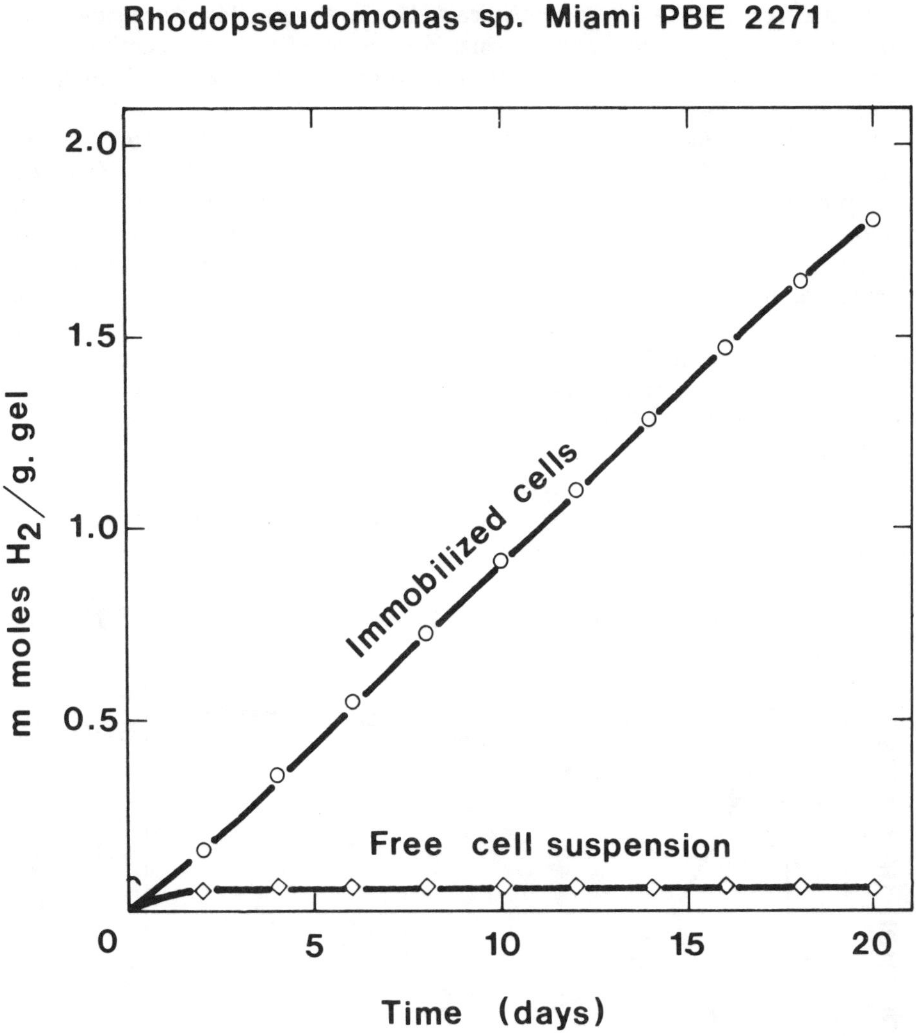

FIG. 2. Continuous hydrogen photoproduction from malate by immobilized nonsulfur photosynthetic bacteria, *Rhodopseudomonas* sp. Miami PBE 2271.

PBE 2271. Immobilized Miami PBE 2271 produced hydrogen at high rates (more than 3 μmol/mg protein/h) from fumarate, lactate, malate, pyruvate, succinate, acetate, dextrin, glucose, fructose, maltose, and mannitol. β-Glycerophosphate, butyrate, arabinose, xylose, pectin, and cellobiose were also used for hydrogen production. Therefore, the marine photosynthetic bacterium Miami PBE 2271 would be suitable for hydrogen production from various kinds of organic wastes.

Hydrogen Production from Orange-Processing Waste Water by Immobilized Nonsulfur Photosynthetic Bacteria

Orange processing is Florida's number one agricultural industry and the industry uses in excess of 90% of the total orange crop. In the process of orange juice production, various byproducts and wastes are produced. Some of them are utilized as animal feed, molasses, and chemicals. However, uses for processing waste water and evaporator condensate, which are produced in large quantities, have not yet been established. Presently they are transported to fields and spread on the ground.

Hydrogen production rates from various orange-processing byproducts and wastes were tested with immobilized marine photosynthetic bacteria. Hydrogen was produced from animal feed, citrus molasses, processing waste water and evaporator condensate. The highest hydrogen production rate was observed with processing waste water (133 μmol/mg BChl/h). Hydrogen production was increased resulting in the decrease of Total Organic Carbon (TOC). TOC in the processing water was reduced from 2025 ppm to 794 ppm with simultaneous production of hydrogen. Fig. 3 shows the relationship between hydrogen production and Biological Oxygen Demand (BOD). Hydrogen production was associated with decreasing BOD of processing waste water. BOD was decreased from 700 ppm to 50 ppm. These results indicate that processing waste water contains biologically stable compounds (e.g., lignin and cellulose) and that most biologically active substances were removed with the coupling of hydrogen photoproduction.

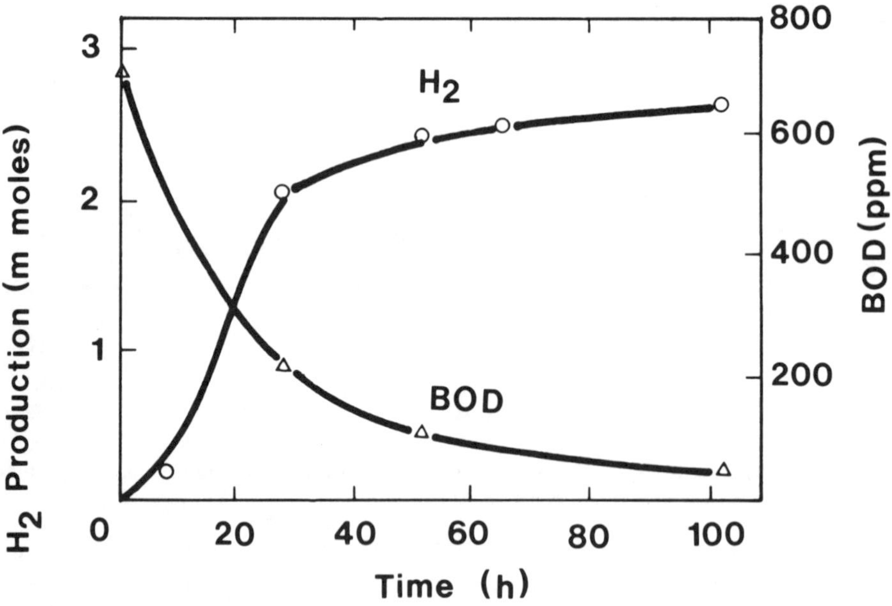

FIG. 3. Relationship between hydrogen photoproduction and BOD decrease from orange-processing waste water by immobilized nonsulfur photosynthetic bacteria. Immobilized cells of *Rhodopseudomonas* sp. Miami PBE 2271 (equivalent to 0.3 g wet wt cells were incubated at 37 C in orange-processing waste water (pH 7.6) at the incident light intensity of 150 μEinstein/m²/s.

Outdoor Hydrogen Production by Immobilized Nonsulfur Photosynthetic Bacteria from Orange-Processing Waste Water

Hydrogen production from orange-processing waste water was also carried out in an outdoor setting. Fig. 4A shows a schematic diagram of the biosolar reactor using immobilized photosynthetic bacteria. Processing waste water was diluted with seawater to give a concentration of 430 ppm total organic carbon (TOC), and 1 liter of the diluted waste water was added to the reactor. The pH of the waste water was adjusted to 7.6. The temperature was not controlled. Fig. 4B shows the time course of hydrogen production, light intensity, and temperature during the outdoor experiment. Hydrogen was continuously produced from 9 a.m. to 4 p.m. Maximum hydrogen production rates were observed between 11 a.m. and 1 p.m.

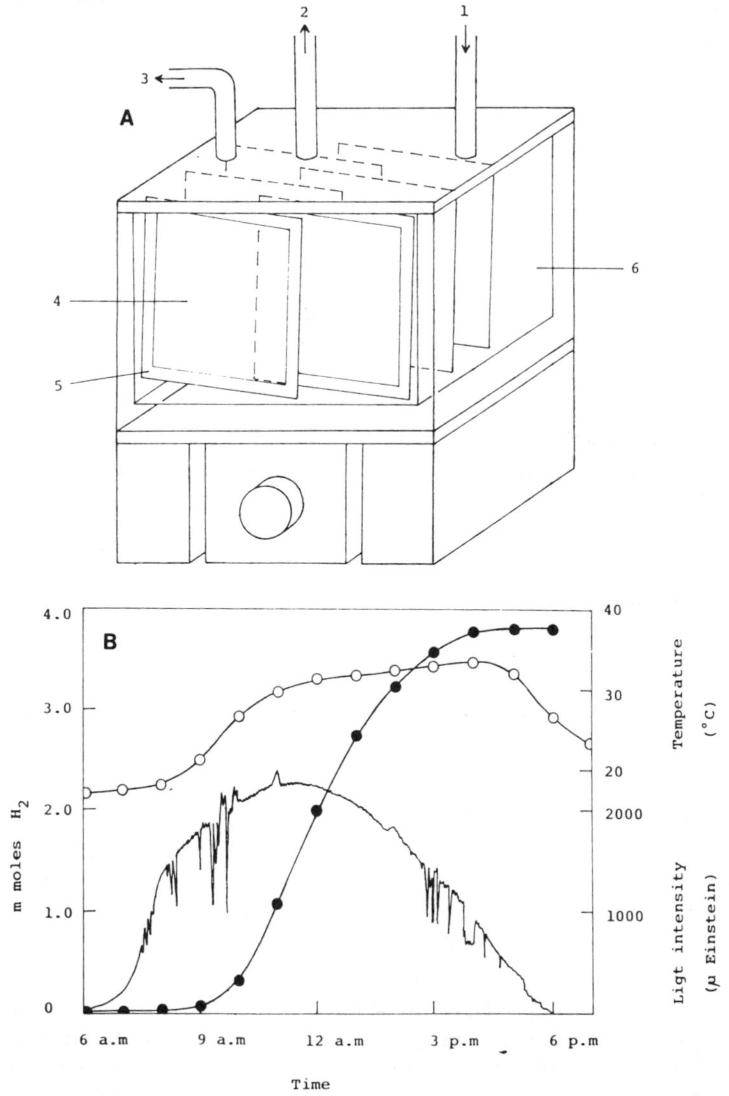

FIG. 4. A. Schematic drawing of biosolar reactor used for outdoor hydrogen photoproduction by immobilized nonsulfur photosynthetic bacteria (500 g gel, 11 mg BChl): 1. Inlet of substrate (orange-processing waste water), 2. Outlet of hydrogen, 3. Outlet of waste water, 4. Immobilized photosynthetic bacteria, 5. Transparent polyester film, and 6. Seawater containing orange-processing waste water. B. Hydrogen production from orange-processing waste water by nonsulfur photosynthetic bacteria in the biosolar reactor under natural sunlight conditions.

when light intensity was very high. Hydrogen was not produced in the dark. The TOC of the waste water was reduced to 270 ppm. The BOD of the waste water was decreased to 40 ppm. These experiments could be repeated for several days when new orange-processing waste water was introduced into the reactor without loss of activity of the immobilized photosynthetic bacteria.

Hydrogen Photoproduction by Sulfur Photosynthetic Bacteria

Among the 225 strains of marine sulfur photosynthetic bacteria tested in this laboratory's collection, 60 strains exhibited high growth rates and biomass yields in culture medium containing sulfide and malate. Fifty-three strains with high hydrogen production ability were tested for growth and hydrogen photoproduction at sulfide concentrations ranging from 2.5 to 20 m*M*. Most strains showed optimum growth at 2.5 to 7.5 m*M* of sulfide, but some strains exhibited high optimum ranges of 10 to 20 m*M* sulfide. Optimum sulfide concentrations for hydrogen production varied widely.

The fact that photosynthetic bacterial strains have different preferences for sulfide concentrations and tolerance to high sulfide concentrations for growth and hydrogen production indicates that proper strains could be chosen for application according to sulfide abundance in industrial wastes and polluted environments.

Most strains tested exhibited uptake hydrogenase activity, which diminishes the accumulation of hydrogen in closed vessels. However, several strains, which apparently lack this activity, showed stable and high rates of hydrogen photoproduction. The fact that many of these strains take up sulfide at high rates and use it for both growth and hydrogen production makes them prime candidates for use in treatment of sulfide wastes.

Fig. 5 shows sulfide removal from medium and simultaneous hydrogen photoproduction by marine *Chromatium* sp. Miami PBS 1071. The sulfide added is completely removed from the medium during the hydrogen production process.

Hydrogen Photoproduction from Sulfide by Immobilized Sulfur Photosynthetic Bacteria

The marine photosynthetic sulfur bacterium, *Chromatium* sp. Miami PBS 1071 was tested in immobilization. Whole cells of this strain were immobilized in an agar gel matrix and grown on either molecular nitrogen or ammonia as the nitrogen source. Immobilized cells grew in the agar gel matrix but not in the surrounding suspension media. High biomass yield was obtained. Hydrogen photoproduction by immobilized cells was carried out by periodic addition of supplemental sulfide in the reaction system. Sulfide was the sole electron donor used for the reaction. CO_2 was not produced and only H_2 was evolved in the system. Hydrogen gas in this immobilized system was continuously produced for several weeks at a constant rate. This is in contrast to free cell systems where the hydrogen production dropped within a few days. The rates of hydrogen photoproduction, on a reaction volume basis, were improved by growing cells in an immobilized system. Immobilized ammonia-grown cells produce hydrogen at the same rates as N_2-grown cells, but a short lag period was observed before initiation of hydrogen production. Rates of hydrogen production up to 6 μmol/g gel/h were observed.

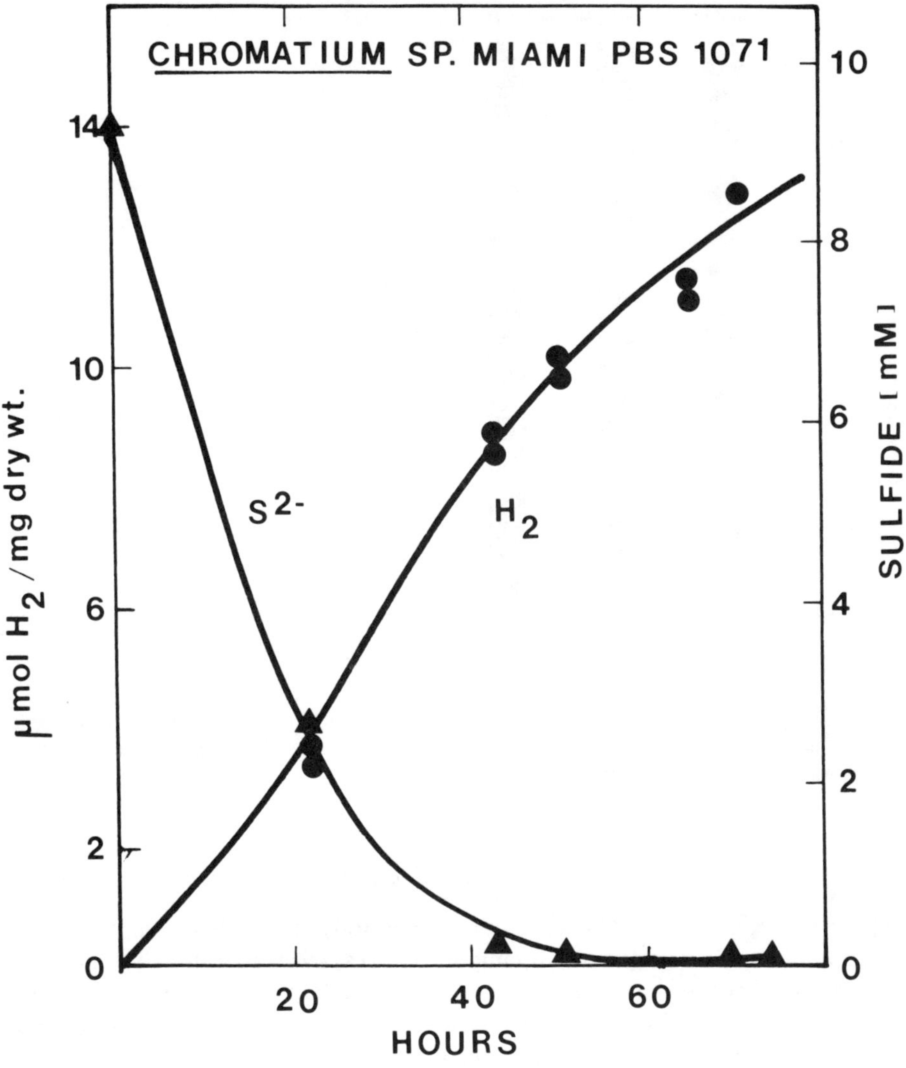

FIG. 5. Removal of sulfide and simultaneous hydrogen photoproduction by marine sulfur photosynthetic bacterium, *Chromatium* sp. Miami PBS 1071; 16 h-old N_2-grown cells were used at pH 7.6 and 30 C.

Hydrogen Photoproduction from Sulfide by Immobilized Sulfur Photosynthetic Bacteria in Outdoor Natural Sunlight Conditions

Fig. 6 shows the hydrogen photoproduction from sulfide in outdoor natural sunlight conditions. Hydrogen was produced during the daytime period with the highest rates of hydrogen photoproduction being observed during the high midday sunlight intensity period (i.e., 2,000 μEinstein/m^2/s). At night there was no hydrogen production. Hydrogen production continued for 4 d, the length of the testing period. Rates of hydrogen production were 0.7–1.5 μmol/g gel/h.

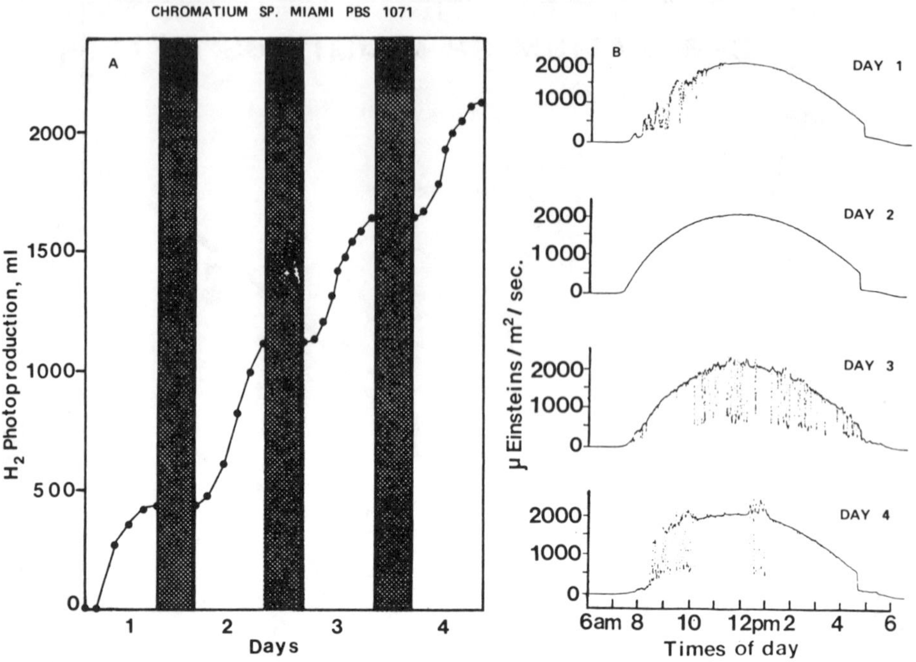

FIG. 6. Hydrogen photoproduction by immobilized sulfur photosynthetic bacteria *Chromatium* sp. Miami PBS 1071 in outdoor natural sunlight conditions. Cells were grown in agar matrix in N_2. N_2 was then changed to Argon gas for H_2 photoproduction.

Discussion

Photosynthetic nonsulfur bacteria have a thermodynamic advantage, in the production of usable energy from organic substances, over other nonphotosynthetic bacteria in that the organic substances are completely decomposed to hydrogen and CO_2 (Gest et al. 1962; Gest 1963; Thauer 1977; Mitsui et al. 1980). In general, nonsulfur photosynthetic bacteria can utilize a wide variety of substrates for growth and hydrogen production (Mitsui et al. 1980; Kumazawa and Mitsui 1982; Mitsui et al. 1983). Substrate specificity differs from strain to strain. Thus the choice of the strain(s) is one of the most important factors in successful waste treatment and hydrogen photoproduction since organic substrates and their concentrations differ in the different wastes that are to be treated. A survey of photosynthetic bacterial strains of differing substrate specificity offers the potential for efficient removal of particular organic substances from particular wastes. Our survey of photosynthetic bacteria from the marine environment, which was for different substrate specificities, may open up opportunities for such future uses. Strains such as *Rhodopseudomonas* sp. Miami PBE 2271, which have wide substrate specificity, may have advantages in that they can be used for a wide variety of wastes.

Sulfur photosynthetic bacteria have a narrower organic substrate specificity (Mitsui et al. 1980; Kumazawa and Mitsui 1982) but are unique in that they efficiently utilize sulfide for hydrogen production. Many strains isolated from the marine environment have high hydrogen photoproduction activity from sulfide, but optimum ranges of sulfide concentration differ from strain to strain (Mitsui et al. 1984). These strains may also have the potential to be used in a variety of wastes and environments.

Although hydrogen photoproduction by marine photosynthetic bacteria appears promising at the moment, there are many problems to be surmounted before practical application is achieved. One of the difficult problems in using photosynthetic bacteria for hydrogen production is the extreme oxygen sensitivity of the hydrogen-producing system, especially with regard to nitrogenase. A small amount of oxygen contamination to the operation system may cause no hydrogen production at all.

Recently, immobilization techniques for enzymes and bacteria have been developed for industrial applications. The immobilization method offers the considerable advantage of continuous use of enzymes and bacteria (reviewed by Zaborsky 1974; Chibata and Tosa 1977; Fukui and Tanaka 1982). Stabilization of hydrogenase and nitrogenase has been successfully achieved by immobilizing living nonphotosynthetic bacteria, *Clostridium butyricum* in polyacrylamide (Karube et al. 1976) and agar (Matsunaga et al. 1980). The hydrogenase and/or nitrogenase of the immobilized bacteria was protected from the deleterious effects of oxygen because the diffusion of oxygen was limited by the gel matrix. Using marine nonsulfur photosynthetic bacterium, *Rhodopseudomonas* sp. Miami PBE 2271, Matsunaga and Mitsui (1982) demonstrated that immobilization dramatically prevented inhibitory effect of oxygen on hydrogen photoproduction. In free cell suspensions 50% inhibition occurred at approximately 0.5% oxygen, whereas with immobilized cells this level of inhibition did not occur until approximately 5% oxygen.

Free cell suspensions of the strain PBE 2271 have some protection mechanism for oxygen that may be attributed to the respiratory activity of the strain. PBE 2271 further established an anaerobic environment around the cells when they were immobilized in the gel matrices, thus protecting them from the deleterious effects of oxygen. A similar effect was observed for inhibition from molecular nitrogen.

In addition, it was shown that immobilized marine photosynthetic bacteria can be used over wide ranges of salinity (Matsunaga and Mitsui 1982). At 10% NaCl, 50% of hydrogen production activity of the maximum rate was retained. Even with freshwater, 93% of activity was retained whereas cell suspensions showed only 52% of the maximum activity. Thus immobilization also protected cells from osmotic stress.

Thus, as shown in this study, hydrogen production by immobilized photosynthetic bacteria was maintained for a prolonged period of time (30 d) with rates of 4,030 ml/g BChl/h (or 99 ml/mg protein/h, or 44.5 ml/g dry wt/h) with periodic addition of the organic substance, malate.

Similar effects of immobilization were also observed in sulfur photosynthetic bacterium *Chromatium* sp. Miami PBS 1071. With periodic addition of sulfide, hydrogen production continued for over 2 wk without any reduction of the production rate and with removal of the sulfide waste (see also Ikemoto and Mitsui 1984).

In addition to the stabilization effect, these results also indicate that immobilized systems using photosynthetic bacteria have the following advantages over the free cell system: (1) cells can grow to high biomass yield in immobilized gel, thus, all harvesting procedures (e.g., centrifugation) are eliminated and hydrogen production can be carried out in the same reactor; and (2) continuous hydrogen production can be carried out without the loss of cells.

Using free living cells, treatment of waste water rich in organic substances or sulfide with photosynthetic bacterial growth has been studied by Kobyashi (1977), but this was not coupled with hydrogen production. Hydrogen photoproduction by free living cells of *Rhodospirillium rubrum* and *Rhodopseudomonas capsulata* from whey, yogurt, or other lactic acid-containing wastes (Zurrer and Bachofen 1981; Jouanneau et al. 1982) have been studied. Using immobilized cells of *Rhodopseudomonas palustris,* hydrogen production from sugar refinery wastes and raw paper mill effluent (Vincenzini et al. 1981) have been studied. We have successfully applied the immobilized photosynthetic bacteria for orange-processing water (or sulfide) treatments and hydrogen production not only in indoor conditions but also in outdoor natural sunlight conditions.

TOC and BOD were significantly reduced from orange-processing waste water, and simultaneously stable and prolonged hydrogen production was associated. Efficient sulfide removal was also coupled to the stable hydrogen photoproduction.

Although the results described here were preliminary and not yet optimized for the system, they are a good indication that the system is applicable for industrial waste treatment and simultaneous clean energy (hydrogen) production. Once the study of the optimization of the small-scale system is completed, a cost analysis should be made to see if a successful system can be established.

Acknowledgments

This work has been supported by the National Science Foundation, The Solar Energy Research Institute Department of Energy, and the National Aeronautics and Space Administration. Any opinions, findings, and conclusions or recommendations expressed in this publication are those of the authors and do not necessarily reflect the view of the granting agencies. The authors thank Mr. R. Cook, Plymouth Citrus Products, Inc., Florida, for providing information and samples of orange-processing products.

Literature Cited

APHA. 1975. Standard Methods, 14th Edition. American Public Health Association, Washington, DC.

Chibata, I., and T. Tosa. 1977. Transformation of organic compounds by immobilized microbial cells. *Adv. Appl. Microbiol.* 21:1–27.

Fukui, S., and A. Tanaka. 1982. Immobilized microbial cells. *Annu. Rev. Microbiol.* 36:145–172.

Gest, H. 1963. Metabolic aspects of bacterial photosynthesis. Pages 129–150 *in* H. Gest, A. San Pietro, and L. P. Vernon, eds., *Bacterial Photosynthesis*. The Antioch Press, Yellow Springs, OH.

Gest, H., and M. D. Kamen. 1949. Photoproduction of molecular hydrogen by *Rhodospirillium rubrum. Science* 109:558–559.

Gest, H., J. G. Ormerod, and K. S. Ormerod. 1962. Photometabolism of *Rhodospirillium rubrum:* Light dependent dissimilation of organic compounds to carbon dioxide and molecular hydrogen by an anaerobic citric acid cycle. *Arch. Biochem.* 97:21–23.

Ikemoto, H., and A. Mitsui. 1984. Continuous hydrogen photoproduction from sulfide by an immobilized marine photosynthetic bacterium, *Chromatium* sp. Miami PBS 1071. Pages 779–782 *in* C. Sybesma Martinus Nijhoff, ed., *Advances in Photosynthesis Research,* Vol. II. Dr. W. Junk Publishers, The Hague, the Netherlands.

Jouanneau, Y., J. W. Willison, A. Colbeau, P. C. Hallenbeck, C. Riolacci, and P. M. Vignais. 1982. Pages 174–179 *in* D. O. Hall and W. Palz, eds., *Photochemical, Photoelectrochemical and Photobiological Processes,* Series D, Vol. I. D. Reidel Publishing Co., Dordrecht, Holland.

Karube, I., T. Matsunaga, S. Tsuru, and S. Suzuki. 1976. Continuous hydrogen production by immobilized cells of *Clostridium butyricum. Biochem. Biophys. Acta* 444:338–343.

Kobayashi, M. 1977. Utilization and disposal of wastes by photosynthetic bacteria. Pages 443–453 *in* H. G. Schlegel and I. Barner, eds., *Microbial Energy Conversion*. Pergamon Press, Oxford and New York.

Kumazawa, S., and A. Mitsui. 1981. Characterization and optimization of hydrogen photoproduction by a salt water blue-green alga, *Oscillatoria* sp. Miami BG7. I. Enhancement through limiting the supply of nitrogen nutrients. *Intl. J. Hydr. Energy* 6:339–348.

______. 1982. Hydrogen metabolism of photosynthetic bacteria and algae. Pages 299–316 *in* A. Mitsui and C. C. Black, eds., *CRC Handbook of Biosolar Resources,* Vol. 1, *Basic Principles.* CRC Press, Boca Raton, FL.

Kumazawa, S., S. Izawa, and A. Mitsui. 1983. Proton efflux coupled in dark H_2 oxidation in whole cells of a marine sulfur photosynthetic bacterium (*Chromatium* sp. strain Miami PBS 1071). *J. Bacteriol.* 154:185–191.

Matsunaga, T., and A. Mitsui. 1982. Sea water based hydrogen production by immobilized marine photosynthetic bacteria. *Biotechnol. Bioeng.* 12:441–450.

Matsunaga, T., I. Kerube, and S. Suzuki. 1980. Some observations on immobilized hydrogen-producing bacteria: Behavior of hydrogen in gel membrane. *Biotechnol. Bioeng.* 22:2607–2615.

Mitsui, A. 1975. Photoproduction of hydrogen via microbial and biochemical processes. Pages 31–48 *in* T. N. Veziroglu, ed., *Symposium Proceedings of Hydrogen Energy Fundamentals.* University of Miami Press, Miami.

______. 1976a. A survey of hydrogen producing photosynthetic organisms in tropical and sub-tropical marine environments. *NSF Annual Report,* pp. 1–68.

______. 1976b. Bioconversion of solar energy in salt water photosynthetic hydrogen production system. Pages 77–99 *in* T. N. Veziroglu, ed., *Proc. of First World Hydr. Energy Conf.: Progress in World Hydrogen Energy Projects and Planning,* Vol. 2, 41B. University of Miami Press, Miami, FL.

______. 1979. Biological and biochemical hydrogen production. Pages 171–191 *in* T. Ohta, ed., *Solar Hydrogen Energy Systems.* Pergamon Press, Oxford and New York.

______. 1980. Saltwater-based biological solar energy conversion for fuel, chemicals, fertilizer, food and feed. Pages 486–491 *in Proceedings of Bio-Energy '80. World Congress and Exposition.* Bio-energy Council, Washington, DC.

______. 1981. Progress report: A study of hydrogen production by tropical marine photosynthetic bacteria for applied systems. (Nov. 1, 1979–Oct. 30, 1980) Pages 1–19 *in Proc. of the Rev. Meeting on Solar Hydr. Prod. Progr.* SERI/SP-624-1095.

Mitsui, A., E. J. Phlips, S. Kumazawa, K. J. Reddy, S. Ramachandran, T. Matsunaga, L. Haynes, and H. Ikemoto. 1983. Progress in research toward outdoor biological hydrogen production using solar energy, sea water, and marine photosynthetic microorganisms. *Biochem. Eng. III, Ann. N.Y. Acad. Sci.* 413:514–530.

Mitsui, A., Y. Ohta, J. Frank, S. Kumazawa, C. Hill, D. Rosner, S. Barciella, J. Greenbaum, L. Haynes, L. Oliva, P. Dalton, J. Radway, and P. Griffard. 1980. Photosynthetic bacteria as alternative energy sources: Overview on hydrogen production research. Pages 3483–3510 *in* T. N. Veziroglu, ed., *Alternative Energy Resources,* Vol 8, *Hydrogen Energy.* Hemisphere Publishing Co., Washington, DC.

Mitsui, A., S. Kumazawa, E. J. Phlips, K. J. Reddy, K. Gill, G. R. Renuka, T. Kusumi, G. Reyes-Vasques, K. Miyazawa, L. Haynes, H. Ikemoto, E. Duerr, C. B. Leon, D. Rosner, R. Sescoa, and E. Moffat. 1984. *In* H. Ghose, ed., *Proc. VIIth Intl. Biotechnol. Symp.* (in press).

Ohta, Y., and A. Mitsui. 1981. Enhancement of hydrogen photoproduction by marine *Chromatium* sp. Miami PBS 1071 grown in molecular nitrogen. Pages 303–307 *in* M. Moo-Young and C. W. Robinson, eds., *Advances in Biotechnology,* Vol II. Pergamon Press. Oxford and New York.

Ohta, Y., J. Frank, and A. Mitsui. 1981. Hydrogen production by marine photosynthetic bacteria. Effect of environmental factors and substrate specificity on growth of a hydrogen-producing marine photosynthetic bacterium, *Chromatium* sp. Miami PBS 1071. *Intl. J. Hydr. Energy* 6:451–460.

Phlips, E. J., and A. Mitsui. 1983. Role of light intensity and temperature in the regulation of hydrogen photoproduction by the marine cyanobacterium *Oscillatoria* sp. strain Miami BG7. *Appl. Environ. Microbiol.* 45:1212–1220.

Thauer, R. 1977. Limitation of microbial H_2-formation via fermentation. Pages 201–204 *in* H. G. Schlegal and J. Barns, *Microbial Energy Conversion.* Pergamon Press, New York.

Vincenzi, M., R. Materassi, M. R. Tredici, and G. Florenzano. 1982. Hydrogen production by immobilized cells. II. H_2-photoevolution and waste water treatment by agar-entrapped cells of *Rhodopseudomonas palustris* and *Rhodospirillium molishianum. Intl. J. Hydr. Energy* 9:725–728.

Zborsky, O. 1974. *Immobilized Enzymes.* CRC Press, Cleveland, OH.

Zurrer, H., and R. Bachofen. 1979. Hydrogen production by the photosynthetic bacterium *Rhodospirillum rubrum. Appl. Environ. Microbiol.* 37:779–783.

CHAPTER **13**

Fermentation of Lignocellulosic Materials Treated by Ammonia Freeze-Explosion

Bruce E. Dale*, Linda L. Henk, and Ming Shiang

Department of Agricultural and Chemical Engineering, Colorado State University, Fort Collins, Colorado 80523

The cost of fermentable sugars represents approximately half of the production costs of medium- to large-scale fermentation chemicals. The future of biotechnology in the production of such commodities by fermentation therefore depends strongly on efficient hydrolysis of lignocellulosic materials to yield inexpensive sugar syrups. This paper reviews briefly the features of cellulose that limit its hydrolysis and the various pretreatments that have been employed to increase cellulose hydrolysis. A new pretreatment technique called ammonia freeze-explosion (AFEX) is analyzed in terms of chemical and physical effects on lignocellulosic materials. Yields of fermentable sugars by enzymatic hydrolysis following ammonia freeze-explosion are typically four to five times greater than those obtained on untreated materials. Similar yield increases are observed when AFEX-treated straws are used as fermentation substrates to produce the cellulase enzymes from *Trichoderma reesei* (Rutgers C30 mutant) and fungal protein by *Chaetomium cellulolyticum*. Ethanol yields approaching theoretical are obtained by fermentation of hydrolyzates derived from alfalfa press cake with *Saccharomyces uvarum*. There is no indication that any fermentation inhibitors or sugar degradation products are formed during the AFEX process. The simplicity and effectiveness of the AFEX process appear to make it one of the more promising cellulose pretreatment technologies.

Introduction

The largest reservoir of potentially fermentable sugars on earth is the lignocellulosic materials, estimated at 100 billion tons annually of renewable resource (Sarbolouki and Moacanin 1980). Perhaps the major problem in utilization of the loignocellulosics as fermentation substrates is the resistance of cellulose to hydrolysis, either by acids or enzymes. Another key problem is the cost of the cellulase enzymes used to hydrolyze cellulose to glucose. The high cost of these cellulase enzymes is also largely due to the resistance of cellulose to hydrolysis. A wide variety of pretreatment techniques have, therefore, been used in an attempt to increase the hydrolysis of cellulose (Millet et al. 1975, 1976). No single technique has yet found widespread commercial application. The major factors influencing cellulose reactivity are believed to be particle size and surface area, lignin content, and cellulose crystallinity.

*To whom correspondence should be directed.

A new technique called ammonia freeze-explosion (AFEX) has recently been developed for increasing cellulose hydrolysis (Dale and Moreira 1982). This technique simultaneously decreases cellulose crystallinity and particle size, thereby increasing the surface area exposed for enzymatic attack. Some apparent decreases in lignin content are also observed. These various effects combine to greatly increase the yield of fermentable sugars obtainable from lignocellulosic materials such as wheat and rice straw and cornstover. Typical sugar yields are four to five times greater following AFEX treatment.

The various milling techniques are perhaps the best-developed pretreatment technologies and do indeed produce a more easily hydrolyzed cellulose. However, the large power requirements for these milling/grinding methods tend to make them quite expensive. The steam explosion approaches (Saddler et al. 1983; Schultz et al. 1983) are also quite well developed but require considerable steam, which cannot be easily recovered. Steam treatments also inevitably degrade some of the sugars at the high temperatures involved. The various chemical treatments are generally hampered by the costly chemicals involved and the difficulty and expense of recovering and recycling these chemicals. Many of the pretreatment chemicals are also toxic or inhibitory to microorganisms so the recovery levels must be very high.

The question of sugar yields from lignocellulosic materials is particularly important when one considers the potential of the new biotechnology in production of medium- to large-scale commodities by fermentation. A recent review paper (Dale and Linden 1984) summarizes data from many sources showing that approximately 40 to 70% of the production costs of commodity fermentation chemicals (ethanol, acetone/butanol, citric acid, glycerol, single cell protein, and the cellulase enzymes) are due to the cost of the fermentable carbohydrates. Therefore, progress in such fermentations at the industrial level is governed more by carbohydrate costs than by sophisticated techniques in genetic manipulation; however, genetic engineering techniques may prove valuable for pharmaceuticals or fine chemicals production by fermentation.

Traditional sources of fermentable carbohydrates (molasses, corn syrups, etc.) are generally too expensive at this time to allow fermentation processes to compete economically with synthetic routes to the same commodities. In the absence of government intervention, competing food and feed uses of starches and sugar are likely to keep the cost of these carbohydrates unacceptably high for commodity fermentations. It appears that only the lignocellulosic materials, particularly the agricultural residues such as straws, can offer much hope for sufficiently low-priced fermentable sugars on which to base a fermentation industry for commodity chemicals. Indeed, some recent studies have projected sugar costs of less than 10 cents per pound by hydrolysis of lignocellulosic materials (Dale and Linden 1984); hence, the importance of pretreatment technologies to improve cellulose hydrolysis and the overall process of renewable resource conversion called "biomass refining."

MATERIALS AND METHODS

AFEX Treatment and Solid Substrate Fermentation

Samples of wheat straw, barley straw, and cornstover were obtained field dry from agricultural areas in northern Colorado. Aspen chips were the gift of Aspenal, Inc., Angora, MN, and the alfalfa press cake was donated by Atlantic Richfield Solar Industries, Dublin, CA. Rice straw was provided by the Western Regional Laboratory of the U.S. Department of Agriculture. All samples were ground to pass a 2-mm screen prior to AFEX treatment and subjected to forage fiber analysis (Goering and Van Soest 1970) and protein determination by the micro-Kjeldahl technique.

The AFEX treatment has been described in detail elsewhere (Dale and Moreira 1982). Briefly, approximately 150 g of the sample of interest are placed in a highly modified 1-gal pressure vessel (Autoclave Engineers). The vessel is then sealed and the desired amount of liquid anhydrous ammonia at a given pressure is added while the straw sample is being stirred. After an appropriate time, a large ball valve connected to the autoclave is opened. The liquid ammonia evaporates violently with the pressure decreasing to near atmospheric in less than a second. The temperature also drops precipitously to near −29 C. The autoclave is then unbolted, and the treated straw is removed and allowed to stand overnight to evaporate most of the residual ammonia. Typically ammonia amounting to 1.0 to 1.5% of the dry weight of the straw remains with the straw even after prolonged airing, giving an increase in the crude protein (N × 6.25) content of the straw of 6 to 9 %. The AFEX-treated samples are used in subsequent experiments for enzymatic hydrolysis and fermentation without further processing.

Wheat straw samples subjected to the AFEX treatment, as well as samples treated by steam, autoclaving, and sodium hydroxide were sterilized and then inoculated with *Chaetomium cellulolyticum* and fermented in a stationary layer fermenter as described elsewhere (Abdullah et al., in press 1984). Changes in the straw composition were determined by forage fiber analysis and changes in protein content by micro-Kjeldahl. Fungal biomass was assumed equal to the increased protein content (N × 6.25) times two (approximately half of the fungus dry weight is protein). The technical parameters for this solid substrate fermentation of wheat straw were then evaluated (Laukevics et al., in press 1984).

Enzymatic Hydrolysis of Cellulosic Materials

Commercial cellulase enzymes were donated by Novo Corporation, Danbury, CT. One part Novozyme (cellobiase) to 10 parts Celluclast 200 L *T. reesei* cellulase complex was prepared in 0.1 *M* citrate buffer, pH 4.8. Enzymatic activity (I.U./ml) was determined by standard filter paper assay (Mandels et al. 1976). An appropriate amount of each enzyme was used to obtain an activity of 80 I.U./g of dry AFEX-treated material. Enzymatic hydrolyses were done routinely using the following conditions:

(1) Enzyme (activity) to fiber ratio = 80 I.U./g dry weight; (2) Enzyme (liquid) to fiber ratio = 20 ml/g dry weight; (3) Temperature 50 C; (4) 90 rpm rotary shaking; (5) Duration of hydrolysis: 3 h, 24 h.

At the end of hydrolysis, fibers were separated by filtration, and the resulting hydrolyzates were analyzed for total DNSA reducing sugars (Miller 1959) and for glucose by glucose oxidase (Beckman glucose analyzer).

Cellulase Production by T. reesei, *Rutgers C30 (new type)*

Trichoderma reesei Rut-C30 new type (obtained from Professor Harvey Blanch,

Berkeley), maintained on potato dextrose agar slants, was grown in Mandels' medium (Mandels and Weber 1969) with 0.1% yeast extract and an initial pH of 5.0. AFEX-treated rice straw and wheat straw were used as a carbon source at 1% (w/v) concentration. Control experiments using untreated rice and wheat straws were conducted simultaneously. These cellulase fermentations were carried out in 500-ml baffled shake flasks using a working volume of 100 ml, 30 C, and 120 rpm for 5 d. Culture filtrates were analyzed by the following assays:

(1) Total reducing sugar (Miller 1959); (2) Forage fiber analysis (Goering and Van Soest 1970); (3) Standard filter paper assay.

Ethanol Production from AFEX-treated Lignocellulosics Using Saccharomyces uvarum

A cellulase preparation was mixed as previously described in 0.1 *M* citrate buffer, pH 4.8. Flasks, each containing 6 g AFEX-treated wheat straw (dry wt) and 200 ml of cellulase solution, were prepared for hydrolysis at 50 C, 150 rpm. At specific times, 5-ml samples were taken for enzyme kinetic studies. These samples were clarified by filtration through a fiberglass prefilter (Boehringer Mannheim Biochemicals) and a 0.45 μm nitrocellulose filter (Schleicher and Schuell).

Enzymatic hydrolysis continued for 24 h, at which time the temperature of the hydrolyzate was lowered from 50 C to 30 C in anticipation of inoculation with a 10% inoculum of *Saccharomyces uvarum,* NRRL Y-1347. No additional nutrient salts were added. An aerobic atmosphere was maintained for the first 3 h and then airlocks were set in place. Samples were taken periodically after inoculation and clarified as previously described.

Total reducing sugars and glucose were assayed as previously described. Ethanol was analyzed using a Varian Model 2400 gas chromatograph with a flame ionization detector. The 6-ft × 1/8-inch column was packed with chromosorb M-AW coated with 10% AT-1000. Similar experiments were performed using untreated and AFEX-treated alfalfa press cake. All conditions for hydrolysis remained the same as those used in the wheat straw experiment except for the following:

(1) Cellulase (liquid) to fiber ratio was 10:1; (2) Tap water was used in place of 0.1 *M* citrate buffer. The pH of the slurry was adjusted to pH 4.8 using 50% H_2SO_4 prior to the addition of the cellulase-cellobiase preparation.

Results and Discussion

Enzymatic Hydrolysis of AFEX-treated Materials

Selected lignocellulosic materials were subjected to AFEX treatment and enzymatic hydrolysis. The best hydrolysis results obtained to date for each of the substrates are summarized in Table 1 and are compared with controls (untreated samples). Total cellulose plus hemicellulose contents for these samples as given by the forage fiber analysis are also summarized in Table 1. In general, total sugar yields by DNSA are four to five times greater for treated materials than for untreated controls. With most substrates tested, we have been able to achieve hydrolysis at well above 80% of theoretical. Both hemicelluloses and cellulose are hydrolyzed to about the same extent. The higher than theoretical result for alfalfa press cake in Table 1 may be attributed to the fact that this material has a substantial soluble sugar content before hydrolysis. Aspen, a woody material, is more resistant to the pretreatment, as expected. All of the lignocellulosics studied to date respond somewhat differently to the treatment, but in general the optimum pretreatment conditions are in the neighborhood of 1 lb of liquid anhydrous ammonia per pound of dry fiber, 250 psig total pressure, 30 min or less contact time with ammonia, and a moisture content above 10% (dry basis). Essentially these conditions were used to prepare samples for the fermentation experiments, which are summarized later. Available evidence indicates that the added moisture allows some ammonium hydroxide to form that hydrolyzes hemicelluloses and thus promotes the overall effect of the AFEX treatment. Hemicelluloses hydrolysis has been shown to increase the subsequent hydrolysis of cellulose (Chen and Anderson 1980).

TABLE 1. *Enzymatic hydrolysis of AFEX-treated lignocellulosics*

Sample	Sample Treatment Number	Cellulose plus Hemicelluloses Content (% of dry wt)	Total Sugars after Hydrolysis (mg/g dry fiber) 3h	24 h	Percent of Theoretical Yield[a] in 24 h
Alfalfa press cake	5-25-83-3	44.9	342	594	119
Alfalfa press cake	untreated	44.9	—	323	65
Aspen chips	6-22-83-2	72.8	124	359	44
Aspen chips	untreated	72.8	—	58	7
Barley straw	1-3-84-3	64.5	278	593	83
Barley straw	untreated	64.5	100	181	25
Corn stover	6-23-83-4	64.3	476	679	95
Corn stover	untreated	64.3	—	201	28
Rice straw	5-18-83-1	56.6	—	584	94
Rice straw	untreated	56.6	—	204	32
Wheat straw	10-26-83-2	64.8	—	637	88
Wheat straw	untreated	64.8	—	150	31

[a] Theoretical yield calculated as total sugars after hydrolysis ÷ (percent cellulose plus hemicelluloses) × 11.1 (to account for water of hydrolysis).

As mentioned previously, cellulose hydrolysis is limited primarily by surface area, crystallinity, and lignin content. Hemicelluloses pre-hydrolysis by ammonium hydroxide has the effect of increasing available surface area. The rapid pressure release also shatters and splits plant material parallel to the fiber axis as demonstrated by numerous microphotographs taken of a variety of AFEX-treated materials. All such microphotographs show the same phenomenon—a large increase in fiber surface exposed parallel to the fiber axis. Physically, the AFEX process seems similar to chopping wood on a microscopic level. Incidentally, creating new surface area parallel to the fiber axis is an efficient way of generating surface. In contrast, grinding techniques that repeatedly cut the ends of the fibers must spend large amounts of energy to create relatively little new surface area.

Liquid ammonia has long been known to be a decrystallizing agent for cellulose (Barry et al. 1936) and indeed, crystallinity of AFEX-treated, pure cellulose samples is substantially decreased as measured by x-ray diffraction methods (Segal et al. 1959). Hemicelluloses pre-hydrolysis, increased fiber surface area, and decreased crystallinity all contribute to the increased sugar yields upon hydrolysis noted in Table 1. In addition, we have occasionally measured apparent decreases in lignin content by the Van Soest procedure on AFEX-treated samples. Presumably this is because of a mild lignin depolymerization caused by ammonia since no wash streams are employed in the process to remove lignin.

The amount of enzyme used in the hydrolysis also strongly influences the sugar yields obtainable. In order to quantitate the effect of the pretreatment on wheat straw hydrolysis at various enzyme levels, we ran parallel hydrolysis experiments on AFEX-treated wheat straw at 10, 20, 40, and 80 I.U. enzyme/g dry fiber with a companion hydrolysis of untreated wheat straw at 80 I.U./g. These results are summarized in Fig. 1. It is obvious that the AFEX treatment is a strong potentiator of hydrolysis. Eight-fold less enzyme (10 I.U./g) will still yield twice as much sugar on AFEX-treated wheat straw as on untreated straw. Not only is the final extent of hydrolysis substantially increased but the rate of hydrolysis increases even more rapidly because of the treatment. For instance, at 80 I.U./g after 5 h of hydrolysis, the AFEX-treated straw yields about 350 mg sugar/g fiber versus less than 70-mg sugar/g fiber for untreated straw. The net effect of the

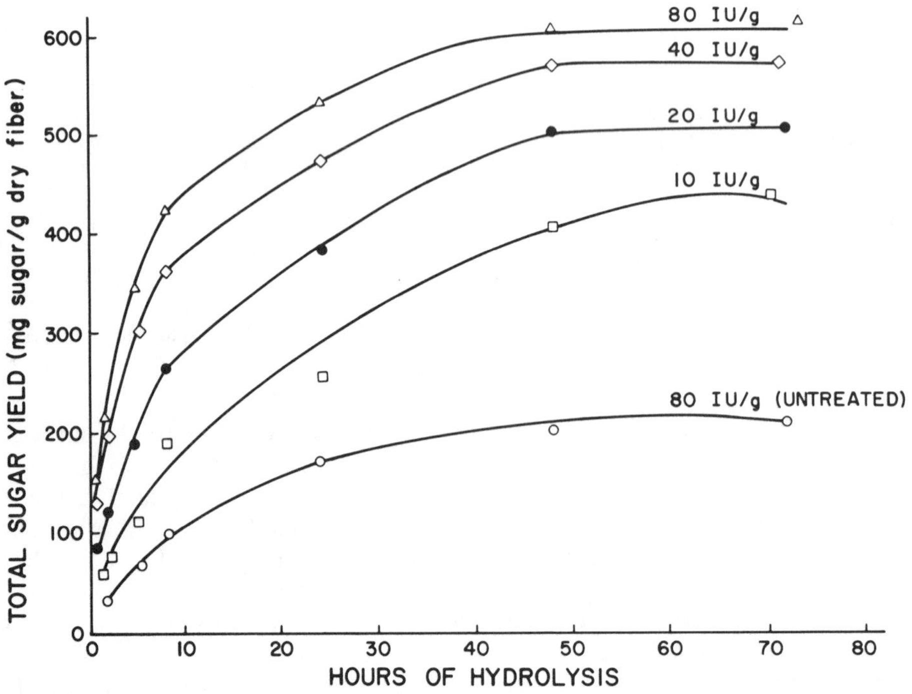

FIG. 1. Effect of enzyme loading on hydrolysis of AFEX-treated wheat straw (12–7–83–1) with time.

AFEX pretreatment is therefore to greatly increase the efficiency of lignocellulose hydrolysis as well as the efficiency with which the hydrolysis reactor is used (overall hydrolysis rate).

Of course, it is one thing to produce sugars that respond positively to the DNSA test, it is another thing to actually be able to ferment such sugars. We now discuss the response of AFEX-treated materials to direct fermentation by *Chaetomium cellulolyticum* and *Trichoderma reesei* as well as fermentation by *Saccharomyces uvarum* of enzymatically hydrolyzed sugar syrups derived from AFEX-treated lignocellulosics.

Solid Substrate Fermentation for Single-Cell Protein

One measure of the effectiveness of a pretreatment is the degree to which the pretreatment makes available fermentable carbohydrates for the growth of cellulolytic fungi. We have conducted some preliminary solid substrate fermentations with *Chaetomium cellulolyticum* using AFEX-treated wheat straw to compare the AFEX treatment with other pretreatments that have received more study (Ulmer et al. 1981). These results are summarized in Table 2. Even without any optimization of the AFEX treatment for this fermentation, AFEX gave technical parameters superior to the other pretreatments. With continued study, it should be possible to improve these parameters even further.

TABLE 2. *Solid substrate fermentation of pretreated wheat straw by* Chaetomium cellulolyticum

Pretreatment	Biomass generated[a] (g/g)	Protein yield[b] (g/g)	Bioconversion efficiency[c] (% w/w)	Protein productivity[d] (g/Lh)
None	0.0088	0.004	2.04	0.009
Steam	0.145	0.073	34.3	0.151
Alkali-washed	0.0688	0.034	16.2	0.072
Alkali-unwashed	0.132	0.066	31.1	0.137
AFEX	0.181	0.091	42.8	0.189

[a] Grams biomass dry weight generated per gram of dry straw fermented.
[b] Grams of fungal protein produced per gram of dry fiber, percent.
[c] Percent conversion of potential carbohydrates to fungal biomass per gram of dry straw based on a theoretical conversion of 0.565 g/g.
[d] Assumes average value of 150 g/L straw concentration during fermentation.

Cellulase Production from Untreated and AFEX-treated Straws

Enzymatic activity derived from the fermentation of AFEX-treated wheat straw and rice straw by *T. reesei* Rut-C30 new type was four times greater than the enzymatic activity derived from untreated wheat and rice straws (Fig. 2). Enzyme activity was measured by standard filter paper assay. Generally speaking, the enzyme activities from a wheat straw fermentation were higher than those from rice straw. AFEX-treated straw apparently provides a suitable substrate for a cellulolytic organism such as *T. reesei.* Work has been done to optimize this fermentation with the addition of corn-steep liquor and a surfactant such as Tween 80 by other workers.

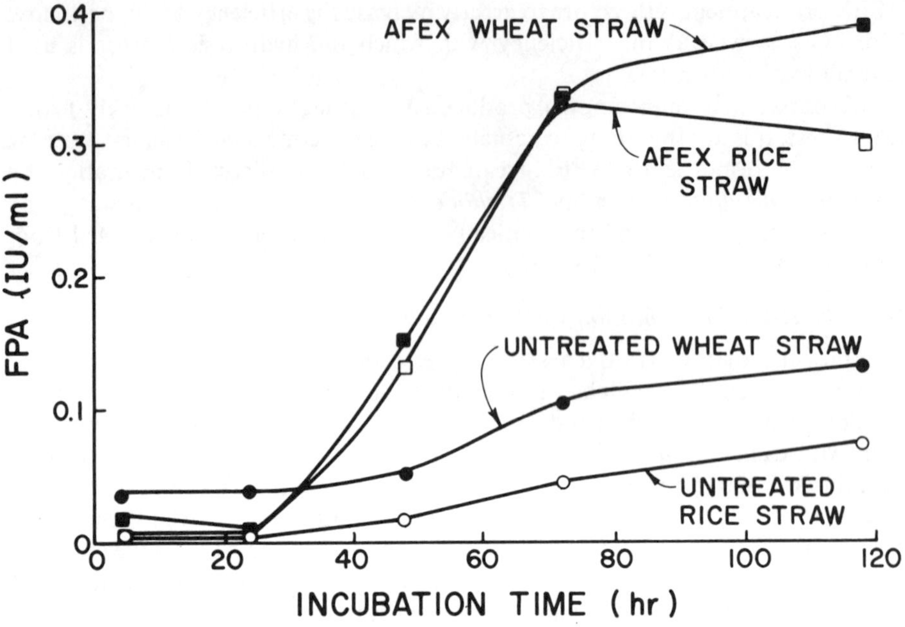

FIG. 2. Cellulase activity (FPA) from untreated and AFEX-treated straws (*T. reesei* Rut C30 new type).

Ethanol Production from AFEX-treated Lignocellulosics

In reviewing the total reducing sugar data (DNSA) after 24 h of enzymatic hydrolysis, 6.0 g of AFEX-treated wheat straw yielded 20.8 mg total reducing sugar/ml (Fig. 3). The total percentage of sugars available for fermentation was 2.1%.

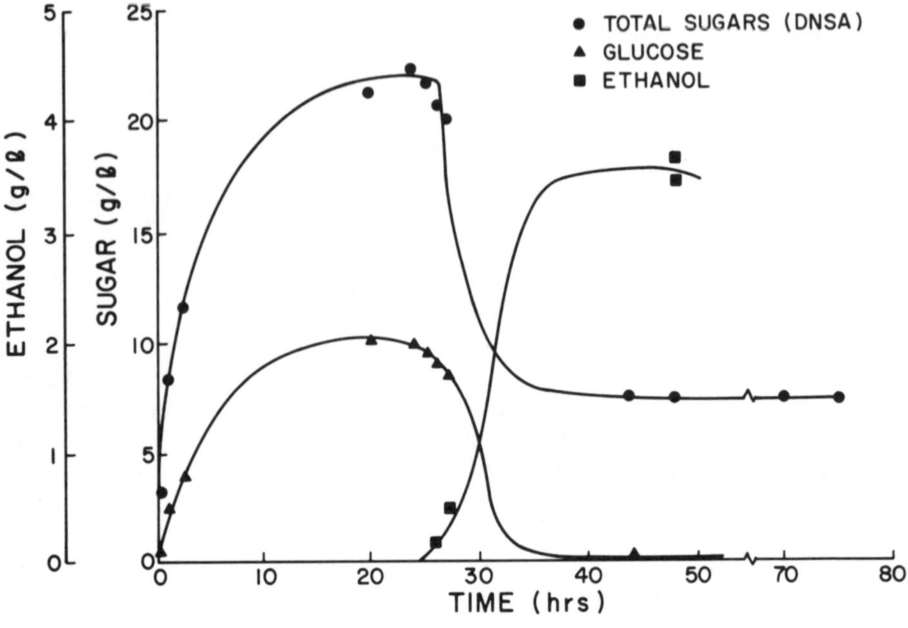

FIG. 3. Enzymatic hydrolysis and ethanol fermentation of AFEX-treated wheat straw (*S. uvarum*).

After analyses were completed, it was apparent the fermentation was essentially complete at 20 h after inoculation with *S. uvarum*. Glucose was depleted, leaving a residual amount of total reducing sugars (8 g/l) that was not fermentable by *S. uvarum*. These sugars are presumably pentoses (xylose) derived from the hemicellulose portion of wheat straw during enzymatic hydrolysis.

Ethanol yields equaled 21% (41% of theoretical) based on total reducing sugar. These low yields are attributed to the initial low concentration of sugar available for fermentation. Apparently, a large percentage of available sugar was used for biomass production (growth) and little was available for actual ethanol production. By establishing a preconcentration step in the fermentation procedure, more sugar from hydrolysis will be available for actual ethanol production. It is encouraging that *S. uvarum* grows readily on the enzymatic hydrolysis product of AFEX-treated wheat straw.

In a similar experiment using AFEX-treated alfalfa press cake, theoretical yields for ethanol production were reached. Once again, *S. uvarum* was the yeast selected for the fermentation. All enzymatic hydrolysis conditions remained constant as previously described in the methods section except for a fiber to enzyme ratio of 1:10.

Total reducing sugars resulting from 24 h of enzymatic hydrolysis of alfalfa press cake equals 32.8 mg/ml. This concentration of total sugars dropped to 13 g/l in 25 h after inoculation with *S. uvarum*. Ten grams per liter of ethanol were produced during this time. Calculating yield as the amount of ethanol produced (g/l) divided by the amount of sugar consumed (g/l) times 100, a yield of 51% (100% of theoretical) was obtained. Untreated alfalfa press cake was used for a control in these experiments. The amount of ethanol produced equaled 6.2 g/l and the amount of sugars consumed (DNSA) equaled 17.8 g/l for a yield of 35% (69% of theoretical) with untreated press cake.

Obviously, AFEX treatment improves the hydrolytic and fermentation capacity of lignocellulosic substrates such as rice straw, wheat straw, and alfalfa press cake. The cellulase, ethanol, and single-cell protein fermentations clearly demonstrate the improvement obtained in product yields when lignocellulosics are pretreated using the AFEX process.

Conclusions

The AFEX treatment substantially increases the yield of sugars from cellulosic materials. The potency of available cellulase enzymes is enhanced many fold. No inhibitors of either hydrolysis or fermentation appear to be formed by the treatment. For each of the fermentations tested to date, both the rate and extent of fermentation are improved by AFEX pretreatment. However, none of these fermentations has been optimized for the AFEX treatment, and further improvements can be expected with optimization.

Whether or not the AFEX technique has industrial potential remains to be determined. In spite of this, technical prospects for its development seem to be favorable since the treatment is reasonably simple and does not require extreme conditions. Preliminary engineering analyses estimate commercial-scale operating costs in the region of $20/ton of straw processed and capital investment costs for the pretreatment of about $80/ton of annual capacity (D. H. Westermann and L. P. Weiner, personal communication). Cellulose pretreatment research is crucial to the development of a commodities biotechnology industry.

ACKNOWLEDGMENTS

The authors express their gratitude to Doug Miller and to Jeff Miller for their technical assistance in AFEX treatment and analysis.

LITERATURE CITED

Barry, A. J., F. C. Peterson, and A. J. King. 1936. X-ray studies of cellulose in non-aqueous systems. I. Interaction of cellulose and liquid ammonia. *J. Am. Chem. Soc.* 58:333–337.

Chen, W. P., and A. W. Anderson. 1980. Extraction of hemicellulose from rye grass straw for the production of glucose isomerase and use of the resulting straw residue for animal feed. *Biotechnol. Bioeng.* 22:519–531.

Dale, B. E., and J. C. Linden. 1984. Fermentation substrates and economics. *Annu. Repts. Ferment. Proc.* 7:107–133.

Dale, B. E., and M. J. Moreira. 1982. A freeze-explosion technique for increasing cellulose hydrolysis. *Biotechnol. Bioeng.* 12:31–43.

Goering, H. K., and P. J. Van Soest. 1970. Forage fiber analyses—apparatus, reagents, procedures and some applications. Agric. Handbook No. 379, Agricultural Research Service, U.S. Department of Agriculture. Jacket Number 387-598. Superintendent of Documents, Washington, DC 20402.

Mandels, M., and J. Weber. 1969. Advanced Chemistry. Ser. 95:391.

Mandels, M., R. Andreotti, and C. Roche. 1976. Measurement of saccharifying cellulase. *Biotechnol. Bioeng.* 6:21–33.

Miller, G. L. 1959. Use of dinitrosalicylic acid reagent for determination of reducing sugar. *Anal. Chem.* 41:426–428.

Millet, M. A., A. J. Baker, and L. D. Satter. 1975. Pretreatments to enhance chemical, enzymatic and microbiological attack of cellulosic materials. *Biotechnol. Bioeng.* 5:193–219.,

________. 1976. Physical and chemical pretreatments for enhancing cellulose saccharification. *Biotechnol. Bioeng.* 6:125–153.

Saddler, J. N., E. K. C. Yu, M. Mes-Hartiee, N. Levitin, and H. H. Brownell. 1983. Utilization of enzymatically hydrolyzed wood hemicelluloses by microorganisms for production of liquid fuels. *Appl. Environ. Microbiol.* 45:153–160.

Sarbolouki, M. N., and J. Moacanin. 1980. Chemicals from biomass—U.S. prospects for the turn of the century. *Solar Energy* 25:303–315.

Schultz, T. P., C. J. Biermann, and G. D. McGinnis. 1983. Steam explosion of mixed hardwood chips as a biomass pretreatment. *I&EC Prod. Res. Dev.* 22:344–348.

Segal, L., J. J. Creely, A. E. Martin, Jr., and C. M. Conrad. 1959. Cellulose crystallinity determination by x-ray diffraction. *Text. Res. J.* 29:786–797.

Ulmer, D. C., R. P. Tengerdy, and V. G. Murphy. 1981. Solid state fermentation of steam treated feedlot waste fibers with *Chaetomium cellulolyticum*. *Biotechnol. Bioeng.* 11:359–369.

CHAPTER **14**

Anaerobic Digestion of Woody Biomass

D. P. CHYNOWETH AND D. E. JERGER

Institute of Gas Technology, Chicago, Illinois 60616

Woody biomass without pretreatment is generally considered to be refractive to anaerobic decomposition. This refractory property is attributed to its low moisture content, crystalline nature of the cellulose, and complex association of the component carbohydrates with lignin. This study investigated the methane fermentation (anaerobic digestion) of various wood species using conventional anaerobic digestion and batch anaerobic biogasification potential (ABP) assays. Most experiments were conducted at 35 C with a particle size in the range of 1–2 mm and with a full complement of inorganic nutrient supplements. Conventional CSTR semicontinuous-feed anaerobic digestion resulted in low methane yields and low conversion (less than 5% organic reduction). Significantly higher conversion (as high as 54%) and higher methane yields (as high as 5.4 SCF/lb VS added) were observed for several hardwood species in ABP assays employing low loading and long residence time (60 d). One softwood (loblolly pine) and eucalyptus were refractory under these conditions. Pretreatments, including particle size reduction and NaOH, increased rates but not total conversion. These results demonstrate that woody biomass can be decomposed by the methane fermentation and can support the potential for development of this process for commercial wood conversion applications.

INTRODUCTION

Woody biomass species are attractive as potential energy crops. It was projected that with well-planned, intensive harvesting, U.S. forests could supply up to 20 quads of energy from woody biomass even after satisfying the needs of lumber and paper industries (Klass 1982). Growth yields are currently as high as 5 to 10 dry tons/acre-year in several different geographic locations within the United States and may be significantly improved through ongoing research efforts. Short-rotation, intensive silviculture has recently received wide attention as a means of increasing growth yields of woody species, including such species as sycamore, cottonwood, and hybrid poplar. For example, yields of 10 to 20 dry tons/acre-year have been achieved for hybrid varieties of poplar (Zavitkovski 1981). Other advantages of woody plants as energy crops are that methods for growth and harvest are commercial and projected costs are lower than those of other biomass species. For example, the cost of wood feedstocks has been estimated in the range of $1 to $4/million Btu (gross energy) (Office of Technology Assessment 1980; Lother and Zavitkovski 1981; Johnson et al. 1980).

Wood is currently the most widely used biomass energy resource in the United States, accounting for over two quads of energy per year. It is burned in numerous types of combustion devices ranging from wood stoves to large industrial cogeneration boiler systems (Johnson et al. 1980). Other thermal processes, such as thermal gasification and pyrolysis, are under investigation for conversion of wood to gaseous and liquid fuels.

Research on the bioconversion of wood has been largely limited to production of ethanol following extensive feedstock pretreatment. Microbial conversion of woody species is generally not considered technically feasible without some type of pretreatment to disassociate the biodegradable components (mainly cellulose and hemicellulose) from lignin and to saccharify the polysaccharides. With few exceptions (Wilson and Pidgen 1964; Colberg et al. 1981) this research has been limited to conversion to liquid fuels. In general, most state-of-the-art pretreatment techniques are expensive and often energy-intensive, and application of this technology to methane production via anaerobic digestion has been minimal.

Institute of Gas Technology (IGT) recently used enrichment techniques to develop the first reported microbiological inoculum that can convert woody biomass to methane without pretreatment (Jerger et al. 1982a,b). Methane yields of 5 to 6 ft^3/lb of wood added were achieved for several hardwoods, corresponding to conversions of about 80%–90% of the structural carbohydrates. The following paper summarizes the results of this work and more recent research on the effects of particle size reduction and pretreatment with NaOH.

Materials and Methods

Feedstock Processing and Analysis

The woody biomass samples screened during this study were obtained from managed, experimental growth stands. The source and growth characteristics of each species were described previously (Jerger et al. 1982a). One 6-month-old sample referred to as "wood grass" was obtained from Dula's Nursery in Oregon. The samples were freshly harvested and shipped to IGT in a frozen state. Upon receipt, they were chopped with a Sears shredder and ground through an 0.8-mm (0.03-in) Urschell mill head (Comitrol 3600). The feed material was mixed to homogeneity, packed in 0.5-gal containers, and frozen.

Proximate and ultimate analyses were performed on the samples to provide data for experimental design and calculation of theoretical yields and material balances for the experiments. Moisture, volatile solids, and ash content of the samples were determined by the methods outlined in *Standard Methods for the Examination of Water and Wastewater* (American Public Health Association 1975). Analytical methods used to perform heating value, C, H, N, P, and S determinations have been discussed in detail in a previous paper (Jerger et al. 1982a).

The lignin content of the woody feeds was determined by solubilizing the lignin with chlorine dioxide. The carbohydrate material is not appreciably attacked under these conditions and is filtered and retained as holocellulose (alpha cellulose and hemicellulose). The holocellulose is then treated with sodium hydroxide and acetic acid following ANSI/ASTM D–1103–60 procedures resulting in the determination of alpha-cellulose concentration.

Anaerobic Biogasification Potential

A serum bottle bioassay procedure, the anaerobic biogasification potential (ABP) assay, was used to determine the biodegradability of the wood species. The protocol for this procedure, discussed in a previous publication (Jerger et al. 1982b), involved addition of a defined complete nutrient growth medium and sewage sludge inoculum to a 250-ml serum bottle containing the woody feedstock. The reactors were incubated from 60 to 90 d; during this period, gas production and gas composition measurements were taken. At the termination of the incubation period, the reactor contents were analyzed for pH, volatile fatty acids, and total and volatile solids (VS) concentrations.

Bench-Scale Digester Studies

The digesters used in the semicontinuous-feed digestion, including baseline digestion, nutrient-effect, and retention time studies were constructed of Plexiglas, contained baffles to enhance mixing, and had several ports for feeding, gas removal, and installation of analytical probes. A culture volume of 1.5 L was used during the experimental studies and the digesters were incubated at 35 C on a gyratory shaker to provide mixing. The start-up, daily operation, and analysis of these digesters has been discussed previously (Chynoweth et al. 1982).

Pretreatment

Physical. The physical pretreatment studies were conducted with hybrid poplar and consisted of mechanical size reduction using an Urschell mill and a wet extrusion milling technique. Wood particles with a diameter of $\leq$0.8 mm (20 mesh) and $\leq$8 mm (2–0.5 mesh) were obtained by passing chips through an Urschell mill (Comitrol 3600) with head sizes of 0.8 and 8 mm. A particle size of approximately 0.003 mm was achieved using a wet grinding procedure developed by Draiswerke, Inc., and demonstrated by Fort Pitt Machine Company, Pittsburgh, PA. A 5% slurry of poplar ($\leq$0.8-mm particles) was used as the starting material.

Chemical. Chemical treatment consisted of treating cottonwood, hybrid poplar, and sycamore with 5% and 50% by weight of NaOH per gram feed volatile solids. Approximately 200 mg of the woody feed was added to a 1 *N* NaOH solution, and the volume was brought up to 10 ml. The reactors were incubated at 100 C for 2.5 h. Alkalinity titrations were performed using pH indicators, methyl orange for standardization of 0.02 *N* H_2SO_4, and phenolphthalien for endpoint titration to pH 8.3, to determine the normality of NaOH before and after wood pretreatment.

Results

Conventional semicontinuous-feed digestion studies confirmed previously reported results that wood is highly refractory to anaerobic decomposition (Table 1) without pretreatment. However, continued operation of these same digesters without feeding in the batch mode resulted in significant conversion with associated increases in methane production. These results were attributed to either enrichment and adaptation of a wood-degrading inoculum or increased organism and particulate matter retention resulting from batch operation.

TABLE 1. *Effect of retention time on biodegradability of woody biomass*

Wood Species	Volatile Solids Reduction, % 15-Day Retention Time[a]	120-Day Retention Time [b]
Black Alder	3	16
Cottonwood	7	32
Eucalyptus	2	11
Hybrid Poplar	6	29
Loblolly Pine	2	3
Sycamore	8	44

[a] 35 C; 15-d HRT; 0.1 lb VS/ft^3-d loading.
[b] 35 C; 120-d incubation; batch-fed.

Batch bioassays were employed to carefully document these findings. In this assay, substrates under study were incubated under batch conditions with excess nutrients and a standardized inoculum. Gas quantity and quality were measured during the experiment, which was continued until no additional gas production was observed. At the completion of the experiment, total gas and methane production and organic matter reduction were determined. The anaerobic biodegradability of several woody species studied is presented in Table 2. The highest methane yields of 5.2 SCF/lb VS added were achieved from hybrid poplar and sycamore. Red and black alder and cottonwood exhibited methane yields of 4.5, 3.9, and 3.6 SCF/lb VS added, respectively, whereas eucalyptus and loblolly pine exhibited poor anaerobic biodegradability with methane yields of 0.23 and 1.0 SCF/lb VS added, respectively.

TABLE 2. *Anaerobic biodegradability of woody species*

Species	Methane Yield, SCF/lb VS added	Methane Yield, SCM/kg VS added	Volatile Solids Reduction, %	Recovery Effluent Volatile Solids, %	Recovery Effluent Carbon
Black Alder	3.9	0.24	32.5	108	b
Cottonwood	3.6	0.22	32.3	106	107
Eucalyptus	0.23	0.014	1.00	107	b
Hybrid Poplar	5.2	0.32	53.8	106	101
Loblolly Pine	1.0	0.063	3.63	103	b
Red Alder[a]	4.5	0.28	48.4	91.5	b
Sycamore	5.2	0.32	56.7	106	98.3

[a] 90-day incubation.
[b] Carbon analyses not conducted on effluent samples.

Volatile solids conversion efficiencies from these experiments confirmed the results obtained from gas production analyses. Volatile solids reductions of

56.7%, 53.8%, 32.5%, and 32.3% were observed for sycamore, hybrid poplar, black alder, and cottonwood, respectively. Loblolly pine and eucalyptus again demonstrated poor biodegradability with volatile solids reductions of only 3.6% and 1.0%, respectively. The validity of these unexpectedly high conversion data is supported by good material balances obtained from all reactors (Table 2).

The effect of harvest age on the anaerobic biogasification potential is shown in Fig. 1. These data show that the conversion rate of the 6-month-old "wood grass" was significantly greater than the 2- or 3-year-old poplar samples. The ultimate methane yields were approximately the same for all three samples.

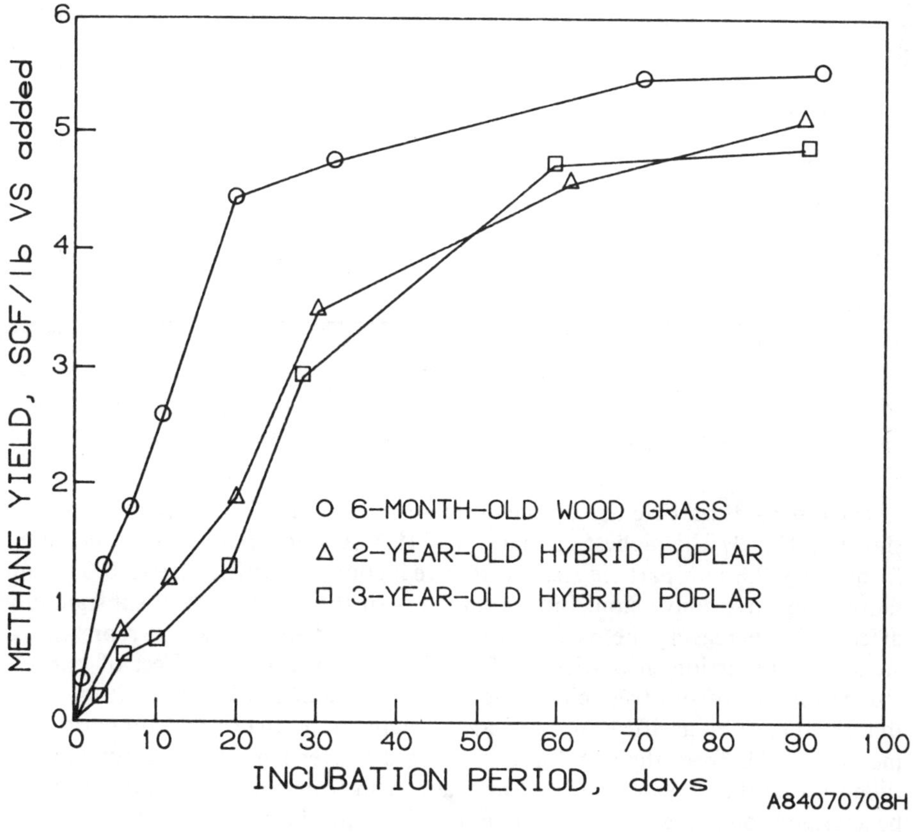

FIG. 1. Anaerobic biogasification potential of Poplar "wood grass" (from Dula's Nursery) and hybrid Poplar (from Pennsylvania State University).

The ABP assay in its standard volume or scaled-up version was employed to evaluate several process variables, such as the change in the major component cellulose during the wood fermentation and the effect of particle size and alkaline pretreatment on the wood fermentation. The results of these studies will be used as a basis for design and operation of larger bench-scale reactors. Data in Fig. 2 show that about 75% reduction in cellulose occurred in the batch fermentation of wood. A close correlation between accumulative methane yields and cellulose reduction was observed. The results of analyses of other organic components were inconclusive.

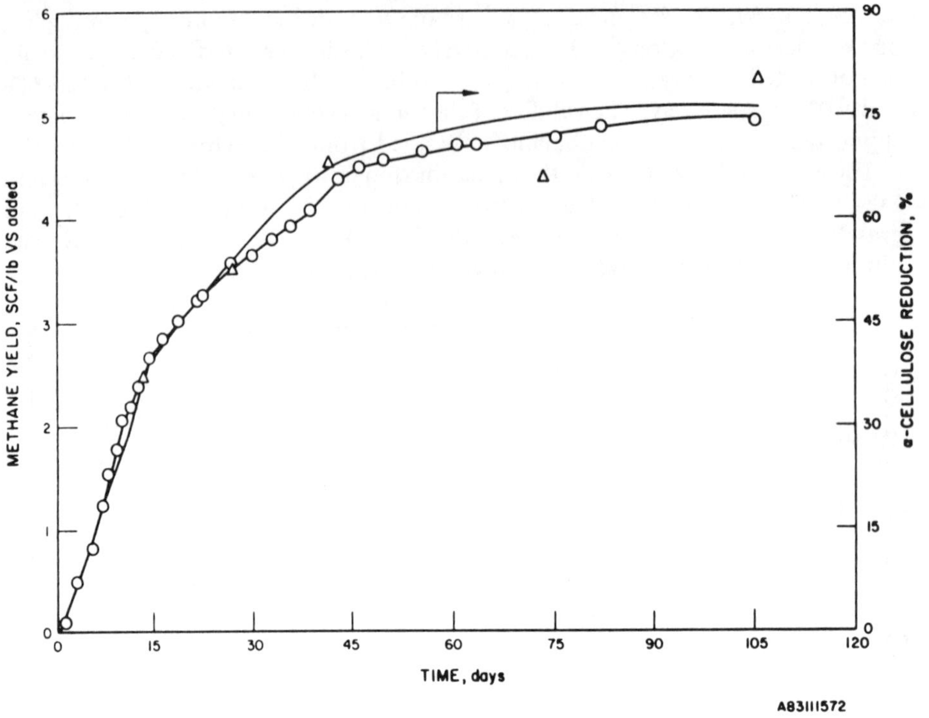

FIG. 2. Methane yield and cellulose reduction during digestion of hybrid Poplar.

The effects of particle size on the ultimate biodegradability of hybrid poplar is shown in Fig. 3. The highest methane yield (5.3 SCF/lb VS added) was achieved from the ≤0.8-mm particle size; further reduction in particle size to 0.003 mm did not improve conversion. Biodegradability of the wood was not substantially affected by increasing the particle size to ≤8-mm diameter, which represents a significant reduction in particle surface-to-volume ratio. The effect of particle size on the rate of substrate biodegradability was significant. Our results indicate that the rate of biogasification is inversely related to particle size. Comparison of methane yields from the 0.003 and ≤8-mm-diameter particles as a function of time showed that 95% of the 60-d methane yield from the larger particles could be achieved following a 20-d incubation period with the small particles. Approximately 90% of the substrate biodegradability achieved from the ≤0.8-mm-diameter particles at 60 d was obtained following a 30-d incubation period from the 0.003-mm particles.

Pretreatment of hybrid poplar with sodium hydroxide led to significant increases in the biodegradation rate, as shown in Fig. 4 but did not affect the ultimate methane yield. Comparable biodegradability of 5% NaOH-treated cottonwood was achieved following a 28-d incubation period versus a 60-d incubation period without treatment. The quantity of sodium hydroxide consumed during pretreatment of hybrid poplar was measured. All of the sodium hydroxide at the 5% (w/w) concentration was consumed during treatment. At the 50% NaOH/VS (w/w) concentration, the sodium hydroxide consumption was approximately 0.14 g/g feed volatile solids.

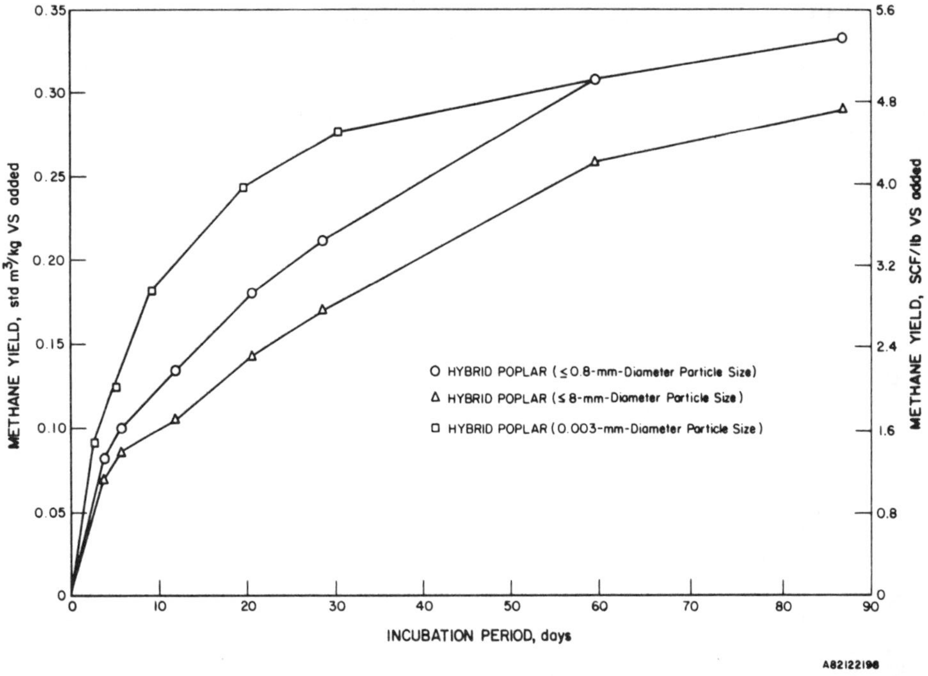

FIG. 3. Methane yields from hybrid Poplar at various particle sizes.

Discussion

Chemical compositional analyses have demonstrated that lignocellulosic materials make up 90% or more of the organic matter of woody biomass. Lignocellulose is composed of three primary constituents: cellulose, hemicellulose, and lignin. Although these constituents differ qualitatively and quantitatively from species to species and within a given species depending upon age and growth conditions, the relative quantities are 40% to 45% cellulose, 20% to 30% hemicellulose, 20% to 30% lignin (Sjostrom 1981). A minor component of wood is the extractables, which include a variety of compounds, such as resins, phenolics, flavenoids, tannins, and stilbenes.

Pure cellulose and hemicellulose have been demonstrated to be biodegradable by mixed and pure cultures under anaerobic conditions (Stafford et al. 1980); however, the degree and rate of cellulose biodegradability reportedly is influenced by factors such as the degree of crystallinity and polymerization of the molecule and capillary structure (Cowling 1975). Although hemicellulose is generally considered more biodegradable than cellulose, both groups become increasingly resistant to microbial attack when associated with lignin in a lignocellulosic matrix (Lother and Zavitkovski 1981). Fan et al. (1981) concluded that the amount of lignin governs the rate of hydrolysis of lignocellulosic materials, whereas reduction in crystallinity and particle size only moderately affects hydrolysis rate. Although lignin is resistant to microbial decomposition it can be metabolized aerobically (Scheffer 1966; Zeikus 1980). However, no

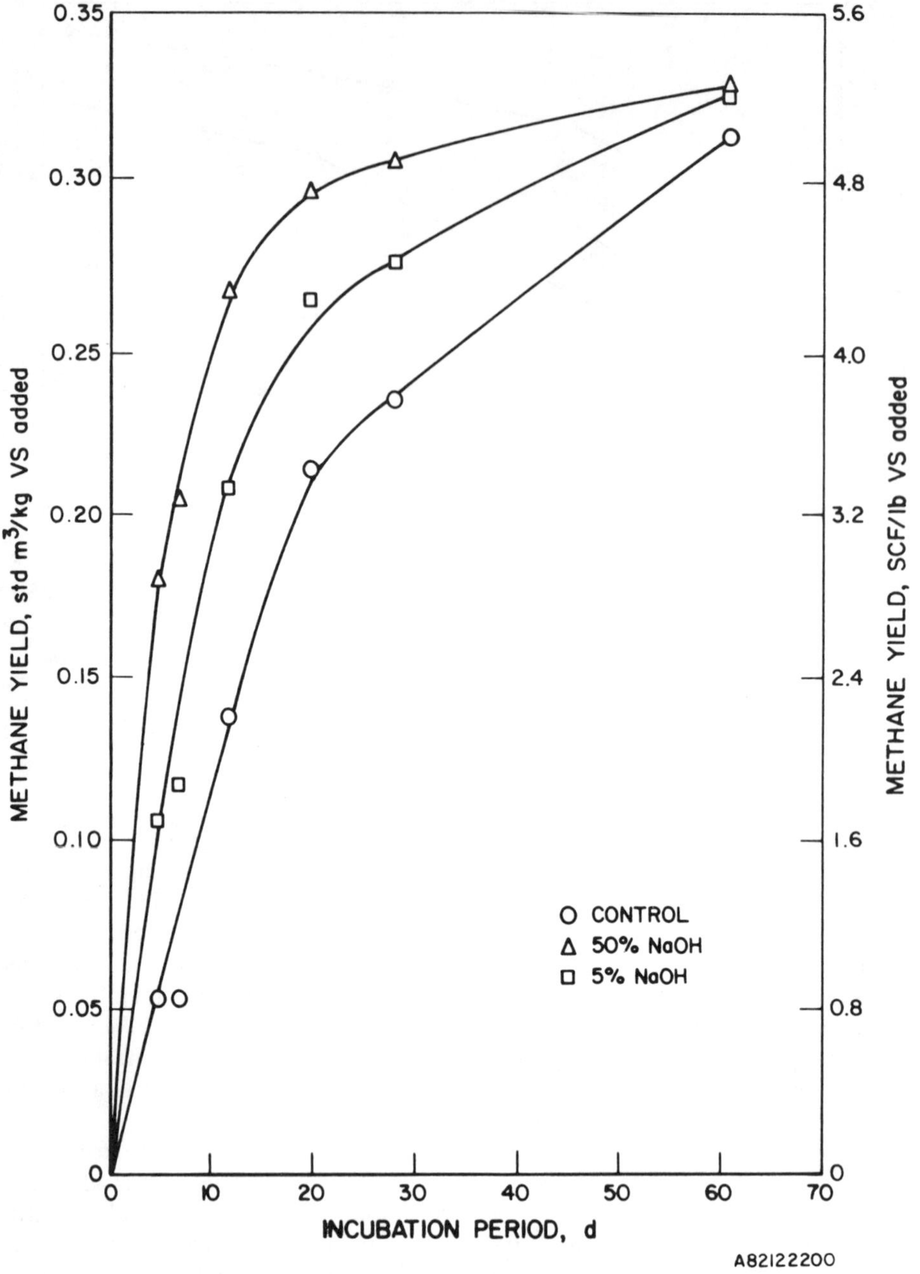

FIG. 4. Methane yields from hybrid Poplar following sodium hydroxide pretreatment.

evidence exists to date to document hydrolysis of the lignin polymers under anaerobic conditions. It has been documented, however, that the structural subunits of lignin, such as vanillic acid, cinnamic acid, catechol, protecatechoic acid, phenol, *p*-hydroxybenzoic acid, syringic acid, and syringaldehyde can be metabolized under anaerobic conditions (Taylor 1982). Thus, anaerobic degradation of lignin appears to be dependent upon a pretreatment step that may include aerobic treatment or some form of physical or chemical treatment. It has generally been thought that the organic matter in wood should be poorly biodegradable under the anaerobic conditions of the methane fermentation because lignin is refractory and retards degradation of cellulose and hemicellulose.

This supposition has been supported by studies on conventional anaerobic digestion of wood. Chandler et al. (1980) demonstrated a linear relationship between volatile solids reduction and lignin content in a study to evaluate the anaerobic biodegradability of several herbaceous and woody species. Using a model developed by these investigators, an organic matter (volatile solids) reduction of only 15% would be predicted for hardwood feed with a lignin content of 25% dry weight. These results are comparable to those observed in conventional semicontinuous-feed digestion studies conducted in our laboratory and those reported by other researchers using *in vitro* digestion methods (Baker et al. 1975). However, batch-fed digester conversion efficiencies of 53.8% and 56.7%, respectively, reported here for hybrid poplar and sycamore represent a drastic increase in comparison to literature values for the anaerobic biodegradability of biomass containing such a high lignin concentration.

The anaerobic conversion efficiencies achieved from woody substrates in this study at long solids retention times and with macro- and micronutrients, suggest that the association between lignin and other cell wall polymers are not as detrimental to microbial enzyme attack as was previously reported (Cowling 1975; Chandler et al. 1980; Zeikus 1980b). These results refute the current theory that wood is refractory to anaerobic decomposition and suggest that biogasification of wood is technically feasible and needs further documentation and study. Studies comparing methane yields from both herbaceous and woody biomass at long solids residence times in our laboratories have not demonstrated a linear relationship between lignin concentration and anaerobic biodegradability; however, the data suggest that lignin concentrations have an effect on the rate of methane fermentation. Preliminary data from our laboratory indicate that the wood lignin is refractory, suggesting that this component could be recovered and used as a chemical feedstock with substantial byproduct credits.

Although current results suggest that wood conversion requires a longer retention time than that required for herbaceous and aquatic species, this limitation is negated by the fact that, at equal loadings, wood retention is longer than that of most other biomass types because of its lower water content. This concept is illustrated in Table 3.

TABLE 3. *Comparison of stirred tank reactor hydraulic retention times at a loading of 0.3 lb VS/ft^3-day for select biomass feeds*

Feed	Total Solids Concentration, %	Hydraulic Retention Time, d
Water Hyacinth	3.5	7
Kelp	7.0	16
Hybrid Poplar	30	70

The biogasification of wood and wood residues as an energy source may be a technically feasible process because high bioconversion efficiencies have been achieved from select hardwood species following reduction to a particle size diameter of $\leq$0.8 mm. This particle size was necessary to provide a feedstock amenable to laboratory-scale operation. The development of an anaerobic digestion process for woody biomass will depend on obtaining these high methane yields using a particle size that can be produced economically.

Numerous studies on the biogasification of biomass and wastes have used hammermilled feedstocks as the standard substrate (Gracheck et al. 1981; Foutch et al. 1981). Feed preparation costs using hammermilling for MSW have been estimated at between $1.15 and $5.80 per dry metric ton depending on the particle size desired (Gracheck et al. 1981). Successful size reduction of hybrid poplar chips by hammermilling was achieved in our laboratory. A wood particle representative of the size range achieved by hammermilling was obtained using an 8-mm Urschell mill head. A methane yield of 4.5 SCF/lb VS added was obtained with this particle size compared with a methane yield of 5.2 SCF/lb VS added from the smaller ($\leq$0.8-mm-diameter) particles. These results are encouraging because the larger particle represents a size that may be achieved more economically at a larger scale.

The high methane yields (4.0 to 5.4 SCF/lb VS added) achieved from the woody species examined in this study were obtained at much more rapid rates following pretreatment with sodium hydroxide. Only 12 to 14 d were necessary to achieve similar bioconversion efficiencies requiring up to 60 d solids retention times without treatment. Approximately 14% (w/w) NaOH was necessary to achieve these increased rates.

Based on a price of $370 per ton for sodium hydroxide, the cost of caustic pretreatment would be approximately $45 per dry ton of hybrid poplar. These costs compare favorably with data presented by Fan et al. (1981) for caustic pretreatment of wheat straw ($36 per dry ton). He concluded, after screening several physical and chemical pretreatment techniques, that caustic pretreatment was one of the most promising candidates for large-scale process development based on cost and effectiveness.

Conclusions

This work demonstrated that the nonlignin fraction of several species of hardwood biomass were decomposed by a mixed population methane fermentation without physical or chemical pretreatment. These results refute previous theory and reports that wood carbohydrates are refractory to anaerobic decomposition because of their complexation with lignin. Pretreatment of wood by particle size reduction or with sodium hydroxide increased conversion rates but did not affect yields. Studies are in progress to employ these findings for development and optimization of this process for commercial production of substitute natural gas.

Acknowledgments

This work was supported by the Gas Research Institute under GRI Contracts No.

5081-323-0463 and 5080-323-4323(F). The authors thank John Conrad and Al Razik for their technical assistance, and General Electric Company and Dula's Nursery for providing the woody biomass samples.

Literature Cited

American Public Health Association. 1975. *Standard Methods for the Examination of Water and Wastewater.* 14th ed., Washington, DC.

Baker, A. J., M. A. Millett, and L. D. Satter. 1975. Wood and wood-based residues in animal feeds. Pages 75–105 *in ACS Symp. Ser.* 10: *Cellulose Technology Research,* American Chemical Society, Washington, DC.

Chandler, J. A., W. J. Jewell, J. M. Gossett, P. J. Van Soest and J. B. Robertson. 1980. Predicting methane fermentation biodegradability. *Biotechnol. Bioeng. Symp.* 10:93–108.

Chynoweth, D. P., S. Ghosh, and M. P. Henry. 1982. Biogasification of blends of water hyacinth and domestic sludge. Pages 742–755 *in Proc. of the 1980 International Gas Research Conference,* Gas Research Institute, Chicago, IL.

Colberg, P. J., K. Baugh, T. Everhart, A. Bachman, D. Harrison, L. Y. Young, and P. L. McCarty. 1981. Heat treatment of organics for increasing anaerobic biodegradability. *In Annu. Progr. Rept.,* SERI/TR-98174-1, 92 pages, March.

Cowling, E. B. 1975. Physical and chemical constraints in the hydrolysis of cellulose and lignocellulosic materials. *Biotechnol. Bioeng. Symp.* 5:163–182.

Office of Technology Assessment, Congress of the United States. 1980. *Energy From Biological Processes,* Vol. I. OTA-E-124, Washington, DC.

Fan, L.T., M. M. Gharpuray, and Y. H. Lee. 1981. Evaluation of pretreatment for enzymatic conversion of agricultural residues. *Biotechnol. Bioeng. Symp.* 11:29–46.

Foutch, G. L., and J. L. Gaddy. 1981. Culture studies on the conversion of cornstover to methane. *Biotechnol. Bioeng. Symp.* 11:249–262.

Gracheck, S. J., D. B. Rivers, L. C. Woodford, K. E. Giddings, and G. H. Emert. 1981. Pretreatment of lignocellulosics to support cellulose production using *Trichoderma reesei* QM 9414. *Biotechnol. Bioeng. Symp.* 11:47–66.

Hewett, C. E., C. J. High, N. Marshall, and R. Wildermuth. 1981. Wood energy in the United States. *Annu. Rev. Energy* 6:139.

Jerger, D. J., J. R. Conrad, K. F. Fannin, and D. P. Chynoweth. 1982a. Biogasification of woody biomass. Pages 341–372 *in Symp. Papers: Energy from biomass and Wastes* VI. Institute of Gas Technology, Chicago, IL.

Jerger, D. J., D. A. Dolenc, and D. P. Chynoweth. 1982b. Bioconversion of woody biomass as a renewable source of energy. *Biotechnol. Bioeng. Symp.* 12:233–248.

Johnson, R. C., R. K. Lay, and L. C. Newman. 1980. *Energy from Biomass: A Technology Assessment of Terrestrial Biomass Systems.* The MITRE Corp., MTR-80W259.

Klass, D. L. 1982. Energy from biomass and wastes: 1981 update. Pages 1-61 *in Symp. Papers, Energy from Biomass and Wastes* VI. Institute of Gas Technology, Chicago, IL.

Lother, D. C., and J. Zavitkovski. 1981. A Financial Analysis of Poplar Intensive Culture in the Lake States. Pages 145–164 *in Symp. Papers: Energy from Biomass and Wastes* V. Institute of Gas Technology, Chicago, IL.

Scheffer, T. C. 1966. Natural resistance of wood to microbial deterioration. *Annu. Rev. Phytopathol.* 4:147–170.

Sjostrom, E. 1981. *Wood Chemistry Fundamentals and Applications.* Academic Press, New York.

Stafford, D. A., D. L. Hawkes, and R. Horton. 1980. *Methane Production from Waste Organic Matter.* CRC Press, Boca Raton, FL.

Taylor, G. T. 1982. The methanogenic bacteria. *Progr. Ind. Microbiol.* 16:231–329.

Wilson, R. K., and W. J. Pidgen. 1964. Effect of sodium hydroxide treatment on the utilization of wheat straw and poplar wood by rumen microorganisms. *Can. J. Anim. Sci.* 44:122–124.

Zavitkovski, J. 1981. Some promising forest ecosystems of the temperate zone for biomass production and energy storage. Pages 129–144 *in Symp. Papers: Energy from Biomass and Wastes* V. Institute of Gas Technology, Chicago, IL.

Zeikus, J. G. 1980a. Chemical and fuel production by anaerobic bacteria. *Annu. Rev. Microbiol.,* pages 423–464.

Zeikus, J. G. 1980b. Fate of Lignin and Related Aromatic Substrates in Anaerobic Environments. Pages 101–110 *in Lignin Biodegradation: Microbiology, Chemistry and Potential Applications.* Vol. I, CRC Press, Boca Raton, FL.

V
SYMPOSIUM: EXTRACELLULAR MICROBIAL POLYSACCHARIDES

A. I. Laskin, *Convener*
Exxon Research and Engineering Company, Annandale, Pennsylvania

PANELISTS

M. C. Cadmus
J. W. Costerton
D. Gutnick
G. Holzwarth
O. Pines
J. Shabtai
M. E. Slodki
W. C. Wernau

CHAPTER 15

The Role of Bacterial Exopolysaccharides in Nature and Disease

J. W. COSTERTON

Department of Biology, University of Calgary, Calgary, Alberta, Canada T2N 1N4

All bacteria growing in natural or pathogenic environments elaborate at their cell surfaces an exopolysaccharide glycocalyx. This consists most often of a highly hydrated matrix of fibrillar polyanionic polymers that constitute an ion exchange resin and surround the cell in a layer. This surface matrix mediates the adhesion of bacterial cells to surfaces and the subsequent formation of surface-associated biofilms in all aquatic systems. The glycocalyx matrix uses its inherent ion exchange properties to trap and concentrate nutrients in the biofilm and this property also protects the cells within the biofilm from chemical antibacterial agents that must first saturate the matrix before reaching the innermost bacterial cells.

INTRODUCTION

The production of exopolysaccharides is a significant energy cost to bacteria, and yet direct observations of bacterial cells in a wide variety of natural and industrial environments show, unequivocally, that all such cells are surrounded by structured exopolysaccharides and that many produce very large amounts of extracellular glycocalyx material. Exopolysaccharides escaped detection by electron microscopy for many years because of their low affinity for heavy metal stains and because of their pronounced tendency to collapse during the dehydration stages of preparation for this type of observation. However, the development of a specific stain (ruthenium red) for organic polyanions (Luft 1971) and the development of stabilizing agents, such as specific antibodies (Bayer and Thurow 1977) and lectins (Birdsell et al. 1975), have allowed us to demonstrate their ubiquity, their remarkable dimensions, and their complex organization (Fig. 1). In this review, we consider the various roles of bacterial glycocalyces in bacterial growth and persistence in natural, medical, and industrial environments. It is useful, in this context, to remember that the only environment in which these structures do not have a positive survival value is in the *in vitro* monospecies culture. When bacteria are shielded from antibacterial agents and competition and are fed a predigested nutrient "compost," they often conserve metabolic energy and dispense with glycocalyx production (Doggett et al. 1964). This profound cell surface difference between cultured cells and wild cells makes extrapolation from *in vitro* data to environmental situations especially hazardous.

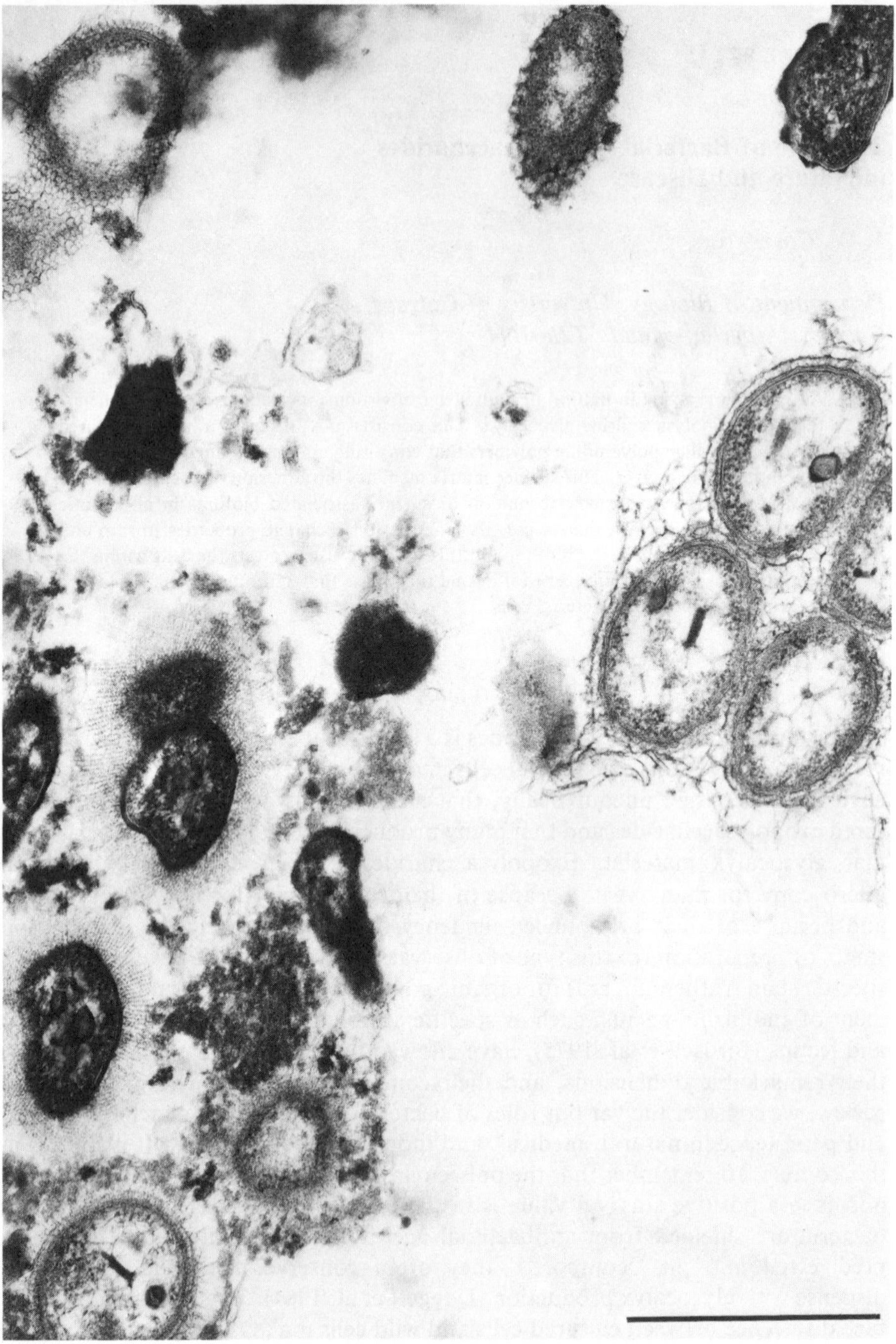

FIG. 1. Transmission electron micrograph (TEM) of a natural aquatic population of bacteria demonstrating their exopolysaccharide structures. Extracellular glycocalyces are seen to surround all of the cells in these mixed microbial populations and a wide variety of glycocalyx morphologies are seen. Bar 1.0 μm.

Discussion

Role of Exopolysaccharides in Ecological Positioning and Nutrition

Perhaps the simplest bacterial strategy for ecological positioning to obtain optimal nutrition is seen in pristine aquatic environments, such as alpine streams (Geesey et al. 1978), where bacteria adhere avidly to available surfaces by means of their glycocalyces. These adherent bacteria rapidly develop adherent biofilms in which they are bound to each other, and to the inert substrate, by a fibrous exopolysaccharide matrix that is usually very extensive (Fig. 2). Bacterial exopolysaccharides are usually composed of uronic acids and other anionic molecules (Sutherland 1977), and they act as ion exchange resins that extract charged organic molecules and inorganic ions from the passing water. This ion exchange effect, added to the natural concentrations of organic molecules at surfaces in aquatic environments (Zobell 1943), produces a favorable concentration of bacterial nutrients within the bacterial biofilm, and even in very oligotrophic waters, the adherent bacterial biofilms thicken to macroscopic dimensions.

Within the microbial biofilm, a complex population depends on many other nutrient factors such as the recycling of the components of cells that die and lyse within the matrix and on association with primary producers such as algae. Locke et al. (1984) have developed a biofilm concept that visualizes the adherent biofilm as a coherent community within which cell-cell association and nutrient cycling are fully operative. In our laboratory, well-developed stable biofilms are "fed" daily for as little as 15 min, and they thrive, using nutrient "trapping" and nutrient recycling. Bacteria usually form metabolically functional consortia in natural environments (Costerton et al. 1981) in which a primary colonizer attaches at an ecologically favorable site and a secondary colonizer often associates, by glycocalyx interaction, with the primary organism (Fig. 3) to begin a functional metabolic consortium that may finally include several species. A consensus is developing that such complex endergonic processes as the production of methane and H_2S actually require bacterial consortia (Hamilton 1984).

In the more complex case in which the substratum is not inert, bacteria show great specificity in their colonization of surfaces, and both starch and cellulose have been shown to be avidly colonized by amylolytic and cellulolytic organisms, respectively (Minato and Suto 1978). Thus, when a nutrient substratum (e.g., a leaf) is introduced into an aquatic environment, it will be colonized by specific organisms (Cheng et al. 1980) that will begin a sequence of monospecies and consortial developments until the nutrients are exhausted or the available space is filled (Fig. 4). Animal tissues may themselves be sources of nutrient for bacteria (e.g., urea, keratin, etc.), and many animal tissues rapidly develop commensal bacterial communities on their surfaces (Marrie et al. 1978) by mutual attraction of their glycocalyces by tissue lectins. In some cases the animal tissue and the bacterial community may develop a physiological cooperation based on shared enzymes and metabolic complementation (Cheng and Costerton 1979). The role of the bacterial glycocalyx is paramount in all of these interactions because this structure constitutes the functional external bacterial surface and provides the matrix that holds the members of the consortia in close juxtaposition.

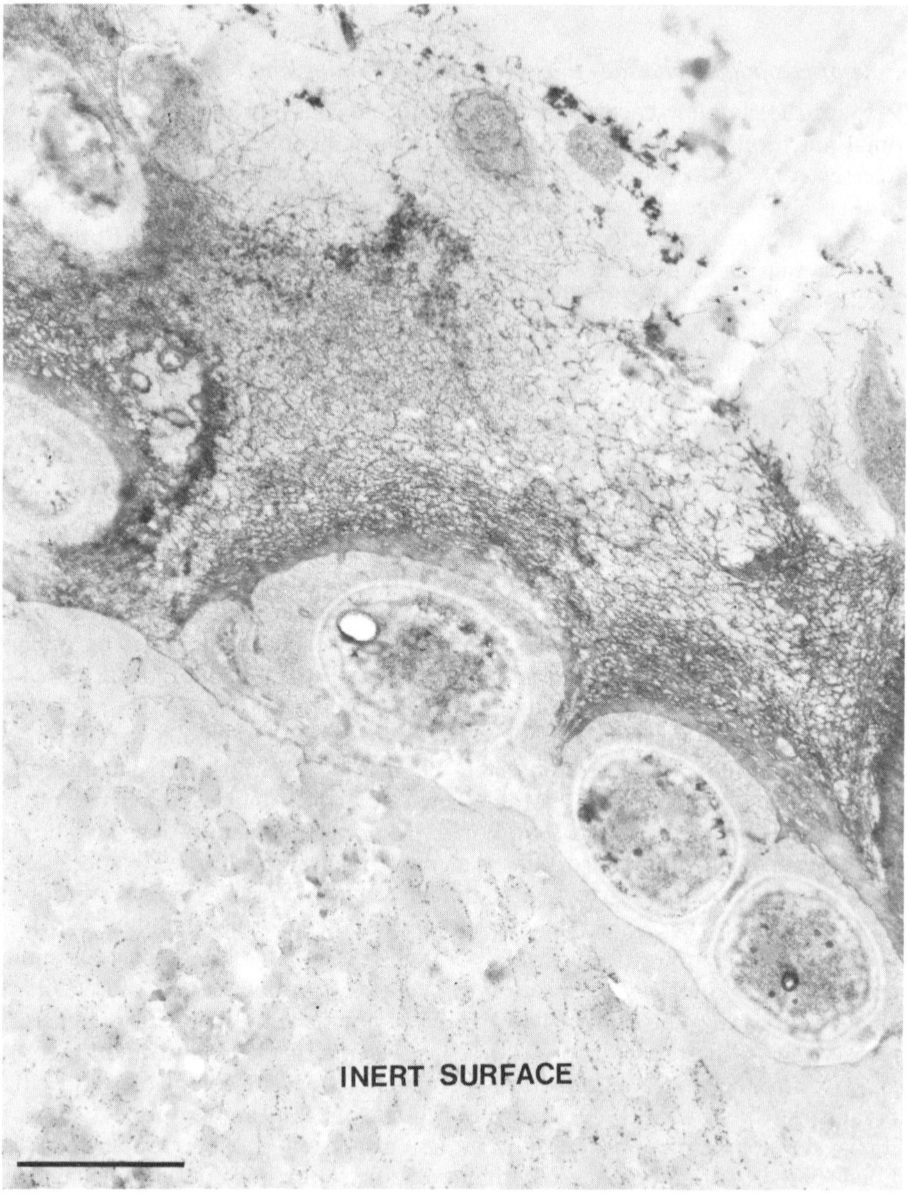

FIG. 2. TEM of the microbial biofilm that developed on a submerged surface in the Athabasca River. The sessile bacterial cells are adherent to the colonized surface and are embedded in an extensive matrix of their own fibrous anionic exopolysaccharide glycocalyces. Bar 1.0 μm.

Role of Exopolysaccharides in Protection from Antibacterial Factors

In any natural environment, bacteria operate at the interface between organic nutrients and the biological food chain. Bacteria are the only organisms capable of using dissolved organic nutrients in low concentrations (Costerton et al. 1981), and their cells constitute the first stage in the natural biological predation series.

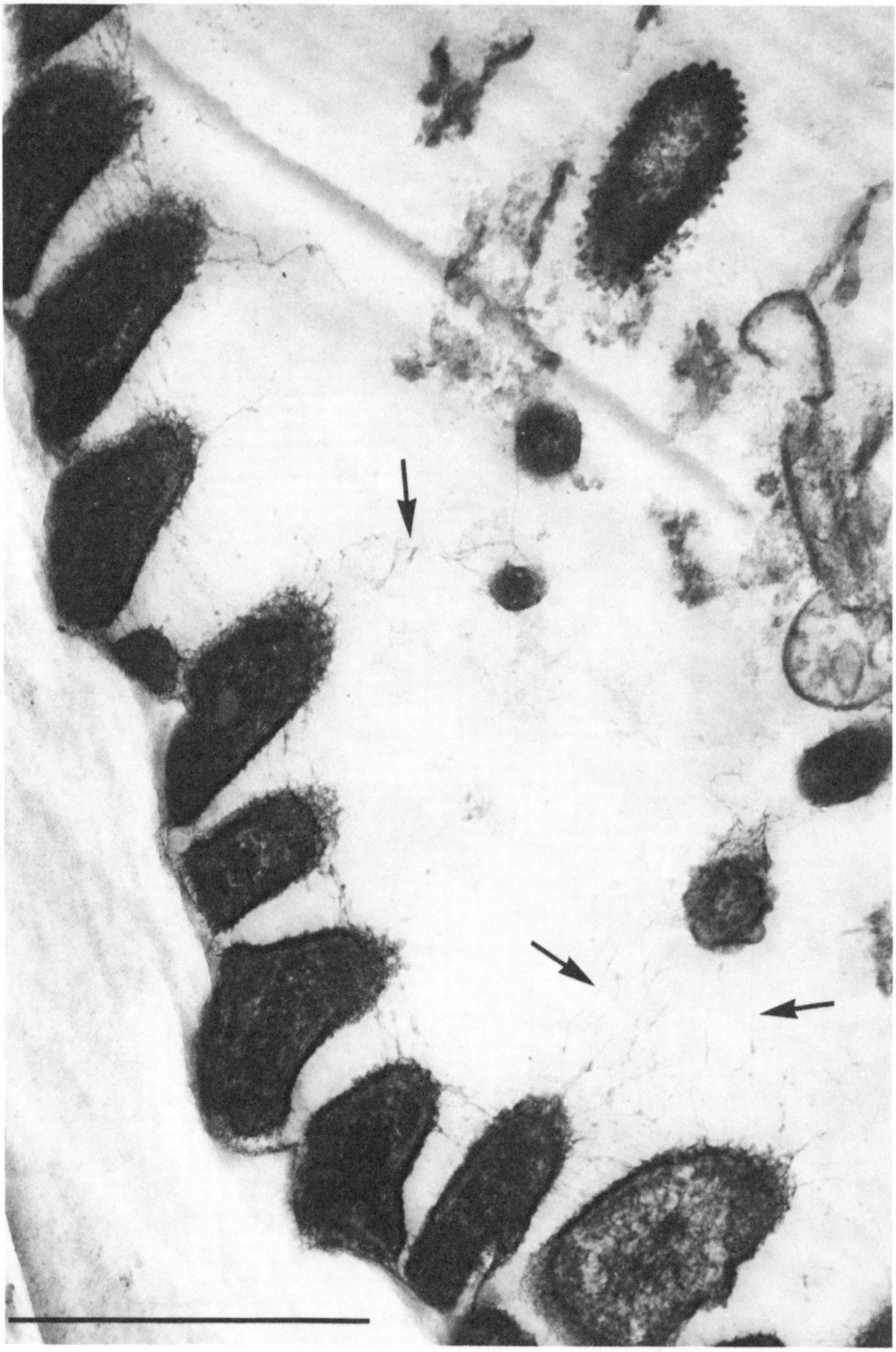

FIG. 3. TEM of a cellulose fragment colonized and partly digested by primary cellulolytic bacteria that adhere to the cellulose and to each other by means of glycocalyx fibers. A functional consortium is set up when the very small secondary colonizers adhere to the layer of primary colonizers, also by means of their fibrous glycocalyces (arrows). Bar 1.0 μm.

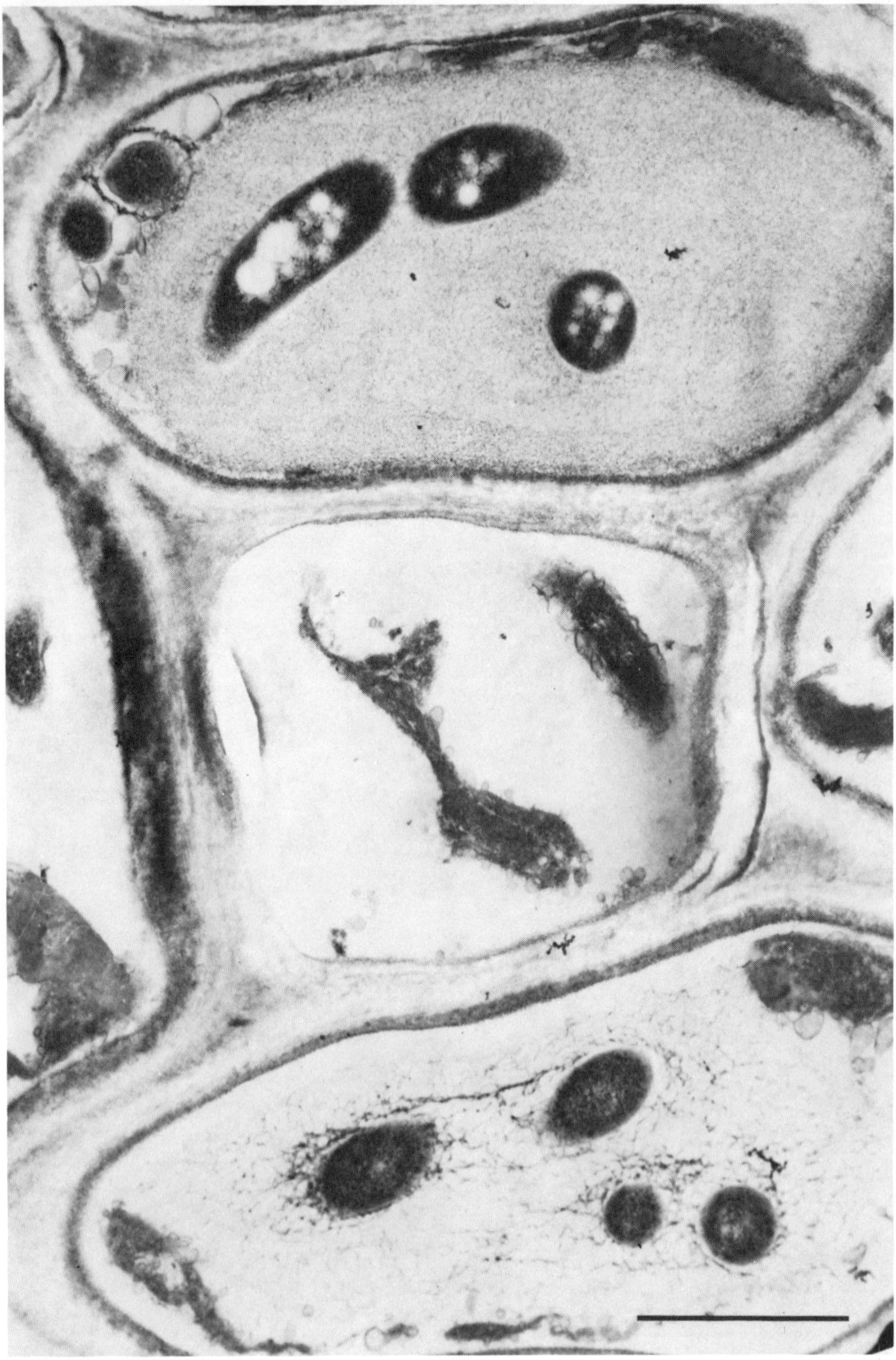

FIG. 4. TEM of a partially digested legume leaf showing that different bacteria have come to predominance in different niches and have filled the available space with their cells and their fibrous glycocalyces. Bar 1.0 μm.

Thus, a very wide variety of eukaryotic organisms seek to consume bacteria by engulfment, sieving, vortex feeding, rasping, etc., and the bacterial counter strategy is the formation of exopolysaccharide-enclosed adherent biofilms (Costerton et al. 1981). Similarly, the defensive phagocytic cells of animals seek to engulf bacteria, following opsonization by specific antibodies, but this attack is frustrated by simple capsule formation (Baltimore and Mitchell 1980) and entirely obviated by adherent biofilm formation (Marrie and Costerton 1983; Mayberry-Carson 1984). The corollary of this bacterial strategy is that these microorganisms are often confined to a dental plaque-like biofilm on a surface, and while they can persist indefinitely, they cannot launch a disseminating invasion of the ecosystem until the predatory cells are weakened or withdrawn. These surface-associated bacteria are very difficult to recover and even more difficult to quantitate by conventional microbiological methods, and they usually remain undetected until the development of the *in situ* biofilm sampler—the Robbins Device.

Biofilm bacteria are also protected from subcellular and molecular antibacterial agents. Adherent matrix-enclosed cells are protected from bacteriophage, specific antibodies (Schwarzmann and Boring 1971), and surfactants (Govan 1975), and all of these agents can be used to favor the growth of biofilm-producing bacteria, even in *in vitro* culture conditions (Govan 1975).

The bacterial glycocalyx that constitutes the matrix of the bacterial biofilms is an anionic ion exchange resin, and one cannot expect that molecular antibacterial agents, such as biocides and antibiotics, would have the same access to a deep biofilm cell as to a free-floating uncoated cell. As a molecular antibacterial agent penetrates a biofilm, it is bound to its target in superficial cells and is ionically bound to the matrix components so that it is depleted as it penetrates. Biofilms that are about 200 cells (1 mm) in thickness are routinely found in medical (Marrie and Costerton 1984) and industrial (Costerton and Lashen 1984) systems, and it is naive to expect an antibacterial agent to penetrate the biofilm and to kill its innermost cells at a concentration that will kill free-floating naked cells in culture. We have shown that biocides must be used in substantially higher concentrations than have heretofore been used (Ruseska et al. 1982) and that a long-term "soaking" dose of biocide yields maximum cost effectiveness. We have also shown that 1,000 μg/ml of tobramycin fails to kill biofilm cells of *Pseudomonas aeruginosa* when the same cells in suspension have an MIC of 0.6 μg/ml (Nickel et al. 1984). Clearly, we cannot predict an effective biofilm-killing dose from *in vitro* test data, because biocides differ markedly in their penetrating power. We must test biocides and antibiotics against sessile bacterial populations, under realistic conditions, in order to determine appropriate doses and dosing patterns for effective control of biofilm bacteria. It is sobering to realize that the bacteria involved in corrosion and in fouling in industrial systems are, by definition, the innermost cells of thick bacterial biofilms. The flow-through Robbins Device (Fig. 5) provides a system in which biofilms can be developed, monitored, and challenged with antibacterial agents to establish effective doses for the control of biofilm bacteria.

Factors Affecting Exopolysaccharide Production

Nutritional conditions have a profound effect on exopolysaccharide production by bacteria and factors such as divalent cation concentration, carbon-nitrogen

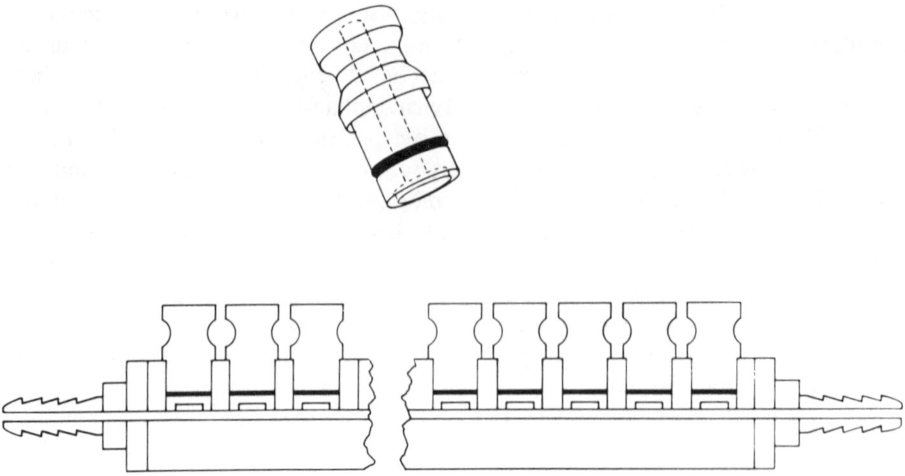

FIG. 5. Diagrammatic representation of the modified Robbins Device in which discs of various materials can be exposed to flowing fluids containing bacteria, and biofilms can be built up and aseptically sampled by removing specimen "plugs."

ratio, and specific substrate (e.g., gluconate) availability have been shown to be important in this regard (Chan et al. 1984). Further, the physical nature of the medium (solid vs. liquid) has been shown to affect exopolysaccharide production in different ways in different strains. Bacteria in batch culture have been shown to continue to produce very large amounts of exopolysaccharide even after the culture had entered the stationary phase of growth (Chan et al. 1984). However, the sine qua non of exopolysaccharide production is the availability of sufficient organic carbon and energy for this very endergonic process. Recent fascinating observations by Amy and Morita (1983) have confirmed that starvation radically reduces bacterial cell size and also reduces exopolysaccharide production. This observation confirms earlier reports (Novitsky and Morita 1976) that deep sea bacteria, most of which are starved ultramicrocells, show a sharply reduced tendency to adhere to available surfaces in comparison to the bacteria in more nutritionally favorable estuarine environments.

These data allow the hypothesis that a point source of organic nutrients in a flowing aquatic system would stimulate the development of especially thick metabolically active biofilms on adjacent submerged surfaces, and that this burgeoning biofilm would trap a large proportion of the organic nutrients for conversion into biomass and exopolysaccharide (Locke et al. 1984). Studies of the sessile and planktonic bacterial populations of receiving streams (Osborne et al. 1983) have supported this hypothesis in that point source nutrient stimulation has been shown to produce thickening and activation of biofilms on adjacent surfaces and very limited downstream effects (Ladd et al. 1979). In contrast to streams, in which bulk flow of water occurs, groundwater is exposed to very large amounts of the surface area of soil particles, and proportionately larger biofilm populations develop on these surfaces and sharply reduce the nutrient content of the flowing water (Ladd et al. 1982). This nutrient withdrawal from flowing groundwater is so effective that both bacterial numbers and total organic content decrease sharply downward through the soil horizons, but starved ultramicrocells are still present in groundwater in deep soil horizons. These small but effective replicating units can be restored to full size, complete metabolic activity, and effective exopolysaccharide production and adhesion by the simple provision of suitable nutrients. When water containing bacteria is flowed through the pore spaces of a solid matrix (such as a sand pack, or sandstone) a similar pattern of inlet face biofilm plugging is seen (Gupta and Swartzendruber 1962; Hart et al. 1960). We have examined bacterial growth in glass bead "cores" (Shaw et al. 1984) that provide a constant pattern and size of pores ($\sim 33 \mu m$), and mimic a very porous sandstone (~ 8 darcys), and we note very extensive adhesion, proliferation, exopolysaccharide formation, and plugging of the inlet face of this solid matrix by bacteria (Fig. 6). This phenomenal biofilm formation sharply reduces both the hydraulic flow and the total organic carbon content of the water. The internal surfaces of the core can be exposed by simple fracturing of the fragile glass structure and show no bacterial adhesion or biofilm formation (Fig. 7). Thus, the basic bacterial "strategy" governing the growth in solid matrices is identical with that seen in bulk flow systems, in that these microorganisms form exopolysaccharide-enclosed biofilms on the first available surfaces, reduce the nutrient content of the percolating water, and penetrate the porous matrix in the form of nonadherent, starved ultramicrocells. The practical sequellae of this basic pattern of bacterial growth in solid matrices are that the bacteria introduced into oil-bearing formations will normally express metabolic activity only very near the injection face, but that metabolic activity (including exopolysaccharide production and pore plugging) can be induced at considerable depth if nutrients can be made to penetrate and to stimulate preexisting ultramicrocells.

CONCLUSIONS

Ecological studies of bacterial growth in natural systems have discovered certain patterns, or "strategies," that are used by bacteria in whatever environment they find themselves. Unlike test tube bacteria, wild strains produce large amounts of exopolysaccharide and they use this external component to colonize specific surfaces, to mediate specific associations with other bacteria and with tissues, to produce the matrix of a protective biofilm, and to trap and concentrate nutrients from flowing fluids. In ecosystems with a finite supply of nutrients, these molecules are adsorbed and metabolized by bacteria within adherent biofilms until starvation overtakes any cells that penetrate past this nutrient-rich zone. These starved cells reduce their cell size and their exopolysaccharide production and remain as minute, nonadherent propagules until they encounter sufficient concentrations of nutrients to resume normal metabolic activity. Bacteria are relatively primitive cells that employ these basic strategies of growth and survival, in all environments, and their growth and distribution in any particular system may be anticipated if we bear in mind their predictable responses of growth and exopolysaccharide production in response to the availability of surfaces and of nutrients.

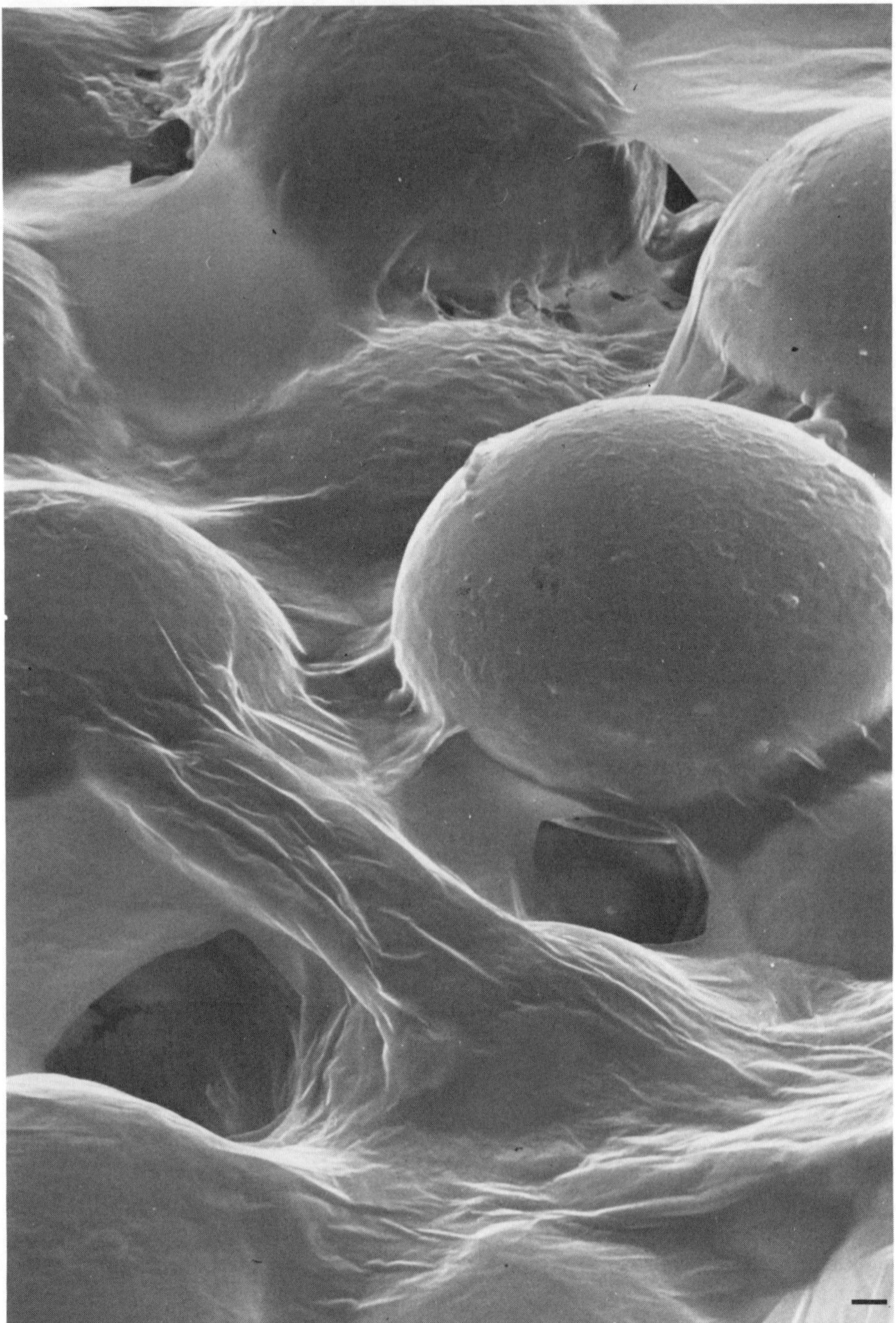

FIG. 6. Scanning electron micrographs (SEM) showing bacterial growth and glycocalyx production that has occluded the inlet surface of a glass bead "core," which mimics an open sandstone rock, with a "skin" of biofilm within which the bacteria are virtually buried. The inlet surface of the core is nutrient-rich and bacteria adhere and proliferate. Bar 5 μm.

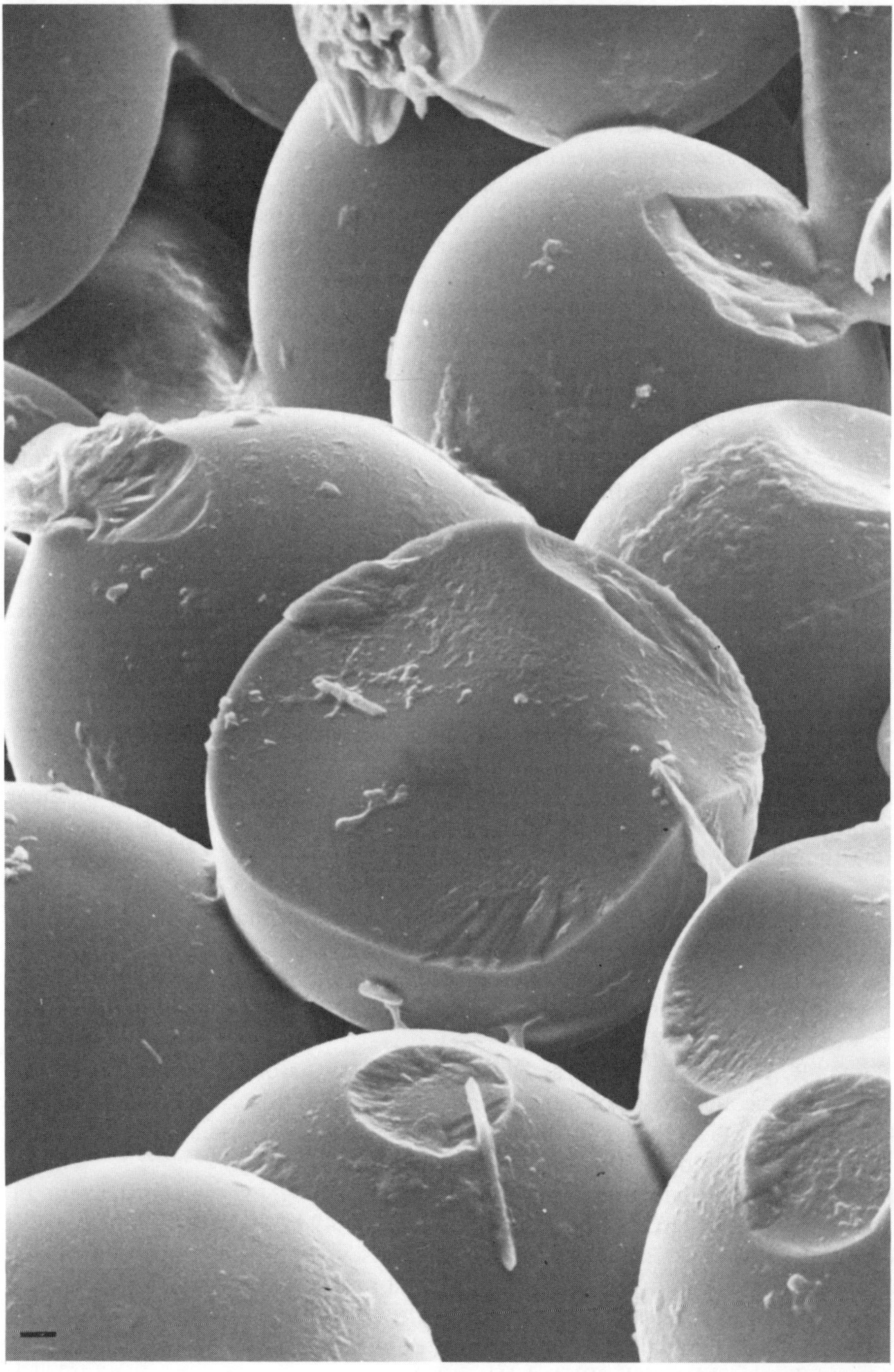

FIG. 7. SEM of the same core seen in Fig. 6 that has been broken 2 cm below the heavily colonized inlet face. Note the absence of adherent bacterial growth in this nutrient-depleted zone of the glass bead core. Bar 5 μm.

LITERATURE CITED

Amy, P. S., and R. Y. Morita. 1983. Starvation-survival patterns of sixteen isolated open ocean bacteria. *Appl. Environ. Microbiol.* 45:1109–1115.

Baltimore, R. S., and M. Mitchell. 1980. Immunologic investigations of mucoid strains of *Pseudomonas aeruginosa:* Comparison of susceptibility to opsonic antibody in mucoid and nonmucoid strains. *J. Infect. Dis.* 141:238–247.

Bayer, M. E., and H. Thurow. 1977. Polysaccharide capsule of *Escherichia coli*: Microscope study of its size, structure and sites of synthesis. *J. Bacteriol.* 130:911–936.

Birdsell, D. C., R. J. Doyle, and M. Morgenstern. 1975. Organization of teichoic acid in the cell wall of *Bacillus subtilis. J. Bacteriol.* 121:726–734.

Chan, R., J. S. Lam, K. Lam, and J. W. Costerton. 1984. Influence of culture conditions on expression of the mucoid mode of growth of *Pseudomonas aeruginosa. J. Clin. Microbiol.* 19:8–16.

Cheng, K. -J., and J. W. Costerton. 1979. Adherent rumen bacteria: Their role in the digestion of plant material, urea, and epithelial cells. Pages 227–250 *in* Y. Ruckebush and P. Thivend, eds., *Digestive Physiology and Metabolism in Ruminants.* MTP Press, Lancaster.

Cheng, K. -J., R. E. Howarth, and J. W. Costerton. 1980. Sequence of events in the digestion of fresh legume leaves by rumen bacteria. *Appl. Environ. Microbiol.* 40:613–625.

Costerton, J. W., and E. S. Lashen. 1984. Influence of biofilm on efficacy of biocides on corrosion-causing bacteria. *Mater. Performance* 23:34–37.

Costerton, J. W., R. T. Irvin, and K. -J. Cheng. 1981. The bacterial glycocalyx in nature and disease. *Annu. Rev. Microbiol.* 35:299–324.

Doggett, R. G., G. M. Harrison, and E. S. Wallis. 1964. Comparison of some properties of *Pseudomonas aeruginosa* isolated from infections of persons with and without cystic fibrosis. *J. Bacteriol.* 87:427–431.

Geesey, G. G., R. Mutch, J. W. Costerton, and R. B. Green. 1978. Sessile bacteria: An important component of the microbial population in small mountain streams. *Limnol. Oceanogr.* 23:1214–1223.

Govan, J. R. W. 1975. Mucoid strains of *Pseudomonas aeruginosa:* The influence of culture medium on the stability of mucus production. *J. Med. Microbiol.* 8:513–522.

Gupta, R. P., and D. Swartzendruber. 1962. Flow-associated reduction in the hydraulic conductivity of quartz sand. *Soil Sci. Soc. Am. Proc.* 126:6–10.

Hamilton, W. A. 1984. The sulphate-reducing bacteria: Their physiology and consequent ecology. *Proc. of FEMS Symp.* pp. 1–5.

Hart, R. T., T. Fekete, and D. L. Flock. 1960. The plugging effect of bacteria in a sandstone system. *Can. Mining Met. Bull.* 53:495–501

Ladd, T. I., J. W. Costerton, and G. G. Geesey. 1979. Determination of the heterotrophic activity of epilithic microbial populations. Pages 180–195 *in* J. W. Costerton and R. R. Colwell, eds., *Native Aquatic Bacteria: Enumeration, Activity and Ecology.* ASTM Technical Publication No. 695. American Society of Testing Materials, Philadelphia, PA.

Ladd, T. I., R. M. Ventullo, P. M. Wallis, and J. W. Costerton. 1982. Heterotrophic activity and biodegradation of labile and refractory compounds by groundwater and stream microbial populations. *Appl. Environ. Microbiol.* 44:321–329.

Locke, M. A., R. R. Wallace, J. W. Costerton, R. M. Ventullo, and S. E. Charlton. 1984. River epilithon: Towards a structural-functional model. *Oikos* 42:10–22.

Luft, J. H. 1971. Ruthenium red and ruthenium violet. I. Chemistry, purification, methods of use for electron microscopy and mechanism of action. *Anat. Rec.* 171:347–368.

Marrie, T. J., and J. W. Costerton. 1983. A scanning and transmission electron microscopic study of the surface of intrauterine contraceptive devices. *Am. J. Obstet. Gynecol.* 146:384–393.

______. 1984. Scanning and transmission electron microscopy of "in situ" bacterial colonization of intravenous and intraarterial catheters. *J. Clin. Microbiol.* 19:687–693.

Marrie, T. J., G. K. M. Harding, and A. R. Ronald. 1978. Anaerobic and aerobic urethral flora in healthy females. *J. Clin. Microbiol.* 8:67–72.

Mayberry-Carson, K. J., B. Tober-Meyer, J. K. Smith, D. W. Lambe, Jr., and J. W. Costerton. 1984. Bacterial adherence and glycocalyx formation in osteomyelitis experimentally induced with *Staphylococcus aureus*. *Infect. Immunol.* 43:825–833.

Minato, H., and T. Suto. 1978. Technique for fractionation of bacteria in rumen microbial ecosystem. II. Attachment of bacteria therefrom. *J. Gen. Appl. Microbiol.* 24:1–16.

Nickel, J. C., I. Ruseska, and J. W. Costerton. 1984. Antibiotic resistance in an "in vitro" catheter-associated infection. Submitted for publication.

Novitsky, J. A., and R. Y. Morita. 1976. Morphological characteristics of small cells resulting from nutrient starvation of a psycrophilic marine vibrio. *Appl. Environ. Microbiol.* 32:617–622.

Osborne, L. L., R. W. Davies, R. M. Ventullo, T. I. Ladd, and J. W. Costerton. 1983. The effects of chlorinated municipal sewage and temperature on the abundance of bacteria in the Sheep River, Alberta. *Can. J. Microbiol.* 29:261–270.

Ruseska, I., J. Robbins, E. S. Lashen, and J. W. Costerton. 1982. Biocide testing against corrosion-causing oil field bacteria helps control plugging. *Oil Gas J.* March 8:253–264.

Schwarzmann, S., and J. R. Boring, III. 1971. Antiphagocytic effect of slime from a mucoid strain of *Pseudomonas aeruginosa*. *Infect. Immunol.* 3:762–767.

Shaw, J. C., B. Bramhill, N. C. Wardlaw, and J. W. Costerton. 1984. Bacterial fouling in a model core system. Submitted for publication.

Sutherland, I. W. 1977. Bacterial exopolysaccharides—their nature and production. Pages 27–96 *in* I. W. Sutherland, ed., *Surface Carbohydrates of the Prokaryotic Cell.* Academic Press, London.

Zobell, C. E. 1943. The effect of solid surfaces on bacterial activity. *J. Bacteriol.* 46:39–56.

CHAPTER 16

Fermentation Methods for the Production of Polysaccharides

W. C. Wernau

Pfizer Central Research, Groton, Connecticut 06340

A number of fermentation-derived bacterial and fungal polysaccharides have sparked commercial interest over the past 20 years. The most successful of the bacterial polysaccharides, xanthan gum, serves to illustrate the array of approaches investigated for the production of these materials. Strain selection, inoculum character, substrates, fermentor designs, and fermentation modes will be discussed and comments offered on differences between bacterial and fungal polysaccharide fermentation.

Introduction

Xanthan gum is the most widely used microbial polysaccharide. It has applications in the food industry, in the oilfield, and in a variety of industrial areas as well. Many of these applications arise from the exceptional viscosity characteristics and suspending abilities of the polymer. Xanthan gum is produced by aerobic fermentation by the bacterium *Xanthomonas campestris*. Discovered at the USDA's Northern Regional Research Laboratories, *Xanthomonas campestris* strain NRRL B-1459 was used initially in commercializing the product. Since that time, other strains of *Xanthomonas campestris* and other species of *Xanthomonas* have been studied as well. In this paper we will illustrate the similarities between xanthan gum fermentation and other traditional fermentations and also point out some of the unique features encountered when fermenting polysaccharides.

Discussion

Strain selection. As in any commercial fermentation, strain selection is extremely important. In the case of xanthan gum, the original NRRL B-1459 strain proved to be difficult to stabilize, breaking down upon subculture and degenerating during continuous fermentation (Lindblom and Patton 1967). Traditional culture preservation techniques were applied, including lyophilization, dried paper storage, and subculturing. These were successful in preserving, but not in stabilizing, the culture (Jeanes et al. 1976). Recently patented are degeneration-resistant strains of *Xanthomonas* derived from semicontinuous and continuous cultures of *Xanthomonas campestris* (Weisrock 1982a; Bauer and Khosrovi 1983). As the patent literature amply illustrates, many other species and strains of *Xanthomonas* have been investigated (Patton and Lindblom 1962). In certain applications, the yellow pigment coproduced during xanthan fermentation is undesirable. Wagner (1978) and others have isolated colorless mutants of *Xanthomonas campestris* without difficulty.

Inoculum character. The fermentation of xanthan gum starts off with inoculum staging. McNeely (1968a) specifies an optimized inoculum for production of biopolymer, as does Colin and Fleury (1971). In the latter case, a very small inoculum (<0.2%) of composition similar to the final fermentation medium is specified. In the former case, as opposed to conventional wisdom, it was found that the fermentation cycle could be shortened considerably by using inoculum in the stationary phase of growth rather than the logarithmic phase. Claimed advantages resided in cost savings and reduced chance of contamination.

Substrates. A wide variety of practical and exotic substrates have been proposed as raw materials for xanthan gum fermentation. These span the range from agricultural waste products to refined commercial carbohydrate sources. Among the cruder substrates can be listed flour and bran, cereal grain hydrolyzate, dry-milled corn starch, kenaf juice, lactoserum and lactoserum plus cereal bran combinations, soybean whey liquid, and defatted soybean powder (McNeely 1969a, 1969b; Miescher 1969; Cadmus et al. 1971; Colin and Merle 1972; Kanda 1977, 1978). The more refined substrates include starch, glucose and cerelose, corn syrup, sucrose, etc. (McNeely 1968b). Generally speaking, cost and the end-use of the product will dictate the choice of substrate, with the cruder materials being perfectly acceptable for drilling muds quality material, for example, but quite unsuitable for some enhanced oil applications (Wernau 1978). Various crude substrates require quite different media to achieve optimized fermentation performance. In a sense this situation differs little from that encountered during antibiotic or other traditional fermentation processes. A balance must be drawn among relevant variables such as nitrogen, carbon, and phosphorus levels and the requirements for yield and productivity. As opposed to traditional processes for antibiotics, cell growth in the xanthan fermentation is an order of magnitude lower in terms of grams dry weight of cells per liter of broth. Consequently, crude substrates with high available nitrogen content can cause excessive growth and are not very suitable unless used in combination with other crude substrates containing high carbon content. It should be clear that productivity is proportional to cellular concentration in the fermentor, but that polysaccharide yield is inversely proportional to cellular concentration.

Other additives in the fermentation medium can affect fermentation performance and broth quality. The nature of the nitrogen source (organic or inorganic) appears to be important in terms of broth quality. Partially digestible crude materials will, of course, leave residues in the broth. In applications where whole broth is used (some enhanced oil applications) these crude materials may be unacceptable, leading to plugging of formations or wellbores (Wernau 1978). Crude materials will, of course, also contribute to recovery purification difficulties where solid product forms are used (e.g., food applications) (McNeely 1968b). Regarding vitality of the cells, a rich nutritious medium can produce very "healthy" cells. However, it is not always the objective of the industrial fermentation practitioner to produce such cells but more often to optimize production of a desired end-product. Medium ingredients should be formulated with this objective in mind. Defined minimal media will be discussed below.

By implication, the carryover of inoculum components can affect yield and performance by affecting growth and product quality. Certain strains of *Xan-*

thomonas may perform better in the presence of certain amino acids (Duc et al. 1978). A water-soluble medium can give high quality broths for enhanced oil recovery applications, and patents cover such media as well as use of certain organic acids and Krebs cycle acids to improve performance (Wernau 1978). In the xanthan fermentation these Krebs cycle acids, and particularly citric acid, serve to solubilize medium components such as calcium and iron salts in the presence of phosphates and to stimulate activity of the Krebs cycle. This leads to a better metabolic balance between carbon flow from glucose through the hexose monophosphate and Entner Doudoroff pathways and final oxidation through the Krebs cycle. Broth ion levels can be critical and a low Ca^{++} xanthan gum can be produced with "smoother" properties for certain applications (Richmon 1983). Certain cations bind quite strongly to xanthan. Ferric ion, for example, has been found to have detrimental effects on injectivity of xanthan during enhanced oil applications (Philips et al. 1982). Yet some iron is needed for growth of *Xanthomonas campestris* during the fermentation. A careful balance must be struck between the needs of the organism and the requirements of the product. Deoxycholate and cholate are reported to improve yield and the use of excess $SO_4^{=}$ is reported to reduce viscosity (Weisrock 1981, 1983a). The use of oxidative phosphorylation uncouplers is also claimed to have a beneficial effect upon yield (Nichiden-Kagaku 1982).

The pyruvic acid content of xanthan gum is a function of fermentation conditions and strain. It is possible to produce polymers containing a range of pyruvate contents from no pyruvic acid to polymers containing a ketal-linked pyruvic acid on almost every sidechain of the molecule (Philips et al. 1982). Pyruvate-free xanthan has been patented as has its deacetylated derivative (Wernau 1981a). These various xanthans appear to possess differences in tolerance for "hardness" salts in brines, viscosity effectiveness, and broth viscosity (Philips et al. 1982). Although these differences in primary structure affect the properties of the polysaccharide, medium composition also influences the result. For example, during continuous fermentation the choice of limiting nutrient influenced the pyruvate content of the polymer (Davidson 1978).

Fermentor design. The application of engineering to the xanthan gum fermentation has led to quite a few interesting fermentor designs. Because of the highly viscous and pseudoplastic character of the broths, xanthan fermentations offer a significant challenge to the biochemical engineer concerned with scale-up and to the laboratory investigator as well. In the traditional Rushton turbine stirred-tank configuration, mixing and mass transfer problems predominate. Mass transfer correlations developed for inviscid systems cannot predict performance under highly viscous conditions. Since shear rate in the tank varies (being highest at the stirrer), the effective broth viscosity also varies. Air bubbles tend to coalesce because of reduced velocity away from the stirrer, and interfacial gas bubble area per unit liquid volume is significantly less than in inviscid systems. In smaller laboratory fermentors, wall effects become significant, and it is not uncommon to see an impeller turning at 1000 rpm and stagnant broth a few inches away. Poor mixing, particularly near the walls of the fermentor, is worsened by the presence in laboratory fermentors of excessive baffling, cooling devices, pH probes, dissolved oxygen probes, thermowells, and sampling lines. At larger scale, cooling coils and baffles can represent impediments to good mixing. Proper design is therefore of great importance in order to balance mixing and oxygen transfer requirements with heat transfer demands of the fermentor. Fortunately, the oxygen demands of the xanthan fermentation are low by industrial standards and so too is the heat of fermentation (W. C. Wernau, unpublished data).

These problems have led to the proposal of alternative fermentor configurations. Since xanthan gum is produced quite nicely on agar plates, Lipps proposed the use of solid surface methods, including a moving belt configuration where inoculation took place at one end and product was removed at the other (Lipps 1966a, 1966b). Emulsion fermentation has also been proposed in which an oil/water emulsion of xanthan is produced, with the oil phase being the proposed continuous phase, leading to a reduction in broth viscosity and facilitated oxygen transfer (Engelskirchen et al. 1984; Maury 1982). Similar techniques for reducing viscosity are employed in oil/water emulsions proposed for pipeline transportation of heavy crude oils, where high viscosities are encountered. The lipopolysaccharide "emulsan" from *Acinetobacter calcoaceticus* has been proposed for use in this latter application (Basta 1983). New stirred-tank designs have also been patented (Hitachi 1981). As a final illustration, a patent has recently been issued for the production of polysaccharides, including xanthan gum, by polysaccharide-producing strains fixed on porous substrates or gel bases with continuous addition of substrate at controlled dilution rate (Kanegafuchi 1983).

Fermentation modes. Not only is fermentor design important but also the physiological state of the cells. In traditional batch fermentation, this state varies as a result of external changes in medium composition as the fermentation proceeds. The earliest patents in this area covered the use of pH control and then moved on to consider continuous fermentation methods, either single or multistage (Moraine and Rogovin 1971; Silman and Rogovin 1970; Lindblom and Patton 1967; Rogovin 1969; Patton and Lindblom 1962). Culture instability was a serious problem with the early continuous processes. One method proposed for stabilizing the culture was to separate the polymer-producing activities of the culture from its growth (Lindblom and Patton 1967). This two-stage process employed a small first stage dedicated to cell growth and a larger second stage where polymer production occurred. A fed-sugar process was patented that gave improved yield and productivity by allowing the engineer to balance the oxygen transfer capabilities of the fermentor with the productivity of the process (Wernau 1981b). In this way, excessive by-product acid formation could be avoided. Recently a series of patents have appeared on semicontinuous and continuous fermentation of xanthan gum in which special, degeneration-resistant cultures are used (Weisrock 1982b, 1983b). The use of simple compounds in the continuous fermentation of xanthan gum was patented in the U.K. and several other countries by the Ministry of Defense (Ellwood et al. 1978). Choice of limiting nutrient in continuous fermentation can be critical as illustrated in work carried out by the Ministry of Defense and by others (Ellwood et al. 1978; Davidson 1978).

Against this background of batch-fed, semicontinuous, and continuous processing, one must still consider traditional batch processing as a viable alternative. Advantages of the latter include reduced contamination risk and less complexity in industrial practice.

Fungal polysaccharides. The xanthan gum fermentation is illustrative of a bacterial fermentation process with mass transfer and mixing difficulties. There are also several fungal polysaccharides of commercial significance. These include scleroglucan and pullulan, among others. Scleroglucan serves to illustrate some additional complications that can be encountered in fungal systems. Scleroglucan is a $\beta=(1\rightarrow3)$glucan with $\beta=(\rightarrow6)$glucose sidechains every third residue. Its backbone is the same as that of the bacterial $\beta=(1\rightarrow3)$glucan known as curdlan (Bluhm et al. 1982). To begin with the scleroglucan polysaccharide produced by *Sclerotium rolfsii* or any of a variety of related organisms is not totally excreted as a freely soluble material as is xanthan gum. Rather, some of the polymer adheres to the mycelium as a gelatinous capsular coating, potentially inhibiting mass transfer of nutrients into the cells and severely limiting final concentration of the polysaccharide. Yield tends to be lower than for xanthan fermentation (Halleck 1967). In addition, the broths tend to be more gelatinous than xanthan broths, further inhibiting mass transfer and mixing. Removing the polymer from the cells requires heating and shearing a diluted polymer broth, with subsequent cell removal by traditional means of DE filtration or by application of microscreening techniques first proposed at Oak Ridge National Laboratories, followed by polishing filtration (Griffith et al. 1979). In some fungal systems, including this one, it has been observed that the producing organisms make enzymes that degrade the polysaccharides, and the broths need to be harvested before these enzymes are induced (Griffith and Compere 1978). Generally speaking, fungal mycelia are easier to remove from diluted polymer broths than are bacteria, but the additional cost of reconcentration imposes a severe economic penalty.

The morphological form of a fungus is also important in polysaccharide fermentations. If grown in pellet form, the fungus is much less productive, with the internal pellet mass transfer resistance being added to the already severe external mass transfer problems. For this reason, dispersed mycelial growth is preferred. The high shear fields present in industrial fermentors are helpful in creating this type of growth. However, the presence of mycelium exacerbates the mixing and mass transfer requirements of the fermentation. It is well known that mycelium itself can greatly increase the effective viscosity of fermentation broths.

Conclusions

Polysaccharide fermentations offer an exciting challenge to the microbiologist and to the biochemical engineer. The unique problems encountered as a result of the highly viscous nature of high molecular weight polymers have engendered creative approaches to improving mass transfer and mixing. In addition, the application of conventional fermentation engineering and science has led to improved processes for the production of commercially significant polysaccharides.

Literature Cited

Basta, N. 1983. New biopolymer vies for many surfactant uses. *Chem. Eng.* May 2nd:20–22.

Bauer, K. A., and B. Khosrovi. 1983. Process of using *Xanthomonas campestris*

NRRL B-12075 and NRRL B-12074 for making heteropolysaccharide U.S. Pat. 4,400,467 Aug.

Bluhm, T. L., Y. Deslandes, R. H. Marchessault, S. Perez, and M. Rinaudo. 1982. Solid-state and solution conformation of scleroglucan. *Carbohyd. Res.* 100:117–130.

Cadmus, M. C., M. O. Bagby, M. Burton, K. A. Burton, and I. Wolff. 1971. Nitrogen source for improved productions of microbial polysaccharides. U.S. Pat. 3,565,763 Feb.

Colin, P., and M. Fleury. 1971. Processes for carrying out polysaccharide-producing fermentations. U.S. Pat. 3,594,280 July.

Colin, P., and R. Merle. 1972. Process for producing polysaccharides by fermentation. U.S. Pat. 3,671,398 June.

Davidson, I. W. 1978. Production of polysaccharide by *Xanthomonas campestris* in continuous culture. FEMS *Microbiol. Lett.* 3:347–349.

Duc, N. C., M. Brehant, J. P. Benoit, and M. H. Sechet. 1978. Process for producing a "*Xanthemonas*-type" [sic] polysaccharide. U.S. Pat. 4,104,123 Aug.

Ellwood, D. C., C. G. T. Evans, and R. G. Yeo. 1978. Production of bacterial polysaccharides. U.K. Pat. 1,512,536 June.

Engelskirchen, K., W. Stein, M. Bahn, L. Schieferstein, J. Schindler, and R. Schmid. 1982. Processes for *Xanthomonas* biopolymers. Can. Patent 1,164,376 Mar.

Griffith, W. L., and A. L. Compere. 1978. Production of a high viscosity glucan by *Sclerotium rolfsii* ATCC 15206. *Dev. Ind. Microbiol.* 19:609–617.

Griffith, W. L., G. B. Tanny, and A. L. Compere. 1979. Evaluation of tangential filtration methods for the recovery of scleroglucan. *Dev. Ind. Microbiol.* 20:743–750.

Halleck, F. E. 1967. Polysaccharides and method for production thereof. U.S. Pat. 3,301,848 Jan.

Hitachi, KK. 1981. Aeration-stirring tank—for contacting gas with viscous liquid used for cultivating microbes. Jap. Pat. J5 6021–635 Feb.

Jeanes, A., P. Rogovin, M. C. Cadmus, R. W. Silman, and C. A. Knutson. 1976. Polysaccharide (xanthan) of *Xanthomonas campestris* NRRL B-1459: Procedures for culture maintenance and polysaccharide production, purification, and analysis. Avail. from Agricultural Research Service, U.S. Dept. of Agriculture as ARS-NC-51.

Kanda, H. 1977. Xanthan gum. Jap. Kokai Pat. 7,710,492 Jan.

______. 1978. Process for producing *Xanthomonas* polysaccharide on a soybean whey medium. U.S. Pat. 4,071,406 Jan.

Kanegafuchi Chem. KK. 1983. Production of polysaccharides of *Xanthane* [sic] genus. Jap. Pat. J5 8146–290A Aug.

Lindblom, G. P., and J. T. Patton. 1967. Heteropolysaccharide fermentation process. U.S. Pat. 3,328,262 June.

Lipps, B. J., Jr., 1966a. Fermentation process for preparing polysaccharides. U.S. Pat. 3,251,749 May.

______. 1966b. Fermentation process for producing a heteropolysaccharide. U.S. Pat. 3,281,329 Oct.

Maury, L. G. 1982. Polysaccharide gum production by fermentation with aqueous nutrient medium dispersed in oil. U.S. Pat. 4,352,882 Oct.

McNeely, W. H. 1968a. Process for producing polysaccharides. U.S. Pat. 3,391,061 July.

______. 1968b. Process for producing a polysaccharide. U.S. Pat. 3,391,060 July.

______. 1969a. Process for preparing polysaccharide. U.S. Pat. 3,427,226 Feb.

______. 1969b. Process for producing a polysaccharide. U.S. Pat. 3,433,708 Mar.

Miescher, G. M. 1969. Process for the production of polysaccharide gum polymers. U.S. Pat. 3,455,786 July.

Moraine, R. A., and P. Rogovin. 1971. Xanthan biopolymer production at increased concentration by pH control. *Biotechnol. Bioeng.* 13:381–391.

Nichiden-Kagaku. 1982. Preparatory method for microbial acidic polysaccharides. Jap. Pat. J5 7146–589 Sept.

Patton, J. T., and G. P. Lindblom. 1962. Process for synthesizing polysaccharides. U.S. Pat. 3,020,206. Feb.

Philips, J. C., J. W. Miller, W. C. Wernau, B. E. Tate, and M. H. Auerbach. 1982. A new high-pyruvate xanthan for enhanced oil recovery. *SPE Sixth International Symp. on Oilfield and Geothermal Chemistry,* Dallas, Texas, Jan. 25–27. SPE paper 10617.

Richmon, J. B. 1983. Process for producing low calcium xanthan gums by fermentation. U.S. Pat. 4,375,512 Mar.

Rogovin, S. P. 1969. Continuous process for producing *Xanthomonas* heteropolysaccharide. U.S. Pat. 3,485,719 Dec.

Silman, R. W., and P. Rogovin. 1970. Continuous fermentation to produce xanthan biopolymer: Laboratory investigation. *Biotechnol. and Bioeng.* 12:75–83.

Wagner, M. 1978. Microbiological colourless polysaccharide production. DL Pat. 129,802 Feb.

Weisrock, W. P. 1981. Method for improving xanthan yield. U.S. Pat. 4,301,247 Nov.

______. 1982a. Semi-continuous method for production of xanthan gum using *Xanthomonas campestris* ATCC 31600 and *Xanthomonas campestris* ATCC 31602. U.S. Pat. 4,328,308 May.

______. 1982b. Semi-continuous method for production of xanthan gum using *Xanthomonas campestris* ATCC 31601. U.S. Pat. 4,328,310 May.

______. 1983a. Method for producing a low viscosity xanthan gum. U.S. Pat. 4,377,637 Mar.

______. 1983b. *Xanthomonas campestris* ATCC 31600 and process for use. U.S. Pat. 4,407,951 Oct.

Wernau, W. C. 1978. Process for producing *Xanthomonas* hydrophilic colloid, product resulting therefrom, and use thereof in displacement of oil from partially depleted reservoirs. U.S. Pat. 4,119,546 Oct.

______. 1981a. *Xanthomonas* biopolymer for use in displacement of oil from partially depleted reservoirs. U.S. Pat. 4,340,678 July.

______. 1981b. Fermentation process for production of xanthan. U.S. Pat. 4,282,321 Aug.

CHAPTER 17

Xanthan and Scleroglucan: Structure and Use in Enhanced Oil Recovery

G. Holzwarth

Department of Physics, Wake Forest University, Winston-Salem, North Carolina 27109

Two biopolymers, xanthan and scleroglucan, are strong candidates for the task of mobility control in chemically enhanced oil recovery. In this application, polymer-thickened brine is used to drive a slug of surfactant through the porous reservoir rock to mobilize residual oil; the polymer prevents fingering of the drive water through the surfactant bank and ensures good areal sweep. The structures of the two biopolymers bear many similarities. Both are polysaccharides and both have a comb-like primary structure, molecular weight of $2\text{–}12 \times 10^6$, and multistranded helical conformations. Xanthan is most probably a double-stranded polymer, although some evidence favors a single-stranded model. For scleroglucan, light scattering and x-ray data support a three-stranded model in solution, with interchain H-bonds. These structural features provide both polymers with a chain stiffness similar to that of DNA. This stiffness gives their aqueous solutions high viscosity at low concentration, even in the presence of dissolved salts, and makes the solutions commercially useful.

Discussion

Chemically Enhanced Oil Recovery

Oil occurs in the pores of reservoir rock. When a well is drilled through the impermeable overburden, some oil may flow to the surface by natural pressure, but this primary production soon subsides. To force more oil from the formation, several injection wells are commonly drilled around each producing well. Water is injected, which forces additional oil toward the producing well. Such secondary production often continues for decades, until the costs of water and electricity for the pumps exceed the value of the produced oil.

About two-thirds of the oil in known reservoirs in the United States, or 300 billion barrels, remains behind in such "watered out" reservoirs, trapped by capillary forces and adsorption in the small pores of the reservoir. Some of this trapped oil can be recovered by tertiary methods, one of which is chemically enhanced oil recovery (CEOR). About 5–12 billion barrels are potentially recoverable by CEOR, according to three estimates made in the late 70s (*Oil and Gas Journal* 1980).

An example of a CEOR process is shown diagrammatically in Fig. 1. A surfactant mixture is slowly injected into the reservoir at the injection well. This specially tailored mixture must create a very low interfacial tension between the resident oil and water at the temperature and salinity of the reservoir. The surfactants are

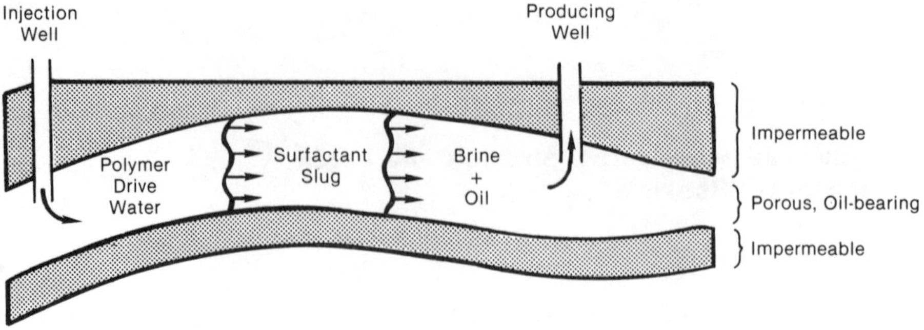

FIG. 1. Schematic representation of the chemically enhanced oil recovery process.

expensive, so one injects only the smallest feasible slug, typically 0.4 pore volume.

Immediately after the surfactant bank has been injected, a polymer-thickened drive water is injected into the formation over a period of many months. This moves the surfactant slug from the injection well to the production well at a frontal advance rate of 1–2 ft/d in a piston-like manner and with good areal sweep. It takes 6 months to several years for the surfactant bank to travel from the injection well to the production well.

Polymer is added to the drive water to control its mobility (Sandvik and Maerker 1977). The mobility of a fluid through a porous medium is the flow rate per unit area divided by the pressure gradient driving the flow. The drive water, the surfactant slug, and the mobilized oil bank each have their own mobilities, designated M_d, M_s, and M_o. If $M_d < M_s < M_o$, the fluids will proceed sequentially through the formation. However, if $M_d > M_s$ or $M_s > M_o$, the fluid of higher mobility will finger through the bank of lower mobility, which is ahead of it in the formation. Such fingering leaves in place large patches of the oil one is attempting to mobilize.

Mobility is the ratio of rock permeability P to fluid viscosity η:

$$M = P/\eta$$

Biopolymers like xanthan or scleroglucan are added to drive water to reduce M_d by increasing η. The viscosity of the drive water without polymer is essentially that of pure water, 1 cP at 20 C. This viscosity must be increased to 10–50 cP at a shear rate of 1 sec^{-1} for good mobility control; 1 sec^{-1} is a rough average of the shear rate in a typical reservoir during the EOR process.

The viscosity of solutions containing 100 to 1000 ppm xanthan is shown as a function of shear rate in Fig. 2. Scleroglucan behaves similarly. Three points are noteworthy in the figure. First, xanthan solutions are non-Newtonian; the viscosity decreases by orders of magnitude as the shear rate increases. This makes xanthan easy to pump. Second, only 600–2000 ppm of polymer is needed. Even so, 1–2 lb of polymer are consumed per barrel of oil produced. Third, the viscosity is highly nonlinear at concentrations greater than 300 ppm. This means that polymer-polymer interactions contribute substantially to the observed viscosity.

Other criteria which the displacement fluid must satisfy to meet this potentially lucrative market, are (a) propagation without plugging or excessive adsorption;

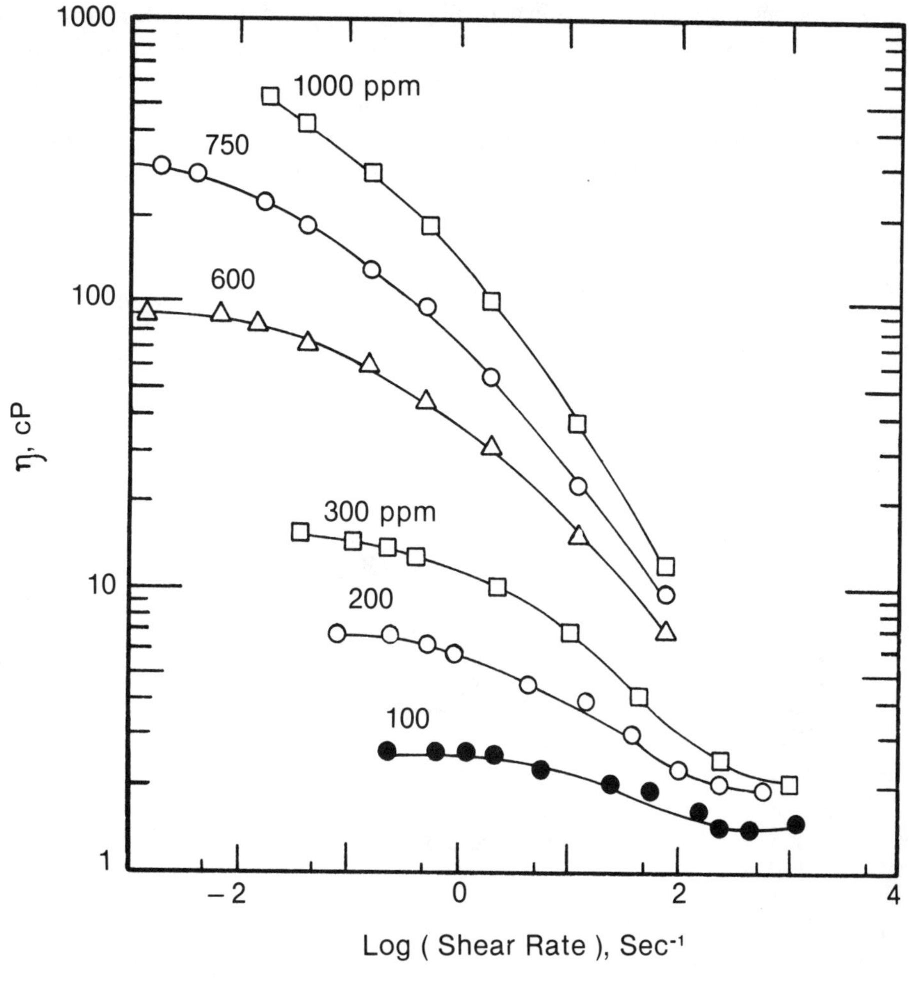

FIG. 2. Viscosity of xanthan at 20 C in 0.8 *M* NaCl, 0.04 *M* PO_4 buffers, pH 7. Data from Holzwarth (1981).

(b) stability to mechanical, chemical, and biological insult for 6–24 months at 20–95 C; (c) compatibility with the surfactant; and (d) low cost (Holstein 1982). Surveys of hundreds of commercial polymers (Szabo 1979; Davison and Mentzer 1980) have identified xanthan and scleroglucan as particularly suitable biopolymers for this application.

Xanthan

Xanthan is the extracellular polysaccharide of the bacterium *Xanthomonas campestris*, a cabbage pathogen. It is produced by submerged aerobic fermentation on glucose and has been a commercial product of Kelco Co. since 1967. Kelco, Pfizer, and Rhone-Poulenc are among the 1984 suppliers. Xanthan was first identified as an unusually effective viscosifier of water and brine by Jeanes et al. (1961).

The primary structure is shown in Fig. 3. It is comb-like, with a pentasaccharide repeating unit (Jansson et al. 1975). The backbone of the comb is poly-β-1,4-D-glucopyranose, as in cellulose, but with a 3-sugar, negatively charged side chain attached at C3 to alternate backbone residues. In addition, about one-third of the side chains bear a charged pyruvate group on the terminal mannose (Sandford et al. 1977; Sutherland 1981). Hence xanthan is a polyelectrolyte.

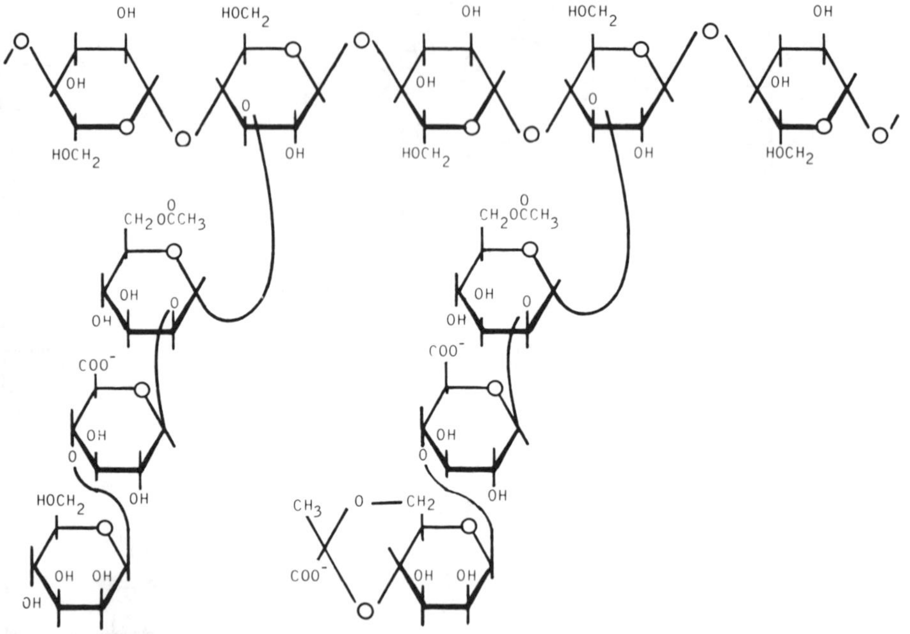

FIG. 3. Primary structure of xanthan as determined by Jansson et al. (1975). Reproduced with permission from Holzwarth (1976).

Typical synthetic polyelectrolytes, such as poly-sodium-acrylate, are highly extended when dissolved in deionized water; this chain expansion is due to Coulombic repulsion between the charged groups of the chain. This makes many polyelectrolytes excellent viscosifiers at low ionic strength, but their solution viscosity decreases sharply upon the addition of salt. Xanthan, by contrast, is highly extended both in the presence and absence of salt. This is a consequence of chain stiffening maintained by structural features that are now at least partially understood. Its intrinsic viscosity $[\eta]_0$ is 8000–12000 ml/g.

The rigid structure of xanthan can be "melted out" like the double-helix of DNA. When a solution of xanthan at low ionic strength is heated, the melting is readily observed by viscometry, optical rotation, circular dichroism, and NMR (Jeanes et al. 1961; Holzwarth 1976; Rinaudo and Milas 1978; Morris et al. 1977). The melting temperature moves systematically to higher values as the ionic strength increases, just as for DNA, and presumably for similar reasons: in the native conformation the charged COO – groups are closer to one another than in the high-temperature denatured form. Repulsion between their charges destabilizes the native conformation with respect to the denatured one. The addition of salt decreases this repulsion.

In the NMR spectrum of native xanthan, each line is greatly broadened because of the slow tumbling time of the stiff chain (Morris et al. 1977). Both the main-chain and side-chain protons are undetectable. This suggests a structure for the native molecule in which the side chains are folded down against the main chain, thereby stiffening it (Morris et al. 1977).

Electron micrographs of native and denatured xanthan show that the native molecule is a 40 Å-wide fiber, whereas the denatured chain is only 20 Å wide (Holzwarth and Prestridge 1977). This suggests that native xanthan is multistranded.

X-ray diffraction studies on oriented xanthan fibers were initially interpreted to originate from a single-stranded helix with five pentasaccharide repeating units per turn (Moorhouse et al. 1977). The proposed structure has a mass/length ratio of 996 daltons/nm. Subsequently, the fiber diffraction data were reexamined to see whether the diffraction pattern and fiber density could be better fitted with multistranded models. The reexamination supports a parallel or antiparallel double-stranded 5_1 helix with the side chains folded down against the main chain (Okuyama et al. 1980). Four intramolecular H-bonds occur in both double-stranded models; the antiparallel model has, in addition, one interchain H-bond. It would be useful to test for these H-bonds by other methods.

The molecular weight of xanthan has been a contentious topic, with values ranging from 2×10^6 to 12×10^6 (Dintzis et al. 1970; Holzwarth 1978; Rinaudo and Milas 1978; Wellington 1981; Paradossi and Brant 1982). The discrepancy is probably due equally to sample variation and methodological discrepancies. Sample variation is apparent in reported measurements of 2×10^6 and 3.6×10^6, both determined by light scattering from a single laboratory (Rinaudo and Milas 1978; Milas and Rinaudo 1979). In my experience at Exxon's Corporate ResearchLaboratory, using band sedimentation and intrinsic viscosity, Kelco xanthans gave $M_w = 8\text{–}12 \times 10^6$ while Pfizer xanthans fell in the $4\text{–}8 \times 10^6$ range. Two-fold variations exist between batches from the same supplier. Such variations may have occurred during biosynthesis or during processing. In a preliminary study at Exxon of changes during gum production, 3-fold variations in M_w were observed between inoculation and harvesting in shake-flasks. Others have suggested that molecular weights greater than 2×10^6 are the result of aggregation (Dintzis et al. 1970; Frangou et al. 1982).

Some initial progress has been made recently in understanding the biosynthesis of xanthan. It is known that the pentasaccharide repeating unit is synthesized first, then polymerized (Ielpi et al. 1981a), and that pyruvate is added at the pentasaccharide stage (Ielpi et al. 1981b). But the location and control of the polymerization reaction are unexplored.

Scleroglucan

A screen of 140 natural and synthetic polymers for use in EOR pointed to scleroglucan as outstandingly stable at high temperature: it maintained its high viscosity for 500 d at 90 C (Davison and Mentzer 1980). The primary structure is shown in Fig. 4. Like that of xanthan, it is comb-like, but with a single

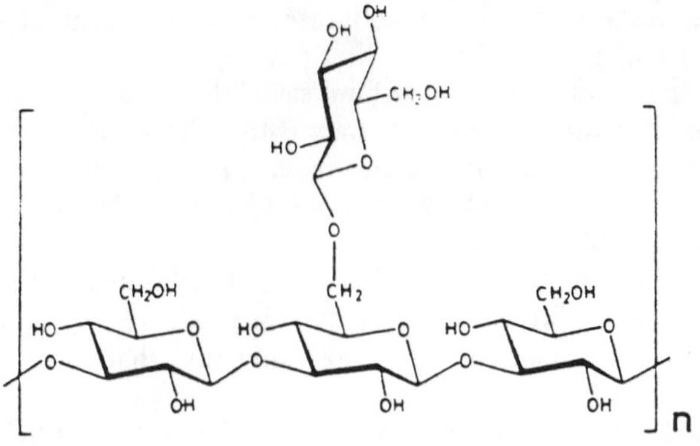

FIG. 4. Primary structure of scleroglucan. Reproduced with permission from Yanaki (1981).

glucopyranose group attached to every third backbone glucopyranose unit (Johnson et al. 1963; Halleck 1967; Sanford 1979; Rinaudo and Vincendon 1982). Schizophyllan has almost identical primary and secondary structure (Norisuye et al. 1980; Yanaki et al. 1981).

When dissolved in water, scleroglucan is an extremely stiff chain, with native intrinsic viscosity at zero shear rate $[\eta]_o = 6600$ ml/g. The value of $[\eta]_o$ drops to 243 ml/g when scleroglucan is dissolved in dimethysulfoxide (DMSO). Fig. 5 is a plot of $[\eta]$ against weight percent water in DMSO:water mixtures. The value of $[\eta]$ remains high as the percent of water decreases, then drops sharply at 13% water. Light scattering measurements show that M_w decreases from 5.4×10^6 in water to 1.4×10^6 in DMSO (Yanaki et al. 1981). These experiments by themselves are strong evidence that native scleroglucan is triple-stranded in solution.

There is good evidence that scleroglucan is also triple-stranded in solid fibers. Fiber x-ray diffraction of β-(1,3)-D-xylan had already in 1969 revealed a three-stranded structure (Atkins and Parker 1969); this particular structure has three interchain H-bonds at the center of the helix. The extra CH_2OH group and the side-chain glucose unit of scleroglucan can be appended to the triple-helical main chain of β-(1,3)-D-xylan without a change in the backbone structure; they fit on the outside (Norisuye et al. 1980). Scleroglucan fibers themselves have low crystallinity and therefore give poor x-ray patterns, but the patterns are similar to those of curdlan, a known triple helix (Marchessault et al. 1977). Conformational analysis further supports the triple-helical model for scleroglucan (Bluhm et al. 1982).

The reason scleroglucan is so effective as a viscosifier of water even at very low polymer concentrations lies in two factors: (a) its molecular weight is high; (b) it is a very stiff chain. Measurements of $[\eta]_o$ for two different molecular weights prepared by sonication show that the persistence length is about the same as that of schizophyllan, 180 nm (Yanaki et al. 1981). This is three times the persistence length of DNA, which is a very stiff molecule.

Like xanthan, scleroglucan is very resistant to shear degradation. One can speculate that the three strands provide three times the tensile strength of an individual strand, much as steel cables gain strength from many fine wires.

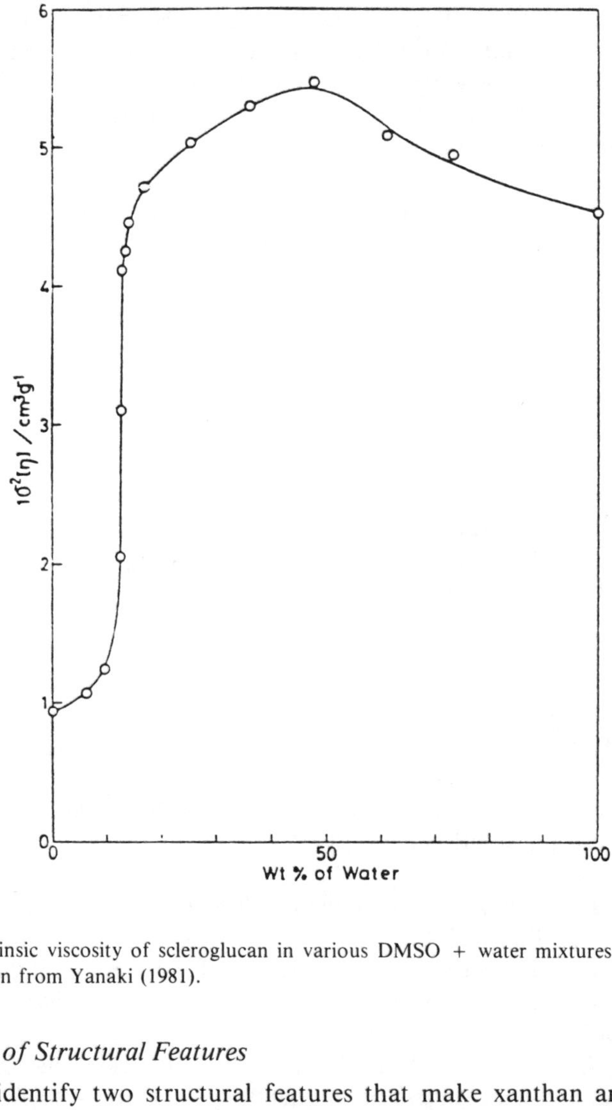

FIG. 5. Intrinsic viscosity of scleroglucan in various DMSO + water mixtures. Reproduced with permission from Yanaki (1981).

Summary of Structural Features

One can identify two structural features that make xanthan and scleroglucan useful for EOR. These are high molecular weight and long persistence length. Persistence length is a measure of chain stiffness for a particular type of polymer; it is independent of M_w. For a flexible polymer like polyethylene it is 0.2–0.5 nm; for a rigid rod like TMV it exceeds 1000 nm. The persistence length of xanthan has been estimated as 60 nm (Holzwarth 1978); that of scleroglucan is 180 nm (Yanaki et al. 1981). In both cases the contour length is about 15 times as long as the persistence length. This means that neither polymer is truly a rigid rod, like a matchstick, but is instead a very stiff coil or worm-like chain.

Both M_w and persistence length contribute to $[\eta]$. For a series of molecular weights of a given polymer, $[\eta]$ is proportional to M^b; for a given molecular weight the exponent b is a measure of persistence length. For rigid rods, $b = 1.8$, whereas for random-flight coils $b = 0.5$–0.8. For xanthan, $b = 1.35$ for a sonicated xanthan with $M_w = 400{,}000$; b still remains quite high, 0.96, for the native molecule (Holzwarth 1978). For scleroglucan, b = 1.21 (data of Yanaki et al. 1981). By contrast, b = 0.68 for polyacrylamide in 0.2 *M* NaCl (Kulicke et al. 1982) and b = 0.83 for 30% hydrolyzed polyacrylamide in 0.5 *M* NaCl (Klein and Conrad 1978). Thus, in brine, both biopolymers are stiffer than the synthetic polymers, which are their primary EOR competitors. As a consequence, although their moloecular weights are not higher than those of some polyacrylamides, their viscosities are favorable. This, combined with resistance to shear degradation and thermal degradation, makes these polymers good candidates for use in EOR.

Acknowledgments

I thank Wake Forest University and Chevron Research Corporation for research support, the American Chemical Society for permission to reproduce Fig. 3, and Dr. T. Yanaki and The Polymer Society of Japan for permission to reproduce Figs. 4 and 5. Most of my own work on xanthan was done while I was employed at Exxon Research and Engineering Co. Some of the data in Fig. 2 were provided by Pat Whitcomb of Henckel (Minneapolis) and Walt Gale of Exxon Production Research Co. (Houston).

Literature Cited

Atkins, E. D. T., and K. D. Parker. 1969. The helical structure of a β-D-1,3-xylan. *J. Polym. Sci.* 28C:69–81.

Bluhm, T. L., Y. Deslandes, R. H. Marchessault, S. Perez, and M. Rinaudo. 1982. Solid-state and solution conformation of scleroglucan. *Carbohydr. Res.* 100:117–130.

Davison, P., and E. Mentzer. 1980. Polymer flooding in North Sea reservoirs. *Soc. Petrol. Eng.,* Paper 9300.

Dintzis, F. R., G. E. Babcock, and R. Tobin. 1970. Studies on dilute solutions and dispersions of the polysaccharide from *Xanthomonas campestris* NRRL B-1459. *Carbohydr. Res.* 13:257–267.

Frangou, S. A., E. R. Morris, D. A. Rees, R. K. Richardson, and S. B. Ross-Murphy. 1982. Molecular origin of xanthan solution rheology: Effect of urea on chain conformation and interactions. *J. Polym. Sci. Polym. Lett. Ed.* 20:531–538.

Halleck, F. E. 1967. Polysaccharides and methods for production thereof. U.S. Pat. 3,301,848 Jan.

Holstein, E. D. 1982. Future of EOR by chemical flooding looks promising. *World Oil.* July:133–144.

Holzwarth, G. 1976. Conformation of the extracellular polysaccharide of *Xanthomonas campestris. Biochemistry* 15:4333–4338.

______. 1978. Molecular weight of xanthan polysaccharide. *Carbohydr. Res.* 66:173–186.

______. 1981. Is Xanthan a wormlike chain or a rigid rod? Pages 15–23 *in* D. A. Brant, ed., *Solution Properties of Polysaccharides,* Amer. Chem. Soc. Symp. Ser. Vol. 150.

Holzwarth, G., and E. B. Prestridge. 1977. Multistranded helix in xanthan polysaccharide. *Science* 197:757–759.

Ielpi, L., R. Couso, and M. Dankert. 1981a. Lipid-linked Intermediates in the Biosynthesis of xanthan gum. *FEBS Lett.* 130:253–256.

______. 1981b. Xanthan gum biosynthesis. Pyruvic acid acetal residues. *Biochem. Biophys. Res. Commun.* 102:1400–1408.

Jansson, P. E., L. Kenne, and B. Lindberg. 1975. Structure of the extracellular polysaccharide from *Xanthomonas campestris. Carbohydr. Res.* 45:275–282.

Jeanes, A., J. E. Pittsley, and F. R. Senti. 1961. Polysaccharide B-1459. A new hydrocolloid polyelectrolyte produced from glucose by bacterial fermentation. *J. Appl. Polym. Sci.* 5:519–526.

Johnson, J., S. Kirkwood, A. Misaki, T. E. Nelson, J. V. Scletti, and F. Smith. 1963. Structure of a glucan. *Chem. Ind.* :820.

Klein, J., and K. D. Conrad. 1978. Molecular weight determination of poly-(acrylamide) and poly (acrylamide-co-sodium acrylate) *Makromol. Chem.* 179:1635–1638.

Kulicke, W. M., R. Kniewske, and J. Klein. 1982. Preparation, characterization, solution properties and rheological behavior of polyacrylamide. *Prog. Polym. Sci.* 8:373–468.

Marchessault, R. H., Y. Deslandes, K. Ogawa, and P. R. Sundararajan. 1977. X-ray diffraction data for β-(1,3)-D-glucan. *Can. J. Chem.* 55:300–303.

Milas, M., and M. Rinaudo. 1979. Conformational investigation of the bacterial polysaccharide xanthan. *Carbohydr. Res.* 76:189–196.

Moorhouse, R., M. D. Walkinshaw, and S. Arnott. 1977. Xanthan gum—molecular conformation and interactions. Pages 90–102 *in* P. A. Sanford and A. I. Laskin, eds., *Extracellular Microbial Polysaccharides,* Amer. Chem. Soc. Symp. Ser., Vol. 45.

Morris, E. R., D. A. Rees, G. Young, M. D. Walkinshaw, and A. Darke. 1977. Order-disorder transition for a bacterial polysaccharide in solution. *J. Mol. Biol.* 110:1–16.

Norisuye, T., T. Yanaki, and H. Fujita. 1980. Triple helix of a *Schizophyllum commune* polysaccharide in aqueous solution. *J. Polym. Sci. Polym. Phys. Ed.* 18:547–558.

Oil and Gas Journal. 1980. DOE report examines enhanced oil recovery constraints. April 28, 1980:105–114.

Okuyama, K., S. Arnott, R. Moorhouse, M. O. Walkinshaw, E. D. Atkins, and C. Wolf-Ullish. 1980. Fiber diffraction studies of bacterial polysaccharides. Pages 411–427 *in* A. D. French and K. H. Gardner, eds., *Fiber Diffraction Methods,* Amer. Chem. Soc. Symp. Ser., Vol. 141.

Paradossi, G., and D. A. Brant. 1982. Light scattering study of a series of xanthan fractions in aqueous solution. *Macromolecules* 15:874–879.

Rinaudo, M., and M. Milas. 1978. Polyelectrolyte behavior of a bacterial polysaccharide from *Xanthomonas campestris*: Comparison with carboxycellulose. *Biopolymers* 17:2663–2678.

Rinaudo, M., and M. Vincendon. 1982. ^{13}C nmr structural investigation of scleroglucan. *Carbohydr. Polym.* 2:135–144.

Sanford, P. A. 1979. Exocellular microbial polysaccharides. *Adv. Carbohydr. Chem. Biochem.* 36:265–313.

Sandford, P. A., J. E. Pittsley, C. A. Knutson, P. R. Watson, M. C. Cadmus, and A. Jeanes. 1977. Variation in *Xanthomonas campestris* NRRL B-1459: Characterization of xanthan products of differing pyruvic acid content. Pages 192–210 *in* P. A. Sanford and A. Laskin, eds., *Extracellular Microbial Polysaccharides.* Amer. Chem. Soc. Symp. Ser., Vol. 45.

Sandvik, E. I., and J. M. Maerker. 1977. Application of xanthan gum for enhanced oil recovery. Pages 242–264 *in* P. A. Sandford and A. Laskin, eds., *Extracellular Microbial Polysaccharides.* Amer. Chem. Soc. Symp. Ser., Vol. 45.

Sutherland, I. W. 1981. *Xanthomonas* polysaccharides—improved methods for their comparison. *Carbohydr. Polym.* 1:107–115.

Szabo, M. T. 1979. An evaluation of water-soluble polymers for secondary oil recovery. *J. Pet. Technol.* 1979:553–570.

Wellington, S. L. 1981. Xanthan gum molecular size distribution and configuration. *Polymer Preprints 22,* Amer. Chem. Soc. Meet., Aug. 27, 1981.

Yanaki, T., T. Kojima, and T. Norisuye. 1981. Triple helix of scleroglucan in dilute aqueous sodium hydroxide. *Polymer J.* 13:1135–1143.

CHAPTER **18**

Enzymic Breakage of Xanthan Gum Solution Viscosity in the Presence of Salts

M. C. Cadmus and M. E. Slodki

Northern Regional Research Center, Agricultural Research Service, U.S. Department of Agriculture, Peoria, Illinois*

A major potential use for the extracellular heteropolysaccharide of *Xanthomonas campestris* NRRL B-1459 is as a suspending agent for proppage in subterranean, hydraulic, fracture fluids. After fracture, it is necessary to reduce the viscosity of the fluids to permit stimulated flow of oil or gas. Because chemical viscosity breakers are inefficient, enzymic depolymerases have been sought. None, however, have been active in the presence of brines, which must be used to maintain the porosity of underground formations. By soil-enrichment culture in the presence of 4% NaCl, we have isolated a *Bacillus* sp. that produces a xanthan gum-degrading enzyme complex active over the pH range 5.5–8.0 in the presence of salts. A notable feature of this complex is significant stability to inactivation by heat when salts are present. For example, 1–6% concentrations of Na or K chlorides completely protect activity for 20 min at 44 C; 50% of control activity (0.3% buffer, pH 5.4) is lost under these conditions. At 46 C, the control is inactive after 20 min, while 75% of activity remains if salts are present. Lower concentrations of Mg and Ca chlorides also are protective. The enzyme complex degrades trisaccharide side chains from native xanthan gum with release of the constituent sugars, including 6-*O*-acetyl-D-mannose.

Introduction

The unique physical properties of xanthan gum have made it a useful thickening and suspending agent, initially as a food additive, and now increasingly in petroleum drilling and waterflooding operations. Commercial success of this polysaccharide is because of a remarkable combination of useful properties: high viscosity of dilute aqueous solutions; resistance to degradation by shear; insensitivity of viscosity, over a broad range of pH, to the presence of salts and divalent cations; good thermal stability, which is enhanced by salts; and pseudoplastic rheological behavior.

Xanthan gum structurally consists of a linear backbone of β-(1→4) linked D-glucosyl residues, which has three-unit-long side chains appended on alternate glucosyls (Fig. 1). In these side chains, D-mannose residues that are appended directly to the backbone bear *O*-acetyl substituents on C-6 (Jansson et al. 1975; Melton et al. 1976). Pyruvic acetal substituents are on the terminal D-mannosyl

*The mention of firm names or trade products does not imply that they are endorsed or recommended by the U.S. Department of Agriculture over other firms or similar products not mentioned.

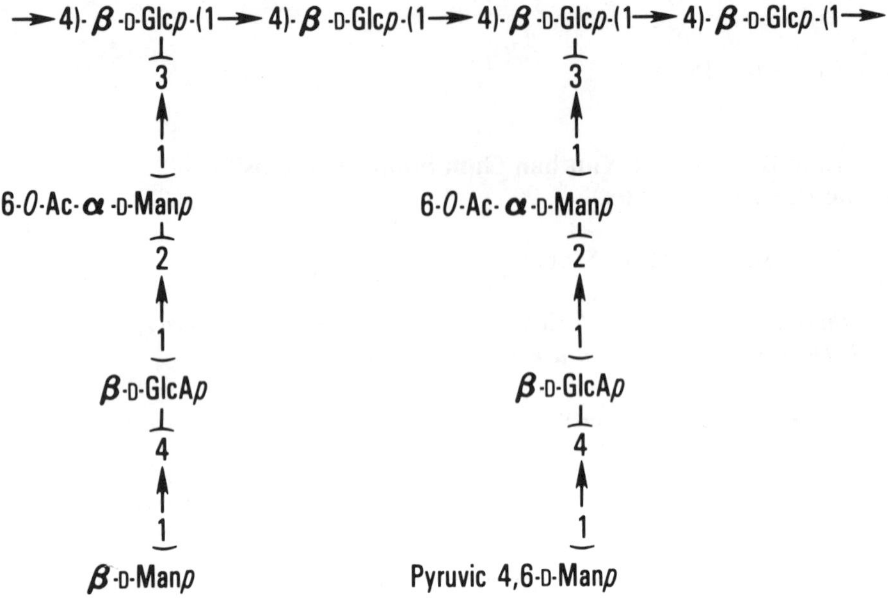

FIG. 1. Structure of xanthan gum.

residues of some side chains; their occurrence may vary depending on the strain of *Xanthomonas campestris* that produces the gum and on the fermentative conditions (Sandford et al. 1977; Cadmus et al. 1978).

The apparent remarkable stability of this gum to microbial attack (nonsterile solutions were stored many years at room temperature without loss of viscosity) prompted a search for enzymes that could alter xanthan for subsequent chemical or biological modification. Secondarily, we sought a means to break the viscosities of xanthan gum-based hydraulic fracture fluids that could be used to stimulate the flow of natural gas and petroleum from tight underground rock formations. For this purpose, a xanthanase must be reasonably stable to thermal inactivation over a broad range of pH and not be inactivated by high concentrations of salts. This latter attribute is important because brines are employed in fracture fluids in order to maintain the stability of clay formations.

Materials and Methods

Enrichment. A salt-tolerant strain of *Bacillus* sp. was isolated from a soil enrichment culture (Cadmus et al. 1982) containing 4% NaCl; the strain was designated NRRL B-4529 (K11). A second microorganism was isolated (*Flavobacterium* sp. K17, NRRL B-14010) that enhanced growth, stabilized stock cultures, and increased enzyme productivity of the *Bacillus* strain, even though as a pure culture it produced no xanthan-degrading enzymes. When cultivated as a mixed culture, the organisms were designated 11 + 17 (NRRL B-4530).

Enrichment broth (EB) consisted of xanthan, 0.25%; $(NH_4)_2SO_4$, 0.05%; NaCl, 4%; yeast extract, 0.05%; and 0.03 *M* K phosphate buffer (pH 6.8). The soil samples were added to broth in Erlenmeyer flasks; these were shaken at 30 C for 3 to 4 wk. Biodegradation was easily observed by loss of viscosity along with abundant cell growth. Positive samples were streaked on an agar medium (KGM) similar to EB except that small amounts of glucose and mannose (0.1% each) were added and the 4% NaCl was omitted. Because of slow growth of the *Bacillus* and its apparent affinity for other microorganisms, several series of plates were streaked before the xanthanase producer could be isolated in pure culture. The enzyme-producing strain is an aerobic, gram-positive, spore-forming, motile rod that forms short and long chains of random curvature.

Production. Fresh stock cultures of the *Bacillus* strains were maintained on KGM agar slants. Culture broths (EB + 4% NaCl, 10 ml/50 ml-Erlenmeyer flask) were inoculated with two loopfuls of cells from a mixed-stock culture (11 + 17). Flasks were incubated 4 to 5 d until viscosity diminished and cell growth was evident. A second-stage flask containing the same medium was inoculated (10% v/v) from the first stage. After 48 h, 20 ml of the second stage was inoculated into each of six Fernbach flasks containing 500 ml of test medium that consisted of xanthan gum, 0.2%; $(NH_4)_2SO_4$, 0.05%; tryptone, 0.18%; KH_2PO_4, 0.15%; K_2HPO_4, 0.07%; NaCl, 4%; and Speakman salt solution B (Snell and Strong 1939), 0.25% v/v. All flasks were shaken (200 rpm) at 30 C. Maximum production of enzyme was achieved in 72 h.

Enzyme. Fermentation broth (3000 ml) from the mixed culture was cooled overnight at 4 C, then centrifuged in the cold for 20 min at 20,000 x *g* to remove microbial cells. The supernatant beer was concentrated to about 100 ml in a membrane filtration apparatus (30,000 molecular weight cut-off) and then recentrifuged to remove any remaining cells or particles. The enzyme concentrate was dialyzed against 0.05 *M* sodium acetate buffer, pH 5.4, for 5 d (5–10 C). Activity of the supernatant beer was 2 units/ml (a unit of xanthanase = 1 μmol of apparent mannose liberated per min at 42 C and pH 5.4); after concentration the activity was 75 units/ml. Concentrates can be stored at −20 C without significant loss of activity for at least 9 yr.

Procedures. Enzyme activity was measured using a buffered-substrate solution containing 0.05 *M* sodium acetate (pH 5.4); xanthan gum, 0.15%; $MgSO_4$, 0.01%; and $MnSO_4$, 0.002%. Test salts were added to this stock solution. Enzyme (0.1 ml) and stock solution (2.4 ml) were equilibrated at 42 C and then mixed; after 20 min, the mixture was placed in a 100 C water bath for 5 min to inactivate the enzyme. Activity was measured as the amount of reducing sugar liberated.

For evaluation of heat stability in the presence of salts, the enzyme concentrate was mixed with one part distilled water. Salt solutions were prepared in 2-fold concentration; 1.0-ml portions were mixed with equal volumes of diluted enzyme. The salt-enzyme mixtures were exposed to various temperatures for 20 min, then quickly cooled in an ice bath. After cooling, the salt-enzyme mixture was diluted with 2.0 ml of 0.025 *M* sodium acetate buffer (pH 5.4) and stored at 4 C until ready for use; 0.1-ml portions were used to assay activity.

Viscosities were measured at 35 C with a Brookfield viscometer (Model LVF) at

30 rpm; reducing sugars were determined in an automatic analyzer by the potassium ferricyanide method (Hoffman 1937).

Polysaccharide was prepared in Fernbach flasks using a distillers solubles medium (Jeanes et al. 1976).

Results and Discussion

Products. The nature of the xanthanase we have described is important to any consideration of activity under various conditions. The same high- (HMWF) and low-molecular weight (LMWF) products are formed from the action of enzymes excreted by salt-sensitive and salt-tolerant *Bacilli* (Cadmus et al. 1982). These major fractions were separated with a membrane filter having a 10,000 molecular weight cut-off. The components of the LMWF were separated by chromatographic methods into four separate products. These were determined to be D-glucuronic acid, D-mannose, pyruvylated hexose, and *O*-acetylated hexose. A key identification was mass-spectrometric confirmation of *O*-acetyl substitution on the C-6 of a mannose residue. This evidence was obtained by use of hexadeuterioacetic anhydride in place of acetic anhydride in the derivatization step to form the per-*O*-acylated aldononitrile used for g.l.c.-m.s. By both ion impact and chemical ionization modes of mass spectrometry, it was easy to distinguish mannose derived from the terminal side-chain residue (total substitution by trideuterioacetyls) from the interior side-chain residue having a natural *O*-acetyl group on C-6. The latter residue could not have been liberated by chemical means.

The HMWF was that portion of the enzyme-hydrolyzed xanthan that did not pass through the membrane. Structure determination of the HMWF by methylation analysis (procedure of Seymour et al. 1976) indicated considerable cleavage of side chains from the backbone; approximately 90% of the residues were D-glucose in $(1 \rightarrow 4)$ linkage. This information, along with the absence of glucose in the LMWF, showed that this xanthanase removes the individual side-chain residues and likely is a complex of at least three enzymes.

Stabilization by salts. We previously reported that xanthanase was a reasonably stable enzyme (Cadmus et al. 1982): solutions could be stored for long periods in the cold; lyophilized preparations stored at room temperature remained active for at least 1 yr; and solutions were stable from pH 5.5 to 8. Heat inactivation studies, however, showed that activity began to diminish at 40 C and was completely lost at 46 C.

In this study, we showed that salts stabilize the xanthanase complex from the salt-tolerant organism to thermal inactivation. Fig. 2 illustrates this stabilization over a range of NaCl concentrations. In the absence of added salt, the enzyme retains only 44% of its original activity after being heated for 20 min at 44 C and is completely inactivated at 46 C. However, when the NaCl is increased to an optimal level of 4%, 65% of the activity remains at 48 C, and total inactivation does not occur until 50 C. Lower amounts of salt afforded lesser degrees of stabilization.

Other salts, including KCl, $MgCl_2$ and $CaCl_2$, were also used. A comparison of 2% concentrations of NaCl, $MgCl_2$ and $CaCl_2$ is illustrated in Fig. 3. Magnesium chloride gave slightly better protection to thermal inactivation than did NaCl, but

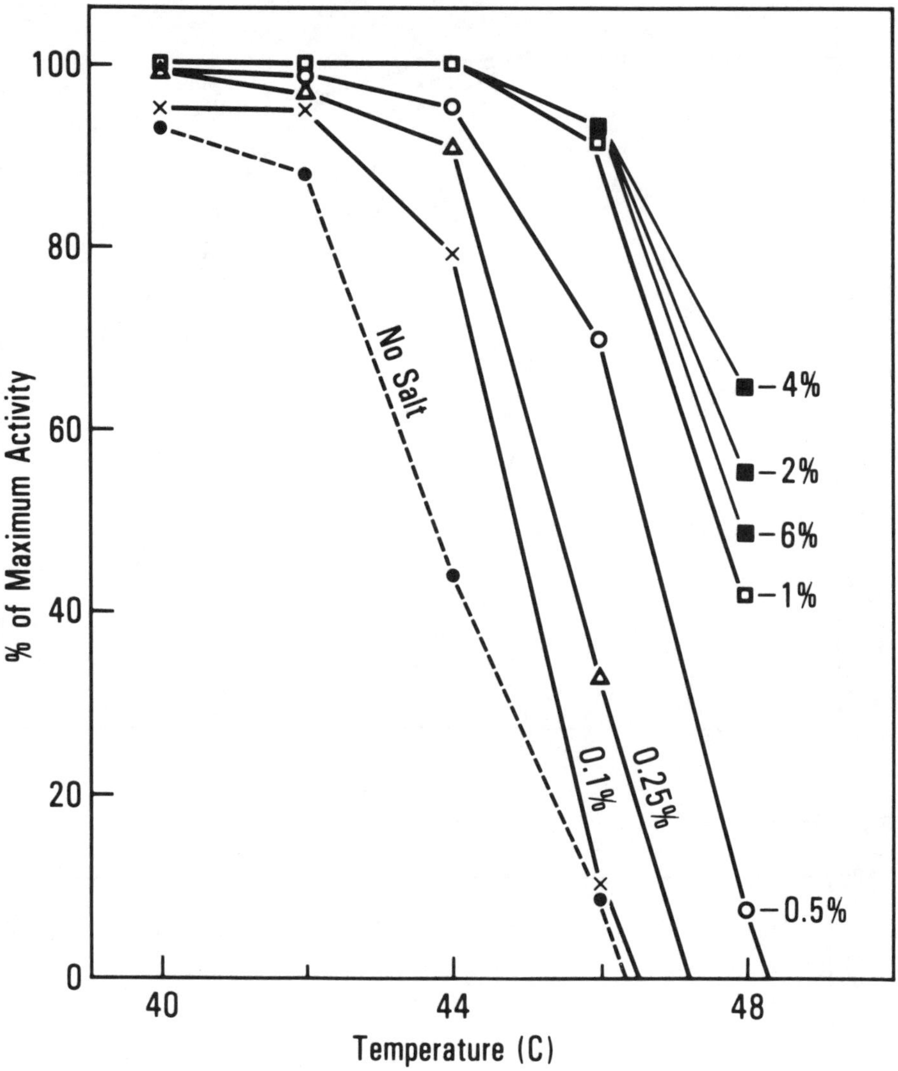

FIG. 2. Stability of xanthanase heated 20 min in the presence of NaCl (maximum activity is obtained by assay of salt-free enzyme not subjected to thermal denaturation).

xanthanase in the presence of $CaCl_2$ retained 76% of its original activity after exposure for 20 min at 50 C; inactivation occurred at 52 C. Concentrations of $MgCl_2$ and $CaCl_2$ above 2% (not shown) in the thermal stabilization studies gave significantly less protection to the enzyme. All our experiments with KCl gave nearly identical results to those done with NaCl.

Activities in the presence of salts. As explained earlier, practical use of xanthanase requires compatability with brines. Fig. 4 shows the effects of up to 8% concentrations of NaCl on activity. In a 60-min assay (42 C, pH 5.4), half of the activity measured in the absence of added salts is available in 4% salt; in 8% NaCl, one-third of the activity remains.

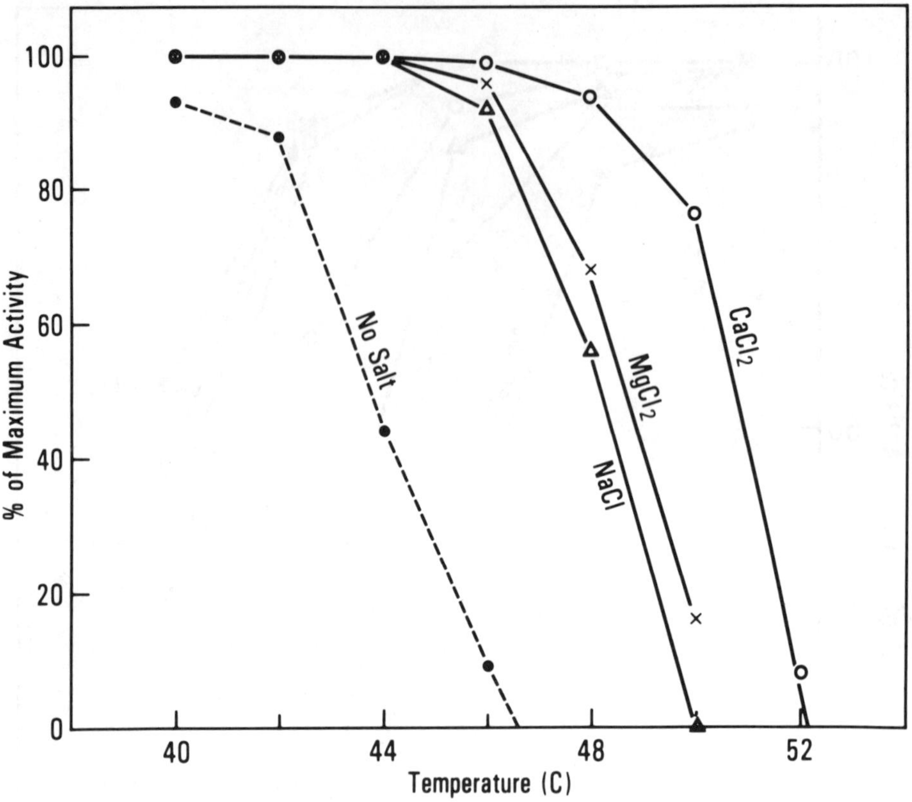

FIG. 3. Stability of xanthanase in the presence of various salts (2% salts; heated 20 min).

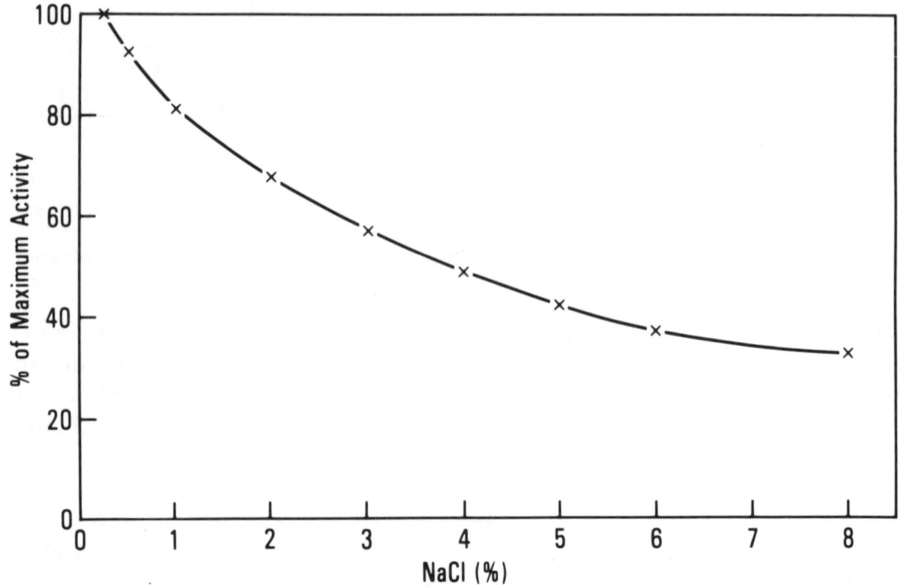

FIG. 4. Activity of xanthanase in the presence of NaCl (42 C; 0.025 *M* sodium acetate buffer, pH 5.4; 60 min assay).

In another experiment (Fig. 5), Na, K, Mg, and Ca chlorides (0–4%) were incorporated in xanthanase reaction mixtures incubated 60 min at 42 C (pH 5.3). Sodium and potassium chlorides gave similar results; i.e., 55% of the salt-free activity at 4% salt concentration. However, the divalent cation salts reduced activities about 70% at the concentration. It appears that reduction in activity is roughly proportional to the amount of salt present.

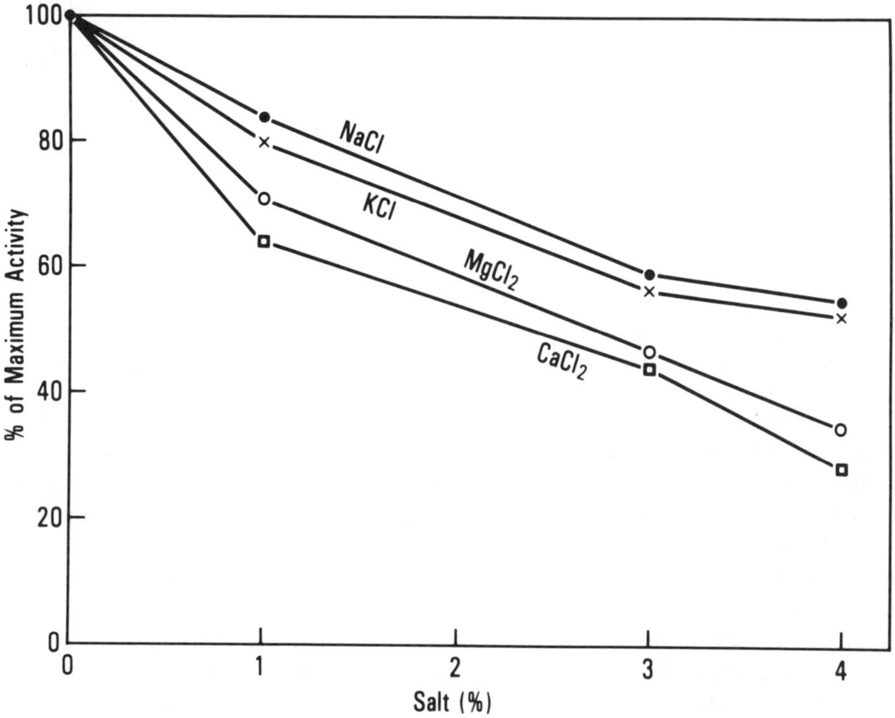

FIG. 5. Activity of xanthanase in the presence of various salts (42 C; 0.025 *M* sodium acetate buffer, pH 5.4).

Fig. 6 gives the relationship between reducing sugar release and viscosity reduction of a 100-ml mixture of 0.5% xanthan, 4% NaCl, and 10 units of xanthanase. Over the 48-h incubation period (37 C, pH 5.4), the increase in reducing sugar mirrors the loss of viscosity; initially activity is rapid, then diminishes with time. Viscosity can be reduced further and reducing sugar liberation increased by the addition of more enzyme. In a second batch (not shown) viscosity was reduced from 764 to 168 mPa's over the same period by use of 20 units of xanthanase.

From the foregoing results it is clear that xanthanase activity in the presence of salts can be considerably less than in their absence. Even so, it should be remembered that other xanthanases (Cripps et al. 1981; Sutherland 1981), which attack the β-(1→4)-linked glucosidic backbone, are inactive in the presence of moderate concentrations of salts. The native, rod-like helical conformation of xanthan gum is promoted by the presence of salts (Holzwarth 1976). Consequently, it is noteworthy that the xanthanase complex attacks this highly ordered, hydrogen-bonded structure that must exist in 8% brines.

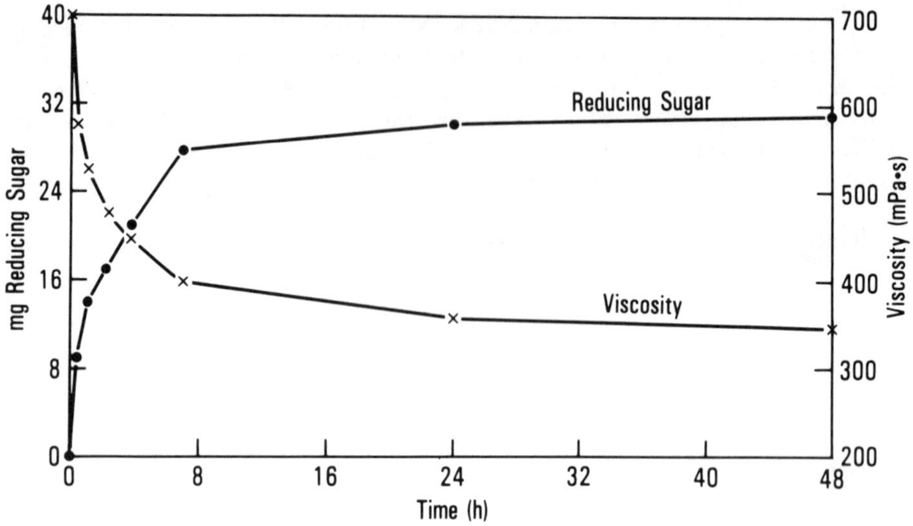

FIG. 6. Relationship between reducing sugar liberation and viscosity reduction in the presence of 4% NaCl (0.5% xanthan gum; 10 units xanthanase; 0.05 *M* sodium acetate buffer, pH 5.4; 37 C).

Xanthan gum has been widely investigated with regard to the contributions of molecular conformation and substituents (*O*-acetyl and pyruvic acetal) to the unusual properties of its aqueous dispersions and solutions. To date, the roles of the various structural components are not understood completely. We hope that when these factors are understood, the principles involved also will help explain the behavior of other polysaccharides and provide rational bases for genetic modification of structures in order to obtain desirable properties. Enzymes capable of removing side-chain components discretely from xanthan, particularly in saline solutions wherein the ordered conformation is favored, could become useful tools toward this objective. For these reasons, we can look forward to their separation and description.

Literature Cited

Cadmus, M. C., C. A. Knutson, A. A. Lagoda, J. E. Pittsley, and K. A. Burton. 1978. Synthetic media for production of quality xanthan gum in 20-liter fermentors. *Biotechnol. Bioeng.* 20:1003–1014.

Cadmus, M. C., L. K. Jackson, K. A. Burton, R. D. Plattner, and M. E. Slodki. 1982. Biodegradation of xanthan gum by *Bacillus* sp. *Appl. Environ. Microbiol.* 44:5–11.

Cripps, R. E., H. J. Sommerville, and M. S. Holt. 1981. Xanthanase enzyme. *Eur. Pat. Appl.* 30:393.

Hoffman, W. S. 1937. A rapid photoelectric method for the determination of glucose in blood and urine. *J. Biol. Chem.* 120:51–55.

Holzwarth, G. 1976. Conformation of the extracellular polysaccharide of *Xanthomonas campestris*. *Biochemistry* 15:4333–4339.

Jansson, P-E., L. Kenne, and B. Lindberg. 1975. Structure of the extracellular polysaccharide from *Xanthomonas campestris*. *Carbohydr. Res.* 45:275–282.

Jeanes, A., P. Rogovin, M. C. Cadmus, R. W. Silman, and C. A. Knutson. 1976. Polysaccharide (xanthan) of *Xanthomonas campestris* NRRL B-1459; Procedures for culture maintenance and polysaccharide production, purification, and analysis. *ARS-NC-51*. U.S. Agricultural Research Service, North Central Region, U.S. Department of Agriculture, Washington, DC.

Melton, L. D., L. Mindt, D. A. Rees, and G. R. Sanderson. 1976. Covalent structure of the extracellular polysaccharide from *Xanthomonas campestris*: Evidence from partial hydrolysis studies. *Carbohydr. Res.* 46:245–257.

Sandford, P. A., J. E. Pittsley, C. A. Knutson, P. R. Watson, M. C. Cadmus, and A. Jeanes. 1977. Variation in *Xanthomonas campestris* NRRL B-1459: Characterization of xanthan products of differing pyruvic acid content. *ACS Symp.* Ser. 45:192–210.

Seymour, F. R., M. E. Slodki, R. D. Plattner, and R. M. Stodola. 1976. Methylation and acetolysis of extracellular D-mannans from yeast. *Carbohydr. Res.* 48:225–227.

Snell, E. E., and F. M. Strong. 1939. A microbiological assay for riboflavin. *Ind. Eng. Chem. Anal. Ed.* 11:346–350.

Sutherland, I. W. 1981. *Xanthomonas* polysaccharide—improved methods for their comparison. *Carbohydr. Polym.* 1:107–115.

CHAPTER **19**

Emulsan: A Case Study of Microbial Capsules as Industrial Products

JOSEF SHABTAI, OPHRY PINES, AND DAVID GUTNICK

Department of Microbiology, George S. Wise Faculty of Life Sciences
Tel Aviv University, Ramat Aviv 69928 Israel

Emulsan is a polyanionic galactosamine-containing bioemulsifier produced by *Acinetobacter calcoaceticus* RAG-1. In early exponential phase, emulsan is located on the cell surface as a minicapsule comprising up to 20% of the cell dry wt. As the cells approach stationary phase, the interaction of the polymer with the cell surface is weakened through the action of at least one enzyme, an esterase. Possible functions for the cell-associated bioemulsifier include phage receptor, masking polymer of cell-surface hydrophobic structures, and as an exocellular shield that enhances tolerance to toxic cations such as cetyltrimethyl ammonium bromide (CTAB). Mutants resistant to CTAB show enhanced production of emulsan. Experiments using mixed cultures demonstrated that cell-associated emulsan is responsible for CTAB tolerance. The cell-associated emulsan minicapsule confers a distinct advantage for the wild type growing on crude oil in sea water. Reconstitution of a phage receptor at an oil/water interface suggests that the conformation of cell-associated emulsan resembles its conformation at the surface of an oil droplet.

INTRODUCTION

Microbial polysaccharides have received a considerable amount of attention in recent years, both with respect to their biological function as well as for their important industrial and medical applications (Costerton et al. 1981; Deavin et al. 1977; Dudman 1977; Kang and Cottrell 1979; Sutherland 1979, 1982; Sutherland and Ellwood 1979; Gutnick and Rosenberg 1977; Pace 1980). Most of these biopolymers are associated with the outer surface of the cell either as a slime or in a tighter association of the cell surface in the form of a capsule (Sutherland 1982). The term "glycocalyx" has been used to refer to a polysaccharide cell-surface structure that is located external to either the outer membrane of gram-negative or the peptidoglycan layer of gram-positive bacteria (Costerton et al. 1981).

The hydrocarbon-degrading organism *Acinetobacter calcoaceticus* RAG-1 is enveloped by a cell-surface heteropolysaccharide capsule during exponential growth (Goldman et al. 1982; Pines et al. 1983; Rubinovitz et al. 1982). As the cells approach stationary phase the polysaccharide is released into the growth medium to form the potent bioemulsifier-emulsan (Goldman et al. 1982; Rosenberg et al. 1979b). Emulsan forms stable oil-in-water emulsions by forming a film around the surface of an oil droplet, thus preventing coalescence of the droplets (Goldman et al. 1982; Rosenberg et al. 1979a; Zuckerberg et al.

1979). Moreover, as a polyanion emulsan binds both organic and inorganic cations at the oil-water interface (Rosenberg et al. 1979b; Zosim et al. 1982; Zuckerberg et al. 1979). The polysaccharide backbone of emulsan (M.W. 10^6) is composed of D-galactosamine, an aminohexuronic acid, and a third unidentified amino sugar. In addition, fatty acids are covalently linked to the polysaccharide backbone via ester and amide linkages (Belsky et al. 1979; Zuckerberg et al. 1979).

The cell-associated form of the emulsan polymer in exponential phase is visualized in Fig. 1 (top left) as a capsular envelope using specific im-

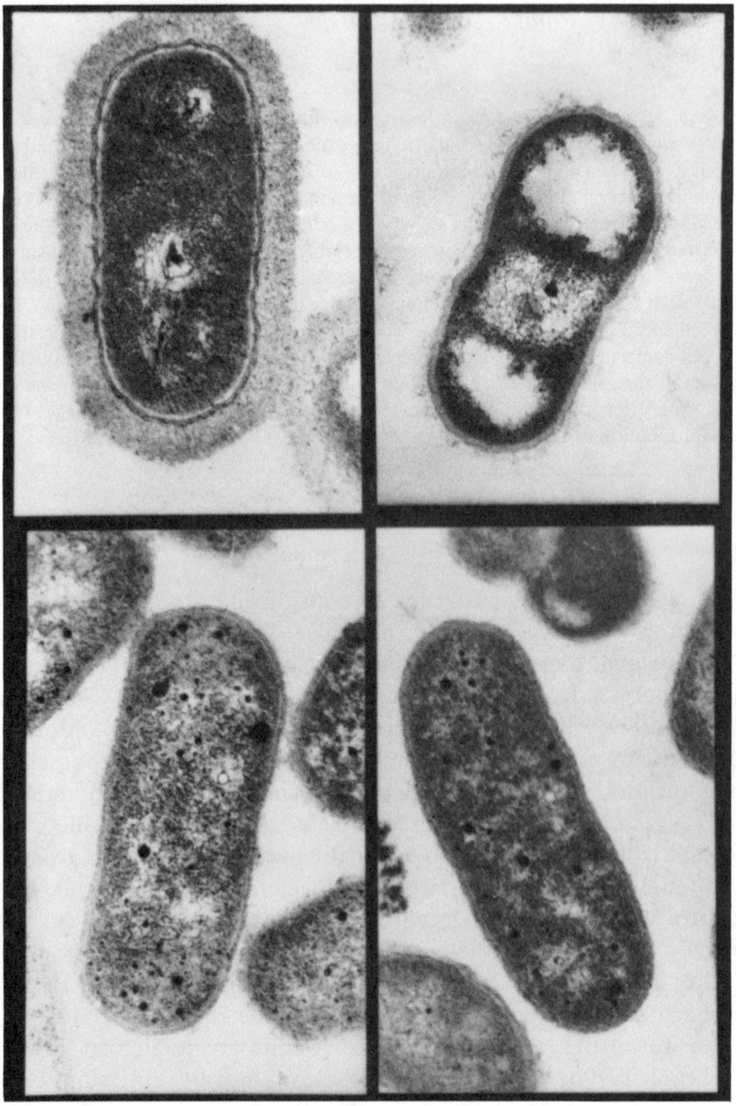

FIG. 1. Transmission electron microscopy of RAG-1 and mutants TR3 and TL4. Upper left—RAG-1 exponential phase; lower left—RAG-1 stationary phase; upper right—TL4; and lower right—TR3. Preparation of cells and immunocytochemical labeling were described previously (Pines and Gutnick 1984).

munocytochemical labeling and electron microscopic techniques (Pines et al. 1983). While tightly bound to the outer surface of the producing cell, in stationary phase the cells are almost completely devoid of capsule (Fig. 1, bottom left). The upper and lower right-hand panels in Fig. 1 are controls using two RAG-1 mutants that do not produce the active emulsifier. Thus, during emulsan production, the outer emulsan layer (which is almost 20% of the cell mass), is somehow destabilized during its conversion to an active bioemulsifier. This alteration in emulsan distribution to release a cell-free polymer was accelerated when exponentially grown cells were exposed to chloramphenicol (CAP).

This report describes several experimental approaches designed to shed more light on the release process and to try to gain some insight into the biological role of the polymer. The results suggest that it is the cell-associated form of the emulsan polymer that serves as a protective shield and is required for growth on hydrocarbons.

Materials and Methods

Bacterial strains and culture conditions. The strains used in these studies were the following: the parent *Acinetobacter calcoaceticus* RAG-1 (ATCC 31012), mutant RAG-10 deficient in emulsan production, and mutant CTR-1049 with enhanced tolerance to cetyltrimethyl ammonium bromide (CTAB).

All strains were cultivated in a minimal ethanol salts medium (ETMS) containing per liter: 22.2 g $K_2HPO_4 \cdot 3H_2O$; 7.26 g KH_2PO_4; 4.0 g $(NH_4)_2SO_4$; 0.2 g $MgSO_4 \cdot 7H_2O$ and 25 ml absolute ethanol. Growth experiments were carried out in shake flasks filled to 20% of their capacity. Cultures were grown in an incubated gyratory shaker at 30 C and 250 rpm. Growth was followed turbidometrically in a Klett-Summerson colorimeter equipped with green filter. One hundred Klett Units correspond to a dry cell weight of 0.36 g/l. Viable counts were determined by counting colonies arising on plates after spreading suitable dilutions of the culture.

Production of emulsan in a 2-liter fermentor system. The fermentation process was carried out in a NBS Multigen 2-liter fermentor, 1.2 liter working volume. The process was pH-controlled using base solution that consisted of a mixture of 2 *M* NaOH and 1 *M* NH_4OH. $FeSO_4$ was added to the medium to final concentration of 1 mg/l. Ethanol was fed into the medium slowly and continuously at a rate of 0.6 g/l-h to a total amount of 30 g/l. The fermentation was run at 30 C with 0.5 vvm aeration and 700 rpm agitation rate. Foaming in the fermentation was controlled automatically using a silicone antifoam emulsion (Dow-Corning).

Emulsan preparation and measurement. Crude emulsan was precipitated from the supernatant fluid in presence of ammonium sulfate (40% saturation) as described by Rosenberg et al. (1979b). Further purification of the polymer was achieved by CTAB precipitation and deproteinization as described by Goldman et al. (1982). Protein content was determined according to Lowry et al. (1951). De-esterified emulsan was prepared by alkaline hydrolysis (Zuckerberg et al.

1979). Emulsan was measured in a standard emulsification assay (Rosenberg et al. 1979b). The assay is based on emulsification of a mixture of an aromatic and an aliphatic hydrocarbon in a buffer solution in the presence of magnesium ion. The apoemulsan (deproteinized) preparation used in this study had a specific emulsifying activity of 155 U/mg dry wt (1 U = 6.5 μg apoemulsan).

Esterase assay. The assay is a slight modification of the assay of Higgins and Lapides (1947) and Krish (1966) and is based on hydrolysis of 4-nitrophenyl acetate to yield 4-nitrophenol (PNP). A sample of the enzyme (0.2 ml) was mixed with 1.7 ml of phosphate buffer (75m*M*, pH 7.0) containing 10 m*M* $MgSO_4$, and the reaction was started by adding 0.1 ml of 100 m*M* PNP-acetate in absolute ethanol. The reaction was run at 30 C and was followed by recording the continuous change in absorbance for 5 min at 405 nm in a Gilford 2400 spectrophotometer using automatic zeroing reference control. The activity is expressed in nmol PNP hydrolyzed from the substrate. One unit of esterase activity was defined as 1 nmol PNP/min. Specific activity is related to protein concentration either of the enzyme solution or in the cell suspension as measured after alkaline pretreatment of the cells (0.2 *N* NaOH, 100 C, 20 min).

Release of emulsan in presence of chloramphenicol (CAP). Cells growing exponentially were harvested by centrifugation, washed twice with phosphate buffer, and resuspended in fresh medium containing 50 μg/ml of CAP. The suspension was incubated for 8 h, during which samples were taken for measurements of growth, emulsan, and the esterase.

CTAB tolerance assay. The assay was based on exposure of cellular samples from the growing bacterial cultures to a series of increasing concentrations of CTAB. Aliquots of 0.1 ml from washed cell-suspensions or whole-culture samples (cells and supernatant) were inoculated into a series of tubes containing 1.9 ml ETMS medium and increasing concentrations of CTAB up to 10 μg/ml at intervals of 0.5 μg/ml. The cells were grown for 24 h with gyrotory shaking at 30 C. CTAB tolerance is the highest concentration of CTAB in μg/ml in which growth occurred.

Protection of emulsan against CTAB. Washed cells of *A. calcoaceticus* RAG-1 were exposed to increasing concentrations of CTAB in the presence of increasing amounts of either pure apoemulsan (deproteinized) or de-esterified (base treated) apoemulsan. The cellular samples were then assayed for their CTAB-tolerance in the standard exposure assay to the toxic detergent.

Protection of emulsan in growing cultures. Independently growing cultures of RAG-1 and RAG-10 and a mixed culture of those that received equal inoculum of each were equally divided after 24 h of growth and CTAB was added to one-half of the cultures to final concentration of 100 μg/ml. Growth of the cells was allowed to proceed, and at various times samples from each of the cultures were removed, and viable counts of RAG-1, RAG-10, and emulsan activity were determined according to the standard described procedures.

Selection and isolation of CTAB resistant mutants. Selection of mutants having elevated tolerance to the toxic cationic detergent was carried out on semisolid medium and in liquid cultures. The selection was made on semisolid medium containing 10 μg/ml CTAB. A cell suspension of 0.1 ml containing 10^5 cells/ml was spread on the agar, followed by placing a crystal of N-methyl-N-nitro-N-nitrosoguanidine at the center of the plate. The plates were incubated at 30 C for 30 h. Colonies that appeared on the agar were isolated and rechecked for their tolerance to CTAB on the same semisolid medium. Selection in liquid cultures was achieved by inoculating a sample of nitrosoguanidine-mutagenized culture into ETMS medium allowing for phenotypic expression and then transferring an aliquot into an ETMS medium containing 100 μg/ml CTAB. The selection was carried out for 6 h and was followed by regrowing the survivors in the regular ETMS medium. Samples were plated on semisolid medium containing 25, 50, 75, or 100 μg/ml CTAB for isolating the CTAB tolerant colonies.

Phage techniques and immunocytochemical labeling. Phage techniques including isolation, propagation, and filtering were as previously described (Pines and Gutnick 1984a; Pines et al. 1983). Antibody isolation (Goldman et al. 1982) and immunocytochemical labeling were previously described (Pines et al. 1983).

Results

Role of Esterase in Emulsan Release

The requirement for both a carbon and nitrogen source in the accelerated release of emulsan from the cell surface in the presence of chloramphenicol suggested that there was an active microbial process in the destabilization of the minicapsule (Rubinovitz et al. 1982). An esterase activity was found to participate in the process. The esterase appeared both in a cell-associated and in a cell-free form (Fig. 2). During the first 12 h there was a rapid release of esterase yet a slower release of total protein into the growth medium, accompanied by a decrease in cell-bound enzyme activity. Cell-free esterase activity continued to rise slowly for the next 36 h. The production of emulsan appeared to be delayed relative to the early appearance of the cell-free enzyme. The production of emulsan began at about 18 h and was accompanied by corresponding release of proteins. By 48 h about 210 U/ml emulsan activity (1.3 g/l) and 400 μg/ml protein were found in the cell-free supernatant. The continuous release of the esterase from the cells was reflected in a corresponding decline in cell-bound activity from about 600 U/mg protein to about 200 U/mg protein throughout the 72 h growth cycle. It should be noted that the specific activity of the cell-bound esterase was almost an order of magnitude higher than that of the cell-free enzyme, even after some of the enzyme was released from the cells. A material balance calculated on the basis of both activity and protein in the supernatant and on the cell surface shows clearly that the majority of the enzymatic activity was cell associated despite the partial loss of the enzyme into the medium.

As mentioned previously, emulsan release from the cells was accelerated in presence of CAP, and the release required the presence of both carbon and nitrogen sources (Rubinovitz et al. 1982). In contrast to the above conditions,

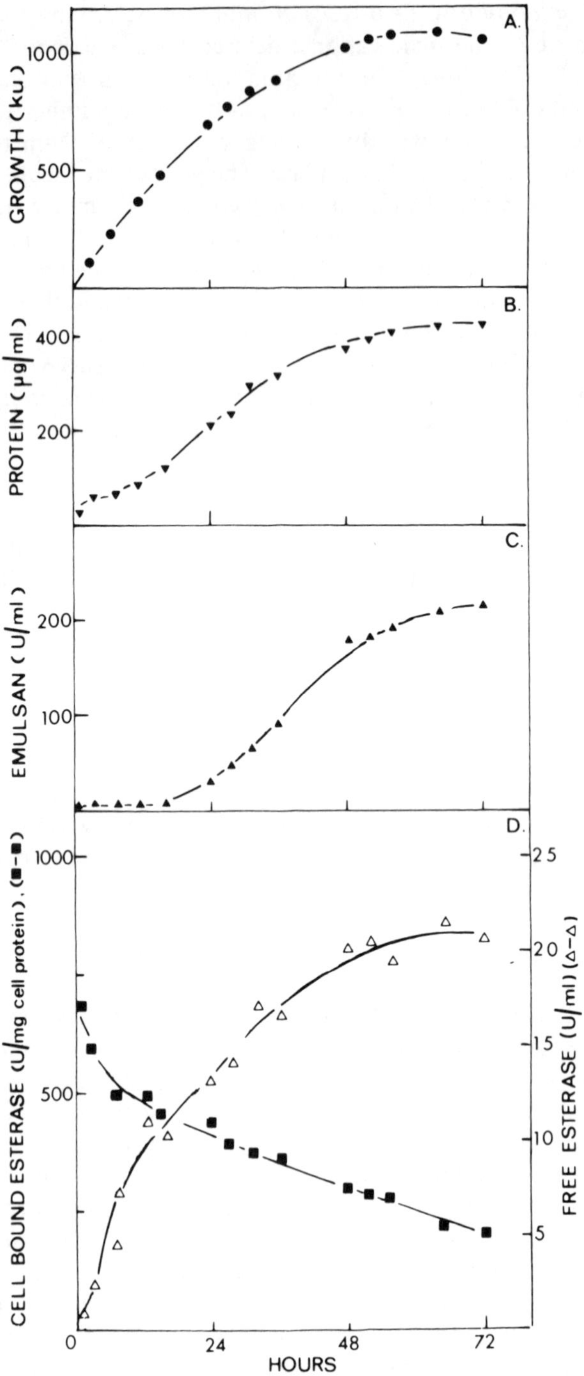

FIG. 2. Esterase distribution during growth of RAG-1. A single colony of RAG-1 was inoculated into 20 ml ETMS medium and allowed to grow overnight. This inoculum was diluted 1:50 into fresh ETMS; samples were removed at the indicated times and assayed for growth (panel A), cell-free protein (panel B), emulsan activity (panel C), cell-free esterase (△ – △ panel D), and cell-bound esterase (■ – ■ panel D).

starvation for carbon alone was sufficient to bring about CAP-mediated esterase release from the cell surface (Fig. 3 B, C). When starved for nitrogen the cells retained the same high level of cell-bound esterase (500 U/mg protein) and no enzyme appeared in the supernatant fluid. Whereas, in the complete system emulsan release proceeded almost linearly and continuously for the entire 4 h treatment, over 90% of the released esterase appeared in the supernatant within the first hour (Fig. 3 B, D). No resynthesis of the enzyme could occur in the presence of CAP (Fig. 3 C), which led to complete cessation of growth (Fig. 3A) and protein synthesis (Rubinovitz et al. 1982). The difference in the requirements for the release of esterase and emulsan suggested a way to study the possible involvement of esterase in emulsan release. It was of interest to determine whether cells from which esterase had been previously released in the presence of CAP, but in the absence of a carbon source, could subsequently release emulsan in a complete CAP-system in which no new esterase could be synthesized. Fig. 3 A′, B′, C′, D′ presents the results of such an experiment. The cells from the five cultures, which were exposed to CAP under different conditions as described above (Fig. 3 A, B, C, D), were washed and resuspended in presence of CAP along with a carbon and nitrogen source. Cells from which about 50% of the cell-bound esterase had been released in the first stage (Fig. 3 B, C) no longer released emulsan in the second stage (Fig. 3 D′). In contrast, cells that retained both their esterase and emulsan during starvation for nitrogen in presence of CAP (Fig. 3 B, C, D) released up to 100 U/ml emulsan within 2 h. The results indicate that an active esterase on the cell surface was required for the release of emulsan. The esterase was found to have a strong affinity for emulsan, as was observed by its specific binding to emulsan immobilized on DEAE-Sephacel beads (Shabtai and Gutnick 1985b).

In addition to de-esterifying the water-soluble substrates, PNP-Ac and triacetin, the RAG-1 esterase also catalyzed the hydrolysis of lipophilic esters, such as PNP-palmitate, but at a rate only about one-tenth of the PNP-Ac rate. In addition, esterase was capable of deacetylating the polysaccharide backbone of emulsan. Hence, cell-associated emulsan may serve as a substrate for the esterase in the release process. Thus, the esterase is likely to be one of the functions that participate in weakening the minicapsular structure of the bacteria facilitating the separation of the surface active form of emulsan from the cells.

Role of Emulsan in Protecting the Cell Against CTAB Toxicity

The polyanionic nature of the emulsan backbone suggested the possibility that the capsule could serve as a cellular protection device against toxic cations. In a series of experiments RAG-1 cells were exposed to the harsh cationic detergent (Ulitzur and Shilo 1970; Hotchkiss 1946; Salton 1951; Chaplin 1951; Salton et al. 1951), cetyltrimethyl ammonium bromide (CTAB), and were shown to be protected in the presence of emulsan. The rationale for this experiment stemmed from the fact that one way of purifying the polymer involves emulsan precipitation in the presence of CTAB. Emulsan at a final concentration of 90 μg/ml enhanced the tolerance of RAG-1 to CTAB from 0.5 μg/ml to about 10 μg/ml. When the de-esterified polymer was used in place of the intact emulsan, the required amounts of the deacylated polymer for neutralizing the same concentration of CTAB were nearly doubled (Shabtai and Gutnick 1985a). The crude

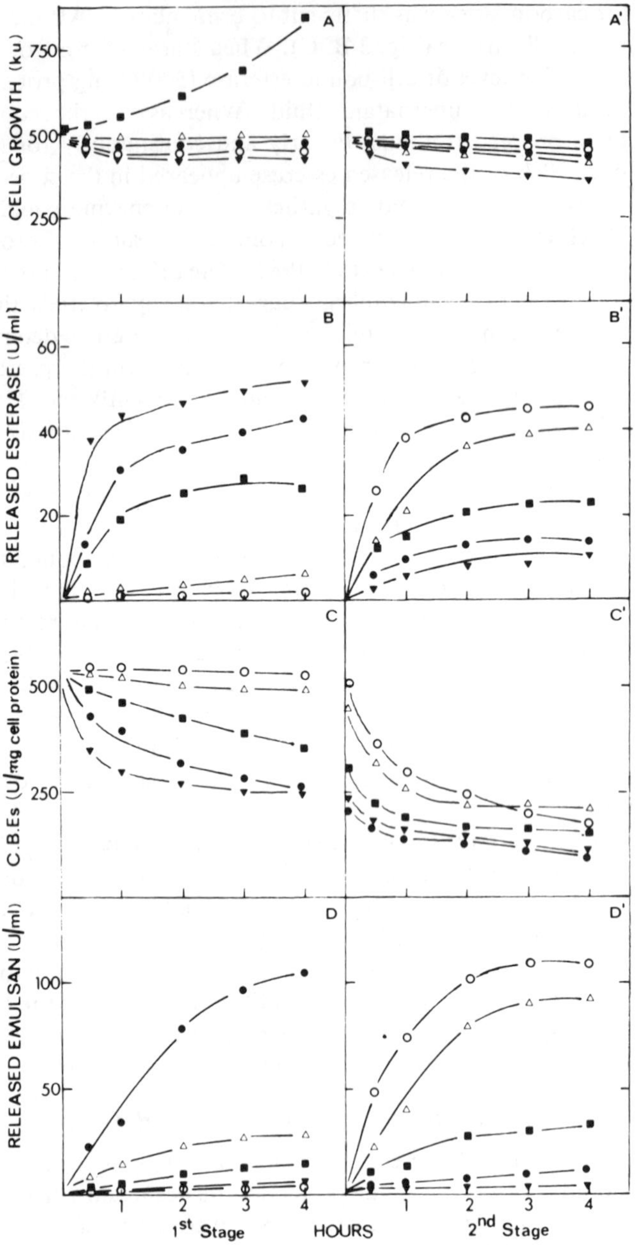

FIG. 3. Release of esterase and emulsan following pretreatment of RAG-1 cells in the presence of CAP. Five exponential cultures of RAG-1 were incubated as described in the legend to Fig. 2. Each culture was pretreated separately in minimal salts media containing ETMS (■ – ■), ETMS + CAP (● – ●), ETMS + CAP – ethanol (▼ – ▼ –), ETMS + CAP – ammonium sulfate (△ – △), and ETMS + CAP – ethanol – ammonium sulfate (○ – ○). At various times samples were removed and assayed for growth (panel A), cell-free esterase (panel B), cell-bound esterase (panel C), and cell-free emulsan (panel D). After 4 h incubation, each culture was harvested and resuspended in complete ETMS medium containing 50 μg/ml CAP (panels A′, B′, C′, D′). The symbols in panels A′, B′, C′, D′ refer to the pretreatment conditions prior to suspending in complete ETMS/CAP medium. At various times samples were removed and assayed.

emulsan in the supernatant of the RAG-1 culture exhibited a similar protective action as did the purified polymer; this was demonstrated by comparison of the tolerance of washed RAG-1 cells with that of whole culture. About 10 μg of active emulsan was required for neutralizing 1 μg of CTAB.

When a mixed culture of RAG-1 and a mutant defective in emulsan production was exposed simultaneously to 100 μg/ml CTAB, a rather surprising fact was discovered (Fig. 4). The emulsan that was produced by the parent strain did not protect the defective mutant in the initial stage of the exposure to CTAB (Fig. 4 B). In the absence of CTAB (Fig. 4 A), equal numbers of the two strains were obtained.

In the presence of CTAB in either the mixed culture (Fig. 4 B) or in the two separate cultures (Fig. 4 D), the emulsan-defective mutants stopped growing immediately. Under these conditions the parent RAG-1 continued to grow, even though the culture contained only about 250 μg/ml *cell-free* emulsan (45 U/ml, Fig. 4 C) at the time of CTAB addition. When the extracellular emulsan activity in the mixed culture reached 75 U/ml (Fig. 4 C), the emulsan-defective mutant began to grow. The results suggest that the cell-associated emulsan is the major factor in protecting RAG-1 cells against CTAB. The close association of emulsan and its producing cells in the form of minicapsular-envelope served as a shield against the external toxic agent.

The fact that the emulsan-minicapsule could protect against CTAB suggested a possible way for selecting mutants with elevated protection against the toxic compound. It appeared likely that among mutants with enhanced tolerance to CTAB one might find those with enhanced emulsan production.

Such CTAB-tolerant mutants were obtained using a direct selection on a semisolid agar medium, which contained elevated levels of the detergent 5-fold higher than the minimal toxic concentration. Similar mutants were obtained by selection in liquid medium following the mutagenesis step. Among the isolated mutants several exhibited enhanced production of emulsan.

One of this group, CTR-1049, exhibited a 2- to 3-fold increase in the yield of emulsan as compared to the parent strain under the same standard conditions of medium and growth. Both strains grew in a similar fashion in the pH-controlled fermentation that was carried out in the 2-liter fermentor. The production of emulsan was faster by the mutant. After about 20 h of fermentation, the mutant CTR-1049 produced about 3-fold more emulsan than the parent RAG-1. The final ratio of emulsan to cell-mass in the fermentation using the mutant was close to 135 U/mg, while the ratio was only 60 U/mg in the case of the parent strain. The overall productivity of the process reached about 15 U emulsan/1-h for RAG-1 compared to 60 U emulsan/1-h for mutant CTR-1049.

Possible Role for Emulsan in Growth of RAG-1 on Crude Oil

It was of interest to determine whether an effect of the mutation of emulsan-deficiency could be observed during growth of mixed cultures in the parent (RAG-1) and a mutant (TR3) on crude oil. A mixed culture of RAG-1 and TR3 inoculated with an equal number of each of the two strains was grown for 120 h as described in Materials and Methods. The enumeration of each strain in the mixed culture was based on (a) the different colony morphology of emulsan-producing (mucoid) and emulsan-deficient (translucent) mutants (Fig. 5), and (b)

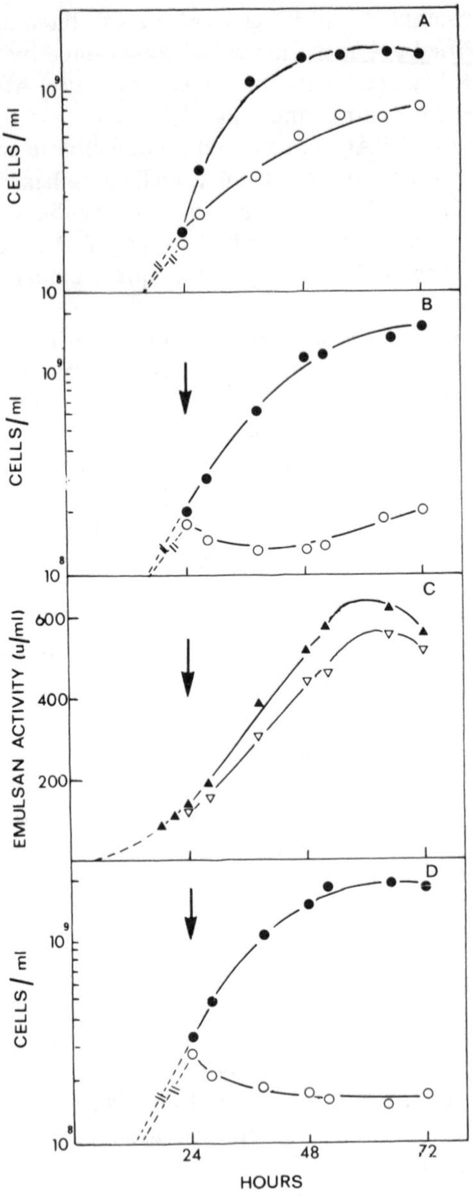

FIG. 4. Effect of CTAB on growth and emulsan production of RAG-1 and RAG-92-10 in mixed cultures. Independently growing cultures of RAG-1 and RAG-92-10 on ETMS were prepared according to Materials and Methods. A mixed culture, which received an equal inoculum of each of the strains, was also prepared on ETMS. All of the three cultures were allowed to grow for 24 h, and the cultures were then equally divided. CTAB was added at the time indicated by arrows to one-half of each of the cultures to a final concentration of 100 μg/ml and growth of the cells allowed to proceed. At various times samples from each of the cultures were removed and viable counts of RAG-1 (● – ●), RAG-92-10 (○ – ○), and emulsan activity were determined. A, mixed culture of RAG-1 and RAG-92-10 without CTAB addition; B, mixed culture of RAG-1 and RAG-92-10 in presence of CTAB; C, emulsan activity in cell-free supernatants of mixed culture in absence (▲ – ▲) and in the presence (▽ – ▽) of CTAB; D, growth of RAG-1 (● – ●) and RAG-92-10 (○ – ○) in separate independent cultures in presence of CTAB.

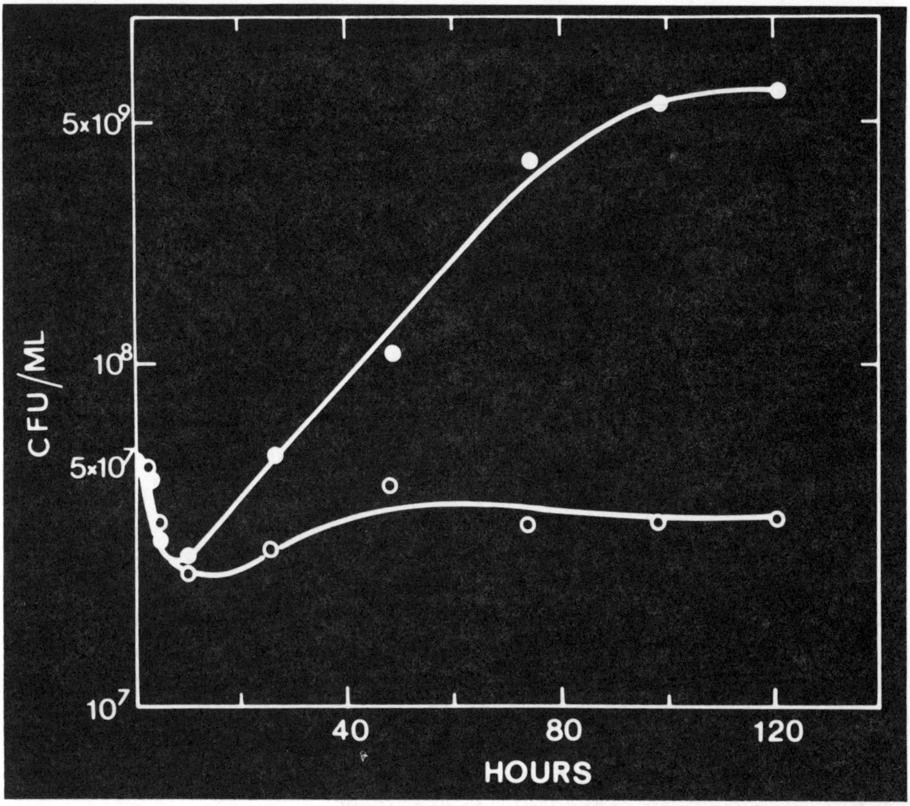

FIG. 5. Growth of a mixed culture of RAG-1 and TR3 on seawater medium (SM) supplemented with crude oil. The SM contained 90% seawater from a Tel Aviv beach previously filtered through a 0.45 μm filter, 10% ETMS, and sufficient ammonium sulfate to make the concentration 2%. Iranian crude oil (Agha Jari obtained from the Haifa refineries) was added to a final concentration of 2% (v/v). Cells were grown in volumes of 4.5 ml medium in acid washed 14-mm diameter tubes, each inoculated with 5×10^7 cfu/ml each of RAG-1 (● – ●) and TR3 (○ – ○). For determination of viable counts (cfu), an aliquot of 500 μg/ml of emulsan was added to a growth tube, the tube was vigorously mixed on a Vortex mixer for 2 min to dislodge adherent cells, and appropriate dilutions plated on minimal plates. RAG-1 cells were counted as white mucoid colonies and TR3 cells as translucent colonies.

the sensitivity of wild type strains to phage ap3 and the sensitivity of emulsan-deficient strains to phage nϕ owing to the presence or absence of emulsan on their cell surface (Pines and Gutnick 1984a; Pines et al. 1983).

Reversion of the Emulsan-Negative Phenotype

In order to test the hypothesis that cell-associated emulsan is involved in growth of RAG-1 on crude oil, it was of interest to examine the growth pattern of emulsan-producing revertants of TR3. Initially it was expected that the isolation of bacteriophage nϕ, which attacks only emulsan-deficient strains (Pines and Gutnick 1984b), and the isolation of mutants of TR3 resistant to nϕ, would provide a selection for emulsan-producing strains. However, all of the nϕ resistant derivatives of mutant TR3 were found to be emulsan-deficient. Since these TR3 derivatives did not adsorb either ap3 or nϕ, it is apparent that the frequency of mutation in the nϕ receptor is much higher than the reversion frequency of the original emulsan-deficient mutation.

Nevertheless, it was expected that growth on crude oil under the specific conditions described above might provide sufficient selective pressure to enrich for emulsan-producing revertants within an emulsan-negative culture. In order to rule out the possibility of contamination, the wild type was replaced with a double auxotroph, RA15, which requires both tryptophan and lysine for growth. The emulsan-negative derivative of strain RA15 (TRA15) was then selected by its resistance to phage ap3 as previously described (Pines and Gutnick 1984b; Pines et al. 1983). TRA15 was defective in emulsan production and required both tryptophan and lysine.

A culture of TRA15 was grown on crude oil, transferred to fresh medium, and grown as described in Materials and Methods. Samples from the aqueous phase were treated with phage nϕ and plated. After 48 h of growth bacteria with wild-type colonies appeared in the transferred culture with a frequency of 8×10^{-5} of the total bacterial population in the aqueous phase and 5×10^{-3} of the nϕ resistant colonies. Whereas stationary phase cultures of mutant TRA15 contained only about 25% of the emulsan activity and 17% of the cell-free hexosamines of the parental strain, the revertant strain, REV15, produced 68% of the emulsan activity and 81% of the cell-free hexosamines of the original parent strain RA15. In addition, washed exponential phase cells of the original strain, RA15, and the revertant, REV15, contained about 40 μg hexosamines per mg cell dry wt. In contrast, the emulsan-deficient mutant, TRA15, had only about 16 μg hexosamines per mg dry wt. Moreover, the change in cell-bound emulsan associated with the reversion of the emulsan deficiency was correlated both with an acquired resistance to phage nϕ and sensitivity to phage ap3.

Discussion

Role of Esterase

The results presented in this report indicate a role for a cell-bound esterase in the release of emulsan from the cell surface of RAG-1. The basis of this conclusion stems from studies in the release of emulsan and esterase in the presence of CAP. In the absence of a carbon source the esterase was released while emulsan was retained on the cell surface. Once esterase was removed, the emulsan was no longer released in the presence of CAP since no new enzyme could be synthesized in the presence of the protein synthesis inhibitor. It is still not clear why the release of esterase requires the presence of a nitrogen source. Such a requirement for bioemulsifier release probably is related to synthesis of amino sugar precursors (Rubinovitz et al. 1982).

A significant amount of active esterase was retained on the cell surface after some of the enzyme was released in the presence of CAP. It was of interest that this residual enzyme was not active in subsequent biopolymer release. One possibility to explain this may be that only a fraction of the RAG-1 esterase is specifically associated with emulsan at the cell surface. An alternative explanation might be that there is more than a single esterase (or lipase) and only one enzyme participates in the release. Support for the first hypothesis comes from the isolation of point mutants of RAG-1 defective in esterase that were also defective in the release of emulsan (unpublished).

The mechanism by which esterase participates in the release process remains unclear. The high affinity of the enzyme for the acylated form of the polymer and the possibility of an enzyme-substrate relationship (Shabtai and Gutnick 1985b) suggests a role for the enzyme in acylation or deacylation of emulsan or other cell-surface polymers. Such reactions would be expected to modify the hydrophobicity of the cell surface (Sutherland 1979, 1982), which in turn could weaken the affinity of the polymer for the surface.

Accelerated release of a lipopolysaccharide-phospholipid protein complex from the outer membrane of *E. coli* and *S. typhimurium* has been reported previously (Jann 1968; Hungerer et al. 1967). This complex appeared to contain only a single protein species. The esterase protein of RAG-1, which was released either in an emulsan-bound form or as a free enzyme, was only one of a number of proteins (10–100,000 M.W.) associated with the cell-free emulsan.

CTAB Protection

Both forms of emulsan, the cell-free as well as the cell-associated, were found to neutralize the toxic effect of the cationic detergent CTAB. Moreover, mutants of RAG-1 defective in emulsan production were found to be less tolerant to the detergent than the wild type strains. Since growth of the parent and an emulsan-deficient mutant in mixed culture in the presence of CTAB did not increase the tolerance of such a mutant to the detergent during the first 24 h (Fig. 4 B), the cell-associated form of the polymer was more active in protection during the early stages of growth. It was only after about 48 h, when sufficient cell-free emulsan appeared in the medium from the parental strain (75 U/ml or 0.5 mg/ml), that the mutant began to grow in the presence of CTAB. Extracellular polyanions have been implicated in the protection of bacteria against various toxic materials including antibiotics (Costerton 1981), surfactants (Govan and Fyfe 1978), or heavy metals (Engel and Owen 1970). In the case of emulsan, it appears that the neutralization of CTAB requires more than electrostatic interaction of the cationic surfactant with the negatively charged uronic acid residues on the emulsan polymer. The requirement for 10 μg emulsan (about 15 nanoequivalents of carboxyl groups) to neutralize 1 μg CTAB (about 2.7 nmol) indicates a rather more complex mechanism of protection. Of interest was the finding that a mutant of RAG-1, which produces a similar biopolymer that is inactive as an emulsifier yet still contains uronic acids, was found to be less tolerant to CTAB than the parent.

Role for Emulsan in Growth on Oil

The competition experiments between RAG-1 and TR3 in mixed cultures growing on crude oil provide evidence that the production of extracellular emulsan by the wild type does not complement an emulsan-deficient mutant. Furthermore, since exogenously added emulsan did not reduce the competitive advantage of the wild type it appears that the biologically active form of the polymer is in fact cell associated. The fact that both strains show similar growth characteristics on water-soluble substrates suggests that emulsan-deficiency specifically affects the interaction of the organism with oil. This defect is most probably a point mutation since (a) spontaneous mutants are isolated at high frequency (in preparation), and (b) the emulsan-deficient phenotype can be reverted. The results suggest that the defect in emulsan production and the defect in growth on crude oil result from the same genetic lesion. It remains to be determined whether the importance of the cell-associated form of emulsan as reflected in mixed culture experiments for both CTAB protection and growth on crude oil reflect similar properties of the cell-bound biopolymer. One observation related to this involves cell-bound emulsan, which serves as a receptor for the bacteriophage ap3 (Pines et al. 1983; Pines and Gutnick 1984a). The cell-free material showed virtually no such activity. It was of interest, therefore, that the ap3 receptor could be reconstituted at an emulsan-stabilized oil/water interface (Pines and Gutnick 1984a). These results suggest that the conformation of the bioemulsifier on the cell surface is similar to its conformation at the oil/water interface. One schematic model describing this similarity is illustrated in Fig. 6. In this model the hydrophobic side chains of the polymer are oriented toward the oil droplet surface or toward a hydrophobic layer on the cell surface. One prediction of this model is that emulsan is associated with some hydrophobic cell-surface component. It has recently been shown that removal of the cell surface emulsan layer by either mutation or enzyme degradation gave rise to a cell with enhanced hydrophobicity (in preparation). The model is currently being tested by isolating and characterizing a variety of mutants altered in the association of emulsan with

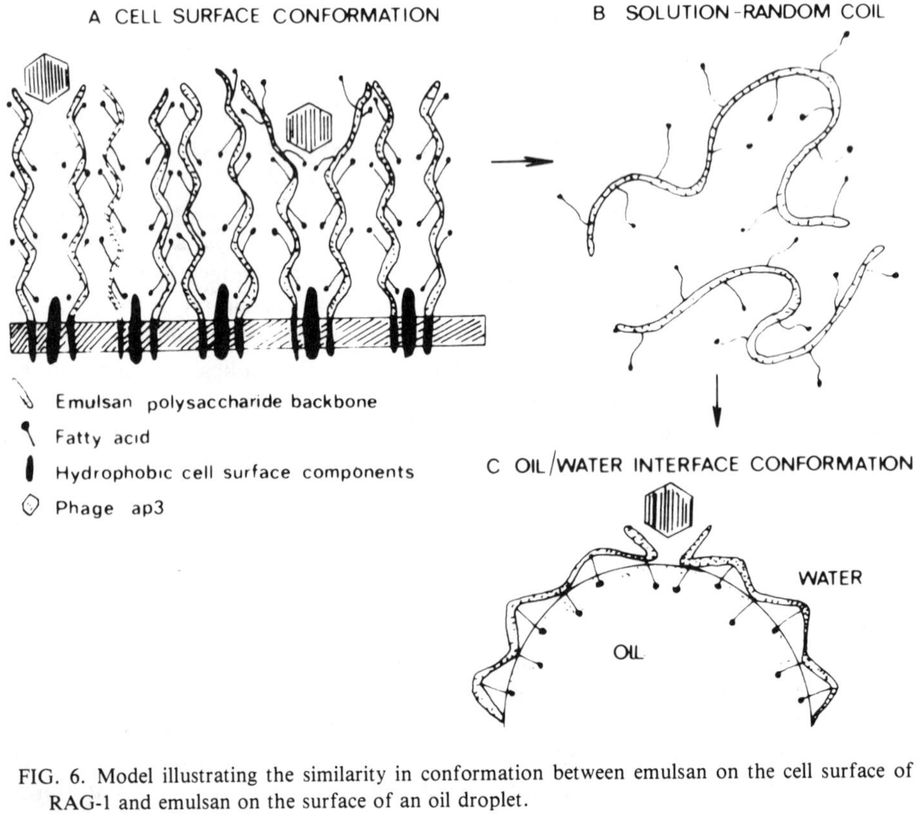

FIG. 6. Model illustrating the similarity in conformation between emulsan on the cell surface of RAG-1 and emulsan on the surface of an oil droplet.

the cell surface. Some of the CTAB resistant strains described above appear to have such characteristics in addition to showing enhanced emulsan production.

Literature Cited

Belsky, I., D. L. Gutnick, and E. Rosenberg. 1979. Emulsifier of Arthrobacter RAG-1: Determination of emulsifier bound fatty acids. *FEBS Lett.* 101:175–178.

Chaplin, C. F. 1951. Observation on quaternary ammonium disinfectants. *Can. J. Bot.* 29:373–382.

Costerton, J. W., R. T. Irvin, and K. J. Chong. 1981. The role of bacterial surface structures in pathogenesis. *CRC Critical Rev. Microbiol.*, pp. 303–338.

Deavin, L., T. R. Jarmain, C. J. Lawson, R. C. Righelato, and S. Solocombe. 1977. Pages 14–20 *in* P. A. Sanford and A. Laskin, eds., *Extracellular Microbial Polysaccharides,* American Chemical Society, Washington, DC.

Dudman, W. F. 1977. The role of surface polysaccharides in natural environments. Pages 357–414 *in* I. Sutherland, ed., *Surface Carbohydrates of the Prokaryotic Cell.* Academic Press, New York.

Engel, W. B., and R. A. Owen. 1970. Metal accumulating properties of fuel utilizing bacteria. *Dev. Ind. Microbiol.* 11:196–209.

Goldman, S., Y. Shabtai, C. Rubinovitz, E. Rosenberg, and D. L. Gutnick. 1982. Emulsan production in *Acinetobacter calcoaceticus:* Distribution of cell-free and cell associated cross-reacting material. *Appl. Environ. Microbiol.* 44:165–170.

Govan, J. R. W., and J. A. M. Fyfe. 1978. Mucoid *Pseudomonas aeruginosa* and cystic fibrosis: Resistance of the mucoid form to carbenicillin, flucloxacillin and tobramycin, and the isolation of mucoid variants in vitro. *J. Antimicrob. Chemother.* 4:233–240.

Gutnick, D. L., and E. Rosenberg. 1977. Oil tankers and pollution: A microbiological approach. *Annu. Rev. Microbiol.* 31:379–396.

Higgins, C., and J. Lapides. 1977. Chromogenic substrates. IV. Acylesters of *p*-nitrophenol as substrates for the colorimetric determination of esterase. *J. Biol. Chem.* 170:467–482.

Hotchkiss, R. D. 1946. The nature of the bacteriocidal action of a surface active agent. *Annu. N.Y. Acad. Sci.* 46:479–483.

Hungerer, D., K. Jann, B. Jann, F. Orskov, and I. Orskov. 1967. Immunochemistry of K-antigens of *Escherichia coli. Eur. J. Biochem.* 2:115–126.

Jann, K. 1968. Immunochemistry of K-antigens of *Escherichia coli. Eur. J. Biochem.* 5:456–465.

Kang, K.S., and I.W. Cottrell. 1979. Polysaccharides. *Microb. Technol.* 1:417–481.

Krish, K. 1966. Reaction of a microsomal esterase from hog liver with diethyl *p*-nitrophenyl phosphate. *Biochim. Biophys. Acta* 122:265–280.

Lowry, O. H., N. J. Rosenbrough, A. L. Farr, and R. J. Randall. 1951. Protein measurement with the Folin phenol reagent. *J. Biol. Chem.* 193:265–275.

Pace, G. W. 1980. Production of extracellular microbial polysaccharides. *Adv. Biochem. Eng.* 15:41–70.

Pines, O., and D. L. Gutnick. 1984a. Specific binding of a bacteriophage at a hydrocarbon-water interface. *J. Bacteriol.* 157:179–183.

Pines, O., and D. L. Gutnick. 1984b. Alternate hydrophobic sites on the cell surface of *Acinetobacter calcoaceticus* RAG-1. *FEMS Microbiol. Lett.* 22:307–311.

Pines, O., E. A. Bayer, and D. L. Gutnick. 1983. Localization of emulsion-like polymers associated with the cell surface of *Acinetobacter calcoaceticus. J. Bacteriol.* 154:893–905.

Rosenberg, E., A. Perry, D. T. Gibson, and D. L. Gutnick. 1979a. Emulsifier of *Arthrobacter* RAG-1: Specificity of hydrocarbon substrate. *Appl. Environ. Microbiol.* 37:409–413.

Rosenberg, E., A. Zuckerberg, C. Rubinovitz, and D. L. Gutnick. 1979b. Emulsifier of *Arthrobacter* RAG-1: Isolation and emulsifying properties. *Appl. Environ. Microbiol.* 37:402–408.

Rubinovitz, C., D. L. Gutnick, and E. Rosenberg. 1982. Emulsan production by *Acinetobacter calcoaceticus* in the presence of chloramphenicol. *J. Bacteriol.* 152:126–132.

Salton, R. J. 1951. The adsorption of cetyltrimethyl ammonium bromide by bacteria, its action in releasing cellular constituents and its bacteriocidal effects. *J. Gen. Microbiol.* 5:391–404.

Salton, R. J., R. W. Horne, and V. E. Cosslett. 1951. Electron microscopy of bacteria treated with cetyltrimethyl ammonium bromide. *J. Gen. Microbiol.* 5:405–407.

Shabtai, Y., and D. L. Gutnick. 1985a. Tolerance of *Acinetobacter calcoaceticus* RAG-1 to the cationic surfactant cetyltrimethyl ammonium bromide (CTAB)—role of the bioemulsifier, emulsan. *Appl. Environ. Microbiol.* 49:192–199.

Shabtai, Y., and D. L. Gutnick. 1985b. Exocellular esterase and emulsan release from the cell surface of *Acinetobacter calcoaceticus* RAG-1. *J. Bacteriol.* 161:1176–1184.

Sutherland, I. W. 1979. Microbial exopolysaccharides: Control of synthesis and acylation. Pages 1–34 *in* R. C. W. Berkeley, G. W. Gooday, and D. C. Ellwood, eds., *Microbial Polysaccharides and Polysaccharases.* Academic Press, New York.

______.1982. Biosynthesis of microbial exopolysaccharides. *Adv. Microb. Physiol.* 23:79–150.

Sutherland, I. W., and D. C. Ellwood. 1979. Microbial exopolysaccharides—industrial polymers of current and future potential. Pages 107–150 *in* A. T. Brill, D. C. Elwood, and C. Ratledge, eds., *Microbial Technology Current State, Future Projects.* Cambridge University Press.

Ulitzur, S., and M. Shilo. 1970. Effect of *Prymnesium pavuum* toxin, cetyltrimethyl ammonium bromide and sodium dodecyl sulfate on bacteria. *J. Gen. Microbiol.* 62:363–370.

Zosim, Z., D. L. Gutnick, and E. Rosenberg. 1982. Properties of hydrocarbon-in-water emulsions stabilized by *Acinetobacter calcoaceticus* RAG-1 emulsan. *Biotechnol. Bioeng.* 24:281–282.

Zuckerberg, A., A. Diver, Z. Peeri, D. L. Gutnick, and E. Rosenberg. 1979. Emulsifier of *Arthrobacter* RAG-1: Chemical and physical properties. *Appl. Environ. Microbiol.* 37:414–420.

VI
SYMPOSIUM: APPLICATIONS OF INDUSTRIAL STARTER CULTURES

G. L. Enders and W. T. H. Chang, *Conveners*
Miles Laboratories, Elkhart, Indiana, and
National Science Council, Taiwan, China

PANELISTS

M. A. Daeschel
G. L. Enders
H. P. Fleming
L. M. Johnson
H. S. Kim
S. W. King
M. Krupa
C. S. McDowell
R. F. McFeeters
C. H. Tseng

CHAPTER 20

Recent Developments of Industrial Malolactic Starter Cultures for the Wine Industry

STEPHEN W. KING

Microlife Technics, Inc., Sarasota, Florida 33578

Two strains of *Leuconostoc oenos* provided as concentrated, frozen starter cultures were used to induce malolactic fermentation (MLF) in wines of the Bordeaux region of France and of the Napa Valley, CA, during the 1983 harvest. Both strains were comparable in their ability to induce and accelerate MLF in winery and laboratory trials. Under cellar conditions in Bordeaux, the cultures accelerated MLF in 52.2% of red wines and 20% of white wines when compared to uninoculated controls. MLF was accelerated in laboratory sublots of each wine nearly 100% of the time, and each sublot generally underwent MLF faster than its winery counterpart. Results were similar in California. Factors that affect the ability of a starter culture to induce and accelerate MLF under winery and laboratory conditions were examined. The economics and convenience of using a starter culture as well as the manner in which it may influence future oenological practices are discussed.

INTRODUCTION

Malolactic fermentation (MLF) is the conversion of L-malic acid to L-lactic acid and carbon dioxide by certain strains of lactic acid bacteria in wine. The fermentation is desirable, particularly in wines from cool climates, since it reduces the titratable acidity of the wine (Kunkee 1967a; Beelman and Gallander 1979). In addition, it is thought to increase the sensory complexity of the wine (Pilone and Kunkee 1965; Fornachon and Lloyd 1965) and render it biologically stable (Kunkee 1974). However, the harsh environment of wine does not always favor the growth of malolactic bacteria. Thus, malolactic fermentation occurs sporadically and is unpredictable. Low pH (Peynaud and Domercq 1967; Bousbouras and Kunkee 1971), the presence of sulfur dioxide (Lafon-Lafourcade 1975; Mayer et al. 1975), ethanol (Peynaud 1967), and the lack of nutrients in new wine (Banner 1973; Beelman et al. 1982) appear to be the most inhibitory factors. The individual and synergistic effect of these compounds have been documented (Beelman et al. 1982; King 1982).

In recent years the use of pure culture inoculation of selected strains of *Leuconostoc oenos*, a lactic acid bacterium indigenous to grapes and wine, has enabled the winemaker to have better control over the malolactic process (Kunkee et al. 1964; Kunkee 1967b; Beelman et al. 1977). This in turn has resulted in the development and use of commercial starter cultures (Beelman 1980; Haymon and Monk 1982; Lafon-Lafourcade 1983).

Although starter cultures greatly improve the chance of malolactic fermentation, they can be inconvenient to use, and malolactic fermentation still may not occur. These cultures inoculate only small quantities of wine, and to inoculate large volumes of wine, bulk starter procedures must be employed. This is often time consuming and occupies large areas of wine cellar space.

Therefore, it would be useful to develop a commercial malolactic starter culture that carries out a clean, rapid, and predictable fermentation requiring little time to prepare for inoculation and requiring no bulk starter procedures. The purpose of this report is to describe an effort to develop such a culture.

Materials and Methods

Isolation and Characterization of Malolactic Bacteria

One red wine and one white wine sample undergoing spontaneous malolactic fermentation were received from Wurzburg, Germany. One ml of each wine was transferred to 10 ml of a nutrient broth (Gestrelius and Kjaer 1983) to which 0.15% (w/v) cyclohexamide had been added to prevent yeast growth. The wine broths were incubated at 24 C for 5 d. One loopful of each broth was streaked onto a 2% agar plate of the same medium. The plates were incubated at 24 C for 5 d in an Anaerobic Jar Gas Pak (Baltimore Biological Lab, Cockeysville, MD). One colony from each plate was transferred back into 10 ml of the nutrient broth which was once again incubated at 24 C for 5 d.

The two isolates were designated SK-4 (red wine) and SWK-2 (white wine) and characterized using the methods of Garvie (1967) and Pilone and Kunkee (1972). *Leuconostoc oenos* strains ML-34 and PSU-1 were used as controls during characterization.

Evaluation of Malolactic Conversion

The two malolactic isolates along with an "in-house" strain of *Leuconostoc oenos* (designated Kli) were evaluated for efficacy of malolactic conversion under simulated wine conditions. Ethanol and sulfur dioxide (added as potassium metabisulfite) were added at concentrations of 12% (v/v) and 50 mg/L, respectively to 90 ml aliquots of Concord grape juice (Welch Foods, Westfield, NY). The pH was adjusted to 3.2 with 85% phosphoric acid, and the juice was placed at 24 C for 24 h to allow the sulfur dioxide to bind to grape juice components and equilibrate.

One tenth (0.1) ml of a stationary phase culture of the two isolates and *L. oenos* Kli were added to the simulated wine to give an initial population of approximately 1×10^6 cfu/ml. Malolactic fermentation was monitored every 5 d by paper chromatography (Kunkee 1974). All evaluations were done in triplicate and incubated at 18 C.

Industrial Preparation of Malolactic Bacteria

L. oenos Kli and SK-4 were grown separately in a nutrient medium in 400-gal stainless steel tanks. The organisms were harvested, concentrated, and quick frozen in 85 g aliquots and stored at -30 C. These preparations were subsequently used in winery tests.

Winery and Wine Laboratory Testing

The industrial preparations were tested in 12 red wines and one white wine of the Napa Valley, CA. In the Bordeaux region of France, 26 red and 10 white wines were tested. California grape varieties were listed in Table 3. Bordeaux red wines were a blend of Cabernet Sauvignon, Cabernet Franc, and Merlot grapes and white wines were usually a blend of Chardonnay, Semillon, and various other varieties. The bacteria were bioactivated before inoculation into wine by adding 85 *g* thawed culture to about 5 gal fresh grape juice diluted 1:1 with water. Yeast extract was added to 0.5% (w/v) and the pH was adjusted to 4.0–4.5 with calcium carbonate. The bioactivation mixture was incubated at 25 C for 24 to 72 h (Lafon-Lafourcade 1983). In the Napa Valley the bacteria were added before or during alcoholic fermentation, whereas in Bordeaux they were added subsequent to alcoholic fermentation. In both locations the bacteria were added at a rate of approximately 1:1000 to give an initial population of 10^6 to 10^7 cfu/ml.

Malolactic fermentation and chemical analysis of the wines were monitored by enzymatic and/or chemical methods at the discretion of the winery or laboratory. In Bordeaux, 3 to 5 gal sublots of each wine inoculated were transported to a laboratory at the Institute d'Oenologie Universite de Bordeaux II for parallel testing.

Results and Discussion

Characterization of Malolactic Isolates

The characteristics of SK-4 and SWK-2 were nearly identical to *L. oenos* ML-34 and PSU-1 (Table 1). However, SK-4 and SWK-2 use mannose, whereas ML-34 and PSU-1 do not. SK-4 and SWK-2 are nearly identical except that SWK-2 gave a variable response to lactose. Much heavier growth of SWK-2 was apparent on esculin compared to SK-4. Microscopic examination of SK-4 and SWK-2 showed cocci that were slightly elongated, occurring mostly in pairs and short chains. Based on these observations and in accordance with the criteria set forth in *Bergey's Manual,* these isolates would be classified as two distinct strains of *Leuconostoc oenos*. The characteristics of *L. oenos* Kli are also shown in Table 1.

Evaluation of Malolactic Conversion

On the average SK-4 and Kli induced MLF in simulated wine in nearly equal time (Table 2). SWK-2 did not initiate MLF in two of the three replications. Direct plating (0.1 ml) methods revealed no viable cells in either of the flasks after 50 d. Based on the results of this and other laboratory tests, it was decided that SK-4 and Kli would be tested under winery conditions.

Wine Cellar and Wine Laboratory Trials

Field testing was done on both a small- and large-scale basis in California (Table 3). Trials no. 1 through 4 ranged from 2.5 to 15 gal, while Trials no. 5 and 6 ranged from 5,000 to 6,000 gal. In Trials no. 1 through 4, SK-4 and Kli completed MLF in 7 to 14 d in all wines and completed MLF before the uninoculated controls. When SK-4 and Kli were compared to *L. oenos* strains ML-34 and

PSU-1 (Trials no. 1 and 4) they performed equally or better than ML-34 and PSU-1.

TABLE 1. *Characteristics of* Leuconostoc oenos *ML-34, PSU-1, SK-4, SWK-2, and Kli*

Characteristics	ML-34	PSU-1	SK-4	SWK-2	Kli
Morphology	Coccus, slightly elongated in pairs and chains				
Gram reaction	+	+	+	+	+
Catalase	−	−	−	−	−
Acid and Gas from Glucose	+	+	+	+	+
Dextran from Sucrose	−	−	−	−	−
Arabinose	−	−	−	−	−
Fructose	+	+	+	+	+
Galactose	−	−	−	−	−
Glucose	+	+	+	+	+
Lactose	−	+/−	−	+/−	−
Maltose	−	+/−	−	−	−
Mannose	−	−	+	+	+
Mannitol	−	−	−	−	+
Raffinose	−	−	−	−	−
Rhamnose	−	−	−	•	•
Ribose	+	+	+	+	+
Salicin	+	+	+	+	+
Sucrose	−	−	−	−	−
Trehalose	+	+	+	+	+
Xylose	−	−	−	−	−
Esculin	+	+	+	++	+
Melibiose	+/−		+	•	•
Cellobiose	+	+	+	•	•
Citric acid	+	+	+	+	+

TABLE 2. *Evaluation of* Leuconostoc oenos *SK-4, SWK-2, and Kli in simulated wine*

Strain	Replication	MLF Completed by (days)
SK-4	1	15
	2	30
	3	25
SWK-2	1	—
	2	—
	3	25
Kli	1	10
	2	35
	3	35

In large-scale trials (no. 5 and 6) SK-4 and Kli were not as successful at inducing MLF. SK-4 induced MLF in two out of four wines in which it was inoculated, and Kli induced MLF in one out of two wines into which it was inoculated. Overall, in both large- and small-scale trials, SK-4 and Kli appeared to be equal in the time they took to complete MLF.

These data demonstrated that MLF can be easily controlled on a small-scale basis, but in large-scale operations it is still unpredictable. Several explanations may be offered for these differences. Trials no. 1 through 4 were conducted in

TABLE 3. *Field trial results in the Napa Valley, CA*

Trial No.	Volume (gal)	Varietal	pH	SO_2(mg/1)	Strain	MLF Completed by (days)
1	2.5	Pinot Noir	3.62	0	SK-4	7
	2.5	Pinot Noir	3.69	0	Kli	9
	2.5	Pinot Noir	3.55	0	ML-34	5
	2.5	Pinot Noir	3.78	0	PSU-1	18
	2.5	Pinot Noir	3.82	0	Control	18
2	15	Pinot Noir	3.66	0	SK-4	11
	15	Pinot Noir	3.66	0	Kli	13
	15	Pinot Noir	3.66	0	Control	[a]
3	10	Chardonnay	3.47	24	SK-4	14
	10	Chardonnay	3.47	24	Kli	14
	10	Chardonnay	3.47	24	Control	34
4	5	Ruby Cabernet	3.59	0	SK-4	7
	5	Ruby Cabernet	3.59	0	Kli	7
	5	Ruby Cabernet	3.59	0	ML-34	36
	5	Ruby Cabernet	3.59	0	Control	36
5	5000	Cabernet Sauvignon	3.30	53	SK-4	[a]
	5000	Cabernet Sauvignon	3.30	66	SK-4	54
	5000	Cabernet Sauvignon	3.30	40	Kli	30
	5000	Cabernet Sauvignon	3.30	82	Kli	[a]
	5000	Cabernet Sauvignon	3.30	—	Control	[a]
6	6000	Cabernet Sauvignon	3.40	—	SK-4	53
	6000	Gamay Beaujolais	3.40	54	SK-4	[a]
	6500	Gamay Beaujolais	3.40	—	SK-4	[a]

[a] MLF never completed.

laboratories with wineries where environmental conditions could be carefully controlled, whereas in a production cellar, the winemaker does not always have control over all environmental factors.

One striking difference between the two sets of trials is that no sulfur dioxide was added to small batches of wine except in Trial no. 3. The large volumes of wine contained copious yet acceptable concentrations of sulfur dioxide ranging from 40 to 82 mg/L. Sulfur dioxide is added as a preservative to grape juice and wine to prevent growth of and spoilage by undesirable microorganisms. Sulfur dioxide exists in free and bound form in wine. Burroughs and Sparks (1964) identified many sulfur dioxide binding compounds in juice and cider, of which acetaldehyde seemed to be the most important binding agent (Fornachon 1963). For years it was believed that only the free form was inhibitory to bacterial growth. However, Fornachon (1963) found that acetaldehyde-bound sulfur dioxide inhibited the growth of malolactic organisms at 100 mg/L. Lafon-Lafourcade (1975) demonstrated that in red wine 10 mg/L acetaldehyde-bound sulfur dioxide was inhibitory to the growth of *L. oenos* but not to its malolactic activity; 20 mg/L-bound sulfur dioxide was bacteriostatic, and malolactic activity was decreased by 38%, by 50% at 50 mg/L, and totally inhibited at 100 mg/L. Thus, in Trials no. 5 and 6 data for sulfur dioxide appear to support earlier findings, especially in Trial no. 5 where one wine contained 82 mg/L sulfur dioxide and

MLF never occurred. In Trial no. 3 sulfur dioxide was present at 24 mg/L, and MLF took nearly twice as long to induce compared to Trials no. 1, 2, and 4, where sulfur dioxide was not used.

The effect of pH may be another factor contributing to the difference in results between small- and large-scale trials. The pH of the wine in Trials no. 5 and 6 was 3.3 and 3.4, respectively. In Trials no. 1 through 4 the lowest pH was 3.47 (Trial no. 3), and the highest pH was 3.82 (Trial no. 1). Bousbouras and Kunkee (1971) investigated the relationship between initial pH and MLF in wine. They found that at seven different pH values between 3.1 and 3.8, the resulting MLF rates were directly proportional to these initial pH values, i.e., the lower the initial pH, the slower the MLF.

The practices of pumping over and racking may also affect MLF. Red wines are traditionally fermented with the grape skins, pulp, and seed. These grape components are brought to the surface of the wine tank during alcoholic fermentation by the buoyant effect of carbon dioxide to form a thick cap. In order to promote color extraction and tannins from the skins and seeds, the cap and fermenting juice were mixed daily. This is done by drawing the juice from the bottom of the tank and spraying it over the entire surface of the cap (Amerine et al. 1980). In our tests, this was done once or twice daily for 60 to 90 min each time. This in effect aerated the fermenting juice. Since *L. oenos* is microaerophilic, this aeration may have slowed its growth, hence delaying MLF.

When the alcoholic fermentation was completed, the wine was drawn off the cap and lees (sedimented yeast and grape particulate) at which time it was once again aerated. It is possible that many of the malolactic bacteria were left behind in the lees and cap that were discarded, thus reducing the overall bacterial population of the wine. Unfortunately, no study was conducted during the trials to see if the lees or cap contained a substantial number of bacteria. Pumping over and racking are practiced frequently in large-scale operations perhaps contributing to delayed MLF. However, they are common and necessary steps in winemaking, and it would be difficult to have wineries change these practices.

Fourteen and nine Bordeaux red wines were inoculated with SK-4 and Kli, respectively (Table 4). SK-4 was successful at inducing MLF in Trials no. 1 through 9, and in nearly every case it completed MLF before the uninoculated controls. In cases where laboratory sublots of each wine were inoculated, the wines always completed MLF before the analogous wines in the field. In trials no. 10 through 14, SK-4 appeared to induce MLF. However, the uninoculated controls in the field and laboratories underwent MLF in the same time. Therefore, malolactic conversions were probably due to spontaneous growth of indigenous bacteria. Kli was able to induce MLF in four red wine trials (no. 14 through 17). In Trials no. 18 and 19, and 21 through 23, MLF was most likely the result of growth of indigenous bacteria.

SK-4 and Kli were used in five trials each in Bordeaux white wines (Table 5). SK-4 failed to complete MLF in any of the wines inoculated in the field. Spontaneous MLF occurred in Trial no. 1, and in only one case (Trial no. 2) MLF was induced in the laboratory sublot.

Inoculation of white wines with Kli proved to be successful in only one case (Trial no. 6) when compared to the uninoculated control. Kli also induced MLF in Trial no. 9 laboratory sublot.

TABLE 4. *Field trial results in Bordeaux red wines*

Trial No.	Strain	pH	MLF Completed in the Field by (days)	MLF Completed in the Lab by (days)
1	SK-4	3.50	21	15
	Control	3.50	a	a
2	SK-4	3.64	40	25
	Control	3.64	a	a
3	SK-4	3.77	12	12
	Control	3.77	a	a
4	SK-4	3.70	19	11
	Control	3.70	a	a
5	SK-4	3.67	22	12
	Control	3.67	28	a
6	SK-4	3.60	35	b
	Control	3.60	a	b
7	SK-4	3.55	40	b
	Control	3.55	a	b
8	SK-4	3.65	12	b
	Control	3.65	12	b
9	SK-4	3.60	23	14
	Control	3.60	a	a
10	SK-4	—	4(c)	4(c)
	Control	—	4(c)	4(c)
11	SK-4	3.64	21(c)	21(c)
	Control	3.64	21(c)	21(c)
12	SK-4	3.76	20(c)	20(c)
	Control	3.76	20(c)	20(c)
13	SK-4	3.68	40(c)	40(c)
	Control	3.68	40(c)	40(c)
14	SK-4	—	30(c)	30(c)
	Control	—	30(c)	30(c)
15	Kli	3.60	34	34
	Control	3.60	a	a
16	Kli	3.52	20	b
	Control	3.52	39	b
17	Kli4	3.55	19	11
	Control	3.55	a	a
18	Kli	3.65	13(c)	10(c)
	Control	3.65	13(c)	10(c)
19	Kli	3.62	11(c)	b
	Control	3.62	11(c)	b
20	Kli	—	10	b
	Control	—	a	b
21	Kli	3.65	8(c)	b
	Control	3.65	8(c)	b
22	Kli	3.70	6(c)	6(c)
	Control	3.70	6(c)	6(c)
23	Kli	3.58	21(c)	21(c)
	Control	3.58	21(c)	21(c)

[a] Did not complete MLF.
[b] No laboratory sublot.
[c] Spontaneous MLF due to high population of indigenous bacteria.

TABLE 4. *Field trial results in Bordeaux red wines (contd)*

Trial No.	Strain	pH	MLF Completed in the Field by (days)	MLF Completed in the Lab by (days)

TABLE 5. *Field trial results in Bordeaux white wines*

Trial No.	Strain	pH	SO_2(mg/L)	MLF Completed in the Field by (days)	MLF Completed in the Lab by (days)
1	SK-4	—	—	5(c)	5(c)
	Control	—	—	—	—
2	SK-4	3.25	9.6	—	22
	Control	3.25	9.6	—	[a]
3	SK-4	—	12.0	[a]	[a]
	Control	—	12.0	[a]	[a]
4	SK-4	—	10.0	[a]	[a]
	Control	—	10.0	[a]	[a]
5	SK-4	3.15	—	[a]	[a]
	Control	3.15	—	[a]	[a]
6	Kli	3.55	4.0	23	[b]
	Control	3.55	4.0	[a]	[b]
7	Kli	3.45	16.0	[a]	45
	Control	3.45	16.0	[a]	[a]
8	Kli	3.37	12.8	31	7
	Control	3.37	12.8	—	—
9	Kli	3.30	0	—	31
	Control	3.30	0	—	[a]
10	Kli	3.51	15.0	[a]	[a]
	Control	3.51	15.0	[a]	[a]

[a] Did not complete MLF.
[b] No laboratory sublot.

Both Kli and SK-4 performed well in Bordeaux red wines. However, the extremely warm weather of 1983 provided grapes of low acidity and high pH. Coupled with the fact that no sulfur dioxide was added to these wines initially, conditions were very favorable for MLF. This was demonstrated by the 10 spontaneous malolactic conversions in the red wines. Therefore, these trials were not really a good test of the organisms under harsh wine conditions.

The Bordeaux white wines had very low initial levels of sulfur dioxide and fairly high pH values. However, personal interviews with individual winemakers revealed that sulfur dioxide was added subsequent to inoculation of the cultures, most likely leading to death of the bacteria.

A striking similarity between results in Bordeaux and California was observed. In trials where the cultures were successful at inducing MLF, small-volume laboratory trials always completed MLF more rapidly than large-production trials, once again demonstrating that conditions were more easily controlled in the laboratory. Also, frequent racking of the Bordeaux wines appeared to delay or inhibit MLF as it may have in California.

Difference Between Bordeaux and California in Time of Culture Inoculation

California winemakers prefer a simultaneous alcoholic/malolactic fermentation and inoculate bacterial cultures with the yeast. This facilitates rapid finishing, aging, and bottling of the wine. Conversely, Bordeaux winemakers insist that MLF takes place subsequent to alcoholic fermentation because they experience an increase in volatile acidity (acetic acid) when a simultaneous alcoholic/malolactic fermentation occurs (Brechot et al. 1974). Other investigators have been unable to reproduce these results in American wines (Kunkee and Beelman, personal communication 1983) and the phenomenon is unknown in other wine growing regions of France (Lafon-Lafourcade, personal communication 1983).

In theory, inoculating the bacteria before, rather than after, alcoholic fermentation should give them an advantage in that they can adapt to the ethanol being produced and have more nutrients available. Gallander (1979), however, found little difference in MLF rates when bacterial cultures were added during or after alcoholic fermentation; he also found that inoculation before alcoholic fermentation was not stimulatory to MLF.

Conclusions

The concept of controlling MLF by induction with selected strains of malolactic bacteria has been in existence for nearly 25 yr. Kunkee et al. (1964), Beelman et al. (1977), and Silver and Leighton (1981) discussed the use of various strains of malolactic bacteria for inducing MLF in wine. Beelman (1980) reviewed the development and use of a lyophilized *L. oenos* PSU-1 culture for inducing MLF in Pennsylvania red wines, and Haymon and Monk (1982) demonstrated starter culture preparation of a French commercial culture for use in Australian wines.

As of this writing, only three commercial malolactic starter cultures were available in the United States. A commercial starter culture has the advantage of convenience over traditional winery methods of growing malolactic bacteria from stock cultures using stepwise bulk procedures. This is time consuming and often populations of bacteria are insufficient for inoculation into fermenting juice or wine without employing large volumes of inoculum.

The culture described herein is unique in that bulk starter procedures are not required. During the bioactivation process there is no significant population increase. Instead, a stationary phase culture is created and the enzyme systems of the organism are induced. The bioactivation time spans only 48 h compared to one to two months it may require a winery to grow its own malolactic culture. Thus, man-hours of preparation time are greatly reduced and the use of large tanks for inoculum is eliminated.

Literature Cited

Amerine, M. A., H. W. Berg, R. E. Kunkee, C. S. Ough, V. L. Singleton, and A. D. Webb, 1980. Pages 359–380 *in Technology of Wine Making*. AVI Publishing Company, Inc., Westport, CT.

Banner, M. J. 1973. Relationships between wine yeast and malolactic bacterial growth with regard to changes in the amino acid composition of a synthetic "model-wine" medium. M.S. Thesis. The Pennsylvania State University, University Park, PA.

Beelman, R. B. 1980. Development and utilization of starter cultures to induce malolactic fermentation in red table wines. Pages 109–117 *in* A. D. Webb, ed., *Grape and Wine Cent. Symp. Proc.* University of California, Davis, CA.

Beelman, R. B., and J. F. Gallander. 1979. Wine Deacidification. *Adv. Food Res.* 24:1–53.

Beelman, R. B., A. Gavin III, and R. M. Keen. 1977. A new strain of *Leuconostoc oenos* for induced malolactic fermentation in Eastern wines. *Am. J. Enol. Vitic.* 28:159–165.

Beelman, R. B., R. M. Keen, M. J. Banner, and S. W. King. 1982. Interactions between wine yeast and malolactic bacteria under wine conditions. *Dev. Ind. Microbiol.* 23:107–121.

Bousbouras, G. E., and R. E. Kunkee. 1971. Effect of pH on malolactic fermentation in wine. *Am. J. Enol. Vitic.* 22:121–126.

Brechot, P., J. Chauvet, and M. Croson. 1974. *Annu. Technol. Agric.* 23:411. Pages 81–163 *in* G. Reed, ed., *Biotechnology, Volume 5: Food and Feed Production with Microorganisms.* 1983. Verlag Chemie GmbH, Weinheim, West Germany.

Burroughs, L. F., and A. H. Sparks. 1964. The identification of sulfur dioxide-binding compounds in apple juices and ciders. *J. Sci. Food Agric.* 15:176–185.

Fornachon, J. C. M. 1963. Inhibition of certain lactic acid bacteria by free and bound sulphur dioxide. *J. Sci. Food Agric.* 14:857–862.

Fornachon, J. C. M., and B. Lloyd. 1965. Bacterial production of diacetyl and acetoin in wine. *J. Sci. Food Agric.* 16:710–716.

Gallander, J. F. 1979. Effect of time of bacterial inoculation on the stimulation of malo-lactic fermentation. *Am. J. Enol. Vitic.* 30:157–159.

Garvie, E. I. 1967. *Leuconostoc oenos* sp. nov. *J. Gen. Microbiol.* 48:431–438.

Gestrelius, S. M., and J. H. Kjaer. 1983. Method of deacidifying wine and composition thereof. U.S. Pat. 4,380,522.

Haymon, D. C., and P. R. Monk. 1982. Starter culture preparation for the induction of malolactic fermentation in wine. *Food Technol. Aust.* 34:14–18.

King, S. W. 1982. Metabolic interactions between *Saccharomyces cerevisiae* Monrachet No. 522 and *Leuconostoc oenos* PSU-1 in a model grape juice/wine system. M.S. Thesis. The Pennsylvania State University, University Park, PA.

Kunkee, R. E. 1967a. Control of malo-lactic fermentation induced by *Leuconostoc citrovorum. Am. J. Enol. Vitic.* 18:71–77.

______. 1967b. Malolactic fermentation. *Adv. Appl. Microbiol.* 9:235–279.

______. 1974. Malo-lactic fermentation and winemaking. Pages 151–170 *in* A. D. Webb, ed., *Advances in Chemistry Series* Volume 37.

Kunkee, R. E., C. S. Ough, and M. A. Amerine. 1964. Induction of malo-lactic fermentation by inoculation of must and wine with bacteria. *Am. J. Enol. Vitic.* 15:178–183.

Lafon-Lafourcade, S. 1975. Factors of the malo-lactic fermentation in wines. Pages 43–53 *in* J. G. Carr, C. V. Cutting, and G. C. Whiting, eds., *Lactic Acid Bacteria in Beverages and Food.* Academic Press, Inc., New York.

______. 1983. Wine and Brandy. Pages 81–163 *in* G. Reed, ed., *Biotechnology Volume 5: Food and Feed Production with Microorganisms*. Verlag Chemie GmbH, Weinheim, West Germany.

Mayer, K., U. Vetsch, and G. Pause. 1975. Hemmung des biologischen Sauerbaus durch gebundene schweflege Saure. *Schweiz. Z. Obst-Weinbau.* III:590–596.

Peynaud, E. 1967. Recent studies on the lactic acid bacteria of wine. *Ferment. Vinification.* 1:219:256.

Peynaud, E., and S. Domercq. 1967. Etude de bacilles homolactiques isoles de vins. *Arch. Mikrobiol.* 57:255–270.

Pilone, G. J., and R. E. Kunkee. 1965. Sensory characterization of wines fermented with several malolactic strains of bacteria. *Am. J. Enol. Vitic.* 16:224–230.

______. 1972. Characterization and energetics of *Leuconostoc oenos* ML-34. *Am. J. Enol. Vitic.* 23:61–70.

Silver, J., and T. Leighton. 1981. Control of malolactic fermentation in wine. 2. Isolation and characterization of a new malolactic organism. *Am. J. Enol. Vitic.* 32:64–72.

CHAPTER **21**

Applications of Starter Cultures in the Dairy Industry

CHU H. TZENG

Chr. Hansen's Laboratory, Inc., 9015 West Maple Street, Milwaukee, Wisconsin 53214

Starter cultures are used currently for the production of billions of dollars worth of products in the United States. This report focuses on the review of the applications of lactic acid-producing cultures in the production of dairy-related products. The historical development of starter culture systems and their present state of industrialization are discussed. The discussions also cover characteristic features of lactic acid-producing cultures in commercial use and the impact of biotechnology on the use of these cultures.

INTRODUCTION

Milk and dairy products are among the most favored foods in the American diet. According to the report of U.S. Manufactured Dairy Products as shown in Table 1, buttermilk production increased from 0.89 billion in 1981 to 0.98 billion lb in 1983, approximately 10% increase in 2 yrs. Hard cheeses increased even more than 10% in the same period. Italian type cheese consumption has been escalating at approximately 10% per year in the United States. The consumption of dairy products is almost 36 lb per capita. This industry certainly can contribute significantly to the economy of the country.

TABLE 1. *Production of dairy products in USA 1981–1983*[a]

Hard Cheeses	1981	1982	1983
Cheddar Type	2.64 billion[b]	2.75 billion[b]	2.93 billion[b]
Italian Type	0.99 billion	1.09 billion	1.20 billion
Other Types	0.64 billion	0.70 billion	0.69 billion
TOTAL	4.27 billion	4.54 billion	4.82 billion
Cottage Cheeses			
Direct Acidification	0.25 billion	0.24 billion	0.24 billion
Cultured Cheeses	0.73 billion	0.74 billion	0.72 billion
TOTAL	0.98 billion	0.98 billion	0.96 billion
Cultured Products			
Buttermilk	0.89 billion	0.93 billion	0.98 billion
Sour Cream and Dips	0.43 billion	0.46 billion	0.50 billion
Yogurt	0.58 billion	0.64 billion	0.67 billion
TOTAL	1.90 billion	2.03 billion	2.15 billion

[a] U.S. Manufactured Dairy Products (USDA).
[b] Figures are in pounds.

The production of dairy products involves fermentation. From the standpoint of industrial microbiology, it is interesting to see the existence of significant differences between fermentation in the dairy industry and the conventional fermentation industry as indicated in Fig. 1.

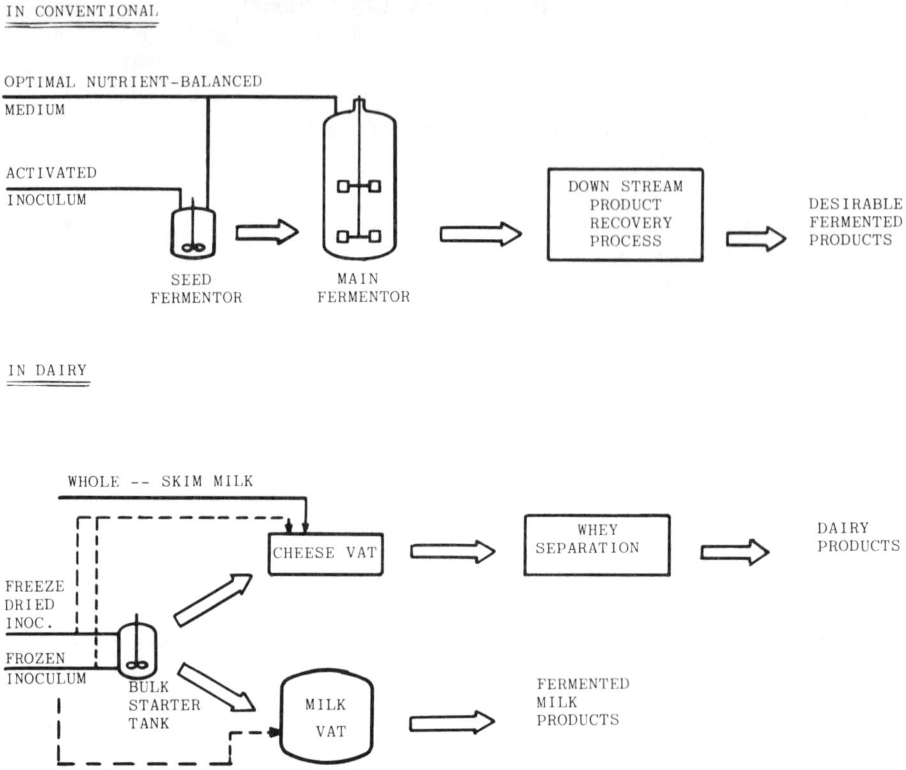

FIG. 1. Comparison of conventional fermentation industry with fermentation in dairy industry.

For instance, in the conventional fermentation industry, the media used in production can be formulated and optimized to the desirable composition. In contrast, milk, almost a fixed substrate, is used for the dairy fermentation. No matter whether it is skim milk, whole milk, or milk with standardized fat content, the composition may be subject to seasonal and geographical variations. Also, additives are not permitted to fortify the milk. This prohibits the addition of compounds that can stimulate the repair of cellular injuries incurred by freezing and lyophilization when a direct vat set of frozen or freeze-dried cells are used.

Even though milk may be only heat-treated (145 F, 60 s), most milk usually is pasteurized (162 F, 16 s) and not sterilized. Natural milk enzymes (e.g., plasmin and lipases), which are not completely inactivated by pasteurization, impose significant changes in flavor, body, and texture developments during cheesemaking (Babel 1984).

Fluid milk dairy fermentations involve submerged stirred culture fermentations in the initial phase followed by static semisolid fermentation when casein molecules reach their isoelectric point and form the coagulated casein particles.

In spite of distinct differences, some advanced technologies developed in the conventional fermentation industry are applicable to the fermentation of starter cultures in the dairy industry.

Discussion

Historical Developments

Chr. Hansen's Laboratory began providing starter cultures to the dairy industry in the late 19th century. However, in the early 1900's to 1930's, many small cheese factories practiced the method of back-slopping a single mother culture, or they borrowed starters from their neighbors. In the late 1940's the use of milk to grow starters was hindered by the presence of antibiotics used for treatment of mastitis in cows. In the 1950's pretested nonfat dry milk (NFDM) was commonly used. Although freeze-dried cultures were introduced to alleviate some cheesemakers' operation problems, many dead cheese vats still sporadically occurred. Whitehead and Cox (1935) discovered that bacteriophage was the agent responsible for the variation of starter culture performance. Some control measures were implemented (Hull 1977, 1978). However, sanitation of equipment, use of enclosed vats, exchange of fresh air in the make room, or even aseptic propagation of starters still could not eliminate phage outbreaks.

The development of a phage-inhibitory medium by Christensen in the early 1960's minimized phage incidence, but the problem still existed (Reddy 1984). The use of modern technology to produce frozen concentrates of starters or lyophilized cultures for direct bulk-starter tank inoculation also gave some advantages to cheese manufacturers (Porubcan and Sellars 1979). The phage problem still continued, even with the introduction of culture rotations of phage-unrelated cultures with defined and undefined strains.

The struggle to develop phage-resistant cultures with desirable functionality led to the use of defined resistant strains in pairs or as cocktail cultures (Richardson et al. 1981; Sandine and Ayres 1983). Some cheese plants have used single strains successfully.

In the Italian cheese industry, nonfat dry milk was used to grow thermophilic cultures. In the 1970's, the industry began to use phage inhibitory media, as shown in Table 2 (Reddy 1984).

In order to obtain consistent, active cultures for optimal productivity in cheese manufacture, the concept of an integrated control system was implemented in the late 1970's. It appears to be gaining more acceptance in the 1980's. The use of this integrated system package to ensure desirable starter culture performance with phage resistance and characteristic functionalities will likely become a trend in the coming years.

Starter Culture Properties and Uses

Properties of lactic starter bacteria and their uses for the manufacture of dairy products have been extensively documented in the literature (Heap and Lawrence 1976; Cox and Stanley 1978; Diebel and Seeley 1978; Sellars and Babel 1978; Kosikowski 1980; Davies and Gasson 1981; Daly 1983b).

There is a wide variety of industrial starter cultures available for the production

TABLE 2. *Significant developments in historical use of starter cultures*

Year	Culture	Media	Fermentation System
to 1930's	"Mother" Culture	Milk	A sterile bottle
	Single-strain cheese starter in New Zealand		Milked the cow into bottle and inoculated culture from back-slop or borrowed from neighbors
	Multiple mixed-strain cultures in rotation		Small-scale cheese vat
1940's	One or more strains of *S. lactis* and/or *L. cremoris*	NFDM (Nonfat dry milk)	
1950's	Mixed Starter (still unknown composition)		In 4-oz bottle to propagate bulk starter for inoculating vats
1960's	Frozen cultures with or without concentration of fermentate (mesophilic)	Phage inhibitory media with phosphate or other suitable chelating agents	Bulk tank Multiple fill cheese vats with time-controlled schedules
	Italian Cultures *S. thermophilus,* *L. bulgaricus* or *L. helveticus*	Whey base or NFDM	Bulk Tank
		NFDM	Multiple
1970's	Multiple mixed strains Multiple strains Mixed strains Paired strains Single strains Cocktail cultures	Start phage inhibitory media for Italian	Bulk tank with natural symbiotic associated growth
1980's	As above and also strains with more phage resistance	Buffered media	Start external pH control fermentor system w/ammonia or neutralizer with various degrees of automation Integrated system
		(Internal pH control)	Constant agitated tank
		Phosphate media	One-step neutralization

of various cheeses and fermented milk products, as shown in Table 3. Throughout much of the dairy world, mesophilic lactic bacteria, which grow well at room temperature, are used as mixed-strain starter cultures of unknown strain composition. Acid producers are always present and, depending on the identity of the flavor-producing bacteria (Daly 1983a,b), these cultures may be separated into four groups as follows: (1) BD cultures (also called D/L type); (2) B cultures (also called L type); (3) O cultures that do not contain diacetyl flavor producing organisms; (4) D cultures contain *S. lactis* subsp. *diacetylactis* as the diacetyl producing organism.

Thermophilic cultures mainly include *Streptococcus thermophilus, Lactobacillus bulgaricus,* and *Lactobacillus helveticus.* They are primarily used to produce Italian cheeses and yogurt. *S. thermophilus* also can be applied for manufacturing cheeses or dairy products that use high cooking or processing temperatures (e.g., >40 C). *L. helveticus* alone can also be used in the same application for making Emmenthaler and Gruyere type cheeses.

TABLE 3. *Dairy cultures made by international Chr. Hansen's Laboratory, Inc.*

Name	Purpose	Temp (°C) for Propagation
Group 1: BD Cultures (also called D/L) Mixed multiple-strain cultures	Cultured butter, cultured milk products, continental cheese and other cheeses with holes (eyes) originating from gas production	19–23
Group 2: B Cultures (also called L) Mixed multiple-strain cultures (bacteriophage unrelated)	Cottage cheese, cream cheese, cheddar cheese, and cheese with only a few or no holes (eyes)	19–23
Group 3: O Cultures Multiple strain cultures (bacteriophage unrelated)	Cheddar cheese and similar types of cheese without holes (eyes)	19–23
Lactobacillus acidophilus *(Lac. acidoph.)*	Cultured milk products	37–40
Lactobacillus bulgaricus *(Lac. bulg.)*	Yogurt, and some Italian cheese types	37–45
Lactobacillus helveticus *(Lac. helv.)*	Highly scalded cheeses, such as Gruyere and Emmenthaler	37–40
Streptococcus cremoris *(St. cremoris)*	Butter, cheese, and cultured milk products	25–30
Streptococcus diacetylactis *(St. diacetyl)*	Butter, cheese, and cultured milk products	20–25
Streptococcus durans *(St. durans)*	Cheese of Cheddar type	37–40
Streptococcus lactis *(St. lactis)*	Cheese	30–37
Streptococcus thermophilus *(St. thermoph.)*	Highly scalded cheeses. Also together with *Lac. bulg.* for Yogurt	37–45
Betacoccus cremoris *(Leuconostoc cremoris)*	Butter and cheese	25–30
Proplonic acid bacteria	Various types of cheese such as Emmenthaler	
Penicillium roqueforti	Various types of blue veined cheeses such as Danablue (Danish Blue)	
Red smear culture	Surface treatment of the cheese	
Gorgonzola culture *(Cult. Gorg.)*	Gorgonzola, Mycella, a.o.	30–35
Casein lactic ferment culture *(Cult. H)*	Casein	37–40
Gruyere	Gruyere/Emmenthaler	37–45
Yogurt	Yogurt	37–45

The use of these cultures is primarily based on the characteristic properties summarized in Table 4 (Law and Sharpe 1978; Auclair and Accolas 1983; Diebel and Seeley 1978).

The terms describing the strains of various types of lactic starter cultures are often confusing. When the term "single strain starter" is used, strains can be derived from three species. Multiple-strain starters may contain the same species as mixed-strain starters according to the definition. The former are composed of defined mixtures of two or more strains, and the latter are composed of undefined numbers and ratios of strains as shown in Table 5 (Cox and Stanley 1978; Cogan 1980; Daniell and Sandine 1981; and Thunell et al. 1984).

In order to provide industrial starter cultures with desirable activities, a variety of bulk starter-culture media have been developed to accommodate their propagation as shown in Table 6. Biolac carries Stimilac and Biolac 50:50; Dairyland Food Labs, Cadilac; Pfizer, Autostat series; Chr. Hansen's, Americana premium, Americana 50:50, Americana complete, Americana Concentrate, and Italiano series.

The Role of Industrial Microbiology

The use of cultures for the production of fermented products has evolved from the inoculation of substrates with uncontrolled wild type culture(s) to the present practice of relying on purified-controllable cultures. A typical example is illustrated in winemaking. In the past, the natural flora occurring on crushed grapes fermented the juice into wine. However, in today's modern industry pure cultures are selected and used. Fermentations in the dairy industry are experiencing similar changes.

In order to produce desirable industrial products via fermentation, both essential genotype and phenotype characteristics of cultures are required. Genes for desired key functions of lactic starter cultures including lactose use, protease production, inorganic salt and antibiotic resistance, citrate use, and host restriction modification system for bacteriophage may be plasmid associated. Various degrees of probabilities exist for losing this extra-chromosomal genetic DNA material from cells during cheese curing or during culture propagations (McKay 1983, 1984, and personal communication).

Intensive discussions are in progress in the dairy industry as to whether single strains, paired strains, multiple strains, or mixed strains should be used. From the standpoint of industrial microbiology, the most important question is whether there exists only one strain that possesses all of the desirable genes that can be properly expressed to give all of the desired characteristics.

Is there sufficient work done on natural selection to find this strain? or Will we need to rely on genetic engineering research to create a "superbug?" Is the use of a mixed-culture essential? Will the creation of a single genetically engineered superbug that can provide all functionalities, become a reality? or Is it a fantasy?

When the dairy industry begins to use defined strains, this will certainly assist in the use of lactic cultures in a more controllable manner. However, when discussions linger over whether single, pair, or multiple strains should be used, are we simplifying or complicating the development of technical know-how?

Some dairy plants have been successfully using a single-strain culture for daily operations. This indicates, however, that this kind of application may be merely

TABLE 4. *Main characteristics and functions of mesophilic and thermophilic lactic starter bacteria*[a]

Properties	*S. lactis*	*S. cremoris*	*S. lactis* subsp. *diacetylactis*	*S. thermophilus*	*L. bulgaricus*	*L. helveticus*	*L. cremoris*
Growth at 10 C	+	+		–	–	–	
40 C	+	–	+	+	+	+	–
45 C–50 C	–	–	–	+	+	+	–
Growth in litmus milk	RAC[b]	RAC	RAC				
Acid % in milk	≤1.8%				≤1.8%	generally >2%	NC
Sugars fermented							
Glucose	+			+	±	+	
Galactose	+			±	±	+	
Fructose	+			+	+	±	
Lactose	+			+	+	+	
Maltose	+			–	–	+	
Sucrose	+			+	–	–	
Catalase	–	–	–	–	–	–	–
Ammonia from Arginine	+	–	+	–			–
Citrate use and production of CO_2 and diacetyl	–	–	+	–	–	–	+
% Salt inhibiting growth	4.0–6.5	2.0–4.0	4.0–6.5	<2.0			<2.0
Serological group	N	N	N	–	E	A	
Lactic acid isomers	L (+)	L (+)	L (+)	L (+)	D (–)	DL	D (–)

[a] From Law and Sharpe (1978); Diebel et al. (1978); Auclair and Accolas (1983).
[b] R = reduction; A = acid; C = clot; NC = no change.

TABLE 5. *Composition of starter culture for hard and cottage cheese*[a]

Type	Species	Characteristics and Method of Use	Cheese Variety/Location
Single-strain Starters	*Streptococcus cremoris* *S. lactis* *S. lactis* subsp. *diacetylactis*	Single or paired	Cheddar in New Zealand, Australia, and selected United Kingdom creameries
Multiple-strain Starters	*S. cremoris* *S. lactis* *S. lactis* subsp. *diacetylactis*	Defined mixtures of two or more strains (may be used in pairs)	Hard cheese varieties in the U.S. and certain U.K. creameries
	Leuconostoc species		Cottage cheese
Mixed-strain Starters	*S. cremoris* *S. lactis* *S. lactis* subsp. *diacetylactis* Leuconostoc species	Unknown proportions of different strains that can vary upon subculture (may be used in pairs)	Most hard cheese varieties in the U.K. and Europe

[a] From CHL Culture Brochure and Cox and Stanley (1978); Cogan (1980); Daniell and Sandine (1981); and Thunell et al. (1984).

TABLE 6. *Starter media*

Biolac	Dairyland Food Lab.	Hansen's	Marschall's	Pfizer
—	—	Americana Premium	Marstar Funnel Grade	—
—	—	Americana 50:50	One-2-One	Autostat M50
—	—	Americana Premium	Marstar MSM	Autostat 200
—	Cadilac	Americana Complete	Co-Star	Autostat 100
Biolac 50:50	—	Americana 50:50	CFS-HT	—
—	—	Americana Premium PHansen Control	CFS-HT Complete	—
—	—	—	Marlac	—
—	—	—	Phase 4	—
—	—	Italiano	Thermostar	—
—	—	Italiano	CR Medium	—
—	—	Italiano	Italian Actilac	—
—	—	Italiano	412-A	—
Stimilac (for lactics)	—	Americana Concentrate (for lactics)	pH Supplement (for lactics)	
—		Italiano	—	—

"product" or "site" (e.g., plant) specific. In order to prevent phage attacks causing dead vats in cheesemaking, a culture containing more than one strain may be necessary. Currently, starter-cultures for dairy products that are requested by the industry not only should produce good quality products but also should be resistant to phage attack. Strain development for phage-resistant cultures will gradually become essential.

A program for improving or developing a desired strain is outlined in Table 7, which includes a series of well-proven techniques (Heap and Lawrence 1976; Lawrence et al. 1976; Marshall and Berridge 1976; Limsowtin et al. 1977; Huggins and Sandine 1979; Thunell et al. 1981; Hurley et al. 1982; Kondo and McKay 1982; West 1982; Hickey et al. 1983; Palmer 1983; Richardson et al. 1983; Sandine and Ayres 1983; Wright et al. 1983).

TABLE 7. *Strain improvement and development program*

Strain Source
Natural Selection or Mutation (Spontaneous and/or Genetic Manipulation)
Phage Resistance Screening Acid-producing Activity
Characterization of Plasmid DNA
Growth in Media with or without Phosphate
Milk Activity and Flavor Tests Characterization and Determination of Key Enzymes including Protease/Peptidase Activities
Compatibility of Cultures (for Culture Blends)
Cheesemaking Trials
Culture Bank
Phage-resistant Cultures

Another extensive discussion in the industry is focused on whether a single, stable, phage-insensitive culture can be established to ensure consistent process control during manufacture and to eliminate variations in product quality. The development of know-how is based on two different philosophies. One can readily illustrate these differences by the following example: Some believe that advanced technologies will enable us to build a super aircraft that can safely fly in any weather regardless of thunderstorms or even hurricanes; on the contrary, others believe that it is safer to develop better detection methods and more accurate weather forecasting in the hopes of preventing tragedy. The dairy industry has made a great deal of effort in identifying problem sources and in developing control measures. These are outlined in Table 8. A continued effort on the part of both starter culture suppliers and users will be necessary to keep the problems under control.

TABLE 8. *Starter culture phage problems and control*

Problem Sources:
Phage contamination from the environment and phage-carrying cultures
Phage mutation
Bacteria cell surface altered and sensitized to phage attack
Changes in host restriction/modification system
Control Measures:
Maintain sanitized conditions of equipment and facilities
Cultures in rotation
The use of slow growing strains
Phage inhibitory media
Use of phage insensitive (resistant) strains or mutants
Determination of the use of specific culture via rapid phage pretest kit

In order to obtain the desired phenotypic characteristics, it may be necessary to rely on the regulation of physiological processes of cultures by their environmental parameters. For example, on the basis of Jacob and Monod's principle, milk protein solids and other protein hydrolysates can regulate proteolytic activities of lactic cultures in the fermentor or starter bulk tank. Many researchers have observed that in the absence of casein or hydrolyzed milk casein, the starter culture may have a weaker proteolytic activity. However, the supply of too many peptides and amino acids from protein hydrolysates also may result in cultures of weak proteolytic activities when they are inoculated into the milk vat for cheesemaking (Babel 1984). The biosynthesis of casein-degrading enzymes is induced by casein components, which function to inactivate repressors. RNA polymerase is then activated for transcription from structural DNA and to make desirable proteases. When too much hydrolysate is supplied, levels of amino acids exceeding cellular metabolic pool capacities inside the cell, they function as corepressors to react with aporepressor. This subsequently blocks the function of RNA polymerase, and the biosyntheses of proteases are then suppressed.

There are many critical parameters that can affect the performance of starter cultures from the stage of cell propagation through cheesemaking and aging. In this regard, performance of lactic cultures is related to the role and functionality of the cell wall and cell membrane. This is becoming well documented. The proteolytic activities of lactic streptococci are strongly associated with the cell envelope on the bacteria. Proteinases and endopeptidases from lactic cultures are particularly important in the flavor and body development of cheeses. Four different types of peptidases have been identified by ultrasonic treatment and another four by detergent treatment (Law and Sharpe 1978 and personal communication). The peptidases localize as periphery enzymes of membrane and as membrane-bound enzymes.

The cell surface also plays a key role in transporting sugar and nitrogeneous compounds. Phage receptor sites and sensitivity and resistance to freezing and lyophilization are associated with the cell wall and cell membrane. Their tolerance to osmotic pressure is particularly important in the application of UF or RO milk.

Much research is still needed for further clarifying the key roles and reactions involving the cell walls and cell membranes. It is necessary to develop a profound knowledge of biochemical and genetic regulation of individual bacterial processes and, consequently, to formulate exactly the defined relationships for these processes. Possibly a system could then be designed for their control.

Value-Added Technologies

Consistency of product quality becomes a key concern of all manufacturers. Product consistency should start with the consistency of raw material. Milk kept under refrigerated conditions deteriorates because contaminated psychrotrophs that produce proteinase and lipase enzymes, which are usually heat resistant under pasteurization conditions, continue to degrade milk components.

Some research institutions have developed systems for controlling the psychrotrophic bacteria in the raw milk supply by lowering refrigerated temperature or by purging carbon dioxide into the storage vessels. The inoculation of lactic bacteria into raw milk is an effective way to suppress the development of psychrotrophic bacteria (Albert et al. 1984). This results in many advantages, such as (Sellars 1984): (a) stimulation of starter activity, (b) reduced "set time," (c) reduced unseparated whey fat losses, and (d) improved quality and yields. One should note that not all industrial lactic starter-cultures give the same effectiveness. As shown in Fig. 2, bacterial counts of raw milk escalated during the storage period when lactic starter bacteria were not inoculated into the milk. However, even when culture No. 0416 was used, the suppression of contaminant growth was less effective, compared with Hansen INOC 1158. Commercial practice will thus determine the degree of effectiveness in controlling psychrotroph development.

Is this practice economical? Theoretical yield loss of approximately 0.05% is shown in Table 9. This is reflected with the increase of nonprotein nitrogen content of whey from 29.8 mg/100 *g* to 42.1 mg/100 *g* during the 5 d storage. During this same period of storage, the psychrotrophic population increased by two logarithmic cycles.

As shown in Table 10, in a cheese plant processing 100,000 lb of milk per day, if the yield difference is 0.025 and 0.05%, respectively, total difference in cheese yield of 37% moisture cheddar cheese per year is shown to be 6,500 and 13,000 lb. In a plant processing 1 million lb of milk per day, this would increase to a difference of 91,250 and 182,500 lb, respectively. At a cost of $6.80 per unit for starter culture inoculation at the rate of 1 can per 50,000 lb, the net gain could be up to $73,547 and $196,735, respectively.

Another example of a practical commercial study showing the economic benefits of the "raw milk inoculation" procedure for improved cheese yield is shown in Table 11 (Sellars 1984). Five thousand gal of grade A milk was inoculated with 1 can (360 ml) of DVS cultures and held at 40–45 F for 12 to 16 h. The results from 110 24,000 lb vats were analyzed for microbiological as well as chemical composition, and cheese yields on each vat. In this study Mozzarella

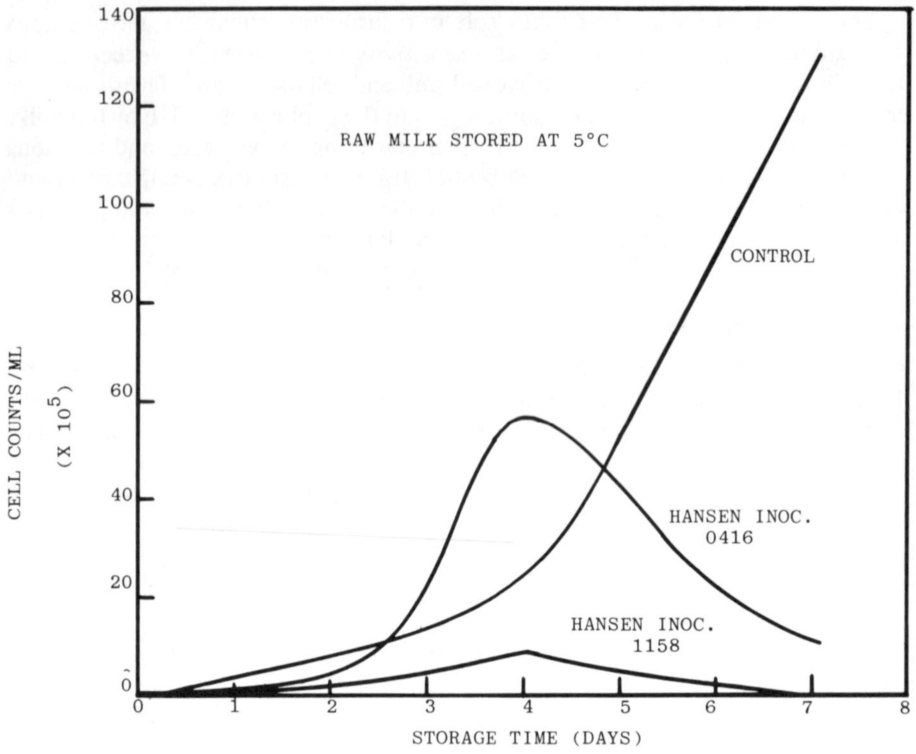

FIG. 2. The inhibition of psychrotrophs by lactic starter cultures.

TABLE 9. *Increase of soluble nitrogen in cheddar cheese whey. Milk was held at 40 F and a portion used to make cheese at 0, 3, and 5 d*[a]

Holding Time of Milk	Nonprotein Nitrogen Content of Whey	Bacterial Count
0 d	29.8 mg/100 g	32,000
3 d	34.3	610,000
5 d	42.1[a]	3,200,000

[a] Theoretical yield loss = ca. 0.05%.
Theoretical yields of 37% moisture cheddar:

$$\text{lb Cheese/100 lb milk} = \frac{(0.93\ \text{Fat} \times \text{Casein} - 0.1) + 1.09)}{\text{T.S.}}$$

cheese was found to have an increase in yield of 0.07% when the raw milk was inoculated with DVS cultures before pasteurization. Economically, the added value of the raw milk inoculation procedure was calculated per each 5,000 gal of milk used. At the rate of DVS cost, $6.85 per unit, the net savings would have been approximately $3,000 had all 110 vats used milk that had been previously inoculated before pasteurization.

Cheese manufacturers have long been searching for technology for the use of whey in order to gain more profit from this byproduct of the cheese industry.

TABLE 10. *Gross difference in 37% moisture cheddar (lb cheese) at different % yield values @ $1.35/lb*

	Extra Cheese when % Yield Difference is			Total/Yr at (%)	
lb Milk/Day	0.025	0.05	Days/Yr	0.025	0.05
100,000	25	50	260	6,500	13,000
500,000	125	250	312	39,000	78,000
1,000,000	250	500	365	91,250	182,500

	Total $ Extra at (%)		Cost[a]	Net Gain $ at (%)	
lb Milk/Day	0.025	0.05	Units/Day	0.025	0.05
100,000	8,775	17,550	3,536	5,239	14,014
500,000	52,650	105,300	21,216	31,434	84,084
1,000,000	123,187	246,375	49,640	73,547	196,735

[a] Cost calculated basis 1 can/50,000 lb @ $6.80 per unit.

TABLE 11. *Economy and yield difference mozzarella cheese inoculated vs. uninoculated*

Parameters:	1 can DVS 360 ml/5,000 gal Grade A Milk		
	40–45 F overnight (12–16 h)		
	vs.		
	Uninoculated Milk (control)		
	Casein, solids, starter, make the same		
	110 24,000 lb vats studied		
Results:	Yield —	Inoculated	8.70%
		Control	8.63%
		Difference	.07%
Added Value	— $52.68/5,000 gal		
Cost DVS	— $ 6.85		
Net Gain	— $45.83/5,000 gal		

Membrane processing is one approach (Amundson 1984). There are currently two categories of products made through commercial processes. The nonfermented products include condensed whey, demineralized whey, whey protein concentrate, lactose, delactosed whey, and reuse of whey as an ingredient into dairy products. The fermented products include whey alcohol for beverages such as Irish cream, for distilled spirits such as vodka, for blending with premium gasoline, for conversion of whey alcohol into acetic acid, or as a direct fermentation into lactic acid for use as an acidulant or a plasticizer precursor. Whey is neutralized with ammonia to make ammonium lactate for cattle feed, for the production of hydrolyzed lactose (whey permeate) in confectionery products, or as substrate for manufacturing baker's yeast, etc. Each of these is "site" and "size" specific in terms of maintaining economical viability because of the limitation of their market potentials. Using *Propionibacterium* cultures, a fermented propionate whey material is now produced as "milk replacer" in the marketplace.

Lactic starter cultures are versatile and research is developing new horizons to the benefit of the dairy industry.

Acknowledgments

The author wishes to express his appreciation to all his colleagues in Research and Development at Chr. Hansen's Laboratory, Inc. (CHL) Research Center in Milwaukee, as well as those in Copenhagen, Denmark, and Arpagon, France, for their efforts and contributions to technology development. The author also would like to acknowledge the commitment of CHL management to technology development. Appreciation is also extended to Dr. Robert L. Sellars for his review of this manuscript.

Literature Cited

Albert, K. A., D. R. McCoy, C. H. Tzeng, and R. L. Sellars. 1984. The use of specially selected strains of lactic acid bacteria for raw milk inoculation and final assay method of inoculated milk. *79th Annu. Meet. Am. Dairy Sci. Assoc.* June 24–27, 1984.

Amundson, C. 1984. Ultrafiltration of raw milk and improved whey processing. Chr. Hansen's Lab. Inc., *Cheesemaking Symp. '84,* Stevens Point, WI.

Auclair, J., and J. P. Accolas. 1983. Use of thermophilic lactic starters in the dairy industry. *Antonie van Leeuwenhoek J. Microbiol. Serol.* 49:313–326.

Babel, F. J. 1984. The proper applications of enzymes for optimum body texture and flavor in cheese. Chr. Hansen's Lab. Inc. *Cheesemaking Symp. '84,* Stevens Point, WI.

Cogan, T. M. 1980. Microbiologie et industrie alimentaire, *Annales du congres international,* Paris, France V4:5.

Cox, W. A., and G. Stanley. 1978. Starters: Purpose, production and problems. Page 279 *in* F. A. Skinner and L. B. Quesnel, eds., *Streptococci.* Academic Press, London.

Daly, C. 1983a. The use of mesophilic cultures in the dairy industry. *Antonie van Leeuwenhoek J. Microbiol. Serol.* 49:297–318.

______. 1983b. Starter culture developments in Ireland. *Irish J. Food Sci. Technol.* 7:39–46.

Daniell, S. D., and W. D. Sandine. 1981. Development and commercial use of a multiple strain starter. *J. Dairy Sci.* 64:407–415.

Davies, F. L., and M. J. Gasson. 1981. Reviews of the progress of dairy science: Genetics of lactic acid bacteria. *J. Dairy Res.* 48:363–376.

Diebel, R. H., and H. W. Seeley, Jr. 1978. Family II. *Streptococcaceae in "Bergey's Manual of Determinative Bacteriology,"* 8th ed., R. E. Buchanan and N. E. Gibbons, eds., p. 490. Williams and Wilkins Co., Baltimore, MD.

Heap, H. A., and R. C. Lawrence. 1976. The selection of starter strains for cheesemaking. *N.Z. J. Dairy Sci. Technol.* 11:16–20.

Hickey, M. W., A. J. Hillier, and G. R. Jago. 1983. Peptidase activities in Lactobacilli. *Aust. J. Dairy Technol.,* Vol. 38, No. 3, pp. 118–123.

Huggins, A. R., and W. E. Sandine. 1979. Selection and characterization of phage-insensitive lactic streptococci. *J. Dairy Sci.* 62:70–76.

Hull, R. R. 1977. Control of bacteriophage in cheese factories. *Aust. J. Dairy Technol.* 32:65–66.

______. 1978. Factory derived cheese starters. *Proc. of the CSIRO and Aust.*

Dairy Corp. Meet. Oct. 4–6, Highett, Australia.

Hurley, M., P. Timmons, F. Drihan, T. Logan, and C. Daly. 1982. Selection and use of multiple strain starters for cheddar cheese manufacture. *Irish J. Food Sci. Technol.* 6:210–214.

Kondo, J. K., and L. L. McKay. 1982. Transformation of *Streptococcus lactis* protoplasts by plasmid DNA. *Appl. Environ. Microbiol.* 43:1213–1219.

Kosikowski, F. 1980. *Cheese and Fermented Milk Foods.* Kosikowski and Associates. Brooktondale, NY.

Law, B. A., and M. E. Sharpe. 1978. Streptococci in the dairy industry. Page 263 *in* F. A. Skinner and L. B. Quesnel, eds., *Streptococci.* Academic Press, London.

Lawrence, R. C., T. D. Thomas, and B. E. Terzaghi. 1976. Review of the progress of dairy science: Cheese starters. *J. Dairy Res.* 43:141–193.

Limsowtin, G. K. Y., H. A. Heap, and R. C. Lawrence. 1977. A multiple starter concept for cheesemaking. *N.Z. J. Dairy Sci. Technol.* 12:101–106.

Marshall, R. J., and N. J. Berridge. 1976. Selection and some properties of phage-resistant starters for cheesemaking. *J. Dairy Res.* 43:449–458.

McKay, L. L. 1983. Functional properties of plasmids in lactic streptococci. *Antonie von Leeuwenhoek J. Microbiol. Serol.* 49:259–265.

______. 1984. Lactic Starter Cultures—what we have learned, and where research is heading. Chr. Hansen's Lab., Inc., *Cheesemaking Symp. '84,* Stevens Point, WI.

Palmer, M. 1983. Defined cultures and guesswork. *Dairy Field* 166 (8):22–26.

Porubcan, R. S., and R. L. Sellars. 1979. Lactic culture concentrates. Page 59 *in* D. Perlman, ed., *Microbial Technology,* 2nd ed., Vol. 1, Academic Press, New York.

Reddy, M. 1984. Review of starter systems. Chr. Hansen's Lab., Inc., *Cheesemaking Symp. '84,* Stevens Point, WI.

Richardson, G. H., C. A. Ernstrom, and G. L. Hong. 1981. The use of defined starter strains and culture neutralization in U.S. cheese plants. *Cult. Dairy Prod. J.* 16:11–14.

Richardson, G. H., C. A. Ernstrom, J. M. Kim, and C. Daly. 1983. Proteinase negative variants of *Streptococcus cremoris* for cheese starters. *J. Dairy Sci.* 66:2278–2286.

Sandine, W. E., and J. W. Ayres. 1983. Method and starter compositions for the growth of acid producing bacteria and bacterial composition produced thereby. U.S. Pat. 4,382,965, May 10.

Sellars, R. L. 1981. Fermented dairy foods. *J. Dairy Sci.* 64:1070–1076.

______. 1984. Improved cheese yield and quality through raw milk inoculation. Chr. Hansen's Lab., Inc., *Cheesemaking Symp. '84,* Stevens Point, WI.

Sellars, R. L., and F. J. Babel. 1978. *Cultures for the Manufacture of Dairy Products.* Chr. Hansen's Lab., Inc., Milwaukee, WI.

Thunell, R. K., F. W. Bodyfelt, and W. E. Sandine. 1984. Defined strains and phage-insensitive mutants for commercial manufacture of cottage cheese and cultured buttermilk. *J. Dairy Sci.* In press.

Thunell, R. K., W. E. Sandine, and F. W. Bodyfelt. 1981. Phage-insensitive multiple-strain starter approach to cheddar cheesemaking. *J. Dairy Sci.* 64:2270–2277.

West, R. A. 1982. The selection and modification of lactic starter culture bacteria for cheese manufacture. Paper presented at the Marschall Int. Dairy Symp., U.K., London, England, Nov. 25.

Whitehead, H. R., and G. A. Cox. 1935. The occurrençe of bacteriophage in cultures of lactic streptococci. *N.Z. J. Dairy Sci. Technol.* 16:319–327.

Wright, S. L., M. A. Shelaih, S. A. Winkel, J. M. Kim, R. J. Brown, and G. H. Richardson. 1983. Characterization of proteinase positive and negative lactic streptococci by growth and acid production in milk. *J. Dairy Sci.* 66 (Suppl. 1):57–66.

CHAPTER 22

Modification of Lactic Acid Bacteria for Cucumber Fermentations: Elimination of Carbon Dioxide Production from Malate[a]

MARK A. DAESCHEL, ROGER F. MCFEETERS, AND HENRY P. FLEMING

Food Fermentation Laboratory, U. S. Department of Agriculture, Agricultural Research Service, and North Carolina Agricultural Research Service, Department of Food Science, North Carolina State University, Raleigh, North Carolina 27695–7624

Malic acid, a natural constituent of cucumbers, undergoes decarboxylation (malate → lactate + CO_2) by naturally occurring lactic acid bacteria and commercial lactic starter cultures. The CO_2 produced by this reaction is of sufficient quantity to cause the fermenting cucumbers to bloat (hollow fruit), resulting in an economic loss. Research is under way to develop lactic starter cultures for cucumber fermentations that lack the ability to decarboxylate malic acid but retain desirable fermentation traits. A selection system has been developed to isolate *Lactobacillus plantarum* mutants that have lost the ability to decarboxylate malic acid. Mutant cultures obtained by this system are presently being evaluated in experimental cucumber fermentations for use as starter cultures.

INTRODUCTION

Malic acid, a dicarboxylic organic acid, is present in many types of plant material, including apples, grapes, cereal grains, legumes, and grasses. Malolactic enzyme, demonstrated to be present in most lactic acid bacteria (Caspritz and Radler 1983) but not in other bacteria, catalyzes the reaction: 1 malate + H^+ → 1 lactate + 1 CO_2. The importance of this reaction during the processing of wine has been well documented (Kunkee 1967). Decarboxylation of malate in wines of high acidity is generally desirable in order to reduce the acidity. In cucumber fermentations, decarboxylation of malate is not desirable because the carbon dioxide produced can cause cucumber bloater damage.

Bloating of cucumbers was a major defect in commercially fermented cucumbers for many years (Jones et al. 1941). Studies of the bloating process pointed to the fact that it was associated with the production of CO_2 gas by microorganisms in the fermentation brines. Early work pointed to yeasts as the

---

[a] Paper no. 9465 of the Journal Series of the North Carolina Agricultural Research Service, Raleigh, NC. Mention of a trademark or proprietary product does not constitute a guarantee or warranty of the product by the U. S. Department of Agriculture or North Carolina Agricultural Research Service nor does it imply approval to the exclusion of other products that may be suitable. This paper was prepared by U. S. government employees as part of their official duties and legally cannot be copyrighted.

* USDA-ARS, Box 7624, NCSU, Raleigh, NC 27695–7624.

major source of CO_2 production (Etchells and Bell 1950). Later evidence showed that in low-salt conditions the heterofermentative lactic acid bacterium, *Lactobacillus brevis,* could also produce sufficient CO_2 to cause bloating (Etchells et al. 1968).

The results of those studies suggested that bloating could be prevented by limiting the microflora in cucumber fermentations to homofermentative lactic acid bacteria, such as *Lactobacillus plantarum* and *Pediococcus pentosaceus,* which were thought to be unable to produce significant amounts of CO_2. However, Fleming et al. (1973a; 1973b) showed that CO_2 was produced by both the brined cucumbers themselves and by the homofermentative lactic acid bacteria used in controlled cucumber fermentations. The combination of these two sources of CO_2 was sufficient to cause significant cucumber bloating damage. These results led to successful efforts to control the CO_2 concentrations in cucumber fermentation by gas purging the brines to remove enough of the CO_2 produced by the fermentation to prevent cucumber bloating. Nitrogen purging of fermentation tanks (Etchells et al. 1973; Fleming 1979) has been widely adopted in the pickling industry and has greatly reduced the incidence of commercially significant bloating.

Recent work in this laboratory has led to the recognition that the major mechanism for CO_2 production by homofermentative lactic acid bacteria in cucumber fermentation is the decarboxylation of malic acid to lactic acid and CO_2. The intent of this paper is to review recent research conducted by our laboratory as it relates to the significance of malic acid in cucumber fermentations and current efforts to develop starter cultures that lack the ability to decarboxylate malic acid.

Discussion

Malic acid and CO_2 production. McFeeters et al. (1982a) found that malic acid is the major organic acid present in the immature cucumber fruit used for commercial processing. The concentration of malic acid in six different cultivars ranged from 14.2 to 23.4 m*M*. In cucumber juice fermentations with *L. plantarum* (McFeeters et al. 1982b), the CO_2 production was directly proportional to the amount of malic acid that was added to the juice (Fig. 1). The relationship between malic acid decarboxylation, CO_2 production, and bloating in fermented cucumbers was demonstrated using strains that do and do not decarboxylate malate (McFeeters et al. 1984). Fig. 2 illustrates this relationship. CO_2 production by the brined cucumbers, not related to malate decarboxylation, was 12.5 m*M* when fermented by *L. plantarum* 965, a nonmalolactic strain. Fermentations conducted with *L. plantarum* WSO, a decarboxylating strain, developed increasing concentrations of CO_2 when additional amounts of malate were added to the fermentations. Fig. 3 shows the relationship of CO_2 production to bloating. Bloater index values indicate the percentage of product damage that renders the pickle unusable for processed dill slices (Fleming et al. 1977). The cucumbers fermented with strain 965 showed only slight bloating. The data indicated that if malic acid decarboxylation could be prevented, bloating could almost be eliminated in pure culture fermentations using homolactic starter cultures. This

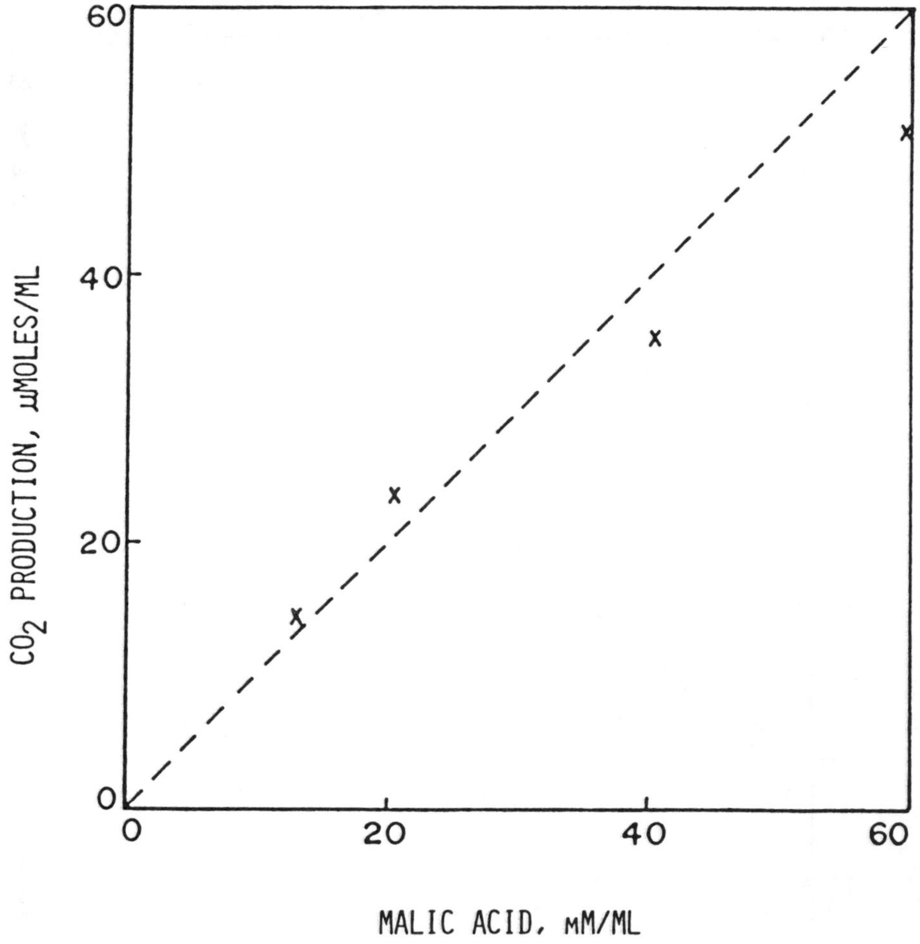

FIG. 1. CO_2 production in cucumber juice supplemented with 6.0% NaCl and malic acid. Determinations were made after incubation at 30 C for 7 d. The dashed line shows the expected CO_2 production if 1 mol of CO_2 were produced per mol of malic acid. From McFeeters et al. (1982b).

may have significant economic advantages since mechanical purging of dissolved CO_2 in commercial brine tanks currently is necessary to ensure bloater-free fermentations. Elimination of purging requirements would result in savings to the industry.

Nonmalolactic cultures. The inability of lactic acid bacteria to decarboxylate malate is rare, as indicated by the surveys of McFeeters et al. (1984) and Caspritz and Radler (1983). With this in mind, Daeschel et al. (1984a) developed nonmalolactic cultures by mutating existing strains of *L. plantarum* containing desirable fermentation characteristics so they no longer possessed malolactic activity. A differential medium designated "MD medium" (Table 1) containing malic acid, glucose, and pH indicator (bromocresol green) as key components was developed to distinguish between malolactic and nonmalolactic strains of lac-

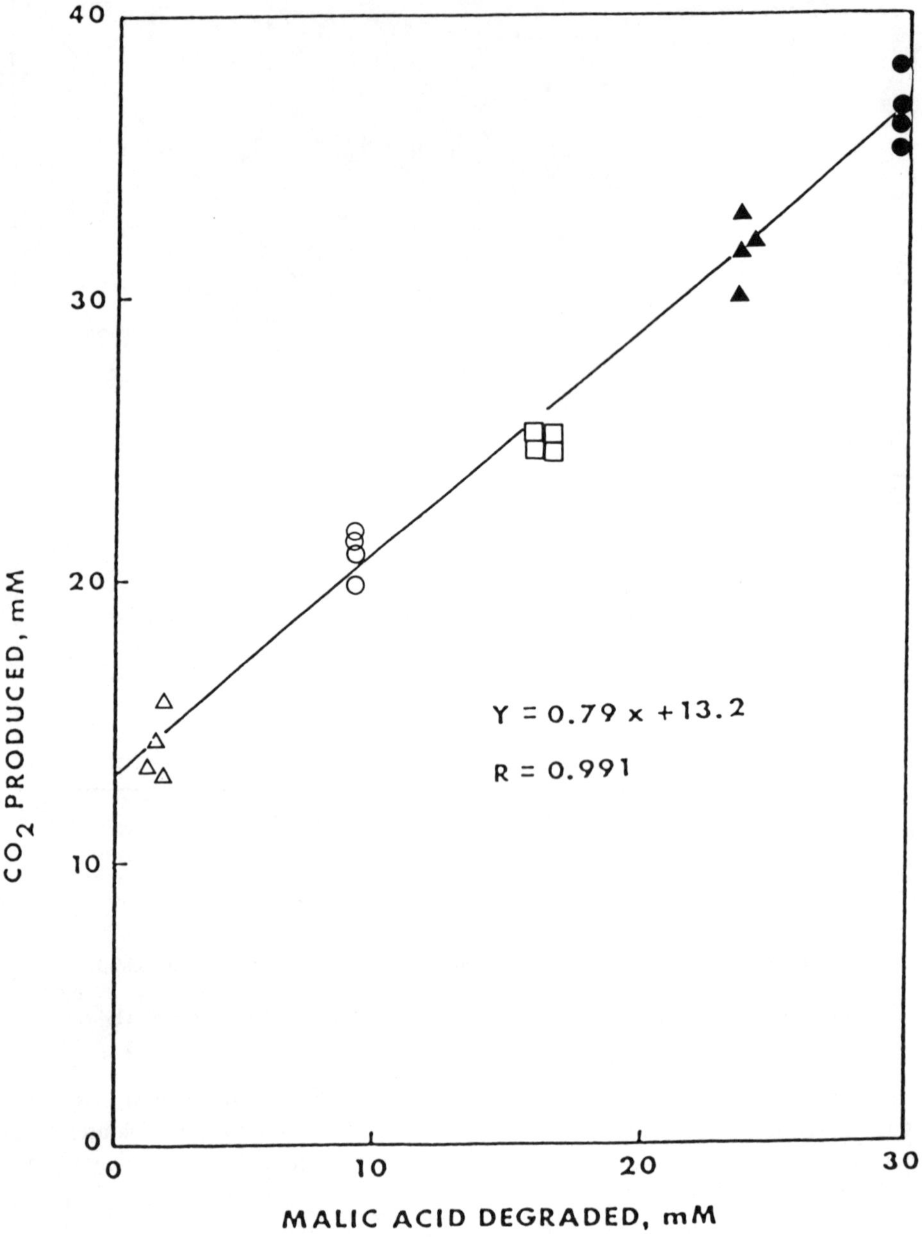

FIG. 2. Relationship between malic acid degradation and CO_2 formation in cucumber fermentations. Cucumbers without added malic acid were fermented with *L. plantarum* 965 (△) and WSO (○). Cucumbers supplemented with 7 (□), 14 (▲), and 21 m*M* (●) malic acid were fermented with *L. plantarum* WSO only. From McFeeters et al. (1984).

tic acid bacteria (Table 2). The differential ability of the medium is based upon the fact that when malate is decarboxylated, there is uptake of a proton. Thus, a lactic acid bacterium that decarboxylates malate will neutralize the lactic acid

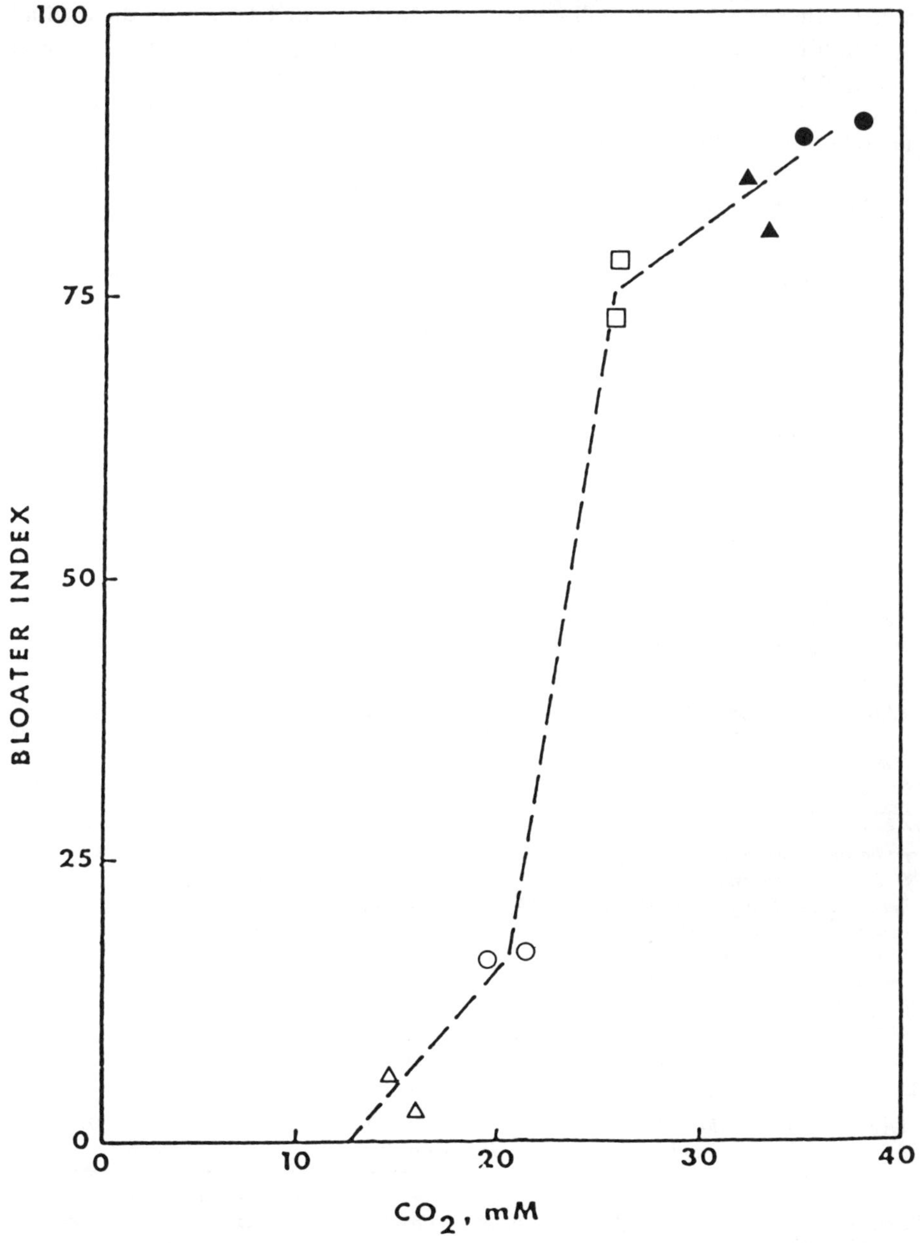

FIG. 3. Relationship between CO_2 production and bloating. Cucumbers without added malic acid were fermented with *L. plantarum* 965 (△) and WSO (○). Cucumbers supplemented with 7 (□), 14 (▲), and 21 m*M* (●) malic acid were fermented with *L. plantarum* WSO only. From McFeeters et al. (1984).

produced from glucose so the pH will not decrease and the medium will remain dark blue. Strains that do not decarboxylate malate will decrease the pH of the medium because the lactic acid from glucose will not be neutralized. The effect will be that the medium will become yellow-green in color. This differential system was used to detect nonmalolactic mutants of *L. plantarum* that had been mutagenized with N-methyl-N′-nitro-N-nitrosoguanidine. The strains giving acid reactions (Table 3) did not produce significant amounts of CO_2, indicating a loss in ability to decarboxylate malate.

TABLE 1. *Formulation and preparation of "MD medium" for the detection of strains and mutants of lactic acid bacteria that do and do not decarboxylate malic acid*[a]

Component	Manufacturer	Amount/Liter
L-Malic acid	Sigma	20 g
Trypticase	BBL	10 g
D(+)Glucose	Sigma	5 g
Casamino acids	Difco	3 g
Phytone	BBL	1.5 g
Yeast extract	Difco	1 g
Tween 80	Atlas	1 g
Bromocresol green	Fisher	20 ml[b]
Agar (when desired)	Difco	20 g

[a] Adjust pH to 7.0 with 10 N KOH. Autoclave at 15 psi for 15 min. Can be stored at room temperature. From Daeschel et al. (1984a).
[b] Stock solution (solubilize 0.1 g in 30 ml of 0.01 N NaOH).

TABLE 2. *Reaction of lactic acid bacteria in MD medium after incubation at 30 C for 1 wk*

Bacterial Strain	pH Reaction	Color	Ability to Decarboxylate Malic Acid
Pediococcus cerevisiae 61	8.58	Blue	+
Pediococcus cerevisiae 39	8.48	Blue	+
Leuconostoc paramesenteroides NCDO 803	8.34	Blue	+
Leuconostoc mesenteroides 43	5.53	Green	–
Leuconostoc mesenteroides LC-33	5.54	Green	–
Leuconostoc dextranicum ATCC 19255	5.52	Green	–
Leuconostoc oenos PSU-1	6.82	Blue	+

From Daeschel et al. (1984b).

TABLE 3. *pH Reaction, CO_2 produced and reducing sugar concentration in MD broth fermented by strains and mutants of* L. plantarum *at 30 C for 7 d*

Strains	pH	CO_2 (mg/100 ml)	% w/v Reducing Sugar	Broth Color
965	5.19	16.9	0.02	Lt. green
WSO	6.82	596.6	0.02	Dk. blue
WSO-M-34[a]	5.28	25.3	0.02	Lt. green
WSO-M-35[a]	5.23	18.9	0.02	Lt. green
Uninoculated control	6.96	15.5	0.48	Dk. blue

From Daeschel et al. (1984a).
[a] Mutants of WSO.

Although mutagenesis is nonselective in its action, we were able to obtain mutants that retained the vigor and desirable fermentation characteristics of the parent and that have not reverted back to the malolactic phenotype. Mutant strains were tested (Daeschel et al. 1984b) for their ability to produce CO_2 from cucumber juice containing the natural malate concentration or additional malate (Table 4). The mutant *L. plantarum* WSO M35 did not produce significant amounts of CO_2 from the cucumber juice with or without additional malate.

TABLE 4. *CO_2 produced in filter-sterilized cucumber juices fermented by strains and mutants of* L. plantarum *at 30 C for 7 d*

Strains	CO_2 mg/100 ml	
	Cucumber Juice[a]	Cucumber Juice with Added Malic Acid[b]
965	22.96	22.85
WSO	98.91	294.5
WSO-M-35	35.23	33.58
Uninoculated control	19.85	6.31

From Daeschel et al. (1984b).
[a] Contains 20 m*M* malic acid.
[b] Additional malic acid added to give a 70 m*M* concentration.

Conclusions

An understanding of the mechanisms of cucumber bloater damage has provided additional parameters on which to base the development of starter cultures for brined cucumbers. Prevention of malate decarboxylation during fermentation by using selected starter strains should reduce the need for mechanical purging systems now employed commercially to prevent bloater damage. Current research (USDA, North Carolina) is directed toward the pilot-scale level using whole cucumbers to determine if the nonmalolactic mutants can successfully mediate bloater-free fermentations.

Acknowledgments

This investigation was supported in part by a research grant from Pickle Packers International, Inc., St. Charles, IL.

Literature Cited

Caspritz, G., and F. Radler. 1983. Malolactic enzyme of *Lactobacillus plantarum*. Purification, properties and distribution among bacteria. *J. Biol. Chem.* 258:4907–4910.

Daeschel, M.A., R. F. McFeeters, H. P. Fleming, T. R. Klaenhammer, and R. B. Sanozky. 1984a. Mutation and selection of *Lactobacillus plantarum* strains that do not produce carbon dioxide from malate. *Appl. Environ. Microbiol.* 47:419–420.

_______. 1984b. Method for differentiating between lactic acid bacteria which do and do not decarboxylate malic acid. U. S. Pat. Appl. 539,028.

Etchells, J. L., and T. A. Bell. 1950. Classification of yeasts from the fermentation of commercially brined cucumbers. *Farlowia* 4:87–112.

Etchells, J. L., A. F. Borg, and T. A. Bell. 1968. Bloater formation by gas-forming lactic acid bacteria cucumber fermentations. *Appl. Microbiol.* 16:1029–1035.

Etchells, J. L., T. A. Bell, H. P. Fleming, R. E. Kelling, and R. L. Thompson. 1973. Suggested procedure for the controlled fermentation of commercially brined pickling cucumbers—the use of starter cultures and reduction of carbon dioxide accumulation. *Pickle Pak Sci.* 3:4–14.

Fleming, H. P. 1979. Purging carbon dioxide from cucumber brines to prevent bloater damage—a review. *Pickle Pak Sci.* 6:8–22.

Fleming, H. P., R. L. Thompson, T. A. Bell, and R. J. Monroe. 1977. Effect of brine depth on physical properties of brine-stock cucumbers. *J. Food Sci.* 42:1464–1470.

Fleming, H. P., R. L. Thompson, J. L. Etchells, R. E. Kelling, and T. A. Bell. 1973a. Bloater formation in brined cucumbers fermented by *Lactobacillus plantarum*. *J. Food Sci.* 38:499–503.

_______. 1973b. Carbon dioxide production in the fermentation of brined cucumbers. *J. Food Sci.* 38:504–506.

Jones, I. D., J. L. Etchells, O. Veerhoff, and M. K. Veldhuis. 1941. Observations on bloater formation in cucumber fermentation. *Fruit Prod. J.* 20:202–206, 219–220.

Kunkee, R. E. 1967. Malolactic fermentation. *Adv. Appl. Microbiol.* 9:235–279.

McFeeters, R. F., H. P. Fleming, and M. A. Daeschel. 1984. Malic acid degradation and brined cucumber bloating. *J. Food Sci.* 49:999–1002.

McFeeters, R. F., H. P. Fleming, and R. L. Thompson. 1982a. Malic and citric acids in pickling cucumbers. *J. Food Sci.* 47:1859–1861, 1865.

_______. 1982b. Malic acid as a source of carbon dioxide in cucumber juice fermentations. *J. Food Sci.* 47:1862–1865.

CHAPTER 23

AgriCultures: Beneficial Applications for Crops and Animals

G. L. ENDERS, JR., AND H. S. KIM

Miles Laboratories, Inc., Elkhart, Indiana 46515

Microorganisms contribute significantly to many, if not all, of the major conversion processes that occur in the agricultural environment. This paper reviews several of the areas to which microorganisms have been applied to elicit beneficial effects in the agricultural field. The roles of bacteria as the key elements for effecting the preservation of forage crops and improving animal health and/or feed use are discussed. Selected topics in the use of biological systems for controlling pests and diseases comprise the third area of this paper. The theoretical aspects of the development of additive products, as well as the advantages and disadvantages of their use, are also reviewed. Commercial products are considered with regard to the types of organisms used and their development. Future prospects for research in these agricultural applications and the impact of biotechnology advances are discussed.

INTRODUCTION

Microorganisms are ubiquitous in the world and play major roles in agriculture. They are an integral part of any food chain. Several of the ways that microorganisms contribute include fixing nitrogen to improve crop nutrition (Dobereiner and Day 1976; Burns and Hardy 1975; Hollaender 1977); controlling pests, plant diseases, and weeds to increase yields (Baker and Cook 1974; Deacon 1983; Schroth and Hancock 1981); supplementing animal feed to improve performance and minimize stress effects (Kinsey 1980; Speck 1981); preserving the nutritional value of forages for winter feeding (McCullough 1978; McCullough and Bolsen 1984; McDonald 1981); and facilitating waste treatment and use (Callander and Barford 1983; Shuler 1980; Gilles 1978). The reader is referred to the above and other extensive reviews that have been written on each of these areas within the last 10 yr. This paper will update some recent findings, but the primary intent is to create an awareness of three areas of lesser developed applications of cultures with respect to silage, probiotics, and biological control agents.

DISCUSSION

Silage Additives

Preserving feed crops in the form of silage has been an established practice in most countries because of the advantages realized by achieving better preservation and maintaining the nutritional value of crops during the winter season.

Despite the long historical use of the ensiling process, better methods for fermentation control are still needed to reduce forage losses. Many parameters affect the quality of the final silage, such as weather conditions, type and condition of the structure containing the crop, stage of crop maturity, and moisture content of ensiled forage. Although some of these parameters are uncontrollable, good management practices can minimize the effects of the others. The natural microbial flora is known to vary both quantitatively and qualitatively during the growing season as well as from year to year. An obvious effect of these fluctuations can be manifested by different rates of fermentation.

To aid silage fermentation, four different general classes of silage additives have been used, which act either as inhibitors for undesirable microorganisms or as fermentation stimulators. The first class, chemical additives, includes those that inhibit the growth of spoilage organisms or reduce the activity of plant enzymes by the action of direct acidification or preservation (Yahara and Nishibe 1975; Beck 1978; McDonald 1981; Bolsen and Hinds 1984; Goering and Gordon 1973). Nutrient sources comprise the second class and serve to stimulate silage fermentation (Shirley et al. 1972; McDonald 1981; Ely 1978; Grieve et al. 1982). These materials are used as energy sources for lactic acid bacteria metabolism (Lindgren et al. 1983). In some instances the additives are intended to increase the nitrogen content of the crop. In the third class, enzymes such as cellulase and/or amylase are applied to break down complex plant constituents in order to make nutrients more available to naturally occurring organisms and to improve the digestibility of organic materials (Owens 1962; Autrey et al. 1975; Bolsen et al. 1981). The fourth class, microbial cultures, is for the purpose of supplementing natural beneficial bacterial populations, thereby stimulating the rapid production of lactic and acetic acids. This results in the inhibition of spoilage organisms and improves aerobic stability. Various lactic acid bacteria, yeast, mold, and spore-forming bacteria have been tried to achieve these goals (Enders et al. 1983a; Moon et al. 1980; Bolsen and Ilg 1981).

A number of recent studies demonstrated the benefits of microbial inoculants on the silage fermentation (Lindgren et al. 1983; O'Leary and Hemken 1982; Enders et al. 1983a, 1983b; Kung et al. 1984; McDonald et al. 1983). Furthermore, several other reports showed beneficial effects on animal performance and feed efficiency by feeding culture-treated silages (Krause and Clanton 1977; Bolsen and Ilg 1981; Bolsen et al. 1981; Grieve et al. 1982).

Improvement of the quality of silage treated with lactic acid bacteria is proposed to be the result of (1) lower pH; (2) inhibition of yeast and mold growth by the production of lactic and/or acetic acids as well as other secondary metabolites; and (3) retardation of the growth of clostridia because of competition with and/or the possible inhibition effect of lactic acid bacteria (Lindgren et al. 1983; Moon et al. 1980; O'Leary and Hemken 1982).

The advantages of microbial inoculants over other additives are that they are (1) not corrosive; (2) cost effective or cheaper than some additives; and (3) easy to apply and handle because they are nontoxic. The primary disadvantage is the susceptibility of the cultures to environmental stresses, which reduce viability. This is especially true during shipping and storage. Culture shipment with dry ice and refrigerated storage are the commercial approaches to rectify this problem.

In the development of microbial additives, three general philosophies have been applied. One approach is based on the different chemical composition of the crops and the use of specifically selected cultures that grow best with each type of crop. The growth of specific lactic acid bacteria can vary because of the types and amounts of carbohydrates in the crops; microbial growth can also be influenced by the crop maturity stage. Thus far, only one major manufacturer has pursued this philosophy to the marketplace. Another approach is the application of a broad spectrum of cultures to all crops to ensure the desired action of at least one organism on each crop. The third approach is the selection of a single nonspecific culture, which can work on any crop. This approach necessitates the use of cultures with a broad capacity for using different crop nutrients for growth. Both of these latter philosophies have been the premise for the majority of products in the market.

Table 1 contains a summary of some currently available culture products for silage fermentation. Although the list is extensive, it is not intended to be all inclusive. The microorganisms listed as active ingredients vary widely. Most of the products contain lactic acid bacteria or fermentation products as the ingredients.

TABLE 1. *Microbial products for silage additives*[a]

Company	Trade Name	Cultures and Enzymes Used[b]
Agriaid Div. Agrimerica, Inc.	Ensila Plus	cellulase, carbohydrase, lactobacilli
AgriTech, Inc.	Sila-Prime	*L. casei, L. plantarum, S. lactis, A. oryzae*
Anchor Laboratories, Inc.	Sila-lator, Silage Inoculant 160 X Conc.	*S. faecium, S. cremoris, S. diacetylactis, L. plantarum, L. acidophilus, P. cerevisiae,* enzymes
Chr. Hansen's Laboratory, Inc.	Biomax SI	*L. plantarum*
Furst-McNess Co.	McNess Silage Starter	*L. plantarum, L. brevis, P. acidilactici,* protease, amylase, gumase
Great Lakes Biochemical Co.	SIC Concentrate	streptococci, lactobacilli
Kent Feeds, Inc.	Silage Supreme	*L. plantarum*
Marschall Div. Miles Laboratories, Inc.	AgMaster Alfalfa	*L. plantarum, P. acidilactici*
	AgMaster Corn	*L. xylosus, P. acidilactici*
	AgMaster High Moisture Corn Inoculants	*P. acidilactici, L. curvatus, S. faecium*
M&M Livestock Product Co.	Si-Lo-Fame	*B. subtilis, A. oryzae*
Pioneer Hi-Bred International, Inc.	Silage Inoculant 1177 and 1174	*L. plantarum, S. faecium*
Triple "F" Feeds	H/M Plus Inoculant	*S. faecium, L. plantarum, B. subtilis, Pediococcus* sp.
Vigortone Products	Vigortone	*L. plantarum*

[a] Includes some of the major manufacturers; many regional products have not been included.
[b] Other ingredients, e.g., carbohydrates, antioxidants, and minerals, have been omitted.

Most commercialized cultures are homofermentative. Heterofermentative lactic acid bacteria generally have been avoided because their metabolism results in dry matter losses because of gas production. The major microorganisms in the commercial products are lactobacilli, especially *L. plantarum*. Fermentation products of yeast and mold are used quite often in combination with lactic acid bacteria. It should be noted that most of the genera represented in the product formulations can be found as natural inhabitants of the crops. However, cultures such as *L. acidophilus,* a normal inhabitant of the intestinal tract, are also used. The function of this specific culture in the silage fermentation and its survival in an atypical environment merit further investigation.

Claims made by manufacturers regarding culture additives include (1) lowered ensiling temperature; (2) improved dry matter and neutral detergent fiber recovery; (3) prolonged bunk life; (4) greater lactic acid content and commensurate lower pH; and (5) improved protein retention in the crop. Many manufacturers also claim benefits to animal performance: (1) a greater fat content in the milk; (2) more milk production; (3) less weight loss during early lactation; and (4) improved silage palatability. When crops have been treated with sufficient quantities of viable organisms, investigators have demonstrated definitive effects on the fermentation characteristics. The inconsistencies in fermentation sometimes observed can be attributed to crop variation, lack of uniform application, degree of anaerobiosis attained in the structure, differences in chopping and loading, and amount of viable inoculum added. Data supporting the animal performance claims have evolved slower. This is in part due to animal handling, environmental factors, and the large variability among animals, thus requiring larger groups of animals for statistical analysis. Further research is required to clarify the mechanism(s) for modifying animal performance that could be related to the surviving cultures in the silage or to the improved chemical composition of the silage or both.

Most products are sold as dry culture blended with carriers in 50 lb bags. Some companies market cultures in dry form with a high viable count in small packages or as frozen liquid concentrates. Many of the dry products are applied as dry additives using a metering applicator; others are rehydrated and applied using a spray applicator.

Application rates are generally 1 lb or 1 gal per ton of crop for alfalfa and corn. Except for one major manufacturer, all recommend twice the usage rate for high moisture corn. Manufacturers recommend inoculation levels ranging from 1×10^4 to 3×10^5 colony forming units (CFU) per gram of crop. In order to ensure culture viability and to meet the recommended inoculation level of each product, careful handling and storage of the products are essential. The use of stabilizers or unique growth conditions during manufacturing of the cultures are employed by several producers to extend the product shelf life. O'Leary and Hemken (1984) evaluated several commercial silage products to determine storage stability, use stability, and sensitivity to abuse. Two products were found to be stable during storage at recommended and elevated temperatures whereas the others tested were not stable under normal (recommended) storage conditions and were consequently not as effective in aiding silage fermentation. In our laboratory, several silage inoculants were assayed for total viable organisms and the results are illustrated in Table 2. Many companies do not specify the viable microbial count in their product, thus it can only be assumed that the low counts observed were either because of the low level initially in the manufactured product or the poor stability during shipping and storage under stress conditions.

TABLE 2. *Microbial content and calculated application rate for several commercial silage inoculants*

Product	Product Form	Application Method	Recommended Storage	Total CFU/g[a] Purchased Product	CFU/g Crop Based on Recommended Application Rate[b]
A	dry	dry	cool, dry place out of direct sunlight	2.3×10^{7}	1.2×10^{4}
B	dry	liquid	Frozen	3.8×10^{10}	2.0×10^{5}
C	dry	liquid	cool, dry place out of sunlight	2.4×10^{7}	4.0×10^{3}
D	dry	liquid or dry	15.5 C or below	4.9×10^{9}	5.4×10^{4}
E	dry	liquid	refrigeration	9.8×10^{9}	1.1×10^{5}
F	dry	dry	not mentioned	7.6×10^{4}	3.8×10^{1}
G	dry	dry	cool, dry place out of sunlight	3.2×10^{5}	1.6×10^{2}
H	dry	dry	cool, dry place	2.6×10^{6}	1.3×10^{3}
I	dry	liquid	frozen	5.4×10^{9}	1.7×10^{4}
J	dry	dry	not mentioned	6.5×10^{4}	3.3×10^{1}
K	dry	dry	not mentioned	6.7×10^{5}	3.4×10^{2}

[a] Assayed using Difco Eugon and/or Lactobacillus MRS media.
[b] Determined using the actual assayed total CFU/g product and calculated based on haylage (50–65% dry matter) application if specified.

In the future, the production of the appropriate balance of organic acids (lactic and acetic acids primarily), other secondary metabolites (bacteriocins, organic byproducts, etc.), as well as bacteriophages should be further investigated for the purpose of improving silage quality. Selection of microorganisms that produce beneficial effects should be continued as well as the isolation and characterization of the factors responsible for producing the desired effects.

Probiotics

Feed supplements have been used by farmers to maximize animal performance by improving animal health, increasing feed efficiency, obtaining better weight gain, and enhancing milk production. The term "feed supplements" includes any feed material that can stimulate animal growth when mixed with a feedstuff (Ensminger and Olentine 1978). Probiotics, defined as live microbial food additives that can aid in the maintenance of good health by establishing a desirable balance of gastrointestinal organisms, have recently been spotlighted as feed supplements.

Interest in using *L. acidophilus* cultures as probiotics has increased since Metchnikoff's first mention of the desirability of consuming lactobacilli capable of living in the intestinal tract (Metchnikoff 1908). The use of lactobacilli, including *L. acidophilus*, for animals under stresses such as breeding, birth, weaning, post antibiotic treatments, shipping, and environmental changes are considered prime target areas for probiotics (Crawford 1979). Despite the complexity in the etiology of scours, reduced mortality and decreased severity of diarrhea were observed with lactobacilli-treated pigs (Muralidhara et al. 1977; Parker and Crawford 1978; Maxwell et al. 1983). Other beneficial influences such as improved feed efficiency and weight gain in broilers (Fuller 1977; Francis et al. 1978; Potter et al. 1979), turkeys (Potter et al. 1979), and growing-finishing pigs (Maxwell et al. 1983; Olsson 1967), were reported. Several studies also demonstrated strong evidence for the beneficial effects of *L. acidophilus* with humans (Brown 1977; Tramer 1973; Rettger et al. 1935). Gilliland (1979) mentioned several desirable characteristics of lactobacilli that should be present when considering cultures as dietary adjuncts. Such lactobacilli must be capable of surviving and growing in the intestine. They also must be resistant to acid in the stomach and bile salts in the intestine. Two other desirable properties of a probiotic are the ability to exert antagonistic effects toward undesirable microorganisms in the intestinal tract (Mitchell and Kenworthy 1976; Sandine et al. 1972; Shahani and Ayebo 1980) and the ability to stimulate growth of other types of bacteria that are beneficial to the host (Tannock and Savage 1974). Savage et al. (1968) showed that lactobacilli stimulated action of the intestinal villi, which aids the extraction of nutrients from feedstuffs during digestion.

The advantages of using probiotics include the following: (1) the microorganisms are typical of the natural flora in the gastrointestinal tract; (2) they colonize in the intestine and thus minimize the establishment of enteropathogens; (3) no residues are found in the tissues and thus no withdrawal periods are required as is true for antibiotics; and (4) they facilitate the normalization of the gut flora after stress or therapy and thereby aid in improving digestion.

Some difficulties in the development of probiotics are (1) selecting organisms that meet all the criteria required to survive and grow in the intestines; (2) determining the significance of host specificity of the cultures; (3) identifying cultures that are able to adhere to the intestinal wall; and (4) developing suitable model systems to investigate the in vivo changes in metabolism.

Table 3 contains a partial summary of probiotic products currently on the market. The most common active ingredient of these probiotic products is *L. acidophilus*. The major claims of these products are that they (1) provide the necessary balance of natural microflora in the intestinal tract; (2) have antagonistic effects on disease-producing organisms; (3) favor good digestion and absorption of nutrients; and (4) provide growth factors. However, published information on lactobacilli used as probiotics is limited and the results of the studies vary (Miles et al. 1981; Damron et al. 1981). This variability must be minimized in future studies by applying sufficient amounts of viable culture, controlling environmental conditions, and increasing the number of experimental animals.

Most probiotic products do not specify the viable count but merely designate the quantity to be added to the feed. Although most products assayed in our laboratory had viable counts of less than 10^7 organisms per gram, one product did conform to its recommended dosages ranging from 1×10^9 to 5.6×10^{10} CFU per animal per day depending on the type of animal. A major concern is re-

TABLE 3. *Probiotic products*[a]

Company	Trade Name	Cultures and Enzymes Used[b]
AgriTech Inc.	Lacto-Grass	*L. acidophilus, B. subtilis, L. plantarum*
Anchor Laboratories, Inc.	Feed Mate, Feed Mate Ruminant, Feed Mate PM, Feed Mate calf/veal, Feed Mate Crumbles	*B. subtilis, L. acidophilus, A. oryzae, S. faecium*
Bio-Ceutic Labs	MicroVet VMS, Ruminant, Crumbles	*S. faecium, L. acidophilus, B. subtilis, L. plantarum, A. oryzae*
Fermented Products Inc.	Gold Seal Acidolac	*L. acidophilus*
Great Lakes Biochemical Co., Inc.	Lact-A-Bac-AF	*L. acidophilus*
Chr. Hansen's Laboratory, Inc.	Biomax BC Biomax 10 and 40, Biomax D, Biopak, Biomax FG	*L. acidophilus, L. lactis, B. subtilis*
Pioneer Hi-Bred International, Inc.	Probios, oral paste and granules	*L. acidophilus*
Polybac Corp.	AGTX-10	*L. acidophilus, B. subtilis, A. oryzae, A. niger,* Torula yeast, marine algae
Prairie States Enterprises Star Labs	Primalac Probiotic	*L. acidophilus, L. casei, B. bifidus, A. oryzae, Torulopsis*

[a] Includes some of the major manufacturers; many regional products have not been included.
[b] Other ingredients, e.g., vitamins, minerals, and carbohydrates, have been omitted.

taining viability of the culture until the time of application. Daily inoculation levels should be calculated based on the viability of the culture. In the selection of cultures for products, care must be taken with regard to the source of *L. acidophilus*. Several articles have suggested the importance of host specificity, such as using *L. acidophilus* of human origin as a dietary adjunct for humans and not other animals to obtain best performance (Gilliland et al. 1975; Mitsuoka 1969). Morishita et al. (1971) further reported that *L. acidophilus* isolated from a human intestinal tract could not be implanted in the intestinal tract of a chicken. Most available products are sold either as powders, pastes, or liquids. The major concern in the selection of the product form should be related to the quality and stability of the culture. Problems of low activity similar to those illustrated in Table 2 for silage additives are observed with probiotics. Conclusions concerning the efficacy of these products often rely on trends rather than statistically significant data. In the future, it might be possible to better design studies for statistical analysis using single-parent, cloned embryo implants to control and minimize these animal variations.

Biological Controls for Crops

The ever-increasing need for food and greater crop production relies on the technological advances of applied biology with respect to improved crop varieties (e.g., higher yield, drought resistance, etc.), and insect, disease, and weed control. In the latter areas, chemical insecticides and herbicides have made significant contributions. The major disadvantages of the chemicals, however, have been the development of resistance by the pest or weed and the undesirable residues in the environment. As a result, in the last 5 yr, there has been a significant surge of interest in the applications for biological controls.

Although various definitions have been proposed for biological control (Baker and Cook 1974; Garrett 1970), a simple composite is proposed for this review. Biological control is a means by which undesirable organisms (or their effects) are inhibited, partially or totally, by another organism, the host, or the pest itself. Biological control agents are generally considered safe compared to chemicals, although the safety testing required to register potential organisms often poses a significant barrier to product development. While not universally true, many biological control agents produce longer lasting effects. In addition, they have, at most, slight effects on the ecological balances in nature. If narrow host range agents are selected, then natural predators are not inhibited as could be true for indiscriminant chemicals.

Despite the logical advantages of this type of control, progress has been slow because of a variety of factors, including the following: lack of adequate funding, which results in intermittent peaks of activity; absence of quantitative data partially because of the slower advances and greater time requirements, especially related to the interactions in the ecosystem; taxonomic characterization of many new isolates; knowledge of the physiology of antagonistic microflora; and the need for the development of model systems that can better simulate real environmental situations.

Many organisms have been characterized with regard to inhibitory properties, but few have had extensive basic research performed related to their mechanisms of inhibition. Table 4 lists some of the inhibitory microorganisms reported in the literature as well as their use or potential applications as biological control agents. The scope of biological control agents appropriate for both insect pests and plant diseases includes bacteria, viruses, and fungi (Deacon 1983). Other, as yet untapped, sources of organisms include parasitic nematodes and protozoa. While many bacteria (besides those shown in Table 4) have the potential to infect and kill insects: the two most widely commercialized are *Bacillus thuringiensis* and *B. popillae* (Falcon 1971). These bacteria have been demonstrated to be safe for man and other vertebrates and are compatible with other chemical insecticides; *B. thuringiensis* shows a very broad spectrum of activity. The *B. popillae* species has not been widely marketed because of both a narrower spectrum of activity and the high cost of production. Viruses similarly pose production problems and are costly to manufacture in large quantities. Their use has been limited and slow to develop because of public and regulatory concern for safety. The most appropriate viruses for exploitation are those currently used belonging to the group of baculoviruses. Unlike the *B. thuringiensis* bacteria, most viruses have very restricted host ranges. They are capable of causing disease in all developmental stages of the insect, and infection usually occurs through the gut of the host. A detailed discussion of the infectious process was reviewed by Tinsley (1979). The disadvantage of baculovirus sensitivity to ultraviolet (UV) inactivation is partially offset by their persistence in soils; thus they have the potential for long-term control applications. Biological control using fungi has been applied mostly in the Eastern European countries. These organisms, like viruses, infect all insect developmental stages, survive well in the soil environment, and are rapidly inactivated by UV radiation. In addition, they require high relative humidity conditions to effect initiation of infection. Natural populations of *Entomorphthora* sp. have been used to control aphid populations (Baird 1958; Ferron 1979; Wilding and Perry 1980). If the culture production technology and strain development work permit cost-effective products to be manufactured that function under practical field conditions, then perhaps the natural control organisms can be supplemented by improved strains. Soil-borne pests, like corn root worm and Colorado potato beetles, require agents that can be concentrated in the soils. In this instance, parasitic nematodes found in the soils could be applicable (Brogna 1981). The major limitations of this group are the lack of growth and media formulation technology and the need to develop preservative systems to provide adequate shelf life. Parasitic protozoa face the same drawbacks as nematodes but additionally require a host or tissue culture system to grow. The development of suitable tissue culture methods should open up this area for commercialization.

Microbial control of plant diseases is hindered by the nature of the infection, namely, because the pathogen is inside and thereby protected by host tissues. Thus, control must be exercised at the early stages of insect development or during the pathogen's dormant stages in the environment. Understanding the mechanism of action will necessitate a knowledge of natural community interactions, pathogen-control agent interactions, as well as specific alterations of host metabolism. Inhibition of *Agrobacterium tumefaciens* by the *Agrobacterium radiobacter* bacteriocin to prevent crown gall tumors is probably one of the best-known examples of plant disease control (Kerr and Htay 1974; Moore 1979; Süle and Kado 1980). Many other examples of cross protection using related or attenuated organisms have been reported for plant diseases (Chamberlain et al. 1964; Chanon et al. 1978; Ikediugwu and Webster 1970).

The use of biological agents for weed control is another application illustrated in Table 4. Classically, the control agent is imported into a new geographical area where the host is the pest. The special problems associated with this approach include the following: (1) sufficient control must be exercised so as not to import other pests; and (2) the weed pathogens must be highly selective for the weed host and not damaging to other economic crops. The use of mycoherbicides has been extensively reviewed in several reports (Templeton and Smith 1977; Templeton et al. 1979). To date, only a few of these microbial controls for weeds have been commercialized. This area offers significant potential for commercial application but will require basic research into the mechanisms of action and technology development for large-scale production.

By comparison, the use of microbial cultures as biological control agents, thus far, is less developed than the areas of silage and probiotic applications. A few of the major producers currently active include Sandoz, Inc.; Biochem Products; Fairfax Biological Laboratory, Inc.; Reuter Laboratories, Inc.; Abbott Laboratories; Battelle; the World Health Organization; and both the U.S. and Canadian

TABLE 4. *Some commercial and potential biological control agents*[a]

Category	Biological Agent	Applications	Remarks and Companies/Products
Bacteria	*Bacillus thuringiensis*	registered for use with more than 20 crops, trees, and ornamentals.	
		active against 100 *Lepidoptera* (moths) species, alfalfa caterpillar, cabbage looper, white butterfly, tobacco budworm, and European corn borer.	Sandoz, Inc., three strains marketed; Biochem Products, new strains in development; Abbott Labs., 'Dipel.'
	Bacillus thuringiensis Israelensis	combats mosquitoes (esp. Aedes) and black flies.	Promoted by the World Health Organization; Biochem Products, 'Bactimos.'
	Bacillus sphaericus	mosquito control (better against Anopheles and Culex).	
	Bacillus popilliae	Japanese beetle control in grass turf; milky disease of *Scarabaeidae* beetles.	Fairfax Biological Laboratory, Inc., 'Doom' and 'Japidemic'; Reuter Laboratories, Inc., 'Milky Spore.'
	Agrobacterium radiobacter	protects possibly via bacteriocin production, against *Agrobacterium tumefaciens,* which causes crown gall.	AgBiochem and Nortel Laboratories market products.
	Bacillus popillae	Japanese beetle control in grass turf; milky disease of *Scarabaeidae* beetles.	Fairfax Biological Laboratory, Inc., 'Doom' and 'Japidemic'; Reuter Laboratories, Inc., 'Milky Spore.'
	Pseudomonas syringae, Erwinia herbicola	potential aids in preventing frost damage to plants; control of Dutch elm disease.	
Viruses	Baculoviruses (nuclear polyhedrosia, granulosis and cytoplasmic polyhedrosis viruses)	species of group infect cotton bollworm and tobacco budworm; gypsy moth control; Douglas fir tussock moth; against coddling moth in apple, pear, and walnut groves; control of European pine sawfly and spruce budworm.	Exempted from EPA regulation of pesticides. Sandoz, Inc., 'Elcar,' 'Gypsy Check,' and 'Biocontrol I.' U.S. and Canadian Forest Services; U. of Wisconsin.

Viruses *(Contd)*	attenuated virus strains	cross protection used with tobacco mosaic virus to protect tomato crops, with tristeza virus in citric crops.	
Fungi	*Beauveria bassiana*	controls aphids; controls Colorado potato beetle.	Used in Soviet Union; 'Boverin.'
	Verticillium lecanii	Control of aphids and scale insects, coffee green bug, aphids in greenhouse chrysanthemums.	Commercialized on small scale. Applied in India; Tate & LyLein, England, 'Vertalec.'
	Metarrhizium anisopliae	controls froghopper in sugar cane plantations; acts against rhinoceros beetle.	Used in Brazil 'Metaquino.'
	Peniophora gigantea	natural inhibitory organism active against *Heterobasidion annosum,* a pathogen infecting coniferous tree roots; mechanism of action unknown.	Used in England; Bio-Basic Ltd., markets in U.S.
	Cercospora rodmanii	used to control water hyacinths.	Abbott Labs., studying application.
	Trichoderma viridae	potential control agent related to apple canker, and silver leaf disease of plum and other fruit trees.	
	Colletotrichum gloeosporioides	acts against northern joint-vetch in rice fields.	Applied in Arkansas, Mississippi, and Louisiana.
Nematodes	Unspecified	corn root worms; Colorado potato beetles.	Several small companies interested.
Protozoa	Unspecified	most insect species susceptible to some protozoan infections.	Very new area; a lot of interest. generated.
	Nosema locustae	pathogen of grasshoppers.	EPA registered; commercial interest.

[a] Not intended to be all inclusive.

Forestry Services. The major emphasis to date has been the commercial availability of *B. thuringiensis*.

Impact of Biotechnology

In the preceding sections the applications of microorganisms to preserve forage quality during ensiling, to aid animal performance, and to control pests and plant diseases were briefly reviewed. An underlying principle common to all of these uses is antibiosis. Antibiosis can be defined as an interaction among organisms where the metabolic byproducts of one or more organisms are lethal or inhibitory to others. Food microbiologists for years have recognized and used the selective effects of physical, chemical, and biological pressures on spoilage microflora. The agricultural environment can be considered analogous to foods except that we do not yet fully understand the complexity of the communities that must be manipulated to obtain the more desired types and densities of microorganisms. Besides these interactions, introduced microflora can produce beneficial products or can aid host metabolism. A prerequisite for developing commercial products in these areas is basic research of the microbial attributes responsible for the effects. Although much work has been done, where specific factors (organic acids produced, toxins elaborated, amino acids secreted) have been defined, the molecular genetics of the producing organism often are not yet developed.

Recently the transfer of genetic elements and the characterization of plasmids (mostly cryptic) have been studied for the lactic acid bacteria (Stadhouders 1983). If desirable functions can be linked to these plasmids and transfer vectors defined, then the functional traits could be selectively amplified, perhaps even in more desirable, yet unrelated, genera. The areas of biological pest and plant disease control could be major beneficiaries of this technology. An example would be the *B. thuringiensis* crystal production which, if plasmid linked, could be transferred to another microorganism, thereby eliminating the need for an insect host for large-scale production. This example also exposes another aspect related to silage additives, probiotics, and biological control agents: the need to further develop fermentation technology. New endeavors in the agricultural area will necessitate novel nutrient formulations, capabilities to scale-up anaerobic cultures, tissue culture systems, and even simultaneous mixed culture growth systems. As biotechnology yields new engineered strains, laboratory model systems will become essential for performance and safety validation before any field testing is conducted. While many systems have been tried for simulating silage fermentations, rumen microbial interactions, and intestinal flora metabolism, none have yet been able to adequately or totally mimic the natural situations. A similar problem is faced by researchers investigating bioinsecticides and bioherbicides.

In summary, the current status of agriCultures in agriculture is one of great potential. With biotechnology advances providing the impetus, companies both large and small, as well as venture capital organizations, have initiated research programs. But in order to succeed, these programs will require coordination with basic research conducted at academic communities. An integrated interdisciplinary approach could yield major new agriCultures during the next decade.

Literature Cited

Autrey, K. M., T. A. McCaskey, and J. A. Little. 1975. Cellulose digestibility of fibrous materials treated with *Trichoderma viridae* cellulase. *J. Dairy Sci.* 58:67–71.

Baird, R. B. 1958. The artificial control of insects by means of entomogenous fungi: A compilation of references with abstracts. Entomology Laboratory, Belleville, Ontario.

Baker, K. F., and R. J. Cook. 1974. Biological control of plant pathogens. Freeman, San Francisco, CA, 433 p.

Beck, Th. 1978. The microbiology of silage fermentation. Pages 61–115 *in* M. E. McCullough, ed., *Fermentation of Silage—A Review.* NFIA, West Des Moines, IA.

Bolsen, K. K., and M. A. Hinds. 1984. The role of fermentation aids in silage management. Pages 79–112 *in* M. E. McCullough and K. K. Bolsen, eds., *Silage Management.* NFIA, West Des Moines, IA.

Bolsen, K. K., and H. J. Ilg. 1981. Inoculant, enzyme, NPN, or NaOH additives for corn and sorghum silage: Cattle performance. *J. Anim. Sci.* 53(suppl. 1):382.

Bolsen, K. K., H. J. Ilg, and M. A. Hinds. 1981. Inoculant and enzyme additives for alfalfa silage and corn grain supplementation for steers. *J. Anim. Sci.* 53(suppl. 1):383.

Brogna, C. 1981. Bioinsecticides. *Biotechnol. News* 21:5–7.

Brown, J. P. 1977. Role of gut bacterial flora in nutrition and health: A review of recent advances in bacteriological techniques, metabolism, and factors affecting flora composition. *Crit. Rev. Food Sci. Nutr.* 8:229–336.

Burns, R. C., and R. W. F. Hardy. 1975. Nitrogen fixation in bacteria and Higher Plants. Springer-Verlag, Berlin, Heidelberg, New York. 189 p.

Callander, I. J., and J. P. Barford. 1983. Recent advances in anaerobic digestion technology. *Process Biochem.* 18(4):24–30.

Chanon, A. G., N. J. Cheffins, G. M. Hitchon, and J. Barker. 1978. The effect of inoculation with an attenuated mutant strain of tobacco mosaic virus on the growth and yield of early glasshouse tomato crops. *Ann. Appl. Biol.* 88:121–129.

Chamberlain, E. E., J. D. Atkinson, and J. A. Hunter 1964. Cross protection between strains of apple mosaic virus. *N. Z. J. Agric. Res.* 7:480–490.

Crawford, J. S. 1979. Pages 45–55 *in* Probiotics in animal nutrition proceedings. Arkansas Nutrition Conference, Sept. 27–28.

Damron, B. L., H. R. Wilson, R. A. Voitle, and R. H. Harms. 1981. A mixed Lactobacillus culture in the diet of broad breasted large white turkey hens. *Poult. Sci.* 60:1350–1351.

Deacon, J. W. 1983. Microbial control of plant pests and diseases. American Society for Microbiology, Washington, DC, 88 p.

Dobereiner, J., and J. M. Day. 1976. Pages 518–538 *in* W. E. Newton and C. J. Nyman, eds., *Proc. Int. Symp. on Nitrogen Fixation,* 1st. Washington State University Press, Pullman, WA.

Ely, L. O. 1978. The use of added feedstuffs in silage production. Pages 233–280 *in* M. E. McCullough, ed., *Fermentation of silage—A Review*. NFIA, West Des Moines, IA.

Enders, Jr., G. L., S. L. Hoover, and H. S. Kim. 1983a. Development of a selected pair of lactic acid producing bacteria as an inoculant for alfalfa silage. *J. Dairy Sci.* 66(suppl. 1):184–185.

Enders, Jr., G. L., H. S. Kim, and S. L. Hoover. 1983b. Evaluation of culture additives for enhancing cut and shell corn fermentation. *J. Anim. Sci.* 57(suppl. 1):282.

Ensminger, M. E., and C. G. Olentine, Jr. 1978. Feeds and Nutrition. Ensminger Publishing Co., Clovis, CA, pp. 411–436.

Falcon, L. A. 1971. Use of bacteria for microbial control of insects. *In* H. D. Burges and N. W. Hussey, eds., *Microbial Control of Insects and Mites.* Academic Press, London.

Ferron, P. 1979. Biological control of insect pests by entomogenous fungi. *Annu. Rev. Entomol.* 23:409–442.

Francis, C., D. M. Janky, A. S. Arafa, and R. H. Harms. 1978. Interrelationship of Lactobacillus and zinc bacitracin in the diets of turkey poultry. *Poult. Sci.* 57:1687–1689.

Fuller, R. 1977. The importance of lactobacilli in maintaining normal microbial balance in the crop. *Br. Poult. Sci.* 18:85–94.

Garrett, S. D. 1970. Pathogenic root-infecting fungi. University Press, Cambridge, England, 294 p.

Gilles, M. T. 1978. Animal feeds from waste materials. Noyes Publications, Park Ridge, NJ, 347 p.

Gilliland, S. E. 1979. Beneficial interrelationships between certain microorganisms and humans: Candidate organisms for use as dietary adjuncts. *J. Food Prot.* 42:164–167.

Gilliland, S. E., M. L. Speck, and C. G. Morgan. 1975. Detection of *Lactobacillus acidophilus* in feces of humans, pigs, and chickens. *Appl. Microbiol.* 30:541–545.

Goering, H. K., and C. H. Gordon. 1973. Chemical aids to preservation of high moisture feeds. *J. Dairy Sci.* 56:1347–1351.

Grieve, D. B., K. R. Ahrens, J. W. Thomas, and J. T. Huber. 1982. Production of lactating cows and growing steers fed alfalfa haylage treated with ammonia or a microbial inoculant. *J. Dairy Sci.* 65(suppl. 1):143.

Hollaender, A. 1977. Genetic engineering for nitrogen fixation. Plenum Press, New York and London. 538 p.

Ikediugwu, F. E. O., and J. Webster. 1970. Antagonism between *Coprinus hemptemerus* and other coprophilus fungi. *Trans. Br. Mycol. Soc.* 54: 181–204.

Kerr, A., and K. Htay. 1974. Biological control of crown gall through bacteriocin production. *Physiol. Plant Pathol.* 4:37–44.

Kinsey, C. M. 1980. Use of microbial additives in feed. A literature review. Pages 25–30 *in 40th Semiannual Am. Feed Manufacturers Assoc. Meet. Proc.* Hyson, Westcott and Dunning, Inc., Baltimore, MD.

Krause, V., and D. C. Clanton, 1977. Preservatives for silage. Nebraska Beef Cattle Report, EC77–218.

Kung, Jr., L., L. D. Satter, G. L. Enders, Jr., H. S. Kim, R. S. Porubcan, and S. H. Gehrman. 1984. Microbial inoculation of high dry matter alfalfa silage. *J. Dairy Sci.* 67(suppl. 1):134.

Lindgren, S., P. Lingvall, A. Kaspersson, A. deKartzow, and E. Rydberg. 1983. Effect of inoculants, grain and formic acid on silage fermentation. *Swedish J. Agric. Res.* 13:91–100.

Maxwell, C. V., D. S. Buchanan, F. N. Owens, S. E. Gilliland, W. G. Luce, and R. Vencl. 1983. Effect of probiotic supplementation on performance, fecal parameters and digestibility in growing-finishing swine. *Okla. Agr. Exp. St. Anim. Sci. Res. Rep.*, MP114:157–161.

McCullough, M. E. 1978. Fermentation of silage—A Review. National Feed Ingredients Association, West Des Moines, IA, 332 p.

McCullough, M. E., and K. K. Bolsen. 1984. Silage Management. National Feed Ingredients Association, West Des Moines, IA, 275 p.

McDonald, P. 1981. *The Biochemistry of Silage.* John Wiley and Sons, NY, 226 p.

McDonald, P., M. J. Proven, and A. R. Henderson. 1983. The effect of some pre-ensiling treatments on silage composition and nitrogen disappearance in the rumen. *Anim. Feed Sci. Technol.* 8(4):259–270.

Metchnikoff, E. 1908. The prolongation of life. G. P. Putnam's Sons, New York.

Miles, R. D., H. R. Wilson, A. S. Arafa, E. C. Coligado, and D. R. Ingram. 1981. The performance of bobwhite quail fed diets containing lactobacilli. *Poult. Sci.* 60:894–896.

Mitchell, I., and G. D. Kenworthy. 1976. Investigations on a metabolite from *Lactobacillus bulgaricus* which neutralizes the effect of enterotoxin from *Escherichia coli* pathogenic for pigs. *J. Appl. Bacteriol.* 41:163–174.

Mitsuoka, T. 1969. Vergleichende Untersuchungen uber die Laktobazillen aus den Faeces von Menschen, Schweinen und Huhnern. *Zentralbl. Bakteriol. Paraskitenkd. Infektionskr. Hyg. Abt. 1 Orig.* 210:32–51.

Moon, N. J., L. O. Ely, and E. M. Sudweeks. 1980. Aerobic deterioration of wheat, lucerne and maize silages prepared with *Lactobacillus acidophilus* and a *Candida* sp. *J. Appl. Bacteriol.* 49:75–87.

Moore, L. W. 1979. Practical use and success of *Agrobacterium radiobacter* strain 84 for crown gall control. *In* B. Schippers and W. Gams, eds., *Soil-Borne Plant Pathogens.* Academic Press, London.

Morishita, Y., T. Mitsuoka, C. Kaneuchi, S. Yamamoto, and M. Ogata. 1971. Specific establishment of lactobacilli in the digestive tract of germ-free chickens. *Japan J. Microbiol.* 15:531–538.

Muralidhara, K. S., G. G. Sheggeby, P. R. Elliker, D. C. England, and W. E. Sandine. 1977. Effect of feeding lactobacillus on the coliform and lactobacillus flora of the intestinal tissue and feces from piglets. *J. Food Prot.* 40:288.

O'Leary, J., and R. W. Hemken. 1982. Improvement in alfalfa fermentation and feed intake with a mixed frozen concentrated culture. *J. Dairy Sci.* 65(suppl. 1):142.

______. 1984. Evaluation of inoculants for silage. *J. Dairy Sci.* 67(suppl. 1):136.

Olsson, T. 1967. Intestinal disorders in pigs: Prophylaxis and therapy with *Lactobacillus acidophilus. Svensh Veterinartidining.* 18:353.

Owens, F. G. 1962. Effect of enzymes and bacitracin on silage quality. *J. Dairy Sci.* 45:934–936.

Parker, R. B., and J. S. Crawford. 1978. Alternative to antibiotics. *Proc. of 13th Annu. Pacific Northwest Anim. Nutr. Conf.* Nov. 1–2, Vancouver, British Columbia.

Potter, L. M., L. A. Newbern, C. M. Parson, and J. R. Shelton. 1979. Effect of protein, poultry by-product meal, and dry *Lactobacillus acidophilus* culture addition to diets of growing turkeys. *Poult. Sci.* 58:1095.

Rettger, L. G., M. N. Levy, L. Weinstein, and J. E. Weiss. 1935. *Lactobacillus acidophilus* and its therapeutic application. Yale University Press, New Haven, CT.

Sandine, W. E., K. S. Muralidhara, P. R. Elliker, and D. C. England. 1972. Lactic acid bacteria in food and health: A review with special reference to enteropathogenic *Escherichia coli* as well as certain enteric diseases and their treatment with antibiotics and lactobacilli. *J. Milk Food Technol.* 35:691–702.

Savage, D. C., R. Dubos, and R. W. Schaedler. 1968. The gastrointestinal epithelium and its autochthonous flora. *J. Exp. Med.* 127:67–76.

Schroth, M. N., and J. G. Hancock. 1981. Selected topics in biological control. *Annu. Rev. Microbiol.* 35:453–476.

Shahani, K. M., and A. D. Ayebo. 1980. Role of dietary lactobacilli in gastrointestinal microecology. *Amer. J. Clin. Nutr.* 33:2448–2457.

Shirley, J. E., L. D. Brown, F. R. Toman, and W. H. Stroube. 1972. Influence of varying amounts of urea on the fermentation pattern and nutritive value of corn silage. *J. Dairy Sci.* 55:805–810.

Shuler, M. L. 1980. The utilization and recycle of agricultural wastes and residues. CRC Press, Inc., Boca Raton, FL, 304 p.

Speck, M. 1981. Use of microbial cultures: Dietary products. *Food Technol.* 35:71–73.

Stadhouders, J. 1983. Lactic acid bacteria in foods: Genetics, metabolism and applications, Abstracts, Netherlands Society for Microbiology, Wageningen, Netherlands.

Sule, S., and C. I. Kado. 1980. Agrocin resistance in virulent derivatives in *Agrobacterium tumefaciens* harboring the pTi plasmid. *Physiol. Plant Pathol.* 17:347–356.

Tannock, G. W., and D. C. Savage. 1974. Influences of dietary and environmental stress on microbial populations in the murine gastrointestinal tract. *Infect. Immunol.* 9:591–598.

Tinsley, T. W. 1979. The potential of insect pathogenic viruses as pesticidal agents. *Annu. Rev. Entomol.* 24:63–87.

Templeton, G. E., and R. J. Smith, Jr. 1977. Pages 167–176 *in* J. G. Horsfall and E. B. Cowling, eds., *Plant Disease: An Advanced Treatise.* Academic Press, New York.

Templeton, G. E., D. O. TeBeest, and R. J. Smith. 1979. Biological weed control with mycoherbicides. *Annu. Rev. Phytopathol.* 17:301–310.

Tramer, J. 1973. Yogurt culture. *J. Soc. Dairy Technol.* 26:16–21.

Wilding, N., and J. N. Perry. 1980. Studies on *Entomophthora* in populations of *Aphis fabae* on field beans. *Ann. Appl. Biol.* 94:367-378.

Yahara, N., and S. Nishibe. 1975. A comparative effect of four organic acids on the silage fermentation of direct-cut alfalfa. Res. Bul. Hokkaido. *Natl. Agric. Exp. St.* 111:103.

CHAPTER 24

Microbiology in Pollution Control: From Bugs to Biotechnology

LAYNE M. JOHNSON, CURTIS S. MCDOWELL, AND MARK KRUPKA

Polybac Corporation, Allentown, Pennsylvania 18103

Microorganisms have been used in various aspects of pollution control for the past 15 to 20 yr. Because of the advances made in research and development and in application techniques during the past few years, pollution control microbiology has progressed from an art of using "bugs" to the science of biotechnology. This paper serves to illustrate how this science has developed by presenting data from various cleanup jobs including an oil spill and a formaldehyde spill. The recent developments in pollution control are discussed and results obtained from the treatment of complex xenobiotic wastes are presented. In addition, the development of new microbial systems for use in aerobic and anaerobic processes are presented. The potential of pollution control microorganisms and their relation to the advances of microbial genetics are presented to provide an understanding of the role that these organisms play as pollution control biotechnology matures.

INTRODUCTION

Microorganisms are being used to solve problems in the areas of commercial and industrial wastewater treatment, municipal wastewater treatment, and hazardous materials handling. The use of microorganisms in pollution control has been demonstrated in a wide variety of instances (Crueger and Crueger 1982; Omenn and Hollaender 1983). Many waste problems are encountered in several domestic and industrial situations and some of these have been solved by the application of microorganisms. In some cases, the results have been dramatic, leading several persons to believe that the potential of pollution control biotechnology has only begun to be realized (Powledge 1983).

The history of pollution control biotechnology is difficult to trace. Most information regarding the subject is typically considered proprietary and has not been recorded in scholastic journals. Early production of "bugs" can be traced to the 1940s, however. The use of organisms for pollution control is thought to be a direct result of still earlier work in the areas of alcohol production and acetone and butanol fermentations. Some of the methods for producing pollution control microorganisms were developed while microbes were being cultivated for use as animal feed supplements.

Early pioneers in this field of microbiology recognized microorganisms as good sources of enzymes that could be used to solve solids accumulation problems. In the early to mid-1950s, metal pipes were widely used and problems of plugging were not easily corrected by the use of corrosive chemicals. An alternative method relied on the use of lipase-producing microorganisms to dissolve fats and other solids. These organisms were capable of catalyzing the removal of fats without harming pipes.

The use of the organisms was extended and subsequently applied to solving problems in anaerobic digestion. In the early 1950s, microorganisms were used to accelerate the aging process in anaerobic digesters. Such commercial use of microorganisms can be traced to the Milwaukee and Chicago areas.

Further development of pollution control microbiology occurred in the early 1960s when it was shown that microorganisms displayed a propensity for metabolizing certain herbicides and phenolic compounds in industrial waste streams. During this period of time, great impetus was given to the development of this field of microbiology when a commercially available household product was introduced that could be used by the consumer. The product was designed to aid in the efficient operation of septic tanks.

Many of the microorganisms that were used during the late 1940s, 1950s, and early 1960s were isolated from composting operations (James Tobey, Jr., personal communication). Commercial literature and patents described these cultures, and the organisms that were selected for use were chosen because of their ability to degrade solids. Some highly active microbes were isolated and developed and many of these are still being used today.

During the 1960s and 1970s, cell-free products containing enzymes were developed. These products were not directly applied to solving problems of pollution, but rather, they were used in detergents. During the early 1970s an approach designed to increase the efficiency of sewage treatment was adopted. The success of the approach relied upon the selection of microorganisms that could be used to enhance the operation of particular sewage treatment processes. Extensions of this technology included the use of microorganisms to treat exotic industrial wastes or to treat accidental spills of xenobiotic chemicals (Johnson and Thomas 1983).

Dart and Stretton (1980) reported that they observed no "spectacular advances in the use of microorganisms in pollution control" during the period between 1977 and 1980. But after 1980, significant advances in microbial genetics occurred and these began to shed new light on pollution control microbiology. In fact, so many changes have occurred since the pollution control biotechnology industry was born, that today, an annual market of $7–10 million exists. One source reports that the potential total market value has been estimated at $200 million (Crueger and Crueger 1982).

Three major companies, including Polybac Corporation, Allentown, PA (a subsidiary of Cytox Corporation), Sybron Biochemical, Birmingham, NJ (a division of Sybron Corporation), and Environmental Cultures Division of Flow Laboratories, Inc., Inglewood, CA (a subsidiary of Flow General) have developed markets in pollution control biotechnology. Many of the interests of these companies include various microbial processes that are not directly related to pollution control, and these applications include metal recovery, agriculture, agronomy, and food processing.

Discussion

Typically, microorganisms that metabolize complex organic compounds are developed by enrichment and selection techniques commonly used in microbiology (Aaronson 1970). Environmental samples are collected from sites where natural or man-made pressures have been selected for organisms of interest. These microorganisms are further selected by various enrichment and isolation techniques. In many cases, organisms are placed in media where an organic compound may serve as a sole source of carbon and energy or as a source of nitrogen or other essential nutrient. Sometimes it is necessary to employ cometabolism to ensure the isolation of microorganisms (Johnson and Williams 1982). Essentially microorganisms are chosen on the basis of their capacity to multiply within a short generation time and to metabolize organic compounds. Further adaptation and mutation techniques are employed and these vary according to the end use of the organism or to the particular company employing them. One series of adaptation and mutation steps is shown in Fig. 1. Typically, wild type strains that degrade a specific organic chemical or functional group are exposed to increasing concentrations of that compound. Those strains that grow most rapidly and are least inhibited by high concentrations of the chemical are further adapted and isolated. In some instances, those microorganisms that exhibit the greatest poten-

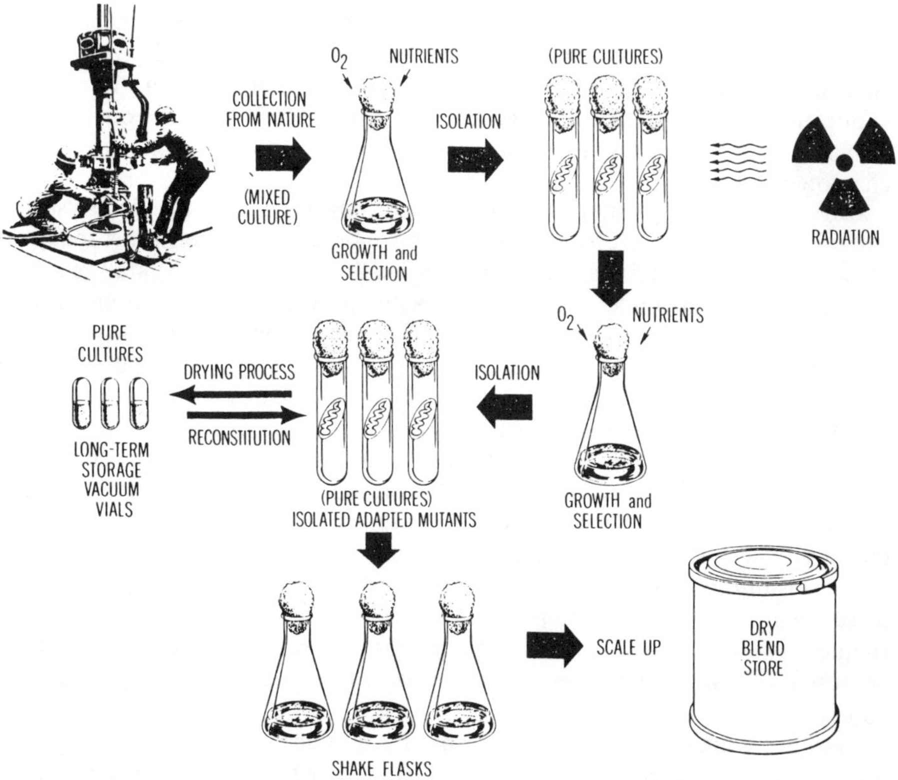

FIG. 1. Simplified selective and adaptation/mutation process. This scheme outlines the steps that are typically undertaken during the development of microbial pollution control products.

tial for end use are developed further by mutagenesis. Various mutagenic procedures may be used to induce genetic changes, but typically cells are irradiated. These genetic alterations can increase growth rate and enhance the biochemical traits desired.

Manufacturing Products Containing Pollution-Control Microorganisms

Cultures that are developed for pollution control purposes are manufactured according to a variety of methods. Typically, pure cultures of microorganisms are grown in liquid media contained in fermenters. In some cases, it is most desirable to manufacture mixed cultures of organisms under conditions of continuous flow. At appropriate times, these organisms may be harvested by filtration or centrifugation. At this point, microorganisms will be handled according to their end use. In instances where insoluble, dry products can be used by the consumer, further manufacturing of harvested cells can include additional fermentation using solid state methods. These organisms are then preserved by one of several methods, including lyophilization, air-drying, thermal air drying, and freezing at −20 C or −180 C.

Where soluble products are required, manufacturing can proceed via several avenues; the net result is that once the organisms are harvested, they are placed in high- or low-moisture matrix. It is often necessary to store and ship high-moisture products at refrigeration temperatures. The major recent developments in the preservation of pollution control microorganisms are being made with soluble types of products. Microorganisms preserved in liquid or gel form offer ease of application and enhanced viability and longevity. Most recently, our work at Polybac Corporation has led us to develop soluble products that are strictly anaerobic.

Once microorganisms have been preserved, they are usually blended to suit an end use. In general, microbial pollution control products on the market today are blends of several organisms that are metabolically related and can catalyze particular bioconversions under the existing uncontrolled conditions present in most bioreactions. For example, the treatment of petrochemical effluents includes using products containing types of microorganisms that degrade hydrocarbons. The ratios of these organisms can be varied to meet the demands of a unique application. In some instances, it is necessary to use more than one product. For example, a program can be established whereby one product containing heterotropic bacteria is used to pretreat a waste system. These bacteria biologically remove compounds that are toxic toward sensitive autotrophic organisms contained in a second product. Once toxicity has been significantly reduced by the organisms in the first product, the second product can be successfully used, and these organisms are allowed to function efficiently. Such synergistic programs are frequently used to mediate successful ammonia removal from waste systems because nitrifying bacteria tend to be subject to the toxic effects of many compounds.

Once pollution control microorganisms are produced, preserved, and blended for use, they are typically stored in containers ranging from a few ounces to hundreds of pounds. The manufacturing and formulation processes are best understood in terms of the application of the products, and these are described below.

Application of Pollution-Control Microorganisms

In most cases, the application of pollution control microorganisms merely begins with the microbiologist. As Kobayashi (1983) pointed out, the "application can only become a reality through the interdisciplinary activity of the genetic engineer with microbiologists of appropriate specialties and with environmental engineers well versed in biotechnology." In order to approach the reality of application, it is necessary to target specific markets. Because of the vast spectrum of man-made organic compounds and manufacturing processes, it is difficult for the pollution control microbiology industries to prepare themselves adequately to solve the variety of problems encountered. Therefore, it is necessary to develop organisms that display a wide variety of biodegradative traits that can be used in many markets.

When it is determined that a microbial product can theoretically solve a pollution control problem, it is necessary for a specialist to consider the engineering aspects of the particular system to be treated. Polybac Corporation has coined the term "Biomass Engineering® " to meet this end. Engineering factors that must be considered are shown in Table 1. These factors include the parameters that are also considered for growing organisms in the laboratory, only on a large scale the economic feasibility for employing pollution control microorganisms must be seriously considered. All these factors must be considered by a competent engineer to ensure the successful use of pollution control microorganisms.

TABLE 1. *Engineering factors that are critical to the adequate and stable operation of biological wastewater treatment systems*

Adequacy of system design
Dissolved oxygen [aerobic systems]
pH conditions
Nutrient availability
Temperature
Waste toxicity and concentrations

If all of the above criteria are met, several additional factors must be considered. Marketing a "high-technology" concept that is based on the use of microorganisms is no easy task. One reason is that the advertising media usually depict microorganisms as germs and the causative agents of disease. Since most actual end users of pollution control biological products are plant operators or other individuals, they are generally not well versed in microbiology. Acceptance of an entity that is not visible to the naked eye is difficult and sometimes requires the personification of the lowly microbe. When microbial products are acceptable to the end user, product application is not usually identified as being free of risks. Therefore, it is not atypical to have difficulty gaining the acceptance of regulatory agencies. In this regard, opinions vary from country to country, from state to state, and even from municipality to municipality. To convince regulatory agencies that microbial pollution control products are safe and effective requires sound, high-quality research. The question of risk has been addressed by the Office of Technology Assessment (OTA) for genetically manipulated microorganisms (Office of Technology Assessment 1981). Although the OTA recognizes that harmful events are "unlikely," it has been agreed that society must ultimately determine what level of uncertainty it is willing to accept.

After theoretical, engineering, marketing, and regulatory factors have been reviewed, one final aspect of application must also be considered. The successful use of pollution control microorganisms typically requires a maintenance program that consists of multiple inoculations, as opposed to a "one shot," starter culture approach. Because so many waste problems occur in continuous flow environments, it is necessary for the engineer to adequately assess each situation and propose a dosage schedule that is economical and effective. Only in very rare situations does a single inoculation suffice. In landfarming operations, multiple inoculations supplemented with nutrient additions are also required.

Much pollution control biotechnology has been applied to solving wastewater treatment system problems. Fewer cases of application exist for handling hazardous materials. Occasionally, accidents occur during manufacturing, storage, or transportation and these often necessitate the implementation of emergency or remedial methods for hazardous waste treatment.

Initially, an emergency response to a hazardous materials spill is undertaken to alleviate any contamination problems that might ensue. Such measures include containment, removal of excess, highly concentrated wastes, or neutralization. After these measures have been taken, remedial action based upon microbiological pollution control technology can sometimes be instituted. A competent engineer who has had experience in solving related problems must evaluate the situation to assess whether or not *in situ* biodegradation is feasible. Several factors are critical to engineering biodegradation of a land-based spill. Growth conditions, such as oxygen requirements, temperature, moisture, and pH effects must be evaluated. Environmental contamination of the area and site geology must also be considered, and the solubility and biodegradability of the material must be ascertained. In most cases, biodegradability is determined by studying samples that have been sent to the laboratory. Most commercial pollution control microbiology laboratories determine biodegradability, toxicity, and feasibility for treatment by performing a battery of tests. These may include the use of electrolytic respirometers, proprietary toxicity tests, shake flask studies or soil degradation studies. In some cases, anaerobic as well as aerobic studies are performed. All of the tests are performed to establish biodegradation kinetics, potential for inhibition at various concentrations, oxygen and nutrient requirements, and temperature effects. Existing, commercially available organisms as well as organisms isolated from the sites may be used to design an effective product. Qualitative and quantitative analytical procedures relying on the use of high-performance, liquid chromatography and gas chromatography are employed. If biodegradability is deemed feasible, then the biomass engineer can design an inoculation program that can be used to ensure successful cleanup. The process of determining feasibility usually entails from 4 to 8 wk of laboratory investigations. In rare cases, where much is known about biodegradability of a compound, an immediate decision regarding *in situ* treatment can be made. This is due to the fact that reference to the biochemical capabilities of microbial strains catalogued in computer files makes a simple task of choosing the proper microbial formulation.

Petrochemical waste treatment systems have existed as prime targets for marketing microbial pollution control products. These systems are typically adequately designed and operate well and usually contain compounds that are relatively biodegradable.

The operation of petrochemical waste treatment systems can become expensive because of excessive costs for chemicals, field labor, or off-site disposal (Zitrides 1982). One specific example of such a system was an oxygen-activated sludge system used to treat wastes containing variable loadings of insoluble, emulsified, and soluble hydrocarbons. Several parameters routinely exceeded the design conditions of this system. High levels of alcohols and insoluble and emulsified oils were present in the influent and these produced foaming and solids-stabilized emulsions. A thick blanket of these emulsions accumulated on the clarifier surface and eventually washed over the effluent weirs, resulting in the deterioration of effluent quality. This solid-liquid separation problem was not successfully solved by expensive, chemical methods. The solids residence time of the system was increased from 7 to 10 d but this attempt created equipment failure. An alternative solution was employed and included the addition of Phenobac® . This product is a dry, insoluble powder that contains mutant bacteria that degrade hydrocarbons. The inoculation program resulted in significant increases in waste removal and process stability as demonstrated by total oxygen demand exerted by the system (Fig. 2). Tertiary butyl alcohol (TBOH) removal was also significantly

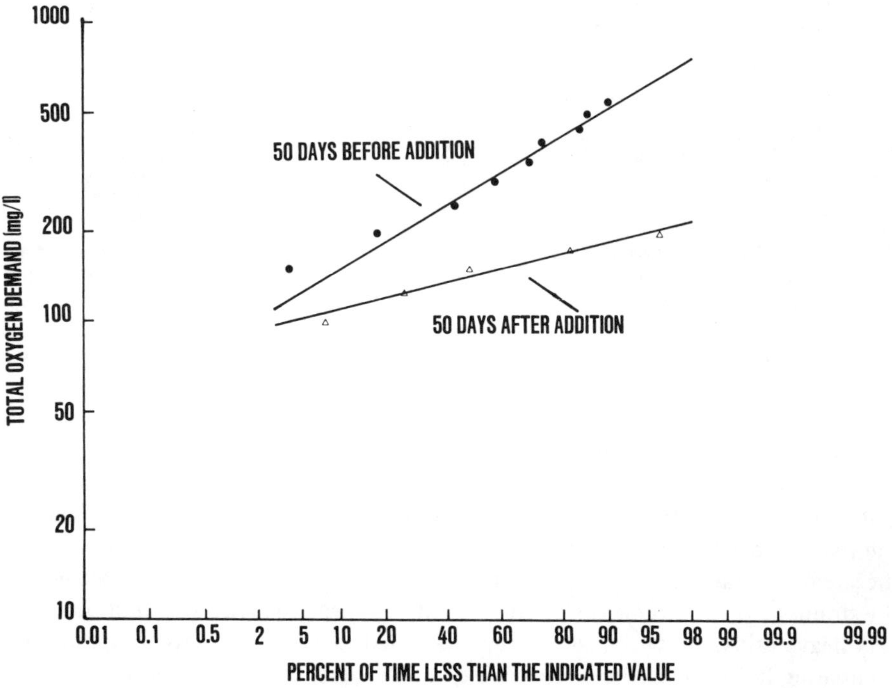

FIG. 2. Comparison of effluent Total Oxygen Demand data before and after the addition of microbial pollution control product. The application reveals significant improvement in the system performance.

enhanced (Fig. 3). As a result of this Biomass Engineering®, a reduction in the use of antifoam agents and coagulation agents occurred. This reduction more than paid for the cost of a continual use of the microbial product.

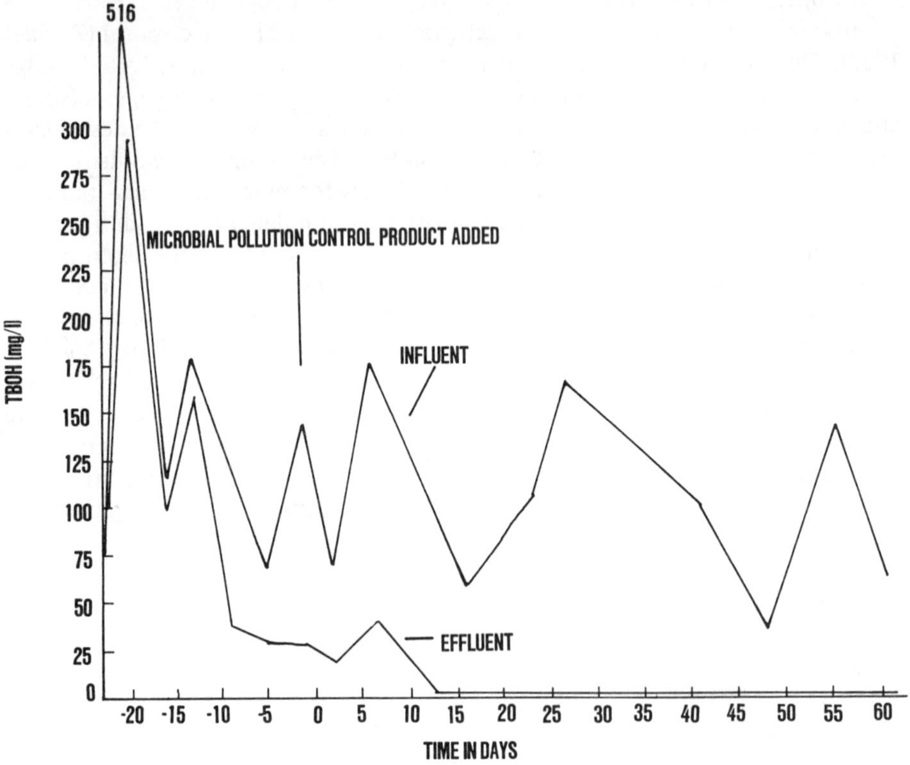

FIG. 3. Biodegradability of tertiary butanol in a petrochemical wastewater process. The data indicate that the levels of available tertiary butanol were essentially zero after addition of microbial pollution control product.

Formaldehyde Spill Case Study

In one accident, the contents of a 21,000 gal railroad tank car containing 50% formaldehyde solution spilled onto a railroad ballast and into a storm drain. The solution emptied into a ditch, which flowed to a tributary of a river that supplied drinking water for 250,000 residents. About 7,000 gal of formaldehyde reached the river before being discovered. A major spill control contractor was called to the site by the railroad company. Initially, an earthen dam was constructed across the drainage ditch. Removal of the remaining formaldehyde was hampered by very heavy rains, but within 24 h, crews had used vacuum trucks to pump the contaminating liquid from the ditch. These liquids were transferred to railroad tank cars and tank trucks and ultimately transported to a secure hazardous waste site.

Two options were considered for the cleanup of the residual formaldehyde; these included biological methods and removal of the ballast followed by transport to a secure landfill. Removal of the track was estimated to cost $50,000, and the resulting delay in train operations was estimated at \$80,000 per hour. Placing the contaminated ballast in a landfill would have added a cost of \$75,000 for a total estimated ballast decontamination of approximately \$445,000.

Alternatively, biodecontamination was undertaken in a cooperative program between the railroad, the spill contractors, and Polybac Corporation. A system was designed and installed whereby formaldehyde contained in a sealed drainage ditch was pumped to an aerated holding tank (Fig. 4). The tank was inoculated

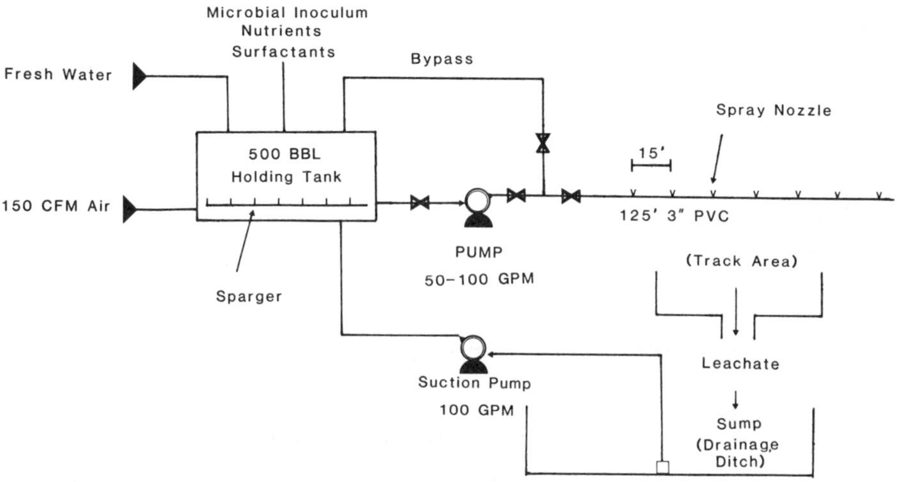

FIG. 4. *In situ* ballast decontamination system. Microbial pollution control products were added to the 500-barrel holding tank where they were mixed with fresh water, nutrients, surfactants, and formaldehyde solution. The contents of the tank were aerated and pumped to PVC pipes, where the contents were sprayed over the ballast. The leachate was collected in a drainage ditch and pumped back to the holding tank.

with organisms capable of mediating formaldehyde breakdown, and the contents of the tank were redistributed over the rail ballast at a rate of 50 to 100 gpm. Microorganisms were continuously cycled through the system and fresh water, nutrients, microorganisms, and surfactant were added daily. This system allowed the microorganisms to reduce the level of formaldehyde from 1,400 mg/l to 1 mg/l within 18 d (Fig. 5). The cost for this treatment was approximately \$45,000. During the treatment period, about 1.7×10^{14} microorganisms were added to the system.

Other emergency spill problems have been solved by using microorganisms. Landfarming operations have also relied upon the use of microorganisms. Some compounds that have been successfully treated in this manner include pentachlorophenol, crude oil, creosote, and many other compounds.

Conclusions

During the past 40 yr, the development of microorganisms for solving problems associated with pollution has progressed from a very narrow range of applica-

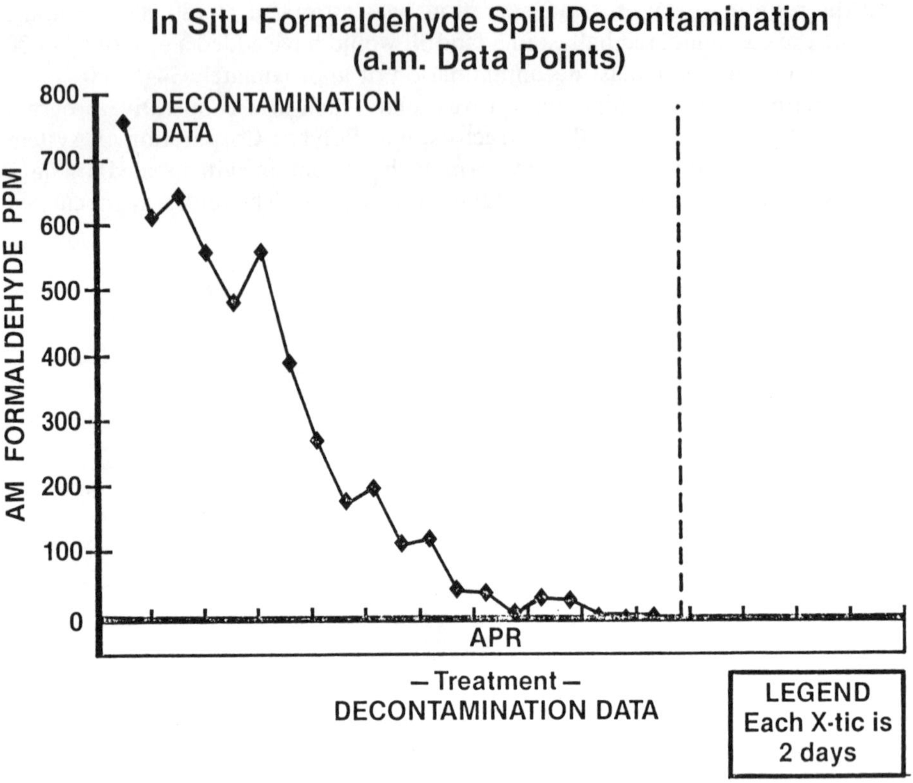

FIG. 5. Decontamination of formaldehyde in an *in situ* spill. This figure presents data to support that formaldehyde concentrations were reduced from 750 ppm to less than 1 ppm in approximately 21 days.

tions to a multimillion dollar business. During the initial stages of development, pollution control microbiology relied upon a handful of bacterial and fungal strains. These cultures were often grown under crude conditions, and the markets that were developed for their use allowed them to fit into a convenient niche that still exists today. But, these microorganisms merely laid the groundwork for a concept that continues to gain wider recognition and acceptance. Today, U.S. manufacturers supply a multitude of organisms to all corners of the globe. Yet, some basic conditions for successful market growth must continue to be met.

In order for pollution control microbiology to successfully continue its sojourn from bugs to biotechnology, it must be realized that such success can only be achieved by gaining answers through quality research. In most commercial operations, manufacturing processes have been mastered, market shares have been identified, sales crews have demonstrated success, and new applications are being developed on a daily basis. In addition, engineering approaches have developed dramatically during the past half decade. But, even if all of these aspects are adequately developed, what purpose can they serve in pollution control microbiology if microbiology is disregarded?

Strong research programs in microbial ecology, physiology, and genetics geared to solving environmental problems will allow pollution control microbiology to continue to develop, most likely at a pace that is more rapid than that experienced during the past 40 yr. This development will require the search for new microorganisms that can benefit the environment, agriculture, and other industrial processes. It will also require an understanding of how these microorganisms survive in the environments into which they are placed. Strong research must also depend upon looking at known microorganisms to see how their metabolic capabilities might be harnessed and put to work to help protect our environment.

One additional aspect of pollution control microbiology is embodied by genetic engineering. Many scientists and businessmen believe that the role of recombinant DNA technology in this field of microbiology is severely limited because of restrictions that have been imposed upon the release of such organisms into our environment. Such restrictions may affect the total pollution control market, but it is the responsibility of the commercial pollution control microbiologist to approach this problem with creativity. The restricted use of whole cells that have been genetically modified may be a reality that society must face. But there are many cell products, including hydrolase and laccase enzymes that have exhibited great potential for large-scale use (Johnson and Talbot 1983). The detoxification enzymes will become more interesting as economical means of producing them become a reality through the use of genetic modification. Such work is already in progress by many groups, and it is merely a matter of time before it can become a commercially viable concept.

Acknowledgments

The authors extend their gratitude to Mr. James F. Tobey, Jr., George A. Jeffrerys and Co., Inc., Salem, VA; Mr. John Biesz and Mr. Thomas Zitrides, Polybac Corporation, Allentown, PA for their insights into the history of pollution control microbiology and for information regarding case studies. Our appreciation is also extended to Kathy Dilcher for her patience and assistance during the preparation of this manuscript.

Literature Cited

Aaronson, S. 1970. *Experimental Microbial Ecology.* Academic Press, NewYork. 231 p.

Crueger, W., and A. Crueger. 1982. *Biotechnology. A Textbook of Industrial Microbiology.* Science Tech, Inc., Madison, WI. p. 276.

Dart, R. K., and R. J. Stretton. 1980. *Microbiological Aspects of Pollution Control.* Elsevier Scientific Publishing Company, New York.

Johnson, L. M., and H. W. Talbot, Jr. 1983. Detoxification of pesticides by microbial enzymes. *Experientia* 39:1236–1246.

Johnson, L. M., and J. M. Thomas. 1983. Biodecontamination of a full-scale formaldehyde spill. Page 372 *in* G. S. Omenn and A. Hollaender, eds., *Genetic Control Environmental Pollutants.* Plenum Press, New York.

Johnson, L. M., and F. D. Williams. 1982. Effect of substrate concentration on the cometabolism of *m*-chlorobenzoate by *Pseudomonas fluorescens. Bull. Environ. Contam. Toxicol.* 29:447–454.

Kobayashi, H. A. 1983. Application of genetic engineering to industrial waste/wastewater treatment. Pages 195–214 *in* G. S. Omenn and A. Hollaender, eds., *Genetic Control of Environmental Pollutants.* Plenum Press, New York.

Office of Technology Assessment. 1981. The question of risk. Pages 197–207 *in Impacts of Applied Genetics. Microorganisms, Plants and Animals.* OTA-HR-132. Washington, DC.

Omenn, G. S., and A. Hollaender, eds. 1983. *Genetic Control of Environmental Pollutants.* Plenum Press, New York. 408 pp.

Powledge, T. M. 1983. Prospects for pollution control with microbes. *Biotechnology* 1:743–755.

Zitrides, T. G. 1982. Full-scale applications of mixtures specialized microbes in spill site decontamination and wastewater treatment. *Impact of Appl. Genet. in Pollut. Control—A Symp.* South Bend, IN.

VII
SYMPOSIUM: CRYOPRESERVATION IN INDUSTRY AND BIOTECHNOLOGY

F. Simione, *Convener*
American Type Culture Collection, Rockville, Maryland

PANELISTS

J. M. Barbaree
T. Feldblyum
R. J. Heckly
F. V. LeSane
C. Y. Neuland
W. C. Nierman
A. Sanchez
G. N. Sanden
D. M. Strong

CHAPTER 25

Principles of Preserving Bacteria by Freeze-Drying

ROBERT J. HECKLY

Naval Biosciences Laboratory, School of Public Health,
University of California, Naval Supply Center,
Oakland, California 94625

Freezing and drying can damage bacteria in many ways. In the absence of protective substances, slow cooling produces solute effects and rapid cooling results in intracellular ice. Both can be lethal. Freezing can affect membrane permeability by damaging lipoproteins, and mutations may result from nucleic acid damage. Free radicals are produced when dried cells are exposed to oxygen, but the evidence that they kill or damage bacteria is not overwhelming. However, addition of sucrose, or similar substances, prevents free radical formation and markedly increases cell survival. Factors affecting survival include age of culture, nutrition, suspending menstrum, rate of cooling, rate and extent of drying, temperature and atmosphere during storage, and rehydration procedure. As a generalization, high survival of bacterial cultures can be obtained if an equal volume of 12% sucrose is added to a mature culture, and the mixture is frozen slowly, dried at 0.1 torr or less, sealed under original vacuum, and stored at or below 4 C. If the culture is reconstituted as soon as possible after the ampule is opened oxygen damage is minimized. Highest viability is usually obtained with complete growth medium.

INTRODUCTION

Hammer (1911) was probably the first to apply freeze-drying to the preservation of bacteria when he dried *Escherichia coli, Staphylococcus aureus* and *Pseudomonas aeruginosa.* The term lyophile was used by Flosdorf and Mudd (1935) to describe the "water loving" properties of their dried biological materials. Since then, lyophilization has come to denote the process of drying any material by sublimation of ice from frozen preparations. However, the descriptive term, freeze-drying, seems to be more commonly used today.

METHODS

Before dry ice or small, low temperature refrigeration systems were generally available, Flosdorf and Mudd (1938) introduced the cryochem process, which used anhydrous calcium sulfate (Drierite) to remove water vapor. Instead of shell-freezing the materials, they used self-freezing. This process is convenient in that the filled ampules are attached to the manifold without freezing. The vacuum is applied slowly and the samples freeze as the rate of evaporation becomes high enough. The main disadvantage is that the self-freezing procedure requires careful control of the vacuum, because material may froth or "bump" out of the container when the suspension becomes superheated. With increasing availability of dry ice, the cryochem process was soon discontinued.

Early investigators, such as Proom and Hemmons (1949) believed that cells were killed solely by penetration of the cell wall by ice crystals. Thus, since it was well established that crystal size is inversely related to cooling rate, few if any cells would be killed if they could be cooled rapidly enough. To this end, ultrarapid cooling methods were devised, such as spraying the culture into liquid nitrogen through a fine needle. Cooling rates were estimated to be about 1,000 C/s. In marked contrast, slow cooling, in the order of 1 C/min, actually yields higher viability. Despite the higher viability of slowly cooled cultures, shell-freezing is commonly used today, particularly as the volume of culture approaches 50% of the total container volume.

Until about 1940, most of the research on freeze-drying was concerned with equipment and methodology. Since then, interest shifted to studies of other factors, particularly the suspending media. Thus, a large number of different mixtures have been concocted for protecting freeze-dried bacteria. At least 30 different mixtures have been used by the various investigators to preserve viability of freeze-dried organisms with varying degrees of success (Heckly 1961, 1978). These suspending media ranged from solutions of glucose, sucrose, or lactose to more complex mixtures, such as those advocated by Naylor and Smith (1946), which contained 0.5% ascorbic acid, 0.5% thiourea, 0.5% NH_4Cl, and 2% dextrin. Subsequently, Fry and Greaves (1951) published on good results obtained with "mist. desicans," which consisted of 3 parts serum and 1 part broth with 7.5% glucose. Despite the large amount of effort directed toward improving the suspending fluid, it does not appear that any of the complex mixtures are in general use today. Instead, sugar solutions, or skim milk, are the most commonly used additives for freeze-drying bacteria.

Discussion

Viability Factors

Many factors affect survival of freeze-dried bacteria. In addition to variations from one species or strain to another, they include the following easily identified variables: age of the culture, nutrition, suspending menstrum, cell concentration, rate of cooling, rate and extent of drying, temperature and atmosphere during storage, and rehydration procedures. Understanding the effects of each of these factors is extremely difficult because, as will be discussed, they are so interrelated.

Resistance to freeze-drying appears to be a genetically stable trait, and some strains of a given species may be more resistant than others. Steel and Ross (1963), as well as others, have observed that as a group gram-positive cells survive better than gram-negative cells when freeze-dried. Spores, on the other hand, are inherently so resistant to all forms of stress, and especially drying, that they present no preservation problem.

Nutrition, as an independent variable, has been little studied, possibly because it is so difficult to separate this variable from the physiological age effects. Many have studied the effect of culture age as a factor, and it was generally concluded that young cultures were more sensitive to the effects of freeze-drying than were cultures in the late growth or early stationary phase. Proom and Hemmons (1949) and Lingg et al. (1967) were among the few who found young cultures to survive better than older cultures.

A demonstration of the close interrelationships of factors was illustrated by the results of an experiment in which samples of a growing culture were removed at intervals and frozen slowly as well as rapidly. As is shown in Fig. 1, cells from young cultures survived freeze-drying best when frozen slowly. However, survival of rapidly cooled samples increased as the culture aged, until at 24 h the survival significantly surpassed that of slowly cooled samples. Thus, it is evident that even small differences in the age of a culture could be responsible for some of the discrepancies reported in the literature regarding the effects of cooling rate or culture age.

Cell concentration is also tied closely to culture age. The percentage of cells surviving freeze-drying was directly correlated with cell concentration (Otten 1930; Fry and Greaves 1951; Benedict et al. 1958). It has been postulated that high survival is obtained because the majority of cells are protected by substances released by lysis of a few cells. These substances must be rather special because a positive correlation between percentage surviving and cell density was observed in a rich medium containing 8% sucrose, 5% skim milk, and 1.5% gelatin (Damjanovic and Radulovic 1967). Perhaps cell lysis releases an antioxidant.

A little-recognized variable concerns the temporal relationships of treatments. It has been demonstrated that contact time of additives (before freezing) can have an unexpected effect on viability (Heckly et al. 1967; Heckly and Dimatteo 1975). Fig. 2 illustrates how resistance of *Serratia marcescens* varied after a culture was mixed with propyl gallate. Note that when cells were stored either 1 or 7 d increasing the time of contact before freezing from 2 min to 5 min increased survival of cells over 1,000-fold. Unfortunately, the exact temporal relationships were not reproducible, even though the general rhythmic response was reproduced in many other experiments.

Although the composition of suspending fluids have been studied in great detail, few individuals have considered pH as a variable. In fact, few publications even indicate the pH of the cell suspension at the time of freezing. As shown in Fig. 3, the rate of pH change may be more important than pH of the suspension at the time of freezing. The results of this experiment show that survival of *S. marcescens* freeze-dried from concentrated cell suspensions was scarcely affected by pH between pH 6 and pH 10 if the sodium hydroxide was added slowly (0.03 pH units per min). However, if the pH was adjusted rapidly (0.2 pH unit/min), the number of viable cells 24 h after drying fluctuated by several logs. Despite the fact that pH can be critical under these specific conditions, it is usually not essential to adjust pH of bacterial cultures prior to freeze-drying.

As indicated earlier, cooling rate is a critical factor, and it is not generally recognized that there is an optimum cooling rate for each system. As discussed by Mazur (1966, 1970, 1977), there is an optimum because of the interactions of two opposing effects. At slow cooling rates, cells become dehydrated because of extracellular freezing, and solutes become concentrated within the cells. In contrast, rapid cooling results in intracellular freezing. This is not lethal per se, since Mazur (1966) showed that high survival can be obtained if these cells are thawed

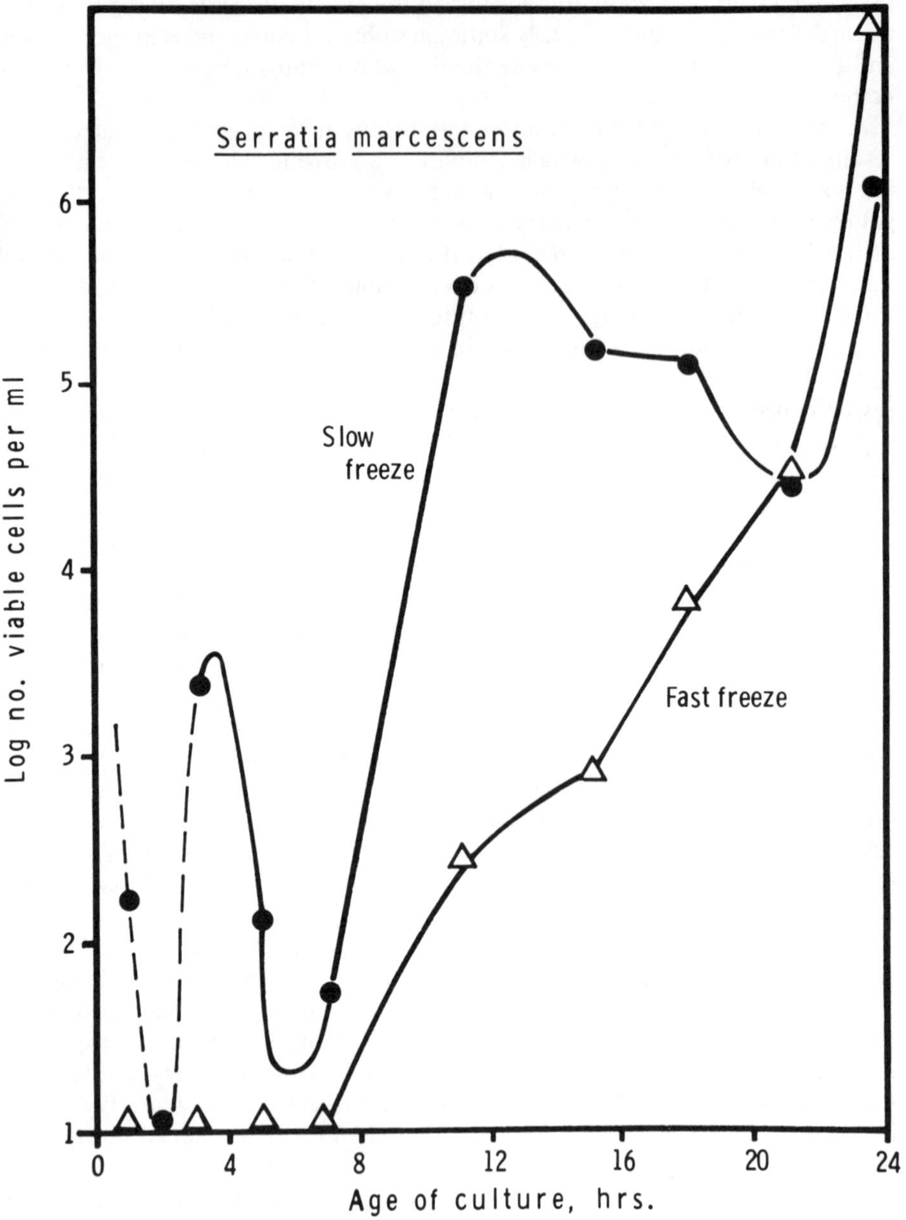

FIG. 1. Effect of culture age and freezing rate on survival of freeze-dried *Serratia marcescens*. Organisms were grown in chemically defined medium at 30 C with shaking. At intervals, samples were removed and frozen rapidly by immersion in a dry ice-ethanol bath or frozen slowly in a −20 C freezer chest. All samples were dried overnight and reconstituted with distilled water for viability assays (from Heckly 1978, with permission).

rapidly enough. Mazur (1977) suggested that injury resulted from intracellular ice only during its recrystallization on warming and that the ice crystals disrupted intracellular membranes. Recrystallization can be prevented if the temperature is kept below about −30 C, as was done for preparing specimens for electron

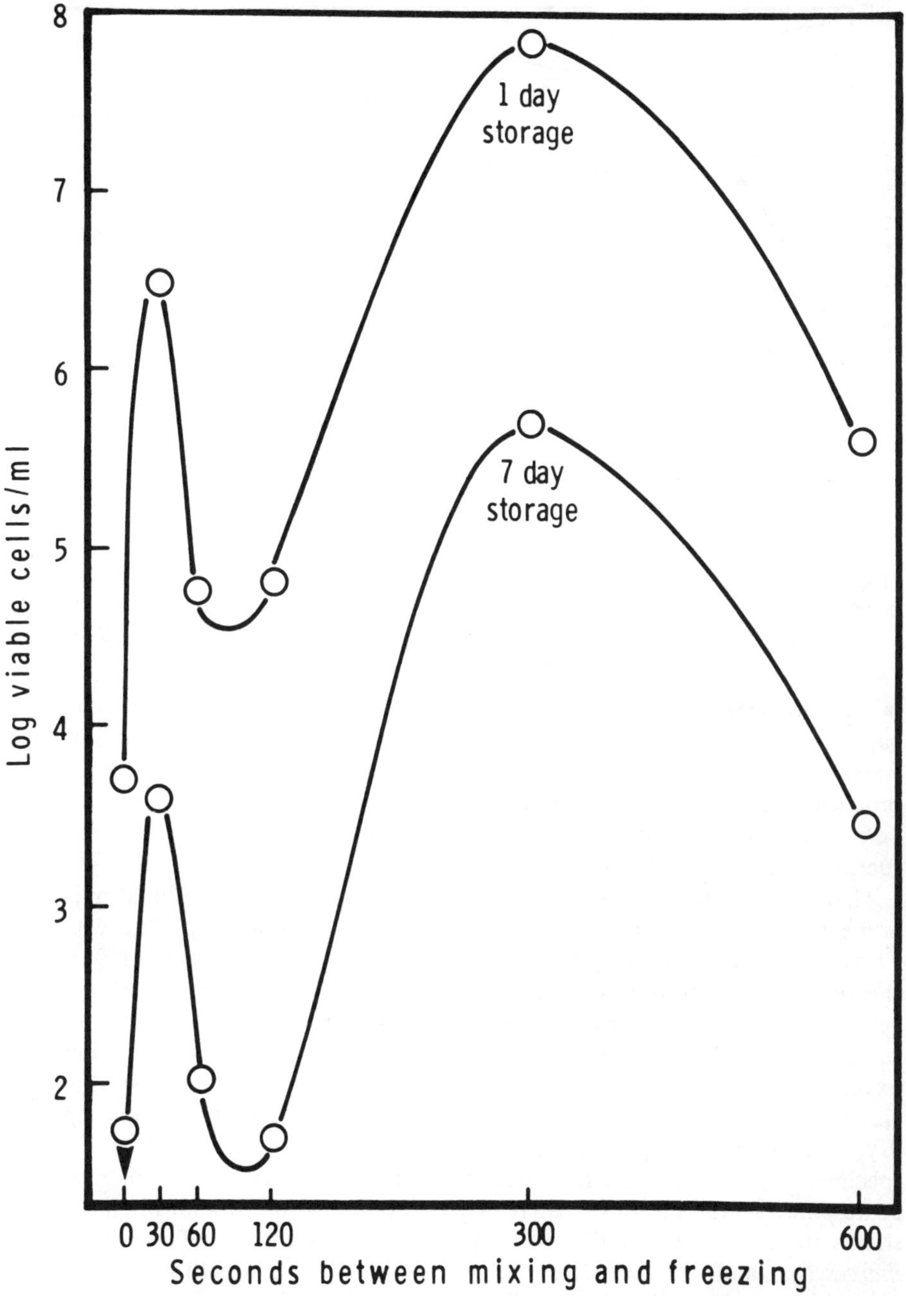

FIG. 2. Effects of propyl gallate on viability of freeze-dried *Serratia marcescens*. Centrifuged cells were resuspended in distilled water and mixed with an equal volume of 0.4% propyl gallate. At the indicated times 0.5 ml samples were removed and rapidly frozen by dispensing into bottles cooled in dry ice. After all samples were collected, they were freeze-dried. Half of the samples were rehydrated after 1-d storage and the other half after 7 d at room temperature (from Heckly et al. 1967, with permission).

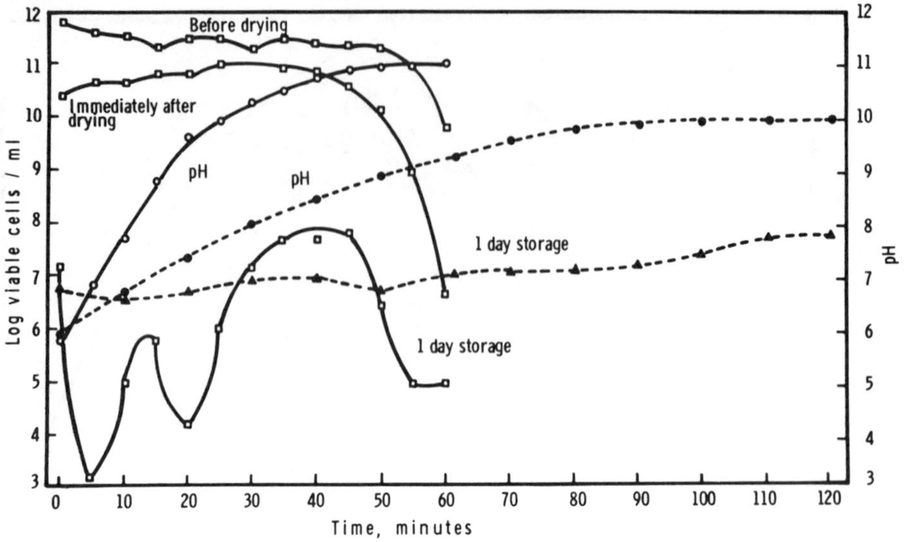

FIG. 3. Effects of pH, before freeze-drying, on viability of freeze-dried *Serratia marcescens*. Broken lines represent data from an experiment in which the pH was increased slowly, whereas the solid lines represent another experiment in which the pH was increased about four times as rapidly (from Heckly et al. 1967, with permission).

microscopy (Heckly and Skilling 1979). Unfortunately, at that temperature the vapor pressure is so low that the time required to sublime even a few milliliters of water becomes impractical for culture preservation. Therefore, since it is rarely practical to determine the optimum cooling rate for each system, bacteria should be cooled slowly, about 1 C/min, with cryoprotective agents, such as lactose or sucrose, to minimize deleterious solution effects.

There is considerable controversy regarding the effects of residual moisture. Some have argued that complete removal of water is incompatible with life. There are a number of reports, similar to that of Fry and Greaves (1951) who showed that prolonged drying, 336 vs. 18 h, decreased the percentage of surviving cells. Such long drying may not necessarily result in lower moisture content; instead, the decreased viability may be due to increased exposure of dried cells to oxygen. As long as there is ice in the sample, movement of the water vapor excludes all oxygen from the dried portions, but as soon as the sample is completely dry, air entering through small leaks in the system can diffuse freely to the dried specimen. Even at 0.05 torr the partial pressure of oxygen is significant and could be responsible for the observed reduced viability. Many organisms have been shown to survive best at the lowest moisture content tested. These include *S. marcescens* and *E. coli* (Dewald et al. 1967), phytopathogenic bacteria (Samosudova 1965), *Bacillus popilliae* (Lingg and McMahon 1969), and *Shigella* (Damjanovic 1974). There is no good evidence that any but the lowest moisture level is advantageous for preserving viability of bacteria.

Over 70 yr ago Rogers (1914) found that survival was highest in cultures stored under vacuum and the lowest in those stored in air or oxygen. Subsequent studies have extended these observations, and essentially confirmed the original observations and quantitated the effects of oxygen. For example, when sealed at 0.6 to

0.7 torr, 3% of *S. marcescens* and 41% of *S. aureus* were viable, whereas only about 1 and 7% survived when the respective cultures were sealed at about 0.1 torr (Christian and Stockton 1956). Lion and Avi-Dor (1963) reported that oxygen induced the inactivation of NADH-oxidase, not only in dried *E. coli* but in cell-free extracts containing enzyme. Furthermore, they showed that potassium nitrate protected oxidase activity and also viability of cells exposed to air. However, potassium nitrate does not appear to be in common usage today.

Dewald (1966) showed that the action of oxygen is temperature dependent and an Arrhenius plot of his data (inactivation rate of dried *S. marcescens* in air vs. 1/T) yielded a straight line. Subsequently, Heckly and Dimmick (1968) showed that viability after 7 d at room temperature was inversely proportional to the log of the oxygen pressure. When washed *S. marcescens* cells, with no additives, were freeze-dried and reconstituted without admitting air, virtually 100% of the cells were found to be viable (Heckly and Dimmick 1968). Fig. 4 shows the arrangement used to rehydrate cells without admitting air. The inner ampule contained degassed distilled water, which was sealed under vacuum. Rehydration was accomplished by shaking the tube to break the delicate tip of the inner ampule. When air was admitted before rehydration, viability was lost rapidly. As is shown in Fig. 5, when the ampules were opened shortly after drying a plot of the log of number of viable cells vs. the square root of the time exposed to air yielded essentially a straight line, similar to that reported by Dewald (1966). However, if the dried cultures were stored in the sealed glass ampules under vacuum for 2 wk, the cells became even more sensitive to oxygen. As is shown in Fig. 5, after as little as 2 min exposure to air the cells aged in vacuum were reduced in viability by almost 3 log. These effects were observed only with washed cell suspensions, but it serves

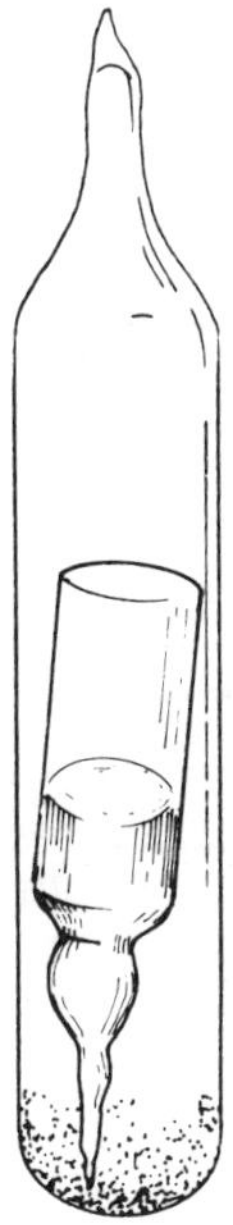

FIG. 4. Diagram of double ampule system used to rehydrate dried materials without exposure to air.

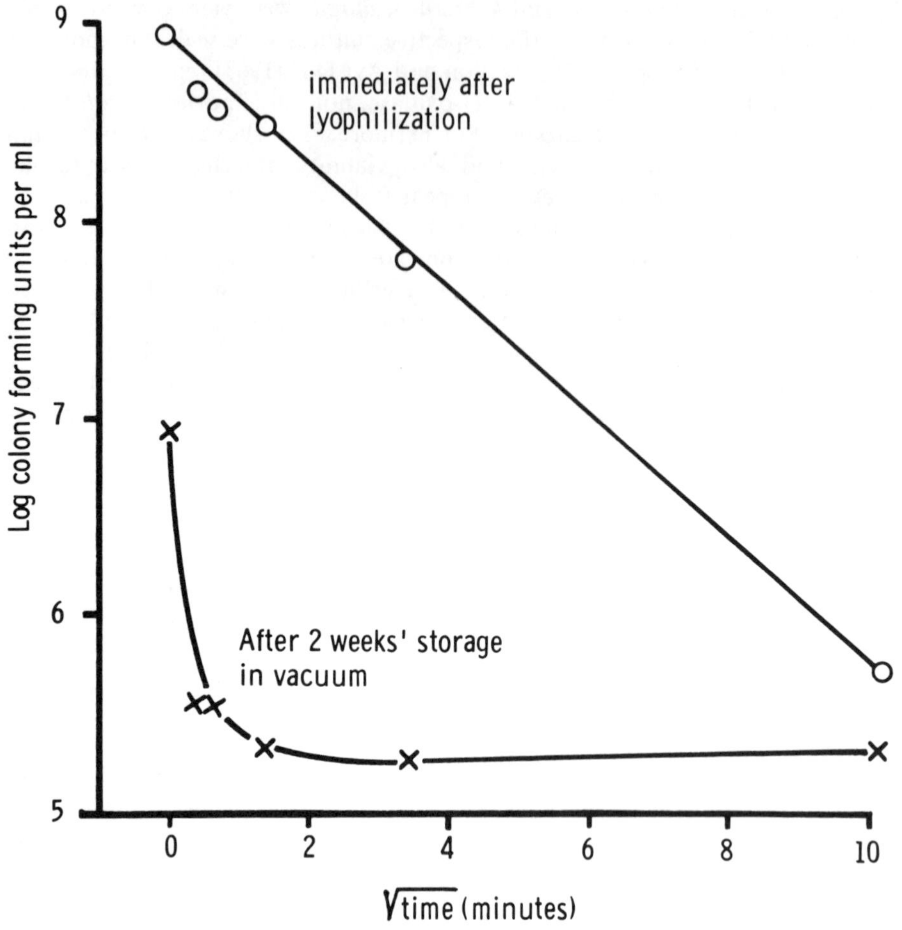

FIG. 5. Effect of exposure to air at 1 atm of pressure on viability of *Serratia marcescens*. The freeze-dried samples were rehydrated by adding distilled water after the indicated periods of exposure to air (from Heckly and Dimmick 1968, with permission).

to illustrate again the interrelationships of the various factors affecting viability. The effect of oxygen on dried organisms, including sublethal effects, were reviewed in detail a few years ago (Heckly 1978).

Wasserman and Hopkins (1957) were among the few investigators who studied the effect of temperature or composition of the reconstitution fluid. They found that 0.05 *M* sodium malate at pH 6 was particularly effective for reconstitution of *S. marcescens*. Furthermore, they reported that 96% of the cells that developed into colonies when rehydrated at 30 C failed to grow when rehydrated with the same solution at 5 C. This was confirmed by Leach and Scott (1959), who extended the studies to show that there is a significant difference between organisms with respect to the optimum temperature of reconstitution. The entire subject of reconstitution, temperature, composition, and volume of reconstitution fluid has been reviewed in detail (Heckly 1961). Evidence is not convincing enough to change the general practice of using either distilled water or growth media at room temperature for reconstitution. Although, in general, the effect of reconstitution fluid temperature may be trivial, manufacturers still recommend activation of commercial yeast with warm water (40–46 C).

Nature of Cell Damage

In attempts to determine the cause of death, many investigators have studied injury caused by freezing and drying. When held in the proper environment, bacteria are able to repair the damage. The time required for cells to recover has been reported to vary from about 30 min to approximately 24 h. Injury has been characterized by increased permeability of leakiness of the cells, inability to grow on selective media, loss of virulence, or mutations. Many attempts have been made to identify the nature of the damage caused by freeze-drying, but it appears that any one of a number of different sites may be affected. A variety of approaches has been used to determine the nature of the damage or injury as a consequence of freezing and drying.

Israeli et al. (1974) reported that freeze-drying of *E. coli* damaged the membrane transport system for o-nitrophenol beta thiogalactopyranoside and of potassium ions and that if oxygen were excluded the damage could be partially repaired after rehydration. Calcott and MacLoed (1975a) demonstrated that the release of cellular constituents (ultraviolet absorbing material, potassium, and beta-galactosidase) was correlated with loss of viability. In a subsequent report, based on electron microscope evidence, Calcott and MacLoed (1975b) showed that both membrane and cell wall were damaged. Glycerol prevented both types of damage, whereas Tween 80 prevented only membrane damage. Only the membrane damage was lethal.

Bretz and Kocka (1967) report that freezing and thawing increased cell permeability, as evidenced by increased sensitivity of *E. coli* to actinomycin D, and that the cells would recover their resistance to the antibiotic within 30 min after thawing. Freeze-dried *E. coli* also have been shown to be similarly damaged (Sinskey and Silverman 1970), but at least 5 h were required to recover antibiotic resistance. This indicates that drying does damage in addition to that caused by the freezing.

Gomez et al. (1973) reported repair of damage to be complete in about 8 h. As a measure of injury, they used the difference in number of *E. coli* or *Salmonella typhimurium* growing on minimal agar vs. the complete trypticase soy medium. Ray et al. (1972) believed that freezing affected the lipopolysaccharide (LPS) of the cell wall. They showed that the rate of repair was temperature dependent and did not require synthesis of protein, ribonucleic acid, or cell wall mucopeptide, but did require energy in the form of adenosine triphosphate (ATP). Subsequently, Kempler and Ray (1978) demonstrated that, at least with *E. coli*, structural damage to the LPS was in the heptose region, possibly with loss of heptose.

A different type of membrane damage, reported by Israeli et al. (1974), indicates that freeze-drying affected the bacterial membrane and involved the DNA initiation complex. The repair process required protein synthesis. They also noted that on exposure of dried bacteria to oxygen, the injury became irreversible and bacteria no longer could divide and form colonies.

Several studies have shown that freezing also may damage deoxyribonucleic acid (DNA). For example, Ohnishi et al. (1977) reported on DNA strand breakage in *E. coli*; they also reported that after rehydration only the radiation-resistant strain (*E. coli* B/r) could repair the damaged DNA. More recently, Tanaka et al. (1979) reported on reversion of *E. coli* induced by freeze-drying. They found a greater mutation frequency in the strain that could repair DNA than in the strain that lacked the repair system. It is logical that genetic changes can occur only if there is repair of DNA; if cells fail to repair DNA, they cannot reproduce. They also concluded that damage was more like that induced by X-ray than by ultraviolet damage.

Attempts to use the Ames strains of *Salmonella typhimurium* (Ames et al. 1975) to identify the mutagenic effects of freeze-drying were not conclusive (Heckly and Quay 1981). Unless survival is very high it is difficult to differentiate between mutagenic effects of freeze-drying and the selection of spontaneous revertants present in the culture prior to freezing. This would be particularly significant if, as Webb (1969) had reported, auxotrophs do not survive drying as well as prototrophs. Freeze-drying of strains TA 100 and TA 1535 appeared to increase the number of revertants, whereas the frame shift mutants, TA 98, TA 1536, TA 1537, and TA 1538 were not significantly affected by freezing and drying. This would confirm the findings of Ohnishi et al. (1977) that freeze-drying can damage or kill cells by strand breakage of DNA.

Yersinia pestis survives well in the dried state, but the virulence of freeze-dried cells, particularly if they are stored for any length of time, is markedly reduced (Heckly and Blank 1980). The loss of virulence for mice was progressive, and after 25 yr, as many as 4,000 viable cells constituted an LD_{50} when assayed shortly after rehydration. Fig. 6 summarizes the results of four experiments showing that the increase in virulence (a reduction of LD_{50} values) was essentially logarithmic as a function of holding time between reconstitution and testing of cultures. Apparently most of the cells were so damaged by storage in the dry state for 25 yr that they were unable to reproduce and cause disease when injected into mice, even though they were able to grow and form colonies on agar. If cells were suspended in the usual growth medium (heart infusion broth) instead of water with which they were reconstituted, virulence was always fully restored after 24 h incubation at 31 C (Heckly and Blank 1980). Repair in reconstituted culture appeared to be temperature dependent.

Recently, Yoshida and Ohtomo (1984) reported that the capsular type of *S. aureus* was altered as a result of freeze-drying and storage for 10 yr at room temperature. This is the first report of antigenic type changes as a result of freeze-drying. Yoshida and Ohtomo suggested that genetic mechanisms were probably involved in cellular injuries due to the treatment. Apparently, capsular type A strains developed capsular type B antigens and, conversely, type B strains developed reactivity to type A antiserum.

When lyophilized bacteria are exposed to oxygen, there is an accumulation of free radicals (Dimmick et al. 1961; Lion et al. 1961) and, furthermore, there appears to be a correlation between the log of the number of cells killed and the free radical concentration at the time of rehydration (Heckly and Dimmick 1968). Although free radicals are known to be very reactive entities, they have not been demonstrated to be the cause of death of bacteria. The work of Cox and Heckly (1973) indicates that free radicals were not involved in early death of washed and lyophilized bacteria when exposed to oxygen. However, this would not rule out free radicals as a factor because in no instance have free radicals been demonstrated without great loss of viability. The free radical in the bacterial system has not been identified, but some properties of the radical have been characterized

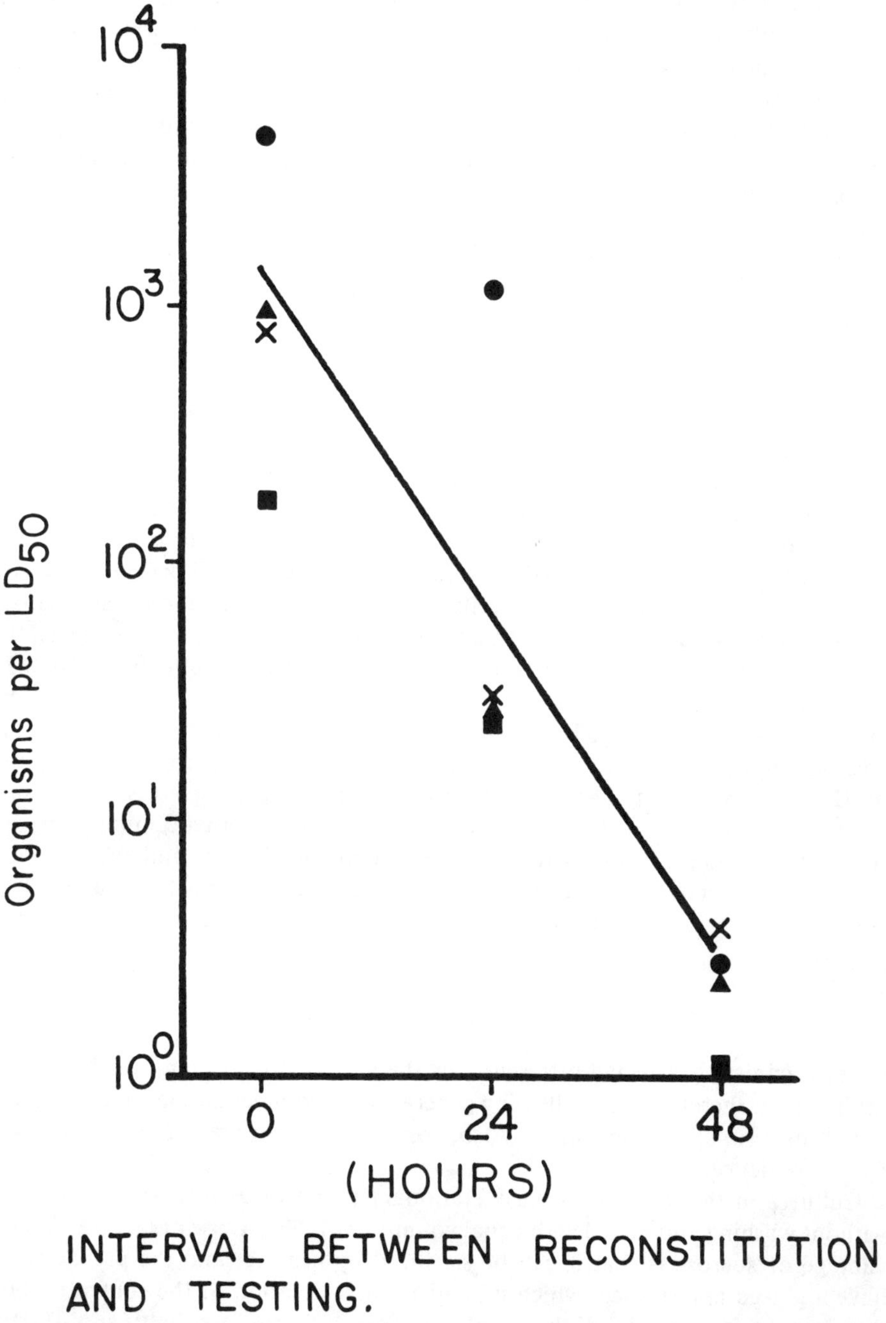

FIG. 6. Effect of interval between reconstitution and testing of freeze-dried *Yersinia pestis* stored for 25 yr at 4 C under vacuum in glass ampules. The different symbols represent results obtained from individual ampules (from Heckly and Quay 1981, with permission).

and compared with free radicals produced in other anhydrous systems. Lyophilized mixtures of casein and antioxidants, such as propyl gallate or ascorbate, have been shown to produce free radicals when exposed to oxygen (Heckly 1976). More recently, Heckly and Quay (1983) reported on free radical formation by exposure to air of freeze-dried mixtures of any one of a number of nitrogenous compounds, such as adenine, histidine, tryptophan, or phenyl alanine, and an antioxidant such as ascorbate or propyl gallate. Of these, histidine and propyl gallate mixtures (His-PG) produced the highest free radical concentration. In this simple system, histidine appears to act as a catalyst to produce free radicals by oxidation of the propyl gallate, which otherwise is stable in air and does not produce free radicals. As shown in Fig. 7, the free radical concentration was proportional to the log of the oxygen partial pressure. Although the bacterial system is much more complex than the His-PG, free radical production in both systems are similarly affected by oxygen, sugars, moisture, and temperature. Since the g-value of the free radical observed in bacterial systems is significantly different from that of the His-PG, or any of the other simple systems tested (Heckly and Quay 1983), it remains unidentified.

The results of adding propyl gallate and sucrose in various proportions on viability of *S. marcescens* are summarized in Fig. 8. The synergistic effect of sucrose and propyl gallate was observed over a broad concentration range. It is suggested that sucrose protected cells during the freezing and drying phase and that propyl gallate acted to protect cells against oxygen during the storage period. Furthermore, since sugars inhibit free radical production in these dry systems, the deleterious effect of free radicals at the instant of rehydration is minimized. In the absence of sucrose, propyl gallate had a protective effect, but even at the highest concentration, which is nearly saturated, fewer than 1% of the cells survived freeze-drying (Fig. 8). In other experiments with high sucrose concentration (6–12%), propyl gallate had no significant effect, because with sugar alone survival approached 100%. It is of interest that dextran or methyl cellulose reduced free radical concentration in the His-PG system, but lactose and sucrose were much more effective. Similarly, lactose and sucrose were much more effective than the higher molecular weight substances in preserving viability of dried bacteria.

Conclusions

Every organism is affected differently by the various stresses imposed by freeze-drying, and, therefore, the following generalized recommendations may not yield optimum results in all instances. Also, some procedures may compromise yield for convenience.

Cultures in the late log or early stationary phase of growth should be used without washing and mixed with equal volumes of 12% sucrose or lactose. An advantage of sucrose is that it can be sterilized by heating, and the preparations have a glazed appearance, which minimizes dissemination of the contents when the ampule is opened. With lactose the preparations are light and fluffy. However, both sugars yield comparably high viability.

The cell suspension should be cooled slowly, preferably by self-freezing, and dried at less than 0.1 torr with a condenser temperature below -50 C for at least

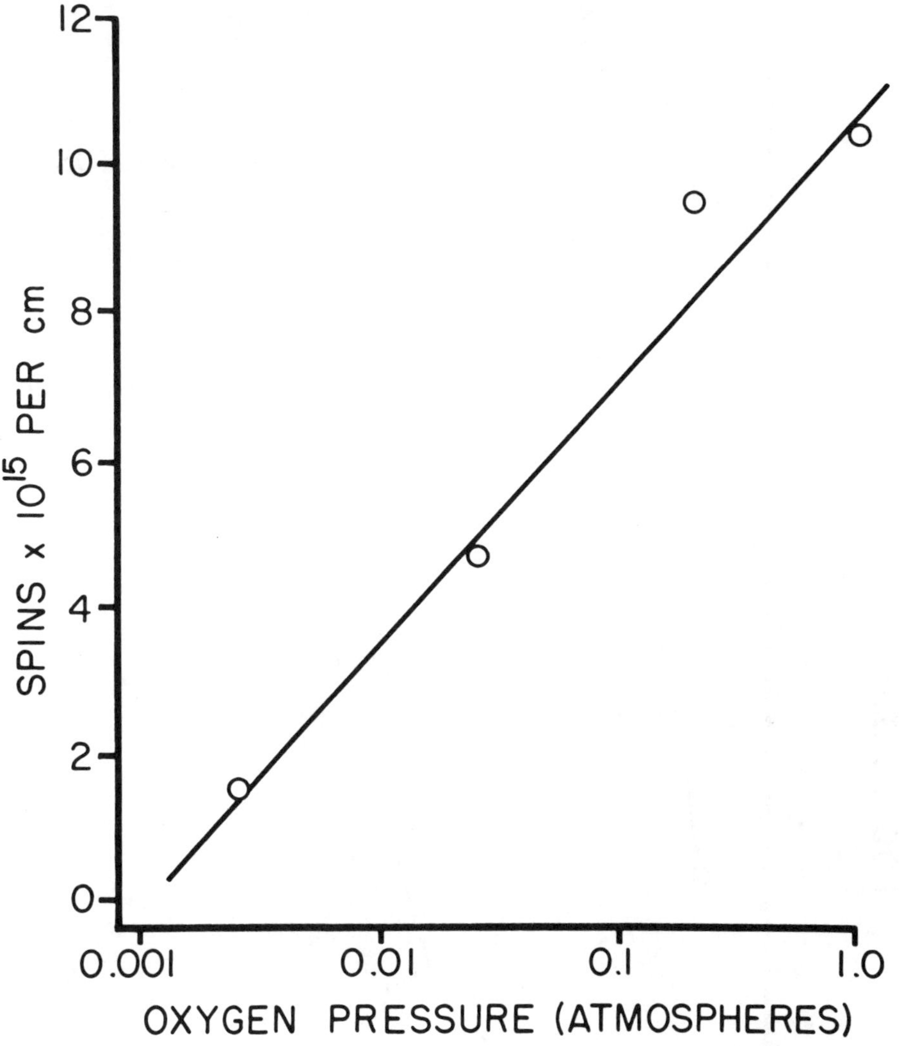

FIG. 7. Effect of oxygen pressure on free radical production by freeze-dried mixtures of propyl gallate and histidine. A solution containing 0.04 *M* of each was adjusted to pH 7 with NaOH before freezing and drying. The dried powder was transferred to EPR tubes, which were evacuated for at least 30 min prior to filling and sealed at the indicated partial pressures with either air or pure oxygen. Data obtained after 15 d at room temperature (from Heckly and Quay 1983, with permission).

twice as long as is required to eliminate frost on the ampules; usually 4–6 h is adequate for 0.2 to 1 ml, but an overnight schedule is most convenient since there is no evidence that over-drying is detrimental and this ensures complete dehydration. The ampules should be sealed under original vacuum and stored at or below 4 C. For maximum survival, the culture should be reconstituted with distilled water as soon as possible after air is admitted to the ampule to minimize oxygen damage to the dried cells. A complete, nonselective, growth medium should be

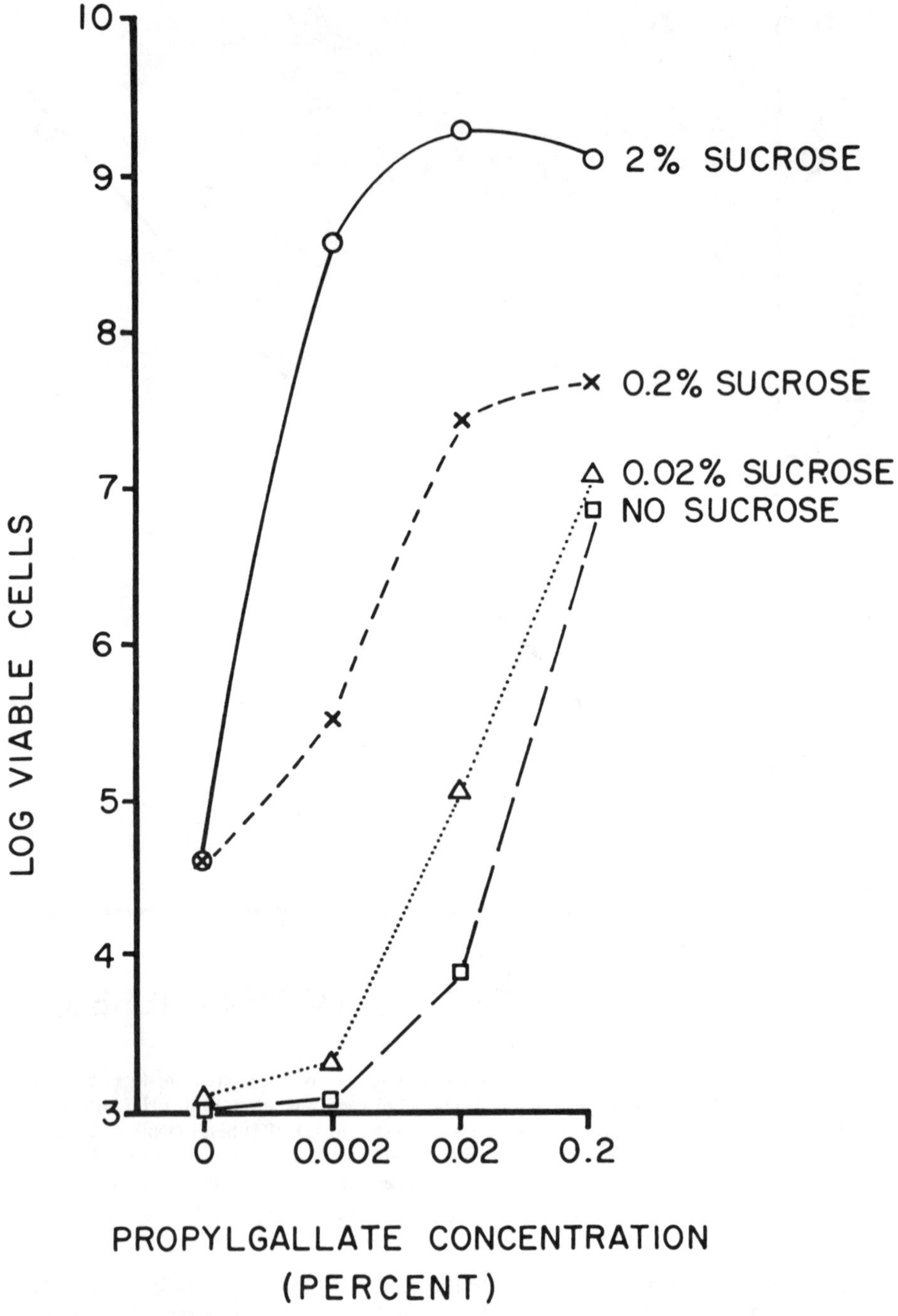

FIG. 8. Synergistic effect of propyl gallate and sucrose on viability of freeze-dried *Serratia marcescens*. Percentages indicated are the concentrations in the cell suspension before freezing. After freeze-drying, dry air was admitted and bottles were sealed with butyl rubber stoppers. Cultures were reconstituted 11 d after they were dried (from Heckly and Quay 1983, with permission).

used for the initial growth and plated to determine how well the preparations are surviving. Viability assays should be determined immediately after drying and at intervals during the storage period to ascertain whether the entire procedure is adequate for preserving the particular strain.

Acknowledgments

This work was supported by the Office of Naval Research, Microbiology Program, Naval Biology Project, under Contract N00014-81-C-0570, between the Office of Naval Research and the University of California.

Literature Cited

Ames, B. N., J. McCann, and E. Yamasaki. 1975. Methods for detecting carcinogens and mutations with the Salmonella/mammalian-microsome mutagenicity test. *Mutat. Res.* 31:347–364.

Benedict, R. G., J. Corman, E. S. Sharpe, C. E. Kemp, H. H. Hall, and R. W. Jackson. 1958. Preservation of microorganisms by freeze-drying I. Cell supernatant, Naylor-Smith solution, and salts of various acids as stabilizers for *Serratia marcescens. Appl. Microbiol.* 6:401–407.

Bretz, H. W., and F. E. Kocka. 1967. Resistance of actinomycin D of *Escherichia coli* after frozen storage. *Can. J. Microbiol.* 13:914–917.

Calcott, P. H., and R. A. McLoed. 1975a. The survival of *Escherichia coli* from freeze-thaw damage: Permeability barrier damage and viability. *Can. J. Microbiol.* 21:1724–1732.

______. 1975b. The survival of *Escherichia coli* from freeze-thaw damage: The relative importance of wall and membrane damage. *Can. J. Microbiol.* 21:1960–1968.

Christian, R. T., and J. J. Stockton. 1956. The influence of sealing pressure on survival of *Serratia marcescens* and *Micrococcus pyogenes* var. aureus desiccated from the frozen state. *Appl. Microbiol.* 4:88–90.

Cox, C. S., and R. J. Heckly. 1973. Effects of oxygen upon freeze-dried and freeze-thawed bacteria: Viability and free radical studies. *Can. J. Microbiol.* 19:189–194.

Damjanovic, V. 1974. Kinetics of thermal death and prediction of the stabilities of freeze-dried Streptomycin-dependent live Shigella vaccines. *J. Biol. Standardiz.* 2:297–311.

Damjanovic, V., and V. Radulovic. 1967. Survival of *Lactobacillus bifidus* after freeze-drying. *Cryobiology* 4:30–32.

Dewald, R. R. 1966. Kinetic studies of the destructive action of oxygen on lyophilized *Serratia marcescens. Appl. Microbiol.* 14:568–572.

Dewald, R. R., K. W. Browall, L. M. Schaefer, and A. Messer. 1967. The effect of water vapor on lyophilized *Serratia marcescens* and *Escherichia coli. Appl. Microbiol.* 15:1299–1302.

Dimmick, R. L., R. J. Heckly, and D. P. Hollis. 1961. Free-radical formation during storage of freeze-dried *Serratia marcescens. Nature* 192:776–777.

Flosdorf, E. W., and S. Mudd. 1935. Procedure and apparatus for preservation in "lyophile" form of serum and other biological substances. *J. Immunol.* 29:389–425.

______. 1938. An improved procedure and apparatus for preservation of sera, microorganisms and other substances—the cryochem-process. *J. Immunol.* 34:469–490.

Fry, R. M., and R. I. N. Greaves. 1951. The survival of bacteria during and after drying. *J. Hyg.* 49:220–246.

Gomez, R., M. Takano, and A. J. Sinskey. 1973. Characteristics of freeze-dried cells. *Cryobiology* 10:368–374.

Hammer, B. W. 1911. A note on the vacuum desiccation of bacteria. *J. Med. Res.* 24:527–530.

Heckly, R. J. 1961. Preservation of bacteria by lyophilization. *Adv. Appl. Microbiol.* 3:1–76.

______. 1976. Free radicals in dry biological systems. Pages 135–158 *in* W. S. Pryor, ed., *Free Radicals in Biology,* Vol. 2, Academic Press, New York.

______. 1978. Preservation of microorganisms. *Adv. Appl. Microbiol.* 24:1–53.

Heckly, R. J., and H. Blank. 1980. Virulence and viability of *Yersinia pestis* 25 years after lyophilization. *Appl. Environ. Microbiol.* 39:541–543.

Heckly, R. J., and J. Dimatteo. 1975. Rhythmic changes in dry heat resistance of *Bacillus subtilis* spores after rapid changes in pH. *Appl. Microbiol.* 29:565–566.

Heckly, R. J., and R. L. Dimmick. 1968. Correlations between free radical production and viability of lyophilized bacteria. *Appl. Microbiol.* 16:1081–1085.

Heckly, R. J., and J. Quay. 1981. A brief review of lyophilization damage and repair in bacterial preparations. *Cryobiology* 18:592–597.

______. 1983. Adventitious chemistry at reduced water activities: Free radicals and polyhydroxy agents. *Cryobiology* 20:613–624.

Heckly, R. J., and D. Skilling. 1979. Device to facilitate freeze-drying specimens for scanning electron microscopy. *Cryobiology* 16:196–199.

Heckly, R. J., R. L. Dimmick, and N. Guard. 1967. Studies on survival of bacteria: rhythmic response of microorganisms to freeze-drying additives. *Appl. Microbiol.* 15:1235–1239.

Israeli, E., E. Giberman, and A. Kohn. 1974. Membrane malfunctions in freeze-dried *Escherichia coli. Cryobiology* 11:473–477.

Kempler, G., and B. Ray. 1978. Nature of freezing damage on the lipopolysaccharide molecule of *Escherichia coli* B. *Cryobiology* 15:578–584.

Leach, R. H., and W. J. Scott. 1959. The influence of rehydration on the viability of dried micro-organisms. *J. Gen. Microbiol.* 21:295–307.

Lingg, A. J., and K. J. McMahon. 1969. Survival of lyophilized *Bacillus popilliae* in soil. *Appl. Microbiol.* 17:718–720.

Lingg, A. J., K. J. McMahon, and C. Herzmann. 1967. Viability of *Bacillus popilliae* after lyophilization of liquid nitrogen frozen cells. *Appl. Microbiol.* 15:163–165.

Lion, M. B., and Y. Avi-Dor. 1963. Oxygen-induced inactivation of NADH-oxidase in lyophilized cells of *Escherichia coli. Isr. J. Chem.* 4:374–378.

Mazur, P. 1966. Theoretical and experimental effects of cooling and warming velocity on the survival of frozen and thawed cells. *Cryobiology* 2:181–192.

______. 1970. Cryobiology: The freezing of biological systems. *Science* 168:939–949.

______. 1977. The role of intracellular freezing in the death of cells cooled at supraoptimal rates. *Cryobiology* 14:251–272.

Naylor, H. B., and P. A. Smith. 1946. Factors affecting the viability of *Serratia marcescens* during dehydration and storage. *J. Bacteriol.* 52:565–573.

Ohnishi, T., Y. Tanaka, M. You, Y. Takeda, and T. Miwatani. 1977. Deoxyribonucleic acid strand breaks during freeze-drying and their repair in *Escherichia coli*. *J. Bacteriol.* 130:1393–1396.

Otten, L. 1930. Die trockenkonservierung von pathogenen bakterien. *Zentralbl. Bakteriol. Parasitenkd.* Abt. I Orig. 116:199–210.

Proom, H., and L. M. Hemmons. 1949. The drying and preservation of bacterial cultures. *J. Gen. Microbiol.* 3:7–18.

Ray, B., D. W. Jansen, and F. F. Busta. 1972. Characterization of the repair of injury induced by freezing *Salmonella anatum*. *Appl. Microbiol.* 23:803–809.

Rogers, L. A. 1914. The preparation of dried cultures. *J. Infect. Dis.* 14:100–123.

Samosudova, E. V. 1965. Lyophilization of phytopathogenic bacteria. *Mikrobiologiya* 34:551–555.

Sinskey, T. J., and G. J. Silverman. 1970. Characterization of injury incurred by *Escherichia coli* upon freeze-drying. *J. Bacteriol.* 101:429–437.

Steel, K. J., and H. E. Ross. 1963. Survival of freeze-dried bacterial cultures. *J. Appl. Bacteriol.* 26:370–375.

Tanaka, Y., M. Yoh, Y. Takeda, and T. Miwatani. 1979. Induction of Mutation in *Escherichia coli* by freeze-drying. *Appl. Environ. Microbiol.* 37:369–372.

Wasserman, A. E., and W. J. Hopkins. 1957. Studies in the recovery of viable cells of freeze-dried *Serratia marcescens*. *Appl. Microbiol.* 5:295–300.

Webb, S. J. 1969. Some effects of dehydration on the genetics of microorganisms. Pages 153–167 *in* T. Nei, ed., *Freezing and Drying of Microorganisms*, University Park Press, Baltimore, MD.

Yoshida, K., and T. Ohtomo. 1984. Alteration of capsular type of encapsulated strains of *Staphylococcus aureus* during freeze-drying and storage. *Experientia* 40:93–94.

——. 1970. [illegible]: the freezing of biological systems. Science 168: 939–949.

——. 1977. The role of intracellular freezing in the death of cells cooled at supraoptimal rates. Cryobiology 14: 251–[illegible].

[illegible], J. F., and [illegible] A. Sommer. 1980. Factors affecting the viability of [illegible] during [illegible] storage. [illegible] 3: [illegible].

[illegible], T. [illegible], [illegible] Tanaka, [illegible], [illegible] Lindsay, and [illegible]. 19[illegible]. [illegible] between [illegible] and [illegible] and their [illegible]. [illegible] 50: [illegible]–1396.

[illegible], K. 1976. [illegible] nitrogen [illegible]. [illegible] Journal [illegible] 1(3): [illegible].

[illegible]

CHAPTER 26

Problems in Freeze-Drying: I. Stability in Glass-Sealed and Rubber-Stoppered Vials

J. M. BARBAREE*, A. SANCHEZ**, AND G. N. SANDEN***

*Centers for Disease Control, Center for Infectious Diseases, *Division of Bacterial Diseases, **Division of Viral Diseases, and ***Biological Products Program, Atlanta, Georgia 30333*

The use of flame-sealed vials with internal rubber stoppers (FSRS) was compared with conventional rubber-stoppered vials for lyophilized bacteria. *Legionella bozemanii* and *L. gormanii* suspensions were dispensed into both types of vials, lyophilized, sealed under vacuum, and subjected to thermal degradation studies. We made first order least squares regression analysis plots, estimated time required to degrade to 100 colony-forming units/ml at three temperatures, and found that the FSRS vials showed no apparent advantage over the conventional rubber-stoppered tubing vial for stability and efficacy. This information should be considered in the acceptance or rejection of rubber-stoppered vials for lyophilized biologic reference or standard materials.

INTRODUCTION

Glass-sealed ampules of freeze-dried products have been accepted as having a more permanent seal than vials with rubber stoppers. World Health Organization (WHO) (1978) guidelines for producing freeze-dried international biologic standards recommend the use of glass-sealed vials over rubber-stoppered ones. The physical and chemical formulations of rubber stoppers yield porosity and permeability characteristics that may lead to shortening the life of freeze-dried products (Amoignon 1975). On the other hand, Greiff et al. (1975) reported the presence of a microscopic pore in the seal area of each glass-sealed vial that he examined. Conversely, McKenzie et al. (1977) disagreed on draw-sealed vials. When compared to flame-sealed vials, the rubber-stoppered vials are more convenient because rehydration is easier and the vials are less fragile and easier to handle during and after processing.

Spaun and Lyng (1970) described a vial that allowed stoppering with a rubber stopper below a flame seal. In contrast to other flame-sealed vials, this vial reduced exposure of the sample to heat from the flame-sealing process, eliminated contamination due to glass splinters falling into the vial while opening it, and prevented exposure of the sample to the environment when the vial was opened. Also, these vials had a thick wall in the flame-seal section (1.4 mm), enabling storage in liquid nitrogen vapors, which in some cases can result in the loss of the seal (Barbaree and Smith 1981).

At the Centers for Disease Control, we investigated the possible use of this container system for WHO reference and standard materials. Historically, our lyophilized reference preparations have been vacuum sealed in rubber-stoppered vials. For comparison, suspensions of *Legionella bozemanii* and *L. gormanii* were used in the containers. This report gives our results and conclusions.

MATERIALS AND METHODS

Suspending media. Newborn calf serum containing 5% (w/v) concentrations of either myo-inositol or D-trehalose were used as menstrua for the suspension and lyophilization of *Legionella bozemanii* and *L. gormanii*, respectively. Suspending media were prepared by mixing nine parts of serum with one part of a 50% (w/v) solution of the carbohydrate in glass-distilled water. The inositol solution (SI) was warmed to make the solution. The SI or serum-trehalose (ST) solutions were prepared and sterilized by filtration the day before usage.

Vials and closures. All vials used were Type 1, Class A, borosilicate glass. Fig. 1 shows three stages of sealing for the custom-made FSRS vials, i.e., sealing with the rubber stopper via use of the plunger, constriction of the tubing before the flame seal, and after the flame and rubber stopper seals have been made. The 3-ml (practical fill capacity) tubing vial used for comparison with the FSRS vial closely resembled the dimensions of the bottom portion of the FSRS vial. The

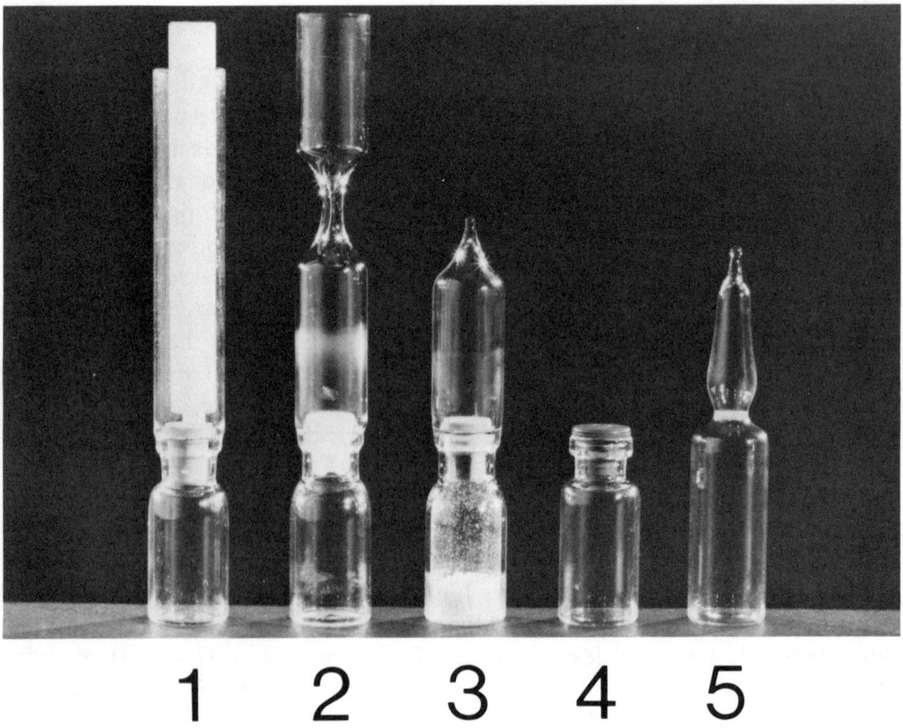

FIG. 1. Vials and stages of stoppering. 1 = FSRS vial with plunger on top and stopper inserted; 2 = FSRS vial with constriction; 3 = Sealed FSRS vial containing lyophilized product; 4 = Conventional RS vial; 5 = Flame-sealed ampule.

rubber stopper in the FSRS vial was an IGLU butyl rubber stopper, and a 13-mm (diameter) split, gray, siliconized, butyl rubber closure was used for the 3-ml tubing vial.

Organisms and culture conditions. Legionella bozemanii strain WIGA and *Legionella gormanii* strain LS-13 were obtained from stock cultures maintained at the Centers for Disease Control, Atlanta, GA. Growth of *L. bozemanii* was initiated from frozen egg-yolk-sac cultures inoculated with infected guinea pig tissues. Primary growth from yolk-sac cultures was accomplished on buffered charcoal-yeast-extract (BCYE) agar for 1 wk at 35 C and in a moist atmosphere with 2.5% (w/v) CO_2. Growth from BCYE plates was suspended in aqueous 5% D-trehalose solutions (w/v) dispensed in 1.0-ml volumes into screw-capped tubes and frozen in a −70 C freezer for later use in lyophilization experiments. Subsequent culturing of organisms was performed solely on charcoal-yeast-extract (CYE) agar at 35 C in a moist atmosphere.

Harvesting and dispensing. Aseptic techniques were used throughout the study. Kolle flasks containing approximately 50 ml of CYE agar were used to culture *Legionella* organisms. Cultures were prepared by inoculating flasks with actively growing cells suspended in distilled water, and incubating for 4 d. Surface growth was removed by adding approximately 10 ml of suspending medium to each flask and spinning sterile paper clips over the surfaces with a magnetic stirrer. The suspensions were then transferred to a flask containing additional menstruum and a stirring bar, and stirred during dispensing.

An LKB model 2075 dispenser with an elongated tip was used to dispense 1-ml aliquots into the FSRS vials. We used an adjusted 2-ml Cornwall dispenser to dispense 1-ml aliquots into the 3-ml tubing vials. In all cases, the vials were placed in metal lyophilization trays with removable liners used to bring the vials into immediate direct contact with the precooled (−50 C) lyophilizer shelves. After dispensing and partially stoppering, the plungers were placed on top of the stoppers in the FSRS vials just before loading the freeze-dryer.

Lyophilization. Lyophilization was performed in a Hull model 652M3FX20 lyophilizer with two shelves (3 ft^2/shelf). One tray of vials per shelf (2 shelves) was frozen at −50 C in the lyophilizer chamber for $\geq$3 h, after which the shelf temperature was changed to −40 C, and primary drying was initiated. Drying continued at −40 C for 48 h, then the shelf was changed to 25 C, and the vials were dried an additional 24 h. This cycle for freeze-drying bacteria was reported earlier by Barbaree (1977). The vials were then sealed by insertion of the rubber stoppers under vacuum, removed from the shelves, machine capped or flame sealed, and checked for vacuum with a high frequency generator (sparker) before being stored in a −70 C freezer until usage. The vials were rechecked for vacuum before usage. Less than 1% of them lost vacuum while in storage.

Flame-sealing. Flame-sealing of the stoppered and lyophilized vials required the initial formation of a thick-walled constriction in the upper portion of the glass-sealed vials. This was accomplished by placing vials on a motor driven roller assembly in which a gas-oxygen torch could be inserted in and withdrawn from under a rotating vial. It was necessary to have the glass thick at the constriction (1.0–1.5 mm) in order for the glass to withstand the vacuum during the final sealing process. This was done by heating the flame-seal area with the torch until the glass softened, slightly stretching the glass from the ends (approximately 1 cm), then allowing this area to fall back in on itself as it rotated, resulting in a thickening of the walls around the area of the constriction. By doing this once or several times, depending on the skill of the operator, a final constriction was made by removing the flame and stretching the glass into the hour-glass configuration required for final flame sealing under vacuum.

Vials with constrictions were fitted onto a vertical manifold of a Virtis freezemobile, whose condenser was cooled to a temperature of −60 C. A vacuum of 10–20 μm was pulled on each vial, followed by flame sealing with a hand-held gas-oxygen torch with dual tips.

Accelerated thermal degradation studies. Lyophilized vials of *L. bozemanii* and *L. gormanii* were placed in 25, 35, 45, and 55 C incubators. At designated times, duplicate vials were retrieved from the incubators and placed at −70 C until viability assays could be performed.

Viability was compared using regression lines of survival curves as reported by Barbaree et al. (1982) and Damjanovic (1974). Logarithms of the colony-forming units (CFU's) were plotted against time of exposure at a specific temperature. A least squares regression analysis for the responses at each temperature was made, and estimates of times for degradation to 100 CFU's at specific temperatures were calculated.

Viability assays. We performed colony-forming unit (CFU) determinations on lyophilized cultures in the following manner: (1) The stoppers were removed from the vials; (2) the cultures were quickly rehydrated with 1.0 ml glass-distilled water and agitated for a few seconds; (3) after about a minute, the cell suspensions were agitated again; (4) 10- and 100-fold dilutions in physiologic saline were made; and (5) 0.1 ml of each dilution was plated on CYE agar and incubated at least 1 wk at 35 C or until a countable population (30–300 CFU) developed.

We gained entry into the flame-sealed vial by scoring the vial just below the tapering of the flame seal with a file or a diamond-tipped pen, heating only the scored area with a Bunsen Burner until an orange flame flickered from the glass surface, then quickly dipping the score in and out of a beaker of cool water. This caused a cracking of the glass, which radiated from the score. The top of the flame-sealed top could then be removed by tapping the glass with a metal rod, and a syringe needle could be inserted into the rubber stopper for reconstituting the dried cells.

Residual moisture analysis. Percent residual moisture (total weight) was determined using the Karl-Fischer method, by titrating the cell suspensions in vials (12 vials per lot) using a dual Methrohm Dosimat E3535 modified to inject standardized methanol and Karl-Fischer reagents into the vials. Visual endpoints were used in the standardization of the reagents and in all determinations on the lyophilized cultures; average values for percent residual moisture were recorded.

Vacuum tests. An Electro Technic BD10-A high-frequency generator (sparker) was used to detect the presence or absence of vacuum after the final seals were made. The FSRS vials were checked for vacuum above and below the internal stopper. Vials without vacuum in one or both areas were rejected.

Results

As shown in Table 1, the residual moistures in the lyophilized FSRS and RS vials with *L. bozemanii* were 1.22% and 1.28%, respectively, and 0.71% and 0.83% in those with *L. gormanii*, respectively.

No effect on viability rates after restoring vials at −70 C was noticed.

Figs. 2 and 3 (first order plots) present the thermal degradation data graphi-

TABLE 1. *Residual moisture analyses of lyophilized bacterial suspensions in flame-sealed and rubber-stoppered vials*

Bacteria	Menstrua	Residual Moisture (%)			
		FSRS Vials[a]		RS Vials	
		Average	Range	Average	Range
Legionella gormanii	Serum-5% Trehalose	[b]0.71	0.57–0.82	[b]0.83	0.74–0.97
L. bozemanii	Serum-5% Inositol	[c]1.22	0.86–1.69	[b]1.28	1.11–1.38

[a] FSRS = Flame-sealed and also rubber-stoppered inside the vial; RS = Rubber-stoppered.
[b] Results with 12 vials were used to determine the average.
[c] Results with 11 vials were used to determine the average.

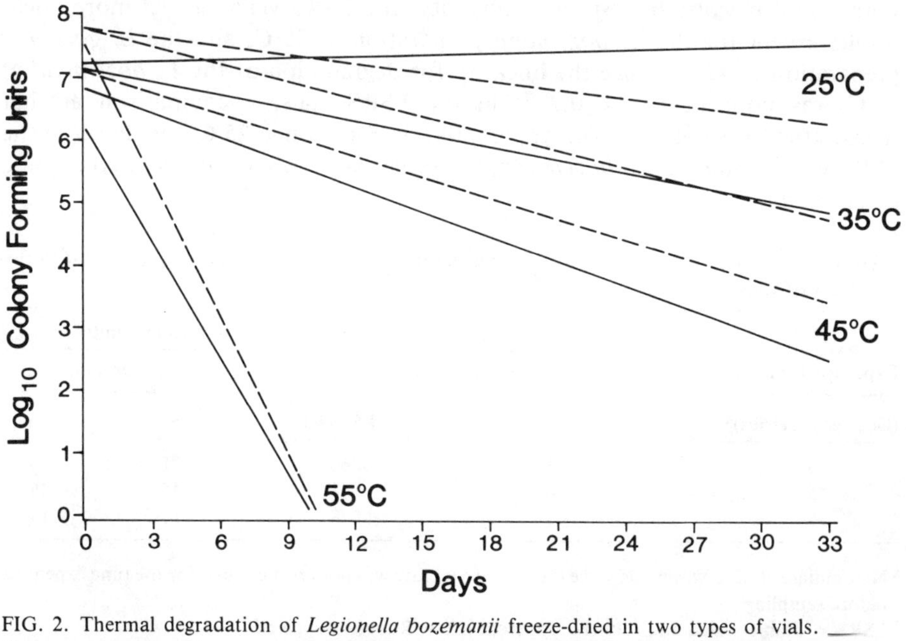

FIG. 2. Thermal degradation of *Legionella bozemanii* freeze-dried in two types of vials. ___ Flame-sealed and rubber-stoppered vials; __ __ Rubber-stoppered vials.

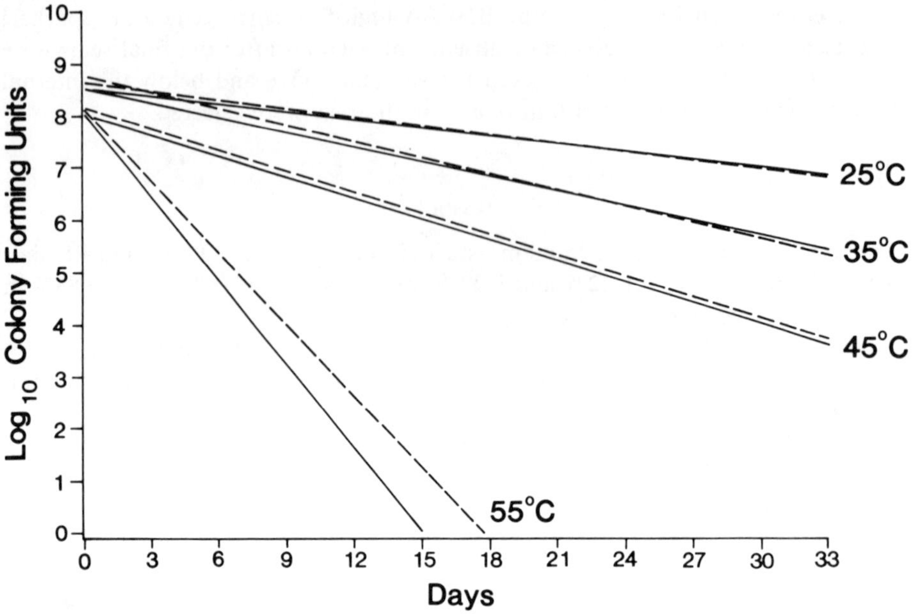

FIG. 3. Thermal degradation of *Legionella gormanii* freeze-dried in two types of vials. ____ Flame-sealed and rubber-stoppered vials; __ __ Rubber-stoppered vials.

cally as least squares regression curves. *L. bozemanii* showed slightly higher loss of viability rates at 25 C, 35 C, and 45 C in the RS vials. *L. gormanii* had slightly higher rates at 25 C and 35 C in the RS vials, but higher at 45 C and 55 C in the FSRS containers. No significant differences were apparent between results using both types of vials. In respect to linearity, the FSRS vials yielded more linear results except for the *L. bozemanii* preparation at 25 C, and the *L. gormanii* preparation at 45 C. Since the linearity for degradation of the *L. bozemanii* at 25 C was poor (C.D. = 0.272) in the FSRS vials, a comparison at that temperature was not possible. Apparently, degradation at 25 C was slower in the FSRS vials containing *L. bozemanii*, but essentially the same in those containing *L. gormanii*.

TABLE 2. *Estimates of degradation of* Legionella bozemanii *and* L. gormanii *freeze-dried in two types of vials*

Exposure Temperature	Days to degrade to 100 colony-forming units			
	L. bozemanii		*L. gormanii*	
(Degrees—Celsius)[a]	FSRS vials[b]	RS vials	FSRS vials	RS vials
35	61	66	71	66
45	36	48	45	46
55	7	8	11	14

[a] No estimate at 25 C was made since the loss of viability was not great enough for the time expended before sampling.

[b] FSRS = Flame-sealed and rubber-stoppered; RS = Rubber-stoppered.

Table 2 shows estimates of degradation to 100 CFU using the slope values from the least squares regression curve for each organism, container, and temperature. Stability of *L. gormanii* was better in the FSRS vials at 35 C compared with results at 45 C and 55 C. Results with *L. bozemanii* showed that the stability period was slightly longer at the three elevated temperatures.

When the overall processing time and loss of vials during processing were considered, RS vials gave the best results. Approximately 12% of the FSRS vials were rejected from the first lyophilization run with suspensions of *L. bozemanii* because of the absence of vacuum or an ineffective seal caused by faulty manipulation of the vials while making the constriction or flame seal and by not positioning the stopper and plunger appropriately before lyophilization. Fewer FSRS vials (about 10%) were rejected on the second run in which suspensions of *L. gormanii* were freeze-dried in both types of vials. Mistakes were minimized because of the experience gained from previous errors made in sealing the FSRS vials. At least 2 to 4 min per vial were needed to make the constrictions and final flame seals; therefore, much more time was consumed processing the FSRS vials compared with the conventional rubber-stoppered containers. Less than 1% of the RS vials were not acceptable owing to defective vials.

Discussion and Conclusions

Apparently, the use of different types of vials did not cause a difference in residual moisture in the final product, but the suspending medium did.

The question of which type of seal on a container is best for lyophilized materials is controversial, and the answer must be based on the features that are most important to the scientist or organization preserving the material. Features for three kinds of containers, the FSRS vial, the flame-sealed ampule, and the conventional RS vial are listed in Table 3. Use of either of the two flame-sealing methods is more expensive, mainly because of the time, personnel, and care required for processing. Rehydration by injection through a rubber stopper is more exacting and convenient. Ampules are more fragile. Dispensing into the conventional vials is much easier and may be important in maintaining reproducibility. When stability is considered, it must be noted that no great differences between the types of vials were found in this study. If storage for $\geq$15–20 yr is probable, deterioration of the elastomeric material in the rubber stopper and the tightness of the stopper-vial fit could become a factor, although we have not presented data to support a difference.

In conclusion, we did not show an important difference in the stability of cultures lyophilized in the FSRS and RS vials. The double security of having two seals, glass and rubber, is more reassuring to some people; however, the convenience, saving, and easy manipulation with a conventional rubber-stoppered vial that appears to have equal stability are factors that cannot be ignored. If the selection of the container system is based on all of these factors, including accelerated degradation studies, the rubber-stoppered vial can be at least equal for international biologic standards.

Table 3. *Comparison of the advantages and disadvantages of types of containers for lyophilization*

Feature	FSRS[a]	FS[b]	RS[c]
Expense	Most expensive in cost for container and processing time	Expensive because of time and equipment needed for sealing	Least expensive in time and direct cost
Rehydration	No loss of product during rehydration	Possible loss of product when seal is broken	No loss of product during rehydration
Sturdiness	Sturdy	Fragile	Sturdy
Sealing	More difficult and time consuming	Time consuming	Easy and not time consuming
Ease in Dispensing	Long tubes require long tips on dispenser and more time for dispensing	Same as for flame-sealed and rubber-stoppered vial	Easy to use
Stability	Should be more stable, but accelerated temperature studies do not show it	Usually accepted as being more stable	Comparable stability except possible long-term deterioration of stopper

[a] Flame-sealed and rubber-stoppered vial.
[b] Flame-sealed ampule.
[c] Rubber-stoppered vial.

Literature Cited

Amoignon, J. 1975. Physico-chemical contamination of freeze-dried pharmaceutical and biological products. Pages 445–460 *in* S. A. Goldblith, L. Rey, and W. W. Rothmayr, eds., *Freeze Drying and Advanced Food Technology.* Academic Press, New York.

Barbaree, J. M. 1977. Methods for determining favorable conditions for freeze-drying biological reference materials. *Dev. Biol. Stand.* 36:335–341.

Barbaree, J. M., and S. J. Smith. 1981. Loss of vacuum in rubber stoppered vials stored in a liquid nitrogen vapor phase freezer. *Cryobiology* 18:528–531.

Barbaree, J. M., F. Thompson, and S. J. Smith. 1982. Use of thermal stability studies to compare *Bacteroides fragilis* lyophilized in skim milk and Polyvinylpyrrolidone solutions. *Cryobiology* 19:92–98.

Damjanovic, V. 1974. Kinetics of thermal death and prediction of the stabilities of freeze-dried streptomycin dependent live *Shigella* vaccines. *J. Biol. Stand.* 2:297–311.

Greiff, D., H. Melton, and T. Rowe. 1975. On the sealing of gas filled ampoules. *Cryobiology* 12:1–14.

McKenzie, A. P., D. G. Welkie, M. G. Lagally, M. Pace, and F. I. Elliott. 1977. On the adequacy of the draw-sealing of gas-filled ampoules. *Dev. Biol. Stand.* 36:151–160.

Spaun, J., and J. Lyng. 1970. Ampoule for freeze-drying of pharmaceuticals with stopper inserted below flame sealing. *Dan. Tidsskr. Farm.* 44:292–300.

WHO Expert Committee on Biological Standardization. 1978. Guidelines for the preparation and establishment of reference materials and reference reagents for biological substances. *World Health Org. Tech. Rep. Ser.,* No. 626.

CHAPTER 27

Problems in Freeze-drying: II. Cross-Contamination During Lyophilization

J. M. BARBAREE*, A. SANCHEZ**, AND G. N. SANDEN***

Centers for Disease Control, Center for Infectious Diseases,
**Division of Bacterial Diseases, **Division of Viral Diseases, and*
****Biological Products Program, Atlanta, Georgia 30333*

Vial-to-vial contamination (cross-contamination) during freeze-drying was shown in three different freeze-dryers. Vials containing pure culture suspensions of *Proteus mirabillis* and *Pseudomonas aeruginosa* and having unique antibiotic susceptibilities to neomycin and colistin, respectively, were clustered in the chambers of the freeze-dryers on the sides of a shelf, in the middle of a shelf, and on two different shelves. Intrashelf cross-contamination (12%) occurred in a freeze-dryer with the external condenser port below the shelf; no intershelf cross-contamination occurred in a lyophilizer with an external condenser port in the rear chamber wall. Intrashelf (22%) and intershelf vial-to-vial contamination (80%) occurred in a lyophilizer with an internal condenser. These studies show that vial-to-vial contamination can occur under the right conditions during lyophilization, and products such as bacterial reference cultures should not be mixed in the same run.

INTRODUCTION

Diagnostic, research, and small production laboratories that lyophilize reagents and reference cultures often face the dilemma of using a chambered freeze-dryer to preserve small batches of products without wasting the space on the lyophilizer shelf or shelves. Efficient use of space often dictates the lyophilization of more than one product per run. If cross-contamination occurs in products such as key reference strains of bacteria, the results can be detrimental to the attainment of valid test results or good quality products.

The transfer of microorganisms such as *Bacillus* sp., *Salmonella* sp., and certain viruses by sublimation to a freeze-dryer condenser when these organisms were freeze-dried was reported by Stein and Rogers (1950). This report established the migration of microorganisms from vials to the condenser and showed the possibility for cross-contamination.

We undertook a study to demonstrate vial-to-vial contamination (cross-contamination) during lyophilization in three of our small production lyophilizers. We previously reported some of our results showing cross-contamination during lyophilization (Barbaree and Sanchez 1982). Subsequently, we have followed up that report and compiled more data. Our results are given in this paper.

Methods and Materials

Tests for cross-contamination. We used the procedure for checking cross-contamination reported by Barbaree and Sanchez (1982). In this method, two separate bacterial cultures, *Pseudomonas aeruginosa* (ATCC culture 27853) and *Proteus mirabilis* (CDC culture from Vitek study 3025–1) were suspended in a serum-inositol medium (SI) and used as test organisms to indicate vial-to-vial contamination. These bacteria are susceptible to colistin and neomycin, respectively. The bacterial suspensions were adjusted to $>10^9$ bacteria/ml, dispensed carefully into 5-ml (practical fill) wide-mouth (20-mm) tubing vials held in trays with removable bottoms, partially stoppered with butyl rubber split stoppers, arranged in various patterns on the freeze-dryer shelves, frozen on the shelf, lyophilized, and sealed under vacuum. Then the lyophilized vials were rehydrated with sterile distilled water, plated onto Mueller-Hinton agar (M-H) in quadrant plates containing two sections each with 8 μg colistin and 8 μg neomycin, respectively. Inoculated plates were incubated for 2–4 d at 35 C and observed for growth on the antimicrobial-containing medium. Cross-contamination was confirmed by bacteriologic identification.

Lyophilizers. We used three types of freeze-dryers (A,B,C) in the experiment. Freeze-dryer A was a Virtis freeze-mobile 10–145–MRBA[1] with a No. 10–MRSA chamber. Lyophilizer B was a modified Hull research model 651–6FX20 freeze-dryer. Freeze-dryer C was almost identical to B except it contained two internal condenser plates, each vertically parallel to the right and left sides of the chamber.

Results and Discussion

Table 1 gives the results for 11 lyophilization runs. Only $\leq$3.3% cross-contamination occurred in freeze-dryer A when the vials containing a different test organism were clustered on each side of the shelf (runs 1–3). However, when vials containing *Pseudomonas aeruginosa* were surrounded by vials with *Proteus mirabilis,* 8.3% to 13.3% cross-contamination occurred (runs 4–7) in the same lyophilizer. In freeze-dryer B, only 1.9% intrashelf cross-contamination occurred (run 9). No cross-contamination occurred shelf-to-shelf in this lyophilizer. High rates of cross-contamination occurred in freeze-dryer C, which contained the same size of chamber as B, and had two internal condenser plates. Intershelf contamination was 80%, and intrashelf was 22.2%.

We found two main factors for the likelihood of vial-to-vial contamination, the port for the external condenser and the presence of internal condenser plates. Cross-contamination in freeze-dryer A indicated that the deposition of bacteria occurred in an umbrella-like pattern as the water migrated to the condenser port below the shelf. Freeze-dryer B yielded minimal cross-contamination with the external condenser port in the center of the rear chamber wall. Results with freeze-dryer C overwhelmingly indicated that the bacteria were dispersed easily among

[1] Use of trade names is for identification only and does not imply endorsement by the Public Health Service or by the U.S. Department of Health and Human Services.

TABLE 1. *Results of cross-contamination during 11 lyophilization runs*

Run No.[a]	Vial array[b]	Freeze-dryer	Condenser location	Cross-contamination: Vials contaminated	%
1[c,d]	Left side of tray with *Proteus mirabilis* and right side with *Pseudomonas aeruginosa*	A	External, port centered below shelf	1/106	0.9
2				4/120	3.3
3				1/120	0.8
4	*Pseudomonas aeruginosa* surrounded by *Proteus mirabilis*			13/120	10.8
5				10/120	8.3
6				16/120	13.3
7				11/120	11.7
8	*Proteus mirabilis* on top shelf and *Pseudomonas aeruginosa* on bottom shelf	B	External, port in center of rear chamber wall	0/506	0
9	As in runs 4–7			5/270	1.9
10	As in run 8	C	Internal, port in center of rear chamber wall	221/276	80.0
11	As in runs 4–7			60/270	22.2

[a] Runs 1–8 were reported by Barbaree et al. (1982).

[b] The postlyophilization colony forming units for either type of bacteria ranged from 1×10^9/ml to 9.4×10^9/ml.

[c] The SI suspending medium was used in all runs, but was diluted 1/10 for run number 3.

[d] The cycle for all runs except No. 1 was freezing on shelf at ≤ -40 C for ≥ 2 h, setting the shelf temperature at +25 C when lyophilization was started, and sealing under 13 to 20 μm vacuum after 48 h of lyophilization. In run No. 1, the shelf temperature remained at −40 C for 46 h of lyophilization and changed to 25 C for the final 24 h before sealing under vacuum.

other vials due to the sublimation pattern in the presence of the condenser plates within the chamber. Other factors such as the lyophilization cycle and suspending media may contribute to cross-contamination, but the evidence suggests that the physical characteristics of the freeze-dryer may dictate the mechanism for vial-to-vial contamination.

Perhaps the lyophilization of some reagents, such as types of antisera, may be done in freeze-dryer B on different shelves without much or any contamination. Since subtle changes in lyophilization conditions may occur run-to-run, the mixture of two types of one reagent in the same run is not recommended regardless of the data from run 8 (Table 1).

Literature Cited

Barbaree, J. M., and A. Sanchez. 1982. Cross-contamination during lyophilization. *Cryobiology* 19:443–447.

Stein, C. D., and H. Rogers. 1950. Recovery of viable microorganisms and viruses from vapors removed from frozen suspensions of biologic material during lyophilization. *Am. J. Vet. Res.* 11:339–344.

CHAPTER **28**

Cryopreservation of Lymphocytes and Lymphoid Clones

DOUGLAS M. STRONG,* FRANCES V. LESANE, AND CAROLYN Y. NEULAND

Uniformed Services University of the Health Sciences, Bethesda, Maryland 20814-4799

The advent of techniques for the clonal expansion of antibody-producing hybridomas and other lymphoid clones has made a significant impact on both research and clinical medicine. The use of monoclonal antibodies and cloned T cells that recognize unique antigenic determinants has generated the need for more stringent quality control, using frozen cellular reagents. Experiments were carried out using commercial monoclonal antibodies against lymphocyte subpopulations to determine the effects of freezing on lymphocyte subsets. In addition, various freezing protocols were tested to determine the optimal procedure for cryopreservation of lymphoid clones. Little or no difference was found in the T-cell subpopulations following freezing; however, a significant increase in monocytes from a mean of 6.4% ± 1.7 to 15.1% ± 3.3 was seen by cell sorter analysis. Longitudinal studies performed on aliquots of cryopreserved mononuclear cells demonstrated the utility of this procedure for the quality control of monoclonal reagents, FACS analysis, and technical procedures. Experiments on the cryopreservation of T- and B-cell lymphoid clones demonstrated that B-cell clones required slower cooling rates (0.5 C/min) for optimal recovery following freezing. Cryopreservation serves as an important tool for the long-term preservation of valuable cell lines as well as for ensuring optimal quality control of both assays and reagents.

INTRODUCTION

The report by Polge et al. (1949), of the cryoprotective properties of glycerol, presented cellular biologists and scientists from a variety of disciplines with an important tool for the long-term preservation of many different kinds of cells. Immunologists have used cryopreservation to preserve mononuclear cells from a variety of animal as well as human sources for use in immunological assays, both for the convenience and for quality control. The recent development of hybridoma technology and cloning procedures has made cryopreservation even more critical for preserving valuable cell lines that have been cloned and retain a defined function and specificity (T cells, B cells, and monocytes).

An important application of cryopreservation is the use of mononuclear cells with defined cell surface determinants for the quality control of monoclonal antibody reagents. This is particularly critical in the clinical diagnosis laboratory, where control of day-to-day variation is important in longitudinal studies of patients on specific therapies (Strong et al. 1975).

This report describes the results of comparisons of fresh and frozen-thawed samples of human peripheral blood mononuclear cells for reactivity with several

*Present address: Genetics Systems Corporation, Seattle, WA 98121.

common monoclonal reagents. In addition, a cloned T-cell line and an EBV-transformed B-cell line from the same donor were tested for recovery following freezing, using several different protocols to ascertain an optimal procedure for each.

Materials and Methods

Peripheral Blood Cell Preparation, Freezing, and Thawing

Peripheral blood mononuclear cells (PBMC) were separated from heparinized venous blood (Boyum 1968) by Ficoll-Hypaque density gradient centrifugation (LSM, Litton Bionetics) then washed three times in RPMI 1640 medium (GIBCO, Grand Island, NY) supplemented with 20% Fetal Bovine Serum (Biofluids, Rockville, MD), 0.1% gentamicin (M.A. Bioproducts, Walkersville, MD), 1% L-glutamine (GIBCO), and 25 m*M* Hepes buffer (GIBCO). One-half of the PBMC preparation was then resuspended to 20×10^6 cells/ml in complete RPMI 1640 medium and diluted to 10×10^6 cells/ml by the slow addition of precooled 15% dimethyl sulfoxide (Me_2SO) in complete RPMI 1640 medium. One milliliter aliquots were dispersed into Nunc vials (Cooke, Alexandria, VA), which were then capped and placed into the precooled freezing chamber of a CryoMed liquid nitrogen controlled-rate freezer. The PBMC aliquots were first cooled to −2 C for 10 min then frozen at a rate of 1 C/min to −30 C and 5 C/min to −80 C, as previously described (Strong et al. 1982). Immediately following freezing, the vials were transferred until needed to the gaseous phase of a liquid nitrogen storage freezer.

Frozen cells were thawed rapidly by agitation in a 37 C water bath. When the last ice crystal dissolved, the cells were transferred immediately to a 15-ml plastic tube and the Me_2SO slowly diluted with room temperature RPMI 1640 medium containing 10% FBS. The cells were then centrifuged ($150–200 \times g$) for 10 min, gently resuspended in fresh complete RPMI medium, and counted. Viability by trypan blue dye exclusion was usually >90%, while cell number recovery ranged from 80% to 90%.

The other half of the PBMC preparation, which was not frozen, was held at 22 C until the freezing was complete, at which time aliquots of both fresh and frozen-thawed samples were analyzed for surface antigen expression, as described in the following section.

Flow Cytofluorographic Analysis: Staining Method

To ascertain the relative distribution of mononuclear cell subpopulations in fresh and frozen-thawed PBMC preparations, aliquots of cells (0.5 to 1×10^6 cells) were incubated with a pretested dilution of monoclonal antibody for 30 min at 4 C in RPMI 1640 medium containing 2.5% bovine serum albumin (BSA), and 0.1% NaN_3. Cells were then washed in cold medium and counterlabeled with fluorescein-conjugated $F(ab')_2$ fragments of goat anti-mouse IgG (Cappel Laboratories) for 30 min at 4 C. Following this second incubation, the cells were washed and resuspended to a volume of 0.7 ml with cold phosphate buffered saline (PBS) containing BSA and NaN_3. All reagents were ultracentrifuged at $100{,}000 \times g$ for 30 min to remove aggregates and immune complexes prior to use.

Monoclonal Reagents

OKT3, a pan-T-cell reagent from Ortho Pharmaceutical Co. (Kung et al. 1979; Reinherz et al. 1980); OKT4, a monoclonal that detects the inducer/helper subpopulation of human T lymphocytes (Kung et al. 1979; Reinherz et al. 1979, 1980; Thomas et al. 1980) (Ortho Pharmaceutical Co.); OKT8, a monoclonal antibody that recognizes the cytotoxic/suppressor T-cell subpopulation (Reinherz et al. 1980; Thomas et al. 1980) (Ortho Pharmaceutical Co.); 7.2, a monoclonal reagent from New England Nuclear that identifies a determinant on the framework region of the human Ia-like antigen (Hansen et al. 1980); 20.2, a monoclonal antibody that recognizes monocytes, granulocytes, and histiocytes, was a gift from Dr. John Hansen (Kamoun et al. 1983).

The negative control reagent used was a supernatant fluid from the $P3 \times 63Ag^8$ myeloma cell line containing a nonspecific IgG_1, immunoglobulin. All samples were kept on ice until analyzed on the FACS.

Flow Cytometric Analysis

The quantitative measure of immunofluorescence of labeled PBMC was determined using a FACS II (Becton-Dickinson, Sunnyvale, CA) and a PDP 11/34 computer from Digital Equipment Corporation. All samples were analyzed using a single argon laser with a forward scatter detector and right-angle PMT detecting green fluorescence. A lower level scatter window was set to exclude electronic noise, debris, and damaged or drying cells. Fluorescence staining was determined on all cells above this window, i.e., all peripheral blood mononuclear cells. Fluorescence profiles were presented as linear or logarithmic histograms from which the percentage of positive cells was calculated by subtracting the negative control, P3, staining. Normally, 10,000 cells were analyzed.

Growth and Characteristics of Established HTLV Cell Lines, T and B Pairs

Hut 102–2B was established, as previously described (Poiesz et al. 1980), from a patient with mycosis fungoides and was shown to be infected with HTLV. It was grown in RPMI 1640 medium supplemented with 10% Fetal Bovine Serum; 1% L-glutamine; 0.1% gentamicin; and 2% Hepes buffer.

Rob-B was established from the peripheral blood of the same patient mentioned above following transformation with EBV. This B-cell line was grown in RPMI 1640 medium and 10% Fetal Bovine Serum.

T- and B-cell lines were cultured for prolonged periods of time in RPMI 1640 supplemented medium. The cultures were passed and refed with fresh medium every 2 to 3 d, checking for cell concentration and viability. Both lines were cloned using limiting dilution techniques and regrown in bulk cultures from isolated clones.

Cultured T- and B-Cell Line Cryopreservation

Cultured cells were removed from their flask on the day after feeding, spun down at 1,400 rpm for 10 min; supernatants were discarded and the cells were washed in fresh medium. Each culture was then diluted to a known volume, a small aliquot was removed, and a cell count and viability done in trypan blue on a hemocytometer.

Cell cultures were then resuspended in medium and 20% Fetal Bovine Serum (FBS) and divided into three separate tubes with equal amounts of cells/ml to which different concentrations of dimethyl sulfoxide (Me_2SO) were added dropwise. From each concentration of Me_2SO-suspended cells, 1 ml was transferred to a 2-ml freezing vial (Nunc, Arthur Thomas Co.), capped, and placed in a precooled chamber of a liquid nitrogen controlled-rate freezer (CryoMed, Mt. Clemens, MI) applying cooling rates of 0.5 C, 1 C or 5 C/min to −30 C and then 5 C/min to −80 C. The remaining cells were centrifuged at 1,000 rpm for 10 min and resuspended in 10% FBS.

Cell lines were thawed and washed as described before. All samples, fresh (unfrozen) and frozen, were counted using a hemocytometer. Viability was examined by the trypan blue dye exclusion test.

Lymphoproliferation Assay

Functional viability of cell lines was determined by DNA synthesis as measured by uptake of ^{3}H-thymidine (^{3}HTdr). Several concentrations of each cell line were tested at 24, 48, and 72 h, noting concentrations that provided a linear curve.

For the assay, 0.25, 0.125, 0.06, and 0.03×10^6 cells/ml were used as responders in 10% FBS. One-tenth milliliter of cell suspensions was added to wells of a Linbro round-bottom microtiter plate in triplicate (Flow Laboratories, Inc., McLean, VA), plus 0.1 ml of 10% FBS medium and then incubated at 37 C in a humidified incubator (Bellco Glass, Inc., Vineland, NJ) with 5% CO_2. The cultures were pulsed 18 h prior to harvesting with 1 μCi/well of ^{3}HTdr (Schwarz/Mann, Cambridge, MA; specific activity 6 Ci/mM) and harvested onto glass fiber filter paper with a multiple sample harvester (M-24, Brandel, Gaithersburg, MD). The discs were placed in biovials containing 1 ml Ready-Solv solution (Beckman, Columbia, MD) and counted in a liquid scintillation counter (Packard, Downers Grove, IL). Results were calculated as counts per minute (CPM).

Results and Discussion

Monoclonal Antibody Markers on PBMC Following Freezing

Experiments were carried out to determine whether freezing and thawing caused significant differences in mononuclear cell populations from peripheral blood as determined by monoclonal antibody markers and FACS analysis. PBMC were divided into two aliquots: one diluted in Me_2SO and frozen to −80 C, thawed and washed, and the other held at room temperature. The cells were then labeled with monoclonal antibodies as previously described. When analyzing cells on a flow cytometer, it is first necessary to select the population to be analyzed, based on their light scatter properties. When light scatter histograms are examined, a heterogeneous distribution is normally seen, with subcellular debris, smaller cells (such as erythrocytes), and damaged or dying cells on the left-hand side of the curve, monocytes on the far right, and lymphocytes (the larger peak) in the center. Usually, when cells are frozen and thawed, an increase of 10 to 20% is found in the subcellular debris range; i.e., particles smaller than the lymphocytes. The light scatter profile between individuals may vary slightly in both the fresh and frozen samples. The smaller particles were excluded from analysis by the FACS and all forward scatter; i.e., cells in the lymphocyte and monocyte ranges were examined with the monoclonal reagents.

The effects of freezing on PBMC subpopulation expression can be seen for six subjects tested (Table 1). The mean ± S.E. has been calculated for the fresh and frozen samples and statistical differences determined using the paired student *t* test.

A slight decrease was noted in the percentage of T-cells and T-cell subsets following freezing; however, this decrease was not statistically significant. The T-helper to T-suppressor ratios for the tested individuals were also calculated and no difference was seen (1.56 vs. 1.54). A significant increase in cells bearing the 7.2 and 20.2 markers was noted. The monoclonal antibody 20.2 stains monocytes and an increase in the mean percent positive of 6.4% to 15.1% was seen.

Although the majority of reports in the literature on the recovery of lymphoid function following freezing suggests that functional activity is retained, there is a difference of opinion as regards certain cell surface characteristics. Jewett et al. (1976) reported an increase in E-rosette positive cells (T-cells) and a decrease in surface immunoglobulin (sIg) positive cells (B-cells) following freezing. On the other hand, Callery et al. (1980) found the reverse to be true. The majority of studies, however, conclude that under optimal conditions, lymphocyte populations are quite stable and only minor changes are normally observed, as determined by surface markers (Sears et al. 1978; Polizzi et al. 1978; Strong et al. 1975). In fact, Slease et al. (1980) demonstrated no significant differences in sIg positive cells following freezing using the more sensitive FACS analysis of lymphoid cells. The important issue here, however, is that shifts in cell populations can occur for a variety of reasons, and care must be taken if accurate and reproducible results are required.

An increase in monocytes as determined by the 20.2 marker was observed in this study. Van Der Meulen et al. (1981) and Weiner and Norman (1981) have reported inhibition of monocyte function following freezing. Dean and Strong (1977), on the other hand, have reported an increase in specific functional responses using monocyte chemotaxis. FACS analysis of mixed lymphoid/monocyte populations resulted in increased levels of nonspecific fluorescence usually attributed to monocyte Fc receptor binding of fluorescent antibodies. This binding can often be greater for the specific reagents than for negative controls resulting in shifts of the curves for "background" fluorescence. This seems to be particularly true for frozen-thawed cells, where this background is increased. Thus increases in percent positive cells could be due to this shift in intensity of fluorescence rather than to actual increases in the numbers of cells.

Longitudinal Studies

Alterations by freezing in mononuclear populations, as measured by surface markers, are of concern when contemplating longitudinal studies of patients using cryopreserved cells. The use of cryopreservation for sequential analysis has been recommended by several authors as a means of decreasing the assay-to-assay variability, which can be immense in many immunological determinations (Weiner et al. 1973; Sears and Rosenberg 1975; Jewett et al. 1976). An equally important application of cryopreservation is the use of multiple aliquots of cells

TABLE 1. *Effect of freezing on PBMC subpopulation expression*

Sample	OKT3		OKT4		OKT8		OKT4/OKT8		7.2		20.2	
	Fresh	Frozen	Fresh	Frozen	Fresh	Frozen	Fresh	Frozen	Fresh	Frozen	Fresh	Frozen
F3646	72.4	73.0	36.1	32.1	29.8	28.4	1.21	1.13	8.2	16.8	1.1	11.1
F3647	62.4	62.5	28.2	32.7	25.6	25.8	1.10	1.27	18.1	25.1	5.0	8.8
F3648	63.4	56.5	36.7	35.5	19.9	19.7	1.84	1.80	18.6	26.3	9.5	11.4
F3649	64.3	55.6	26.7	30.6	21.8	19.1	1.22	1.60	18.1	26.7	2.7	8.5
F4061	65.7	71.9	33.4	27.7	21.6	18.1	1.55	1.53	20.6	31.5	9.2	23.7
F4062	75.0	62.2	49.8	37.0	20.5	19.6	2.43	1.89	22.4	36.4	11.1	27.2
Mean	67.2	63.6	35.2	32.6	23.2	21.8	1.56	1.54	17.7	27.1[a]	6.4	15.1[a]
S.E.	2.1	3.0	3.4	1.4	1.6	1.7	0.21	0.12	2.0	2.7	1.7	3.3

[a] $P \leq 0.05$

from a single normal donor for the quality control of reagents, FACS, and technical procedures.

An example of this approach is shown in Fig. 1. A leukocyte rich fraction was obtained by leukophoresis of a single donor on a Haemonetics-30 continuous-flow cell separator. Following standard gradient separation, several hundred ali-

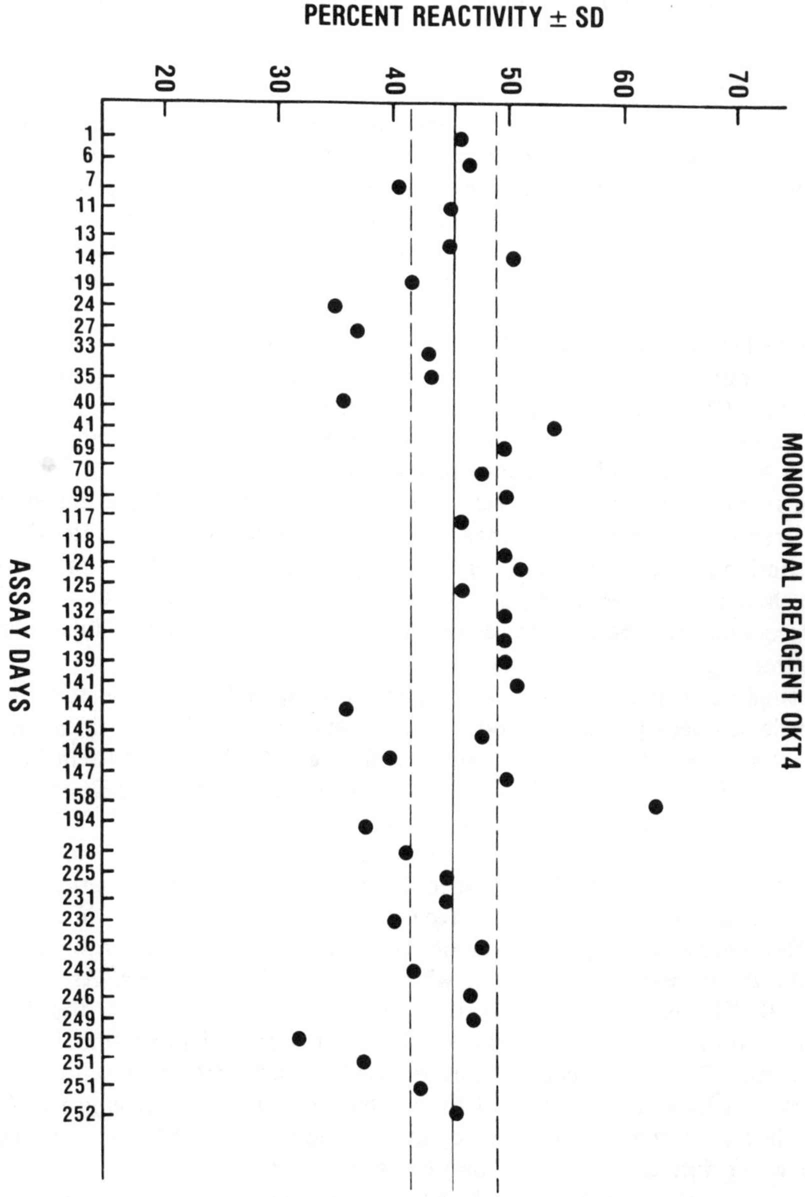

FIG. 1. Day-to-day variability in percent reactivity of monoclonal reagent OKT4 on normal leukophoresed, cryopreserved cell sample, F1515.

quots of cells were cryopreserved and stored in the gaseous phase of liquid nitrogen. Individual aliquots were then assayed daily using the standard battery of monoclonal reagents. The day-to-day variability of the FACS analysis for OKT4 is shown in Fig. 1. Using this quality control procedure, the accuracy and precision of this system can be easily determined. For example, the employment of a new technician starting at day 150, resulted in consistently lower values. These subtle shifts can be determined quickly using these types of standards for cellular assays.

Factors Affecting Cell Recovery

A great deal of investigation has gone into the determination of the mechanisms of freezing injury. Cells that are cooled too slowly to temperatures below freezing are damaged by the resulting increase in salt concentration and cellular shrinkage that occurs as water is removed during the formation of ice (Lovelock 1953a). Conversely, if cooling is too rapid, a new mechanism is invoked in which shrinkage no longer occurs but the cell is damaged by the formation of intracellular ice, either during freezing or upon thawing (Mazur et al. 1969). Cryoprotectants, such as glycerol or dimethylsulphoxide, reduce the amount of ice present during freezing and reduce solute concentration thus reducing ionic disturbances (Polge et al. 1949; Mazur et al. 1969; Farrant et al. 1974; Lovelock 1953b). However, these compounds can themselves cause osmotic injury since they are hypertonic and can cause damage during their addition or removal (Strong 1976). Optimum cooling rates vary from cell type to cell type depending on differences in membrane permeability and intracellular water, which is removed during the dehydration phase of "slow" cooling and extracellular ice formation (Mazur et al. 1969). In general, the larger the cell volume, the slower the rate of cooling must be to allow equilibration of intra- and extracellular water during freezing.

Although optimal conditions for lymphoid cells have been known for sometime, little has been published on differences between T- and B-cell lines. Since the description of cell fusion hybridoma technology (Kohler and Milstein 1975), greater attention has been focused on the need to preserve valuable T- or B-cell clones with well-defined specificities.

A series of experiments was carried out to determine the optimal cooling rate and Me_2SO concentrations for a cloned T- and B-cell pair from a patient with mycosis fungoides. These two lines have very different characteristics in culture. Hut 102–2B grows in singlets and doublets primarily and is considerably larger than Rob-B, the B-cell counterpart, which grows in large clusters. As seen in Table 2, the slower cooling rate of 0.5 C/min gave better results for the B-cell line. The optimal cooling rate for the T-cell line was greater than 1 C/min but less than 5 C/min. These differences can be ascribed to the differences in cell size and water content as well as the size of the clusters in which these cells grow. The larger cells and clusters required more time for equilibration of intra- and extracellular water during the dehydration process of freezing.

The concentration of Me_2SO played an important role both in terms of cryopreservation as well as in contributing to the osmotic stress on the cell membrane. This osmotic fragility can be mimicked by treatment with 0.5–3.0 × isotonic saline or simple dilution at 4 C from the hypertonic conditions (1,800

TABLE 2. *Effects of different cooling rates and Me_2SO concentrations on the recovery of cloned lymphoid lines following freezing*[a]

		Cooling Rate C/min		
		0.5 C	1.0 C	5.0 C
Me_2SO	Cell Line[b]	3H-Thymidine Uptake (CPM ± S.E.		
10.0%	Hut 102-2B	1606 ± 41	5837 ± 490	2845 ± 266
	Rob-B	8020 ± 469	3628 ± 424	2307 ± 171
7.5%	Hut 102-2B	1806 ± 87	3786 ± 529	3159 ± 274
	Rob-B	7821 ± 319	1892 ± 83	2405 ± 199
5.0%	Hut 102-2B	2037 ± 118	2855 ± 229	1500 ± 17
	Rob-B	8492 ± 413	2747 ± 314	2241 ± 17

[a] Hut 102-2B is an HTLV-infected T-cell line established from a patient with mycosis fungoides (Poiesz et al. 1980). Rob-B is an EBV-transformed B-cell line from the same donor. Both lines were cloned by limiting dilution. Lines were cooled to − 80 C, thawed, washed, and plated at 2.5×10^5 cells/well with $_3$H-thymidine and harvested after 24 h.

[b] 2.5×10^5 cells/culture harvested at 24 h following thawing.

mOsm) of Me_2SO (Strong 1976). Thus one important factor in recovery of frozen-thawed cells is dilution with warm as opposed to 4 C tissue culture medium in a stepwise procedure to reduce osmotic shock (Weiner 1976; Strong et al. 1975; Birkeland 1980).

Conclusions

The handling of cells prior to and following freezing and thawing is at least as important as the freezing itself. The following precautions are considered important in assuring good recovery of mononuclear cells and lymphoid lines following cryopreservation: obtain PBMC free of platelet and granulocyte contamination; harvest cell lines during log phase of growth when viability is maximal; freeze only small aliquots (0.2–2.0 ml) containing a minimum of 10×10^5 cells/ml; maintain cell suspensions in serum-containing medium at all times; control the cooling rate (1 C/min for lymphoid cells, 0.5 C/min for B-cell lines) to − 30 C, and 5 C/min to − 80 C; store in the vapor phase of liquid nitrogen; thaw rapidly in a 37 C water bath with mixing; dilute the cell suspension slowly with room temperature or warm medium containing 10% serum, or add Me_2SO dropwise to allow for osmotic equilibrium; handle carefully, centrifuge slowly, and resuspend cells slowly prior to dilution and assay or growth.

Acknowledgments

The authors would like to thank Cindy Pittenger for excellent technical and editorial assistance, and Charles Thomas and Stephen Trost for FACS technical assistance. This work was supported by contract YO1-CP-30500, Public Health Service from the Division of Cancer Cause and Prevention, National Cancer Institute.

Literature Cited

Birkeland, S. A. 1980. Cryopreservation of human lymphocytes for sequential testing of immune competence. *J. Immunol. Meth.* 35:57–67.

Boyum, A. 1968. A one-stage procedure for isolation of granulocytes and lymphocytes from human blood. *Scand. J. Clin. Lab. Invest.* 21:51.

Callery, C. D., M. Golightly, N. Sidele, and S. H. Golub. 1980. Lymphocyte surface markers and cytotoxicity following cryopreservation. *J. Immunol. Meth.* 35:213–223.

Dean, D. A., and D. M. Strong. 1977. Improved assay for monocyte chemotaxis using frozen stored responder cells. *J. Immunol. Meth.* 14:65–72.

Farrant, J., S. C. Knight, L. E. McGann, and J. O'Brien. 1974. Optimal recovery of lymphocytes and tissue culture cells following rapid cooling. *Nature* 249:452–453.

Hansen, J. A., P. J. Martin, and R. C. Nowinski. 1980. Monoclonal antibodies identifying novel T-cell antigens and Ia antigens of human lymphocytes. *Immunogenetics* 10:247–260.

Jewett, M. A. S., S. Gupta, J. A. Hansen, S. Cunningham-Kundles, F. P. Siegel, R. A. Good, and B. Dupont. 1976. The use of cryopreserved lymphocytes for longitudinal studies of immune function and enumeration of subpopulations. *Clin. Exp. Immunol.* 25:449–454.

Kamoun, M., P. Martin, M. E. Kadin, L. G. Lum, and J. A. Hansen. 1983. Human monocyte-histiocyte differentiation antigens defined by monoclonal antibodies. *Clin. Immunol. Immunopath.* 29:181–195.

Kohler, G., and C. Milstein. 1975. Continuous cultures of fused cells secreting antibody of predefined specificity. *Nature* 256:495.

Kung, P. C., G. Goldstein, E. L. Reinherz, and S. F. Schlossman. 1979. Monoclonal antibodies defining distinctive human T-cell surface antigens. *Science* 206:347–349.

Lovelock, J. E. 1953a. The mechanism of the protective action of glycerol against haemolysis by freezing and thawing. *Biochim. Biophys. Acta* 11:28–36.

______. 1953b. The haemolysis of human blood cells by freezing and thawing. *Biochim. Biophys. Acta* 10:414–426.

Mazur, P., J. Farrant, S. P. Leibo, and E. H. Y. Chu. 1969. Survival of hamster tissue culture cells after freezing and thawing. Interactions between protective solutes and cooling and warming rates. *Cryobiology* 6:1–9.

Poiesz, B. J., F. W. Ruscetti, A. F. Gazdar, P. A. Bunn, J. D. Minna, and R. C. Gallo. 1980. Detection and isolation of type C retrovirus particles from fresh and cultured lymphocytes of a patient with cutaneous T-cell lymphoma. *Proc. Natl. Acad. Sci. USA* 77:7415–7419.

Polge, C., A. U. Smith, and A. S. Parkes. 1949. Revival of spermatozoa after vitrification and dehydration at low temperatures. *Nature* 164:666.

Polizzi, G. J., E. Cohen, and S. G. Gregory. 1978. E-rosette formation of fresh versus frozen-thawed lymphocytes from leukemia patients and normal individuals. *Cryobiology* 15:67–72.

Reinherz, E. L., P. C. Kung, G. Goldstein, and S. F. Schlossman. 1979. Separation of functional subsets of human T-cells by a monoclonal antibody. *Proc. Natl. Acad. Sci. USA* 76:4061–4065.

Reinherz, E. L., P. C. Kung, G. Goldstein, R. H. Levey, and S. F. Schlossman. 1980. Discrete states of human intrathymic differentiation: Analysis of normal thymocytes and leukemic lymphoblasts of T-cells lineage. *Proc. Natl. Acad. Sci. USA* 77:1588–1592.

Sears, H. F., and S. A. Rosenberg. 1975. Advantages of cryopreserved lymphocytes for sequential evaluation of human immune competence. I. Mitogen stimulation. *J. Natl. Cancer Inst.* 58:183–187.

Sears, H. F., S. C. Tondreau, and S. A. Rosenberg. 1978. Advantages of cryopreserved lymphocytes for sequential evaluation of human immune competence. II. Mixed lymphocyte cultures and mononuclear cell subpopulations. *J. Natl. Cancer Inst.* 61:1011–1016.

Slease, R. B., D. M. Strong, R. Wistar, Jr., R. E. Budd, and I. Scher. 1980. Determination of surface immunoglobulin density on frozen-thawed mononuclear cells analyzed by the fluorescence activated cell sorter. *Cryobiology* 17:523–529.

Strong, D. M. 1976. Cryobiological approaches to the recovery of immunological responsiveness to murine and human mononuclear cells. *Transplant. Proc.* 8:203–208.

Strong, D. M., J. N. Woody, M. A. Factor, A. Ahmed, and K. W. Sell. 1975. Immunological responsiveness of frozen-thawed human lymphocytes. *Clin. Exp. Immunol.* 21:442–435.

Strong, D. M., J. R. Ortaldo, F. Pondolfi, A. Maluish, and R. B. Herberman. 1982. Cryopreservation of human mononuclear cells for quality control in clinical immunology. I. Correlations in recovery of k- and NK-cell functions, surface markers, and morphology. *Clin. Immunol.* 2:214–221.

Thomas, Y., J. Sosman, O. Irigoyen, S. F. Friedman, P. C. Kung, G. Goldstein, and L. Chess. 1980. Functional analysis of human T-cell subsets defined by monoclonal antibodies. I. Collaborative T-T interactions in the immunoregulation of B-cell differentiation. *J. Immunol.* 125:2402–2408.

Van Der Meulen, F. W., M. Reiss, E. A. M. Stricker, E. Van Elven, and A. E. G. Van Dem Borne. 1981. Cryopreservation of human monocytes. *Cryobiology* 18:337–343.

Weiner, R. S., J. Breard, and C. O'Brien. 1973. Cryopreserved lymphocytes in sequential studies of immune responsiveness: Problems and prospects. Preliminary report, Pages 117–131 *in* R. S. Weiner, R. K. Oldham, and L. Schwartzenberg, eds., *Cryopreservation of Normal and Neoplastic Cells.* Paris, INSERM, pp. 117–131.

Weiner, R. S. 1976. Cryopreservation of lymphocytes for use in *In Vitro* assays of cellular immunity. *J. Immunol. Meth.* 10:49–60.

Weiner, R. S., and S. J. Norman. 1981. Functional integrity of cryopreserved human monocytes. *J. Natl. Cancer Inst.* 66:255–260.

CHAPTER **29**

Cryopreservation of Cultures That Contain Plasmids

WILLIAM C. NIERMAN AND TAMARA FELDBLYUM

American Type Culture Collection, 12301 Parklawn Drive, Rockville, Maryland 20852

Cultures of *Escherichia coli, Bacillus subtilis,* and *Saccharomyces cerevisiae* that contained plasmids were monitored through cryopreservation by lyophilization and freezing in liquid nitrogen to determine viability and plasmid retention. Plasmids were selected to evaluate a range of plasmid sizes, copy numbers, replication systems, and selectable genetic markers. Preservation of any of the selected cultures by freezing in liquid nitrogen resulted in neither loss of plasmid nor viability. Lyophilization of plasmid-containing *E. coli* cultures yielded similar results. Excessive plasmid loss was observed in the instance of a particularly unstable plasmid. Lyophilization of vegetative *B. subtilis* resulted in a loss in viability of about an order of magnitude. The lyophilized *B. subtilis* required at least 5 h of rehydration before the full establishment of the plasmid coded antibiotic resistance. The viable cells did retain the plasmids. However, no viability loss was observed when freeze-drying *B. subtilis* spores. In addition, the dried spores did not require an adaptation period before challenge with antibiotic. Lyophilization of *S. cerevisiae* resulted in a loss of viability of two to three orders of magnitude. For the survivors, the fraction containing plasmids was the same before and after lyophilization.

INTRODUCTION

Recombinant DNA technology has provided a powerful method for isolation and preparation of specific genes, for the study of their structure and function, for their transfer to various species, and for efficient expression of their products. This technology provides new approaches to basic and applied biological sciences including advances in medicine and food production, improvements in various industrial processes, and restoration of the environment. A large number of recombinant DNA host-vector systems have been developed for the application of this technology (for reviews see Bolivar and Beckman 1979; Rodriguez and Tait 1983). The organisms most frequently used in recombinant DNA research are those with the best understood genetics and/or those with a historical application to industrial processes, i.e., *Escherichia coli, Saccharomyces cerevisiae,* and *Bacillus subtilis.* The application of this technology to applied and basic research problems has resulted in the widespread maintenance of stock cultures of plasmid-containing organisms in university and industrial laboratories and in culture collections. However, there have been very few data presented in the scientific literature on the suitability of standard preservation methods for the maintenance of plasmids and the stability of plasmid structure in cultures of *E. coli, B. subtilis,* and *S. cerevisiae.*

At the American Type Culture Collection (ATCC) two general methods are used for the preservation of cultures that contain plasmids. They are (1) lyophilization and storage at 4 C and (2) freezing and storage in liquid nitrogen. For discussion of the methods see Gherna (1981). Both of these methods are satisfactory for preserving a broad range of bacterial species for periods of more than 30 yr. In this report, we present the results of our study of the preservation of plasmid-containing cultures of *Escherichia coli, Bacillus subtilis,* and *Saccharomyces cerevisiae* by both lyophilization and freezing in liquid nitrogen.

Materials and Methods

Cells and media. All bacteria, plasmid, and yeast strains were obtained from the ATCC collection. Plasmids are listed in Table 1. The plasmid host strains employed were *Escherichia coli* RR1 ATCC 31343 (Rodriguez and Tait 1983) (F-*hsd*520 *ara*-14 *leu*B6 *pro*A2 *lac*Y1 *galK*2 *rpsL*20 *xyl*-5 *mtl*-1 *supE*44λ^-), *Bacillus subtilis,* BD170 ATCC 33608 (Gryczan et al. 1978) *trp*C2 *his*B2), and *Saccharomyces cerevisiae* SHY1 ATCC 44769 (Botstein et al. 1979) (α*ste*-VC9 *ura*3-52 *trp*1-289 *leu*2-3 *leu*2-112 *his*2-1 *adel*-101). Rich, minimal, and sporulation media were prepared as described in the indicated references: *E. coli* LB medium (Maniatis et al. 1982), GB medium (Zacher et al. 1980), *B. subtilis* VY broth (Gryczan et al. 1978) and sporulation medium (Donnellan et al. 1964), *S. cerevisiae* YM, and Yeast Minimal Media (Wickerham 1951; Sherman et al. 1981). Antibiotics and amino acids were purchased from Sigma and used at the indicated concentrations: ampicillin, 50 μg/ml; tetracyline, 20 μg/ml; kanamycin, 25 μg/ml; streptomycin, 5 μg/ml; chloramphenicol, 25 μg/ml; trimethoprim, 1 μg/ml; neomycin, 5 μg/ml; and amino acids, 40 μg/ml.

TABLE 1. *Plasmids*

Escherichia coli	ATCC No.
pBR322	37017
pGV1106	37169
pKC16	37085
pLG338	37130
pRK290	37168
R6K	37120
pSC101	37032
pSE150	37182
Bacillus subtilis	
pBD8	37097
pC194	37034
pUB110	37015
pTL12	37075
Saccharomyces cerevisiae	
YEp13	37115
YRp7	37060

Cryopreservation. Cells were prepared for freezing by adding an equal volume of 20% glycerol to a liquid culture followed by freezing as described in Gherna

(1981). Cells were prepared for lyophilization from liquid cultures by suspending a cell pellet after centrifugation in 20% skim milk and freeze-drying by the double vial method as described in Gherna (1981).

DNA manipulations. The methods for bacterial and yeast transformation, and small-scale plasmid purification have been described (Maniatis et al. 1982; Rodriguez and Tait 1983; Sherman et al. 1981). Restriction enzymes were purchased from the Bethesda Research Laboratories and used in accordance with the manufacturer's directions.

Results and Discussion

Cryopreservation of Plasmid-Containing Cultures of Escherichia coli

In order to test adequately the procedures of freezing in liquid nitrogen and lyophilization for the preservation of plasmid-containing *Escherichia coli*, a diverse set of plasmids were selected for evaluation. These are presented in Table 2. The plasmids included provide an assortment of replication systems, a range of sizes from 38 to 4.3 kilobase pairs of DNA, six different antibiotic resistance genes, and a temperature-sensitive bacteriophage lambda repressor gene. In addition, they provide a range of copy numbers, i.e., the number of plasmid molecules present in the cell per host chromosome. Two of them can replicate in a broad range of gram-negative host bacteria. To provide a standard host for the plasmids, they were all transformed into the same *E. coli* host, RR1 (Rodriguez and Tait 1983).

Overnight cultures of plasmid-containing *E. coli* RR1 were grown under antibiotic selection and prepared for freezing in liquid nitrogen and lyophilization as described in Material and Methods. Before freezing and lyophilization, an aliquot of the bacteria was diluted, spread to plates with and without antibiotic selection, incubated overnight at 37 C, and the resultant colonies counted. This process was repeated with vials immediately after freezing and lyophilization and again after a storage period of 10 months. Lyophilized bacteria were rehydrated by the addition of LB medium prior to dilution and plating. When plasmid-containing *E. coli* cultures were preserved by freezing in liquid nitrogen, no loss in viability or loss of plasmid from the viable cells was detected. Similar results were obtained when plasmid-containing *E. coli* were lyophilized, as can be observed in the data presented in Table 3. Only in the case of pGV1106 was any loss of the plasmid detected. Plasmid pGV1106 was selected for this study because it was previously determined to be unstable. This plasmid is derived from the W incompatibility group plasmid Sa and has a very broad host range among gram-negative bacteria. When *E. coli* containing the plasmid pGV1106 were grown overnight in nonselective medium, more than 90% of the cells lost the plasmid. When such cells were plated on solid medium without antibiotic, plasmid-containing cells could be obtained from only about half of the colonies (M. Hyde, unpublished observations). Thus pGV1106 was readily lost from *E. coli* when grown without antibiotic selection. In addition to its inherent instability when grown without selection, the data in Table 3 demonstrate that pGV1106 was also readily lost during lyophilization.

TABLE 2. Escherichia coli *plasmids*

Plasmid	Size (kb)[a]	Replication	Genetic Markers[b]	Copy No.	Broad Host Range	Reference
pBR322	4.3	ColE1	Tc^r Ap^r	10–200	No	Bolivar et al. 1977
pGV1106	8.6	Sa	Km^r Sm^r Sp^r	3–5	Yes	Leemans et al. 1982
pKC16	11.0	ColE1	Ap^r λimm		No	Rao and Rogers 1978
pLG338	7.3	pSC101	Km^r Tc^r	6	No	Stoker et al. 1982
pRK290	20.0	RK2	Tc^r	8	Yes	Ditta et al. 1980
R6K	38.0	R6K	Ap^r Sm^r	10–15	No	Kontomichalou et al. 1970
pSC101	8.9	pSC101	Tc^r	6	No	Cohen and Chang 1977
pSE150	7.7		Cm^r Sp^r	1	No	Pang and Walker 1983

[a] kb = kilobase pairs. [b] Tc^r = tetracycline resistance, Sm^r = streptomycin resistance, Sp^r = spectinomycin resistance, Ap^r = ampicillin resistance, Km^r = kanamycin resistance, Cm^r = chloramphenicol resistance, λimm = immunity to bacteriophage λ.

TABLE 3. *Lyophilization of* Escherichia coli *plasmids*

		Cell Density (cells/ml)					
		Before Lyophilization Antibiotic Selection		After Lyophilization Antibiotic Selection		10 Months After Lyophilization Antibiotic Selection	
Strain Designation	Selective Antibiotic	*No*	*Yes*	*No*	*Yes*	*No*	*Yes*
RR1		10^{15}		10^{15}		10^{15}	
RR1(pBR322)	ApTc	10^{12}	10^{12}	10^{12}	10^{12}	10^{12}	10^{12}
RR1(pGV1106)	Km	10^{10}	10^{10}	10^{10}	10^{9}	10^{10}	10^{9}
RR1(pKC16)	Ap	10^{11}	10^{11}	10^{11}	10^{11}	10^{11}	10^{11}
RR1(pLG338)	TcKm	10^{14}	10^{14}	10^{14}	10^{14}	10^{14}	10^{14}
RR1(pRK290)	Tc	10^{14}	10^{14}	10^{14}	10^{14}	10^{14}	10^{14}
RR1(R6K)	ApSm	10^{12}	10^{12}	10^{12}	10^{12}	10^{12}	10^{12}
RR1(pSC101)	Tc	10^{12}	10^{12}	10^{12}	10^{12}	10^{12}	10^{12}
RR1(pSE150)	Cm	10^{15}	10^{15}	10^{15}	10^{15}	10^{15}	10^{15}

The structural integrity of plasmids obtained after freeze-drying was checked by purifying the plasmid from small liquid cultures and analyzing by digestion with the restriction endonucleases. No evidence for any alteration of plasmid structure by lyophilization was obtained by this method.

In general the preservation data for plasmid-containing cultures of *E. coli* indicate that they can be satisfactorily preserved by either freezing in liquid nitrogen or by lyophilization. Plasmids that are readily lost when grown nonselectively can also be lost when cultures are lyophilized. In the case of pGV1106, the plasmid loss was only an order of magnitude. Plasmid-containing cultures of pGV1106 can be readily recovered after lyophilization by rehydration and growth in a selective medium.

Cryopreservation of Plasmid-Containing Cultures of Bacillus subtilis

Many plasmids that replicate in *Bacillus subtilis* were originally isolated from *Staphylococcus aureus* or are derivatives of *S. aureus* plasmids. Three such plasmids were included in this study: pBD8, pC194, and pUB110 (see Table 4). They represented a range of plasmid sizes and antibiotic resistance markers. The fourth plasmid, pTL12, is indigenous to *B. subtilis*. The genetic markers in this plasmid, *leuA, leuB,* and tmp^r were all derived from *B. subtilis* 168 (Tanaka and Kawano 1980), and the replicator originated from a *B. subtilis* (natto) plasmid (Tanaka and Sakaguchi 1978).

TABLE 4. Bacillus subtilis *plasmids*

Plasmid	Size(Kb)	Genetic Markers[a]	Reference
pBD8	9.0	KmSmCm	Gryczan et al. 1980
pC194	3.0	Cm	Gryczan et al. 1978
pUB110	4.4	Km	Gryczan et al. 1978
pTL12	9.6	Tmp	Tanaka and Kawano 1980

[a] Km = kanamycin resistant, Sm = streptomycin resistant, Cm = chloramphenicol resistant. Km = kanamycin resistant, Tmp = trimethoprim resistant.

Preservation of Vegetative B. subtilis

Our initial preservation studies were done using vegetative plasmid-containing *B. subtilis* cells. The cells were grown overnight, concentrated as described in Materials and Methods, and preserved by lyophilization and freezing in liquid nitrogen. To determine viability and plasmid retention through the freeze-drying processes, the freeze-dried bacterial cultures were rehydrated with an appropriate liquid medium, diluted, and plated with and without antibiotic to select for total viability and the presence of the plasmid. With freeze-dried vegetative *B. subtilis*, about 5 h of rehydration without antibiotic selection were required for the plasmid-containing cells to develop resistance to the antibiotic. During this rehydration period little growth occurred but the density of antibiotic resistant cells increased by a factor of 10 (data not shown). The required rehydration period was growth stage independent. The same lag in the development of resistance to the antibiotic was observed whether the cells were freeze-dried from early log phase or from stationary phase. Thus a long rehydration period without antibiotic was employed when determining the extent of plasmid retention after lyophilization of vegetative *B. subtilis* cells.

As with *E. coli, B. subtilis* vegetative cells preserved by freezing in liquid nitrogen showed no loss in viability or detectable loss of plasmid. When vegetative *B. subtilis* cells were lyophilized some losses occurred. As the data in Table 5 show, upon freeze-drying, the viability of vegetative *B. subtilis* was reduced by a factor of 10–100 but for all plasmids except pTL12 the plasmid was retained in all viable cells. In the case of pTL12, there was a 10-fold loss of plasmid from the viable cells. The most serious problem in the freeze-drying of vegetative *B. subtilis* appeared after storage at 4 C for 10 months. As Table 5 reveals, there was a considerable loss of viability during storage. We have no data as to how rapidly the viability decline occurred or as to whether the decline will continue beyond 10 months. Even with this extensive reduction in viability, plasmids can be readily recovered and propagated. However, with the survivors of lyophilization and storage comprising such a small fraction of the original population, it is conceivable that some genetic variation might occur.

Preservation of Plasmid-Containing B subtilis *Spores*

The suitability of preserving the plasmid in *B. subtilis* spores was studied using *B. subtilis* BD170(pC194). Cells were grown on solid sporulation medium at 37 C for 24–48 h. The resultant cells and spores (sporulation was >75%) were harvested, heat treated to kill vegetative cells, and preserved by freezing and lyophilization as described in Materials and Methods. The data presented in Table 6 show that no loss was observed in viability or plasmid retention for at least 4 months when spores were preserved by these methods. In addition, no rehydration period in the absence of antibiotic was required when rehydrating the spores. Presumably antibiotic resistance was activated on a time course similar to the return to vegetative growth. These data indicated that lyophilization of *B. subtilis* spores was superior to the lyophilization of the vegetative cells

Cryopreservation of Plasmid-Containing Cultures of Saccharomyces cerevisiae

The plasmids that are commonly employed in *Saccharomyces cerevisiae* host-vector systems are in fact *S. cerevisiae/E. coli* shuttle vectors. They are chimeric plasmids that contain a replication and selection system functional in *E. coli* and, in general, a replication and selection system functional in *S. cerevisiae.* Manipulations involving plasmid preparation and the alteration of plasmid structure are more readily carried out in *E. coli.* The plasmids are then transformed into *S. cerevisiae* to test the plasmid construction in the *S. cerevisiae* system. In contrast to the use of antibiotic resistance for selection in bacteria, the routine method of selecting transformed yeast cells is by complementing a host auxotrophic mutant with a wild-type gene carried on the incoming plasmid. Some of the wild-type genes that are used as selectable traits in yeast vectors are *LEU*2, *URA*3, and *TRP*1 (Hinnen et al. 1978; Struhl et al. 1979; Stinchcomb et al. 1979; Tschumper and Carbon 1980). Plasmids that are maintained extrachromosomally in yeast contain yeast DNA sequences, which function in plasmid replication. The plasmids contain either a replication origin from 2 micron circle DNA, an en-

TABLE 5. *Lyophilization of vegetative plasmid-containing* Bacillus subtilis

Strain Designation	Selective Antibiotic	Cell Density (cells/ml): Before Lyophilization Antibiotic Selection — *No*	Before Lyophilization — *Yes*	After Lyophilization Antibiotic Selection — *No*	After Lyophilization — *Yes*	10 Months After Lyophilization Antibiotic Selection — *No*	10 Months After Lyophilization — *Yes*
BD170		10^{14}		10^{14}		10^{9}	
BL170(pBD8)	KmSmCm	10^{12}	10^{12}	10^{10}	10^{10}	10^{6}	10^{6}
BD170(pC194)	Cm	10^{13}	10^{13}	10^{11}	10^{11}		
BD170(pUB110)	Km	10^{13}	10^{13}	10^{12}	10^{12}	10^{11}	10^{6}
BD170(pTL12)	Tmp	10^{12}	10^{12}	10^{11}	10^{10}	10^{6}	10^{6}

TABLE 6. *Preservation of spores of* Bacillus subtilis *BD170(pC194)*

	Lyophilized		Liquid Nitrogen	
	Total Viability (spores/ml)	Cm Resistant (spores/ml)	Total Viability (spores/ml)	Cm Resistant (spores/ml)
Before preservation	10^{14}	10^{14}	10^{14}	10^{14}
Immediately after preservation	10^{14}	10^{14}	10^{14}	10^{14}
4 Months after preservation	10^{14}	10^{14}	10^{14}	10^{14}

dogenous yeast plasmid (Hicks et al. 1979); an autonomously replicating yeast chromosomal sequence or ars element (Stinchcomb et al. 1979; Tschumper and Carbon 1980); or a yeast chromosome centromere or CEN sequence in conjunction with an ars element (Clarke and Carbon 1980).

Two *S. cerevisiae/E. coli* shuttle vectors were selected for the cryopreservation study. YEp13 (Broach et al. 1979) confers ampicillin and tetracycline resistance in *E. coli* and contains the yeast wild-type *LEU*2 gene. Episomal replication is maintained through a 2 micron circle DNA replication origin. YRp7 (Botstein et al. 1979) confers ampicillin and tetracycline resistance in *E. coli* and contains the yeast *TRP*1 gene. It is replicated via a chromosomal ars sequence in the plasmid. The plasmids were purified from their *E. coli* host and transformed into spheroplasts of *S. cerevisiae* SHY1 (*trp*^{1}–289 *leu* 2–3 *leu*2–112). Cultures of SHY1, and SHY1 transformed with each plasmid, were grown under selective conditions and preserved by freezing in liquid nitrogen and by lyophilization. Prior to lyophilization, cultures were either frozen rapidly by plunging into liquid nitrogen or slowly by placing in a −70 C freezer. The cell densities and extent of plasmid retention before and after preservation were determined by dilution and plating on selective and nonselective media. Early work revealed that rapidly freezing *S. cerevisiae* in liquid nitrogen prior to freeze-drying reduced the viability by a factor of 10^{6}, whereas slow freezing had no effect. Thus the cells were routinely frozen in a −70 C freezer for an hour before drying.

Yeast 2 micron circle DNA vectors and ars vectors are mitotically unstable. Even under selective growth conditions, only a minority of the cells in the culture population retains the plasmids. When plasmid-containing *S. cerevisiae* were preserved by freezing in liquid nitrogen, no loss of viability was observed and the fraction of plasmid-containing cells was the same before and after freezing.

The data for a typical freeze-drying experiment of plasmid-containing *S. cerevisiae* are presented in Table 7. These data demonstrate what we observe to be a general variable loss in viability of *S. cerevisiae* upon lyophilization, which is independent of the presence or absence of a plasmid. In this experiment the loss of viability under nonselective conditions varied from a factor of 10 for SHY1 to 100 for SHY1(YRp7). The experiments with freezing rate prior to freeze-drying suggested that the variability was because of differences in freezing rates upon placement of the culture vials in the −70 C freezer.

The fraction of viable yeast cells containing the plasmid remained approximately the same before and immediately after lyophilization. The data showed that while YEp13 was more stable than YRp7, the stabilities were not altered by

TABLE 7. *Lyophilization of* Saccharomyces cerevisiae *plasmids*

Strain Designation	Selective Marker	Cell Density (cells/ml)					
		Before Lyophilization Selection		After Lyophilization Selection		9 Months After Lyophilization Selection	
		No	*Yes*	*No*	*Yes*	*No*	*Yes*
SHY1		2×10^9		2×10^8		2×10^8	
SHY1(YEp13)	LEU2	3×10^{10}	1×10^{10}	4×10^9	1×10^9	1×10^9	2×10^6
SHY1(YR7)	TRP1	1×10^9	6×10^7	5×10^7	4×10^6	3×10^6	5×10^5

freeze-drying. However, storing the freeze-dried material at 4 C appeared to bring about a decrease in viability, and for SHY1(YEp13) a decrease in the fraction of viable cells containing the plasmid. The storage period tested thus far has only been 9 months. Longer storage periods are required to determine if freeze-drying is a suitable long-term method for preserving plasmid-containing cultures of *S. cerevisiae*. It was clear that freezing in liquid nitrogen was the safest method for preservation of the cultures.

Conclusions

The data presented in this report indicated that the preferred methods for preserving plasmid-containing cultures of *Escherichia coli, Bacillus subtilis,* and *Saccharomyces cerevisiae* are either freezing in liquid nitrogen or freeze-drying. Both methods worked equally well with *E. coli* cultures that contained plasmids. For *B. subtilis* vegetative cells, liquid nitrogen freezing is the method of choice. If *B. subtilis* spores are to be preserved, liquid nitrogen and freeze-drying are equally satisfactory methods. *S. cerevisiae/E. coli* shuttle plasmids are better preserved in *E. coli*. If plasmid preservation in *S. cerevisiae* is required, then freezing in liquid nitrogen is the better procedure.

Acknowledgment

The authors express appreciation to Ellen Baque for assistance in preparation of the manuscript.

Literature Cited

Bolivar, F., and K. Beckman. 1979. Plasmids of *Escherichia coli* as cloning vectors. Pages 245–280 *in* R. Wu, ed., *Methods in Enzymology,* Vol. 68, Academic Press, New York.

Bolivar, F., R. L. Rodriguez, P. J. Greene, M. C. Betlach, H. L. Heyneker, and H. W. Boyer. 1977. Construction and characterization of new cloning vehicles II, a multipurpose cloning system. *Gene* 2:95–113.

Botstein, D., S. L. Falco, S. E. Stewart, M. Brennan, S. Scherer, D. T. Stinchcomb, K. Struhl, and R. W. Davis. 1979. Sterile host yeasts (SHY): A eukaryotic system of biological containment for recombinant DNA experiments. *Gene* 8:17–24.

Broach, J. R., J. N. Strathern, and J. B. Hicks. 1979. Transformation in yeast: Development of a hybrid cloning vector and isolation of the CANl gene. *Gene* 8:121–133.

Clarke, L., and J. Carbon. 1980. Isolation of a yeast centromere and construction of functional small circular chromosomes. *Nature* 287:504–509.

Cohen, S. N. and A. C. Y. Chang. 1977. Revised interpretation of the origin of the pSC101 plasmid. *J. Bacteriol.* 132:734–737.

Ditta, G., S. Stanfield, D. Corbin, and D. R. Helinski. 1980. Broad host range DNA cloning system for gram-negative bacteria: Construction of a gene bank of *Rhizobium meliloti. Proc. Natl. Acad. Sci. USA* 77:7347–7351.

Donnellan, J. E., Jr., E. H. Nags, and H. S. Levinson. 1964. Chemically defined, synthetic media for sporulation and for germination and growth of *Bacillus subtilis*. *J. Bacteriol.* 87:332–336.

Gherna, R. L. 1981. Preservation. Pages 208–217 *in* P. Gerhardt et al. eds., *Manual of Methods for General Bacteriology,* American Society for Microbiology, Washington, DC.

Gryczan, T. J., S. Contente, and D. Dubnau. 1978. Characterization of *Staphylococcus aureus* plasmids introduced by transformation into *Bacillus subtilis*. *J. Bacteriol.* 134:318–329.

Gryczan, T., A. G. Shivakumar, and D. Dubnau. 1980. Characterization of chimeric plasmid cloning vehicles in *Bacillus subtilis*. *J. Bacteriol.* 141:246–253.

Hicks, J. B., A. Hinnen, and G. R. Fink. 1979. Properties of yeast transformation. Pages 1305–1313 *in Cold Spring Harbor Symp. on Quant. Biol.* Vol. 43. Cold Spring Harbor Laboratory, Cold Spring Harbor, New York.

Hinnen, A., J. B. Hicks, and G. R. Fink. 1978. Transformation of yeast. *Proc. Natl. Acad. Sci. USA* 75:1929–1933.

Kontomichalou, P., M. Mitani, and R. C. Clowes. 1970. Circular R-factor molecules controlling penicillin synthesis, replicating in *Escherichia coli* under either relaxed or stringent control. *J. Bacteriol.* 104:34–44.

Leemans, J., J. Langenakens, H. De Greve, R. Dubnau, M. Van Montagu, and J. Schell. 1982. Broad-host-range cloning vectors derived from the W-plasmid Sa. *Gene* 19:361–364.

Maniatis, T., E. F. Fritsch, and J. Sambrook. 1982. *Molecular Cloning: A Laboratory Manual,* page 440. Cold Spring Harbor Laboratory, Cold Spring Harbor, New York.

Pang, P., and G. C. Walker. 1983. Identification of the *uvrD* gene product of *Salmonella typhimurium* LT2. *J. Bacteriol.* 153:1172–1179.

Rao, R. N., and S. G. Rogers, 1978. A thermoinducible lambda phage-ColE1 plasmid chimera for the overproduction of gene products from cloned DNA segments. *Gene* 3:247–263.

Rodriguez, R. L., and R. C. Tait. 1983. Pages 17–19 *in Recombinant DNA Techniques: An Introduction.* Addison-Wesley Publishing Company, Reading, MA.

Sherman, F., G. R. Fink, and J. B. Hicks. 1981. *Methods in Yeast Genetics,* page 114. Cold Spring Harbor Laboratory, Cold Spring Harbor, New York.

Stinchcomb, D. T., K. Struhl, and R. W. Davis. 1979. Isolation and characterization of a yeast chromosomal replicator. *Nature* 282:39–43.

Stoker, N. G., N. F. Fairweather, and B. G. Sprat. 1982. Versatile low-copy-number plasmid vectors for cloning in *Escherichia coli*. *Gene* 18:335–341.

Struhl, K., D. T. Stinchcomb, S. Scherer, and R. W. Davis. 1977. Functional genetic expression of eukaryotic DNA in *Escherichia coli*. *Proc. Natl. Acad. Sci. USA* 76:1035–1039.

Tanaka, T., and N. Kawano. 1980. Cloning vehicles for the homologous *Bacillus subtilis* host-vector system. *Gene* 10:131–136.

Tanaka, T., and K. Sakaguchi. 1978. Construction of a recombinant plasmid composed of *B. subtilis* leucine genes and a *B. subtilis* (natto) plasmid: Its use as cloning vehicle in *B. subtilis* 168. *Mol. Gen. Genet.* 165:269–276.

Tschumper, G., and J. Carbon. 1980. Sequence of a yeast DNA fragment containing a chromosomal replicator and the TRP2 gene. *Gene* 20:157–166.

Wickerham, L. J. 1951. Taxonomy of yeast. Page 2 *in Technical Bulletin No. 1029.* Northern Regional Research Laboratory, Bureau of Agricultural and Industrial Chemistry. U.S. Department of Agriculture, Washington, DC.

Zacher, A. N. III, C. A. Stock, J. W. Golden II, and G. P. Smith. 1980. A new filamentous phage cloning vector: fd-tet. *Gene* 9:127–140.

VIII
CONTRIBUTED PAPERS

CHAPTER 30

Overproduction and Purification of the Three Enzymes Constituting the *Escherichia coli* Proline Biosynthetic Pathway

A. H. Deutch*, C. J. Smith**, and K. E. Rushlow***

Bethesda Research Laboratories, Inc.
Gaithersburg, Maryland 20877

Proline is synthesized from glutamate through a series of three enzymatic reactions catalyzed by the gene products of the *pro*B, *pro*A, and *pro*C loci. The *Escherichia coli (E. coli)* genes encoding the enzymes of the proline biosynthetic pathway were subcloned onto an expression plasmid carrying both the bacteriophage lambda P_L promoter and the lambda gene encoding a temperature-sensitive cI repressor protein. One plasmid construct contained a tandem repeat of a 2.9 kb *E. coli* DNA insert. Depression of the P_L promoter by thermal inactivation of the cI repressor protein resulted in the simultaneous overproduction of the *pro*B and *pro*A gene products (7% and 15% of the total soluble protein, respectively). Thus, the *pro*B and *pro*A loci are closely linked and transcribed in the same direction. A second plasmid construct contained a 1.2 kb *E. coli* DNA fragment and allowed the overproduction of the *pro*C gene product (5% of the total soluble protein). The high expression levels obtained facilitated the purification to homogeneity of the gene products of the *pro*B, *pro*A, and *pro*C loci.

Introduction

Proline is a key amino acid, from an industrial point of view, based upon its high cost, its use as a component in parenteral and enteral nutrition formulations for humans, its use as a pharmaceutical intermediate for the synthesis of antihypertensive drugs, and its potential role in agriculture as an osmoprotectant. For the past several years, this laboratory has been studying the molecular biology of the proline biosynthetic pathway in the gram-negative bacterium *E. coli* with the objective of genetically engineering *E. coli* to overproduce proline.

Proline is synthesized from glutamate by the action of three enzymes: γ-glutamyl kinase (GK), γ-glutamyl phosphate reductase (GPR), and Δ'-pyrroline-5-carboxylate reductase (P5CR), the gene products of the *pro*B, *pro*A, and *pro*C loci, respectively (Hayzer and Leisinger 1980; Baich and Pierson 1965). GK [ATP:L-glutamate 5-phosphotransferase; EC 2.7.2.11] catalyzes the ATP-dependent γ-phosphorylation of glutamate to form γ-glutamyl phosphate and ADP (Baich 1969). GPR [L-glutamate 5-semialdehyde:$NADP^+$ oxidoreduc-

*Present Address: Washington Research Center, W. R. Grace & Co., 7379 Route 32, Columbia, MD 21044.

**Present Address: Laboratory of Molecular Microbiology, NIAID, Building 550, Ft. Detrick, Frederick, MD 21701.

***Present Address: Syngene Products & Research, 225 Commerce Drive, P.O. Box 2211, Ft. Collins, CO 80522.

tase (phosphorylating); EC 1.2.1.41] catalyzes the NADPH-dependent reduction of γ-glutamyl phosphate to L-glutamate 5-semialdehyde and $NADP^+$ (Baich 1969). Due to the extreme liability of γ-glutamyl phosphate, it has been proposed that GK functions in a complex with GPR (Baich 1969). L-glutamate 5-semialdehyde undergoes a spontaneous cyclization to form Δ′-pyrroline-5-carboxylate, which is then converted to proline through the action of P5CR [L-proline:NADP(P)$^+$ 5-oxidoreductase; EC 1.5.12] (Vogel and Davis 1952; Rossi et al. 1977a).

The overproduction of amino acids via fermentation is partially dependent upon overcoming the normal cellular regulatory circuits such as feedback inhibition and repression. Regulation of proline biosynthesis is thought to be exerted primarily through feedback inhibition of GK by proline (Baich 1969; Baich and Pierson 1965), although some regulation of proline biosynthesis may also be exerted through P5CR (Rossi et al. 1977a, 1977b). *E. coli* mutants that excrete proline can be selected by resistance to the toxic proline analogue 3, 4-dehydroproline; these strains possess GK activity with decreased sensitivity to feedback inhibition by proline (Baich and Pierson 1965; Tristram and Thurston 1966).

GPR has previously been purified to homogeneity and characterized (Hayzer and Leisinger 1982); however, P5CR has been only partially purified (Rossi et al. 1977a), and GK has only been studied in cell-free extracts or crude preparations (Baich 1969; Hayzer and Leisinger 1980).

We have constructed two plasmids, which facilitate the production and purification of the enzymes of the proline biosynthetic pathway (Deutch et al. 1982; Smith et al. 1984; Deutch et al. 1984). In this report we describe the production and purification of GK (resistant to feedback inhibition by proline), GPR, and P5CR, and we present evidence that the *pro*B and *pro*A loci are closely linked and transcribed in the same direction.

Materials and Methods

Enzymes. Restriction endonucleases and T_4 DNA ligase were supplied by Bethesda Research Laboratories, Inc. and were used as suggested by the manufacturer.

Bacterial strains and plasmids. BRL 1945 is a *rec*A derivative of *E. coli* K-12 strain K802 (*hsd*R2 *sup*E *lac*Y *gal*K *met*A) constructed in our laboratory by F. Bloom. Plasmid pAD-13 contains the *pro*BA locus from a 3, 4-dehydroproline-resistant *E. coli* strain, cloned onto the expression plasmid pGW7. Plasmid pGW7-*pro*C contains the *pro*C gene cloned onto pGW7. Plasmid pGW7 is a derivative of pBR322 and contains the bacteriophage λP_L promoter and the lambda gene encoding a temperature-sensitive cI repressor protein. The *pro* genes were cloned onto pGW7 in such a manner as to be under the control of the temperature-inducible λP_L promoter (Deutch et al. 1982; Smith et al. 1984; Deutch et al. 1984).

Enzyme overproduction—Analytical scale. Strain K802 with and without pAD-13 was grown at 30 C in a complex medium (YET broth; containing 10 g Bacto-Tryptone, 5 g yeast extract, 5 g NaCl per liter) with and without 100 μg/ml ampicillin, respectively. At an optical density of 0.65–0.80, at 660 nm, 230-ml portions of each culture were incubated at 30 C and 42 C with vigorous shaking for a period of 2 h. Strain BRL 1945 containing pGW7 or pGW7-*pro*C was grown at 30 C in YET broth containing ampicillin. After reaching an optical density of 0.7–1.0, 230-ml portions of each culture were incubated as described above.

Enzyme overproduction—Preparative scale. Four one-liter cultures of strain K802/pAD-13 were grown at 30 C, in YET broth supplemented with ampicillin, to an optical density of 1.0, at 660 nm. The culture temperature was then increased to 42 C and incubation was continued with vigorous shaking for 2 h. Strain BRL 1945/pGW7-*pro*C was used to inoculate 450 ml of YET broth supplemented with ampicillin, and the culture was grown as described above.

Extract preparation, enzyme assay, and enzyme purification. Details on the preparation of enzyme extracts, enzyme assays, and enzyme purifications have been given previously (Deutch et al. 1982; Smith et al. 1984).

SDS-polyacrylamide gel electrophoretic analysis of proteins accumulated during induction of the λP_L promoter. Following 2 h of incubation with shaking at 30 C and 42 C (see above), 0.1-ml aliquots of each culture were subjected to centrifugation in an Eppendorf microfuge. The cell pellets were resuspended in electrophoresis buffer, boiled for 5 min, and either electrophoresed on a 10% polyacylamide slab gel (P5CR), or on a 7.5%–20% polyacrylamide gradient slab gel (GK/GPR), containing 0.1% SDS (Laemmli 1970). After electrophoresis the protein bands were visualized by staining with Coomassie Brilliant Blue R-250.

Results and Discussion

Overproduction of GK, GPR, and P5CR

Recombinant plasmids, containing either the *pro*BA locus or the *pro*C gene subcloned onto the expression plasmid pGW7, were constructed to facilitate the production and purification of the three enzymes comprising the proline biosynthetic pathway. Plasmid pAD-13 contains a tandem repeat of a 2.9 kb DNA fragment, oriented as shown in Fig. 1, encoding both GK and GPR (Smith et al. 1984; Deutch et al. 1984). Plasmid pGW7-*pro*C contains a single copy of a 1.2 kb DNA fragment, oriented as shown in Fig. 1, encoding P5CR (Deutch et al. 1982).

The effect of temperature on the specific activities of GK and GPR, in cells either containing no plasmid or pAD-13 or, in the case of P5CR, cells either containing pGW7 or pGW7-*pro*C, is shown in Table 1.

A 38-fold and 60-fold increase (compared to wild-type levels, 30 C) was observed in the specific activities of GK and GPR, respectively, in cell harboring pAD-13 and grown under conditions where the λP_L promoter was repressed (30 C). An additional 2.2-fold and 4.8-fold increase, respectively, in the specific activities of GK and GPR was obtained by growing the culture at 42 C, thermally inactivating the repressor of the P_L promoter. At 42 C, strain K802/pAD-13 showed a 94-fold and 483-fold increase in the specific activities of GK and GPR, respectively, in comparison to wild-type cells (K802, 42 C). The temperature-

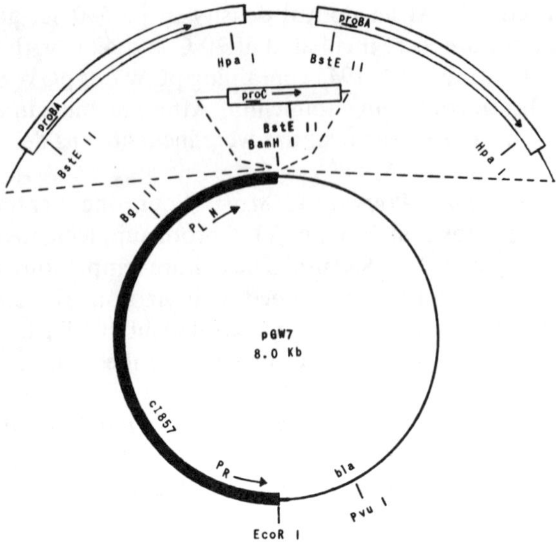

FIG. 1. Schematic presentation of plasmids pAD-13 and pGW7-*pro*C. Plasmid pAD-13 contains a tandem repeat of a 2.9 kb *E. coli* DNA fragment, within which is encoded the *pro*B (GK) and *pro*A (GPR) gene products, inserted at the *Bam*Hl site of the expression plasmid pGW7. Plasmid pGW7-*pro*C contains a 1.2 kb *E. coli* DNA fragment, within which is encoded the *pro*C (P5CR) gene product, inserted at the *Bam*Hl site of pGW7. Plasmid pGW7 is comprised of DNAs derived from phage λ (thick line) and from pBR322 (thin line). The direction of transcription is indicated by the arrows.

TABLE 1. *Relative specific activities of GK, GPR and P5CR as a function of depression of the* λP_L *promoter*[a]

		Relative Specific Activity[b]		
Strain	Temperature	GK	GPR	P5CR
K802	30 C	1.0	1.0	n.d.[c]
	42 C	0.9	0.6	n.d.
K802/pAD-13	30 C	38	60	n.d.
	42 C	85	290	n.d.
BRL 1945/pGW7	30 C	n.d.	n.d.	1.0
	42 C	n.d.	n.d.	1.0
BRL 1945/pGW7-*pro*C	30 C	n.d.	n.d.	2.2
	42 C	n.d.	n.d.	190

[a] Culture growth conditions and strains are described in Materials and Methods.

[b] Data are the average of two separate experiments. One relative specific activity unit for GK = 0.592 U, where 1 U = 1.0 nmol product formed per min per mg protein; for GPR = 2.09 U, where 1 U = 1.0 nmol NADP reduced per min per mg protein; and for P5CR = 0.634 U, where 1 U = 1 μmol NADPH oxidized per min per mg protein.

[c] n.d. = not determined.

dependent, simultaneous overproduction of GK and GPR is attributed to an increase in transcriptional activity originating from the derepressed P_L promoter, since this over production is dependent on the orientation of the *pro*BA DNA with respect to the P_L promoter. These experiments demonstrate that the *pro*B and *pro*A genes are closely linked (see also Le Rudulier and Valentine 1982;

Hayzer 1983; Mahan and Csonka 1983), are transcribed in the same direction, and possibly comprise a polycistronic operon.

The specific activity of P5CR in strain BRL 1945/pGW7-*pro*C was elevated only 2.2-fold as compared to enzyme activity isolated from cells that contained the expression plasmid lacking DNA encoding *pro*C. This attenuated gene dosage effect has been observed with other plasmids carrying the 1.2 kb DNA fragment encoding *pro*C and is not the result of low plasmid copy number (data not shown). This observation is also independent of the orientation of the 1.2 kb fragment in pGW7 (for cells grown at 30 C) and is independent of temperature when the 1.2 kb DNA fragment is subcloned onto pGW7 in an orientation opposite to that required for transcription from the P_L promoter (Deutch et al. 1982). Furthermore, the structural gene for P5CR is located well within the boundaries of the 1.2 kb DNA fragment and thus probably contains all required regulatory regions (Deutch et al. 1982). The lack of a strong gene dosage effect on P5CR activity might be an indication of either autogenous regulation of the *pro*C gene by the *pro*C gene product or, of the requirement for positive activation of the *pro*C gene. Additional evidence that regulation of proline biosynthesis may occur at *pro*C is provided by the observation that a class of proline-excreting mutants, *arg*D, possess elevated levels of P5CR activity (Rossi et al. 1977b).

Derepression of the P_L promoter resulted in an additional 86-fold increase in the specific activity of P5CR, and an overall increase of 190-fold over that observed in cells possessing the expression vector without the 1.2 kb DNA fragment insert.

Total soluble protein obtained from cells grown at 30 C or 42 C was fractionated by SDS-polyacrylamide gel electrophoresis and the abundant proteins were detected by staining with Coomassie Brilliant Blue (Fig. 2). Analysis of protein prepared from cells containing pAD-13 revealed the plasmid-specified accumulation of two major protein species of molecular weight 37 k and 41.5 k.

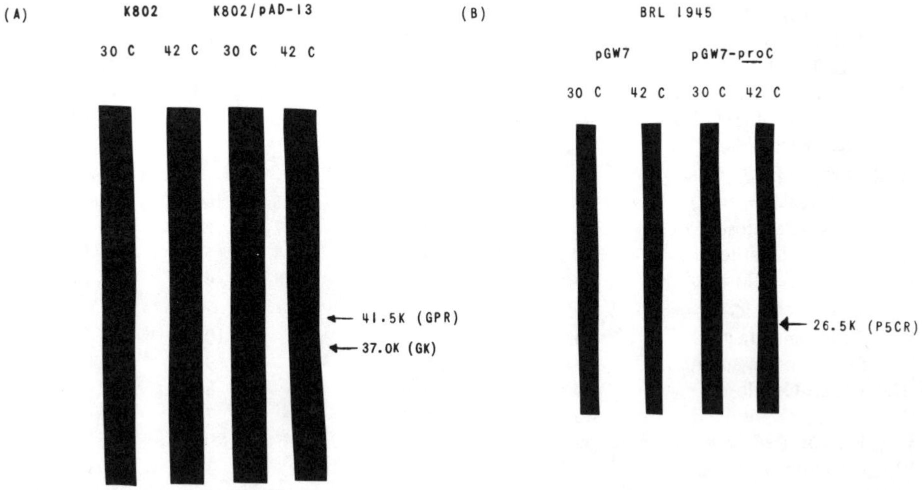

FIG. 2. SDS-polyacrylamide gel electrophoretic analysis of proteins accumulated during induction of the λP_L promoter. Overproduction of (A) GK and GPR; (B) P5CR. Bacterial strains and growth conditions are described in Materials and Methods.

These protein bands were prominent even when protein was examined from cells grown at 30 C, and the intensity of staining increased further when the temperature of the culture was raised to 42 C. Densitometric analysis of the Coomassie stained gel indicated that the 37 k and the 41.5 k proteins together represented 20–25% of the total soluble protein, with the 41.5 k protein occurring in a 2- to 3-fold molar excess over the 37 k protein.

A new major protein band was evident in temperature-induced cells harboring pGW7-*pro*C (Fig. 2). The 26.5 k protein represented approximately 5% of the total soluble protein.

Purification of GK, GPR, and P5CR

GK was purified to homogeneity, in 20% yield, from strain BRL 1945/pAD-13 (Table 2). Analysis of purified GK by SDS-polyacrylamide gel electrophoresis showed a single protein band, which comigrated with the temperature-inducible 37 k protein. A coupled enzyme assay, employing GPR, was required in order to detect GK activity in other than crude preparations (Smith et al. 1984), providing evidence in support of the suggestion that an interaction between these two enzymes is necessary for GK activity (Baich 1969).

It is interesting that the use of Procion Red Agarose in the purification of GK and P5CR resulted in the recovery of greater enzyme activity than that applied to the columns (Table 2): a stimulation of enzyme activity that has yet to be explained.

TABLE 2. *Purification of GK, GPR and P5CR*

Enzyme fraction	Volume (ml)	Activity[d] Total units	Activity[d] U mg^{-1}	Yield	Fold Purification
GK (K802/pAD-13)[a]					
I. Cell-free extract	53	1,065	0.58	100	1
II. DEAE-Cellulose	160	814	0.94	76	1.6
III. Procion Red Agarose	54	980	11.8	92	20.2
IV. Cibacron Blue Agarose	41	375	11.5	35	19.6
V. Hydroxylapatite	29.7	206	41	19.5	70.7
GPR (K802/pAD-13)[b]					
I. Cell-free extract	38	810	0.6	100	1
II. DEAE-Cellulose	240	525	3.7	65	5.9
III. Hydroxylapatite	370	480	23.4	59	38
IV. DEAE-Cellulose	24.6	381	28.8	47	46
P5CR(BRL 1945/pGW7-proC)[c]					
I. Cell-free extract	8.0	9,216	87	100	1
II. Streptomycin sulfate	8.8	8,756	95	95	1.1
III. DEAE-Cellulose	34.5	7,900	317	85.7	3.6
IV. $(NH_4)_2SO_4$	10.0	6,820	524	74	6.0
V. Procion Red	33.0	8,448	2,910	91.6	33.0
VI. Cibacron Blue	3.2	3,228	1,530	35	17.5

[a] Starting material was six 1-liter cultures.
[b] Starting material was four 1-liter cultures.
[c] Starting material was a 450-ml culture.
[d] See Table 1.

GPR was purified to homogeneity, in nearly 50% yield, from BRL 1945/pAD-13 (Table 2), and was used as a reagent in the coupled enzyme assay for the detection of purified GK. The purified GPR was essentially free of contaminating protein and appeared on a SDS-polyacrylamide gel as a single band, which comigrated with the temperature-inducible 41.5 k protein.

P5CR was purified to homogeneity, in 35% yield, from K802/pGW7-*pro*C (Table 2). The purified protein was free of contaminating protein as judged by SDS-polyacrylamide gel electrophoresis, and comigrated with the temperature-inducible protein of 26.5 k.

Conclusions

In wild-type *E. coli* both GK and GPR exhibit very low specific activities, making their measurement difficult. The high expression levels obtained by cloning the *pro*BA locus onto the expression plasmid pGW7 facilitated the purification of these enzymes in high yield. A second important factor in the purification of GK was the use of a coupled assay for detection of GK activity in purified preparations. The use of GPR as a reagent in the coupled assay required large amounts of the enzyme, and was facilitated by the construction of a plasmid capable of directing the overproduction of GPR. The purification procedures outlined for GK, GPR, and P5CR, yielded apparently homogeneous proteins.

Literature Cited

Baich, A. 1969. Proline synthesis in *Escherichia coli:* A proline-inhibitable glutamic acid kinase. *Biochim. Biophys. Acta* 192:462–467.

Baich, A., and D. J. Pierson. 1965. Control of proline biosynthesis in *Escherichia coli. Biochim. Biophys. Acta* 104:397–404.

Deutch, A. H., K. E. Rushlow, and C. J. Smith. 1984. Analysis of the *Escherichia coli pro*BA locus by DNA and protein sequencing. *Nucleic Acids Res.* 12:6337–6355.

Deutch, A. H., C. J. Smith, K. E. Rushlow, and P. J. Kretschmer. 1982. *Escherichia coli* Δ^1-pyrroline-5-carboxylate reductase: Gene sequence, protein overproduction and purification. *Nucleic Acids Res.* 10:7701–7714.

Hayzer, D. J. 1983. Sub-cloning of the wild-type *pro*AB region of the *Escherichia coli* genome. *J. Gen. Microbiol.* 129:3215–3225.

Hayzer, D. J., and T. Leisinger. 1980. The gene-enzyme relationships of proline biosynthesis in *Escherichia coli. J. Gen. Microbiol.* 118:287–293.

Hayzer, D. J., and T. Leisinger. 1982. Proline biosynthesis in *Escherichia coli:* Purification and characterization of glutamate-semialdehyde dehydrogenase. *Eur. J. Biochem.* 121:561–565.

Laemmli, U. K. 1970. Cleavage of structural proteins during the assembly of the head of the bacteriophage T_4. *Nature* 227:680–685.

Le Rudulier, D., and R. C. Valentine. 1982. Genetic engineering in agriculture: Osmoregulation. *Trends in Biochem. Sci.* 7:431–433.

Mahan, M. J., and L. N. Csonka. 1983. Genetic analysis of the *pro*BA genes of *Salmonella typhimurium:* Physical and genetic analyses of the cloned *pro*B$^+$

A+ genes of *Escherichia coli* and of a mutant allele that confers proline overproduction and enhanced osmotolerance. *J. Bacteriol.* 156:1249–1262.

Rossi, J. J., J. Vender, C. M. Berg, and W. H. Coleman. 1977a. Partial purification and some properties of Δ^1-pyrroline-5-carboxylate reductase from *Escherichia coli. J. Bacteriol.* 129:108–114

Rossi, J. J., J. Vender, C. M. Berg, and W. H. Coleman. 1977b. Proline excretion in *Escherichia coli:* A comparison of an *arg*D+ strain and a proline-excreting *arg*D derivative. *Biochem. Gen.* 15:287–296.

Smith, C. J., A. H. Deutch, and K. E. Rushlow. 1984. Purification and characteristics of a γ-glutamyl kinase involved in *Escherichia coli* proline biosynthesis. *J. Bacteriol.* 157:545–551.

Tristram, H., and C. F. Thurston. 1966. Control of proline biosynthesis by proline and proline analogues. *Nature* (London) 212:74–75.

Vogel, H. J., and B. D. Davis. 1982. Glutamic γ-semialdehyde and Δ^1-pyrroline-5-carboxylic acid, intermediates in the biosynthesis of proline. *J. Am. Chem. Soc.* 74:109–112.

CHAPTER **31**

Antibiotic Production by *Streptomyces cinnamonensis* ATCC 12308

DONALD W. THAYER[1], CARYL E. HEINTZ[2], JOHN N. MARX[2], DAVID E. COX[2], AND ROBERT HUFF[2]

Eastern Regional Research Center, ARS, USDA, Philadelphia, Pennsylvania 19118[1], and Texas Tech University, Lubbock, Texas 79409[2]

The ability of *S. cinnamonensis* ATCC 12308 to produce polyether type antibiotics was investigated. This organism was found to produce both whole-broth antibiotic activity and methanol-extractable antibiotics. The addition of Fe_2SO_4 increased antibiotic production. Methyl propionate and ethyl acetate enhanced antibiotic production. The methanol extracts were active against *Staphylococcus aureus, Micrococcus luteus, Streptococcus faecalis, Bacillus megaterium, Alcaligenes faecalis, Pseudomonas aeruginosa,* and *Pseudomonas maltophilia.* Thin-layer chromatographic results indicated that the active component in the methanol extracts differs from several authentic samples of polyether antibiotics.

INTRODUCTION

Polyether antibiotics have proven to be of great value for the prevention of coccidiosis in poultry and other animals (Reid et al. 1977; Reid et al. 1975) and the increase of feed-efficiency ratios in poultry and beef cattle. These antibiotics are monocarboxylic acid ionophores with furan-pyran polyether structures. In the crystalline state the two ends of the antibiotic molecule are held together by a hydrogen bond between the carboxyl group and a tertiary hydroxyl on the terminal tetrahydropyran ring to form a cyclic structure. Because all of the oxygen functions are concentrated in the center of the molecule and the hydrophobic alkyl groups are on the surface, the polyether antibiotics are almost insoluble in water and soluble in organic solvents in either the salt or free acid forms (Westley 1977). Biologically, the polyether antibiotics have broad spectrum activity against gram-positive bacteria, yeasts, fungi, and protozoa but are too toxic for most therapeutic uses. The inclusion of small amounts of some of the polyether antibiotics, such as monesin in the forage fed to steers, results in an increase in feed efficiency. It was discovered by Dinius et al. (1976) that this increase resulted from an increase in the ratio of propionic acid to acetic acid in the ruminal content.

At least nine species of *Streptomyces* produce polyether antibiotics and appear to comprise a class of fairly closely related organisms. *Streptomyces cinnamonensis* (ATCC 15413), which produces monensins A, B, C, and D (Haney and Hoehn 1968), was isolated from soil and classified as a strain of *Streptomyces cin-*

namonensis Okami, NRRL B1588 by C. E. Higgons (Stark et al. 1968). The spore chain morphology of ATCC 15413 was described (Haney et al. 1964) as having straight to flexuous sporophores and spore chains with neither spirals nor hooks present, which differs from the description by Pridham and Tresner (1974), who described *S. cinnamonensis* spore chains as forming hooks, open loops, or greatly expanded coils. The literature is further confused by the existence of several species with very similar names and descriptions, such as *Streptomyces cinnamoneus*, which is now classed as *Streptoverticillium cinnamoneum* (Baldacci and Locci 1974) and *Streptoverticillium cinnamoneum* forma *azacoluta*. The ambiguities concerning the possible coidentity of *S. cinnamonensis* ATCC 15413 and *S. cinnamonensis* ATCC 12308 was the subject of this study. Classic taxonomical approaches were planned for one study and in the other the ability of the two strains to produce polyether antibiotics would be compared. This paper reports the results of study of antibiotic production by *S. cinnamonensis* ATCC 12308.

Materials and Methods

Cultures. Streptomyces cinnamonensis ATCC 12308 and *Streptomyces cinnamonensis* ATCC 15413 were obtained from the American Type Culture Collection, Rockville, MD. Slant cultures were grown for 5 d on the agar medium of Stark et al. (1968) consisting of glucose, 10 g; Soytone (Difco), 10 g; agar, 25 g; and distilled water 1,000 ml. The pH was adjusted to 7.3 with 1 *N* NaOH before sterilization by autoclaving. Spores were harvested from the slants to make suspensions for inoculation of the culture media.

Culture media. The media used in this study were based on those of Stark et al. (1968) and are described in Table 1.

TABLE 1. *Basal media used during fermentation studies*

Medium component	Concentration (g/liter)		
	1	2	3
Glucose	5.0	30.0	10.0
Nutrisoy flour	15.0		
Soytone (Difco)			15.0
Soy grits		25.0	
Refined soybean oil		20.0	
Methyl oleate		20.0	
Dextrin	20.0		20.0
Yeast extract (Difco)	2.5		2.5
$CaCO_3$	1.0	1.0	1.0
KCl	0.10	0.1	0.1
K_2HPO_4	0.10	0.1	0.1
$MnCl_2 \cdot 4H_2O$		0.6	0.6
$Fe_2(SO_4)_3$		0.3	0.3
pH	8.0	8.0	7.0

Vegetative inoculum. Vegetative inocula were grown as described by Stark et al. (1968).

Growth. Mycelial growth was measured as packed cell volume by centrifuging a 10-ml sample of whole-broth culture for 30 min at 500 × g. Growth was recorded as percentage of solids (v/v).

Comparison of antibiotic production by S. cinnamonensis ATCC 15413 and S. cinnamonensis ATCC 12308. Cultures of *S. cinnamonensis* ATCC 15413 and *S. cinnamonensis* ATCC 12308 were grown in the Stark et al. (1968) production medium (Medium 2, Table 1) using a 2% inoculum from the vegetative culture. The cultures were incubated at 32 C for periods of several days. Values for pH, packed cell volume, viable cell number, whole-broth antibiotic activity, and methanol extractable antibiotic activity were measured on at least a daily basis. The methanol extractable activities reached maxima after approximately 4 d incubation. This incubation period was, therefore, selected as the standard for comparison and the results that are reported and discussed in the results section are those obtained after 4 d incubation.

Assay. Antibiotic titers in fermentation broth samples were measured by microbiological assay on Tryptic Soy Agar (TSA, Difco). Antibiotic titers were calculated as micrograms per milliliter assayed against monensin standards. Five replicate assays were made of each sample using 12.5-mm diameter Whatman No. 1 filter paper discs. *Bacillus megaterium* ATCC endospore suspensions (2×10^9 per ml) were used as lawns on TSA, by the Kirby-Bauer technique, 40 ml per 15 × 150-mm Petri dish. Each disc received exactly 0.10 ml of the mixture to be assayed. After they had dried, five discs for each assay were placed on the inoculated assay plate. The diameters of the inhibition zones were measured with calipers after incubation of the plates for 16 h at 30 C.

Assay samples were prepared as described by Stark et al. (1968). Methanol extracts were prepared by adding 10 ml of methanol to 2 ml of fermentation medium containing both cell mass and product, referred to as whole-broth in the remainder of the paper. Stark et al. (1968) established that this technique reproducibly extracted 90% of monensin activity. After extraction for approximately 30 min, the cells were removed by centrifugation. Assay samples were diluted, when necessary, with 25% (v/v) methanol in distilled water.

Identification of methanol extractable components. Authentic samples of monensin Na, A204 A, and dianemycin were provided by Lilly Laboratories. Authentic samples of lasalocid Na, X-205 Na, and nigericin were obtained from Hoffman-La Roche Inc. In addition, a sample of rumensin, which is the commercial product intended as a ruminant feed additive and contains monensins A and B, was obtained from a commercial source. The antibiotic(s) were extracted and crystallized from rumensin as described by Haney and Hoehn (1968).

The methanol extracts were analyzed by thin-layer chromatography on Eastman Chromogram silica gel, without fluorescent indicator, thin-layer chromatography sheets 20 × 20 cm. Samples were applied 2 cm from the bottom of the plates and developed with ethyl acetate. Whatman 3MM, 20 × 20-cm filter paper sheets were also used as chromatographic supports and were developed with ethyl acetate as described above. Following chromatographic development the dry plates were sprayed with 3% vanillin in 1.5% ethanolic sulfuric acid

(Haney and Hoehn 1968) or 3% H_2SO_4 and heated in an oven for 5 min at 100 C. Several other reagents including ninhydrin, *p*-dimethylaminobenzaldehyde/antimony trichloride, iodine/sodium azide, nitroprusside, isatin, sulfanilic acid, dinitrosalicylic acid, alkaline silver oxide, aniline/xylose, acridine, and dinitrophenylhydrazine were used to develop the chromatograms and identify the chemical properties of any of the solutes contained in the methanol extracts. There were no ninhydrin or isatin reacting compounds, presumably amino acids or proteins, in the extracts. Dinitrosalicylic acid and alkaline silver oxide, which are tests for sugars, dinitrophenylhydrazine as a test for keto acids, and aniline/xylose and acridine as tests for organic acids all failed to react. The exception was the Ehrlich *p*-dimethylaminobenzaldehyde/HCl reagent (Smith 1969), which was more sensitive than the vanillin reagent in reacting with known polyether antibiotics.

Antibiotic stabilities. Aliquots of 50 ml each from ATCC 12308 fermentation broth containing antibiotic activity received the following treatments: none, 50 C for 30 min, 75 C for 30 min, 100 C for 30 min, adjust to pH 2.0 with HCl, adjust to pH 10 with NaOH, and adjust to 0.g *M* with trichloracetic acid. Each aliquot was then assayed for whole-broth and methanol extractable activity.

Solubility of methanol extractable antibiotics. Aliquots of 100 ml each of ATCC 12308 fermentation broth were adjusted to pH 9.0, mixed with 3 g of Avicel, and filtered. Each filter cake was extracted four times with 25 ml of one of the solvents.

Isolation procedure. Fermentation broth (500 ml) containing active antibiotic activity produced by ATCC 12308 was adjusted to pH 9.0 with NaOH, mixed with 15 g of Avicel, and filtered. The filter cake was repeatedly extracted with $CHCl_3$ (totalling 200 ml). The residual filter cake following extraction with $CHCl_3$ was extracted with 150 ml of CH_3OH. The original 500 ml of filtrate was adjusted to pH 9.0 and extracted with 200 ml of $CHCl_3$ in a continuous extractor for 24 h. This extract was concentrated *in vacuo* to 25 ml. Each of the three extracts had biological activity.

In vitro antimicrobial spectrum. The methanol extract of an active ATCC 12308 culture was placed on either 0.1 or 0.2 ml amounts on 13-mm filter paper discs. The Kirby-Bauer technique was used to inoculate TSA in a 150 × 15-mm Petri dish with each culture. Five antibiotic discs and one 3 μg control of monensin Na were placed on each plate and incubated 24 h at 30 C.

Effect of medium composition on antibiotic production. The effect of medium component on the production of whole-broth and methanol extractable antibiotics was investigated by omitting one component at a time from medium 3 in Table 1. The effect of adding 10.0 g of one of the following to medium 3 was determined in a separate study: refined soybean oil (Leveland Vegetable Oil Co.), ethyl acetate (J. T. Baker Co.), methyl deconate (Eastman), methyl formate (Eastman), methyl hexanoate (Eastman), methyl laurate (Eastman), methyl myristate (Eastman), methyl oleate (Eastman), methyl octanoate (Eastman), methyl palmitate (Eastman), methyl propionate (Eastman), and methyl stearate (Eastman).

A volume of 100 ml of medium was used per 500-ml baffled Erlenmeyer flask. Each medium was adjusted to pH 7.0 and two flasks per medium were used in the study. The inoculum was 2.0 ml from a 3 d culture of ATCC 12308 in medium 1. The cultures were incubated at 32 C with agitation at 300 rpm (2.5 cm radius). Several preliminary studies indicated that methanol extractable activity peaked in medium 3 after 4 d incubation, and, subsequently, all cultures were harvested at 96 h for this study.

Results

The principal questions to be answered in this study were (1) does *S. cinnamonensis* ATCC 12308 produce both whole-broth and methanol soluble antibiotics, and (2) if antibiotics are produced, are they the same as those produced by *S. cinnamonensis* ATCC 15413? Both ATCC 12308 and ATCC 15413 produced both whole-broth antibiotic and methanol soluble antibiotic activities in the Stark et al. (1968) production medium. When concentrated methanol extracts of ATCC 15413 cultures grown in medium 2 were subjected to thin-layer chromatography on silica gel, the major peaks corresponded to those produced by monensin Na or $CHCl_3$ extracts of rumensin (Table 2). The reaction of these chromatographic zones to vanillin was similar to that of authentic samples of monensin Na or of extracts of rumensin, which produced two chromatographic zones that turned red following reaction with vanillin. Typically, the methanol extracts of ATCC 12308 cultures contained two to three solutes with R_F values of 0.08–0.19, 0.36–0.45, and 0.93–0.97 (minor component). The second solute migrated, typically, just below the lower chromatographic zone produced by monensin Na; but unlike monensin or extracts from ATCC 15413 cultures, it produced a blue reaction color with vanillin rather than a red color. Bioassay of the chromatogram indicated that both the lower purple zone and the middle blue zone were highly active against *B. megaterium*. The tentative conclusion of these studies was that the two strains of *S. cinnamonensis* do not produce the same antibiotics, though both produced methanol soluble antibiotics, which based on their solubility and reaction with vanillin were presumptively of the polyether type.

The methanol extract of ATCC 12308 cultures had antimicrobial activity against *Alcaligenes faecalis* ATCC 19018, *Bacillus megaterium* ATCC 14581, *Pseudomonas aeruginosa, Pseudomonas maltophilia* ATCC 13048, *Staphylococcus aureus*, and *Streptococcus faecalis* ATCC 19433. The crude methanol extract was inactive against *Acinetobacter lwoffi* ATCC 7976, *Bacillus cereus* ATCC 14579, *Bacillus subtilis* W23, *Bordetella bronchicamis* ATCC 10580, *Enterobacter aerogenes* ATCC 13048, *Escherichia coli* K-12, *Flavobacterium meningosepticum, Moraxella osloensis* ATCC 19976, *Pseudomonas aeruginosa* ATCC 10145, *Sarcina lutea, Serratia marcescens* ATCC 13880, *Shigella sonnei* ATCC 10580, and *Staphylococcus epidermidis* ATCC 14990. The monensin Na control was active against *Alcaligenes faecalis* ATCC 19018, all of the *Bacillus* species, *Moraxella osloensis* ATCC 19976, and *Pseudomonas aeruginosa, Sarcina lutea,* and *Staphylococcus aureus*. These results differ from those reported

TABLE 2. *Results of silica gel thin-layer chromatography of polyether antibiotics and extracts of cultures of* S. cinnamonensis *ATCC 15413 and* S. cinnamonensis *ATCC 12308*

Sample	Amount	Vanillin reaction color	Ehrlich reaction	Zone cm	R_F
CH_3OH EXT	10 μl			0.6–3.1	0.11
ATCC 15413				6.1–7.4	0.51
Medium 1					
CH_3OH EXT	10 μl			0.5–4.0	0.17
ATCC 15413		red		5.3–5.7	0.42
Medium 2		red		6.7–7.5	0.55
				12.2–12.8	0.96
CH_3OH EXT	10 μl	purple		1.4–2.9	0.16
ATCC 12308		blue		5.1–5.9	0.42
Whole broth					
Medium 1					
CH_3OH EXT	5 μl	purple	purple	1.9–3.0	0.19
ATCC 12308		blue	blue	5.0–6.5	0.44
Whole broth			red	5.8–6.3	0.47
Medium 2			lt blue	7.0–7.8	0.57
$CHCl_3$ EXT	5 μl	trace	purple	2.0–2.8	0.18
ATCC 12308		blue	lt pink	4.6–6.5	0.42
Mycelium			blue	7.0–7.6	0.56
		lt blue	white	7.8–9.6	0.66
		lt green	grey/green	8.8–10.1	0.73
		blue/grey	white	10.9–12.6	0.90
CH_3OH EXT	5 μl	blue		0.3–0.6	0.04
ATCC 12308		purple	purple	1.8–2.7	0.17
$CHCl_3$ EXT		blue	pink	5.2–6.5	0.45
Mycelium		lt purple	pink	6.7–7.7	0.55
$CHCl_3$ EXT	13.7 μg	red	green	5.6–7.1	0.49
Rumensin		red	green	7.1–8.8	0.61
Monensin Na	5.2 μg	red		5.4–5.8	0.43
		red		6.8–7.8	0.57
A 204 A	5.8 μg	purple		9.6–10.8	0.79
Dianemycin	17.1 μg	lavender		2.8–3.9	0.26
		lavender		5.5–6.9	0.48
		red		7.3–7.8	0.59
Lasalocid Na	19.9 μg	blue-grey		10.6–11.7	0.86
X-205 Na	8.6 μg	violet		7.4–8.3	0.61
Nigericin	17.8 μg	purple		5.0–6.2	0.43
		purple		6.5–7.0	0.52

by Haney and Hoehn (1968) in that the methanol extract was active against gram-negative bacteria as well as gram-positive species. However, until the antibiotics(s) responsible for this activity are purified these results can only be considered as tentative.

Based on the biological activities of the extracts, the mycelial antibiotic(s) produced by ATCC 12308 were not soluble in water but were soluble in the organic solvents in order of decreasing solubility: methanol, *n*-butanol, chloroform, acetone, dichlorethane. The antimicrobial component of the methanol extract retained 96% of its activity after the broth was heated to either 50 or 75 C and 92% after being heated to 100 C. Acidifying the broth to pH 2.0 before extraction with methanol resulted in the extract having a slightly greater activity than when the broth was extracted at pH 8. When NaOH was added to the fermentation liquor adjusting to pH 10.0 before extraction with methanol, there was 8% lower activity in the resulting extract. Haney and Hoehn (1968) reported that monensin was stable as the sodium salt but degraded rapidly under acidic conditions. Temperature stability values were not reported.

Comparison of the methanol extracts by paper and thin-layer chromatography (Table 2) of fermentation broth and mycelium from ATCC 12308 and ATCC 15413 cultures combined with the color reactions of the chromatographic zones indicates that the two strains do not produce identical methanol soluble substances. The results of silica gel thin-layer chromatography are reported in Table 2. Of the authentic samples of polyether antibiotics, nigericin was the closest in R_F values and color reaction with vanillin to those of the methanol extracts of ATCC 12308. Bioassay of the methanol soluble components initially failed to resolve any of the zones, that is, inhibition of *B. megaterium* occurred from the spot of application to the solvent front on silica-gel thin-layer chromatograms. Later bioassays of sufficiently diluted samples obtained by extracting the mycelium of ATCC 12308 with chloroform and then by methanol indicated that the bioactive components of ATCC 12308 migrated on Whatman No. 3MM chromatography paper when developed by ethyl acetate at an average R_F value of 0.11 and 0.46 whereas the rumensin bioinhibition occurred at an average R_F value of 0.84. The results obtained with bioassay of silica gel thin-layer chromatograms of either of these extracts of ATCC 12308 produced inhibition at an average R_F value of 0.12 (0–0.23) and rumensin R_F 0.52 (0.30–0.72). Thus the bioactive component(s) of the methanol extracts of the ATCC 12308 culture is different from that of rumensin.

By varying the inoculum level (0.5, 1.0, 2.0, 5.0, and 10.0% v/v) and measuring the production of both whole-broth and methanol soluble antibiotic activities in the Stark et al. (1968) production medium, an inoculum of 2.0% was optimum for the production of antibiotic activities by ATCC 12308 (data not shown).

The results of a study in which components of medium 3 were sequentially eliminated are presented in Table 3. The deletion of Soytone from the medium strongly enhanced production of methanol extractable antibiotic activity by ATCC 12308 even though the packed cell volume was reduced by 76%. The deletion of any of the minerals but especially $CaCO_3$, K_2HPO_4, and $Fe_2(SO_4)_3$ inhibited both growth and production of methanol extractable antibiotic(s). The elimination of glucose enhanced production of methanol extractable activity whereas the elimination of dextrin from the medium resulted in neither whole-broth nor methanol extractable activity being produced. The deletion of yeast extracts from medium 3 resulted in enhanced whole broth activity by ATCC 12308. The results indicated that the whole broth and methanol soluble activities are affected differently by the medium components and, thus, there are presumably antibiotics other than the polyether antibiotics present in the whole broth.

Stark et al. (1968) reported a marked increase in the titer of monensin when the refined soybean oil was supplemented by the addition of methyl oleate. In this

TABLE 3. *Effect of deletion of components of medium 3 on growth, pH, whole-broth, and methanol-extractable antibiotic activities produced by ATCC 12308*

Component eliminated	Final pH	Packed cell volume %	Whole-broth relative activity	Methanol extractable relative activity
None	8.5	16.5	1.0	1.0
Glucose	8.8	10.0	1.1	1.3
Soytone	7.0	4.0	2.6	1.9
Dextrin	9.0	10.5	0.0	0.0
Yeast extract	7.0	12.5	1.7	0.9
$CaCO_3$	8.2	13.0	0.4	0.6
KCl	8.6	7.2	0.8	0.7
K_2HPO_4	8.2	9.5	0.9	0.6
$MnCl_2$	8.3	10.5	1.0	0.9
$Fe_2(SO_4)_3$	8.4	9.8	0.8	0.7

study the addition of soybean oil or a fatty acid ester supplementation of a basal medium was investigated with differing results (Table 4). Methyl oleate, at least by itself, did not increase the production of methanol extractable antibiotics. The supplementation of the basal medium in which ATCC 12308 was growing with ethyl acetate was markedly stimulatory to production of the presumably polyether type antibiotics in the methanol extracts. Acetate was not especially stimulatory to the production of monensin.

TABLE 4. *Effect of supplementation of medium 3 with 1.0% soybean oil or fatty acid esters on antibiotic production by ATCC 12308*

Supplement added	Final pH	Packed cell volume %	Whole-broth relative activity	Methanol extractable relative activity
None	8.4	9.0	1.0	1.0
Soybean oil	8.0	18.0	1.0	0.6
Ethyl acetate	8.5	10.4	0.8	2.5
Methyl decanoate	7.7	1.8	0.0	0.0
Methyl formate	5.2	1.8	0.0	0.0
Methyl hexanoate	8.7	6.6	1.1	1.1
Methyl laurate	8.2	9.4	0.2	0.0
Methyl myristate	7.9	11.0	0.6	0.2
Methyl octanoate	6.9	6.4	0.6	0.0
Methyl oleate	7.7	13.0	1.6	0.6
Methyl palmitate	7.6	13.5	1.1	0.5
Methyl propionate	8.4	9.2	0.8	0.4
Methyl stearate	7.7	16.5	0.7	0.6

DISCUSSION

The chromatographic studies described above indicated that *S. cinnamonensis* ATCC 12308 does not produce the two major components of monesin Na that are produced by *S. cinnamonensis* ATCC 15413. Under the conditions of our experiments ATCC 15413 did not produce the methanol extractable substances synthesized by ATCC 12308. This does not indicate that ATCC 15413 cannot produce these compounds, only that it did not do so under the conditions we employed. The two major components in the methanol extracts of ATCC 12308 cultures appear to be similar to nigericin.

The results of the physiological study of *S. cinnamonensis* ATCC 12308 were markedly different from those reported by Stark et al. (1968) for *S. cinnamonensis* ATCC 15413. The production of monensin was not reported to be sensitive to substitutions or variations of concentrations of glucose in the fermentation medium. The elimination of dextrin in this study markedly lowered antibiotic titers. The deletion of glucose actually produced a small increase of antibiotic titer when dextrin remained. The increase in methanol extractable antibiotic titer when Soytone was eliminated also differs from the results reported for ATCC 15413.

Both cultures of *S. cinnamonensis* apparently are sensitive to the presence of minerals such as iron and calcium for the production of polyether type antibiotics. The requirements of ATCC 12308 were not quantified. Stark et al. (1968) reported an optimum concentration of 0.3 g/liter of $Fe_2(SO_4)_3$ for the production of monensin by ATCC 15413.

The observation that ethyl acetate but not methyl oleate was stimulatory to the production of polyether type antibiotics by ATCC 12308 differs markedly from the results reported by Stark et al. (1968) for ATCC 15413.

The results presented in this study establish that *S. cinnamonensis* ATCC 12308 and *S. cinnamonensis* ATCC 15413 are physiologically different.

Acknowledgments

This research was completed at Texas Tech University and was supported by a grant from Micro Chemical, Inc., Amarillo, Texas.

Literature Cited

Baldacci, E., and R. Locci. 1974. Genus II. Streptoverticillium. Pages 829–842 *in* R. E. Buchanan and N. E. Gibbons, eds., *Bergey's Manual of Determinative Bacteriology*. 8th ed. The Williams and Wilkins Company, Baltimore, MD.

Dinius, D. A., M. E. Simpson, and P. B. Marsh. 1976. Effect of monensin fed with forage on digestion and the ruminal ecosystem of steers. *J. Anim. Sci.* 42:229–234.

Haney, M. E., Jr., and M. M. Hoehn. 1968. Monensin a new biologically active compound I. Discovery and isolation. *Antimicrob. Agents Chemother.* 1967:349–352.

Haney, M. E., Jr., M. M. Hoehn, and J. M. McGuire. 1964. A3823 novel antibiotic. U.S. Pat. Appl. No. 399706.

Pridham, T. G., and H. D. Tresner. 1974. Genus I. Streptomyces. Waksman and Henrici 1943, 339. Pages 748–829 *in* R. E. Buchanan and N. E. Gibbons, eds., *Bergey's Manual of Determinative Bacteriology*. 8th ed. Williams and Wilkins Co., Baltimore, MD.

Reid, W. M., J. Johnson, and J. Dick. 1975. Anticoccidial activity of lasalocid in control of moderate and severe coccidiosis. *Avian Dis.* 19:12–18.

Reid, W. M., J. Dick, J. Rice, and F. Stino. 1977. Effects of monensin-feeding regimens on flock immunity of coccidiosis. *Poult. Sci.* 56:66–71.

Smith, I. 1969. *Chromatographic and Electrophoretic Techniques. Vol. I. Chromatography.* Interscience Publishers, New York.

Stark, W. M., N. G. Knox, and J. E. Westhead. 1968. Monensin a new biologically active compound. II. Fermentation studies. *Antimicrob. Agents Chemother.* 1967:353–358.

Westley, J. W. 1977. Polyether antibiotics: Versatile carboxylic acid ionophores produced by streptomyces. Pages 177–223 *in* D. Perlman, ed., *Advances in Applied Microbiology,* Vol. 22, Academic Press, New York.

CHAPTER **32**

Isolation of Cephalosporin C from Fermentation Broths Using Membrane Systems and High-Performance Liquid Chromatography

M. Kalyanpur, W. Skea, and M. Siwak

Millipore Corporation, Bedford, Massachusetts

Cross-flow filtration and preparative high-performance liquid chromatography were used to isolate cephalosporin C from fermentation broths. *Acremonium chrysogenum* (ATCC 48272) was fermented in a complex fermentation medium containing coarse ingredients, such as soy flour, starch, corn steep liquor, and calcium carbonate. The viscous fermentation broth was initially clarified through a cross-flow filtration device with a 0.45 μm microporous membrane. The clarified broth was then filtered through another cross-flow filtration device with a spiral-wound ultrafiltration membrane with a nominal molecular weight limit of 10,000. This step removed most of the precipitate formed in the clarified broth during storage at low temperatures and some soluble proteins and part of the dark reddish brown pigment. The permeates were then concentrated 6-fold using a spiral-wound reverse osmosis module. The permeate containing the antibiotic was loaded onto a large-scale, high-performance liquid chromatograph (KiloPrep™). Cephalosporin C recovered from this system was compared for bioactivity with a sample of the pure antibiotic. Our product was 93% pure.

Introduction

The high dollar value of antibiotics and the level of purity expected of the final product by the pharmaceutical industry have resulted in much attention being paid to the entire process for their manufacture. The array of life-saving drugs available today is a result of concerted efforts on the part of microbiologists, biochemists, chemists, and chemical engineers. They have made significant contributions to strain, media, and process improvements. All this has been accomplished bearing in mind that the development of a good process for producing high yields of an antibiotic in the fermentor should be matched by an equally good process to recover most of that antibiotic in a very pure form.

Usually, the first step in the recovery of an antibiotic is the separation of the microorganisms and other solids in the fermentation broth. The clarified broth is then subjected to extraction with a suitable organic solvent or a series of chromatographic procedures where adsorption on carbon, silica, or either ionic or nonionic resins is followed by selective elution. The final steps are concentration and precipitation or crystallization until the desired level of purity is attained. The selection of the appropriate process depends on the properties of the antibiotic.

The clarification of fermentation broths is usually done with rotary vacuum filters precoated with filter aids, such as diatomaceous earth or in plate and frame filter devices. The separation of solids from the broth can also be accomplished by centrifugation and is often the equipment of choice for separating single-cell organisms like bacteria and yeasts. However, these methods suffer from certain drawbacks. Precoated filters tend to adsorb some of the antibiotics in the broth and are not easily eluted. Such losses can be significant when the product is either produced in low yields or is of a high dollar value. The performance of centrifugal separations is hard to predict because of the wide variation of percentage of solids, particle size or viscosity of the liquid in different fermentation beers.

When solvent extraction is used to remove antibiotics from the aqueous broth, the solvent extracts have to be concentrated by distillation *in vacuo* prior to precipitation or crystallization. The concentration process can take a long time and, even under the mild conditions, one can lose some portion of heat-labile antibiotics. Moreover, this is an energy-intensive process and poses problems of explosion and health hazards, which make it necessary to carry out this operation in specially designed areas. The last, but not the least-important, problem is the cost of disposal of waste solvents.

At Millipore Corporation, we are developing novel separation techniques for the pharmaceutical industry that could replace some of the classical methods described above. We chose cephalosporin C as a model compound to work with because it is presently isolated from fermentation broths by some of these classical methods. Cephalosporin C is a valuable antibiotic as a source of 7-aminocephalosporanic acid which, in turn, is the starting material for a number of semisynthetic cephalosporins; new improved methods for cephalosporin C's separation and purification would be attractive to the industry. The techniques we are developing employ cross-flow filtration devices with microporous, ultrafiltration, reverse osmosis membranes, and process scale high-performance liquid chromatography.

This paper describes the use of these techniques for the recovery of cephalosporin C from fermentation broths. Ultrafiltration is finding increasing application in the pharmaceutical industry, especially in the recovery of products made by bacteria in relatively simple fermentation media. The system we used is under development at Millipore and makes clarification of the viscous cephalosporin C fermentation broths relatively easier than was possible with older systems. Datta et al. (1977) and Trieber et al. (1981) have studied the feasibility of using reverse osmosis for concentrating antibiotics, and our results confirm that this method offers the inherent advantage of excellent recovery of the antibiotic. The use of the KiloPrep™ System for the final recovery of cephalosporin C from the concentrate by reverse phase chromatography completes our aim to use Millipore separation systems in every phase of downstream processing of an important pharmaceutical product.

Materials and Methods

Media. Agar slant medium contained sucrose, 2.0%; K_2HPO_4, 0.05%; KH_2PO_4, 0.05%; KCl, 0.05%; $MgSO_4 \cdot 7H_2O$, 0.05%; $FeSO_4 \cdot 7H_2O$, 0.001%; $NaNO_3$, 0.3%; yeast extract, 0.4%; peptone, 0.4%; agar, 2.0%. The medium was sterilized at 121 C for 20 min. The seed medium contained cornstarch, 4.0%; corn steep liquor, 3.0%; soybean meal, 1.0%; ammonium sulfate, 0.1%; calcium carbonate, 0.3%; and methyl oleate, 2.4%. This medium was adjusted to pH 7.0

before sterilization at 121 C for 45 min. The fermentation medium contained sucrose, 2.0%; cornstarch, 3.0%; beet molasses, 5.0%; soybean meal, 6.0%; ammonium acetate, 0.8%; calcium sulfate, 1.25%; calcium carbonate, 0.5%; and methyl oleate, 3.0%. The medium was adjusted to pH 6.4 before sterilizing at 121 C for 45 min.

Cephalosporin C fermentation. Fermentations were carried out with a strain of *Acremonium chrysogenum,* ATCC 48272, maintained on agar slant medium. The slants were incubated at 25 C for 7 d, and the slant growth suspension in sterile water was used to inoculate 50 ml of the seed medium in 500-ml Erlenmeyer flasks. The flasks were incubated on a rotary shaker at 25 C for 54 h. For fermentations, 11 liters of fermentation medium in NBS Microgen Fermentors was inoculated with 250 ml of a mature seed of the antibiotic-producing organism and the fermentation started at 25 C with 0.5 vvm of air and 500 rpm agitation. The fermentation medium became very viscous after day 1 and the air-agitation rates were increased to 0.7 vvm and 900 rpm, respectively. The pH of the fermentation medium was controlled at 6.0 from day 3 onward, and samples of fermentation broths were tested each day for antibiotic production. The fermentors were harvested on day 6, at which point the assay for cephalosporin C was at its peak value.

Ultrafiltration of fermentation beers. Fig. 1 shows a complete process schematic for the recovery of cephalosporin C. The fermentation beers were initially clarified through a Millipore tangential flow filtration device. The filtration module was of an open channel design that is capable of accepting highly viscous fluids containing high concentrations of particulate matter. The system contained Millipore 0.2 μm microporous GVLP membranes. The fermentation broth was processed at an average pressure of 30 psi and the temperature was maintained at 20 to 25 C. A clear filtrate was obtained, leaving behind a highly viscous cell paste that contained the mycelium and particulates from the fermentation medium.

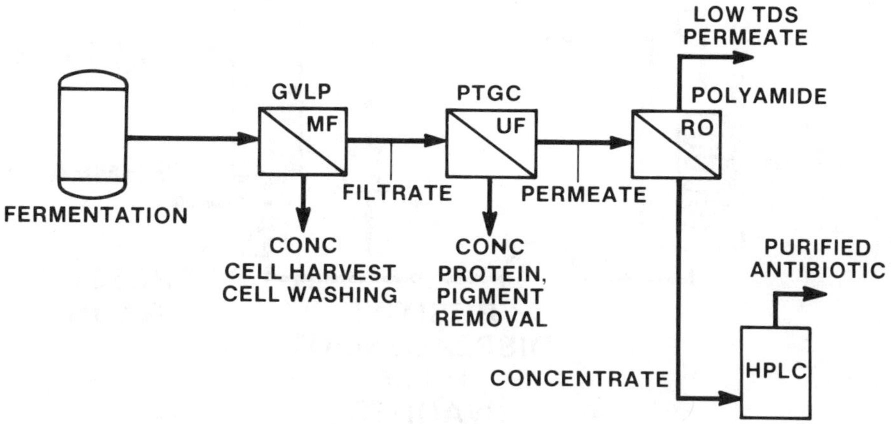

FIG. 1. Schematic of the process for recovery of cephalosporin C.

The clear filtrate was then processed through another Process Ultra Filtration System (Millipore Corporation) with a spiral-wound membrane to eliminate high molecular weight proteins and polysaccharides from the filtrate. The PTGC membrane used in this step had a nominal molecular weight limit of 10,000. The filtration was carried out at 15 C with a recirculation rate of 8 gal/min through a 15 ft² membrane area. More antibiotic was recovered from the retentate by adding 4 liters of water to it and reprocessing this volume through the system. This step helped to remove not only the large molecular weight compounds from the filtrate but also a considerable amount of the dark brown pigment associated with the fermentation product.

Concentration of clarified filtrate by reverse osmosis. The filtrate, or permeate, from the previous step was next processed by reverse osmosis to concentrate the solution for further purification by high-performance liquid chromatography. The equipment used in this step is illustrated in Fig. 2. Concentration was accomplished by recirculating the broth through a reverse osmosis spiral module and removing permeate from the system. The filtration was run at a transmembrane pressure of 500 psi and a recirculation rate of 2.8 gal/min at 20–25 C. The spiral-wound module was of a polyamide composite type membrane with 15 ft² of membrane area.

At volumetric concentration factors of 1, 2, 3, 4.8, and 6, pressure excursions were performed to generate data on flux vs. applied pressure. Samples of the con-

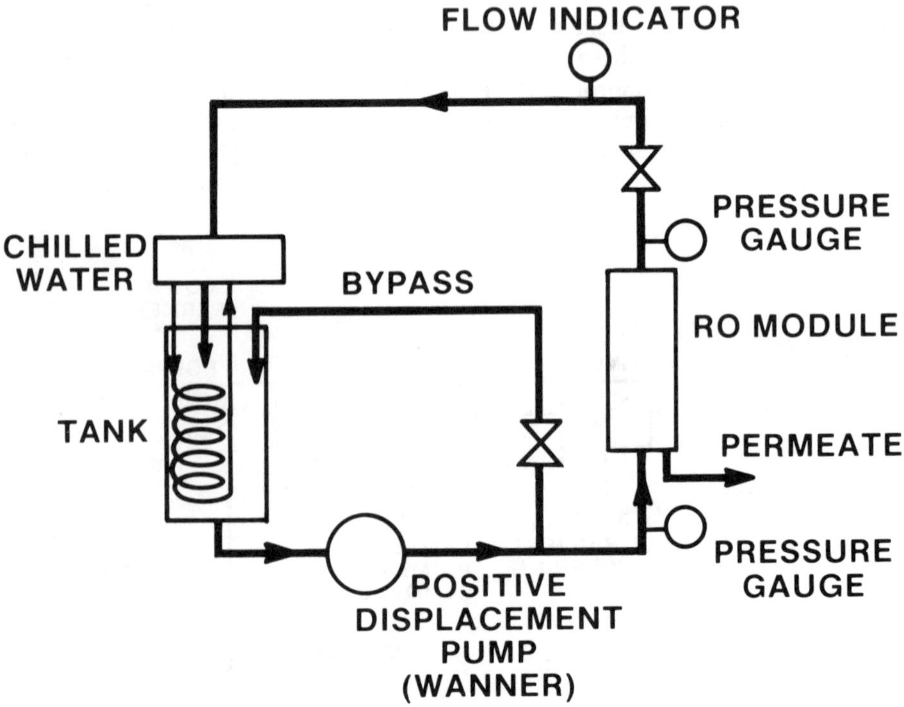

FIG. 2. Schematic of the system for concentration of cephalosporin C by reverse osmosis.

centrate and permeates taken at different concentration factors were analyzed for cephalosporin C content.

Recovery and purification of cephalosporin C from the RO concentrate. High-performance liquid chromatography (HPLC) was used to isolate cephalosporin C from the cell-free clarified broth. Initial experiments focused on the optimization of the separation via analytical HPLC, followed by scale-up to preparative and process scale isolations.

Analytical chromatography was performed with a Waters Associates HPLC system consisting of a model 6000 A pump, a model U6K manual injector, a model 440 UV detector, a μ Bondapak C_{18} column (10 μm, 4.8 mm i.d. × 30 cm), a custom-packed Radial-Pak Prep C_{18} (55–105 μm, 8 mm i.d. × 10 cm) column, and a model 730 data module. A mobile phase of 20 m*M* NH_4OAc, pH 5.2, buffer was used to isolate the component of interest.

Once the optimum analytical conditions were established, studies were conducted to determine the maximum amount of broth that could be loaded onto a column to produce acceptable purity and yield of the cephalosporin C. This was accomplished with the above system using a custom-packed Radial-PAK Prep C_{18} (55–105 μm, 8 mm i.d. × 10 cm) cartridge in a Z-Module™ Radial Compression Separation System (Waters Associates). Having determined the optimum load of cephalosporin C per g of column packing, preparative chromatography was carried out with an Auto 500™ Prep LC system using a Prep PAK C_{18} cartridge (55–105 μm) (Waters Associates). The mobile phase for these isolations consisted of 97.5% 10 m*M* NH_4OAc, pH 5.2, buffer with 2.5% CH_3CN to elute the cephalosporin C within a reasonable time.

A large-scale isolation of cephalosporin C from clarified broth was accomplished using a KiloPrep™ HPLC (Fig. 3) (Millipore Corporation) with a 15 cm × 60 cm column segment packed with Prep PAK C_{18} material (55–105μm). The mobile phase was 97.5% 20 m*M* NH_4OAc, pH 5.2, buffer with 2.5% CH_3CN.

Ammonium acetate and acetonitrile were HPLC grade (J. T. Baker Chemical Co.). Reagent grade water (18 megohm-cm resistivity at 25 C) was obtained from a Milli-Q Water System (Millipore Corporation). Cephalosporin C standard was obtained from Sigma Chemical Co.

Results and Discussion

The strain of *Acremonium chrysogenum* was fermented in the complex production medium containing the coarse ingredients one usually encounters in large-scale antibiotic fermentations in the pharmaceutical industry (Hollander et al. 1984). Such a medium was used to ensure that the broths processed had viscosity, cell biomass, and particulate content similar to fermentations in the industry. At the end of the 6 d fermentations, the broth was bioassayed for cephalosporin C against a strain of *E. coli* using a sample of pure antibiotic supplied by Sigma Chemical Co. (St. Louis, MO).

In the first ultrafiltration step, the open channel microporous module showed high flux rates, which declined as gradual concentration was accomplished. Fig. 4

FIG. 3. Large-scale preparative HPLC system, KiloPrep™ (Millipore Corporation).

shows the drop in flux dependent upon concentration factor. The curve for flux versus concentration factor for the second filtration step is shown in Fig. 5. Classical gel polarization theory predicts that flux rate is described by the following relationship:

$$J = k \cdot \log \frac{C_w}{C_b} \quad (1)$$

where, J = a process flux, k = a constant related to fluid characteristics, C_w = concentration of retained solids at membrane interface, and C_b = concentration of retained solids in bulk solution.

The data show that, in this step, concentration of fermentation broths to 8-fold or greater is possible with the system employed, unless limited to viscosity constraints. Such constraints are likely to be more significant in fermentations of filamentous organisms like *A. chrysogenum* than with bacterial fermentations in simple, defined media.

Flux data for the ultrafiltration of the clarified broth through the spiral-wound system show that a concentration of 20-fold can be achieved and indicate that the levels of retainable proteins, polysaccharides, or any other high molecular weight

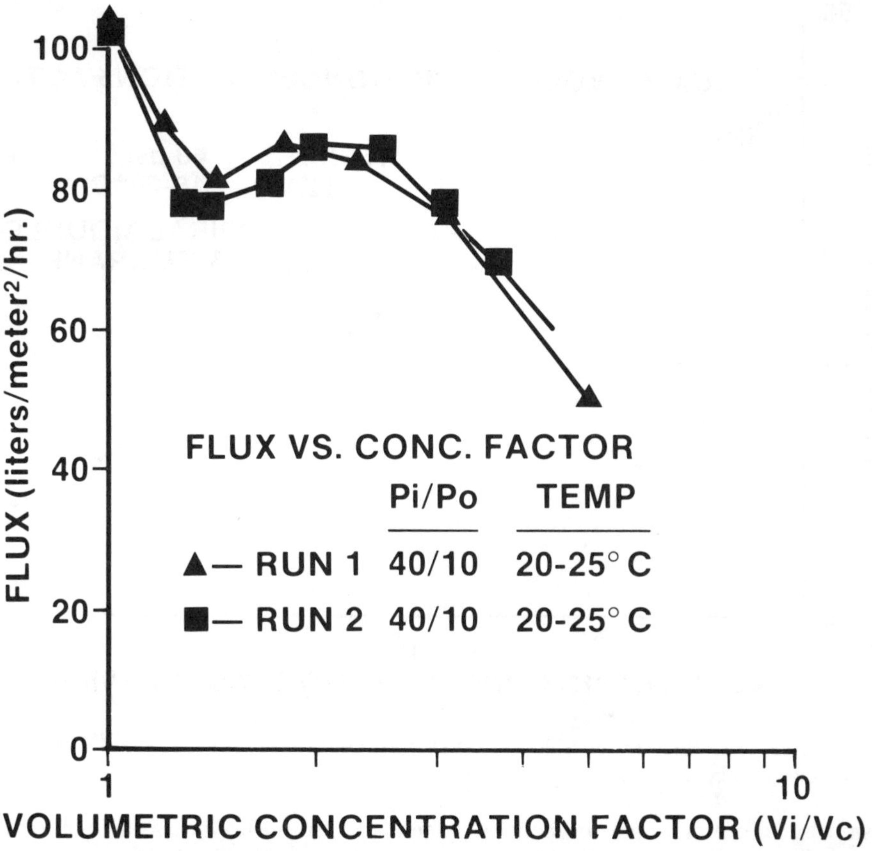

FIG. 4. Flux decay curve for ultrafiltration of fermentation beers through microporous membranes.

products in the filtered broth are probably low. Here it is important to use the minimum levels of nutrients, especially soybean meal and starch, needed to support a high-yielding fermentation. Unnecessarily high levels of nutrients that are not really needed in the fermentation process can only cause problems in downstream processing, especially when membrane systems are used.

The ultrafiltration of the broths through the 10,000 NMWL membranes also removed some of the pigments from the clarified beers. The pigments in microbial fermentations are usually low molecular weight compounds, but perhaps they bind some of the proteins in the broth and were removed along with these proteins. Such a step of cleaning fermentation broths definitely extends column life in the chromatographic procedures and often leads to better resolution of desired compounds.

The recoveries of the antibiotic in the first two ultrafiltration steps were 74% and 80%, respectively. Use of a more efficient cooling system for the concentrate and permeate may have yielded more cephalosporin C. Also, diafiltration of the retentate in the first step could have released more antibiotic from the highly viscous cell paste.

During the concentration of the clarified fermentation broth by reverse

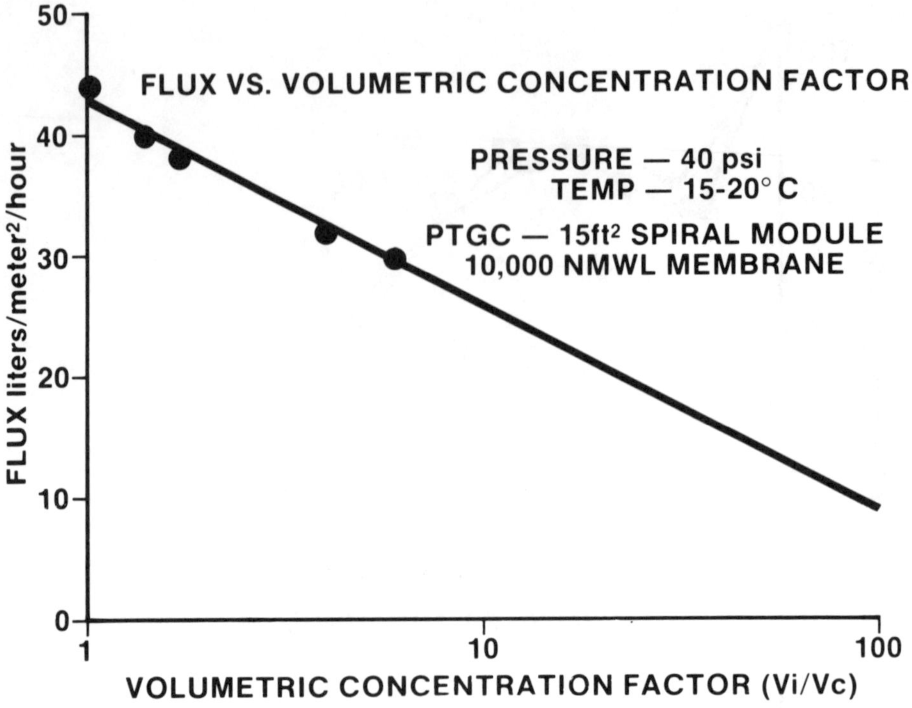

FIG. 5. Flux decay curve for ultrafiltration of clarified broths through the 10,000 NMWL ultrafiltration membrane.

osmosis, volumetric measurements were made to study flux rate as a function of concentration factor and the data are presented in Fig. 6. Volumetric concentration factor (VCF) is defined as:

$$VCF = \frac{V_i - V_p}{V_i} = \frac{V_i}{V_c} \qquad (2)$$

where, V_i = initial volume, V_p = volume of permeate removed, V_c = volume of concentrate.

These results show that flux rate and extent of concentration are a function of operating pressure. Flux rate in a nonpolarising reverse osmosis system is described by the following relationship, where osmotic pressure becomes the flux-controlling factor.

$$\text{Flux} = A\,(\Delta P - \Delta\pi) \qquad (3)$$

where, A = membrane constant, ΔP = transmembrane pressure differential, and $\Delta\pi$ = transmembrane osmotic pressure differential.

At a constant applied pressure (ΔP), the flux rate is controlled by the osmotic pressure of the solution. As $\Delta\pi$ increases due to increased solution concentration, the flux rate drops accordingly. When the osmotic pressure equals the applied pressure, membrane flux will be zero.

The major contribution to osmotic pressure is from low molecular weight compounds like salts, amino acids, residual sugars, and other primary metabolites of the fermentation. Compared to the effect of these compounds, the antibiotic that is present in relatively low concentrations contributes little to the osmotic

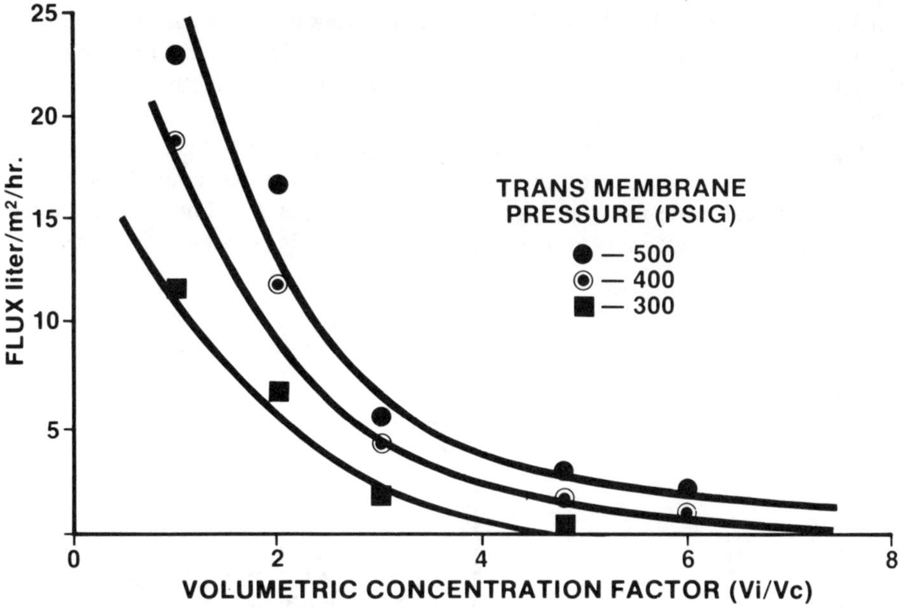

FIG. 6. Filtration characteristics in the concentration of cephalosporin C by reverse osmosis.

pressure. Consequently, antibiotic level has little impact on flux rates and the degree of concentration. Table 1 shows the antibiotic concentration in the permeates and concentrates at various concentration factors. Also included is the antibiotic retention, calculated from the equation:

$$\% \text{ Retention} = 1 - \frac{C_p}{C_r} \times 100 \qquad (4)$$

where, C_p = concentration of antibiotic in the permeate, and C_r = concentration of antibiotic in the retentate.

TABLE 1. *Antibiotic recovery during concentration by reverse osmosis*

Volumetric Concentration Factor	Antibiotic Concentration		Percent Retention
	Retentate g/l	Permeate g/l	
1	2.29	0.0	100.0
2	4.45	0.003	99.9
3	6.90	0.17	99.8
4.8	10.47	0.039	99.6
6	13.71	0.046	99.7

Antibiotic retention by the membrane is very high at >99.6% and almost constant throughout the entire concentration process. Analysis of the composite permeate that represents the total loss of antibiotic through the membrane is 0.009 g/l. Mass balance calculations on antibiotic for this 6-fold concentration yields antibiotic recovery of 99.7%, or a 0.3% loss. The high yield of cephalosporin C demonstrates that reverse osmosis is a very effective process in which degradation of highly labile compounds like antibiotics can be avoided.

The isolation of cephalosporin C was initially carried out by reverse phase analytical HLPC. Fig. 7 shows the antibiotic well resolved from other constituents. Once chromatographic conditions were established on an analytical scale, the development of a scale-up procedure was undertaken. This developmental process involved the determination of the maximum load that could be achieved at the analytical scale using preparative packing materials. The data obtained from these experiments served as the basis for predicting scale-up criteria.

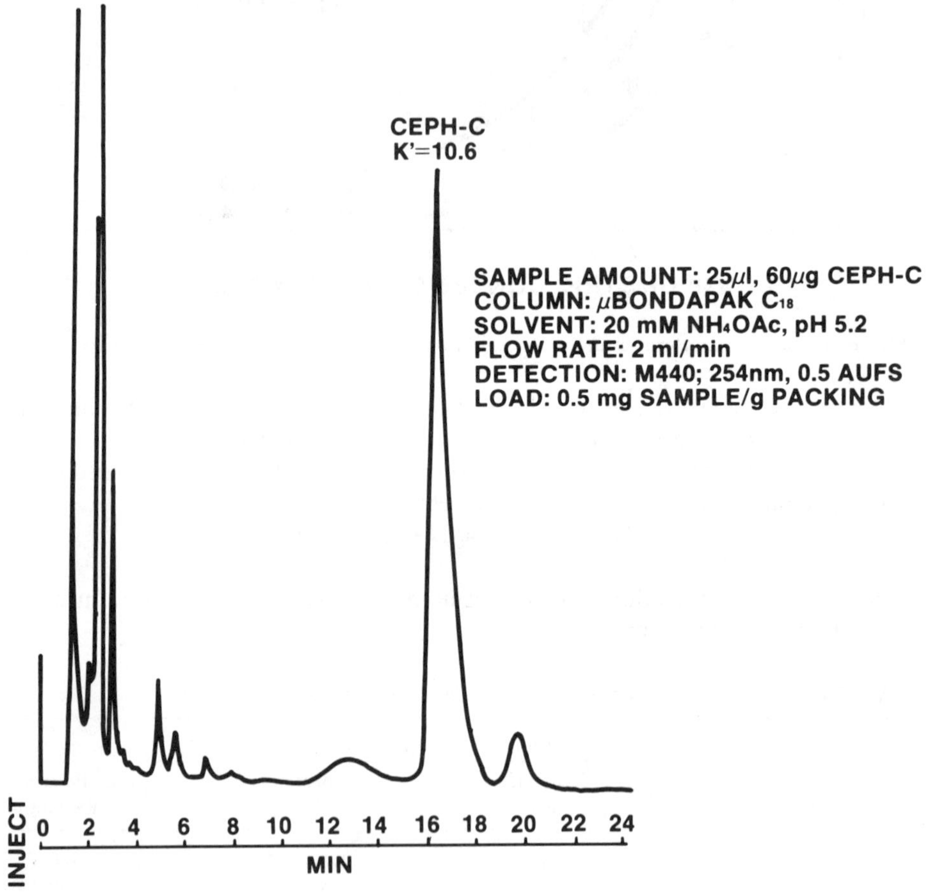

FIG. 7. Analytical HPLC separation of cephalosporin C from cell-free clarified broth.

Having settled on analytical conditions with 10 μm packing to obtain optimum resolution of the antibiotic from other components, the scaling process was started. To determine the maximum load or the volume of broth that could be loaded per g of packing before the packing no longer retained the antibiotic, preparative reverse phase packing was used. The preparative packings, 55–105 μm diameter, were packed into 8 mm × 10 cm Radial-PAK™ cartridges. As can be seen in Fig. 8, the resolution decreases as the particle size increases and as more sample is injected into the system. In addition, cephalosporin C is eluting very

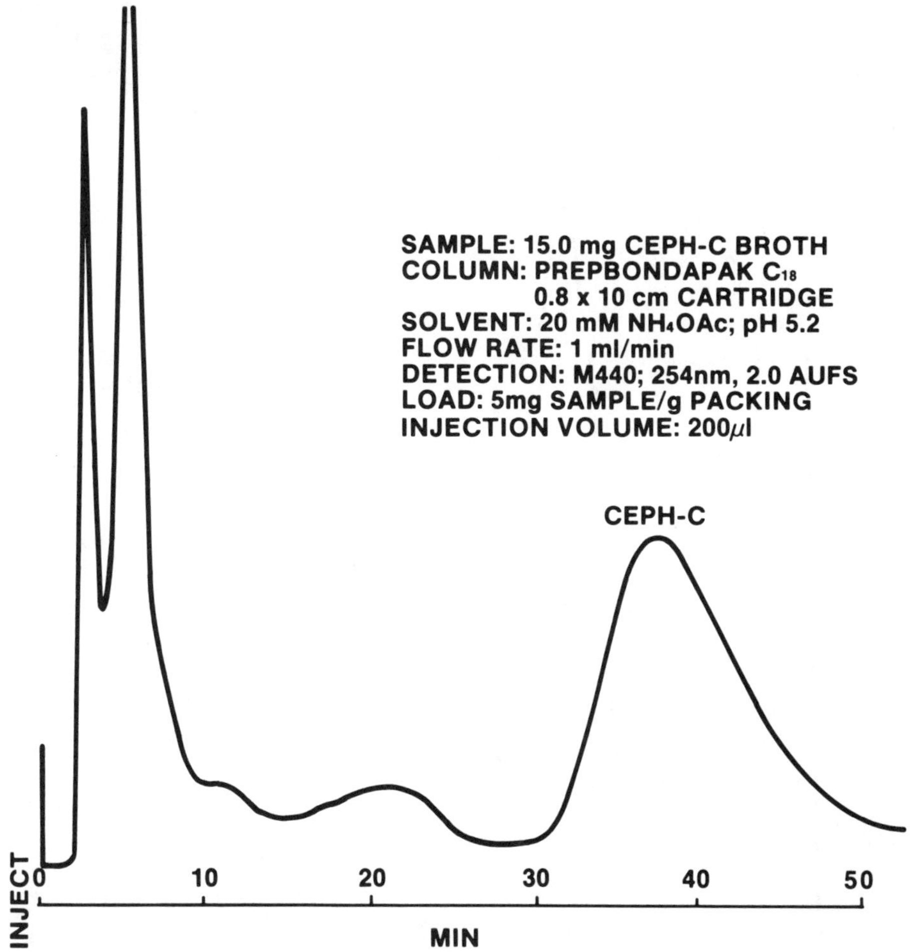

FIG. 8. Analytical loading study using preparative reverse phase C_{18} packing (55–105 μm).

late due to slight differences in chemistry between the analytical and preparative packing materials. This last phenomenon is easily overcome by the addition of small amounts of organic modifier in the mobile phase, as is seen in Fig. 9. The process of loading in mass and volume was continued until the preparative packing could no longer retain the antibiotic, i.e., cephalosporin C could be found eluting in the void volume. Data obtained from these studies were used to calculate the mass load that could be achieved on the preparative scale using the following relationship (Burgoyne et al. 1984).

$$Cm = \frac{R_2^2}{R_1} \times \frac{L_2}{L_1} \tag{5}$$

where, R_1 = radius of the analytical column, R_2 = radius of the preparative column, L_1 = length of the analytical column, and L_2 = length of the preparative column.

In this study, approximately 120 mg of crude sample could be loaded onto one

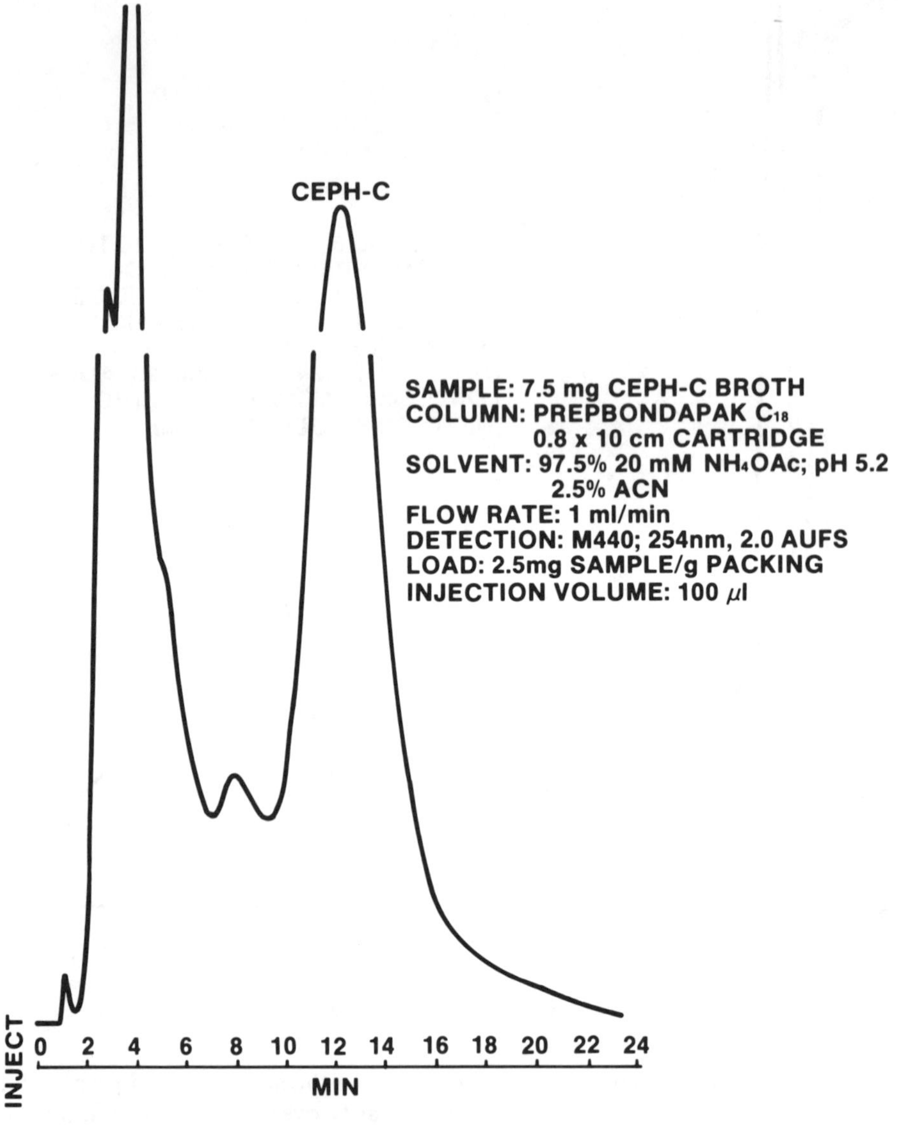

FIG. 9. Analytical loading study using preparative reverse phase C_{18} packing with the addition of an organic modifier to the mobile phase.

g of packing. Fig. 10 shows 1,100 ml (45.6 g) of crude broth loaded onto a single cartridge preparative chromatograph. Fractions were taken as indicated and examined by analytical HPLC (Figs. 11–13). Analysis of these fractions indicated that it was possible to obtain 100% recovery of the antibiotic with a purity of 93%.

Scaling to a much larger system, the KiloPrep™ (Fig. 3) was a very straightforward process after having optimized the mass load with analytical chromatography and determined the recovery/purity data from preparative

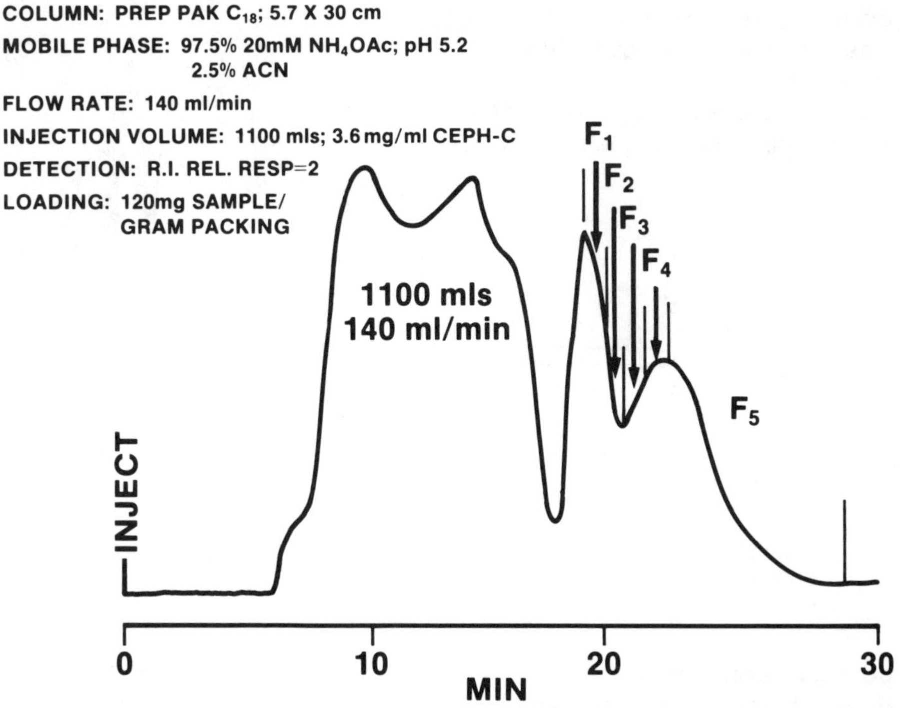

FIG. 10. Preparative isolation of cephalosporin C from cell-free clarified broth.

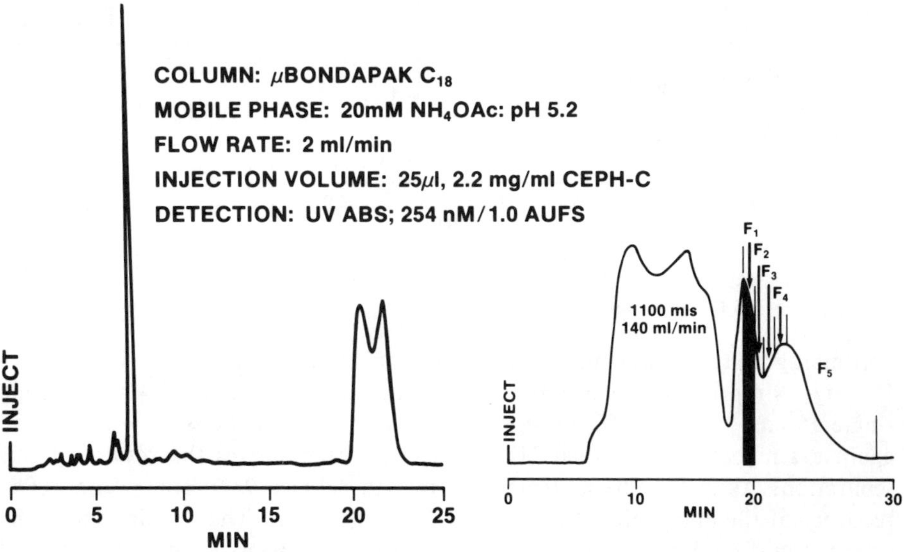

FIG. 11. Analytical HPLC analysis of fraction 1 from a preparative isolation of cephalosporin C. Cephalosporin C is approximately 46% pure.

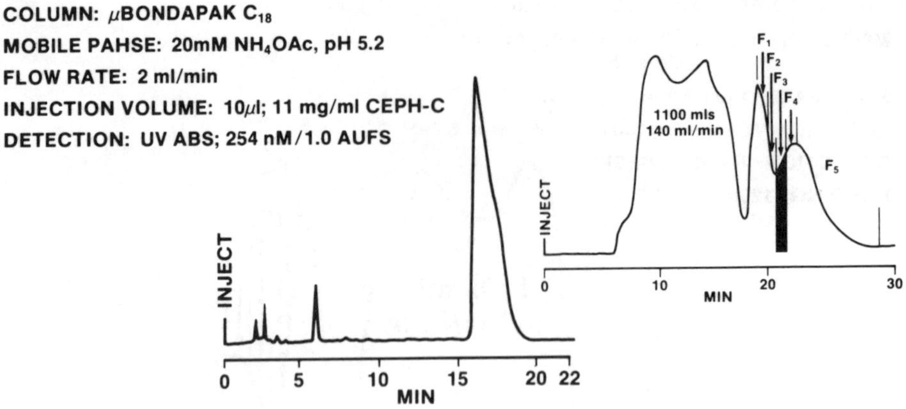

FIG. 12. Analytical HPLC analysis of fraction 3 from a preparative isolation of cephalosporin C. The purity of cephalosporin C is in the 80–85% range. A coeluting peak is evident.

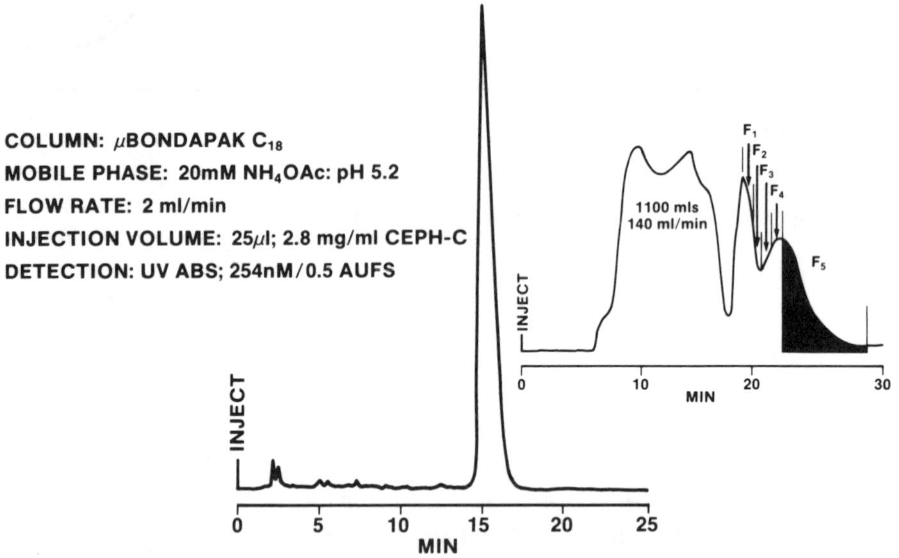

FIG. 13. Analytical HPLC analysis of fraction 5 from a preparative isolation of cephalosporin C. Cephalosporin C is 96% pure.

chromatography. Following the same relationship as stated earlier in equation (5), Fig. 14 represents the chromatographic profile of a 16-liter (554 g) load on a single 15 cm × 60 cm KiloPrep™ column segment. Fractions were taken and again examined via analytical HPLC. Table 2 summarizes the results of this separation. As is indicated, it was possible to achieve 93% purity and 100% recovery of the antibiotic in a very short period of time. The throughput of the system could be dramatically increased by using 20 cm × 60 cm segments and operating the system in a semicontinuous mode. Bioassay of the chromatographically purified antibiotic (Fig. 15) showed that it was as pure and as active as a standard obtained by conventional methods.

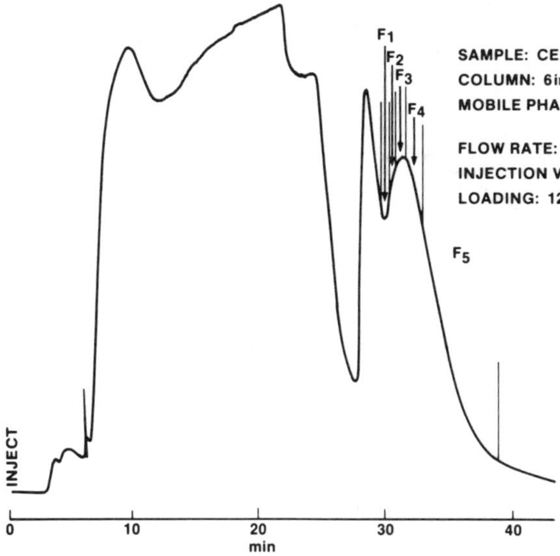

FIG. 14. KiloPrep™ isolation of cephalosporin C from cell-free clarified broth.

TABLE 2. *Summary of analytical HPLC assays of KiloPrep™ purified cephalosporin C. Injection volume: 16 liters. Flow rate: 1 liter per minute*

Fraction Number	Volume (ml)	Cephalosporin C Concentrate (mg/ml)	Cephalosporin C Recovered (grams)
1	450	2.98	1.30
2	800	8.94	7.15
3	900	14.23	12.81
4	1200	16.27	19.52
5	5800	5.7	33.06
Total	9150		

Amount loaded: 65 g cephalosporin C/576 g total mass amount recovered: 73.8 g cephalosporin C. Purity: 93%

Conclusions

A novel process stream for the isolation and purification of the antibiotic, cephalosporin C, from crude fermentation broths is described. A combination of three membrane-based ultrafiltration and reverse osmosis steps followed by large-scale preparative high-performance liquid chromatography was employed. The first two steps in the process showed some antibiotic losses, but almost 100% recovery of the product was possible in the reverse osmosis and chromatographic procedures. The probable reasons for antibiotic losses in the first two steps are discussed. A process stream such as the one described offers several inherent advantages over conventional antibiotic recovery methods in the pharmaceutical industry.

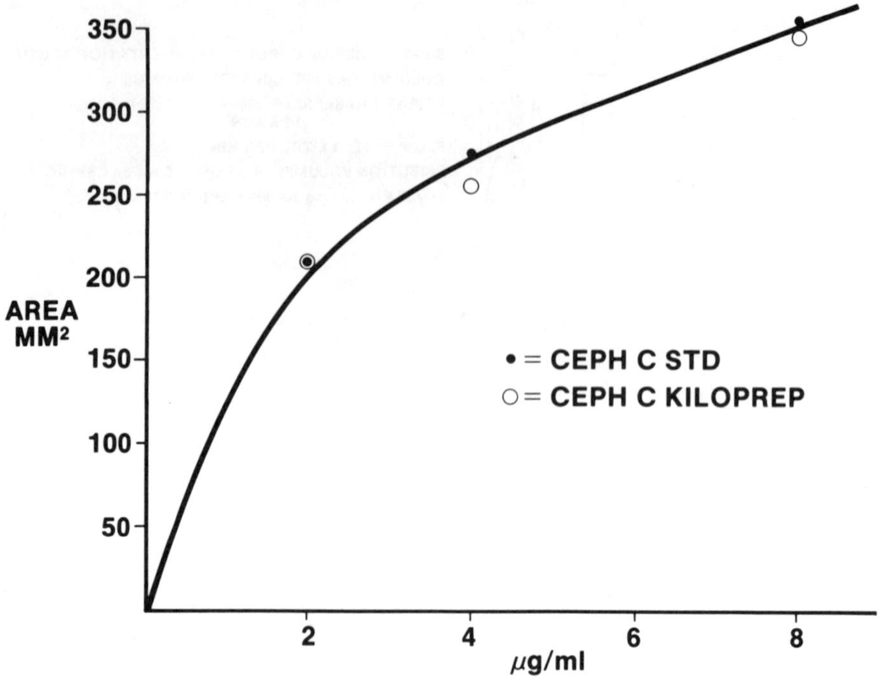

FIG. 15. Biossay of cephalosporin C purified on the KiloPrep™.

Acknowledgments

We thank Professor Arnold Demain for providing media formulations for the cephalosporin C fermentation. We acknowledge the skillful technical assistance of Christina Findeisen, Paul Hatch, and Donna Case.

Literature Cited

Burgoyne, R. F., D. K. Bowles, and A. Heckendorf. 1984. Rapid scale-up of peptide purifications by preparative HPLC. American Biotechnology Laboratory, March 1984, 38–45.

Datta, R., L. Fries, and G. T. Wildman. 1977. Concentration of antibiotics by reverse osmosis. *Biotechnol. Bioeng.* 19:1419–1429.

Hollander, I. J., Y. -Q. Shen, J. Hein, and A. L. Demain. 1984. A pure enzyme catalyzing penicillin biosynthesis. *Science* 224:610–612.

Trieber, L. R., V. P. Gullo, and I. Putter. 1981. Procedure for isolation of thienamycin from fermentation broths. *Biotechnol. Bioeng.* 23:1255–1265.

CHAPTER 33

The Fungus of Dutch Elm Disease and Antibiotics

H. M. MAZZONE

U.S. Department of Agriculture, Forest Service,
51 Mill Pond Road, Hamden, Connecticut 06514

The Dutch elm disease fungus was screened against a number of antibiotics to measure their effectiveness. Plate and liquid cultures were used to determine at what concentration the antibiotics were fungicidal. Antibiotics were tested singly and in combinations. Cerulenin, clotrimazole, polymyxin B, polyoxin D, stendomycin, and tropolone, each tested singly, were lethal to the fungus. Their efficacy, in terms of concentration required, was increased when the antibiotics were used in combinations. This was also true for antibiotics, such as puromycin, which when tested singly were not effective against the fungus. Antibiotics in combination may be increasing the permeability of the fungal wall making the organism more susceptable to the action of these agents. This study demonstrated that the Dutch elm disease fungus can be made sensitive to a variety of antibiotics employed at the μg level. This consideration should be of importance in field trials, which will be even more promising because of the following: (1) the discovery of new antifungal antibiotics; (2) breaking antibiotic resistance of the Dutch elm disease fungus through genetic engineering; (3) the improvement of drug delivery techniques, a factor in the systemic injection of antibiotics into elms.

INTRODUCTION

Dutch elm disease was clearly evident in Europe toward the end of World War I. The causal agent was determined to be an ascomycete, later designated as *Ceratocystis ulmi* (Buisman) C. Moreau. The disease first appeared in the United States at Cleveland, OH, in 1930. Its origin in this country has been attributed to the import of diseased elm logs from Europe. The terrible impact of this disease is well known, and after more than 60 yr of devastation of elms throughout Europe and North America, there is still no workable control procedure, other than that modulated by nature at various periods (Sinclair and Campana 1978).

Soon after their importance as research probes became apparent, antibiotics were tested against the Dutch elm disease fungus. In two laboratory studies Waksman and coworkers screened *C. ulmi* against a number of antibiotics. Clavacin and actinomycin were observed to be highly effective in suppressing the growth of *C. ulmi* in nutrient broth. However, the antifungal activity of these antibiotics, and especially that of actinomycin, could be partly overcome by certain nutrients in the medium, especially peptone (Waksman and Bugie 1943). Candicidin, an antibiotic isolated by the Waksman group was found to be strongly fungicidal to the fungus (Lechevalier et al. 1953). The results of antibiotics in retarding Dutch elm disease in field trials have not produced optimism. Holmes (1955) observed that pleocidin, patulin, aureomycin, candicidin, and polymyxin exhibited some inhibition against *C. ulmi* in plate cultures, but did not prevent the course of the disease when injected into trees. Campana (1977), in field trials, tested nystatin combined with solubilized benomyl, a chemical fungicide to which ascomycetes are particularly sensitive (Delp and Klopping 1968). Some arresting of the symptoms of the disease was apparent.

The search for antibiotics effective against the Dutch elm disease fungus is continuing. Recently, Nickerson et al. (1982) employed cerulenin to study the yeast-mycelial dimorphism in *C. ulmi*. The dimorphic potential of conidiospores was totally inhibited by the antibiotic at a concentration of 8 μg/ml, whereas the blastospores were not inhibited at a concentration of 100 μg/ml.

The present report on the efficacy of antibiotics against the Dutch elm disease fungus has as its basis previous screening studies undertaken in our laboratory (Mazzone et al. 1981a, b). These studies, including the present one, were induced by the great number of antibiotics now available for testing, and by the vigorous research being conducted with antibiotics in fungal research, notably by industry.

Materials and Methods

Fungus. C. ulmi (Alabama), an aggressive strain, was obtained from Dr. P. L. Pusey, Department of Plant Pathology, The Ohio Agricultural Research and Development Center, Wooster, OH. Plate cultures were grown in 100 × 15 mm plastic Petri plates containing 35 ml of potato dextrose agar. Liquid cultures were grown in test tubes containing potato dextrose broth or trypticase soy broth. Plate and liquid cultures of the fungus were grown at room temperature.

Antibiotics. Cerulenin was obtained from CalBiochem-Behring Corp., La Jolla, CA, and as a gift from Makor Chemicals, Ltd., Jerusalem, Israel; clotrimazole was a gift from the Shering Corp., Bloomfield, NJ; 4-desoxypyridoxine HCl was obtained from the Sigma Chemical Co., St. Louis, MO; polymyxin B sulfate was obtained from ICN Pharmaceuticals, Inc., Cleveland, OH, and as a gift from the Pfizer Chemical Co., New York, NY; polyoxin D was obtained from CalBiochem-Behring Corp,; puromycin dihydrochloride was obtained from the Sigma Chemical Co.; stendomycin salicylate was a gift from the Lilly Research Laboratories, Indianapolis, IN; tropolone was obtained from the Aldrich Chemical Co., Milwaukee WI.

Each of the antibiotics was dissolved using the following solvents: water for 4-desoxypyridoxine, polymyxin B, polyoxin D, puromycin, and tropolone; ethanol for cerulenin, clotrimazone, and stendomycin. The antibiotic solutions were sterilized by passage through Millipore filters. Binary mixtures of the antibiotics were also prepared. The solvent used for antibiotic combinations was water unless one of the components was water insoluble (cerulenin, clotrimazole, stendomycin). For combinations involving water-insoluble antibiotics, one or both components, the solvent was ethanol.

Testing of antibiotics against C. ulmi. When plate cultures were used for screening, 5 ml of agar overlays, with fungus alone, or with fungus and antibiotic, were added to the potato dextrose agar plates to give a final volume of 40 ml. The agar overlays were of the following composition: the control plate contained 4.5 ml of

0.7% agar plus 0.5 ml of a stock fungus suspension; the antibiotic plate contained 4.0 ml of 0.7% agar plus 0.5 ml of the stock fungus suspension plus 0.5 ml of antibiotic solution.

Liquid test cultures contained, generally, 5 ml of medium containing 0.5 ml of fungus suspension (control tube), or the same volume of medium containing 0.5 ml fungus suspension plus 0.5 ml of antibiotic solution. Water-insoluble antibiotics were not tested in liquid cultures.

The stock fungus suspension contained approximately 50,000 spores/ml. Standard antibiotic solutions could contain up to 20 mg/ml, and the maximum concentration of antibiotic tested to obtain fungicidal action was 10 mg/plate or liquid culture. The test period for both plate and liquid cultures containing antibiotics was a minimum of 7 d at room temperature (21 C).

Results

When tested singly in plate cultures, cerulenin, clotrimazole, polymyxin B, polyoxin D, stendomycin, and tropolone were fungicidal to the Dutch elm disease fungus. Puromycin and 4-desoxypyridoxine, tested singly, were not inhibitory to the fungus at a concentration of 10 mg/plate culture (Table 1). All of the antibiotics cited above, which were lethal to *C. ulmi*, showed greater efficacy when used in binary combinations (Table 2), except clotrimazole. The concentration of each component of the lethal combination was lower than that required to pro-

TABLE 1. *Effect of antibiotics on* C. ulmi *in plate cultures*

Antibiotic	Effect[a]
cerulenin	fungicidal at 1 mg
clotrimazole	fungicidal at 1 mg
4-desoxypyridoxine	noninhibitory at 10 mg
polymyxin B	fungicidal at 10 mg
polyoxin D	fungicidal at 10 mg
puromycin	noninhibitory at 10 mg
stendomycin	fungicidal at 1 mg
tropolone	fungicidal at 2 mg

[a] Concentrations are per plate culture (40-ml volume).

TABLE 2. *Combinations of antibiotics producing a fungicidal effect on* C. ulmi *in plate cultures*[a]

tropolone 1 mg	+	cerulenin 0.1 mg
tropolone 1 mg	+	4-desoxypyridoxine 1 mg
tropolone 0.5 mg	+	polymyxin B 0.5 mg
tropolone 0.1 mg	+	polyoxin D 0.1 mg
tropolone 1 mg	+	puromycin 1 mg
tropolone 1 mg	+	stendomycin 0.1 mg
tropolone 0.5 mg	+	stendomycin 0.5 mg
polymyxin B 0.5 mg	+	stendomycin 0.5 mg
stendomycin 0.5 mg	+	cerulenin 0.125 mg
stendomycin 0.25 mg	+	4-desoxypyridoxine 1 mg

[a] Concentrations are per plate culture (40-ml volume).

duce fungicidal action when the antibiotic was used singly. This was not observed to be the case when clotrimazole was used in combination with other antibiotics.

In contrast to their inactivity when used singly, puromycin or 4-desoxypyridoxine in combination with tropolone suppressed the growth of *C. ulmi* (Table 2, Fig. 1), as did 4-desoxypyridoxine in combination with stendomycin (Table 2). Combinations of tropolone with other antibiotics showed a greater frequency of effectiveness than other combinations of antibiotics. In plate culture, the combination of tropolone (0.1 mg/plate) and polyoxin D (0.1 mg/plate) provided the lowest concentration of antibiotics that was lethal to *C. ulmi* (Table 2).

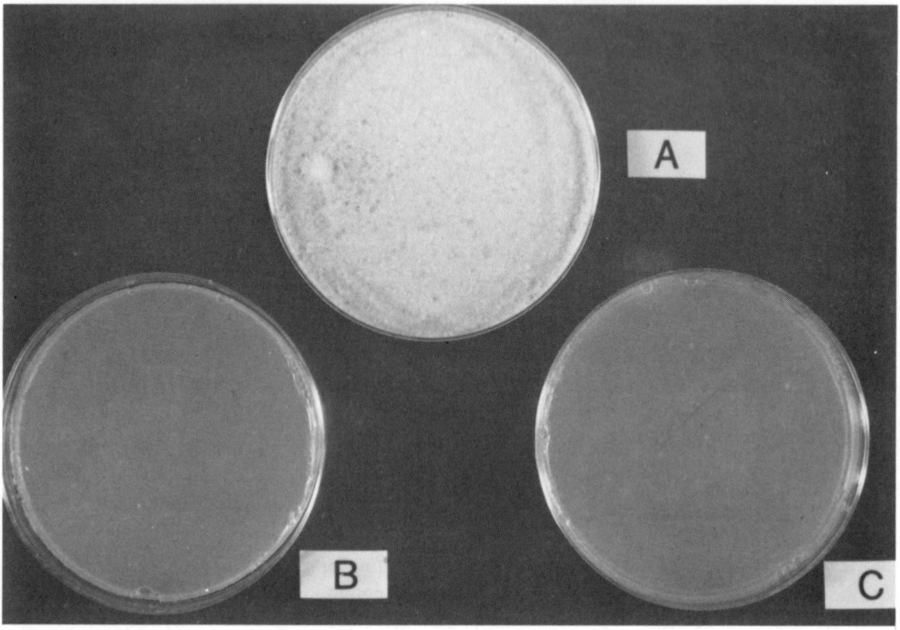

FIG. 1. Fungicidal effect of antibiotic combinations on *C. ulmi*. A. Control: fungus; B. tropolone, 1 mg and puromycin, 1 mg; C. tropolone, 1 mg and 4-desoxypyridoxine, 1 mg.

Liquid cultures of the fungus and antibiotics produced results that complemented those results noted for plate cultures. This is clearly shown for liquid cultures containing a combination of tropolone and polyoxin D, in reducing concentrations, versus *C. ulmi* (Fig. 2). At a total antibiotic concentration of 0.02 mg/ml, the combination of tropolone and polyoxin D was still fungicidal. In liquid cultures this was the lowest concentration of antibiotics that was lethal to *C. ulmi*.

DISCUSSION

An attempt to determine the minimum concentration producing fungicidal action was made by testing the antibiotics used in this study. Obviously, in field trials this factor would be meaningful in terms of providing a maximal effort for the cost of injecting trees with antibiotics. Moreover, the Dutch elm disease problem

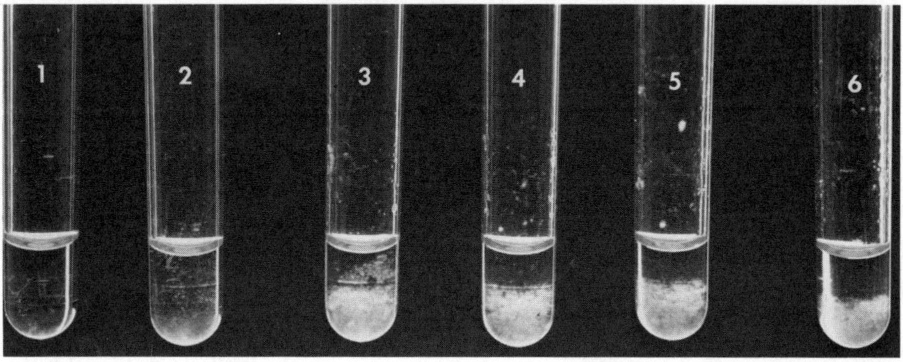

FIG. 2. Effect of tropolone combined with polyoxin D on *C. ulmi*. Tubes: (1) no growth, tropolone 1×10^{-1} mg/ml, polyoxin D 1×10^{-1} mg/ml; (2) no growth, tropolone 1×10^{-2} mg/ml, polyoxin D 1×10^{-2} mg/ml; (3) growth, tropolone 1×10^{-3} mg/ml, polyoxin D 1×10^{-3} mg/ml; (4) growth, tropolone 1×10^{-4} mg/ml, polyoxin D 1×10^{-4} mg/ml; (5) growth, tropolone 1×10^{-5} mg/ml, polyoxin D 1×10^{-5} mg/ml; (6) control culture.

is hardly ameliorated by the application of materials producing fungistatic activity.

When tested singly in a plate culture volume of 40 ml, cerulenin, clotrimazole, and stendomycin, each at a concentration of 1 mg/culture, were fungicidal at the μg level (1 mg per 40 ml = 25 μg/ml). Combinations of antibiotics provided lower concentrations that were lethal to the fungus. The combination of tropolone at 0.1 mg and polyoxin at 0.1 mg gave fungicidal action at 5 μg/ml. Similarly, the combination of stendomycin at 0.5 mg and cerulenin at 0.125 mg was lethal to *C. ulmi* at approximately 16 μg/ml. In liquid cultures, the combination of tropolone and polyoxin D was fungicidal at 20 μg/ml.

In speculating on the mode of action of the antibiotics, especially those used in combinations, it is appropriate to review their pertinent characteristics. Cerulenin is an antifungal antibiotic isolated from *Cephalosporium caerulin* (Hata et al. 1960). Specifically, cerulenin interrupts yeast-type fungal growth by inhibiting the biosynthesis of sterols and fatty acids (Nomura et al. 1972). Clotrimazole is a member of the antifungal group of antibiotics that has imidazole as a key structural moiety (Bowman et al. 1983). 4-desoxypyridoxine may not satisfy the exact definition of an antibiotic in that it is not produced by a microorganism. It is, however, a potent antimetabolite of the pyridoxine compounds involved in vitamin B-6 synthesis (Woolley 1952). Polymyxin B is a peptide antibiotic produced by *Bacillus polymyxa* (Brownlee and Jones 1948). Polymyxin B is believed to exert its effect on microorganisms in the same manner as cationic detergents, i.e., combining with the phosphatidic acids of the cytoplasmic membrane to alter membrane permeability (Rose 1968). Polyoxin D is a nucleotide antibiotic produced by *Streptomyces cacoi* var *asoensis* (Isono et al. 1967). It inhibits cell wall chitin synthesis in fungi (Endo et al. 1970). Puromycin produced by the soil actinomycete *Streptomyces allonigen,* is known to interfere with protein formation by interfering with the function of RNA in the cells involved (Yarmolinsky and de la Haba 1959). Stendomycin is a peptide antifungal antibiotic (Bodanszky and Pearlman 1969). Its precise mode of action is not known. Tropolone, a catechol-like compound exhibiting marked phenolic properties, is a noncompetitive inhibitor of S-adenosylmethionine-catechol methyltransferase (Belleau and Burba 1961). Tropolone has recently been shown to be produced by *Pseudomonas* sp. (Lindberg et al. 1980) and is lethal to fungi (Lindberg 1981).

It is not known in the present study exactly how the antibiotics exerted their effect on *C. ulmi*. However, it seems plausible that in the use of combinations of antibiotics, the fungal wall was made more permeable. Newton (1956) states this mode of action for polymyxin B and in ongoing studies in our laboratory, the alteration of fungal wall permeability also seems to be the mode of action of tropolone. In this connection with tropolone combinations, the associative antibiotic could enter the fungus and exert its action maximally. This is clearly noted in the case of puromycin and 4-desoxypyridoxine, which singly do not inhibit the fungus. However, combinations of each with tropolone, and of 4-desoxypyridoxine with stendomycin, produced a lethal effect on the fungus.

The reason for no observable enhancement of clotrimazole in combination with other antibiotics cannot be explained at this time. Perhaps a coprecipitation of the antibiotics occurred and escaped notice. The water insolubility of cerulenin, clotrimazole, and stendomycin could present a problem in field trials, where the antibiotic would be injected systemically into elms. However, the development of encapsulation techniques providing a slow release of material may permit the use of water-insoluble as well as water-soluble antibiotics.

Perhaps more laboratories, particularly those in industry, will take up the Dutch elm disease challenge. Certainly the production of novel antibiotics and the genetic potential of breaking antibiotic resistance in microorganisms should provide the motivation not to resolve Dutch elm disease by planting other species of trees.

Literature Cited

Belleau, B., and J. Burba. 1961. Tropolones: A unique class of potent noncompetitive inhibitors of S-adenosylmethionine catechol methyl transferase. *Biochim. Biophys. Acta* 54:195–196.

Bodanszky, M., and D. Pearlman. 1969. Peptide antibiotics. *Science* 163:352–358.

Bowman, K., D. Gurwith, M. Gurwith, G. Moorer, and G. Stein. 1983. Comparative study of tioconazole (UK-20,349) and clotrimazole in treatment of vaginal candidiasis. Presented at *23rd Intersci. Conf. Antimicrob. Agents and Chemother.* Las Vegas, NV, October 24–26.

Brownlee, G., and T. S. G. Jones. 1948. A related series of antibiotics derived from *Bacillus polymyxa. Biochem. J.* 43:xxv–xxvi.

Campana, R. J. 1977. Comparative evaluation of solubilized benomyl (MBC-HCl) and nystatin (Ceratocide) injected for arrest of Dutch elm disease following development of symptoms. *Am. Phytopathol. Soc. Proc.* 3 (1976):266 (Abstract).

Delp, C. J., and H. G. Klopping. 1968. Performance attributes of a new fungicide and mite ovicide candidate. *Plant Dis. Rep.* 52:95–99.

Endo, A., K. Kakiki, and T. Misato. 1970. Mechanism of action of the antifungal agent polyoxin D. *J. Bacteriol.* 104:189–196.

Hata, T., Y. Sano, A. Matsumae, Y. Kamio, S. Nomura, and R. Sugawara. 1960. Study of new antifungal antibiotic. *Japan J. Bacteriol.* 15:1075–1077.

Holmes, F. W. 1955. Field and culture tests of antibiotics against *Graphium ulmi*. *Phytopathology* 45:185 (Abstract).

Isono, K., J. Nagatsu, K. Kobinata, K. Sasaki, and S. Suzuki. 1967. Studies on polyoxins, antifungal antibiotics. V. Isolation and characterization of polyoxins C, D, E, F, G, H, and I. *Agric. Biol. Chem.* 31:190–199.

Lechevalier, H., R. F. Acker, C. T. Corke, C. M. Haenseler, and S. A. Waksman. 1953. Candicidin, a new antifungal antibiotic. *Mycologia* 45:155–171.

Lindberg, G.D. 1981. An antibiotic lethal to fungi. *Plant Dis.* 65:680–683.

Lindberg, G. D., H. A. Whaley, and J. M. Larkin. 1980. Production of tropolone by a *Pseudomonas*. *J. Nat. Prod.* 43:592–594.

Mazzone, H. M., G. Wray, and R. Zerillo. 1981a. Studies on the fungal pathogen of Dutch elm disease. Pages 400–401 *in* G. W. Bailey, ed., *Proc. Electron Microsc. Soc. Am. 39th Annu. Meet.*, Atlanta, GA, Claitor's Publishing Div., Baton Rouge, LA.

Mazzone, H. M., J. Kluck, N. P. Dubois, and R. Zerillo. 1981b. Dutch elm disease control with biological agents or their metabolites. Pages 36–45 *in* E. S. Kondo, Y. Hiratsuka, and W. B. C. Denyer, eds., *Proc. Dutch Elm Dis. Symp. and Workshop,* Winnipeg, Manitoba, Canada.

Newton, B. A. 1956. The properties and mode of action of the polymyxins. *Bacteriol. Rev.* 20:14–27.

Nickerson, K. W., D. J. McNeel, and R. K. Kulkarni. 1982. Fungal dimorphism in *Ceratocystis ulmi:* Cerulenin sensitivity and fatty acid synthesis. *FEMS Microbiol. Lett.* 13:21–25.

Nomura, S., T. Horiuchi, S. Omura, and T. Hata. 1972. The action mechanism of cerulenin. *J. Biochem. (Tokyo)* 71:783–786.

Rose, A. M. 1968. *Chemical Microbiology.* 2nd ed., Plenum Press, New York. p. 81.

Sinclair, W. A., and R. J. Campana, eds. 1978. Development and status of Dutch elm disease. Pages 5–6 *in Dutch Elm Disease. Perspectives after 60 Years.* Cornell Agricultural Experiment Station, Ithaca, NY.

Waksman, S., and E. Bugie. 1943. Action of antibiotic substances upon *Ceratostomella ulmi. Proc. Soc. Exp. Biol. Med.* 54:79–82.

Woolley, D. W. 1952. *A Study of Antimetabolites.* John Wiley and Sons, Inc., New York. p. 48.

Yarmolinsky, M. B., and G. L. de la Haba. 1959. Inhibition by puromycin of amino acid incorporation into protein. *Proc. Natl. Acad. Sci. USA* 45:1721–1729.

CHAPTER 34

Comparison of Two Defined Media for Inhibitor and Incorporation Studies of Aflatoxin Biosynthesis

J. W. BENNETT, STANTON KOFSKY, ALAN BULBIN, AND MICHAEL DUTTON*

Department of Biology, Tulane University, New Orleans, Louisiana 70118

Aflatoxins are typical secondary metabolites of *Aspergillus flavus* and *A. parasiticus*. As secondary metabolites, the production of aflatoxins is sensitive to changes in environmental parameters such as substrate, and the kinetics of their production is correlated with the cessation of growth during a special stage called idiophase. We have compared two major approaches to studying aflatoxin biosynthesis at the beginning of idiophase (48 h) using two defined media. In one set of experiments, selected organophosphates were added to the defined media at this time; in the second set of experiments, a blocked aflatoxin mutant was presented with a known late intermediate of aflatoxin biosynthesis and transferred to resting media. Despite the different compositions of the media, comparable results were obtained. Of the 15 organophosphates tested, four exhibited dichlorvos-like activity in both media. Inhibition of aflatoxin production and concomitant accumulation of an orange pigment was observed for ciodrin, phosdrin, hexamethyl phosphorous triamide, and triphenyl phosphate. In the resting cell experiments, a blocked aflatoxin-negative mutant was grown in the two different media and then transferred to a resting cell medium in the presence of sterigmatocystin, a known intermediate in aflatoxin biosynthesis. After additional incubation, both B and G aflatoxins were detected from mycelia pregrown in both media. Moreover, chromatographic profiles of mycelial extracts were also similar. We conclude that the differences in aflatoxin production observed on these two media are more quantitative than qualitative.

INTRODUCTION

Aflatoxins are secondary metabolites produced by certain strains of the common molds *Aspergillus flavus* and *A. parasiticus*. These mycotoxins are acutely toxic and carcinogenic to many animal species, and aflatoxin contamination of foods and feeds is a major problem in agriculture. The aflatoxin literature has been reviewed (Goldblatt 1969; Heathcote and Hibbert 1978; Steyn 1980; Bennett and Christensen 1983).

Various media have been devised for the laboratory production of aflatoxins. Highest yields were obtained with natural substrates, such as peanuts, wheat, or rice, or with complex liquid media containing supplements, such as yeast extract or corn steep liquor (Diener and Davis 1969; Venkitasubramanian 1977). Early work showed that defined media traditionally used for culturing fungi, such as Czapecks-Dox medium and Raulin's medium, were poor substrates for aflatoxin

*Current Address: Department of Biochemistry, University of Natal, Pietermaritzburg, Natal, South Africa.

production (Diener and Davis 1969). Subsequently, two defined media were devised especially for supporting high levels of aflatoxin production. The first of these (AM) was formulated by Adye and Mateles (1964) and has been widely used in studies of aflatoxin biosynthesis (Hsieh and Mateles 1970; Singh and Hsieh 1977; Townsend et al. 1982).

The second (RLSM) was formulated by Reddy et al. (1971) and produced higher yields of total aflatoxin than did the AM medium. Each of these defined media contained a carbon source, NH_4SO_4, $MgSO_4{\cdot}7H_2O$, and various trace elements. However, the proportion of constituents varied between the two media, and RLSM contained 10 g of L-asparagine per liter.

Different laboratories around the world have adopted one or the other of these defined media for studies of aflatoxin biosynthesis, raising the question: Are equivalent results obtained? In this study we have compared the two media for use in two major approaches to the study of aflatoxin biosynthesis: (1) a screening study of organophosphates as possible inhibitors of aflatoxin biosynthesis, and (2) a precursor-feeding study using an aflatoxin-negative mutant and whole cell resting cultures.

Materials and Methods

Fungal strains. The aflatoxigenic wild type strain was *A. parasiticus* (NRRL 5862, SU-1). The aflatoxin-negative strain was a versicolorin-A accumulating mutant of *A. parasiticus* designated *ver-1* (ATCC 36537), which was originally isolated by ultraviolet light irradiation (Lee et al. 1975).

Media and culture conditions. Stock cultures were maintained on potato dextrose agar plus 0.5% yeast extract. The two defined liquid growth media, AM and RLSM, were formulated by Adye and Mateles (1964) and Reddy et al. (1971), respectively. The resting medium (AM-RM) was formulated by Adye and Mateles (1964). AM was composed of the following: sucrose, 50.0 g; $(NH_4)_2SO_4$, 3.0 g; KH_2PO_4, 10.0 g; $MgSO_4{\cdot}7H_2O$, 2.0 g; $Na_2B_4O_7{\cdot}10H_2O$, 0.7 mg; $(NH_4)_6$ $Mo_7O_{24}{\cdot}4H_2O$, 0.5 mg; $Fe_2(SO_4)_3{\cdot}6H_2O$, 10.0 mg; $CuSO_4{\cdot}5H_2O$, 0.11; $ZnSO_4{\cdot}7H_2O$, 17.6 mg; and 1,000 ml deionized H_2O.

RLSM was composed of the following: sucrose, 85.0 g; L-asparagine, 10.0 g; (NH_4SO_4), 3.5 g; KH_2PO_4, 0.75 g; $MgSO_4{\cdot}7H_2O$, 0.35 g; $CaCl_2{\cdot}2H_2O$, 0.75 g; $ZnSO_4{\cdot}7H_2O$, 10.0 mg; $MnCl_2{\cdot}4H_2O$, 5.0 mg; $(NH_4)_2$ $Mo_7O_{24}{\cdot}9H_2O$, 2.0 mg; $Na_2B_4O_7$, 2 mg; $FeSO_4{\cdot}7H_2O$, 2 mg; and 1,000 ml deionized water.

AM-RM was composed of the following: glucose, 15.0 g; KH_2PO_4, 5.0 g; $MgSO_4{\cdot}7H_2O$, 0.5 g; and the same trace minerals as AM.

All liquid cultures were grown at 27–28 C in a gyratory shaker in the dark.

Organophosphates. The following 16 organophosphates were tested: dichlorvos (dimethyl-2,2-dichlorovinyl phosphate), ciodrin [dimethyl-1-methyl-2(1-phenylcarbetho) vinyl phosphate] phosdrin [methyl-3-(dimethoxyphosphinyloxy) crotonate], dimethyl methylphosphonate, triethyl phosphate, triethyl-2-phosphonopropionate, hexamethyl phosphorous triamide, diethyl-2-bromoethyl phosphate, triethyl phosphonocrotonate, trimethyl phosphonoacetate, ethyl-diethoxy-phosphinyl formate, triphenyl phosphate, triethyl phosphonoacetate, tris-(2-chloroethyl) phosphate, methyl-dichlorophosphate and diethyl-chloro-thiophosphate. For each experiment, a freshly prepared 1% stock solution of a given organophosphate was made in acetone. Cultures of wild type *A. parasiticus* SU-1 were grown in 20 ml of AM or RLSM for 48 h, and then the appropriate amount of organophosphate stock solution was added with a micropipet into the culture in order to deliver 20 ppm. Control flasks contained 35 μl acetone alone. Treated cultures were incubated an additional 72 h before extraction.

Presentation of sterigmatocystin to whole cells in resting cell cultures. One hundred ml of AM or RLSM were inoculated with a dense spore suspension of the *ver-1* mutant and incubated in the dark for 48 h on a rotary shaker. The resultant mycelial pellets were collected in cheesecloth and thoroughly washed with AM-RM. Then 1 g (wet wt) of the pellets was added to 9.8 ml of AM-RM in a 50-ml Ehrlenmeyer flask. To this was added 0.65 μmol of sterigmatocystin (Sigma) dissolved in 0.2 ml acetone. Control flasks contained 0.2 ml acetone without the sterigmatocystin. The resting cell cultures were incubated an additional 48 h on the rotary shaker and then extracted.

Mycelial extractions. After the appropriate times of incubation, wet mycelial pellets were filtered through cheesecloth and then soaked in acetone (25–40 ml) for 2 h. The acetone extract was filtered through a Buchner funnel into a side arm flask, diluted to 70% with deionized water, and poured into a separatory funnel. The aqueous acetone mixture was extracted twice with 30 ml of chloroform and evaporated to dryness under a hood. The resultant dried mycelial extract was resuspended in 1 ml of chloroform or acetone prior to chromatographic analysis.

Thin-layer chromatography and identification of metabolites. Mycelial extracts were separated by thin-layer chromatography (tlc) on 250-mm silica gel G plates (Analtech). Extracts and standards were delivered using a glass syringe calibrated in microliters. The silica gel plates were developed in an unlined covered tank containing 100 ml of the developing solvent. The developing solvent for aflatoxins was chloroform:acetone (9:1, v/v); the developing solvent for pigments was toluene:ethyl acetate:acetone:acetic acid (60:20:15:2 v/v/v/v). Developed plates were observed under long wave ultraviolet light. The authentic standards of aflatoxins (B_1, B_2, G_1, and G_2) and of versiconal hemiacetal acetate (VHA) were provided by Mrs. L. S. Lee, Southern Regional Research Center, New Orleans, LA. Aflatoxins were quantified visually according to the method of Pons et al. (1969).

A two-dimensional tlc analysis was performed on mycelial extracts that produced an orange pigment. Aluminum-backed silica gel 60 plates (Merck 5388) cut into 10-cm squares were spotted with 20 μl of extract in the lower left-hand corner. The plates were first developed in chloroform:acetone (9:1, v/v), dried with a hot air blower, and then developed in toluene:ethyl acetate:acetone:acetic acid (60:25:15:2, v/v/v/v). Developed plates were observed under long wave ultraviolet light. Control plates with an authentic standard of VHA were developed in the same tanks as experimental plates.

RESULTS AND DISCUSSION

The results of tlc analysis of mycelial extracts from cultures of wild type *A. parasiticus* grown on AM or RLSM media, with or without 16 organophosphates are presented in Table 1. Controls produced high levels of aflatoxin and no orange pigment. Reduction in aflatoxin production was observed for eight compounds: dichlorvos, ciodrin, diethyl chlorothiophosphate, tris-(2-chlorethyl) phosphate, phosdrin, hexamethyl phosphorous triamide, triethyl 4-phosphocrotonate, and triphenyl phosphate. For five of these compounds (dichlorvos, ciodrin, phosdrin, hexamethyl phosphorous triamide, and triphenyl phosphate) aflatoxin inhibition was correlated with production of an orange pigment. The orange pigment had the same R_f as an authentic sample of VHA in both one-dimensional and two-dimensional tlc.

The first organophosphate to be associated with aflatoxin inhibition was dichlorvos based on experiments using wheat, rice, peanuts, and corn (Rao and Harein 1972). Hsieh (1973) later reported inhibition of aflatoxin production in defined AM medium; subsequently, dichlorvos treatment of aflatoxigenic strains was associated with the accumulation of an orange pigment (Schroeder et al. 1974; Yao and Hsieh 1974) and the characterization of this orange pigment as VHA (Cox et al. 1977; Fitzell et al. 1977). Both C^{14} and C^{13} isotopic-labeling studies indicated that VHA was an intermediate in aflatoxin biosynthesis (Bennett and Christensen 1983). In our screen of organophosphate compounds, four exhibited inhibition of aflatoxin biosynthesis correlated with the accumulation of an orange pigment with the same chromatographic mobility as an authentic standard of VHA. Results were similar in both AM and RLSM. Dutton and Anderson (1980) have previously reported dichlorvos-like activity in RLSM for chlormephos, ciodrin, naled, phosdrin, and trichlorphon.

The insecticidal activity of dichlorvos and other organophosphates is due to their inhibition of cholinesterase activity. This and previous studies (Hsieh 1973; Dutton and Anderson 1980) show that only some organophosphates inhibit aflatoxin production. One possible explanation for the differential activity is that the mechanism of aflatoxin inhibition is not similar to that of the cholinesterase inhibition. In fact, Dutton and Anderson (1980) postulated that dichlorvos and other organophosphates affecting aflatoxin production did so by an inhibition of fungal oxygenase activity. Other explanations for the inactivity of the certain organophosphates is that there is differential solubility or sensitivity to fungal detoxification.

The results of the experiments with resting cell cultures of *ver-1* with and without sterigmatocystin are presented in Table 2. The versicolorin A-accumulating mutant is normally blocked in aflatoxin production. When this mutant was grown for 2 d in AM or RLSM and then transferred to AM-RM in the presence of 0.65 μmol sterigmatocystin, aflatoxins were recovered after 48 h of incubation. Both B and G aflatoxins were detected, with similar concentrations produced by both AM-grown and RLSM-grown cultures. The chromatographic profiles of mycelial extracts were also similar except that AM-grown cultures produced a blue-green compound with an R_f similar to that of aflatoxin B_{2a}. This compound was usually absent in chromatograms of RLSM-grown cultures.

The kinetics of aflatoxin biosynthesis are similar to those of many other fungal

TABLE 1. *Presence of aflatoxins and an orange pigment after thin-layer chromatography of wild type* Aspergillus parasiticus *grown on two defined media, with and without 16 selected organophosphates*

Organophosphate[a]	AM Medium		Reddy Medium	
	Aflatoxins[b]	Orange pigment[c]	Aflatoxins[b]	Orange pigment[c]
Control	++++	–	++++	–
Dichlorvos	++	+	++	+
Ciodrin (33)	++	+	++	+
Dimethyl methylophosphonate (35)	++++	–	++++	–
Triethyl phosphate (37)	++++	–	++++	–
Methyl dichlorophosphate (27)	++++	–	++++	–
Diethyl chlorothiophosphate (34)	+++	–	++	–
Tris(2-chloroethyl)phosphate (29)	++	–	+++	–
Phosdrin (32)	++	+	+++	+
Triethyl phosphonoacetate (35)	++++	–	++++	–
Trimethyl phosphonoacetate (35)	++++	–	++++	–
Hexamethyl phosphorous triamide (45)	+++	+	+++	+
Triethyl 4-phosphonocrotonate (35)	+++	–	+++	–
Triethyl 2-phosphonopropionate (36)	++++	–	++++	–
Ethyl diethoxyphosphinyl formate (36)	++++	–	++++	–
Diethyl 2-bromoethyl phosphonate (35)	++++	–	++++	–
Triphenyl phosphate (35)	+++	+	+++	+

[a] Numbers in parentheses indicate μl of 1% stock solution of compound added to defined media.
[b] Fluorescence under long wave ultraviolet light, with (++++) equivalent to controls.
[c] Orange pigments with same R_f as an authentic standard of versiconal hemiacetal acetate. (+) = presence; (–) = none detected.

TABLE 2. *Recovery of aflatoxins after presentation of 0.65 μmol sterigmatocystin (ST) to a blocked aflatoxin-negative mutant*

Growth Medium[a]	With or Without ST (0.65 μmol)	Aflatoxin (nmol)	
		B_1	G_1
AM	ST	391	Trace[b]
AM	control	ND[c]	ND
RLSM	ST	387	Trace
RLSM	control	ND	ND

[a] See methods for description of media.
[b] Trace = less than 10 nmol.
[c] ND = none detected.

secondary metabolites, i.e., the onset of biosynthesis is usually correlated with the cessation of growth during a special stage called idiophase. The mycelia of our blocked mutant were grown in AM or RLSM and transferred to resting medium at the beginning of idiophase. No aflatoxins were detected in controls, but cultures of both AM-grown and RLSM-grown mycelia produced similar levels of aflatoxin in the presence of sterigmatocystin. Sterigmatocystin is a known precursor of aflatoxin biosynthesis (Hsieh et al. 1973).

Aflatoxins are acetate-malonate derived secondary metabolites. In controlled liquid fermentations, the events of secondary metabolism follow a predictable pattern that can be manipulated by changing the composition of the medium, the temperature of incubation, and other experimental parameters (Foster 1947; Drew and Demain 1977). Since secondary metabolism is so sensitive to changes in the substrate used for growth, there has been difficulty in comparing the results from different laboratories on the events initiating aflatoxin biosynthesis. For example, many of the C^{14}-precursor incorporation studies have been conducted by growing cells in AM medium and transferring to a resting cell culture (Bennett and Christensen 1983); studies on carbohydrate metabolism have involved growth in a peptone medium (Abdollahi and Buchanan 1981); and experiments on the interface between primary and secondary metabolism have employed RLSM (Maggon et al. 1977; Clevstrom et al. 1983).

In summary, we have compared mycelia pregrown in two different defined media for 48 h. In one set of experiments, selected organophosphates were added at the beginning of idiophase; in the second set of experiments, a blocked aflatoxin mutant was presented with a known late intermediate of aflatoxin biosynthesis and transferred to resting media at this time. Despite the different compositions of the media, qualitatively comparable results were obtained in both sets of experiments, although quantitative differences in levels of aflatoxins produced were detected. In the future, workers employing either AM or RLSM may be more confident in making inter-laboratory comparisons of their data on aflatoxin biosynthesis.

Acknowledgments

We thank John Floyd, Louise Lee, and Eivind Lillehoj for their advice and encouragement. This research was supported by a Cooperative Agreement from the U.S. Department of Agriculture (#532569).

Literature Cited

Abdollahi, A., and R. L. Buchanan. 1981. Regulation of aflatoxin biosynthesis: Characterization of glucose as an apparent inducer of aflatoxin production. *J. Food Sci.* 46:143–146.

Adye, J., and R. I. Mateles. 1964. Incorporation of labeled compounds into aflatoxins. *Biochim. Biophys. Acta* 86:418–420.

Bennett, J. W., and S. B. Christensen. 1983. New perspectives on aflatoxin biosynthesis. *Adv. Appl. Microbiol.* 29:53–92.

Clevstrom, G., H. Ljundgren, S. Tegelstrom, and K. Tideman. 1983. Production of aflatoxin by an *Aspergillus flavus* isolate cultured under a limited oxygen supply. *Appl. Environ. Microbiol.* 46:400–405.

Cox, R. H., F. Churchill, R. J. Cole, and J. W. Dorner. 1977. Carbon-13 nuclear magnetic resonance studies of the structure and biosynthesis of versiconal acetate. *J. Am. Chem. Soc.* 99:3159–3169.

Diener, U. L. and N. D. Davis. 1969. Aflatoxin formation by *Aspergillus flavus.* Pages 13–54 *in* L. A. Goldblatt, ed., *Aflatoxins. Scientific Background, Control, and Implications.* Academic Press, New York.

Drew, S. W., and A. L. Demain. 1977. Effect of primary metabolites on secondary metabolism. *Annu. Rev. Microbiol.* 31:343–356.

Dutton, M. F., and M. S. Anderson. 1980. Inhibition of aflatoxin biosynthesis by organophosphorous compounds. *J. Food Prot.* 43:381–384.

Fitzell, D. L., R. Singh, D. P. H. Hsieh, and E. L. Motell. 1977. Nuclear magnetic resonance identification of versiconal hemiacetal acetate as an intermediate in aflatoxin biosynthesis. *Agric. Food Chem.* 25:1193–1197.

Foster, J. W. 1947. Some introspections on mold metabolism. *Bacteriol. Rev.* 11:167–188.

Goldblatt, L. A., ed. 1969. *Aflatoxin: Scientific Background, Control, and Implications.* Academic Press, New York. 472 pp.

Heathcote, J. G., and J. R. Hibbert. 1978. *Aflatoxins: Chemical and Biological Aspects.* Elsevier Science Publishing Co., Amsterdam.

Hsieh, D. P. H. 1973. Inhibition of aflatoxin biosynthesis by dichlorvos. *J. Agric. Food Chem.* 21:468–470.

Hsieh, D. P. H., M. T. Lin, and R. C. Yao. 1973. Conversion of sterigmatocystin to aflatoxin by *Aspergillus parasiticus. Biochem. Biophys. Res. Comm.* 52:992–997.

Hsieh, D. P. H., and R. I. Mateles. 1970. The relative contribution of acetate and glucose to aflatoxin biosynthesis. *Biochim. Biophys. Acta* 208:482–486.

Lee, L. S., J. W. Bennett, A. F. Cucullu, and J. B. Stanley. 1975. Synthesis of versicolorin A by a mutant strain of *Aspergillus parasiticus* deficient in aflatoxin production. *J. Agric. Food Chem.* 23:1132–1139.

Maggon, K. K., S. K. Gupta, and T. A. Venkitasubramanian. 1977. Biosynthesis of aflatoxins. *Bacteriol. Rev.* 41:822–855.

Pons, W. A., A. F. Cucullu, A. O. Franz, and L. A. Goldblatt. 1969. Improved objective flurodensitometric determination of aflatoxins in cottonseed products. *J. Am. Oil. Chem. Soc.* 45:694–699.

Rao, H. R. G., and P. K. Harein. 1972. Dichlorvos as an inhibitor of aflatoxin production on wheat, corn, rice and peanuts. *J. Econ. Entomol.* 65:988–998.

Reddy, T. V., L. Viswanathan, and T. A. Venkitasubramanian. 1971. High aflatoxin production on a chemically defined medium. *Appl. Microbiol.* 22:393–396.

Schroeder, H. W., R. J. Cole, R. D. Grigsby, and H. Hein, Jr. 1974. Inhibition of aflatoxin production and tentative identification of an aflatoxin intermediate "versiconal acetate" from treatment with dichlorvos. *Appl. Microbiol.* 27:394–399.

Singh, R., and D. P. H. Hsieh. 1977. Aflatoxin biosynthetic pathway: Elucidation by using blocked mutants of *Aspergillus parasiticus Arch. Biochem. Biophys.* 178:285–292.

Steyn, P. S. 1980. *The Biosynthesis of Mycotoxins. A Study in Secondary Metabolism.* Academic Press, New York.

Townsend, C. A., S. B. Christensen, and S. G. Davis. 1982. Bisfuran formation in aflatoxin biosynthesis: The fate of the averufin side chain. *J. Am. Chem. Soc.* 104:6152–6153.

Venkitasubramanian, T. A. 1977. Biosynthesis of aflatoxin and its control. Pages 83–98 *in* J. V. Rodricks, C. W. Hesseltine, and M. A. Mehlman, eds., *Mycotoxins in Human and Animal Health.* Pathotox Pubs, Inc., Park Forest, IL.

Yao, R. D., and D. P. H. Hsieh. 1974. Step of dichlorvos inhibition in the pathway of aflatoxin biosynthesis. *Appl. Microbiol.* 28:52–57.

CHAPTER 35

Role of Ammonium Nitrate in Morphological Differentiation of *Aspergillus niger* in a Submerged Culture

JOHN J. JOUNG AND ROBERT J. BLASKOVITZ

Standard Oil Company (Indiana), Naperville, Illinois 60566

We have investigated the effect of ammonium nitrate concentration on morphological development of *Aspergillus niger* (ATCC 9142) in a submerged culture. With a manganese-deficient basal medium, 27 different fermentation media were prepared by varying the concentration of ammonium nitrate, potassium monobasic phosphate, and magnesium sulfate to three different levels. Cultures obtained from these fermentation media were examined for cell morphology, citrate yield, and biomass. The ammonium nitrate concentration in the culture medium dictated morphological differentiation of the *A. niger* culture. Depending on the ammonium nitrate concentration, three different types of morphologies were possible: (1) vegetative mycelia, (2) complete natural morphology comprising fruiting bodies and mycelia, and (3) short, branched, and bent mycelia with bulbous and horny masses. The latter two morphologies were associated with citrate overproduction, but the vegetative mycelia were not efficient for citrate production. The potassium phosphate and magnesium sulfate concentrations did not induce changes in cell differentiation. The nature and concentration of nitrogen source in the culture medium played a major role in cell differentiation of the *A. niger*. Ammonium sulfate could only produce the complete morphology, whereas potassium nitrate produced highly branched mycelia in all of the concentrations tested.

INTRODUCTION

It has been well recognized that a rapid citrate accumulation in a submerged fermentation is associated with certain fungal morphologies. Snell and Schweiger (1949) prepared citrate-producing *A. niger* pellets 0.1 mm in diameter by balancing nutrients in the fermentation medium and limiting the iron concentration below 1 ppm. The cellular morphology was characterized by short, stubby, forked, and bent mycelia with numerous swollen spherical vesicles. These *A. niger* pellets lacked normal fruiting bodies and hyphal stems. Clark (1962) prepared 0.2–0.5 mm fungal pellets by treating the culture medium with potassium ferrocyanide. The inoculum pellets grew further into dense spheres 2 mm in diameter in the fermentation broth, and the citrate productivity of the pellets was very high. Filamentous mycelial pellets that developed in the absence of ferrocyanide did not accumulate citric acid. Later, Clark and his coworkers (1966) showed that manganese deficiency was the cause of the abnormal morphology of citrate-producing cultures.

Impaired protein and lipid synthesis caused the incomplete and abnormal cellular development observed in a manganese-deficient culture. Cycloheximide and excess copper ion had a similar effect as manganese deficiency in *A. niger* cultures (Clark et al. 1966; Wold and Suzuki 1976; Kubicek and Rohr 1977; Orthofer et al. 1978; Kisser et al. 1980; Rohr and Kubicek 1981).

The dependence of citrate overproduction on manganese deficiency is described in the following way: Abnormally high concentrations of ammonium ion and amino acids can be accumulated inside the cells of a manganese-deficient culture during the idiophase of citrate fermentation. Unregulated glycolysis and the subsequent citrate synthesis can be permitted by the antagonistic effect of the ammonium ions to inhibitory metabolites of phosphofructokinase (Habison et al. 1979; Kubicek et al. 1979). A marked decrease of the activities of aconitase and isocitrate dehydrogenase prevents further conversion of citric acid in the tricarboxylic acid cycle resulting in citrate overflow (Ramakrishnan et al. 1955; Ahmed et al. 1972; Kubicek and Rohr 1978; Bowes and Mattey 1979; Szczodrak 1981).

During our previous optimization of the major nutrient concentrations, we observed inconsistent citrate yields in spite of the use of an identical basal medium containing the same amount of trace elements. More systematic studies on yield and morphological effects of the major nutrients were needed to account for the unexpected results. Employing a manganese-deficient medium (0.1 μM), we obtained 27 different nutrient formulations by varying the concentrations of ammonium nitrate, potassium monobasic phosphate, and magnesium sulfate to three different levels. Cultures obtained from these fermentation media were evaluated for citrate yield, cell morphology, and biomass.

Concentration of ammonium nitrate was the only factor for cell differentiation and it significantly influenced the biomass gain of *A. niger* cultures. High concentrations of potassium monobasic phosphate were detrimental to citrate production. Three different types of morphologies were obtained depending on the concentration of ammonium nitrate: (1) complete natural morphology comprising fruiting bodies and mycelia, (2) vegetative mycelia, and (3) highly branched, short, and bent mycelia with bulbous and horny masses. The first and third morphologies were associated with citrate overproduction, but the vegetative mycelia were not efficient for citrate fermentation.

Our results demonstrated a possible regulatory role of ammonium nitrate in morphological differentiation of the fungal culture. Manganese deficiency theory alone could not explain our observations.

Materials and Methods

A. niger strain ATCC 9142 was employed in this study. Conidia for inoculations were obtained by culturing the fungus on potato dextrose agar plates at 25 C. A homogeneous conidial suspension was prepared by gently washing the agar plates with sterile Tween 80 solution (0.8%) and was stored at 4 C until use.

Commercial dextrose derived from corn starch was treated with a cation exchange resin (Dowex 50W–X8). The composition of all minor nutrients and glucose in the media were fixed, but the amounts of the major nutrients (KH_2PO_4, $MgSO_4$, NH_4NO_3) were varied to three different levels in order to construct a (3,3,3) nutrient matrix. In different experiments, ammonium nitrate was replaced with an equivalent amount of ammonium sulfate or potassium nitrate.

The invariant basal medium consisted of, per liter: Glucose, 100 g; NaCl, 0.1 g; $CaCl_2 \cdot 2H_2O$, 0.1 g; Tween 80, 16 mg; $MnSO_4 \cdot 2H_2O$, 15.4 μg; $ZnSO_4 \cdot 7H_2O$, 143.7 μg; $CuSO_4 \cdot 5H_2O$, 249 μg; and $FeSO_4 \cdot 7H_2O$, 556 μg. The amounts of major nutrients in the media are shown in the results section. The initial pH of the media was adjusted to 3.5. All chemicals employed in the experiment were analytical grade.

A 100 ml aliquot of fermentation medium was transferred into a 250-ml Erlenmyer flask and the top was sealed with a cotton ball. The prepared flasks were autoclaved at 126 C for 15 min. The sterile media were treated with 5 mg of $K_4Fe(CN)_6 \cdot 3H_2O$ before the media became cool (70–80 C). Into each flask, 3×10^7 spores were inoculated aseptically, and the inoculated flasks were placed on a New Brunswick G10 gyratory shaker at 150 rpm in a 28 C constant temperature room.

Fungal pellets were obtained from 7 d cultures, and they were examined with an optical microscope. Specimens for scanning electron microscopy were fixed in 3% glutaraldehyde solution for 1 h, and frozen fractures were obtained by submerging the fungal pellets rapidly in liquid nitrogen. The citrate yield shown in the results section was obtained after 4 wk fermentation. Citric acid was tirated with 0.1 *N* KOH solution, and the yield calculation was based on the amount of glucose added to the fermentation medium.

Results and Discussion

Under identical basal nutrient concentrations where manganese was deficient (0.1 μM), the amount of ammonium nitrate in the fermentation medium influenced both citrate yields and cell morphology profoundly. Potassium monobasic phosphate reduced citrate yields significantly at a high concentration (1.5 g/l), but it did not influence cell morphology. The magnesium sulfate had little effect on cell morphology or citrate yields within the concentration ranges studied.

Table 1 shows the nutrient matrix and the respective citrate yields after 4 wk fermentation at 28 C. High citrate yields were associated with both low and high ammonium nitrate concentrations (0.33 g/l and 3 g/l). The yields were always poor with the intermediate concentration of ammonium nitrate (1 g/l) in any combination with the potassium phosphate and magnesium sulfate.

The effect of ammonium nitrate concentration on cell morphology was even more drastic than the effect on citrate yield. All cultures in our experiments produced spherical pellets 0.5 to 2 mm diameter, but cell morphologies were completely different depending on the amount of ammonium nitrate in the medium. Regardless of the amount of magnesium sulfate or of potassium phosphate, flasks containing 0.33 g/l ammonium nitrate all produced pellets of complete natural morphology comprising mycelia, hyphal stems, vesicles, stigmata, and conidial spores. All media containing 1 g/l ammonium nitrate allowed only smooth, long, and filamentous growth. Furthermore, media containing 3 g/l ammonium nitrate developed a similar cell morphology described by Snell and Schweiger (1949). The fungal mycelia were short, branched, and horny with bulbous vesicles. The scanning electron micrographs shown in Figs. 1, 2, and 3 compare these morphological differences.

TABLE 1. *Effect of major nutrients on citrate yield and fungal morphology*

Formulation Number	KH_2PO_4 (g/l)	$MgSO_4 \cdot 7H_2O$ (g/l)	NH_4NO_3 (g/l)	Citric Acid Yield (%)	Morphological Characteristics[a]
1	0.5	0.08	0.33	45	CN
2	0.5	0.08	1.0	17	VM
3	0.5	0.08	3.0	78	BH
4	0.5	0.25	0.33	29	CN
5	0.5	0.25	1.0	13	VM
6	0.5	0.25	3.0	77	BH
7	0.5	0.75	0.33	62	CN
8	0.5	0.75	1.0	54	VM
9	0.5	0.75	3.0	86	BH
10	1.0	0.08	0.33	68	CN
11	1.0	0.08	1.0	13	VM
12	1.0	0.08	3.0	89	BH
13	1.0	0.25	0.33	80	CN
14	1.0	0.25	1.0	23	VM
15	1.0	0.25	3.0	67	BH
16	1.0	0.75	0.33	72	CN
17	1.0	0.75	1.0	13	VM
18	1.0	0.75	3.0	73	BH
19	1.5	0.08	0.33	28	CN
20	1.5	0.08	1.0	7	VM
21	1.5	0.08	3.0	36	BH
22	1.5	0.25	0.33	46	CN
23	1.5	0.25	1.0	6	VM
24	1.5	0.25	3.0	30	BH
25	1.5	0.75	0.33	22	CN
26	1.5	0.75	1.0	4	VM
27	1.5	0.75	3.0	30	BH

[a] CN = complete natural; VM = vegetative mycelia; BH = bulbous and horny.

FIG. 1. Scanning electron micrograph showing complete fruiting bodies obtained from a submerged culture with a low concentration of ammonium nitrate.

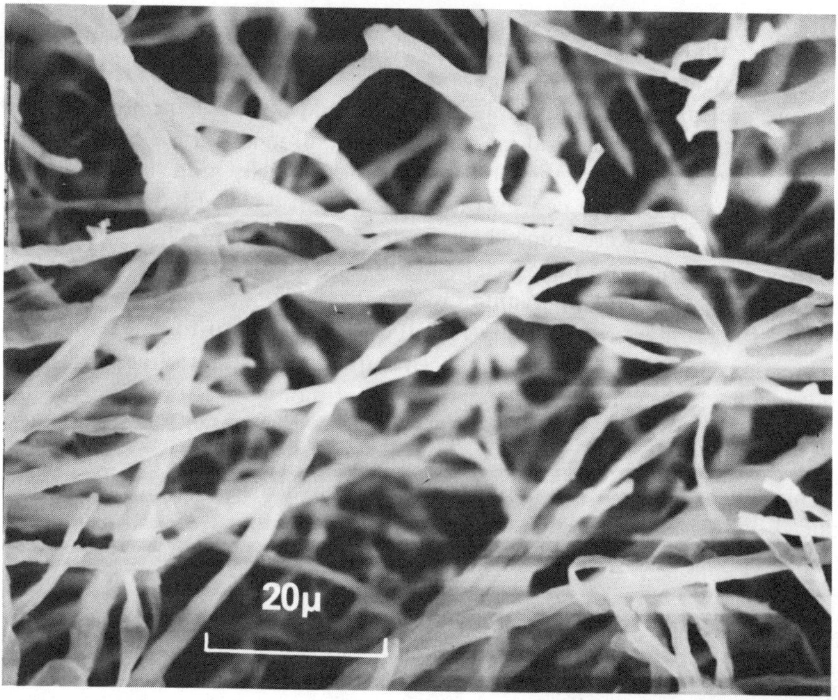

FIG. 2. Vegetative mycelia obtained with an intermediate concentration of ammonium nitrate.

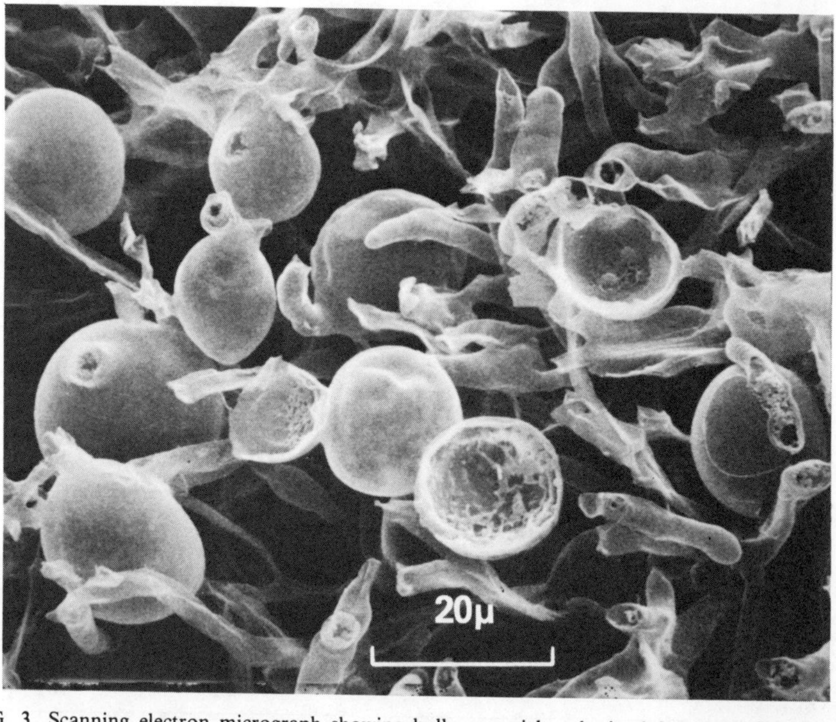

FIG. 3. Scanning electron micrograph showing bulbous vesicles obtained from a submerged culture with a high concentration of ammonium nitrate.

Experiments employing an expanded concentration range of major nutrients resulted in similar effects in cell morphology. The morphological differentiation and biomass yield data for different combinations of the major nutrients after 1 wk culture are summarized in Table 2. Ammonium nitrate concentration of 0.5 g/l resulted in complete natural morphology and all media containing 8 g/l ammonium nitrate produced the short, branched, horny and bulbous fungal biomass. The intermediate concentration (2 g/l) produced filamentous mycelia only, but the staples were somewhat shorter and more branched compared to those of 1 g/l ammonium nitrate. The high biomass yields were associated with filamentous growth, and high ammonium nitrate concentration inhibited the growth of fungal mass significantly for all combinations of the nutrients.

When ammonium nitrate was replaced with ammonium sulfate for its ammonium equivalence, only complete natural morphology was obtained for all 27 different formulations of Table 2. The most extensive growth of hyphal stems was observed with the lowest concentration of ammonium sulfate (0.413 g/l). The replacement with potassium nitrate for its equivalence of nitrate in the media produced only short and highly branched mycelia irrespective to the concentration of the phosphate or magnesium sulfate.

TABLE 2. *Effect of major nutrients on biomass yield and fungal morphology*

Formulation Number	KH_2PO_4 (g/l)	$MgSO_4 \cdot 7H_2O$ (g/l)	NH_4NO_3 (g/l)	Biomass (g/l)	Morphological Characteristics[a]
1	0.2	0.1	0.5	3.5	CN
2	0.2	0.1	2.0	4.4	VM
3	0.2	0.1	8.0	2.3	BH
4	0.2	0.3	0.5	4.1	CN
5	0.2	0.3	2.0	5.2	VM
6	0.2	0.3	8.0	2.8	BH
7	0.2	0.9	0.5	4.2	CN
8	0.2	0.9	2.0	6.1	VM
9	0.2	0.9	8.0	2.8	BH
10	0.8	0.1	0.5	3.5	CN
11	0.8	0.1	2.0	4.1	VM
12	0.8	0.1	8.0	3.1	BH
13	0.8	0.3	0.5	4.8	CN
14	0.8	0.3	2.0	5.9	VM
15	0.8	0.3	8.0	3.2	BH
16	0.8	0.9	0.5	3.2	CN
17	0.8	0.9	2.0	5.0	VM
18	0.8	0.9	8.0	2.4	BH
19	2.4	0.1	0.5	4.8	CN
20	2.4	0.1	2.0	4.8	VM
21	2.4	0.1	8.0	2.4	BH
22	2.4	0.3	0.5	3.7	CN
23	2.4	0.3	2.0	6.1	VM
24	2.4	0.3	8.0	3.0	BH
25	2.4	0.9	0.5	4.6	CN
26	2.4	0.9	2.0	3.5	VM
27	2.4	0.9	8.0	2.5	BH

[a] CN = complete natural; VM = vegetative mycelia; BH = bulbous and horny.

The development and growth of fungal cells were highly specific to the nature and concentration of nitrogen source under our experimental conditions. Manganese or mineral deficiency alone cannot completely explain our results. Ammonium ions and nitrate ions could have directly influenced the gene expression or metabolic regulation for cell differentiation. Yet, there is the possibility that these chemical species may have influenced the availability of mineral nutrients that were treated with potassium ferrocyanide.

It is known that citrate production is stimulated in a nitrogen-limited fermentation, and mycelial growth is associated with a high concentration of nitrogen. But, stimulation of citrate production occurred again at an even higher concentration of ammonium nitrate (3 g/l).

Impaired protein and lipid synthesis along with abnormal morphological development in citrate-producing cultures have been established (Clark 1962; Kubicek and Rohr 1977, 1978; Orthofer et al. 1978; Kisser et al. 1980). Elevated ammonium ion concentration in cells plays a key role in citrate overproduction. Rohr and coworkers described antagonistic effect of ammonium ions to ATP and citrate inhibition of phosphofructokinase allowing unregulated glycolysis of a citrate-producing culture. The origin of an elevated ammonium pool was considered as the results of impaired ammonium fixation and intensified proteolysis in the citrate-producing cells (Mattey 1977; Habison et al. 1979; Kubicek et al. 1979; Rohr and Kubicek 1981).

Nevertheless, our results indicate that the concentration of ammonium nitrate plays a decisive role on cell differentiation and citrate yields in a submerged culture under a nutrient-limiting condition. More studies on cell physiology at a molecular level are needed to account for these unusual, interesting phenomena.

Literature Cited

Ahmed, S. A., J. E. Smith, and J. G. Anderson. 1972. Mitochondrial activity during citric acid production by *Aspergillus niger. Trans. Br. Mycol. Soc.* 59:51–61.

Bowes, I., and M. Mattey. 1979. The effect of manganese and magnesium ions on mitochondrial NADP dependent isocitrate dehydrogenase from *Aspergillus niger. FEMS Microbiol. Lett.* 6:219–222.

Clark, D. S. 1962. Submerged citric acid fermentation of ferrocyanide-treated beet molasses: Morphology of pellets of *Aspergillus niger. Can. J. Microbiol.* 8:133–136.

Clark, D. S., K. Ito, and H. Horitsu. 1966. Effect of manganese and other heavy metals on submerged citric acid fermentation of molasses. *Biotechnol. Bioeng.* 8:465–471.

Habison, A., C. P. Kubicek, and M. Rohr. 1979. Phosphofructokinase as a regulatory enzyme in citrate producing *Aspergillus niger. FEMS Microbiol. Lett.* 5:39–42.

Kisser, M., C. P. Kubicek, and M. Rohr. 1980. Influence of manganese on morphology and cell wall composition of *Aspergillus niger* during citric acid fermentation. *Microbiol.* 128:26–33.

Kubicek, C. P., and M. Rohr. 1977. Influence of manganese on enzyme synthesis and citric acid accumulation in *Aspergillus niger. Eur. J. Appl. Microbiol.* 4:167–175.

______. 1978. The role of the tricarboxylic acid cycle in citric acid accumulation by *Aspergillus niger. Eur. J. Appl. Microbiol. Biotechnol.* 5:263–271.

Kubicek, C. P., W. Hampel, and M. Rohr. 1979. Manganese deficiency leads to elevated amino acid pools in citric acid accumulating *Aspergillus niger. Arch. Microbiol.* 123:73–79.

Mattey, M. 1977. Citrate regulation of citric acid production in *Aspergillus niger. FEMS Microbiol. Lett.* 2:71–74.

Orthofer, R., C. P. Kubicek, and M. Rohr. 1978. Lipid levels and manganese deficiency in citric acid producing strains of *Aspergillus niger. FEMS Microbiol. Lett.* 5:403–406.

Ramakrishnan, C. V., R. Steel, and C. P. Lentz. 1955. Mechanism of citric acid formation and accumulation in *Aspergillus niger. Arch. Biochem. Biophys.* 55:270–273.

Rohr, M., and C. P. Kubicek. 1981. Regulatory aspects of citric acid fermentation by *Aspergillus niger. Proc. Biochem.* June/July:34–38.

Snell, R. L., and L. B. Schweiger. 1949. Production of citric acid by fermentation. U.S. Pat. 2,492,667 Dec.

Szczodrak, J. 1981. Biosynthesis of citric acid in relation to the activity of selected enzymes of the Krebs cycle in *Aspergillus niger* mycelium. *Eur. J. Appl. Microbiol. Biotechnol.* 13:107–112.

Wold, W. S., and I. Suzuki. 1976. The citric acid fermentation by *Aspergillus niger:* Regulation by zinc of growth and acidogenesis. *Can. J. Microbiol.* 22:1083–1092.

CHAPTER **36**

A Simplified Method for the Speciation of Fecal Streptococci

J. Kevin Rumery, Laurie M. Lawrence, and Harold K. Speidel

Northern Arizona University, Department of Biological Sciences, Box 5640, Flagstaff, Arizona 86011

Alternate methods for species differentiation of fecal streptococci were investigated using ATCC reference organisms. Fermentation of various carbohydrates was evaluated as a separation tool by determination of metabolic end products (gas chromatography) and pH (disc assay). Qualitative differences in chromatographic patterns were insufficient to allow separation of the standard organisms. However, there were significant quantitative differences in some metabolites that enabled separation of the species. Determination of carbohydrate fermentation patterns by disc assay proved to be an accurate, rapid, simple technique for speciation. This technique is based upon acid production from six carbohydrates on a phenol red agar and gives more accurate speciation results than Environmental Protection Agency (EPA) identification procedures. Media, time, and labor requirements of this technique are much reduced compared to current EPA procedures.

Introduction

Standard parameters for the measurement of fecal contamination of water are the total coliform (TC), fecal coliform (FC), and fecal streptococcus (FS) (APHA 1975; USEPA 1978). Currently the FC indicator is the method of choice for environmental waters because of the replication and long-term survival of the TC group in natural systems (Mahloch 1974; Van Donsel et al. 1967). This indicator, however, cannot distinguish between different warm-blooded sources of fecal contamination (Geldreich and Kenner 1969).

Although potentially valuable in identifying sources of contamination of water (Geldreich et al. 1968; Geldreich and Kenner 1969; Geldreich 1970; Hartman et al. 1966; Lin 1974; Stuart et al. 1971; Stuart et al. 1976), the species distribution of fecal streptococci has not been used routinely. This is largely because current methodology for speciation is difficult and gives confusing results. The real value of species distributions in identifying sources of contamination cannot be determined until an accurate and workable method has been developed.

In 1978 the United States Environmental Protection Agency (EPA) developed a standard scheme for the verification and speciation of the fecal streptococci based on refinements of previous methods. Although an improvement over previously developed approaches, there are several drawbacks to this method. First, several of the required EPA tests have long time requirements (7–14 d). Since rapid results are important in water quality testing, the requirements reduce the value of speciation as a tool. Second, the number of different media involved makes the logistics of speciation difficult in terms of labor, quality control, and cost. Last, the method has questionable accuracy. Speidel and Hekmati (1982) found problems with the EPA method when testing standard organisms obtained from the American Type Culture Collection (ATCC). In several cases, the EPA scheme failed to identify the ATCC organism. Results obtained in the Speidel study indicated that a more rapid and accurate method might be developed, based on the fermentation of carbohydrates. The current study was designed to investigate two interrelated techniques, analysis of carbohydrate fermentation patterns by disc test and gas chromatography.

Material and Methods

Organisms. Ten standard fecal streptococci cultures were obtained from the American Type Culture Collection (Table 1).

TABLE 1. *Standard ATCC fecal streptococci*

ATCC Number	Name
19433	*S. faecalis*
12958	*S. faecalis* var. *zymogenes*
13398	*S. faecalis* var. *liquefaciens*
19432	*S. faecium*
19434	*S. faecium*
6056	*S. durans*[a]
14025	*S. avium*
15351	*S. bovis*
33317	*S. bovis*
9812	*S. equinus*

[a]Now listed by ATCC as a variant of *S. faecium*.

Verification and speciation of fecal streptococci by standard methods. The standard fecal streptococcal cultures were verified and speciated according to procedures outlined in *Microbial Methods for Monitoring the Environment* (EPA 1978).

Speciation of ATCC cultures by carbohydrate fermentation patterns. The standard organisms were tested using a carbohydrate disc assay. From an initial screen of 21 carbohydrates, six were selected for intensive study (lactose, sucrose, sorbitol, raffinose, arabinose, and inositol). Commercially prepared discs (BBL Minitek) of each of these carbohydrates were used. Plates were prepared using phenol red broth base (Difco) with agar at 20.0 g/liter. The pH of the media was adjusted to 7.4 with 1 *N* NaOH prior to autoclaving. Plates were incubated overnight at 35 C before use to ensure a dry surface and check for contamination.

The carbohydrate fermentation patterns of the standard organisms were determined using the following procedure. Organisms were inoculated into BHI broth and incubated at 35 C for 24 h. Phenol red agar plates were inoculated with five drops of these cultures from a Pasteur pipet. Lawns were formed by spread plate technique using a glass spreading rod. After the organisms were spread on the plate surface, the carbohydrate discs were applied with either a disc dispenser or sterile forceps. The plates were then incubated at 35 C and 45 C under aerobic, anaerobic, and 3–5% carbon dioxide atmospheres. Plates were read at 6, 8, 12, and 24 h. Positive fermentation was recorded if a distinct zone of color change from red to yellow existed around the disc.

Gas chromatographic determination of lactose fermentation products. The standard organisms were screened for qualitative and/or quantitative differences in fermentation products with a Varian Model 3700 Gas Chromatograph equipped with a flame ionization detector. Fermentation products were separated on a column of 15% SP-1220, 1% phosphoric acid on 110/120 Chromosorb W-AW. The carrier gas was nitrogen at a flow rate of 30 ml/min. A temperature program was used with an initial temperature of 50 C for 4 min, increasing at 3 C/min until a final temperature of 150 C was attained. This temperature was held for 30 min. Lactose was the carbohydrate chosen for assay, as this was the only one that all of the standard organisms used. The medium was phenol red broth (Difco) with 1.5% lactose added. The pH was adjusted to 7.3 and medium was dispensed in 100-ml aliquots into 250-ml baffled culture flasks. These were autoclaved at 120 C for 10 min. The standard organisms were incubated in BHI broth at 35 C for 24 h and flasks were inoculated with 0.5 ml of these broth cultures. The flasks were then incubated at 35 C for 24 h. As the time required to run a sample through the instrument was significant, inoculations were staggered by 1 h to ensure uniform culture age. At 24 h, cultures were aseptically sampled for analysis. When immediate injection was not possible, the sample tubes were stored at 4 C to minimize loss of volatile components. A standard sample volume of 5.0 μl was used. Cells were not removed from the cultures before injection. Peak areas were calculated as peak width at half height times peak height.

Results and Discussion

Verification and speciation of standard test organisms. All of the ATCC standard organisms were verified as fecal streptococci when tested by the EPA verification procedures. Using the EPA speciation schemes, only three of the standard cultures were correctly speciated. These were *S. faecalis* (ATCC 19433), *S. faecalis* var *liquefaciens* (ATCC 13398), and *S. avium* (ATCC 14025).

Carbohydrate fermentation patterns. From the 21 carbohydrates originally tested, six gave unique and reproducible patterns for speciation with the ATCC organisms (Table 2). Incubation under anaerobic conditions or 3–5% carbon dioxide yielded variable results. Reproducible patterns were obtained under aerobic conditions at both 35 C and 45 C. Optimum incubation times were 24 h for 35 C and 8 h for 45 C. Incubation beyond these times resulted in zone overlap, weak zones caused by diffusional dilution, and reversal of zone to alkaline conditions.

Gas chromatography. Comparison of chromatograms from all of the ATCC organisms showed no qualitative differences in metabolite patterns. Retention distances for all peaks were identical for all organisms tested. However, signifi-

TABLE 2. *Carbohydrate fermentation patterns obtained using standard ATCC fecal streptococci*

Organism	LAC	IN	RAF	ARA	SUC	SOR
S. faecalis (ATCC-19433)	+	−	−	−	+	+
S. faecalis var *liquefaciens* (ATCC-13398)	+	−	−	−	−	+
S. faecalis var *zymogenes* (ATCC-12958)	+	+	−	−	+	+
S. faecium (ATCC-19434)	+	−	−	+	+	−
S. faecium (ATCC-19432)	+	−	−	−	−	−
S. durans[a] (ATCC-6056)	+	−	−	−	−	−
S. avium (ATCC-14025)	+	−	−	+	+	+
S. bovis (ATCC-33317, ATCC-15351)	+	−	+	−	+	−
S. equinus (ATCC-9812)	+	−	−	−	+	−

[a]Now listed by ATCC as a variant of *S. faecium*.

cant quantitative differences were observed in several peaks. Table 3 gives the retention distances for these peaks. The percentage difference between peak areas of the control (uninoculated, incubated phenol red lactose broth) and each of the tested ATCC organisms is listed in Table 3. Differences in peak areas between organisms that were greater than ±6% were considered to be significant. This value was chosen because it represented the overall accuracy of the peak area determination. Positive values indicate significant increase in area over the control, and negative values indicate significant decrease in area relative to the control. Based on these observations, a tentative flow chart for separation of these standard fecal streptococci was developed (Fig. 1). Comparison of values for peak A allowed separation of both ATCC 33317 and ATCC 13398. The remain-

TABLE 3. *Percent difference from control in major peak areas*

Organism	Peak A[a]	Peak D	Peak E	Peak F	Peak H	Peak I	Peak N
33317	+186	+ 29.1	− 14.7	+34.3	+ 67.5	−29.6	−11.9
13398	+ 46.7	+681	+ 1.7	+43.5	+ 87.5	+20.6	−24.9
19434	+ 22.1	+603	+ 4.2	+35.0	+ 62.5	− 2.6	−21.8
19433	+ 14.9	+603	+19.3	+45.8	− 20.0	+18.3	−22.6
12958	+ 2.6	+681	− 5.3	+37.3	+ 80.0	+29.1	−24.9
14025	− 3.9	+681	+47.2	+30.4	+ 40.0	+72.8	− 8.9
6056	− 23.4	+215	+ 3.2	+40.4	+100	− 6.0	−30.6
19432	− 27.3	+ 92.2	+ 5.3	+21.9	+ 82.5	+72.8	−24.4

[a]retention distances in millimeters: A = 3, D = 6.5, E = 14, F = 33, H = 64, I = 70, N = 112.

ing organisms separated into the following groupings: [12958 and 14025], [19434 and 19433] [6056 and 19432]. Comparison of values for peak B allowed separation of ATCC 6056 and ATCC 19432. Comparison of values for peak E allowed distinct separation of ATCC 12958 and ATCC 14025. The values for this peak allowed separation of ATCC 19434 and ATCC 19433, but not with a large margin. Therefore, the values for these organisms were compared using peak H. This comparison allowed distinct separation between the two organisms.

In addition, comparison of peak areas was made based on direction of change from the control areas. Peak areas that increased over the control were assigned plus (+) designations, those that decreased relative to the control were assigned negative (−) designations, and those without significant change were given zero (0) designations. This approach yielded a set of patterns that also allowed separation of each of the ATCC organisms. These patterns are listed in Table 4.

The discrepancies between ATCC and EPA-derived identifications agree with those reported by Speidel et al. (1982). As indicated in that work, there may be several reasons for these differences. Variations in experimental methods over the years may have led to different specific epithets being attached to the same organism. Identification schemes such as the EPA protocols, if developed with organisms having confusing taxonomy because of these variations, would be ex-

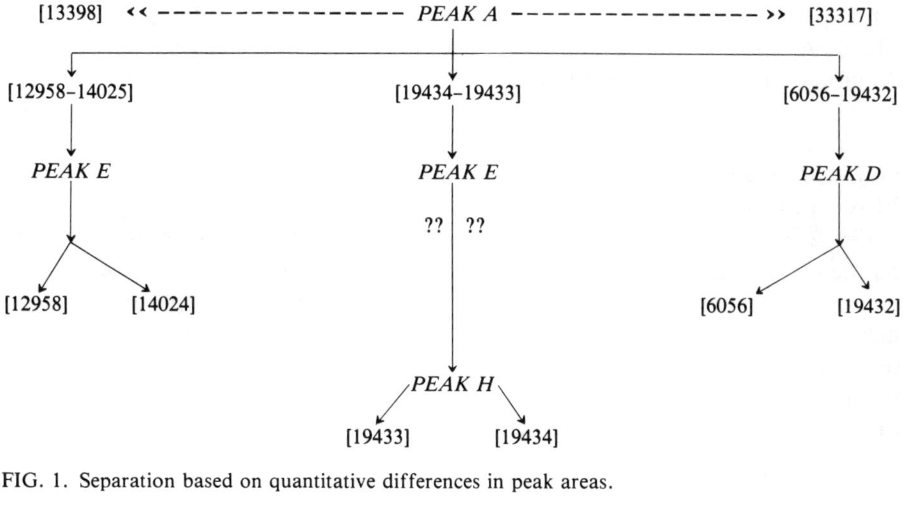

FIG. 1. Separation based on quantitative differences in peak areas.

TABLE 4. *Separation based on direction of change from control*

Organism	Peak A	Peak E	Peak H	Peak I
19433	+	+	−	+
33317	+	−	+	−
19434	+	0	+	0
13398	0	+	+	+
14025	0	+	+	−
12958	0	0	+	+
6056	−	0	+	0
19432	−	0	+	+

pected to give confusing results. This problem, combined with the long incubation times and numerous media required, greatly reduces the validity of the EPA method. In contrast, use of the carbohydrate disc assay (Table 2) and the analysis of metabolites by gas chromatography (Table 3) allows accurate and reproducible separation of the standard ATCC fecal streptococci.

The gas chromatographic approach has high potential but requires further testing. The effects of medium composition, incubation conditions, inoculum size, and different instrument conditions still need to be investigated. Current data represent a small sample size, and it remains to be seen if replicate experiments with the ATCC organisms will produce consistent results. Gas chromatography has some drawbacks. It requires an expensive and sophisticated instrument, which may not be available to many laboratories engaged in pollution assessment. This problem may diminish as more laboratories are equipped with gas chromatographs to perform EPA-required analysis of organics in drinking water. Since simplicity and speed were goals of the current study, complex but potentially more accurate procedures were not used (i.e., extraction of metabolites).

The carbohydrate disk assay appears to be a more viable alternative to the current EPA methodology for the following reasons: (1) Speed. In evaluating water quality and pollution sources, speed is critical. The EPA schemes require up to 14 d for speciation; disc assay can be read in 8 h (45 C) or 24 h (35 C). (2) Simplicity. The EPA schemes require many different media; disc assay requires only one. (3) Accuracy and reliability. EPA speciation results do not correlate with the ATCC designations; results from disc assay correlate exactly. (4) Cost. In terms of media purchase and/or preparation, and the labor involved in quality control, inoculation, and reading of the required tests, the disc assay is much less costly than the EPA schemes.

Fig. 2 shows a proposed scheme for the speciation of fecal streptococci based on the carbohydrate disc assay. Before large-scale application of the method can be considered, some concerns need to be addressed. First, the applicability of the method to field studies needs to be determined. This is being done now by using the disc assay to screen a number of isolates from fecal streptococcus analyses from surface water samples. Second, a large number of fecal streptococci from existing culture collections must be tested by this method. In cases where identifi-

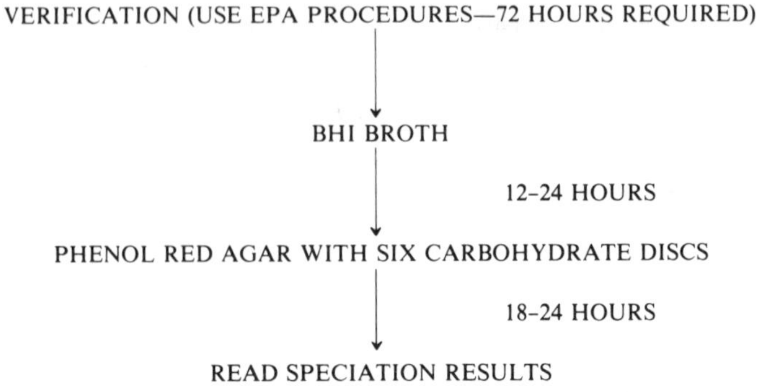

FIG. 2. Tentative speciation procedure based on disc assay.

cations of organisms are contradictory, additional biochemical parameters (including gas chromatography) can be used for clarification. This is necessary to establish a reference data base of fermentation patterns. If these areas are adequately addressed, the proposed procedures may become an accurate and workable method for the speciation of fecal streptococci.

Literature Cited

American Public Health Association. 1975. *Standard Methods for the Examination of Water and Wastewater, 14th ed.* American Public Health Association, Inc., New York.

Geldreich, E. E. 1970. Applying bacteriological parameters to recreational water quality. *J. Amer. Water Works Assoc.* 62:113–120.

Geldreich, E. E., and B. A. Kenner. 1969. Concepts of fecal streptococci in stream pollution. *J. Water Pollut. Contr. Fed.* 41:R336–R352.

Geldreich, E. E., L. C. Best, B. A. Kenner, and D. J. Van Donsel. 1968. The bacteriological aspects of stormwater pollution. *J. Water Pollut. Contr. Fed.* 40:1861–1872.

Hartman, P. A., G. W. Reinbold, and D. S. Saraswat. 1966. Indicator organisms—a review. I. Taxonomy of the fecal streptococci. *Intl. J. Syst. Bacteriol.* 16:197–221.

Lin, S., R. L. Evans, and D. B. Beuscher. 1974. Bacteriological assessment of Spoon River water quality. *Appl. Microbiol.* 28:288–297.

Mahloch, J. L. 1974. Comparative analysis of modeling techniques for coliform organisms in streams. *Appl. Microbiol.* 27:340–345.

Speidel, H. K., and A. M. Hekmati. 1982. Evaluation of current schemes for speciating fecal streptococci. *Dev. Ind. Microbiol.* 23:503–511.

Stuart, D. G., G. K. Bissonnette, T. D. Goodrich, and W. G. Walter. 1971. Effects of multiple use on water quality of high mountain watersheds: Bacteriological investigations of mountain streams. *Appl. Microbiol.* 22:1048–1054.

Stuart, S. A., G. A. McFeters, J. E. Schillinger, and D. G. Stuart. 1976. Aquatic indicator bacteria in the high alpine zone. *Appl. Environ. Microbiol.* 31:163–167.

United States Environmental Protection Agency. 1978. *Manual of Microbiological Methods for Monitoring the Environment, Water and Wastes* EPA 600/8-78-017.

Van Donsel, D. J., E. E. Geldreich, and B. A. Clarke. 1967. Seasonal variations in survival of indicator bacteria in soil and their contribution to storm-water pollution. *Appl. Microbiol.* 15:1362–1370.

CHAPTER 37

Microbiology of the Hands: Factors Affecting the Population

ARTHUR F. PETERSON

Skyland Scientific Services, Inc., Belgrade, Montana 59714

The microflora of the hands is determined by several factors: with hormones, systemic and topical drugs, exposure to chemical agents and certain activities apparently having the greatest effects. In these long-term studies the population is correlated with these factors. The implications, both for studies of skin microbiology and in health care are discussed.

INTRODUCTION

A number of studies on the microbiology of the hands (Peterson 1973, 1974, 1978; Peterson et al. 1978; Rosenberg et al. 1976; Whalen and Peterson 1976) reported the quantitative effects of various antimicrobial agents on the resident and transient microflora. Concurrent with those studies, the quantitative effects of other parameters were examined and qualitative changes were noted. Some of those examinations are reported here.

MATERIALS AND METHODS

Over the course of the period August 1971 to June 1983, 850 volunteers had their hands sampled. These volunteers ranged in age from 18 to 76 yr, were of both sexes, and were varied in racial composition and occupational background. For at least 2 wk prior to and throughout the course of studies, all subjects avoided the use of medicated soaps, shampoos, deodorants, lotions, oral and topical antimicrobials, and avoided contact with solvents, bleaches, detergents, and similar materials. Either prior to their involvement or retrospectively, subjects were interviewed regarding any medication they were using (including oral contraceptives), their activities during the prestudy and study periods, and menstrual status. Demographic data was also recorded.

Following the 2-wk prestudy period, both hands of each subject were sampled by the Standard Glove Fluid Technique on 3 d over a 5- to 7-d period.

One hundred twenty-six volunteers of the original group also had their hands sampled at regular intervals when they were not on the prestudy restrictions. They were interviewed regarding the use of products included in those prestudy restrictions.

The Standard Glove Fluid Technique. Following a thorough rinsing of the hands, sterile surgical gloves with no antimicrobial activity were applied to both hands, and 75 ml of sterile stripping/suspending fluid (0.1% v/v Triton X–100 in 0.1 *M* phosphate buffer, pH 7.8) was instilled into the glove of the hand being sampled. The wrist was secured and the hand massaged through the glove for 1 min. Aliquots were withdrawn, diluted, and plated in triplicate. Trypticase soy broth (TSB) with 1.0% w/v Tween 80 and 0.3% w/v lecithin was used for dilution. Trypticase Soy Agar (TSA) with the same additions was used for all plating.

All plates were incubated at 32–34 C for 48 h. Plates exhibiting between 30 and 300 colonies were counted and populations were determined. In instances where plates exhibited colonial morphology and/or pigmentation differing from normal skin microflora, those colonies were isolated, cultured, Gram stained, and microscopically examined. All data were transformed into base 10 logarithms. Data were then arranged successively into groups to facilitate examination of the following: (1) normal variations among all subjects on prestudy restrictions; (2) normal variations among all males on prestudy restrictions; (3) normal variations among all females on prestudy restrictions: (a) variations with menstrual cycles arranged synchronously; (4) deviation from normal baseline when prestudy restrictions were not followed; and (5) influence of activities that differed from "normal" among subjects on prestudy restrictions. The data are summarized in Tables 1 through 6.

Results and Discussion

Review of the data for all subjects (Table 1) showed a normal distribution of microbial counts on the hands. This same pattern was noted when all males (Table 2) or all females (Table 3) were considered. Apparently the normal microbial population on the hands, both for individuals and for large population, is stable over time, and is not significantly influenced by season, diet, or climate.

TABLE 1. *Means of $\log_{10}$ microbial counts for all subjects on prestudy restrictions (16,240 samples)*

	Dominant Hand	Nondominant Hand
Minimum Count	1.6335	1.7924
Maximum Count	9.7235	9.7937
Mean	6.4260	6.4040
Median	6.1987	6.2201

TABLE 2. *Means of $\log_{10}$ microbial counts for all males on prestudy restrictions (7,460 samples)*

	Dominant Hand	Nondominant Hand
Minimum Count	3.6580	3.4200
Maximum Count	9.7235	9.7937
Mean	6.8579	6.8129
Median	6.2003	6.2316

TABLE 3. *Means of log_{10} microbial counts for all females on prestudy restrictions (8,780 samples)*

	Dominant Hand	Nondominant Hand
Minimum Count	1.6335	1.7924
Maximum Count	7.6532	7.4914
Mean	6.1288	6.1264
Median	6.1996	6.2251

When data were arranged so that menstrual cycles for females coincided, the distribution of counts spread out significantly and the counts were significantly lower than for the same subjects in a postmenstrual period (Table 4). On the other hand, female subjects taking oral contraceptives showed no such population shift. Fluctuations in hormone levels, as with menstrual period females, can have a significant impact on the skin population. Stabilization of hormone levels, as accomplished with oral contraceptives, leads to stabilization of that population.

TABLE 4. *Means of log_{10} microbial counts for females with menstrual arranged synchronously (3,260 samples)*

	Dominant Hand	Nondominant Hand
Menstrual Period[a]		
Minimum Count	1.6335	1.7924
Maximum Count	6.6385	6.5052
Mean	4.3674	4.4920
Median	3.6299	3.6145
Post-Menstrual Period		
Minimum Count	4.3674	4.5119
Maximum Count	7.6532	7.4914
Mean	6.7209	6.7686
Median	6.1980	6.2001

[a] The period from 5 d prior to the menstrual period through the onset.

As would be expected, most agents known or suspected of being effective antimicrobials gave significant reductions of normal hand populations (Table 5) when the hands were exposed to them.

The effect of antimicrobial agents on the hands or in the blood was generally significant, with populations returning to normal 1 to 7 d after use was discontinued. In the case of deodorant soaps, counts increased after 4 wk of use to near initial levels, but qualitative shifts were noted. In some cases, gram-negative organisms had become established on the hands. No single genera predominated, but several subjects had *Pseudomonas* as the only gram-negative.

Many activities outside the normal scope of a volunteer's typical patterns gave significant increases in hand populations, shifts in the qualitative population, or both (Table 6).

Along with antimicrobial agents, certain activities that differed from a subject's typical activity patterns gave rise to increases in populations or qualitative shifts. Occlusion of the hands in a warm, moist environment gave rise to both, with gram-negative organisms and yeasts observed.

TABLE 5. *Effects of agents on the normal population of the hands (all counts as mean* log_{10}*)*

Agent	Normal Population	Post-Use Population
Sodium Hypochlorite (14 subjects)	6.4087	3.7686
Dandruff Shampoo (67 subjects)	6.5670	5.3284
Deodorant Soaps (83 subjects)	6.3685	4.5052[a b] 5.8021[b]
Oral Antibiotics (11 subjects)	6.7348	5.4984[a]
Oral Contraceptives (77 subjects)	6.7356	6.6948
Solvents (26 subjects)	6.6243	5.2131[a]

[a] Statistically significant change.
[b] Samples taken 1 wk and 4 wk after use was begun. Qualitative shifts noted in 47 subjects at 4 wk. Four subjects exhibited *Pseudomonas* as the only gram-negative. 43 had mixed gram-negative populations. Gram-negatives made up 10–40% of the total population of the hands with no pattern evident.

TABLE 6. *Influence of activities on the normal population of the hands (all counts as mean* log_{10}*)*

Activity	Normal Population	Post-Activity Population
Gardening (12 subjects)	5.2409	6.1703[b]
Wound Dressing (6 subjects)	6.2304	6.3600[a]
Food Preparation (11 subjects)	5.8780	6.4303[b]
Dishwashing (with gloves (6 subjects)	6.2101	6.8716[a b]
Handling Animals (17 subjects)	6.1303	6.4211[b]

[a] Statistically significant change.
[b] Establishment of new organisms and some qualitative population shifts noted. Gardening and food preparation gave gram-positive spore formers, yeasts, and molds being found on the hands. Dishwashing gave rise to yeasts and gram-negatives, with *Pseudomonas* predominating.

These studies suggest that certain factors cannot be ignored when studies of skin microbiology are undertaken. The qualitative effects of certain agents and activities certainly merit further study, especially since their potential impact on health care personnel may be significant.

Literature Cited

Peterson, A. F. 1973. The microbiology of the hands: Evaluating the effects of surgical scrubs. *Dev. Ind. Microbiol.* 14:125–130.

______. 1974. The microbiology of the hands: Assessing new scrubbing devices and procedures. *Dev. Ind. Microbiol.* 15:417–419.

______. 1978. The microbiology of the hands: Effects of varying scrub procedures and times. *Dev. Ind. Microbiol.* 19:325–334.

Peterson, A. F., A. Rosenberg, and S. D. Alatary. 1978. Comparative evaluation of surgical scrub preparations. *Surg. Gynecol. Obst.* 146:63–65.

Rosenberg, A., S. D. Alatary, and A. F. Peterson. 1976. Safety and efficacy of the antiseptic chlorhexidine gluconate. *Surg. Gynecol. Obst.* 143:789–792.

Whalen, J. W., and A. F. Peterson. 1976. Iodonium surgical scrub. *Abstr., Annu. Mtg., Am. Soc. Microbiol.*

CHAPTER **38**

Leaching of Pb and Zn from Spent Lubricating Oil

RICHARD W. TRAXLER AND ELEANORE M. WOOD

Department of Plant Pathology/Entomology, University of Rhode Island, Kingston, Rhode Island 02881

Leaching experiments with Pb- and Zn-resistant pseudomonads demonstrated that Pb and Zn were leached from spent lubricating oil into water under both aerobic and anaerobic conditions. Leaching of the metal from oil was more efficient into seawater than into fresh water. Differences in the leaching dynamics of Pb and Zn suggest different mechanisms for these two metals. Chromatographic data indicate that a portion of both the Pb and Zn leached into seawater is present as an organo-metal complex.

INTRODUCTION

Spent lubricating oils, which normally include a mixture of oil grades from automobile engines and transmission fluids, contain large quantities of mercury, cadmium, tin, zinc, chromium, lead, and nickel (Anonymous 1972) as well as other cations such as sodium, magnesium, and potassium. VanVleet and Quinn (1978) demonstrated that sewage plant effluents were a major source of petroleum hydrocarbons found in upper Narragansett Bay sediments. They estimated an annual hydrocarbon input to the bay from the Fields Point sewage plant of 226 tons; analysis showed the hydrocarbons closely resembled a mixture of No. 2 fuel oil, No. 6 fuel oil, and used crankcase oil. Tanacredi (1971) reported used crankcase oil as a major contribution to coastal environments while Porricelli et al. (1971) estimated that 30% of the petroleum in the ocean may be from waste lubricating oil. Hoffman et al. (1981) estimated a hydrocarbon input to Narragansett Bay of 44 metric tons per annum from road run-off during rain storms. A problem of environmental significance is that the spent lube oils thus constitute a reservoir of heavy metals in sediments. An unanswered question concerns the rate and mechanism by which the metals are leached from the oil into the aqueous phase.

Other concerns related to the metal content of spent lube oils are their volatilization if spent oil is burned and their removal for the reprocessing of spent lube oils. Microbial populations that have the capacity for biodegradation of petroleum products exist in soils and sediments as well as fresh and marine water (Atlas 1981). Also, many studies have demonstrated the role of bacteria in metal transformations (Summers and Silver 1978) and Walter and Colwell (1974) have implied the simultaneous biotransformation of both oil and metals in natural environments. Traxler and Wood (1981) demonstrated that many of the bacteria isolated from metal-polluted environments were resistant to those metals causing the pollution, thus a resistant bacterium can possess the genetic information for multiple metal resistance. The biotransformations of metals involve shifts in metal species, which have important environmental implications (Brinckman and Iverson 1975).

This project was initiated as a preliminary area of investigation concerned with the rate and mechanism involved in the release of metals from petroleum hydrocarbons and the role of microorganisms in the processes of metal release and transformation.

Materials and Methods

Bioreactors. The bioreactors were constructed from plexiglass, which reduced metal adsorption to the walls. The reactors were cylinders 135 mm in diameter and 112 mm deep, closed with a bottom plate and a head plate with openings for sampling and other operations. A rubber gasket was fitted between the reactor and headplate, which was secured by threaded plastic bolts. The reactors were charged with 1.5 l of water overlayed with 20 ml of spent lubricating oil. Compressed air or nitrogen were sparged through the water and oil layers then vented to the atmosphere. The reactors were gently agitated with a teflon-coated magnabar at 110 rpm.

Spent lubricating oil. Approximately 20 l of a spent oil was obtained from the drain sump of a local automobile service center. The oil was an unknown mixture of different grades of crankcase oil and transmission fluid. Prior to each use the plastic storage container was shaken to mix the oils and sedimented solids.

Metal analyses. All water samples were acidified with 0.1 *N* HCL to a pH less than 4.0. Metals (Pb, Zn, and Cd) were quantified in a Perkin-Elmer Model 5000 Atomic Absorption Spectrophotometer, and Hg was quantified using the same instrument in the flameless mode with a hydride attachment. Standard curves for each metal were established for each series of samples, and each point on the standard curve and the unknown samples represented the average values from five separate readings.

Microorganisms. The organisms used in this study were *Pseudomonas oleovorans,* a Zn- and Pb-resistant isolate from a marine environment; *Pseudomonas* sp. strain PR72, a Zn- and Pb-resistant isolate from the Providence River; and a *Bacillus subtilis* resistant to Zn and Pb. All organisms were carried on Trypticase Soy Agar (TSA) amended with 50 ppm of Pb as $Pb(NO_3)_2$ and Zn as ($ZnCl_2$) and transferred each month for new working stock cultures. For inoculum, the organisms were grown in TSB-metal medium, recovered by centrifugation, washed three times in sterile distilled water, and resuspended at a density of 1 mg dry wt/ml.

Results and Discussion

The initial leach rate of Pb from spent lubricating oil into fresh seawater is much greater under aerobic conditions than under anaerobic conditions (Fig. 1). In both systems there were initial rates of leaching followed by a secondary rate of leaching. The total Pb leached under aerobic and anaerobic conditions were of the same magnitude. The data in Fig. 2 show the same type of response for the leaching of Zn from the oil into seawater. The quantity of Zn leached, however, was over twice that of the Pb leached from the same oil.

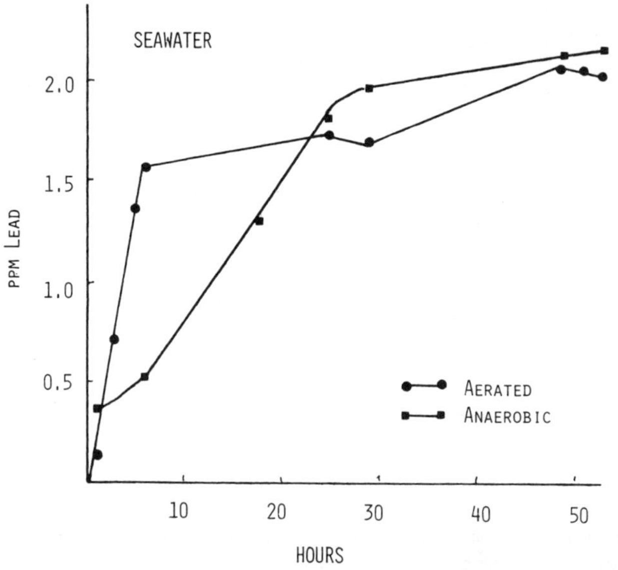

FIG. 1. Leaching of Pb from spent lubricating oil into seawater under ●—● aerobic and ■—■ anaerobic conditions by *Pseudomonas oleovorans*.

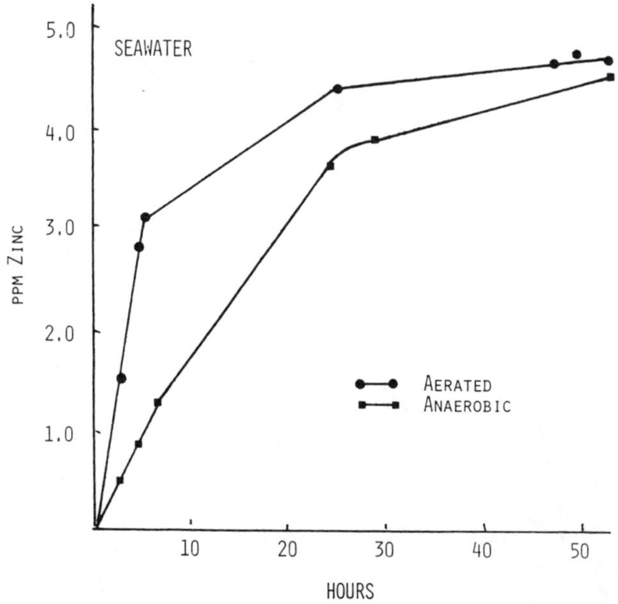

FIG. 2. Leaching of Zn from spent lubricating oil into seawater under ●—● aerobic and ■—■ anaerobic conditions by *Pseudomonas oleovorans*.

Leaching of Zn and Pb from oil into fresh water is shown in Figs. 3 and 4.

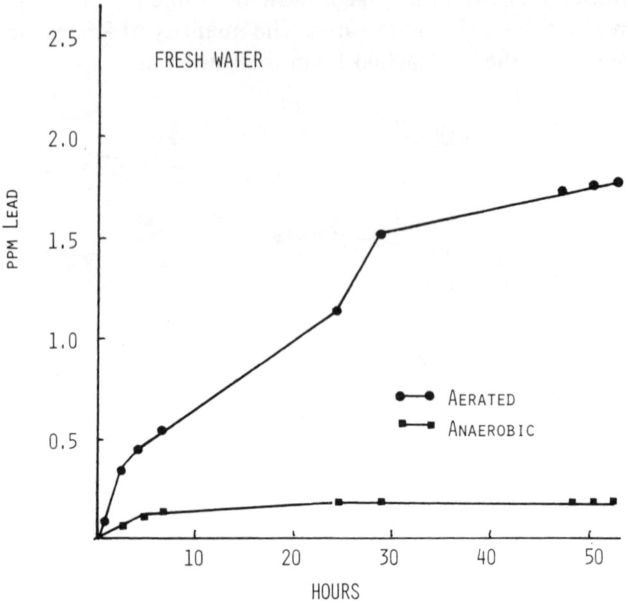

FIG. 3. Leaching of Pb from spent lubricating oil into fresh water under ●—● aerobic and ■—■ anaerobic conditions by *Pseudomonas* sp. PR72.

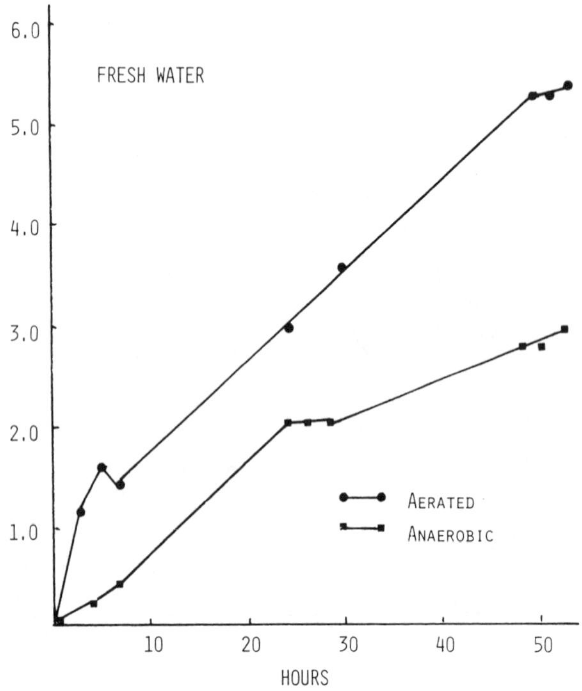

FIG. 4. Leaching of Zn from spent lubricating oil into fresh water under ●—● aerobic and ■—■ anaerobic conditions by *Pseudomonas* sp. PR72.

Under aerobic conditions, Pb leaching did not demonstrate the two leach rates observed with seawater; rather the leaching appears to follow a first-order kinetic. Only a trace of Pb was leached into fresh water under anaerobic conditions. Zn was leached from oil into fresh water under aerobic and anaerobic conditions. Again the high, initial leaching rates followed by a slower secondary rate were not apparent. The total Zn leached under anaerobic conditions did not approach the total Zn leached under aerobic conditions as was found in the seawater system.

These experiments demonstrated a consistent amount of Pb and Zn that could be leached from the oil into the aqueous phase. An experiment was performed to leach the same oil repeatedly with new water and organisms. After 48 h of leaching, the water-organism phase was siphoned from below the oil and replaced with fresh seawater inoculated with *Ps.* PR72. The results (Table 1) showed that with three repeated leaches the final Pb concentrations were in the range of 1.65–1.99 ppm under aerobic conditions. With Zn leaching, the first leach removed a maximum of 4.59 ppm of Zn whereas the secondary and tertiary leaches were 2.4- and 3.4-fold, respectively, less than the primary leach.

TABLE 1. *Repeated leaching of Pb and Zn from spent lube oil into seawater under aerobic conditions*

		PPM	
Leach Number	Time (Hours)	Pb	Zn
1	0	0.30	0
	4	1.36 0.27[a]	2.29 0.57[a]
	24	1.99	4.38
	48	1.92	4.59
2	0	0.19	0.06
	4	0.95 0.19[a]	0.77 0.18[a]
	24	1.32	1.53
	48	1.65	1.88
3	0	0.24	0.08
	4	1.20 0.24[a]	0.52 0.11[a]
	24	1.58	1.04
	48	1.99	1.35

[a] Leach rate mg/h for the first 4 h period.

Examination of the leach rate during the first 4-h portion of the experiment (highest rate) showed that with Pb there was not a significant effect on leach rate by repeated leaches, whereas with Zn the rate of leaching during the initial phase was decreased with each repeated leach.

Oil leached into seawater yielded approximately 1.9 μg/ml of Pb, 4.6 μg/ml of Zn, and 3.8 μg/ml of Cd after 48 h of leaching. If a metal is present in water as an organometallic complex it should be extractable with toluene. A 100-ml portion of 24- and 48-h seawater leachate was extracted with 5 ml of toluene overnight in a continuous liquid-liquid extractor. Duplicate 150 μl aliquots of the toluene were spotted onto silica gel Thin Layer Chromatography (TLC) plates that were developed in hexane:acetone (93:7).

One plate was sprayed with a solution of 10 mg of diphenylthiocarbazone in 20 ml of chloroform to detect areas containing metal. The silica gel was scraped from the second plate in those areas indicating metal by the dithizone reaction on the first plate. Metal was eluted from the silica gel with 2 *N* HCl, filtered and brought to 5 ml with demineralized water for metal analysis.

The leached Pb was not in a toluene-extractable form since only a trace was found at the TLC origin and none was associated with the TLC spots (Table 2). Cd was located in all the reacting areas of the TLC but if present as organo-Cd it was there only in trace amounts. The 100 ml of extracted water contained 380 μg of Cd, whereas the positive areas contained an estimated 1.5 μg at 24 h and 3.75 μg at 48 h on the basis of the extracted 100 ml of water. Somewhat less than 1% of the Cd could be present as an organic complex.

TABLE 2. *TLC Pb and Zn values from* Ps. oleovorans-*leached oil into seawater*

Reactor Time (Hours)	Rf on TLC	Metal eluted from TLC, ppm		
		Pb	Cd	Zn
24	Origin	0.09	0.005	0.178
	0.23	0	0.005	0.224
48	Origin	0.01	0.01	0.306
	0.23	0	0.008	0.137
	0.41	0	0.007	0.079

On the other hand, it is probable that at least two organic forms of leached Zn were represented on the TLC. On the basis of Zn in the 100 ml of water extracted after 24 h of leaching, 13% of the Zn is likely to be organo-Zn and 17% organo-Zn in the 48-h extract.

A similar experiment was performed with a *Bacillus subtilis* isolate that was resistant to 50 ppm of Pb and Zn (Table 3). Included in this aerobic experiment

TABLE 3. *TLC Pb and Zn values for* Bacillus subtilis-*leached oil into seawater*

Sample Time (Hours)	Rf	Metal eluted from TLC, ppm	
		Pb	Zn
A. Seawater-Oil *Bacillus*			
24	Origin (red)	0.07	0.055
	0.18 (yellow)	0.02	0.081
48	Origin (red)	0.04	0.161
	0.18 (yellow)	0.01	0.034
B. Seawater-Oil *Natural populations*			
24	Origin (red)	0.14	0.105
	0.18 (yellow)	0.01	0.044
48	Origin (red)	0.06	0.20
	0.18 (yellow)	0.03	0.067
	0.33 (yellow)	0.02	0.03
	0.91 (blue)	0	0.033

was a bioreactor containing only the microorganism present in seawater. Both reactors were toluene extracted and chromatographed as in the previous experiment. The metal values from the TLC spots were lower than in the experiment reported in Table 2. The results indicate that all the metal-reacting areas on the TLC contained a mixture of Pb and Zn compounds except for the Rf 0.91 spot from the natural population reactor, which contained only Zn.

These data indicate that both Pb and Zn leached into seawater are present as organic-metal complexes. Since metal speciation of the spent lubricating oil was not possible, it is not known if these compounds were initially present in the oil or if they represent microbe-mediated transformation products.

To study the population dynamics and leaching from oil-contaminated sediment we chose to use a sediment from Pawtuxet Cove, which is at the outfall of the Pawtuxet River into Narragansett Bay. This river drains a series of industrial sites, including effluent from metal-working operations. Analysis of the sediment showed values of 5.96 ppm of Zn, 0.18 ppm Cd, 32.9 ppm Pb, 1.5 ppm Sn, and 0.45 ppm Hg. For this experiment, 3 g of sediment were mixed with 20 ml of spent oil and this mixture overlayed with water (pH 6.35, salinity 0.5‰) in an aerobic and in an anaerobic reactor. Each reactor water was sampled for metals and microbial population after 1, 24, 49, and 72 h. The isolation medium was TSA supplemented with 0.1% (w/v) of NH_4NO_3 and the same medium supplemented with 50 ppm of Pb or 50 ppm of Zn. The spread plate dilutions were incubated aerobically at room temperature; a duplicate set was incubated in the anaerobic incubator. Plates were counted after 2 d incubation, and the number of each colony type that developed was determined by visual inspection of the plates.

There was a significant and rapid increase in the microbial numbers in each reactor over the course of the 3-d experiment (Tables 4 and 5) as indicated by the plate count data on nonmetal amended TSA. A selection occurred within the population as indicated by the decreasing number of colony types that grew on

TABLE 4. *Microbial population response in water of an aerobic oil-sediment leach*

Time (Hours) Incubation Conditions	Aerobic Reactor On To		
	TSA	TSA + Pb	TSA + Zn
1			
Aerobic	1.3×10^3(7)[a]	1.8×10^2(6)	5.2×10^2(7)
Anaerobic	3.3×10^2(5)	1.2×10^2(2)	2.1×10^2(6)
24			
Aerobic	$>10^4$(2)	$>10^4$(3)	$>10^4$(2)
Anaerobic	$>10^4$(5)	1.8×10^4(3)	1.4×10^3(4)
48			
Aerobic	$>10^4$(2)	$>10^4$(3)	$>10^4$(2)
Anaerobic	$>10^4$(1)	$>10^4$(1)	$>10^4$(1)
72			
Aerobic	$>10^5$(2)	$>10^5$(3)	$>10^5$(2)
Anaerobic	$>10^5$(3)	$>10^5$(3)	$>10^5$(2)

[a] ( ) = Number colony types represented on the counter plates.

the plating medium. Additionally, under both aerobic and anaerobic conditions a significant portion of the expressed microbial population was resistant to 50 ppm of Pb and Zn (able to develop on the metal supplemented media).

TABLE 5. *Microbial population response in water of an anaerobic oil-sediment leach*

Time (Hours) Incubation Conditions	Aerobic Reactor On To		
	TSA	TSA + Pb	TSA + Zn
1			
Aerobic	5.6×10^2(7)[a]	1.1×10^1(3)	2.7×10^1(7)
Anaerobic	2.6×10^2(6)	1.0×10^1(1)	2.0×10^1(5)
24			
Aerobic	10^4(2)	10^4(3)	10^4(2)
Anaerobic	10^4(5)	1.4×10^3(4)	1.4×10^4(4)
48			
Aerobic	10^4(3)	10^4(3)	10^4(2)
Anaerobic	7.5×10^4(3)	5.6×10^3(2)	8.0×10^3(1)
72			
Aerobic	8.3×10^4(2)	1.5×10^5(3)	6.6×10^4(2)
Anaerobic	1.1×10^5(4)	8.3×10^4(1)	1.4×10^5(3)

[a] (   ) = Number colony types represented on the countable plates.

The quantity of metal detected in the water of both reactors is presented in Table 6. More metal was leached from the oil-sediment into water under aerobic conditions. Zn values were higher than either Pb or Hg values in both reactors. The values are similar to those obtained with the fresh water leaches of oil in the absence of sediment (Figs. 3 and 4).

TABLE 6. *Metal content of leachates from sediment-oil reactors under aerobic and anaerobic conditions.*

Time (Hours)	Aerobic Reactor Water, ppm			Anaerobic Reactor Water, ppm		
	Pb	Zn	Hg	Pb	Zn	Hg
0	0.52	0.02	0.009	0.74	0.02	0.009
24	1.54	5.22	0.008	0.48	2.44	—
48	0.60	4.41	—	0.28	2.39	—
72	2.17	5.59	0.005	0.30	2.65	0.001

These reactors are static systems and cannot be equated directly to events occurring in natural environments. The repeated leach experiments (Table 1) suggest that in a natural system where water replacement occurs there would be continued metal leaching from the oil sediment into water.

These preliminary data indicate that a metal source such as spent lubricating oil is mobilized, and the metal can be transported into water where its toxic effect can be exerted on the biological forms in the water system. Interestingly, the natural microbial flora rapidly develops resistance to these toxic metals and can, by metal transformations, serve as detoxification mechanisms (Schottel 1978). The best-known examples of detoxification mechanisms are the Hg reductase system (Foster 1983; Summer and Silver 1972; Summers and Silver 1978), which converts Hg^{2+} to Hg°, is less toxic, and also is insoluble in water and readily volatilized because of its high vapor pressure. In the case of organomercurials a second enzyme, organomercurial lyase (Tezuka and Tonumra 1978; Schottel 1978), cleaves the C-Hg bonds to release Hg^{2+}, which is reduced by mercuric reductase, resulting in volatilization of the Hg°. Unfortunately, basic knowledge of toxic-detoxification mechanisms for Pb and Zn are unknown.

Conclusions

Leach experiments demonstrated that Pb and Zn were leached from spent lubricating oil into water under both aerobic and anaerobic conditions. Leaching of Pb was more efficient from oil into seawater than into fresh water, particularly under anaerobic conditions. Leaching of Zn was also more efficient into seawater than fresh water.

Differences in the leaching dynamics of Zn and Pb suggest possibly different mechanisms involved in the leaching of these metals from oil or oil entrained in sediment. The experiments in which an oil was repeatedly leached by new water systems might represent the natural environment where an unconfined system is continuously exposed to new water. It appears that in a static system, such as the bioreactors, an equilibrium is established between the two phases of the system, oil-sediment-water or oil-water, so that only a limited amount of a specific metal can leach into the water phase.

Acknowledgments

This work was supported by National Bureau of Standards projects NB79NAA-B-3835, NB81NAA-G-3809, NB81NAA-H-5649, and the Rhode Island Agricultural Experiment Station. We wish to thank William R. Wright and Peter Schauer for the use of their atomic absorption spectrophotometer and Fred Brinckman and Warren Iverson of the National Bureau of Standards for their advice and encouragement on this project. This paper is contribution Number 2200 of the Rhode Island Agricultural Experiment Station.

Literature Cited

Anonymous. 1972. Waste lube oils pose disposal dilemma. *Environ. Sci. Technol.* 6:25.

Atlas, R. M. 1981. Microbial degradation of petroleum hydrocarbons: An environmental perspective. *Microbiol. Rev.* 45:180–209.

Brinckman, F. E., and W. P. Iverson. 1975. Chemical and bacterial cycling of heavy metals in the estuarine system. Page 319 *in* AES Symposium Series, No. 18, *Marine Chemistry in the Coastal Environment.*

Foster, T. J. 1983. Plasmid-determined resistance to antimicrobial drugs and toxic metal ions in bacteria. *Microbiol. Rev.* 47:361–409.

Hoffman, E. J., A. M. Falke, and J. G. Quinn. 1981. Waste lubricating oil disposal practices in Providence, Rhode Island: Potential significance to coastal water quality. *Coastal Zone Manag. J.* 8:337–348.

Porricelli, J. D., V. F. Keith, and R. L. Storch. 1971. Tankers and the ecology. *Trans. Soc. Nav. Arch. Mar. Eng.* 79:169–221.

Schottel, J. L. 1978. The mercuric and organomercurial detoxifying enzymes from a plasmid-bearing strain of *Escherichia coli. J. Biol. Chem.* 253:4341–4349.

Summers, A. O., and S. Silver. 1972. Mercury resistance in plasmid-bearing strains of *Escherichia coli. J. Bacteriol.* 112:1228–1236.

______. 1978. Microbial transformations of metals. *Ann. Rev. Microbiol.* 32:637–672.

Tanacredi, J. T. 1971. Petroleum hydrocarbons from effluents: Detection in marine environments. *J. Water Pollut. Control Fed.* 49:216–226.

Tezuka, T., and K. Tonumra. 1978. Purification and properties of a second enzyme catalyzing the splitting of carbon-mercury linkages from mercury-resistant *Pseudomonas* K-62. *J. Bacteriol.* 135:138–143.

Traxler, R. W., and E. M. Wood. 1981. Multiple metal tolerance of bacterial isolates. *Dev. Ind. Microbiol.* 22:521–528.

VanVleet, E. S., and J. G. Quinn. 1978. Contribution of chronic petroleum inputs to Narragansett Bay and Rhode Island Sound sediments. *J. Fish Res. Board Canada* 35:536–543.

Walker, J. D., and R. R. Colwell. 1974. Mercury-resistant bacteria and petroleum degradation. *Appl. Microbiol.* 27:285–287.

CHAPTER 39

Extractive Fermentation for the Production of Butanol

RICHARD W. TRAXLER, ELEANORE M. WOOD,
JEAN MAYER, AND MASON P. WILSON, JR.

*Center for Biotechnology, College of Resource Development,
University of Rhode Island, Kingston, Rhode Island 02881*

Clostridium acetobutylicum ATCC 824 and several other strains were studied for the fermentation of corn, potato and glucose to butanol and acetone. Butanol was extracted from the medium during fermentation to reduce toxicity and concentrate butanol for more energy-efficient recovery. Dialysis fermentation relieves butanol toxicity with increased yield of product, and solvent extraction can be applied to the nongrowth side of the fermentor for concentration of the butanol.

INTRODUCTION

Zeikus (1980) underlined the need for more intense study and a rebirth of anaerobic fermentation technology as a response to our dwindling petroleum resources. For example, in 1976 the bulk of our acetone (95%) and butanol (90%) were synthesized from petrochemicals rather than by fermentation. Secondary biomass such as crop residues, starchy wastes, and distressed grains are amenable for use as anaerobic fermentation substrates. Butanol blends well with either gasoline or #2 fuel oil at double the alcohol concentration of ethanol and in the presence of traces of water does not phase separate like the ethanol-fuel blends (Noon 1981).

The commercial feasibility of the acetone-butanol fermentation is limited by low yields that are attributed to butanol toxicity and by the energy inefficiency of butanol recovery by distillation from low-yield aqueous beers. In 1980 we proposed the use of extractive fermentation to enhance butanol yield and to improve the energy efficiency of product recovery. Miner and Goma (1982) report a similar approach to overcome ethanol toxicity in *Saccharomyces* fermentations, and Griffith et al. (1983) described a number of solvent systems for the extraction of butanol from fermentation medium.

Other approaches that offer a solution to butanol toxicity are immobilized cell systems (Haggstrom 1981), continuous culture (Leung and Wang 1981), and dialysis fermentation (Schultz and Gerhardt 1969). All of these systems would result in exposure of the cells to lower concentrations of butanol and could be coupled with extraction to achieve solvent concentration. The initial objective of this project was the development of a small-scale on-farm process for enrichment of diesel fuel with butanol. Considering the current value of butanol and its wide industrial use as a chemical feed stock, we have expanded our interest to industrial scale fermentation.

Materials and Methods

Microorganisms. Clostridium acetobutylicum ATCC 824, 10132, NRRL B598, B3179, and *C. butylicum* VPI 2968 were investigated in this study. Spore suspensions were prepared by growth for 4 d at 37 C in a medium of 8% (w/v) stone ground cornmeal in distilled water. The spores were concentrated by centrifugation and 3 ml of the concentrate used to moisten 1 g of sterile sand in tubes with cotton plugs. These were dried in a desiccator and stored in the refrigerator. For outgrowth, one tube of spore-sand mixture was transferred to 25 ml of cornmeal medium, overlayed with CO_2, and incubated at 37 C for 24 h. Inoculum buildup was by transfer of 1 part of stock to 5 parts of fresh medium.

Solvent analysis. Samples were analyzed after clarification using a Hewlett-Packard 5730A gas chromatograph equipped with a FID. A 6-ft × 1/8-inch SS column packed with 15% polypropylene glycol on 80/100 mesh chromosorb W was operated for 3 min at 80 C followed by a 16 C/min program to 150 C final temperature. Each sample was computer integrated against a standard mixture of ethanol, acetone, or isopropanol and butanol containing an internal standard of *n*-propanol. Each sample was spiked with propanol prior to analysis.

Results and Discussion

Five strains of *Clostridium acetobutylicum* and one strain of *C. butylicum* were evaluated for relative yields of butanol and total mixed solvents and the use of 12 carbohydrates as fermentation substrates. *C. acetobutylicum* ATCC 824 and ATCC 10132 and *C. butylicum* VPI 2968 were the three superior strains for solvent production. The preferred sugars for butanol production were glucose, sucrose, cellobiose, fructose, maltose, and mannose. A typical *C. acetobutylicum* ATCC 824 fermentation of glucose is shown in Fig. 1. There is an inoculum-dependent variation in the time at which critical events occur in this fermentation. In general the lag time is about 1 h with maximum acid production between 8–13 h of fermentation. Exponential growth is complete coincident with maximum acid production, which is followed by a 3-h period during which the organism shifts from acid to solvent production. As solvent is produced there is a decrease in extracellular butyrate but not acetate. Maximum solvents are produced by 72 h at a level of 15 mg/ml of butanol and 25 mg/ml of total mixed solvents.

For extractive fermentation the solvent must be immiscible in water, nontoxic to the organisms, have a good affinity for butanol and a low vapor pressure at fermentation temperature. The distribution coefficients were determined for a number of potential extractants (Fig. 2). High molecular weight alcohols are far superior extractants for butanol than hydrocarbons. With all extractants the major portion of the butanol is removed from the aqueous phase within the first few hours of extraction, thus allowing shorter term intermittent extractions. As can be seen in these data, fuel oil is not an efficient extractant but would be of interest for use in a fuel-extender program.

A batch fermentation using a continuous #2 fuel oil extraction after initiation of solvent formation is shown in Table 1. The residual total solvents in the ex-

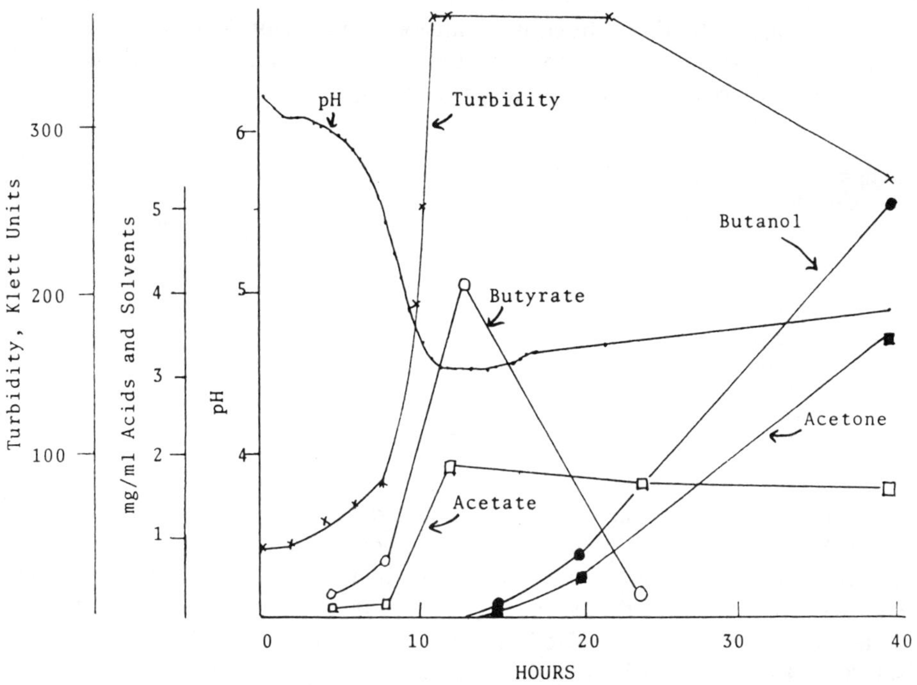

FIG. 1. *C. acetobutylicum* ATCC 824 fermentation of glucose.

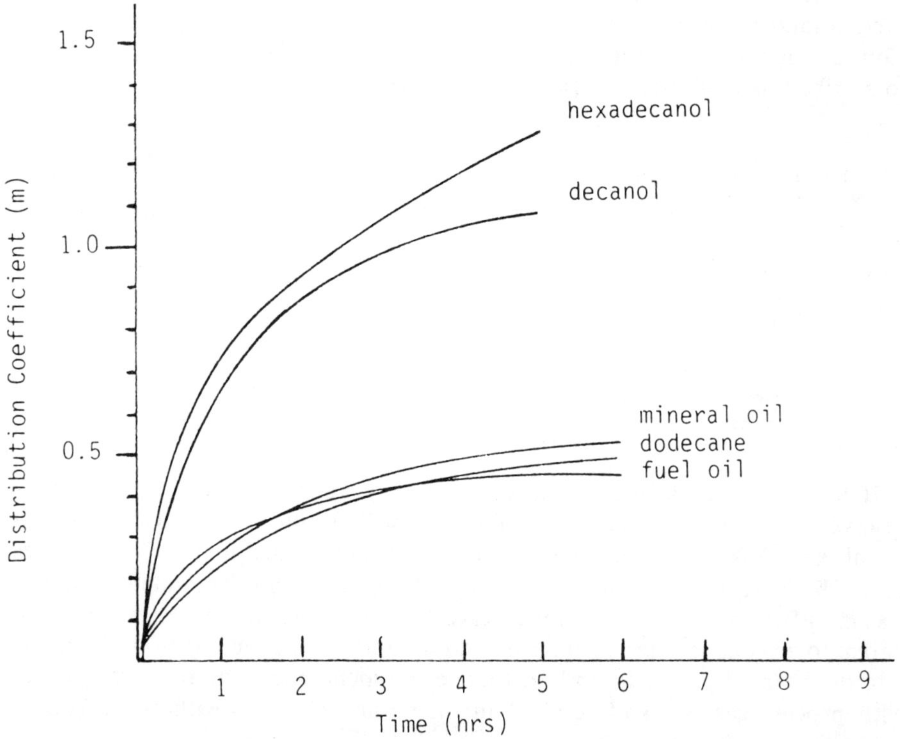

FIG. 2. Distribution coefficients for various solvents with a 1.8% aqueous solution of butanol.

tracted test are 5.2% above the control value with the major increase attributed to acetone. The partition of butanol to fuel oil would account for an additional 5.3 mg/ml of butanol or 20% in butanol production.

TABLE 1. *Solvent production in a #2 fuel oil extracted batch fermentation*

Solvent	Solvent mg/ml	
	Control	Test
Ethanol	1.1	1.6
Acetone	5.9	8.2
Butanol	14.8	13.3
Extracted Butanol		5.3
Total Butanol	14.8	18.6

The data on distribution coefficients (Fig. 2) indicate that aliphatic alcohols are superior butanol extractants and the limited data on solubility of alcohols in water demonstrates insolubility at C-8 and above. Therefore, it seemed logical that octanol and higher alcohols should not be toxic. Miner and Goma (1982) reported toxicity toward *Saccharomyces* up through decanol but not with dodecanol. A series of extractive fermentations (Table 2) with *C. acetobutylicum* ATCC 824 demonstrated increasing toxicity with increasing carbon number of the aliphatic alcohol until hexadecanol, which enhanced solvent production by 8%. The ester, ethylcaproate, also enhanced solvent production, while the cyclohexanols demonstrated the same apparent toxicity as the aliphatic alcohols. Considering the water solubility of octanol, decanol, and dodecanol, it is difficult to ascribe this inhibition to effects on the membrane of the organism.

TABLE 2. *Extractive butanol fermentations*

Extractant	% of Control
Hexanol	− 5%
Octanol	−20%
Dodecanol	−37%
Hexadecanol	+ 8%
Ethylcaproate	+10%
Cyclohexanol	−23%
4-Methylcyclohexanol	−29%

If a semipermeable membrane separates a growth chamber and a medium reservoir, in time an equilibrium of solutes will establish on either side of the membrane. A dialysis fermentation with 8% cornmeal medium (Table 3) resulted in a 12% increase in butanol production. A repeat of this experiment (Table 4) was modified by extraction of the reservoir side with hexanol, resulting in no apparent toxicity as seen in batch fermentations and significant partition of butanol into the hexanol phase. In both of these experiments the membrane was flushed with process gas but still resulted in membrane clogging with the cornmeal medium.

TABLE 3. *Dialysis fermentation with 8% cornmeal*

Time Hrs.	Butanol mg/ml Growth Side	Butanol mg/ml Reservoir Side	Total mg of butanol produced
23.5	10.3	10.8	14,745
47	13.6	13.7	19,105
120	14.7	13.4	19,735
96 control	12.0	—	17,400

TABLE 4. *Four-day dialysis fermentation of 8% cornmeal with hexanol extraction of the reservoir*

Sample Location	mg/ml Butanol	mg/ml Acetone	mg/ml Ethanol	mg/ml Total Solvents
Growth Side	6.2	5.0	1.2	12.4
Reservoir Side	4.9	5.2	1.2	13.9
Hexanol Extract	30.8	7.5	2.9	41.2
Control	13.4	10.1	2.9	26.4

A dialysis experiment was run with 8% glucose medium in the growth chamber and 8% glucose, 0.5% yeast extract, and 0.01% biotin in the reservoir (Fig. 3).

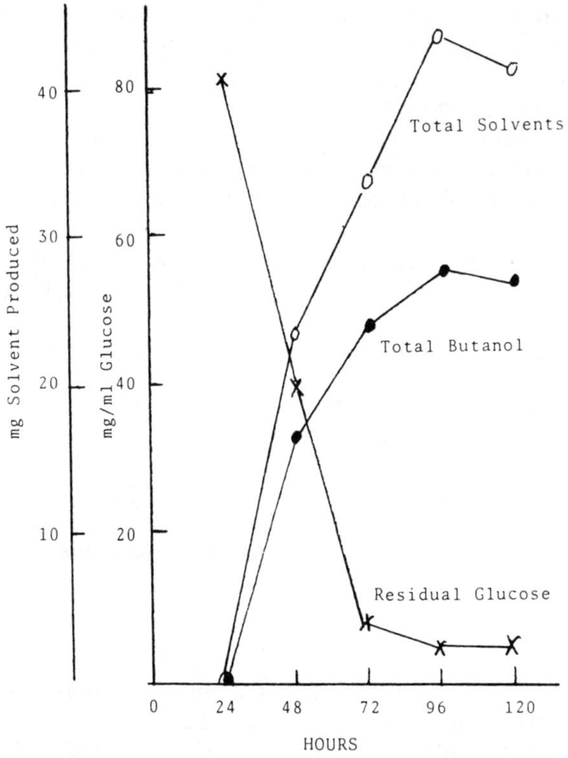

FIG. 3. Solvents and residual glucose in a dialysis fermentation of *C. acetobutylicum* ATCC 824.

Maximum solvents were produced in 4 d. In a batch fermentation, this medium produced 15 mg/ml of butanol and 25 mg/ml of total solvents. This dialysis fermentation increased the butanol yield to 27.7 mg/ml, which is nearly double the batch fermentation. Based on total glucose consumed, the organism used 2.8 mg of glucose to produce each mg of butanol and 1.8 mg of glucose for each mg of total solvents. In a batch fermentation 4 to 4.5 mg of glucose are needed to produce each mg of butanol and 2.1 mg of glucose for each mg of total solvents. Dialysis increased not only solvent yield but also the efficiency of sugar conversion to solvents.

Conclusions

Conventional batch fermentations of *C. acetobutylicum* will yield 15 mg/ml of butanol or about 25 mg/ml of total mixed solvents. Extractive fermentation, and dialysis fermentation with or without extraction, will provide significant increases in butanol yield and also will result in concentration of the butanol in the extractant solvent.

In the conventional butanol fermentation, the aqueous beer is stripped to a distillate of 50% water and 50% mixed solvents, then fractionally distilled to yield anhydrous butanol, acetone, and 95% ethanol. Calculations show that such a system, assuming 90% heat recovery with a counter flow heat exchange, will require 203 calories to recover each gram of butanol, whereas recovery of butanol from a higher boiling extractant with 90% heat recovery will require only 16.5 cal/g. This is 12-fold less recovery energy and provides an energy production factor of 47 vs. 3.9 for the conventional process.

With a semicontinuous fermentation, butanol does not accumulate to toxic levels; thus there would be no yield advantage from extractive fermentation. Extraction of the accumulated process beer would achieve energy savings on product recovery.

Dialysis fermentation provides significant enhancement of solvent yields and more efficient conversion of sugar to solvents. It is possible to recover the solvents from the reservoir side during production for not only yield improvement but also a favorable energy balance in the overall process.

Acknowledgments

This material is based upon work supported by the U.S. Department of Agriculture under Agreement No. 59-2441-1-2-119-0. This paper is contribution number 2227 of the Rhode Island Agricultural Experiment Station.

Literature Cited

Griffith, W. L., A. L. Compere, and J. M. Googin. 1983. Novel neutral solvents fermentations. *Dev. Ind. Microbiol.* 24:347–352.

Haggstrom, L. 1981. Immobilized cells of *Clostridium acetobutylicum* for butanol production. *In* M. Moo-Young and C. W. Robinson, eds., *Advances in*

Biotechnology. Vol. II, Fuels, Chemicals, Foods and Waste Treatment. Pergamon Press, NY (1981).

Leung, J. C. Y., and D. I. C. Wang. 1981. Production of acetone and butanol by *Clostridium acetobutylicum* in continuous culture using free cells and immobilized cells. Pages 348–382 *in Proc. 2nd World Congr. Chem. Eng.,* Montreal.

Miner, M., and G. Goma. 1982. Ethanol production by extractive fermentation. *Biotechnol. Bioeng.* 24:1565–1579.

Noon, R. 1981. A "power-grade" *n*-butanol/acetone recovery system. Kansas Energy Office, Topeka, KS.

Schultz, J. S., and P. Gerhardt. 1969. Dialysis culture of microorganisms: Design, theory and results. *Bacteriol. Rev.* 42:1–47.

Zeikus, J. G. 1980. Chemical and fuel production by anaerobic bacteria. *Annu. Rev. Microbiol.* 34:423–464.

CHAPTER **40**

Studies on Cellobiose Metabolism by Yeasts

A. M. SILLS AND G. G. STEWART

Production Research Department, Labatt Brewing Company Limited, P.O. Box 5050, London, Ontario, Canada N6A 4M3

Direct ethanol production from cellulose in a simultaneous saccharification/fermentation state would reduce the problem of end-product inhibition of cellulases activity by cellobiose and glucose. Of more than 40 yeast species tested for their ability to ferment and/or oxidize cellobiose, only four strains, *Brettanomyces anomalus, Brettanomyces claussenii, Dekkera intermedia,* and *Candida wickerhamii,* were able to ferment cellobiose to ethanol. Significant levels of β-glucosidase activity were obtained under aerobic conditions with *Hansenula* and *Schwanniomyces* species. The production of β-glucosidase activity by *Schwanniomyces castellii* could be enhanced up to 10-fold by increasing the phosphate concentration in the culture medium and up to 6-fold by increasing the rate of aeration/agitation.

INTRODUCTION

Cellulose, a crystalline and insoluble long-chain polysaccharide of β-glucose units linked by β(1,4) bonds, is the world's most abundant renewable biomass resource. It has been estimated that 10^{11} tonnes of this polymer are produced annually worldwide (Ghose 1977).

The utilization of cellulose for ethanol production requires three main steps: pretreatment, saccharification, and fermentation. Pretreatment of cellulosic materials reduces the crystallinity and degree of polymerization of the polysaccharide, thus enhancing the overall enzyme kinetics of the cellulose degradation process using cellulase enzymes. This pretreatment can be accomplished by physical (milling) or chemical (acid or acid-base) means (Tangnu 1982).

Enzymatic saccharification of cellulose to fermentable carbohydrates is normally achieved by the concerted action of the cellulase complex produced by the fungus *Trichoderma reesei* (Wood 1975). At least three different enzymes are involved in this hydrolysis: β(1,4) glucan glucanohydrolase (E.C.3.2.1.4., endoglucanase, Cx-enzyme), which attacks the crystalline cellulose in a random fashion, generating oligosaccharides and "free ends"; β(1,4) glucan cellobiohydrolase (E.C.3.2.1.91, exoglucanase, Cl-enzyme), which removes cellobiose units from the nonreducing ends of the cellulose chains; and β(1,4) glucosidase (E.C.3.2.1.21, cellobiase), which hydrolyzes cellobiose and other cellodextrins into glucose (Shewale 1982).

Trichoderma reesei secretes large amounts of the endo- and exoglucanases while producing small amounts of β(1,4) glucosidase (Freer and Detroy 1983; Gong and Tsao 1979; Sternberg 1976), resulting in a build-up of cellobiose in the saccharification medium. Cellobiose inhibits the action of endo- and exoglucanases, resulting in a decreased rate of cellulose saccharification (Sternberg

1976). To obtain a fast and efficient saccharification of cellulose, β(1,4) glucosidase obtained from *Aspergillus niger* has been used to supplement *Trichoderma* cellulases (Sternberg et al. 1977). Other alternatives have been suggested to prevent cellobiose build-up, such as the use of deregulated mutants of *Trichoderma reesei* with enhanced cellulase production capability (Montenecourt and Eveleigh 1977; Tangnu 1982) and continuous removal of cellobiose by direct alcoholic fermentation (Alexander et al. 1981). The latter alternative appears to be a very convenient process; however, *Saccharomyces* spp., which are the microorganisms utilized in over 95% of the total fermentation ethanol produced worldwide, are unable to ferment cellobiose. Therefore, it would be most advantageous to construct a *Saccharomyces* strain with the capability of fermenting cellobiose directly to ethanol, i.e., able to produce and secrete β(1,4) glucosidase or alternatively able to take up cellobiose followed by intracellular hydrolysis.

In an attempt to select appropriate yeast species that could be used as DNA donors for transformation of *Saccharomyces* spp., several yeast species able to utilize cellobiose in a fermentative or oxidative fashion (Lodder 1970) were compared for their ability to synthesize β(1,4) glucosidase at different times of incubation in a medium containing cellobiose as a sole source of carbon.

Materials and Methods

Microorganisms. The yeast strains employed in this study are referred to by their number in the Labatt culture collection and are listed in Table 1.

Growth media and culture conditions. Yeast strains were cultured in PYNC medium (Ogur and St. John 1956) that contained the following ingredients per liter of distilled water: peptone, 3.6 g; yeast extract, 3.0 g; KH_2PO_4, 2.0 g; $MgSO_4 \cdot 7H_2O$, 1.0 g; $(NH_4)_2SO_4$, 1.0 g; and cellobiose, 20 g. The medium was adjusted to pH 5.7 prior to sterilization by autoclaving. Solid medium for the maintenance of yeast strains was prepared by the further addition of 2% (w/v) Difco Bacto agar.

For inoculation of vegetative growth cultures, cells from 48 h PYNC agar plates were transferred at a density of 10^5 cells per ml to 100 ml volumes PYNC medium in 300-ml Erlenmeyer flasks. Growth culture flasks were incubated at 21 C (unless otherwise specified) in a New Brunswick Model G26 gyratory shaker operated at 150 rpm.

Samples were taken periodically for optical density readings (600 nm) and every 24 h for β-glucosidase activity determination. This enzymatic activity was determined in the cell-free extract (i.e., supernatant after centrifugation) and intracellularly by toluene treatment of the resulting pellet. The pellet was resuspended in the same volume of sterile distilled water. Three drops of toluene were then added, the samples vortexed for 30 s and then incubated at 37 C in a shaker for 40 min. Toluene partially disrupts the cell membrane, allowing *p*-nitrophenyl-β-D-glucoside (PNPG; Boehringer) to diffuse into the cell.

Assay of β-glucosidase activity. β-Glucosidase activity was quantitated by mixing 0.2 ml of the diluted enzyme with 1.8 ml of a 1 m*M* solution of the chromogenic substrate *p*-nitrophenyl-β-D-glucoside (PNPG; Boehringer) in 0.1 *M* citrate phosphate buffer (pH 5.5). The reaction was carried out at 40 C for 10 min and subsequently 1.0 ml of 1 *M* sodium carbonate was added to stop the reaction. The *p*-nitrophenol produced was measured at 400 nm using a spectrophotometer and compared to a series of standards prepared with *p*-nitrophenol (Boehringer) at concentrations ranging from 0.1 to 1.0 m*M*. One international unit (i.u.) is defined as the quantity of enzyme required to release one micromole of *p*-nitrophenol per min under the above experimental conditions.

TABLE 1. *Fermentation and oxidation of cellobiose by several yeast species*

Yeast Species	Code No.	Utilization of Cellobiose[a]	
		Fermentation	Oxidation
Aureobasidium pullulans	1474	0	+ +
Brettanomyces anomalus	256	+	+ +
Brettanomyces claussenii	257	+	+ +
Candida curvata	1269	0	0
Candida flareri	2951	0	+
Candida melinii	609	0	+
Candida museorum	3732	0	0
Candida tenuis	3764	0	+
Candida tropicalis	265	0	+ +
Candida utilis	260	+	+ +
Candida wickerhamii	259	+ +	+ +
Cryptococcus diffluens	748	0	+
Cryptococcus laurentii	1126	0	+ +
Cryptococcus luteolus	1426	0	0
Cryptococcus luteolus	3718	0	+ +
Dekkera intermedia	258	+ +	+ +
Endomyces magnusii	315	0	0
Endomycopsis capsularis	1427	0	+
Endomycopsis fibuligera	240	0	+ +
Hansenula anomala	53	0	+ +
Hansenula saturnus	317	0	+
Hansenula schneggii	899	0	+ +
Hansenula species	337	0	+ +
Kluyveromyces fragilis	102	0	+ +
Kluyveromyces fragilis	103	0	+ +
Kluyveromyces lactis	104	0	+
Kluyveromyces lactis	117	0	0
Kluyveromyces lactis	226	0	0
Kluyveromyces lactis	1437	0	+ +
Pichia polymorpha	223	0	+ +
Rhodosporidium lactosa	3740	0	+
Rhodosporidium marina	3750	0	+
Rhodosporidium rubra	218	0	0
Saccharomyces diastaticus	62	0	0
Saccharomyces diastaticus	1441	0	+
Schwanniomyces castellii	1402	0	+ +
Schwanniomyces castellii	1436	0	+ +
Schwanniomyces occidentalis	1401	0	+ +
Torulopsis sphaerica	1452	0	0
Trichosporon pullulans	727	0	+ +

[a] 0 No growth, + Slight growth, + + Abundant Growth

RESULTS

The capacity of 40 yeast strains to utilize cellobiose as a sole source of carbon under aerobic and anaerobic conditions is presented in Table 1. Only five strains screened, *Brettanomyces anomalus, Brettanomyces claussenii, Candida utilis, Candida wickerhamii* and *Dekkera intermedia* were able to ferment the dissacharide; however, a large number of strains were able to utilize cellobiose under aerobic conditions. The latter strains were compared for their ability to grow in complete medium containing cellobiose (as described in Materials and Methods) and to produce and secrete β-glucosidase activity. A summary of the 10 highest producers of this enzyme is given in Table 2. The three species of the genus *Schwanniomyces* were able to produce and secrete significant levels of β-glucosidase (as determined in cell-free extract assays) and an even higher activity was obtained when the pelleted cells were treated with toluene in order to increase cell permeability. Other strains that produced significant levels of β-glucosidase activity associated with the cell but not in supernatants were *Hansenula spp.* and *Kluyveromyces spp.*

TABLE 2. *Production of β-glucosidase activity by yeasts during growth in complete medium containing 2% (w/v) cellobiose*

Yeast	Code	Maximum β-Glucosidase Activity Detected (i.u./min.)	
		Extracellular	Intracellular
Aureobasidium pullulans	1474	20.4	3.7
Cryptococcus muscorum	3732	6.5	24.7
Endomycopsis fibuligera	240	61.1	62.2
Hansenula schneggii	899	117.9	877.3
Hansenula spp.	337	76.6	214.0
Kluyveromyces lactis	1437	4.4	40.3
Pichia polymorpha	223	19.4	40.7
Schwanniomyces castellii	1402	439.5	769.0
Schwanniomyces castellii	1436	849.2	933.4
Schwanniomyces occidentalis	1401	597.5	1022.1

Hansenula schneggii 899 and *Schwanniomyces castellii 1402* were compared for their ability to produce intracellular and extracellular β-glucosidase activity throughout the growth cycle. *Hansenula schneggii 899* was observed to produce mainly intracellular β-glucosidase activity (Fig. 1A). The highest intracellular activity was detected during late-log phase of growth, and very low levels of extracellular activity were observed throughout the growth cycle. A different pattern was observed with *Schwanniomyces castellii 1402* (Fig. 1B). After 24 h of incubation (mid-log phase), the intracellular β-glucosidase activity was twice as much as the extracellular activity; however, after the culture reached the stationary phase (48 h) most of the activity was present in the cell-free medium. This extracellular activity could not be detected after 72 h of incubation.

The concentration of potassium phosphate monobasic in PYNC medium (see Materials and Methods) was 0.015 *M*. As indicated in Table 3, increasing its concentration in the growth media had a stimulating effect on the production of

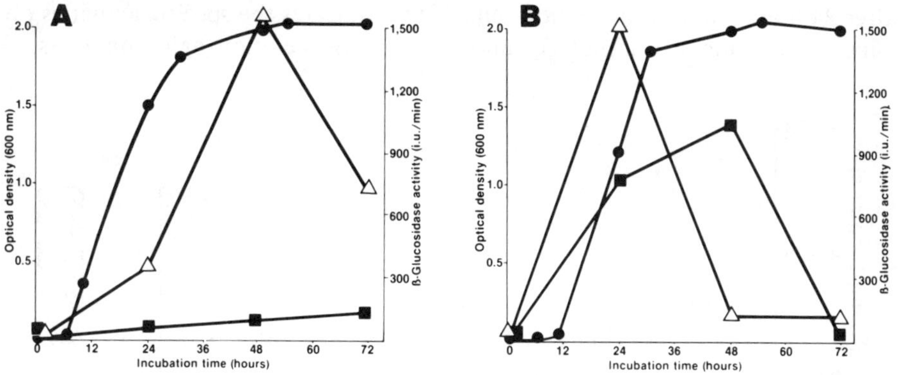

FIG. 1. Production of β-glucosidase activity by *Hansenula schneggii 899* (A) and *Schwanniomyces castellii 1402* (B) during growth in 0.5% (w/v) cellobiose. Optical density (●); internal β-glucosidase activity (△); external β-glucosidase activity (■).

β-glucosidase activity. The concentration of 0.24 *M* appeared to be optimal for β-glucosidase activity, e.g., an increase of 5.4-fold activity extracellularly and 10.0-fold intracellularly after 48 h of incubation over the control. A very significant increase in activity was, however, also detected at 48 h of incubation when the phosphate concentration was increased from 0.015 *M* to 0.060 *M* (4.3-fold extracellularly and 6.7-fold intracellularly). The final biomass, determined after 48 h of incubation was the same (3.5 ± 0.2 g/l) in the different samples. In addition, substituting potassium phosphate dibasic for the monobasic form and adjusting for comparable concentrations of phosphate and potassium (data not shown) indicate that excess phosphate and not potassium was the stimulating factor for β-glucosidase activity.

TABLE 3. *Effect of increasing concentrations of potassium phosphate (monobasic) on the production of* β-glucosidase activity by Schwanniomyces castellii 1402

		β-Glucosidase Activity (i.u./min)			
		24 h		48 h	
$[KH_2PO_4]$ *M*	Generation Time, h	Intracellular	Extracellular	Intracellular	Extracellular
0.015	3.4	346	660	692	388
0.030	3.5	360	1616	1198	605
0.060	3.5	494	2070	2958	2605
0.120	3.9	510	4052	3185	3452
0.240	4.0	533	5211	3750	3891
0.480	5.0	530	5010	3642	3720

The effect of aeration/agitation on β-glucosidase activities produced by *Schwanniomyces castellii 1402* was determined by utilizing flasks with decreasing working volumes including 100 ml (low aeration/agitation), 50 ml (medium aeration/agitation), and 25 ml (high aeration/agitation) in growth medium containing 0.5% (w/v) cellobiose. β-Glucosidase activities and biomass were determined after 24 h and 48 h of incubation, and Fig. 2 indicates the specific activities obtained intra- and extracellularly under increasing aeration/agitation rates. A

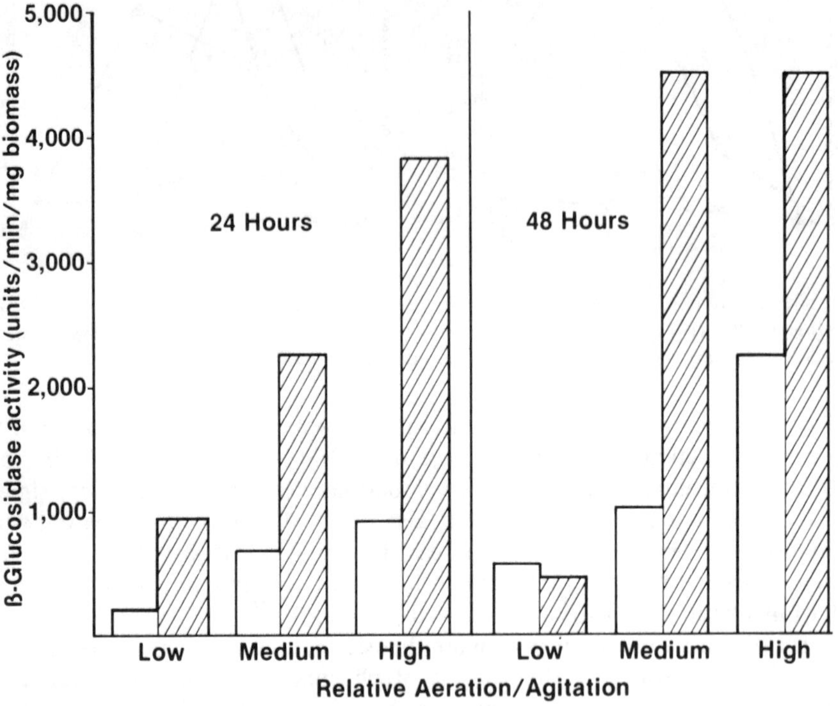

FIG. 2. Effect of aeration/agitation on β-glucosidase production by *Schwanniomyces castellii 1402* in shake flasks. Intracellular activity (▨); extracellular activity (□).

decrease in the working volume contained in shake flasks by one-half (100 ml to 50 ml) increased extracellular activity by 4-fold and intracellular activity by 2.5-fold after 24 h of incubation. A significant increase was also detected on a subsequent decrease in working volume from 50 to 25 ml (medium to high relative aeration/agitation). A similar effect was detected after 48 h of incubation except that the intracellular activity did not vary with medium and high relative aeration/agitation.

Discussion

The major soluble products from the hydrolysis of cellulose by the cellulase enzymes are cellobiose and glucose, and the former must be cleaved by β-glucosidase before it can be fermented to ethanol by *Saccharomyces spp.* Other yeast strains such as *Brettanomyces spp., Dekkera spp.,* and *Candida spp.,* have been reported to ferment cellobiose directly (Blondin et al. 1982; Freer and Detroy 1983; Gonde et al. 1982; Kilian et al. 1983). However, some difficulties were observed with these fermenting strains. For example, *Candida wickerhamii* tolerated only 3.5% (w/v) ethanol (Kilian et al. 1983), and although *Brettanomyces* and *Dekkera spp.* produced up to 7.5% (w/v) ethanol with 80% efficiency from cellobiose, the production of ethanol by a mixture of cellobiose-glucose was hindered considerably by the diauxic effect characteristic of these strains (Gonde et al. 1982; Kilian et al. 1983). As *Saccharomyces spp.* are responsible for most of the fermentation ethanol production due to their high fermentation efficiencies, ethanol tolerance, and osmotic tolerance (Stewart et al. 1984), such a strain with the capability to metabolize cellobiose efficiently would be most useful.

Schwanniomyces castellii produces β-glucosidase activity, which is secreted to the culture medium at significant levels (Fig. 1B). Although *Schwanniomyces castellii* cannot ferment the disaccharide under anaerobic conditions (Table 1), aerobically it produces the highest activity of intracellular and extracellular β-glucosidase activities of all the yeast strains tested in this study. The production of β-glucosidase activity can be enhanced by increasing the concentration of inorganic phosphate (Table 3) and by increasing the relative rate of aeration/agitation (Fig. 2). A high phosphate concentration in the medium might increase ATP formation and lead to a high energy charge in the cell (Demain 1972). Increasing the rate of aeration/agitation can increase the mass transfer of oxygen and substrate thus stimulating the rate of respiration, leading also to a high energy charge in the cell. It appears, therefore, that a high energy charge in the cell is required to obtain significant levels of β-glucosidase activity. The fact that cellobiose is not metabolized anaerobically whereas several other carbohydrates are, such as fructose, glucose, and maltose, might confirm that a high energy state is required for the effective utilization of cellobiose by *Schwanniomyces castellii.*

Currently, studies are being conducted on the regulation of β-glucosidase activity by *Schwanniomyces castellii* and the effect of glucose on cellobiose utilization.

Acknowledgments

The authors express their sincere thanks to A. Wearring and D. Dowhaniuk for technical assistance and to I. Russell, C. J. Panchal, and C. A. Bilinski for critically reading this manuscript. This study was supported in part by Agriculture Canada, contract serial no. OSZ83-00064.

Literature Cited

Alexander, J. K., R. Connors, and N. Yamamoto. 1981. Production of liquid fuels from cellulose by combined saccharification-fermentation or cocultivation of *Clostridia*. Pages 125–130 *in* M. Moo-Young and C. W. Robinson, eds., *Advances in Biotechnology, Vol. II. Fuels, Chemicals, Foods and Waste Treatment,* Pergamon Press, Toronto.

Blondin, B., R. Ratomahenina, A. Arnaud, and P. Galzy. 1982. A study of cellobiose fermentation by a *Dekkera* strain. *Biotechnol. Bioeng.* 24:2031–2037.

Demain, A. L. 1972. Cellular and environmental factors affecting the synthesis and excretion of metabolites. *J. Appl. Chem. Biotechnol.* 22:345–362.

Freer, S. N., and R. W. Detroy. 1983. Characterization of cellobiose fermentations to ethanol by yeasts. *Biotechnol. Bioeng.* 25:541–557.

Ghose, T. K. 1977. Cellulose biosynthesis and hydrolysis cellulosic substances. Pages 39–73 *in* A. Fiechter, ed., *Advances in Biochemical Engineering,* Vol. 6, Springer-Verlag, Berlin.

Gong, C. S., and G. T. Tsao. 1979. Cellulose and biosynthesis regulation. Pages 111–140 *in* D. Perlman, ed., *Annual Reports on Fermentation Processes,* Vol. 3, Academic Press, New York.

Gonde, P., B. Blondin, R. Ratomahenina, A. Arnaud, and P. Galzy. 1982. Selection of yeast strains for cellobiose alcoholic fermentation. *J. Ferment. Technol.* 60:579–584.

Kilian, S. G., B. A. Prior, and P. M. Lategan. 1983. Diauxic utilization of glucose-cellobiose mixture by *Candida wickerhamii. Eur. J. Appl. Microbiol. Biotechnol.* 18:369–373.

Lodder, J. 1970. *The Yeasts, a Taxonomic Study.* North Holland, Amsterdam.

Montenecourt, B. S., and D. E. Eveleigh. 1977. Preparation of mutants of *Trichoderma reesei* with enhanced cellulase production. *Appl. Env. Microbiol.* 34:777–782.

Ogur, M., and R. St. John. 1956. A differential and diagnostic plating method for population studies of respiration deficiency in yeast. *J. Bacteriol.* 92:500–504.

Shewale, J. G. 1982. β-Glucosidase: Its role in cellulase synthesis and hydrolysis of cellulose. *Int. J. Biochem.* 14:435–443.

Sternberg, D. 1976., A method for increasing cellulose production by *Trichoderma viride. Biotechnol. Bioeng.* 18:1751–1760.

Sternberg, D., R. Vijayakumar, and E. T. Reese. 1977. β-Glucosidase: Microbial production and effect on enzymatic hydrolysis of cellulose. *Can. J. Microbiol.* 23:139–147.

Stewart, G. G., C. J. Panchal, I. Russell, and A. M. Sills. 1984. Biology of ethanol-producing microorganisms. *CRC Crit. Rev. Biotechnol.* 1:161–188.

Tangnu, S. K. 1982. Process development for ethanol production based on enzymatic hydrolysis of cellulosic biomass. *Proc. Biochem.* 17(3):36–45, 49.

Wood, T. M. 1975. Properties and mode of action of cellulases. Pages 111–137 *in* C. R. Wilke, ed., *Cellulose as a Chemical and Energy Resource. Biotechnol. Bioeng. Symp.* No. 5, Wiley, New York.

CHAPTER 41

Solvents Production by *Clostridia* as a Function of Wood Stream Organic Toxicant Concentration

A. L. COMPERE, W. L. GRIFFITH, AND J. M. GOOGIN

Oak Ridge National Laboratory, Box X, Oak Ridge, Tennessee 37831

Wood pulping streams are the largest collected source of domestic biomass for fermentation. These streams are generally sterile as a result of their high-temperature digestion and may contain in the neighborhood of 5% fermentable materials, including simple sugars, fatty acids, complex polysaccharides, and sugar acids. Historically, usage of these and other waste biomass streams has been limited by the presence of a number of toxicants, including phenol, cresol, furfural, and cinnamyl alcohol, present at an aggregate concentration of around 0.1 *M*. The response of Northern Regional Research Laboratory (NRRL) industrial *Clostridia* strains B527, B592, B593, and B598 to low concentrations of these materials was measured by gross solvents production. Solvents production by all of the strains tested on all of the materials was effectively terminated at concentrations one to ten thousand times lower than those expected in proposed feedstock streams.

INTRODUCTION

Wood pulp streams, including hydrolyzates, are the largest single source of fermentable U.S. biomass materials. Fermentation of these streams has been constrained by the presence of organic and inorganic toxicants present as a result of either the pulping chemicals or as products of lignin or hemicellulose breakdown. Organic toxicants present in these streams typically include furfural, which is produced by xylose condensation; phenolics, such as phenol or cresol; and lignin breakdown products, such as cinnamic acid or cinnamyl alcohol. Although there is, at present, little ongoing research in the area of neutral solvents production from wood pulp streams, research on the effects of organic toxicants, such as furfural, on neutral solvents fermentations started in the period around World War II. Hydrolyzate streams, which do not generally contain large amounts of salts, provide a good model for the toxic properties of organic materials in other wood streams.

Underkofler et al. (1937) found that use of low-temperature mild acid hydrolysis gave a product from oat hulls that could replace 40 to 50% of the corn mash in a conventional solvents fermentation. More extreme conditions resulted in the production of less effective fermentation feedstocks.

Boehm et al. (1944) found that treatment of wood liquors by solvent extraction followed by hydrated lime and sodium sulfide addition to raise the pH and precipitate acids could provide a good substrate for neutral solvents fermentations. Two groups working for the Department of Agriculture investigated the hydrolysis and fermentation of agricultural residues (Dunning and Lathrop 1945;

and Tsuchiya et al. 1948). They found that the processes were economic, and that only ammonium sulfate and triple superphosphate were required as nutrients for the neutral solvents fermentation of corn cobs, but that control of toxic copper and furfural concentrations were the keys to a good fermentation.

Leonard et al. (1947) found that furfural was the major inhibitor in the neutral solvents fermentations of wood hydrolyzates. When hardwood hydrolyzates were steam stripped to less than 0.05% furfural, it was possible to obtain fermentation of more than 90% of the sugars. Based on their calculations, it was estimated that a ton of maple wood could produce 75 to 100 lb of solvents, 50 lb of furfural, and 120 lb of calcium acetate.

Langlykke et al. (1948) developed a method for hydrolyzing corn cobs to produce a 90% fermentable carbohydrate stream. They found that control of furfural and copper in the stream were crucial to yield. Maddox and Murray (1983) found that use of cation and anion exchange resins to remove furfural (possibly by absorption) and metal salts from acid pine hydrolyzate streams would permit a solvents yield of 17% based on input sugars.

We were interested in determining the effect of furfural, phenol, cresol, and cinnamyl alcohol on solvents production because these organic materials appear to be potentially significant in decreasing the fermentation of wood pulp and hydrolyzate streams. Observation showed that neutralized diluted pulp streams containing volatile acids, simple sugars, or sugar acids are fermentable. However, undiluted, these materials were unable to permit or sustain *Clostridia* growth. Biotoxicity from potentially toxic materials, such as phenol, cresol, furfural, or various sodium salts, was probably responsible.

The amount of organic toxicants in a given pulp or hydrolyzate stream varies considerably as a function of cooking conditions, salts addition, and wood or biomass feedstock. However, as comparison values, Azhar et al. (1982) found that 1 or 2% furfural (0.1 to 0.2 *M*) was not uncommon in wood hydrolyzate streams and amounts up to 3.5% have been reported. Small lignin derivatives, of which phenol, cresol, and cinnamyl alcohol are typical constituents, average around 6% of the total dryweight solids in kraft black liquor or about 1% of material in a weak black liquor stream.

Materials and Methods

Culture maintenance. The *Clostridia* cultures used in this study were obtained as lyophiles from Dr. Larry Nakamura of the Northern Regional Research Laboratory of the U.S. Department of Agriculture, and have been identified as follows: NRRL B527, *Clostridium acetobutylicum;* NRRL B592 and B593, *Clostridium butylicum;* and NRRL B598, *Clostridium pasteurianum*. The cultures were maintained in log phase on chopped liver medium by daily transfer of a 5% inoculum.

Media. The test medium for determining the neutral solvents production from glucose contained, in addition to the toxic materials tested, per liter: 5 *g* Difco Bacto peptone, 0.5 *g* Difco yeast extract, 1 *g* dibasic potassium phosphate, and 27.5 *g* glucose monohydrate. A small marble chip and a small chunk of pork liver were placed in the bottom of a Bellco Hungate tube, medium was added, and the tube was capped prior to sterilization for 25 min at 121 C. After sterilization, tubes were brought to 65 C, aliquotted with separately sterilized solutions of the toxicants tested, and inoculated with 5% of a log phase culture. The tubes were then incubated at 36 C for 72 h. All samples were triplicated, except for unspiked controls. A dozen of each unspiked control were prepared.

Fisher furfural, Mallincrodt phenol, Eastman *m*-cresol, and Aldrich cinnamyl alcohol were used in these experiments. The liquids were autoclaved separately and added to cultures with a sterile micropipette. Phenol was autoclaved and added as a 100% (w/v) solution with a sterile micropipette. The following concentration range was tested in addition to unspiked controls: *m*-cresol, 4 to 19 μM; phenol, 4.3 to 21.2 μM; furfural, 6 μM to 60 μM; and cinnamyl alcohol, 10 μM to 58 μM.

Analytical procedures. Approximately 5 ml liquid samples of cultures were placed in tightly capped test tubes and centrifuged at 4,500 × *g* for 30 min. A 1.3-ml aliquot was removed from each sample, placed in an autosampler vial, spiked with an internal standard of 0.1 ml of 5% (w/v) isobutanol, capped, and placed in the autosampler for gas chromatographic analysis.

Samples were analyzed using a Varian Model 3700 gas chromatograph equipped with a model 8000 autosampler and a CDS 111 autoanalyzer. Standards containing 1-butanol, acetone, isobutanol, and ethanol at levels similar to those expected in samples were used to restandardize between every sample. A 10-ft × 1/8-inch column packed with Waters Porapak Q was operated at 210 C with a flame-ionization detector. Helium at 30 ml/min was used as the carrier gas.

Results

Four organic toxicants found in wood pulp and hydrolyzate streams were tested: furfural, phenol, cresol, and cinnamyl alcohol. Furfural is formed as a xylose condensation product, while the other three are lignin breakdown products. A typical level for such materials in wood pulp or furfural in hydrolyzate streams might be an aggregate total of 0.1 M. Although considerable research on the ability of yeast to tolerate furfural has been performed, and some effort has been made to develop strains adapted to expected furfural concentrations (Azhar et al. 1982), the levels tolerated are still well below those that can occur in hydrolyzate streams.

In contrast to the slow fall in solvents production evinced in the presence of other materials, the decrease in solvents production with increasing furfural concentration was extremely rapid. For example, strain B527 grew readily and produced normal levels of solvents at 6 μM furfural; no growth or solvent production was observed at 12 μM. For strain B592, the cutoff point was at 24 μM, while for strains B593 and B598, solvents production dropped to less than 35% of its normal value between 42 and 48 μM and between 30 and 36 μM, respectively. Microscopic examination showed that the cultures were not motile.

As shown in Fig. 1, the decline in solvents production in response to additions of *m*-cresol was more gradual than that observed for furfural addition. For strains B527, B592, and B593, solvents production decreased above 4 μM *m*-cresol, while with strain B598, solvents production was decreased at 4 μM, the lowest level tested.

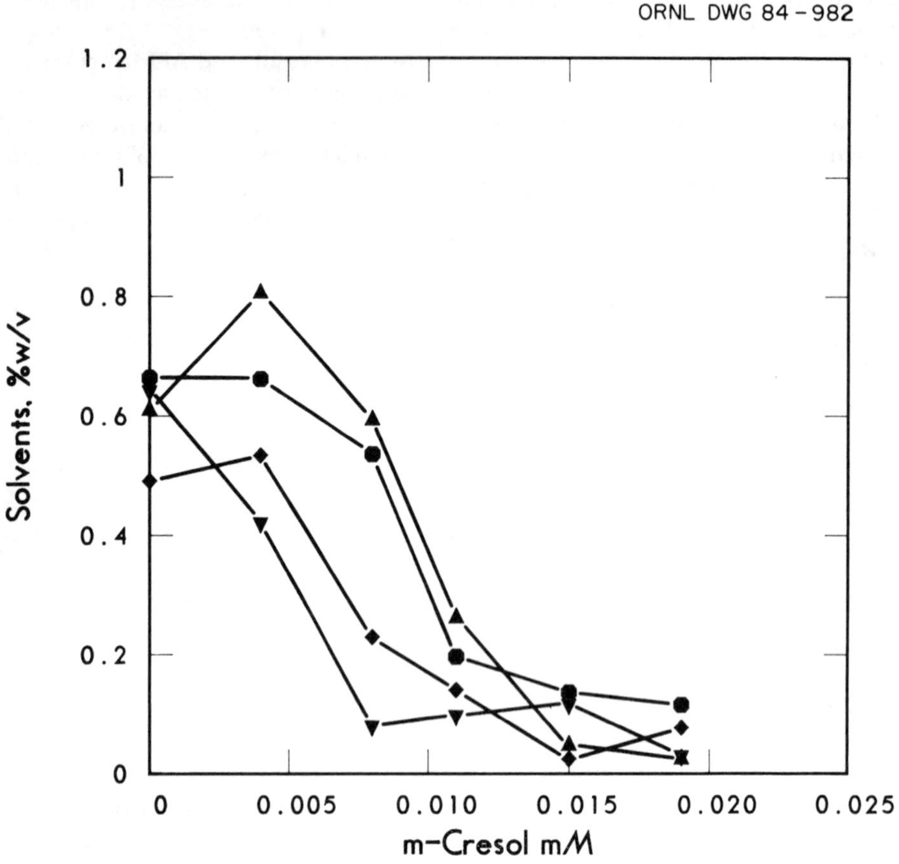

FIG. 1. Total solvents production by strains NRRL B527 (●), NRRL B592 (▲), NRRL B593 (♦), and NRRL B598 (▼) as a function of *m*-cresol concentration.

With phenol, as shown in Fig. 2, a similar pattern occurred. Decreases in solvents production above 4.25 μM phenol were observed for strains B527, B592, and B598. Strain B593 showed a similar drop in solvents production at 8.5 μM, but reached a relatively flat plateau and held solvents production there for concentrations of 12.7, 17, and 21.2 μM phenol. In general, only one of three tubes would grow, and that had a production level only a fraction of normal. It appeared to us that the probability of developing a resistant mutant in one of the three cultures inoculated at a given phenol concentration was relatively high.

Response of the four strains to cinnamyl alcohol addition is shown in Fig. 3. Strain B593 showed a drop in solvents production with cinnamyl alcohol levels above 4 μM. All of the other strains showed a drop in solvents production at 4 μM. However, the drop in solvents production was very gradual, with all strains having substantial solvents production at 19 μM.

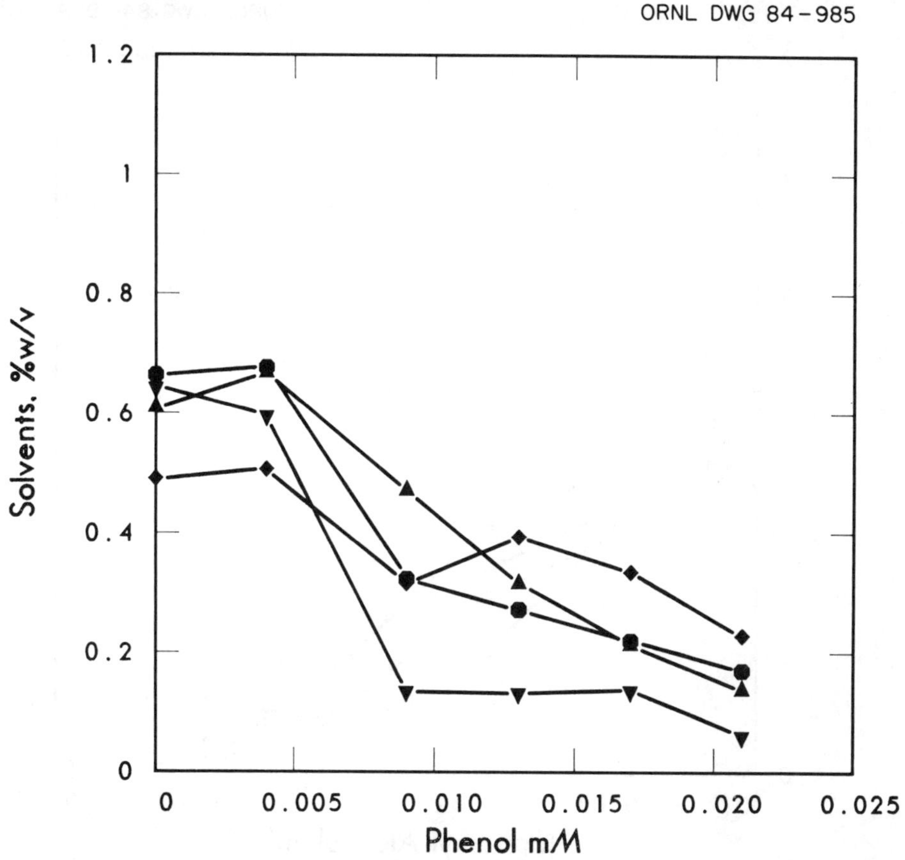

FIG. 2. Total solvents production by strains NRRL B527 (●), NRRL B592 (▲), NRRL B593 (◆), and NRRL B598 (▼) as a function of phenol concentration.

Conclusions

In the case of all of the toxicants studied, the levels required to inhibit solvents production were generally three to four orders of magnitude less than aggregate concentrations of similar materials that might be found in a typical wood pulp or hydrolyzate stream. For such streams to be fermentable will require the development of either significantly more-resistant organisms or effective prefermentation treatment methods.

Acknowledgment

This research was sponsored by the U.S. Department of Energy under Contract DE-AC05-840R21400 with Martin Marietta Energy Systems.

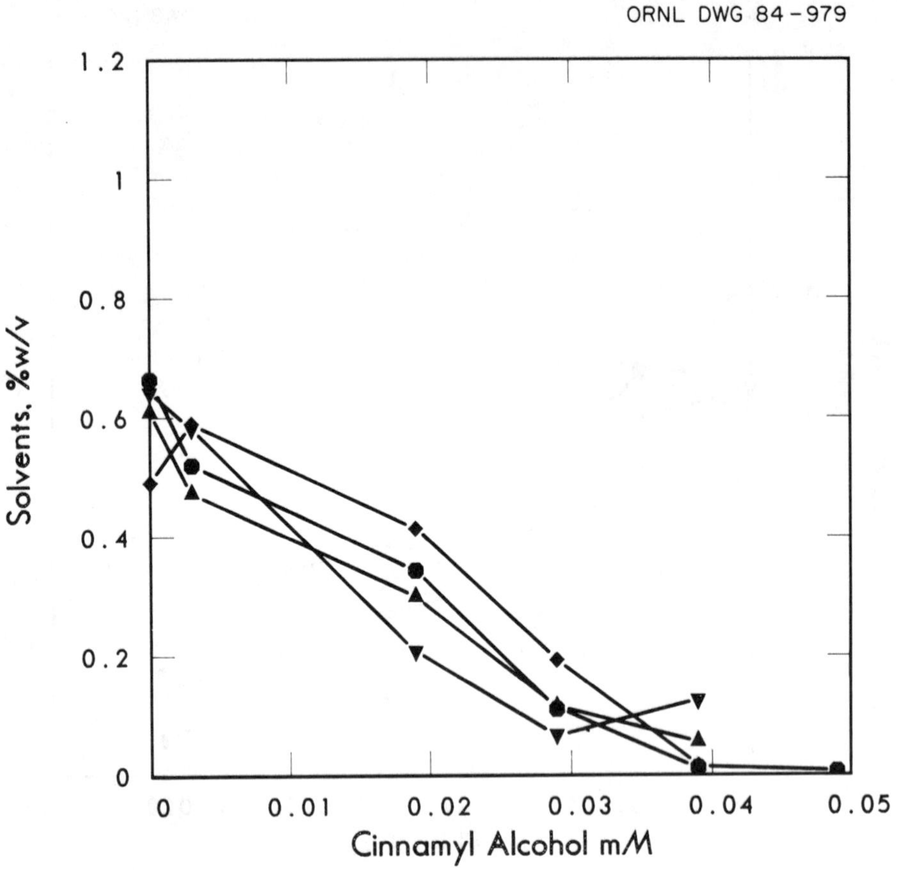

FIG. 3. Total solvents production by strains NRRL B527 (●), NRRL B592 (▲), NRRL B593 (◆), and NRRL B598 (▼) as a function of cinnamyl alcohol concentration.

Literature Cited

Azhar, A. F., M. K. Bery, A. R. Colcord, and R. S. Roberts. 1982. Development of a yeast strain for the efficient ethanol fermentation of wood acid hydrolyzate. *Dev. Ind. Microbiol.* 23:351–360.

Boehm, R. M., H. E. Hall, and J. A. MacDonald. 1944. Preparation of clarified sugar solution for fermentation. U.S. Pat. 2,421,985.

Dunning, J. W., and E. C. Lathrop. 1945. The saccharification of agricultural residues. *Ind. Eng. Chem.* 37:24–29.

Langlykke, A. F., J. M. Van Lanen, and D. R. Fraser. 1948. Butyl alcohol from xylose saccharification liquors from corncobs. *Ind. Eng. Chem.* 49: 1716–1719.

Leonard, R. H., W. H. Peterson, and G. J. Ritter. 1947. Butanol-acetone fermentation of wood sugar. *Ind. Eng. Chem.* 39:1443–1445.

Maddox, I. S., and A. E. Murray. 1983. Production of n-butanol by fermentation of wood hydrolyzate. *Biotechnol. Lett.* 5:175–178.

Tsuchiya, H. M., H. M. Mueller, and V. E. Sohns. 1948. Butanol fermentation of pentosan hydrolyzates obtained from corn cobs. *American Chemical Society 113th Meeting* 10A:18.

Underkofler, L. A., A. I. Fulmer, and M. M. Rayman. 1937. Oat hull utilization by fermentation. Butyl-acetonic fermentation of the acid hydrolyzate. *Ind. Eng. Chem.* 29:1290–1292.

CHAPTER 42

Solvents Production by *Clostridia* as a Function of Sodium Salts Concentration

W. L. Griffith, A. L. Compere, and J. M. Googin

Oak Ridge National Laboratory, Box X,
Oak Ridge, Tennessee 37831

Wood pulping streams are the largest collected source of domestic biomass for fermentation. These streams are generally sterile as a result of their high temperature digestion, and may contain in the neighborhood of 5% fermentable materials, including simple sugars, fatty acids, complex polysaccharides, and sugar acids. Historically, usage of these and other waste biomass streams has been limited by their generally high concentration of sodium salts, including sulfite, sulfate, sulfide, and chloride. The response of Northern Regional Research Laboratory (NRRL) industrial *Clostridia* strains B527, B592, B593, and B598 to low concentrations of these materials was measured by gross solvents production. Kraft black liquor has the highest salts concentrations of the wood pulping liquors, averaging as sodium salts, the following concentrations: bisulfite, 87 m*M* carbonate, 525 m*M* chloride, 3 m*M*; hydroxide, 85 m*M*; sulfide, 62 m*M*; and sulfate, 14 m*M*. Discounting common ion effects and overall osmotic concentration, the levels of sodium chloride and sulfate salts found in black liquor were not sufficient to terminate solvents production by the *Clostridia* strains tested. Sodium sulfide levels in black liquor were around twice the amount readily tolerated by the tested strains. All strains readily tolerated the 10 m*M* sulfite levels that might be expected in weak acid sulfite liquor.

Introduction

Wood product streams are the largest single source of fermentable materials derived from a managed biomass source. Although there is relatively little ongoing research on the fermentation of these materials, there was considerable interest in the use of weak acid sulfite liquor and various hydrolyzates around World War II. Wiley et al. (1941) found that lime pretreatment, which removed furfural and sulfite, permitted roughly 30% yields of solvents, based on reducing substances as glucose, from weak acid sulfite liquor. Mineral salts were used for nitrogen and buffering. Grondal and Berger (1945) reported that feed yeast and neutral solvents could be produced by coculture process from weak acid sulfite liquor. Stark and McGhee (1950) found that sulfite stripping and lime addition permitted solvent yields of 20 to 23% (w/w) based on total sugars, and 33 to 42% (w/w), based on yeast-fermentable sugars. Schroedler (1953) found that neutral solvent fermentation was favored over ethanol fermentation when the feedstock to a weak acid sulfite liquor process was more than half hardwood. Gfeller (1953) found that 75%, and Blinc and Strauch (1958) found that 70%, of the reducing sugars in beech wood acid sulfite liquor could be fermented to neutral solvents. Mueller (1958) found that removal of sulfite and addition of ammonia or organic nitrogen sources increased the amount of solvents that could be produced from both the sugars and the acetic acid present in weak acid sulfite liquor. In all of these cases, removal of toxic salts, primarily sodium sulfite, appeared to be a major factor in the development of successful fermentation conditions.

We were interested in determining the effect of sodium salts on solvents production because these materials are major inorganic constituents of kraft black liquor, the largest wood pulp stream. Kraft liquors contain substantial amounts of the volatile acids that are thought to be neutral solvents precursors (Fricke 1983). Our own observations showed that neutralization followed by several-fold dilution of kraft liquors resulted in their fermentation to neutral solvents, while neutralized weak black liquor did not support the growth of NRRL *Clostridia* strains B527, B592, B593, or B598. We were interested in determining which, if any, sodium salts were responsible for this behavior.

Materials and Methods

Cultures. The *Clostridia* cultures used in this study were obtained as lyophiles from Dr. Larry Nakamura of the Northern Regional Research Laboratory of the U.S. Department of Agriculture and have been identified as follows: NRRL B527, *Clostridium acetobutylicum;* NRRL B592 and B593, *Clostridium butylicum;* and NRRL B598, *Clostridium pasteurianum.* The cultures were maintained in log phase on chopped liver medium by daily transfer of a 5% inoculum.

Media. The test medium for determining the neutral solvents production from glucose contained, in addition to the toxic materials tested, per liter: 5 *g* Difco Bacto peptone, 0.5 *g* Difco yeast extract, 1 *g* dibasic potassium phosphate, and 27.5 *g* glucose monohydrate. A small marble chip and a small chunk of pork liver were placed in the bottom of a Bellco Hungate tube, concentrated medium was added, and the tube was capped prior to sterilization for 25 min at 121 C. After sterilization, tubes were brought to 65 C, aliquotted with separately sterilized solutions of the toxicants tested, and inoculated with 5% of a log phase culture. The tubes were then incubated at 36 C for 72 h.

The sodium chloride, sulfate, and sulfite were Fisher ACS anhydrous reagents, while the Matheson, Coleman, and Bell sodium sulfide was the nonahydrate. Hydrochloric acid was added to decrease the pH of sodium sulfide and sulfite solutions to around 7.5, using phenol red as an indicator.

Analytical procedures. Approximately 5 ml liquid samples of cultures were placed in tightly capped test tubes and centrifuged at 4,500 × *g* for 30 min. A 1.3 ml aliquot was removed from each sample, placed in an autosampler vial, spiked with an internal standard of 0.1 ml of 5% (w/v) isobutanol, capped, and placed in the autosampler for gas chromatographic analysis.

Samples were analyzed using a Varian Model 3700 gas chromatograph equipped with a model 8000 autosampler and a CDS 111 autoanalyzer. Standards containing 1-butanol, acetone, isobutanol, and ethanol at levels similar to those expected in samples were used to restandardize between every sample. A 10-ft × 1/8-inch column packed with Waters Porapak Q was operated at 210 C with a flame-ionization detector. Helium at 30 ml/min was used as the carrier gas.

RESULTS

Of the wood pulping liquors, kraft black liquor contained the highest amounts of sodium salts. This material is produced at pH of 11.6 but can be reduced to pH values near neutrality by carbon dioxide and other available process off gases. Salts in black liquor are primarily sodium, although other bases, such as calcium and magnesium, are used in the cooks. As reported by MacDonald and Franklin (1969), a typical southern pine kraft black liquor might contain these sodium salts: bisulfite, 87 m*M*; carbonate, 525 m*M*; chloride, 3 m*M*; hydroxide, 85 m*M*; sulfide, 62 m*M*; and sulfate, 14 m*M*. Other liquors, such as caustic washes, may contain higher concentrations of chloride and hydrochlorite. Weak acid sulfite liquor, another potential fermentation stream, contains around 10 m*M* sulfite.

As shown in Fig. 1, all strains gave a slight improvement in solvents production

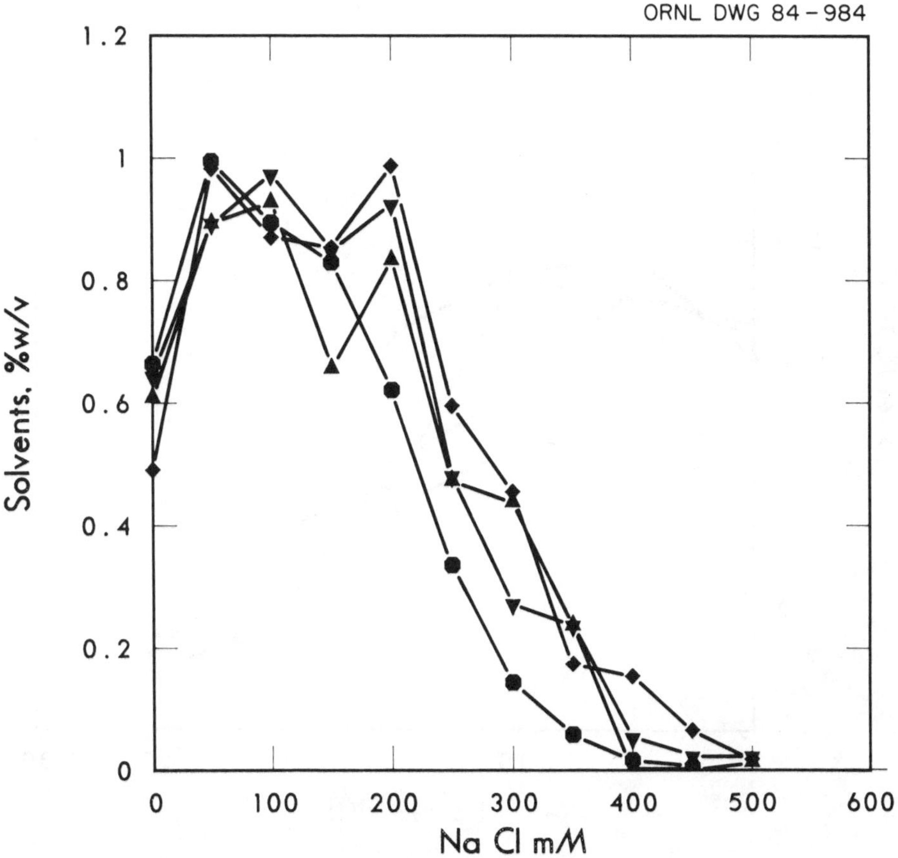

FIG. 1. Total solvents production by strains NRRL B-527 (●), NRRL B-592 (▲), NRRL B-593 (♦), and NRRL B-598 (▼) as a function of NaCl concentration.

with the addition of 50 m*M* NaCl. Above 50 m*M*, solvents production by strain B527 declined gradually, decreasing below 0.1 % (w/v) at around 350 m*M*. Strains B592, B593, and B598 evinced a higher tolerance for NaCl, with a gradual decline between 200 and 400 or 450 m*M*, respectively. The response of strain B592 was somewhat erratic.

As shown in Fig. 2, the strains B527 and B592 responded positively to sodium sulfide levels of 10 m*M* or less, and strain B598, to levels of 5 m*M* or less. Strain B593 did not show much response to concentrations of 10 m*M* or less. At higher concentrations, all strains evinced a decrease in solvents production.

The effect of sodium sulfate at levels less than 200 m*M* was, at most, to reduce solvents production by half. This corresponds to 2.8% (w/v) of sodium sulfate, a level unlikely to be exceeded by conventional pulp streams. Higher sulfate concentrations could conceivably occur in some acid hydrolyzates, but it seems unlikely in terms of process economics.

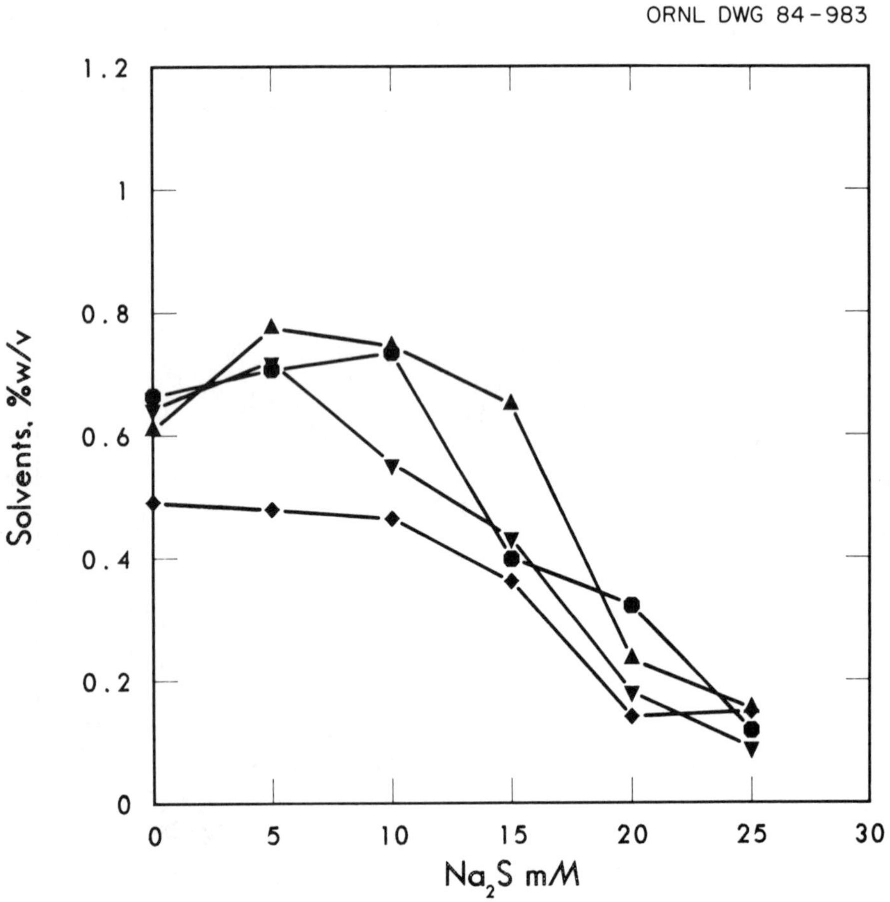

FIG. 2. Total solvents production by strains NRRL B-527 (●), NRRL B-592 (▲), NRRL B-593 (♦), and NRRL B-598 (▼) as a function of Na_2S concentration.

The effect of sodium sulfite addition is shown in Fig. 3. At 15 m*M* or less, solvents production by strains B527, B592, and B593 generally responded positively. A gradual decrease in solvents production occurred above this level. Strain B598 showed an erratic response to sodium sulfite levels of 10 m*M* or less but declined gradually in solvents production at higher levels. At 25 m*M* sodium sulfite, solvents production ceased in all of the strains tested.

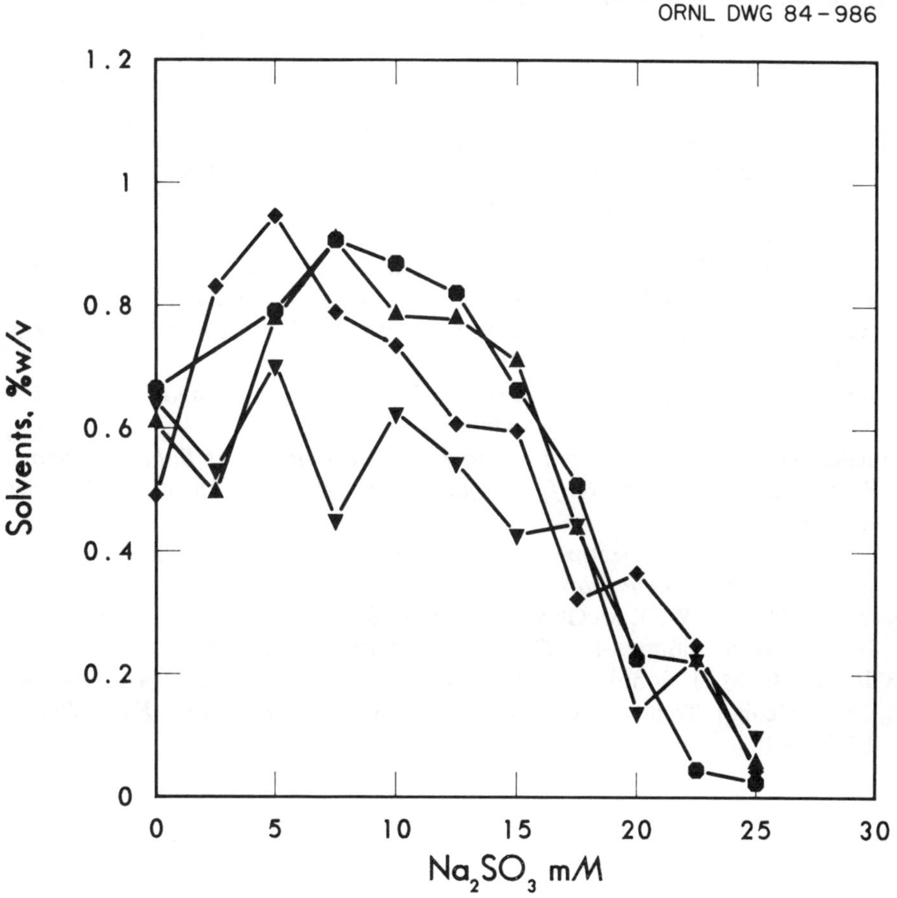

FIG. 3. Total solvents production by strains NRRL B-527 (●), NRRL B-592 (▲), NRRL B-593 (♦), and NRRL B-598 (▼) as a function of Na_2SO_3 concentration.

Conclusions

The levels of sodium chloride and sulfate that permitted solvents production by the *Clostridia* strains tested were above those found in black liquor, the highest salts pulp stream. The levels of sodium sulfide tolerated were around one-half those found in black liquor. All strains tested tolerated the levels of sulfite expected in weak acid sulfite liquor. However, these tests used a single salt rather than the complicated mixtures found in pulp streams, and common ion effects, osmotic pressure, and interaction effects between different toxicants are worthy of further investigation.

ACKNOWLEDGMENT

This research was sponsored by the U.S. Department of Energy under Contract DE-AC-840R21400 with Martin Marietta Energy Systems.

LITERATURE CITED

Blinc, M., and T. Strauch. 1958. Utilization of sulfite waste liquor of beech wood for the production of butanol and acetone. *Nona Proizvodnja* 1958:70–72.

Fricke, A. L. 1983. *Physical Properties of Kraft Black Liquor.* U.S. Department of Energy Report AOE/CE/40606-T1.

Gfeller, V. P. 1953. Zur Kenntnis der Vergarung von Sulfiblaugen durch saccharolytische Clostridien. *Schweiz. Z. All. Pathol. Bakteriol.* 16:867–873.

Grondal, B. L., and H. W. Berger. 1945. Butyl alcohol by fermentation of waste sulfite liquor. *Chem. Eng.* 52:101.

MacDonald, R. G., and J. N. Franklin, eds. 1969. *The pulping of wood.* New York: McGraw-Hill Book Company.

Mueller, H. P. 1958. *Contributions to the Understanding of Butanol Acetone Fermentation in Spent Sulfite Liquor.* Bulletin of the Institute of Paper Chemistry, Zurich.

Schroedler, K. 1953. Utilization of spent sulfite liquor in acetone butanol fermentation. *Wochbl. Papierfabrik* 81:44–46.

Stark, W. H., and W. J. McGhee. 1950. The butyl alcohol acetone fermentation of waste sulfite liquor. *Am. Chem. Soc. 118th Meet.* 22A:56.

Wiley, A. J., M. J. Johnson, E. McCoy, and W. H. Peterson. 1941. Acetone-butyl alcohol fermentation of waste sulfite liquor. *Ind. Eng. Chem.* 33:606–610.

CHAPTER **43**

Importance of Hydrogen Metabolism in Regulation of Solventogenesis by *Clostridium acetobutylicum*

BYUNG HONG KIM AND J. G. ZEIKUS*

Department of Bacteriology, University of Wisconsin-Madison, 1550 Linden Drive, Madison, Wisconsin 53706

Hydrogen metabolism by *Clostridium acetobutylicum* was studied in relation to solvent production. Specific hydrogen production rates and *in vivo* hydrogenase activities were measured during the acidogenic versus solventogenic fermentation time course. The results showed that specific hydrogen production rate decreased as the culture became solventogenic. In pH 5.8 controlled fermentations, the specific activity of *in vivo* hydrogenase remained constant during growth and only acids were produced. In pH 4.5 controlled fermentations, however, the shift from acidogenesis to solventogenesis was accompanied by a corresponding decrease in the specific activity of *in vivo* hydrogenase. These results are discussed in relationship to a proposed model that explains the role of hydrogen metabolism in the bioelectrical regulation of the organism's carbon and electron flow pathways.

INTRODUCTION

Clostridium acetobutylicum ferments sugars to produce three major classes of fermentation end products: (1) acetate and butyrate; (2) butanol, acetone, and ethanol; and (3) CO_2 and H_2. Studies on the course of products formation showed that the neutral solvents production is initiated at the middle of the fermentation, when the pH of the culture decreases to about 4.5 because of the accumulation of the fatty acids (Jones et al. 1982). Since neutral solvents are more reduced than the fatty acids the switch in carbon flow from acids to solvents might be closely related to hydrogen (electron) metabolism. The breakdown of hexose to acetyl-CoA produces 2 mol of NADH and 2 mol of reduced ferredoxin. Petitdemange et al. (1976) demonstrated that NADH-ferredoxin reductase oxidizes NADH to recycle NAD. Hydrogenase (E.C. 1.12.1.1) present in anaerobic bacteria oxidizes reduced ferredoxin to produce molecular hydrogen (Adams et al. 1980). In a previous study, we showed that inhibition of hydrogenase by carbon monoxide increased the solvent yields and decreased acid yields (Kim et al. 1984). From these results, we concluded that electron flow dictates carbon flow under the fermentation conditions employed. The purpose of this present paper is to show that during normal fermentations, in the absence of exogenous modulators of electron flow (i.e., CO), hydrogen metabolism and hydrogenase activity are regulated in relation to solvent production.

*To whom reprint requests should be addressed.

MATERIALS AND METHODS

Chemicals. All chemicals used were reagent grade purchased from Sigma Chemical Co. (St. Louis, MO) and Mallinckrodt (Paris, KY). Gases were purchased from Matheson Scientific (Joliet, IL). Tritiated hydrogen was a gift from Dr. Paul DeLuca, University of Wisconsin-Madison.

Organism maintenance and cultivation. C. acetobutylicum ATCC 4259 was maintained and cultivated as previously described (Kim et al. 1984). A complex medium (CAB) was used in all experiments and contained in grams per liter of distilled water: glucose, 30; yeast extract (Difco), 4; tryptone (Difco), 1; KH_2PO_4, 0.7; K_2HPO_4, 0.7; DL-aspargine, 0.5; $MgSO_4 \cdot 7H_2O$, 0.1; $MnSO_4 \cdot H_2O$, 0.1; $FeSO_4 \cdot 7H_2O$, 0.015; NaCl, 0.1; and 1 ml of 0.2% resazurin. The phosphate-limited chemostat culture utilized a modified CAB medium with 0.1 g KH_2PO_4/L in the place of 0.7 g KH_2PO_4 and 0.7 g K_2HPO_4/L.

Preserved soil culture inocula were revived in anaerobic pressure tubes (Bellco Glass, Inc., Vineland, NJ) containing 10 ml of medium. Heat shock at 80 C for 2 min was applied before the tubes were incubated at 34 C for 2 d. These tubes were used to inoculate 150-ml serum vials containing 70 ml of medium, which were incubated at 34 C for 24 h. This inoculum (5%) was used to initiate the experimental cultures.

Experiments were performed in either 500-ml bottles fitted with anaerobic pressure tube tops with a 70-ml volume of medium or 1-liter Multigen fermentors (New Brunswick Scientific, Edison, NJ) with a 500-ml working volume. The fermentors were equipped with an automatic pH controller (New Brunswick Scientific).

A two-stage chemostat was also used, and it was composed of two Multigen fermentors. The first culture vessel was 1 liter with a working volume of 400 ml and the second 2 liters with a 1250-ml working volume. Constant pH was maintained in both vessels at 5.5–5.8 for the first stage and 4.3–4.5 for the second stage. The flow rate was 80 ml/h to give the dilution rate of 0.2 h^{-1} for the first vessel and 0.064 h^{-1} for the second. Under these conditions, the culture in the first vessel produced acetate and butyrate as fermentation products while only the second stage was solventogenic.

Quantification of growth and fermentation end products. Growth was monitored by reading culture turbidity at 660 nm with a Spectronic 20 spectrophotometer (Bausch and Lomb, Rochester, NY). When the O.D. was over 0.8, the sample was diluted in fresh medium before reading O.D. against a medium blank. The cell concentration was calculated from turbidity using a standard curve. One O.D. unit at 660 nm gave 335 ml/l cell dry weight in both acidogenic and solventogenic cultures.

Fermentation products were analyzed by gas-chromatographic methods previously described (Kim et al. 1984) using a thermal conductivity detector for hydrogen and a flame ionization detector for soluble fermentation products. End products were separated by the following systems: Hydrogen on a stainless steel column packed with Porapak N; ethanol on a Super Q column; and acetate, acetone, butanol, and butyrate on a Chromosorb 101 column.

In vivo hydrogenase activity. Samples of 5 ml were taken from fermentor cultures and placed into anaerobic pressure tubes containing 80% N_2 and 20% H_2. *In vivo* hydrogenase activities were measured by the tritium exchange method as previously described (Kim et al. 1984; Schink et al. 1983). Hydrogenase activity was expressed as dpms ^3H_2O formed/mg cell/min.

Results

Kinetics of hydrogen production. The organism was cultured in a 500-ml bottle, and 0.4-ml gas samples were removed to measure hydrogen production during the fermentation time course. Fig. 1 illustrates the hydrogen production kinetics in relation to growth and the formation of soluble end products. Three distinctly

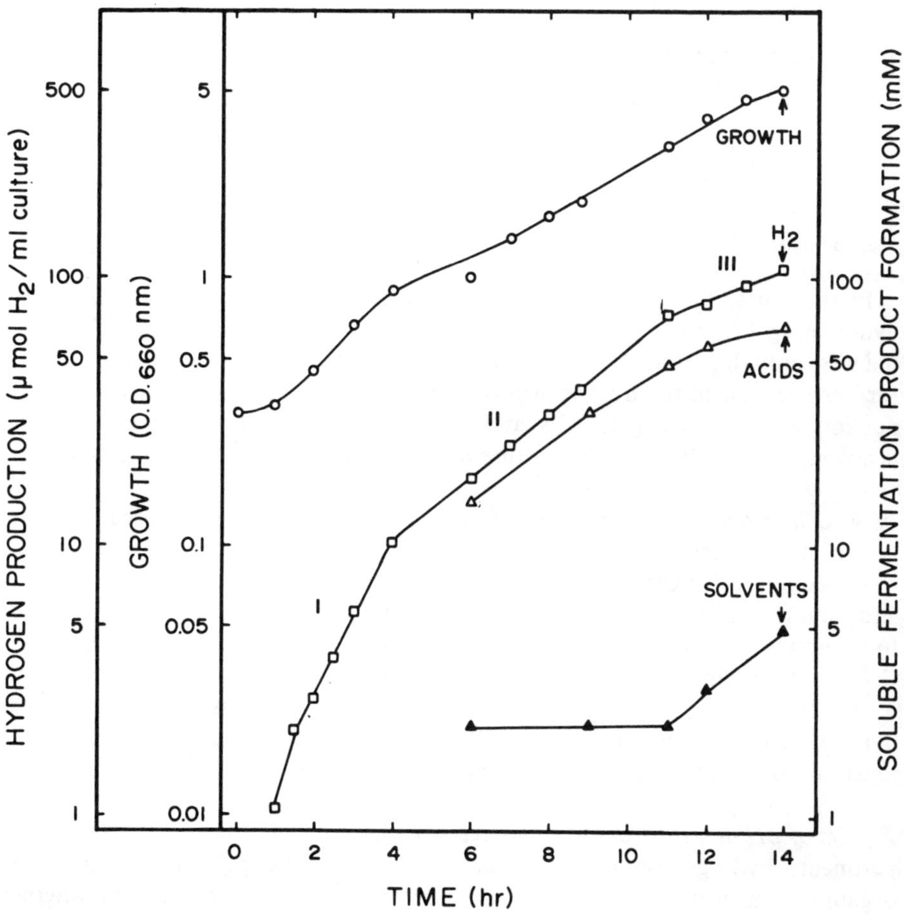

FIG. 1. Relationship between hydrogen production, growth, and initiation of solventogenesis during glucose fermentation by *C. acetobutylicum*. The fermentation was performed in 500-ml bottle cultures fitted with anaerobic pressure tube tops. The initial pH was 5.4, but the fermentation pH was not controlled. Note that solvent levels before 12 h were from the inoculum and that acid and solvents were only monitored after 6 h.

different metabolic phases were observed for hydrogen production. The first decrease of hydrogen production coincided with a decrease in the growth rate and the second decrease coincided with the initiation of solventogenesis. The relationship of hydrogen production to time in each phase was determined as follows:

$$Y = 0.75e^{0.654t} \quad (1.5 \leq t \leq 4)$$

$$Y = 3.47e^{0.272t} \quad (4 \leq t \leq 11)$$

$$Y = 12.68e^{0.156t} \quad (11 \leq t \leq 18)$$

where Y is hydrogen produced by 1 ml of culture (μM H_2/ml culture) at time t (h). The equations were differentiated to calculate the hydrogen production rates and the specific hydrogen production rates (μM H_2/mg cell/h) as shown in Table 1.

TABLE 1. *Relation of incubation time to the specific rate of hydrogen production during glucose fermentations of* C. acetobutylicum *(ATCC 4259)*[a]

Fermentation Time (h)		Specific Hydrogen Production Rate (μM H_2/mg cell/h)
I.	1.5–4	40.86
II.	4–11	17.08
III.	11–18	9.95

[a] See footnote to Fig. 1 for experimental details.

In the initial fast growth period (I), 1 mg of cells produced about 40 μM hydrogen in 1 h, and 17 μM hydrogen were produced by 1 mg cell in 1 h during the next growth period (II). The solventogenic phase (III) was accompanied by a further decrease in the hydrogen production rate. After 18 h the hydrogen production slowed down gradually and hydrogen concentration decreased in the headspace from 276.1 μM/ml culture at 32 h to 267.0 μM/ml culture at 40 h.

In vivo hydrogenase activity as a function of culture age. The organism was grown at a constant pH of 4.5 and 5.8 using Multigen fermentors, and samples were taken at different incubation times to measure *in vivo* hydrogenase activity and soluble fermentation products. Fig. 2 shows the relation between end product formation and *in vivo* hydrogenase activity during pH 4.5 and 5.8 fermentation. Cultures maintained at pH 5.8 only produced acidic soluble end products and had a constant specific hydrogenase activity. However, the specific hydrogenase activity of cultures maintained at pH 4.5 decreased as the culture initiated solventogenesis and continued to do so as more solvents were made.

Effects of pH and fatty acids on in vivo hydrogenase activity. In the previous experiment, hydrogenase activity was measured at the pH under which the organism was grown. The following experiment was performed to test whether the lower hydrogenase activity observed during solventogenic conditions was because of low pH and/or undissociated forms of acetate and butyrate. A special experimental system was used here because the acidogenic phase and solventogenic phase of the culture could be maintained in two different vessels. Culture samples were taken from the 2-stage chemostat maintained at 5.8 and 4.5, placed

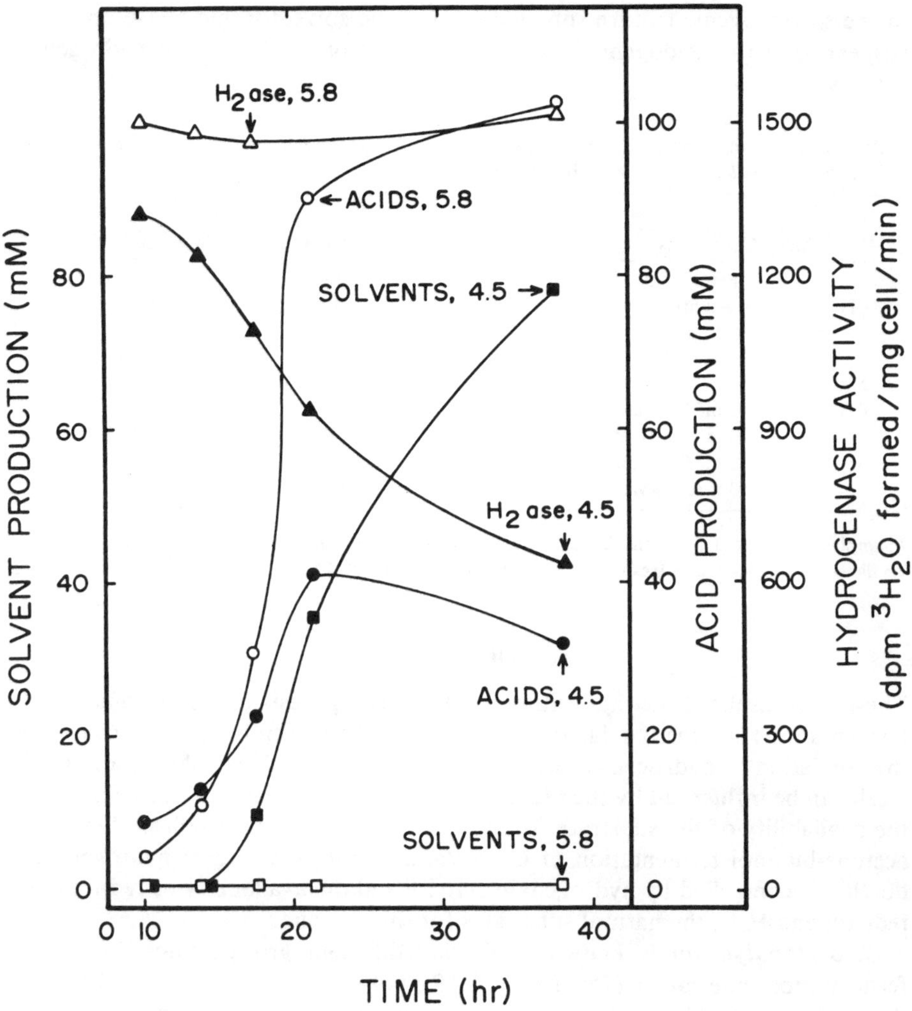

FIG. 2. Relationship between *in vivo* hydrogenase activity and fermentation products formation in fermentors pH controlled at 4.5 and 5.4. The experiments employed Multigen fermentors equipped with automatic pH controller. *In vivo* hydrogenase activities were measured in whole cells directly using samples of the culture.

into anaerobic pressure tubes, and centrifuged. The cells were resuspended in fresh medium at pH 4.5 or 5.8 with or without 30 m*M* acetate and 50 m*M* butyrate prior to measurement of *in vivo* hydrogenase activity. Table 2 shows that acidogenic phase cells grown at a constant pH of 5.8 showed two times as high hydrogenase activity as solventogenic cells grown at pH 4.5, regardless of the assay pH value or the presence of an exogenous fatty acid mixture. Slightly higher hydrogenase activity was obtained when cells were assayed at pH 5.8 than at pH 4.5 with samples of either acidogenic or solventogenic phase cultures. Thus, the decrease in hydrogen metabolism during solvent production is caused by regulation of hydrogenase and not by enzyme inhibition. Similar results were obtained in separate experiments where acidogenic phase cells were resuspended in the solventogenic culture supernatant and the solventogenic phase cells were suspended in the acidogenic culture supernatant prior to assay of hydrogenase activity.

TABLE 2. *Relation of assay conditions to* in vivo *hydrogenase activities of acidogenic and solventogenic phase cultures of* C. acetobutylicum

Assay Conditions[a]		Specific Activity ($dpm\,^3H_2O$ Formed/mg cell/min)	
pH	Fatty Acid Addition	Acidogenic Culture	Solventogenic Culture
4.5	—	1345	648
5.8	—	1577	678
4.5	30 m*M* acetate 50 m*M* butyrate	1286	660
5.8	30 m*M* acetate 50 m*M* butyrate	1532	736

[a] Samples were taken from the 2-stage chemostat and centrifuged. Before assaying activities, the cells were resuspended in fresh medium adjusted to pH 4.5 and 5.8 with or without the fatty acids.

Discussion

These new results show that initiation of solventogenesis by *C. acetobutylicum* fermentations is directly related to a decrease in hydrogen production caused by the regulation of hydrogenase activity. The rate of a biochemical reaction within a cell can be influenced by the regulation of enzyme activity within a cell, and by the availability of the substrate. In the case of hydrogen metabolism during the acetone-butanol fermentation of *C. acetobutylicum*, the rate of hydrogen production is controlled by hydrogenase activity and the availability of reduced ferredoxin and H^+, the natural substrates for this enzyme.

C. acetobutylicum is known to display different growth rates during the fermentation time course (Jones et al. 1982); however, the physiological basis for this phenomenon has not been previously explained. The specific hydrogen production rate of cells decreased during the fermentation time course in association with increased solventogenesis. The initial high hydrogen production rate (initial 4 h) was due to high rates of glucose consumption and growth. Although the specific hydrogenase activity of whole cells did not change during the culture age of 4 to 11 h, the hydrogen production rate of cells was lower at 4–11 h than in the fermentation (17 μM H_2/mg cell/h versus 40.86 μM H_2/mg cell/h). This decrease in hydrogen production coincided with the slower growth and metabolic rates and not regulation of hydrogenase activity. From these observations, it is suggested that the first decrease in the hydrogen production rate (Phase II) was because of the lower availability of reduced ferredoxin caused by slower glucose consumption. Furthermore, the next drop in the hydrogen production rate (Phase III) was the result of lower hydrogen activity and the initiation of solventogenesis triggered by the regulation of enzyme activity.

Fig. 3 presents a hypothetical model to explain in part, the biochemical relationships associated with inhibition of hydrogen production and stimulation of

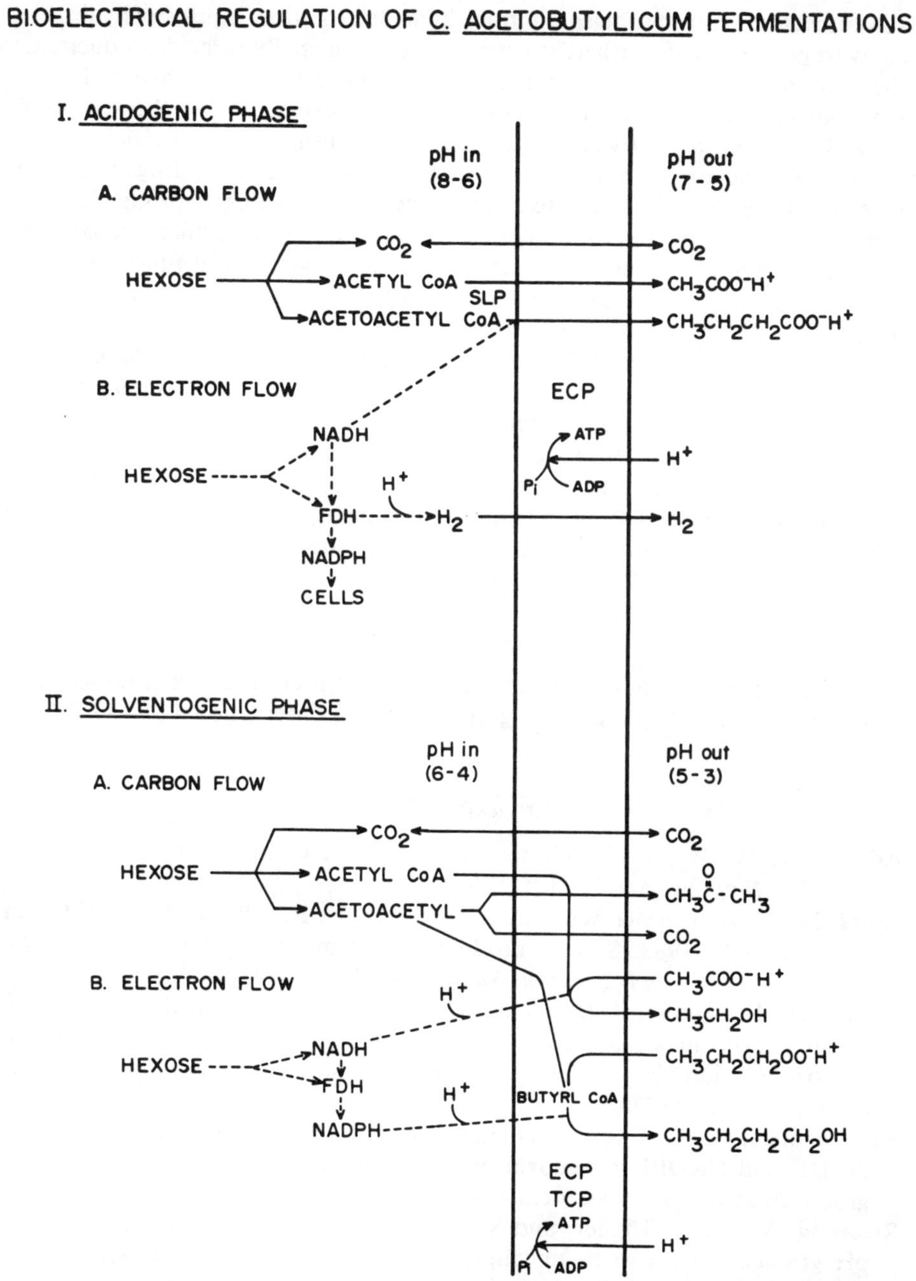

FIG. 3. Model for the hydrogen metabolism and bioelectrical regulation of *C. acetobutylicum*. During the acetogenic Phase (I), electron flow is largely separated from carbon flow because hydrogen production is used to consume internal protons. During the solventogenic Phase (II), carbon and electron flow are both diverted to ethanol and butanol production in order to consume internal protons. In the acidogenic phase, an external pH of 5.4 corresponded to an internal pH of 6.3, whereas in the solventogenic phase, an external pH of 4.0 corresponded with an internal pH of 5.7.

solventogenesis. During Phase I, intracellular electron flow is principally directed to hydrogen production whereas carbon flows principally to acid products. On the other hand, in Phase II, electron flow and carbon flow are diverted to the same solvent end products (i.e., ethanol and butanol). During the acidogenic Phase I, hydrogenase activity is high, and the organism has a very efficient route for the disposal of internal protons by producing gaseous hydrogen. Solventogenesis (Phase II) is associated with decrease in hydrogen production and hydrogenase activity; thus the organism must divert intracellular protons and electrons to neutral solvents in order to maintain physiological function. Based on this argument, both hydrogen production and solventogenesis appear as mechanisms for the maintenance of an internal-alkaline pH gradient, which has been shown to be essential for cell maintenance in other *Clostridium* species (Riebeling et al. 1975); but this has not been associated with conservation of energy (i.e., ATP synthesis via membrane bound ATPase).

Further studies are now needed to demonstrate the exact biochemical mechanisms (i.e., enzymes, electron carriers, internal pH, and membrane components) associated with the regulation of hydrogenase activity in order to understand the importance of this phenomenon in solvent production.

Acknowledgments

This research was supported by the College of Agriculture and Life Sciences, University of Wisconsin, and by a grant from CPC International to J.G.Z.

Literature Cited

Adams, M. W. W., L. E. Mortenson, and J. S. Chen. 1980. Hydrogenase. *Biochim. Biophys. Act* 594:105–176.

Jones, D. T., A. van der Westhuizen, S. Long, E. R. Allock, S. J. Reid, and D. R. Woods. 1982. Solvent production and morphological changes in *C. acetobutylicum*. *Appl. Environ. Microbiol.* 43:1434–1439.

Kim, B. H., P. Bellows, R. Datta, and J. G. Zeikus. 1984. Control of carbon and electron flow in *C. acetobutylicum* fermentations: Utilization of carbon monoxide to inhibit hydrogenase and enhance solvent yields. *Appl. Environ. Microbiol.* 48:764–770.

Petitdemange, H., C. Cherrier, G. Raval, and R. Gay. 1976. Regulation of the NADH and NADPH-ferredoxin oxidoreductases in *Clostridia* of the butyric group. *Biochim. Biophys. Acta* 431:334–347.

Riebeling, V., R. K. Thauer, and K. Jungermann. 1975. The internal-alkaline pH gradient, sensitive to uncoupler and ATPase inhibitor, in growing *C. pasteurianum*. *Eur. J. Biochem.* 55:455–453.

Schink, B., F. S. Jupton, and J. G. Zeikus. 1983. Radioassay for hydrogenase activity in viable cells and documentation of anaerobic hydrogen-consuming bacteria living in extreme environments. *Appl. Environ. Microbiol.* 45:1491–1500.

CHAPTER 44

Relationship Between Lipophilicity and Biodegradation Inhibition of Selected Industrial Chemicals

DINESH D. VAISHNAV AND DEBRA M. LOPAS

University of Wisconsin—Superior, Superior, Wisconsin 54880

Biodegradation of each primary alcohol from methanol ($C_1 \cdot OH$) to 1-dodecanol (1-$C_{12} \cdot OH$) was manometrically measured as a function of alcohol concentration. The study employed resting cells prepared from the growth of a mixed microbial culture on the respective alcohol substrate. Biodegradation rates at each alcohol concentration were calculated in terms of μmol oxygen consumed/μg cell protein/h. Maximum biodegradation rates of all alcohols were correlated with the concentrations of respective alcohols. Gaussian curves were obtained for all alcohols, except 1-undecanol and 1-dodecanol. From the rate inhibition data, concentrations that reduced maximum observed biodegradation rates by 50% (experimental EC_{50}) were statistically derived for 10 alcohols ($C_1 \cdot OH$ to 1-$C_{10} \cdot OH$). The logarithms of 1-octanol/water partition coefficients (Log P) and EC_{50} values of these alcohols were correlated. This relationship was parabolic and defined by the polynomial regression equation: Log EC_{50} (mol/L) = -0.88 (Log P) + 0.07 (Log P)^2 − 0.23. This equation was used for predicting EC_{50} values of 19 narcotic and 2 non-narcotic industrial chemicals based on their Log P values. Experimental EC_{50} values of these test chemicals were derived in the same manner as described for alcohols. The EC_{50} values were compared for each test chemical by taking a ratio of predicted to experimental EC_{50} (mol/L). The mean value of these ratios for the narcotic chemicals was 1.16 ± 0.35. A single structure-toxicity relationship involving only the 1-octanol/water partition coefficient provided a good estimation of concentrations of narcotic chemicals that caused 50% inhibition of their maximum biodegradation rates.

INTRODUCTION

The inventory of industrial chemicals in the United States consists of approximately 30,000 chemicals and is continuously expanding (Veith et al. 1983). Effective hazard assessment methods must rapidly screen chemicals for their biological treatability, environmental persistence, and toxicity to allow testing resources to focus on the potentially hazardous chemicals. One screening technique is to predict biological activities of chemicals from their physicochemical properties. Relationships between biological activities and these properties can be developed around the various types of pharmacologic effects of chemicals (Veith and Konasewich 1975). While predicting the mode of action of chemicals remains a difficult task, it has been shown that gases (Hesser et al. 1978), aliphatic and aromatic hydrocarbons, chlorinated hydrocarbons, alcohols, ethers, ketones, aldehydes, weak acids and bases, and some aliphatic nitrocompounds (Albert 1965; Crisp et al. 1967; Gero 1965; Roth 1980) exhibit narcotic action in vertebrate animals. Narcosis is a reversible state of arrested activity of protoplasmic structures caused by a wide variety of organic chemicals (Veith et al. 1983).

Chemical biodegradation is the microbial activity closely associated with biological treatability and environmental persistence of anthropogenic materials. From literature reports Hansch and Dunn (1972) developed quantitative relationships between lipid solubilities of organic chemicals and inhibition of selected microbial processes by these chemicals. However, systematic data on the effects of very low and high chemical concentrations on their biodegradation are virtually absent from the literature.

This investigation was conducted to (1) determine the possibility of developing a quantitative relationship between the lipophilicities of model narcotic chemicals (primary alcohols) and their biodegradation inhibition, and (2) to test the usefulness of this relationship in predicting concentrations of test narcotic chemicals, which would inhibit their biodegradation.

Materials and Methods

Chemicals. Organic and inorganic chemicals were of analytical and reagent grade, respectively. A mineral salts medium was prepared as described in the *OECD Guidelines for Testing of Chemicals* (1981).

Cultures. Mixed microbial cultures capable of degrading 12 primary alcohols from methanol (C_1•OH) to 1-dodecanol (1-C_{12}•OH) were separately isolated by a conventional enrichment culture technique. An alcohol was added at 100 or 500 ppm in the mineral salts medium inoculated with a wastewater sample. Incubation was conducted on a rotary shaker (150 rpm) at ambient temperature. Additional cultures were similarly obtained that were capable of degrading narcotic and non-narcotic industrial chemicals. All cultures were maintained and individually stored at 4 C in 0.2 *M* phosphate buffer (pH 6.8) containing 100 ppm of an appropriate chemical substrate.

Determining maximum observed biodegradation rates of primary alcohols. Biodegradation of each alcohol (C_1•OH to 1-C_{12}•OH) was studied at various alcohol concentrations employing a manometric technique with resting cells. The cells were prepared according to Lichstein and Oginsky (1965) from 48 to 72-h growth of a preacclimated culture in the mineral salts medium containing a respective alcohol substrate. The final cell suspension was adjusted to contain approximately 2,000 μg cell protein/ml as determined by the modified method of Lowry et al. (Herbert et al. 1971).

All manometric studies were conducted in duplicate on an 18-place Precision Scientific Warburg Apparatus at 30 C using 15-ml single-sidearm flasks as described by Umbreit et al. (1972). Each flask received phosphate buffer (pH 6.8), a desired concentration of alcohol substrate, cell suspension (0.5 ml), and aqueous KOH, to a total of 3.2 ml. Flasks were included for measuring endogenous respiration and abiotic oxidation. Alcohol biodegradation was monitored in terms of oxygen consumption at 15-min time intervals for up to 75 min.

Manometric data were corrected for the endogenous respiration and five biodegradation rates (one for each time interval) at each alcohol concentration were calculated in terms of μmol oxygen consumed/μg cell protein/h. The observed biodegradation rates of an individual alcohol at each concentration were compiled and the maximum rates (one/concentration) were correlated with concentrations.

Relating biodegradation inhibition and lipophilicities of primary alcohols. Relationships were established between alcohol (C_1•OH to 1-C_{12}•OH) concentrations and their biodegradation rates. The rate inhibition data for each alcohol were analyzed by a linear regression technique (Steel and Torrie 1960) to estimate an alcohol concentration that would reduce the maximum observed biodegradation rate of that alcohol by 50% (experimental EC_{50}). Experimental EC_{50} values of 10 alcohols were estimated, but values for 1-undecanol and 1-dodecanol could not be estimated because of the lack of their biodegradation inhibition, even at the highest concentrations (4,816 μmol/3 ml and 4,460 μmol/3 ml, respectively) tested.

The logarithms of 1-octanol/water partition coefficients (Log P) of 10 alcohols were computed by the fragment constant method (Hansch and Leo 1979) and correlated with the experimental EC_{50} values of these alcohols. This relationship was mathematically defined and the resulting algebraic equation was tested for its usefulness in predicting the EC_{50} values of some industrial chemicals.

Comparing predicted and experimental EC_{50} values of test chemicals. Narcotic and non-narcotic test chemicals (total 21) were selected and their Log P values were computed. These values were used in the algebraic equation to calculate predicted EC_{50} values of chemicals. Experimental EC_{50} value of each test chemical was also estimated from observed biodegradation rates. The rate data were obtained and analyzed in the same manner as described earlier for primary alcohols. A comparison of the two EC_{50} values (per test chemical) was made by taking a ratio of predicted to experimental EC_{50} values.

Results and Discussion

Maximum Observed Biodegradation Rates of Primary Alcohols.

Biodegradation of 12 alcohols (C_1•OH to 1-C_{12}•OH) was manometrically followed employing resting cells. Maximum observed biodegradation rates (μmol oxygen consumed/μg cell protein/h) of all alcohols were correlated with concentrations. For each alcohol, except 1-undecanol and 1-dodecanol, a typical Gaussian curve was obtained. The general relationship between alcohol concentrations and corresponding biodegradation rates for 10 alcohols (C_1•OH to 1-C_{10}•OH) is summarized in Fig. 1. There were 10 to 15 concentrations tested for each alcohol ranging from 0.05 μmol for 1-decanol to 24,660 μmol for methanol as added into 3 ml reaction mixtures.

As evident from Fig. 1, biodegradation rates increased with increasing alcohol concentrations (part A), reached and temporarily remained at a maximum with a minimum concentration effect (part B), and gradually declined with further concentration increments (part C). Maximum rates ranged from 0.008 for 1-decanol to 0.018 for methanol, with an average of 0.012 μmol oxygen consumption/μg cell protein/h.

Alcohol concentrations producing maximum biodegradation rates as well as inhibition of these rates decreased with increased Log P values. For example, a maximum biodegradation rate of methanol (Log P = −0.71) was achieved at 2,466 μmol/3 ml while that of 1-decanol (Log P = 4.11) was achieved at 2.6 μmol/3 ml. Similarly, maximum inhibition of methanol biodegradation was at 24,660 μmol/3 ml and that of 1-decanol was at 26 μmol/3 ml. Based on these findings, attempts were made to relate biodegradation inhibition and lipophilicities of alcohols.

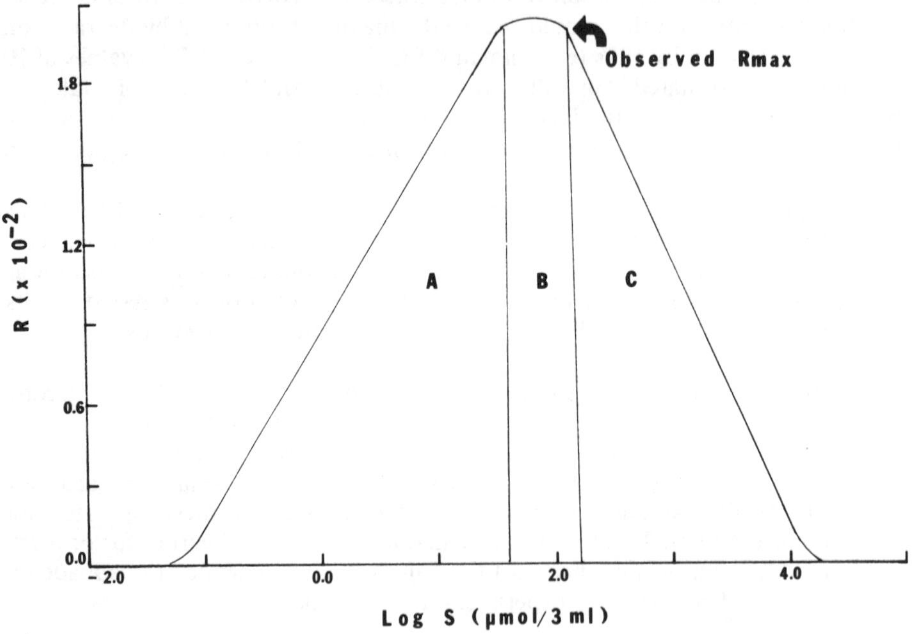

FIG. 1. A summary of concentration effects of 10 primary alcohols (alcohol substrate = S = C_1•OH to 1-C_{10}•OH) on their biodegradation rates (R = μmol O_2 consumed/μg cell protein/h) as manometrically determined employing resting cells of acclimated mixed microbial cultures.

Biodegradation Inhibition and Lipophilicities of Primary Alcohols

A relationship was developed between biodegradation inhibition and the Log P values of 10 alcohols (C_1•OH to 1-C_{10}•OH). First, all concentrations of an individual alcohol causing biodegradation inhibition were identified and their logarithms were related to biodegradation rates (Figs. 2A and 2B). Generally, a linear relationship was observed for each alcohol.

The rate inhibition data for each of the 10 alcohols were then analyzed by a linear regression technique to estimate an experimental EC_{50} value. Correlation coefficients (r) of 10 regression equations ranged from −0.866 for 1-nonanol to −0.999 for methanol, ethanol, 1-propanol, and 1-decanol with a mean value of −0.954 ± 0.046. The experimental EC_{50} values ranged from 0.0028 mol/L for 1-decanol to about 2.8 mol/L for methanol (Table 1). These values decreased as the Log P values of corresponding alcohols increased. A parabolic relationship between the Log P values and logarithms of experimental EC_{50} values for 10

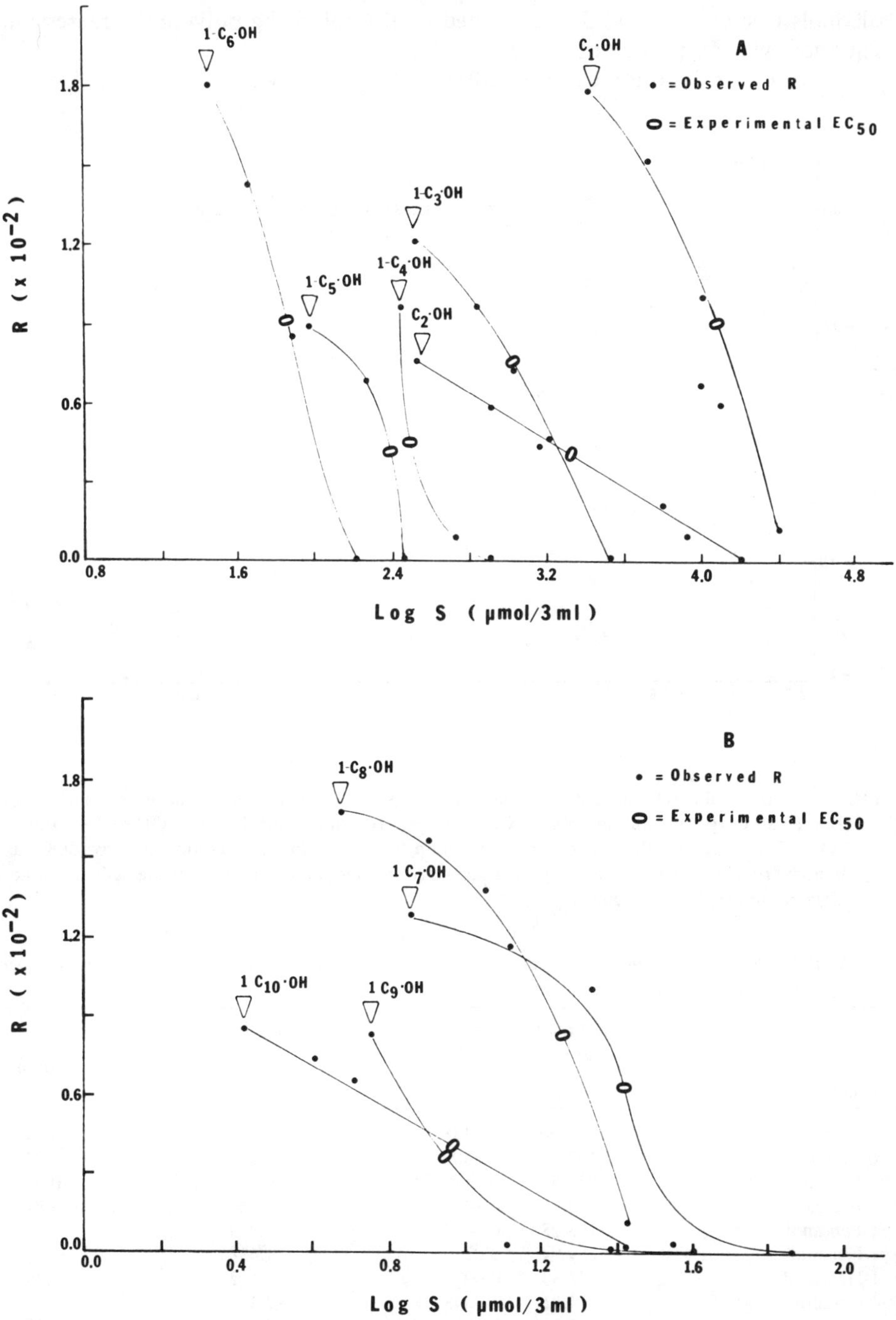

FIG. 2. Inhibitory effects of various concentrations of 10 primary alcohols (alcohol substrate = S = C_1•OH to 1-C_6•OH—Fig. 2A and 1-C_7•OH to 1-C_{10}•OH—Fig. 2B) on their biodegradation rates (R = μmol O_2 consumed/μg cell protein/h) as manometrically determined employing resting cells of acclimated mixed microbial cultures. Note: Concentration of each alcohol was greater than the one producing Rmax for that alcohol.

alcohols is shown in Fig. 3 and defined by the following polynomial regression equation with an r value of −0.997:

$$\text{Log } EC_{50}\ (\text{mol/L}) = -0.88(\text{Log P}) + 0.07(\text{Log P})^2 - 0.23$$

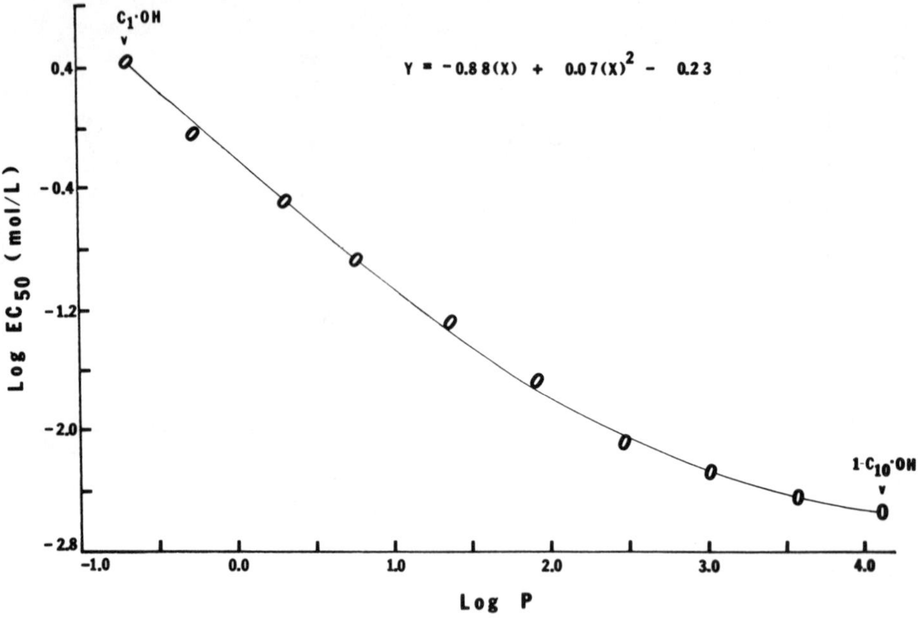

FIG. 3. A parabolic relationship between logarithms of 1-octanol/water partition coefficients (Log P) and experimental microbial EC_{50} values for 10 primary alcohols ($C_1 \cdot OH$ to $1\text{-}C_{10} \cdot OH$). Note: The EC_{50} would reduce the maximum observed biodegradation rate by 50% as manometrically determined employing a respective alcohol substrate and resting cells of an acclimated mixed microbial culture.

TABLE 1. *Primary alcohols, their experimental microbial EC_{50} values*[a] *and logarithms of their 1-octanol/water partition coefficients (Log P)*[b]

Alcohol	Experimental EC_{50} μmol/3 ml	Experimental EC_{50} mol/L	Log EC_{50} mol/L	Computed Log P
Methanol	8,441.00	2.8136	0.449	−0.71
Ethanol	2,192.70	0.7309	−0.136	−0.22
1-Propanol	952.80	0.3176	−0.498	0.33
1-Butanol	429.65	0.1432	−0.844	0.87
1-Pentanol	183.28	0.0610	−1.214	1.41
1-Hexanol	69.04	0.0230	−1.638	1.95
1-Heptanol	23.93	0.0079	−2.102	2.49
1-Octanol	14.51	0.0048	−2.318	3.03
1-Nonanol	10.84	0.0036	−2.443	3.57
1-Decanol	8.62	0.0028	−2.552	4.11

[a] Concentration that would reduce maximum observed biodegradation rate by 50% as manometrically determined employing respective alcohol substrate and resting cells of an acclimated mixed microbial culture.

[b] Computed by fragment constant method of Hansch and Leo (1979).

The finding that microbial toxicity of alcohols was directly related to the Log P values of these alcohols is in agreement with the findings of Hansch and Dunn (1972). The lack of biodegradation inhibition of 1-undecanol and 1-dodecanol could be because of their extremely low water solubilities (49 μmol/L and 10 μmol/L, respectively). In a study with fish, Veith et al. (1983) found that several highly lipophilic chemicals failed to cause acute toxicity via narcosis, and concluded that the toxicities of these chemicals were greater than their water solubilities and could not be measured under equilibrium conditions.

Predicted and Experimental EC_{50} Values of Test Chemicals

The polynomial regression equation as developed based upon alcohol data was used for predicting EC_{50} values of 19 narcotic and 2 non-narcotic chemicals. The latter were ethoxytriglycol and phenol. Selected chemicals were from several chemical classes with Log P values ranging from -1.17 for ethoxytriglycol to 3.54 for 2-decanone (Table 2). All Log P values were computed by the fragment constant method (Hansch and Leo 1979) and substituted in the prediction equation.

As described for the alcohols, experimental EC_{50} values of test chemicals were estimated by a linear regression analysis of the observed biodegradation inhibition rates. The rates were manometrically obtained using resting cells and 5 to 10 inhibitory concentrations of a test chemical. The regression equations for 19 narcotic chemicals had r values ranging from -0.860 for 2-hexanone to -0.999 for acetophenone with a mean value of -0.961 ± 0.042. Experimental EC_{50} values for the narcotic chemicals ranged from 0.005 mol/L for 2-decanone to about 1.2 mol/L for methyl carbitol. These values for ethoxytriglycol and phenol were 0.208 mol/L and 0.004 mol/L, respectively.

Predicted and experimental EC_{50} values of each chemical were compared to assess the prediction accuracy. The EC_{50} values of test chemicals and their comparisons are presented in Table 2. The ratios of predicted to experimental EC_{50} values of narcotic chemicals ranged from 0.44 for naphthalene to 1.60 for benzyl alcohol with a mean value of 1.16 ± 0.35. This ratio was 37.79 for ethoxytriglycol and 9.00 for phenol.

Differences between predicted and experimental EC_{50} values were relatively small for the narcotic chemicals but large for the two non-narcotic chemicals. This may imply that the equation would be only useful in predicting concentrations inhibitory to the biodegradation of chemicals that induce narcosis in vertebrate animals. However, this would include the majority of industrial chemicals (G. D. Veith, personal communication).

CONCLUSIONS

Data presented in this paper have shown that the concentrations of narcotic chemicals that inhibit their biodegradation could be estimated by a single structure-toxicity relationship involving only the 1-octanol/water partition coefficient.

TABLE 2. *Selected industrial chemicals, logarithms of their 1-octanol/water partition coefficients (Log P)*[a], *and their predicted*[b] *and experimental*[c] *microbial EC_{50} values*

Chemical	Computed Log P	EC_{50} (mol/L) Predicted	EC_{50} (mol/L) Experimental	Ratio Predicted / Experimental EC_{50}
Narcotic				
Methyl carbitol	−0.42	1.418	1.254	1.13
Acetone	−0.24	0.966	0.612	1.58
Isopropanol	0.11	0.472	0.514	0.92
2-Butanone	0.30	0.325	0.284	1.14
3-Methyl-2-butanone	0.62	0.178	0.144	1.24
Isobutanol	0.74	0.143	0.197	0.73
3-Pentanone	0.84	0.120	0.092	1.30
Cyclohexanone	0.95	0.099	0.071	1.39
Benzyl alcohol	1.11	0.075	0.047	1.60
2-Pentanol	1.19	0.066	0.053	1.25
Isoamyl alcohol	1.28	0.057	0.046	1.24
2-Hexanone	1.38	0.048	0.055	0.87
Cyclohexanol	1.42	0.045	0.031	1.45
4-Methyl-2-pentanol	1.60	0.034	0.024	1.42
Acetophenone	1.66	0.025	0.016	1.56
2-Octanone	2.46	0.010	0.013	0.77
2-Octanol	2.81	0.007	0.005	1.40
Naphthalene	3.29	0.004	0.009	0.44
2-Decanone	3.54	0.003	0.005	0.60
Non-narcotic				
Ethoxytriglycol	−1.17	7.860	0.208	37.79
Phenol	1.47	0.036	0.004	9.00

[a] Computed by fragment constant method of Hansch and Leo (1979).

[b] Concentration that would reduce maximum observed biodegradation rate by 50% as predicted from Log P value and its following relationship: $\text{Log } EC_{50} \text{ (mol/L)} = -0.88(\text{Log P}) + 0.07(\text{Log P})^2 - 0.23$.

[c] Concentration that would reduce maximum observed biodegradation rate by 50% as manometrically determined employing respective chemical substrate and resting cells of an acclimated mixed microbial culture.

Acknowledgments

This work was supported by Cooperative Agreement CR 809234 between the U.S. EPA and the University of Wisconsin—Superior.

Literature Cited

Albert, A. 1965. *Selective Toxicity*. John Wiley and Sons, New York, 3 p.

Crisp, D. J., A. O. Christie, and A. F. A. Ghobasky. 1967. Narcotic and toxic action of organic compounds on barnacle larvae. *Comp. Biol. Physiol.* 22:629–645.

Gero, A. 1965. Intimate study of drug action. III: Possible mechanism of drug action. Pages 47–49 *in* J. R. DiPalma, ed., *Drill's Pharmacology in Medicine*. McGraw-Hill Book Co., Inc., New York.

Hansch, C., and W. J. Dunn, III. 1972. Linear relationships between lipophilic character and biological activity of drugs. *J. Pharm. Sci.* 61(1):1–19.

Hansch, C., and A. Leo. 1979. *Substituent Constants for Correlation Analysis in Chemistry and Biology.* John Wiley and Sons, New York, 339 p.

Herbert, D., P. J. Phipps, and R. E. Strange. 1971. Chemical analysis of microbial cells. Pages 209–344 *in* J. R. Norris and D. W. Ribbons, ed., *Methods in Microbiology.* Academic Press, New York.

Hesser, C. M., L. Fagraeus, and J. Adolfson. 1978. Roles of nitrogen, oxygen and carbon dioxide in compressed air narcosis. *Undersea Biomed. Res.* 5:391–400.

Lichstein, H. C., and E. L. Oginsky. 1965. *Experimental Microbial Physiology.* W. H. Freeman and Co., San Francisco, CA, 31 p.

OECD Guidelines for Testing of Chemicals. 1981. Section 3—Degradation and accumulation. Organization for Economic Cooperation and Development (OECD) Publications Information Center, 1750 Pennsylvania Ave., Washington, DC, 301(A):9 p.

Roth, S. H. 1980. Membrane and cellular actions of anesthetic agents. *Fed. Proc. Fed. Am. Soc. Exp. Biol.* 39:1595–1599.

Steel, R. G. D., and J. H. Torrie. 1960. *Principals and Procedures of Statistics.* McGraw-Hill Book Co., Inc., New York, 161 p.

Umbreit, W. W., R. H. Burris, and J. F. Stauffer. 1972. Manometric and Biochemical Techniques. Burgess Publishing Co., Minneapolis, MN, 1 p.

Veith, G. D., and D. Konasewich. 1975. Structure-activity correlations in studies of toxicity and bioconcentrations with aquatic organisms. *Symposium Proceedings.* International Joint Commission. Windsor, Ontario, Canada.

Veith, G. D., D. J. Call, and L. T. Brooke. 1983. Structure-toxicity relationships for the fathead minnow, *Pimephales promelas:* Narcotic industrial chemicals. *Can. J. Fish. Aquat. Sci.* 40:743–748.

CHAPTER **45**

Mechanism of Ethylene and Carbon Monoxide Production by *Septoria musiva*

SUSAN K. BROWN-SKROBOT, LEWIS R. BROWN, TED H. FILER, JR.*

*Mississippi State University, Mississippi State, Mississippi 39762, and Southern Forest Experiment Station, P. O. Box 227, Stoneville, Mississippi 38776**

S. musiva, a causative agent of premature defoliation of cottonwood trees, has been shown previously to produce ethylene and carbon monoxide (CO) on media containing glucose, methionine, and iron. Chemical analyses have shown that all three substances are present in the cottonwood leaves. Of seven carbohydrates tested, none supported the production of ethylene and only glucose supported CO production. From 24 amino acids employed as sole carbon sources only methionine and cysteine supported the production of ethylene and CO but four other amino acids supported ethylene production. Metabolic studies have demonstrated that the fungus transaminates methionine to 4-methylmercapto-2-oxobutyric acid, which presumably is acted upon by a peroxide in the presence of iron to form ethylene, formic acid, and dimethyldisulfide. Although the organism grew in a medium containing S-adenosylmethionine, glucose, and iron, no ethylene or CO was produced. Using cell-free preparations, dihydroxyacetone was implicated as a precursor of CO.

INTRODUCTION

S. musiva, a fungal plant pathogen of cottonwood trees *(Populus deltoides)*, is known to produce ethylene and carbon monoxide (Brown-Skrobot et al. 1984) in amounts sufficient to cause premature defoliation. The purpose of this investigation was to shed some light on the pathway(s) by which these gases are produced by the pathogen.

MATERIALS AND METHODS

Cultures and media. The fungus, *Septoria musiva,* isolated from cottonwood *(Populus deltoides)* leaf spots, was received on potato dextrose agar (PDA) from Bernard Smyley of the USDA Forest Service, Stoneville, MS. Additional isolates were obtained in potato dextrose broth (PDB) in 6-oz prescription bottles fitted with serum stoppers. Cultures were maintained on tryptic soy agar (TSA), tryptic soy broth (TSB), and a glucose-peptone broth (G-P broth). Stock cultures were stored at 4 C. Identification of *S. musiva* was by the characteristic hyaline conidia (straight or curved with 1–4 septa) and the pink spore tendrils. All cultures were checked for purity by staining the mycelia and conidia with trypan blue to observe the hyaline conidia. Additionally, the culture was streaked on tryptic soy agar to determine whether bacterial contamination was present. Sections of mycelia also were placed on either potato-dextrose agar (PDA) or rose-bengal agar (RBA) and observed for the presence of fungi other than *S. musiva*. Finally, the *S. musiva* culture was grown in G-P broth. After growth, the culture was examined microscopically for bacterial contamination and streak plates were prepared using nutrient agar (NA) and TSA.

All leaves and stems of cottonwood trees *(Populus deltoides)* were obtained from the Southern Hardwood Laboratories, Stoneville, MS, and were from clones 66 and 261. Both clones are susceptible to infection by *S. musiva*.

Mineral salts broth (MSB) was prepared as described by Brown et al. (1964), and consisted of 1.0 g KNO_3, 0.5 g $K_2HPO_4 \cdot 3H_2O$, 0.2 g $MgSO_4 \cdot 7H_2O$, 0.05 g $FeCl_3 \cdot 6H_2O$ per liter of distilled water. The pH was adjusted to 7.0 using 10% (v/v) hydrochloric acid. Mineral salts agar (MSA) was prepared by adding 1.5% (w/v) Bacto-agar to MSB.

Methionine-glucose broth (M-G broth) consisted of methionine (1.0% w/v), glucose (0.5% w/v), with $K_2HPO_4 \cdot 3H_2O$ (0.05% w/v), and $MgSO_4 \cdot 7H_2O$ (0.02% w/v) dissolved in distilled water. The medium was solidified by the addition of 1.5% (w/v) Bacto-agar (M-G agar).

All carbohydrates and amino acids employed as the sole carbon sources in growth studies were filter-sterilized (0.45 μm filter) and employed at a concentration of 1% (w/v). The remaining constituents of the media were either KNO_3 or NH_4Cl (0.5% w/v) as nitrogen source with $K_2HPO_4 \cdot 3H_2O$ (0.05% w/v), $MgSO_4 \cdot 7H_2O$ (0.02% w/v), and 1.5% (w/v) Bacto-agar with iron powder (100 mg/tube).

All media were either obtained from Difco Laboratories, Inc., or prepared from reagent grade chemicals.

Most experiments were conducted in 16 × 150 mm serum-stoppered test tubes.

Preparation of inoculum. Generally the *S. musiva* isolate was grown in 6-oz bottles containing 50 ml of G-P broth for 14 d, with incubation at room temperature using agitation at 110 rpm/min on a New Brunswick Rotary Shaker. The fungus was collected on a 0.45 μm Millipore filter and washed with 25 m*M* phosphate buffer, removed from the filter, and resuspended in the same solution. The cells were homogenized in a sterile Waring blender for 1 min. The suspension was diluted with buffer to the point where a 1:10 dilution of fungal cells gave a reading of 50% on a Bausch & Lomb Spectronic 20 Spectrophotometer (590 nm).

Resting cell techniques. Cell suspensions were prepared by washing *S. musiva* cells grown in M-G broth and iron with 25 m*M* phosphate buffer (pH 7.0), removing the iron with a magnet, blending in a Waring blender, and resuspending the cells in the phosphate buffer. The fungal cell suspension was then disrupted in an American Instrument Company French Press stainless steel cylinder (3/8 in) (20,000 psi). Tests were performed in serum-stoppered 5-ml serum vials containing 0.5 ml of substrate solution (DHA 100 μmol).

Analytical techniques. All gas-chromatographic (GC) analyses were carried out using a Precision Pressure-Lok syringe and a Fisher Model 1200 Gas Partitioner, using helium as the carrier gas at a flow rate of 30 ml/min, a column temperature of 75 C, and a bridge current of 200 mA. The first column (6.5 ft long by 1/8 inch diameter aluminum) was packed with 800–100 mesh Columpack™ PQ. Column two (11 ft long by 3/16 in diameter aluminum) was packed with 60–80 mesh molecular sieve 13X.

The GC-MS analysis was conducted by the Mississippi State Chemical Laboratory using a Finnigan Model 45–10 GC-MS with either a capillary column (DV-5 liquid phase) for detection of ethylene or a molecular sieve 5A (Linde 6 ft column) with 120–140 mesh for detection of CO.

Hydrolysis of the cottonwood leaves was achieved using sulfuric acid as specified by Browning (1967). Derivitization was achieved using the methodology of Chen and McGinnis (1981). The gas-liquid chromatographic analyses were conducted using a Perkin-Elmer 3920 gas-liquid chromatograph. Nitrogen was employed as the carrier gas with a nickel alloy column packed with 1.0% stabilized diethyl-glycoadipate on 100–120 mesh chromosorb WHP.

The free amino acids were separated from the protein bound amino acids using a solvent extraction system described by Bieleski and Turner (1966). Hydrolysis of the protein bound amino acids was conducted according to Penke et al. (1974). Analysis of cottonwood leaves for free and protein-bound methionine was conducted using thin-layer chromatography according to the method of Pataki (1968). Cottonwood leaves were analyzed for the presence of iron using an Instrumental Laboratories Model 353 Atomic Absorption Spectrophotometer by the Mississippi State Chemical Laboratory.

Analysis for the presence of formic acid was performed using two methods described by Feigl (1958).

Results and Discussion

A previous investigation into the mechanism by which *S. musiva* causes premature defoliation of cottonwood trees revealed that this pathogen produces copious quantities of both ethylene (0.22–0.26 μmol/h/g dry wt mycelium) and carbon monoxide (0.13 μmol/h/g dry wt mycelium) when grown on media containing methionine, glucose, and iron (Brown-Skrobot et al. 1984). Further, the quantities of ethylene and CO produced were sufficient to cause premature leaf fall of *Populus deltoides* (Brown-Skrobot et al. 1984).

All three of the key ingredients of this medium would be expected in cottonwood leaves and chemical analyses proved their presence: methionine (determined by TLC), iron (determined by atomic absorption), and glucose (determined by GLC). In order to identify other carbohydrates that could serve as a substrate for the growth of *S. musiva*, leaves were subjected to acid hydrolysis and gas-liquid chromatographic analysis for aldoses. As expected, glucose was the predominant aldose in both frozen and refrigerated leaves. Other aldoses present in both frozen and refrigerated leaves were D-arabinose, D-mannose, D-rhamnose, D-ribose, and D-xylose. The above analyses were conducted employing leaves of Clone 261, but when refrigerated leaves of Clone 66 were analyzed, galactose was found in addition to the other aldoses present in Clone 261.

Growth on Carbohydrate Media

All aldoses found in the leaves were tested for their ability to serve as substrates for the production of ethylene and/or carbon monoxide. For these experiments, the carbohydrate sources were filter-sterilized into test tubes containing MSB with either KNO_3 or NH_4Cl as the nitrogen source along with iron powder. After inoculation, the tubes were serum-stoppered and incubated at room temperature. After 30 d, gas chromatographic analyses using 1.0-ml samples of the atmospheres from duplicate tubes revealed that none of the carbohydrate sources stimulated the production of ethylene, but carbon monoxide (0.07 μmol) was produced by *S. musiva* growing on glucose with either KNO_3 or NH_4Cl as a nitrogen source. The remaining carbohydrate sources did support growth as indicated by oxygen consumption and carbon dioxide production.

Growth on Amino Acids

Methionine was shown to be a substrate for ethylene production by *S. musiva*. To test the ability of other amino acids to serve as substrates for the production of ethylene and carbon monoxide, growth studies were conducted employing several amino acids as carbon sources. Gas chromatographic analyses of the overlying atmosphere from duplicate samples after 30 d incubation revealed that L-cysteine stimulated the production of an average of 0.09 μmol of ethylene and 0.98 μmol of carbon monoxide, and L-methionine stimulated the production of 0.18 μmol of ethylene and 0.29 μmol of carbon monoxide.

In contrast, L-aspartic acid, DL-ethionine, L-glutamic acid, and DL-homocysteine stimulated the production of ethylene (0.09 μmol) but not carbon monoxide. Amino acids supporting growth but not stimulating the production of either ethylene or carbon monoxide within 30 d were L-alanine, L-arginine, L-asparagine, L-cystine, L-glutamine, L-glycine, L-histidine, L-isoleucine, L-leucine, L-lysine, DL-norleucine, DL-norvaline, L-phenylalanine, L-proline, L-serine, L-threonine, L-tyrosine, and L-valine.

Pathway of Ethylene Production by S. Musiva

Methionine has been shown to be a substrate for the production of ethylene by *S. musiva*. Two possible pathways for the production of ethylene from methionine have been reported in the literature. One pathway proceeds through the formation of S-adenosylmethionine (Fig. 1) while the second pathway proceeds by way of transamination to the alpha-keto acid (Fig. 2). To determine if *S. musiva* could utilize S-adenosylmethionine as a substrate for the production of ethylene, S-adenosylmethionine was substituted for methionine in M-G broth. The organism grew well in this medium but failed to produce either ethylene or CO. Next, *S. musiva* was grown in M-G broth in the presence of isonicotinic acid hydrazide, a known transaminase inhibitor. The lack of ethylene production in the presence of the inhibitor suggests that transamination of methionine was occurring since ethylene was produced in the same system in the absence of the inhibitor. In the transamination pathway, the alpha-keto acid (Fig. 2) formed from the methionine is converted to ethylene via reactions involving peroxides. Since the alpha-keto acid was unavailable and it is known that methional is converted to ethylene via reaction with hydroxyl radicals, an experiment was conducted using disrupted *S. musiva* cells with methional as the substrate. It was found that ethylene was produced from the methional with and without the disrupted cells but the amount of ethylene produced in the presence of disrupted cells was 25% greater than that obtained in the absence of the disrupted cells indicating that

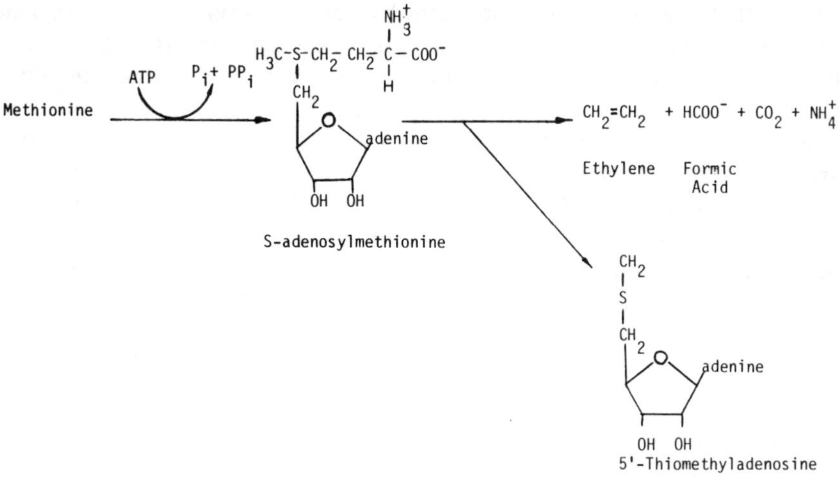

FIG. 1. Pathway for the conversion of methionine to ethylene via S-adenosylmethionine (Metzler 1977).

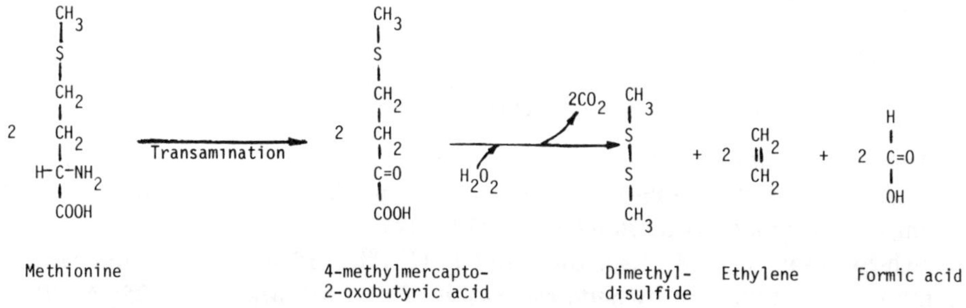

FIG. 2. Pathway for the conversion of methionine to ethylene via transamination. Drawn data are from Metzler (1977) and Hislop et al. (1973).

they stimulated ethylene production, possibly through the formation of hydroxyl radicals. This finding does not prove that the alpha-keto acid is converted to ethyene through the action of peroxide, but it does support the contention that methionine is converted to ethylene via the transamination pathway.

It also may be observed (Fig. 2) that in the transamination pathway, dimethyldisulfide and formic acid are produced along with ethylene. Therefore, the atmospheres from *S. musiva* grown on leaf material and *S. musiva* grown on artificial media were subjected to GC-MS analysis. These analyses showed that dimethyldisulfide was present in the atmospheres of both samples. Establishing that dimethylsulfide was present in the atmosphere overlying the *S. musiva* cultures adds further evidence that the production of ethylene proceeds via transamination of methionine.

When the medium (M-G broth) from a growing *S. musiva* culture producing ethylene was tested for the presence of formic acid by two separate colorimetric tests, both tests were positive. While none of these experiments disprove the existence of the S-adenosylmethionine pathway for the production of ethylene, they do show that the transamination pathway for the production of ethylene is operative in *S. musiva*. Additionally, it should be pointed out that there is no absolute proof that all the ethylene produced by *S. musiva* is by way of methionine.

Other investigations were undertaken using cell-free preparations to determine intermediates in the pathway of carbon monoxide production by *S. musiva*. Disrupted cells of *S. musiva* were offered dihydroxyacetone and produced carbon monoxide (0.18 μmol) and a trace of ethylene. This result was not unexpected since Brown and Brown (1981) showed that carbon monoxide was produced by *Rhizoctonia solani* using dihydroxyacetone; thus, *S. musiva* may produce carbon monoxide by a pathway similar to that employed by *R. solani*.

Conclusions

It is concluded that *S. musiva* can produce ethylene and CO from the glucose, methionine, and iron available within cottonwood trees. Further, metabolic studies have demonstrated that ethylene is probably produced via transamination of methionine to 4-methyl mercapto-2-oxobutyric acid, which reacts with peroxide in the presence of iron to form dimethyldisulfide, ethylene, and formic acid while the CO is produced from DHA. The organism was unable to produce ethylene or CO from S-adenosylmethionine.

Literature Cited

Bieleski, R. L., and N. A. Turner. 1966. Separation and estimation of amino acids in crude plant extracts by thin layer electrophoresis and chromatography. *Anal. Biochem.* 17:278–293.

Brown-Skrobot, S. K., L. R. Brown, and T. H. Filer. 1984. Ethylene and carbon monoxide production by *Septoria musiva. Dev. Ind. Microbiol.* 25:749–755.

Brown, L. R., R. J. Strawinski, and C. S. McCleskey. 1964. The isolation and characterization of *Methanomonas methanooxidans* (Brown and Strawinski). *Can. J. of Microbiol.* 10:792–799.

Brown, S. K., and L. R. Brown. 1981. The production of carbon monoxide and ethylene by *Rhizoctonia solani. Dev. Ind. Microbiol.* 22:725–731.

Browning, B. L. 1967. *Methods of Wood Chemistry,* Vol. II, John Wiley and Sons, Inc., New York, 882 pp.

Chen, C. C., and G. D. McGinnis. 1981. The use of 1-methylimidazole as a solvent and catalyst for the preparation of aldonitrile acetates of aldoses. *Carbohydrate Res.* 90:127–130.

Feigl, R. 1958. *Spot Tests in Organic Analysis.* Elsevier Publishing Company, New York, 600 pp.

Hislop, E. C., G. V. Hoad, and S. A. Archer, 1973. The involvement of ethylene in plant diseases. Pages 87–113 *in* F. Abeles, ed., *Ethylene in Plant Biology.* Academic Press, New York.

Metzler, D. E. 1977. *Biochemistry, The Chemical Reactions of Living Cells.* Academic Press, New York, 1129 pp.

Pataki, G. 1968. *Techniques of Thin-Layer Chromatography.* Ann Arbor Science Publishers, Inc., Ann Arbor, MI, 218 pp.

Penke, B., R. Ferenczi, and K. Kovacs. 1974. New acid hydrolysis method for determining tryptophan in peptides and proteins. *Anal. Biochem.* 60:45–50.

CHAPTER **46**

Ames Tests of Toxic Materials: Pinpoint Colonies Formed in Tests of Chromic Oxide

THOMAS H. UMBREIT, KATHRYN O. COOPER, AND CHARLOTTE M. WITMER

Joint Graduate Program in Toxicology, Rutgers University, Piscataway, New Jersey, and The University of Medicine and Dentistry of New Jersey, Rutgers Medical School, Piscataway, New Jersey 08854

Ames tests on toxic materials can result in formation of pinpoint colonies. Some compounds, such as metal salts, are mutagenic only at high concentrations, and the presence of numerous and large pinpoint colonies may interfere with test results. Replicate plating onto histidine-free medium can distinguish pinpoint colonies from true revertants. We report studies of pinpoint colonies formed in Ames tests of CrO_3, and demonstrate (1) an increase in number and size of pinpoint colonies with increasing CrO_3 concentrations, and (2) the increase in number of pinpoint colonies with decrease in number of lawn colonies. We also demonstrate that pinpoint and lawn colonies require histidine for growth, and thus are not revertants.

INTRODUCTION

The Ames test (Ames et al. 1975) is frequently used as a standard assay for mutagenicity. This test is based on the reversion of mutants unable to synthesize their own histidine to histidine-synthesizing capacity in the presence of a mutagenic chemical. This reversion is determined by the ability of the cells to grow on minimal medium agar plates containing only a trace amount of histidine incorporated into the top agar. The minimal amount of histidine permits a small amount of growth of all cells present during the exposure to the mutagen. The growth is manifested as a faint "lawn" of bacteria on which the revertants stand out as larger colonies. Mutagenic agents are usually somewhat toxic (mutagenesis is usually a lethal event). In the presence of an especially toxic agent, however, many of the lawn cells are killed. In routine assays, the dose of test agent is reduced to overcome this toxicity problem. However, many agents such as metal salts are mutagenic only at toxic levels, and tests carried out at lower levels give false negative results. When toxic doses of mutagen are added to the assay, the lawn is much reduced and very small "pinpoint" colonies appear. These pinpoints may be survivors of the lawn that are able to form colonies by using the histidine present in the top agar (which is now present in a higher amount on a μg/cell basis) or released from the lysed lawn cells (Ames et al. 1975). It becomes difficult to distinguish true revertants from pinpoints on plates when a great deal of lawn death has occurred, and they may cause a false positive result. It is possible to choose colonies for transfer to minimal agar medium that lacks histidine (or test agent) to determine the phenotype of the colony by its ability to grow in the absence of histidine (Cooper et al. 1981). The same result can be obtained by replicate plating onto a minimal medium (Pederson et al. 1982).

Materials and Methods

The Ames test was performed as described by Ames et al. (1975). *Salmonella typhimurium* strains TA 100 and TA 1535 (obtained from Dr. Bruce Ames) were used throughout these studies at approximately 10^8 cells/ml. Chromic oxide (CrO_3, which exists in CrO_4^{2-} in solution at pH 7.4) was used as the test agent at concentrations of 20, 40, 60, and 80 μg/plate. The Vogel-Bonner medium E used in the assay contained no biotin and was designated VB. Biotin was present in the top agar (Ames et al. 1975).

Replicate plating was done with a wooden block and artificial felt-like plush material (Lederberg and Lederberg 1952). Replicates were made of the original Ames assay plates (after 48 h incubation) onto plates of Vogel-Bonner minimal medium E with added biotin (0.45 *μM*. designated VBio), a requirement for growth of *S. typhimurium* strain TA 100. The replicate plates were incubated at 37 C for 48 h prior to counting colonies. Replications were also carried out in series, whereby each plate was used as a template for the next series. After initial replication of the original assay plates, these original plates were stored at room temperature and replicated again at 48, 96, and 144 h. Normal and macroscopic revertant colonies were counted by hand or automatically (New Brunswick Scientific Company Biotran II automatic colony counter), using a low sensitivity setting for counting normal revertants and a high sensitivity setting for detecting macroscopic pinpoints, providing an automatic reproducible size selection. Plates were examined microscopically at 100 times magnification and the number of lawn colonies were counted in one-fourth of a field of view (one field = 2.26 mm^2). The number of microscopic pinpoints were counted in six fields of view. Microscopic pinpoints were counted to exclude lawn colonies and macroscopic pinpoints. All counts were extrapolated to number of colonies per plate, assuming an even distribution over the surface of the plate.

Results

The results of a standard Ames assay of CrO_3 using TA 100 are shown in Table 1. At concentrations of 40 μg/plate, the number of revertants and the number of lawn colonies decline steeply, while the number of macroscopically visible pinpoints greatly increased with increasing CrO_3 concentration. The number of microscopically visible pinpoints also rose but subsequently declined with increasing concentration of CrO_3. As a result of the steep decline in lawn colonies, both the number of macro-pinpoints and micro-pinpoints per lawn colony increased greatly with increasing concentration of the test agent, except for some small declines at the highest concentration tested (Table 2).

Data from replications of these original assay plates are shown in Table 3. For comparison, data from the original plates are included. Macro-pinpoints, micro-pinpoints, and lawn colonies did not replicate onto the minimal agar plates, indicating their inability to synthesize histidine. Replication of original assay plates

TABLE 1. *Ames assay of CrO_3 with TA 100, on VB medium. "Pinpoints," "macro," and "micro" are defined in the text*

Conc μg/pl	Plate	Revertants	Pinpoints Macro	Pinpoints Micro	Lawn × 10^6
0	a	125	12	3045	5.8
	b	115	2	870	6.3
	c	112	9	870	5,6
20	a	216	15	2175	4.8
	b	148	12	3481	6.0
	c	166	17	8702	4.5
40	a	95	96	50472	1.6
	b	20	120	32198	2.3
	c	12	154	40899	1.4
60	a	8	293	21320	.86
	b	14	294	41335	.64
	c	12	265	27412	1.1
80	a	19	667	4786	2.9
	b	45	655	10442	1.8
	c	37	615	6962	3.6

TABLE 2. *Results of the Ames assay of CrO_3 expressed on a per lawn colony basis. All nonstandard terms are defined in the text*

Conc μg/pl	Revertants/ Lawn Colony × 10^{+7}	Macro-Pinpoint/ Lawn Colony × 10^{+7}	Micro-Pinpoint Lawn Colony × 10^{+4}
0	34	63.5	6
20	34.5	68	22.6
40	49.5	250	36
60	180	480	24.5
80	160	380	28

onto trypticase soy agar plates resulted in confluent growth within a 24-h period, demonstrating that the replication transferred lawn cells. Further replications made in a series from each successive template produced similar results: i.e., there were small differences in the numbers of revertants transferred with a consistent lack of growth of lawn, macro-pinpoint and micro-pinpoint colonies on media lacking histidine.

Size changes in micro-pinpoint colonies with increasing concentration of CrO_3 have been documented. Lawn colonies with few pinpoints were observed on the plates containing 0 or 20 μg/plate CrO_3. Pinpoint colonies were seen on plates containing 40 μg/plate CrO_3; the pinpoints were larger at 60 μg/plate CrO_3 and were largest at the maximum concentration tested, 80 μg/plate. Few lawn colonies were seen at 60 or 80 μg/plate.

A nearly identical experiment was done using Ames strain TA 1535, which does not require biotin for growth (data not shown). The number of revertants, macro-pinpoints, and micro-pinpoints generally increased with increasing concentration of CrO_3, while lawn colonies generally decreased in number. Replicate

TABLE 3. *Replicate plating of the Ames assay plates (Table 1) or CrO_3 with TA 100, replicated onto VBio medium*[a]

Conc μg/pl	Plate	Revertants Orig	Revertants 1	Revertants 2	Revertants 3	Pinpoints: Macros Orig	Macros 1	Macros 2	Macros 3	Pinpoints: Micros Orig	Micros 1	Micros 2	Micros 3	Lawn × 10^{-6} Orig
			a b	a b	a b									
0	a	125	140,136	139,138	95, –	12	0	0	0	3045	0	0	0	5.8
	b	115	115,109	98,91	111,81	2	0	0	0	870	0	0	0	6.3
	c	112	114,105	–,–	–,–	9	0	–	–	870	0	–	–	5.6
20	a	216	169,150	136,113	100,78	15	0	0	0	2175	0	0	0	4.8
	b	148	130,111	118,119	82, –	12	0	0	0	3481	0	0	0	6.0
	c	166	155, –	–,–	–,–	17	0	–	–	8702	0	–	–	4.5
40	a	95	43,43	55,56	69,66	96	0	0	0	50472	0	0	0	1.6
	b	20	42,42	69,67	91,96	120	0	0	0	·32198	0	0	0	2.3
	c	12	115, –	–,–	–,–	154	0	–	–	40899	0	–	–	1.4
60	a	8	66,63	73,67	82,47	293	0	0	0	21320	0	0	0	.86
	b	14	39,63	50,52	58,58	294	0	0	0	41335	0	0	0	.64
	c	12	115, –	–,–	–,–	265	0	–	–	27412	0	–	–	1.1
80	a	19	23,20	43,34	64,55	667	0	0	0	4786	0	0	0	2.9
	b	45	27,29	39,35	53,50	655	0	0	0	10442	0	0	0	1.8
	c	37	35,34	–,–	–,–	615	0	–	–	6962	0	–	–	3.6

[a] The original assay plate ("orig") was replicated twice to generate plates 1a and 1b. Plate 1a was replicated to generate plates 2a and 2b, and plate 2a was replicated twice in most cases. No lawn colonies were observed on any replicate plate. "–" = no value obtained. The terms "macros," "micros," and "pinpoints" are defined in the text.

plating of the original Ames assay plates resulted in some increase in the number of revertants per plate. No macro-pinpoint, micro-pinpoint, or lawn colonies were transferred.

A second series of replicate platings made from the original master plates 48 h later gave the same pattern of results; more revertants were obtained, reflecting the longer incubation period of the original plates. Replicate plating onto VB or VBio plates produced no significant differences in the number of revertants obtained.

Discussion

Several groups have shown false positives from toxic chemicals in the Ames assay. Pederson et al. (1982), in a preliminary report, used replicate plating to distinguish "false revertants" from true revertants in assays of $Na_2Cr_2O_7$ and welding fumes. They did not study the relationship between number of lawn cells present and number of pinpoint colonies formed. We originally detected pinpoints with CrO_3 at low concentrations where the morphology was sufficiently different to distinguish from revertants. However, larger pinpoint colonies produced at higher concentrations are difficult to distinguish from revertants. Replicate plating can be used to distinguish revertants from pinpoints.

In this communication, we have used the replicate plating technique to demonstrate (1) the increase in number and size of pinpoint colonies with increasing concentrations of chromic oxide; (2) the increase in number of pinpoints with decrease in the number of lawn colonies; and (3) the requirement of histidine for growth of pinpoint and lawn colonies (and biotin for TA 100).

The technique of replicate plating is affected by several factors that alter the efficiency of replication. One cause of divergent numbers in the tables presented here may be the inclusion of larger pinpoint colonies as revertants in the original assay. The amount of crowding of revertants on the original plate will affect the numbers replicated. Because of the technical requirements of replicate plating, colonies will spread, and overlapping colonies may reduce the accuracy of counts.

Conclusion

Evidence has been presented that pinpoint colonies are cells that require histidine to grow on Vogel-Bonner medium E minimal agar plates. These pinpoint colonies occur in increased numbers and are larger in size with higher concentrations of a toxic agent, and with increasing death of lawn cells. These results support the theory that pinpoint colonies originate as lawn cells that grow on the histidine in the top agar after death of many of the lawn cells. Our results indicate that replicate plating may be a necessary part of the Ames test when toxic agents are employed, as has been suggested previously (deSerres and Shelby 1979). It is possible to distinguish small pinpoints, but larger pinpoints are very difficult to tell from revertants.

Literature Cited

Ames, B. N., J. McCann, and E. Yamasaki. 1975. Methods for detecting carcinogens and mutagens with the Salmonella/mammalian-microsome mutagenicity test. *Mutat. Res.* 31:347–364.

Cooper, K., J. Kelly, and C. Witmer. 1981. Effects on salt concentration on mutagenicity and toxicity with Ames Salmonella strains. *Environ. Mutagen Soc. Annu. Meet.* Abstract.

deSerres, F. J., and M. Shelby. 1979. Recommendations on data production and analysis using the Salmonella/microsome mutagenicity assay. *Mutat. Res.* 64:159–165.

Lederberg, J., and E. M. Lederberg. 1952. Replica plating and indirect selection of bacterial mutants. *J. Bacteriol.* 63:399–406.

Pederson, P., E. Thomsen, and R. M. Stern. 1982. Detection by replicate plating of false revertant colonies induced in the Salmonella/mammalian microsome assay by hexavalent chromium. Pages 18–29 *in* Proc. *2nd Intl. Workshop on* In Vitro *Effects of Mineral Dusts,* Degray Lake, AK.

CHAPTER 47

Microcomputer-Based Gas Control Schemes for Investigating Sulfur Oxidation of *Chlorobium limicola* forma *thiosulfatophilum*.

Jeremy J. Mathers and Douglas J. Cork

Department of Biology, Illinois Institute of Technology, Chicago, Illinois 60616

Tylan mass flow type gas controllers were interfaced to an Apple IIe-ISAAC data acquisition and control system. CO_2, N_2, and H_2S gases were independently controlled and introduced into an illuminated, continuously stirred 800-ml bioreactor containing the green sulfur bacterium *Chlorobium limicola* forma *thiosulfatophilum*. Various real-time monitoring and controlling schemes were developed and incorporated into the software. Emphasis was placed on retarding the oxidative sulfur pathways of the microorganism for the purpose of optimizing product (S^0) recovery.

Introduction

Chlorobium limicola forma *thiosulfatophilum* is a photoautotrophic green sulfur bacterium found in eutrophic lakes and other brackish waters. Van Niel (1931) and Larsen (1952) reported on the unique extracellular sulfur accumulation from pure cultures of this bacterium using sulfide salts and H_2S gas as electron donors. Equations (1) and (2) represent Van Niel and Larsen's reported mass balances for sulfur:

$$2nH_2S + nCO_2 \xrightarrow{h\nu} 2nS^o + n(CH_2O) + nH_2O \tag{1}$$

$$nH_2S + 2nCO_2 + 2nH_2O \xrightarrow{h\nu} nH_2SO_4 + 2n(CH_2O) \tag{2}$$

In later studies (Cork et al. 1982), continuously stirred fed-batch illuminated tank reactors were employed to investigate the unique sulfur metabolism of *C. thiosulfatophilum*. It was found that the illuminated fed-batch reactors were superior to the static culture bottle methods used by Van Niel (1931) and Larsen (1952). It was recognized that hydrogen sulfide space velocity and light were critical factors for the accumulation of elemental sulfur (Cork and Ma 1982). Strict gas control was necessary for such studies, and different real-time methods were required. In order to show effects of space velocities on subsequent sulfur mass balances, computerized hardware and software were developed.

MATERIALS AND METHODS

Fed-Batch CSTR and Computer Hardware

Fig. 1 is a schematic of the reactor used for most of the experimental trials. The reactor contained 800 ml of liquid defined growth medium, and was illuminated at 450 or 2,000 watts/m². The gases used were all 100 mol percent, anaerobic grade H_2S, CO_2, and N_2. The gas feed system consisted of Tylan mass flow-type gas controllers (Tylan Co., Carson, CA) that were interfaced to an Apple IIe-ISAAC microcomputer data acquisition and control package (Cyborg Co., Newton, MA), using 0-5 VDC analog signals to control the flow rate. Details on the interfacing of the valves, the reactor pH and temperature control have been described previously (Cork et al. 1982; Mathers and Cork 1984).

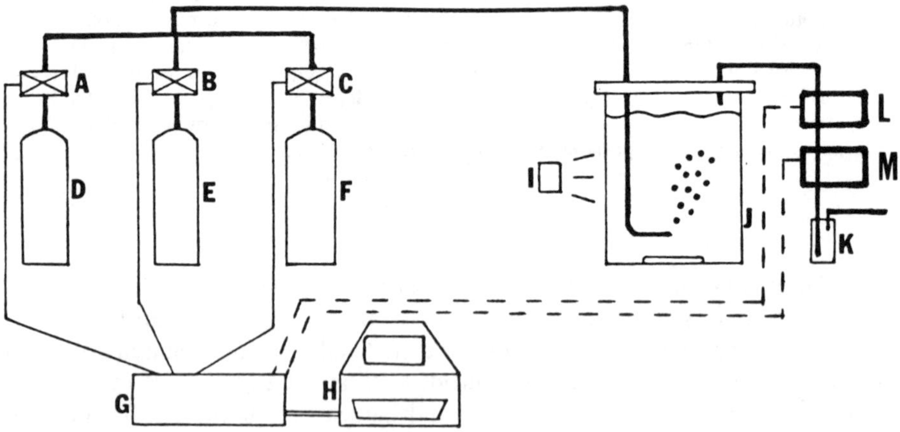

FIG. 1. Laboratory fed-batch reactor. A,B,C: Tylan mass flow type gas controllers. D,E,F: CO_2, N_2, and H_2S gas cylinders. G,H: ISAAC 41A/Apple IIe computer system. I: light source. J: 800-ml liquid volume reactor vessel. K: zinc acetate: trap for H_2S. L: Infrared CO_2 off gas analyzer. M: Electrochemical H_2S off gas analyzer. Dashed lines represent the feedback signals for valve control by the Apple/ISAAC system.

The microcomputer system offered advantages over minicomputer systems in that it offered 12 bit resolution of all signals for the least cost and most space savings. The microcomputer system was set up on an ordinary lab cart next to the reactor. Although the higher priced minicomputer systems offered higher resolution and quicker sampling times, for a small-scale bioreactor these are unnecessary. The proprietary LABSOFT language that is included in the ISAAC package is well documented and is an extension of the Applesoft BASIC.

Experimental Design and Software Implementation

For the investigation of sulfur metabolism, it was known that H_2S space velocity into the reactor was an important factor in the regulation of the oxidative sulfur pathways. H_2S and CO_2 are very soluble gases at a controlled pH of 6.8, used in all trials (N_2 gas is relatively insoluble and was used as a higher velocity, inert carrier gas). The H_2S was completely consumed in all experimental trials, while CO_2 was usually in excess, therefore nonlimiting. All soluble sulfide concentrations were less than 6 m*M*. Concentrations above this level caused low growth rates and sulfide assimilation by the microorganisms. For all experiments, the gas was varied by either a feedback loop or a timed gas flow scheme. Photons were assumed to be limiting due to high bacterial concentrations (ca. 10^9 cells/ml).

Results

Each experiment was blocked off into flow charts in order to translate it to BASIC coding. The code was then written in Applesoft and LABSOFT, and then tested using "dry runs" in order to debug, modify, and insert as many "failsafe" structures as possible to prevent program crashing or dangerously high gas flow rates from occurring. The actual experimental trials were then run, and timely data logs were printed out during the course of each experiment. Figs. 2–4 are the flow charts developed for the controlled-gas experiments; Figs. 2 and 3 show feedback control of the H_2S flow based on detected off gas CO_2 and H_2S levels. Fig. 4 shows a time pulsing of the H_2S, used to show dynamic regulation and deregulation of the oxidative sulfur pathways due to varying soluble $S^=$ levels caused by such pulsing.

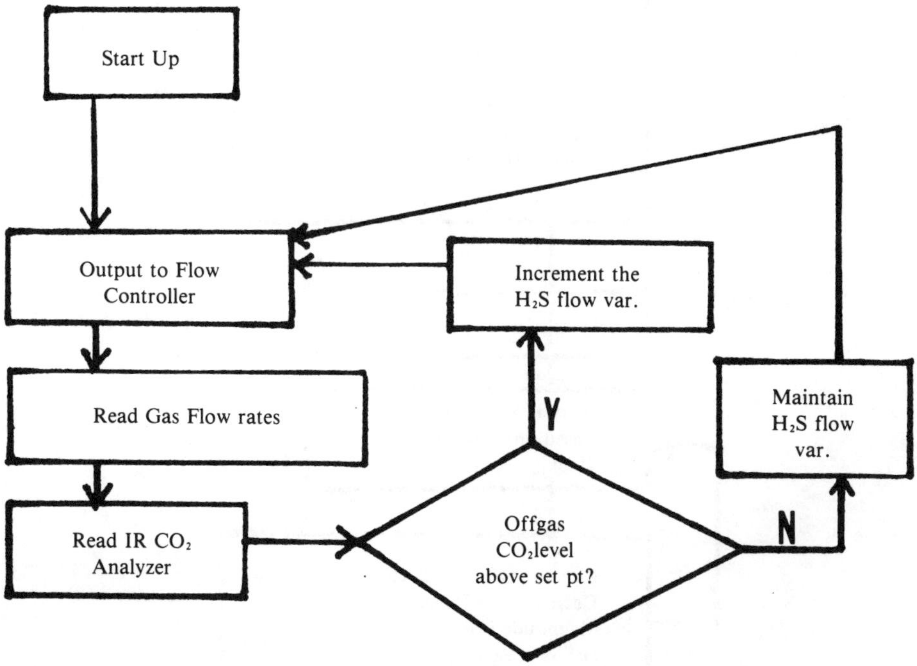

FIG. 2. Algorithm to detect off gas CO_2, with feedback control of H_2S to optimize CO_2 fixation at given H_2S flow rate.

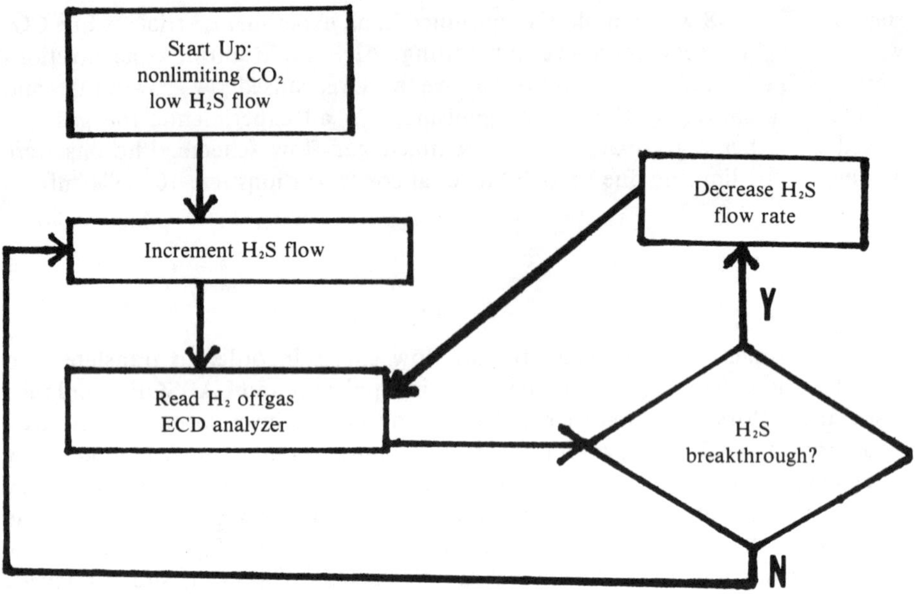

FIG. 3. Algorithm to detect off gas H_2S and adjust H_2S flow to the critical flow rate level required for optimization of S^o formation by *C. thiosulfatophilum*.

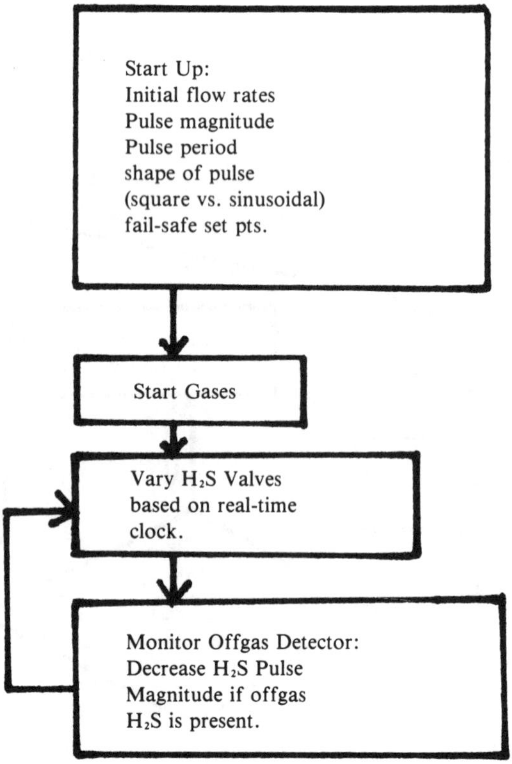

FIG. 4. Algorithm for pulsing H_2S gas. Designed to test the dynamic control of sulfur oxidation by varying the H_2S space velocity.

Discussion

The microcomputer-based gas flow system is a good alternative to larger, more sophisticated computer control systems. It is much more versatile than simple, dedicated boxes in that one may design software around a particular experiment and include integrated real-time and feedback features as well as data logging. Programming is in a higher level language (BASIC), which is easier for beginning or intermediate programmers than assembler code. For our particular application using green sulfur bacteria, strict control of the H_2S, CO_2, and N_2 gases at flow rates of 0.1–10 cc/min was accomplished.

Literature Cited

Cork, D. J. 1982. Acid waste gas bioconversion—an alternative to the claus desulfurization process. *Dev. Ind. Microbiol.* 23:379–387.

Cork, D. J., and S. C. Ma. 1982. Acid gas bioconversion favors sulfur production. *Biotechnol. Bioeng.* 12:285–291.

Cork, D. J., R. Garunas, and A. Sajjad. 1982. *Chlorobium limicola* forma *thiosulfatophilum:* Biocatalyst in the production of sulfur and organic carbon from a gas stream containing H_2S and CO_2. *Appl. Environ. Microbiol.* 45:913–918.

Larsen, H. 1952. On the culture and general physiology of the green sulfur bacteria. *J. Bacteriol.* 64:187–196.

Mathers, J., and D. J. Cork. 1984. Microcomputer interfacing to mass flow type gas controllers for biotechnology applications. *Biotechnol. Lett.* 6(2):87–90.

Van Niel, C. B. 1931. On the morphology and physiology of the purple sulfur bacteria. *Arch. Mikrobiol.* 3:1–8.

CHAPTER **48**

Growth and Biocide Efficacy Studies Using the Iron-Oxidizing Bacterium *Gallionella*

JOHN W. WIREMAN

Biosan Laboratories, Inc., Ferndale, Michigan 48220, and Department of Biological Sciences, Wayne State University, Detroit, Michigan 48202

Procedures for evaluating biocides against the iron-oxidizing bacterium *Gallionella* have been developed. Biocide efficacies were evaluated on a consortium of *Gallionella* and two dominant pseudomonads. The optimum growth conditions for *Gallionella* were found to be an initial pH of 4.2 and a temperature of 15 ± 2 C. The amount of carbon dioxide added was not a critical variable except for the effect of CO_2 on pH. The concentration of heterotrophic pseudomonads also increased approximately 100-fold during a one- to two-week incubation in this minimal salts medium. Quantitation of biocide efficacy on *Gallionella* as well as the pseudomonads was done. The critical variable in biocide evaluations against *Gallionella* involves the time of biocide addition relative to the time of growth initiation. Of the two biocides evaluated, the formaldehyde condensate biocide was more effective against *Gallionella* when added early, whereas the quaternary ammonium biocide was equally effective at all times of addition.

INTRODUCTION

The iron-oxidizing bacterium *Gallionella* sp. is commonly found in iron-rich waters, particularly in cooler environments (Kucera and Wolfe 1957; Hanert 1981). The members of this genus are morphologically very complex with a small kidney-shaped cell attached to a long ribbon-like stalk (Kucera and Wolfe 1957; Hanert 1970). Recent analysis of stalk structure has determined the primary elements to be silicon, aluminum, calcium, and iron (Ridgway et al. 1981).

This organism appears to be one of the few obligate chemolithotrophs that is capable of deriving all of its energy requirements from the oxidation of ferrous iron and deriving its carbon from carbon dioxide (Hanert 1981; Buchanan and Gibbons 1974). The oxidation of ferrous iron to ferric iron results in the precipitation of ferric hydroxide; thus, the growth of *Gallionella* is typified by reddish mats of cells and ferric hydroxide encrusted stalks.

The growth of *Gallionella* in iron pipes is a problem in corrosion as well as clogging owing to the accumulation of cell-stalk masses (Miller and King 1975; Miller 1981; Ridgway et al. 1981). It would be important to determine the requirements for control of *Gallionella* growth in environments where corrosion and clogging are a significant problem (e.g., underground pipes used for oil field injection waters or corrosion of metal surfaces that are difficult or impractical to replace).

This report presents the results of an evaluation of two biocides against *Gallionella* in a first attempt to develop appropriate methodologies for chemical control of this bacterium.

Materials and Methods

Isolation of Gallionella. The strain of *Gallionella* used for these studies was isolated from a dirt-floor basement in Macomb County, Michigan. This basement is approximately one mile from a 10-year-old landfill. The ground water height as well as direction of flow indicate that seepage from the landfill has contaminated the ground water supply and provided a rich and diverse environment for the growth of microorganisms. During the fall of 1983, the ground water level was near the soil surface, and the south wall and the floor of the basement were covered with a reddish layer covering black mud. Microscopic examination revealed that the reddish layer was composed predominantly of *Gallionella*. The ribbon stalks and ferric hydroxide particles were a significant portion of the total mass of the layer. The blackish mud layer contained high levels of metal sulfides, as determined by chemical analysis for sulfide. Samples of the upper reddish layer were used as the primary inoculum for the studies reported here.

Growth of Gallionella. The growth medium for culturing of *Gallionella* was a modification (Hanert 1981) of the medium of Kucera and Wolfe (1957). This modified medium contains 0.005% $K_2HPO_4 \cdot 3H_2O$ rather than the original 0.05% (Hanert 1981). This medium is a minimal salts medium with commercially available granular ferrous sulfide (Baker Chemical Co.). This product was found to be an excellent source of ferrous ions (hydrogen sulfide free) for the growth of *Gallionella*.

Ten ml of modified medium was added to 16-mm, screw-cap tubes and autoclaved. Sterile, filtered carbon dioxide was bubbled through the medium for 5 s to provide carbonate ions in solution. Approximately 0.3 g of ferrous sulfide was then added. The pH of the medium after 5 s of CO_2 addition was approximately 4.3. The tubes were incubated in a tap-water bath at 14–17 C. Flocculent *Gallionella* growth was observed on the sides of the tubes within 4 to 7 d (Fig. 1). The reddish material was stored at 3 C and inoculated into fresh medium at weekly intervals for 12 wk. Even after 12 wk of storage, viable *Gallionella* cells were easily detectable in the stored sample.

Transfer of Gallionella *cultures*. One- to two-week-old cultures of *Gallionella* (Fig. 1B) were used for preparation of test inoculum or for transfer. The tubes were rinsed five times with sterile deionized water and then resuspended in 10 ml of water by scraping the sides of the tubes with a pipet. The standard inoculum was 0.1 ml of such a suspension (Fig. 1B).

Experiments to trace the survival of heterotrophs (Fig. 3) involved a standard plate count of the supernatant of the *Gallionella* culture tube on plate count agar (DIFCO Laboratories) and a count of the attached microorganisms that were removed along with *Gallionella* by scraping the sides of the culture tubes.

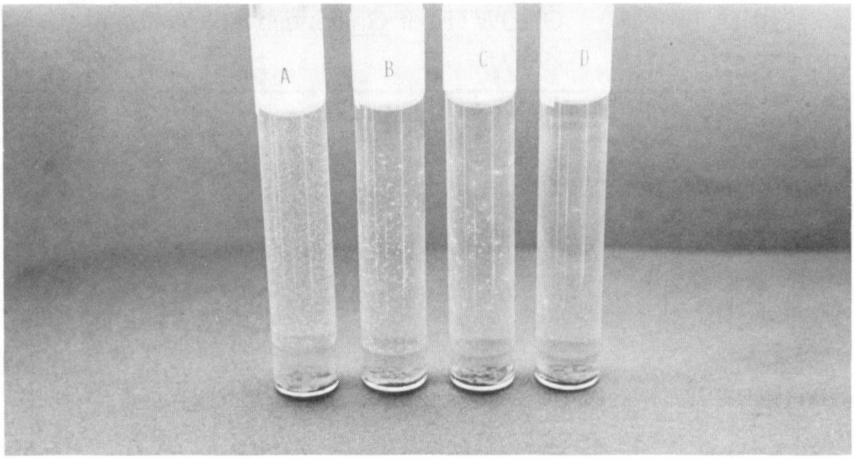

FIG. 1. Culture tubes of *Gallionella* after 10 d of incubation. Tube A was inoculated with 1 ml of a standard 7-d culture, and Tubes B thru D were 1:10 serial dilutions of this inoculum. The standard inoculum used for all testing is equivalent to Tube B. Macroscopic *Gallionella* colonies are visible on the sides of the culture tubes.

Results and Discussion

In order to evaluate the efficacy of biocides against *Gallionella*, the growth conditions were first optimized. The four parameters tested were pH, amount of CO_2 added, amount of FeS added, and temperature.

pH. Previous workers have reported that the growth of *Gallionella* should be optimal at a pH value between 5.5 and 7.6 (Hanert 1981; Kucera and Wolfe 1975; Nunley and Krieg 1968). The results of the experiment shown in Fig. 2 indicate that the strain of *Gallionella* used for these experiments has a broad pH range for growth from $\leq$pH 4.2 to pH 5.2. No growth was observed at pH 6.4, and growth was poor between a pH value of 5.2 and 6.4. The reason for the difference in these observations and previous literature reports is not apparent, but the strain of *Gallionella* used here not only prefers a low pH but also appears to be an acidophile. Bubbling CO_2 from 5 to 15 s into a standard tube results in a pH of 4.4 to 4.1.

Amount of CO_2 added. The amount of CO_2 was varied in several experiments from none added to 15 s of continuous bubbling per 10 ml of medium. Some small *Gallionella* colonies near the bottom of the culture tube were observed in the absence of CO_2 addition if the pH was decreased to 4.5 by the addition of HCl. Strong stimulation of *Gallionella* growth resulted from a small amount of CO_2 addition, but essentially no growth differences were observed between 3 s and 15 s of CO_2 bubbling. The critical variable appeared to be pH (Fig. 2) rather than the amount of CO_2 added. No growth occurred regardless of the amount of CO_2 added if the pH was increased to a value of 6.5 or higher by the addition of NaOH.

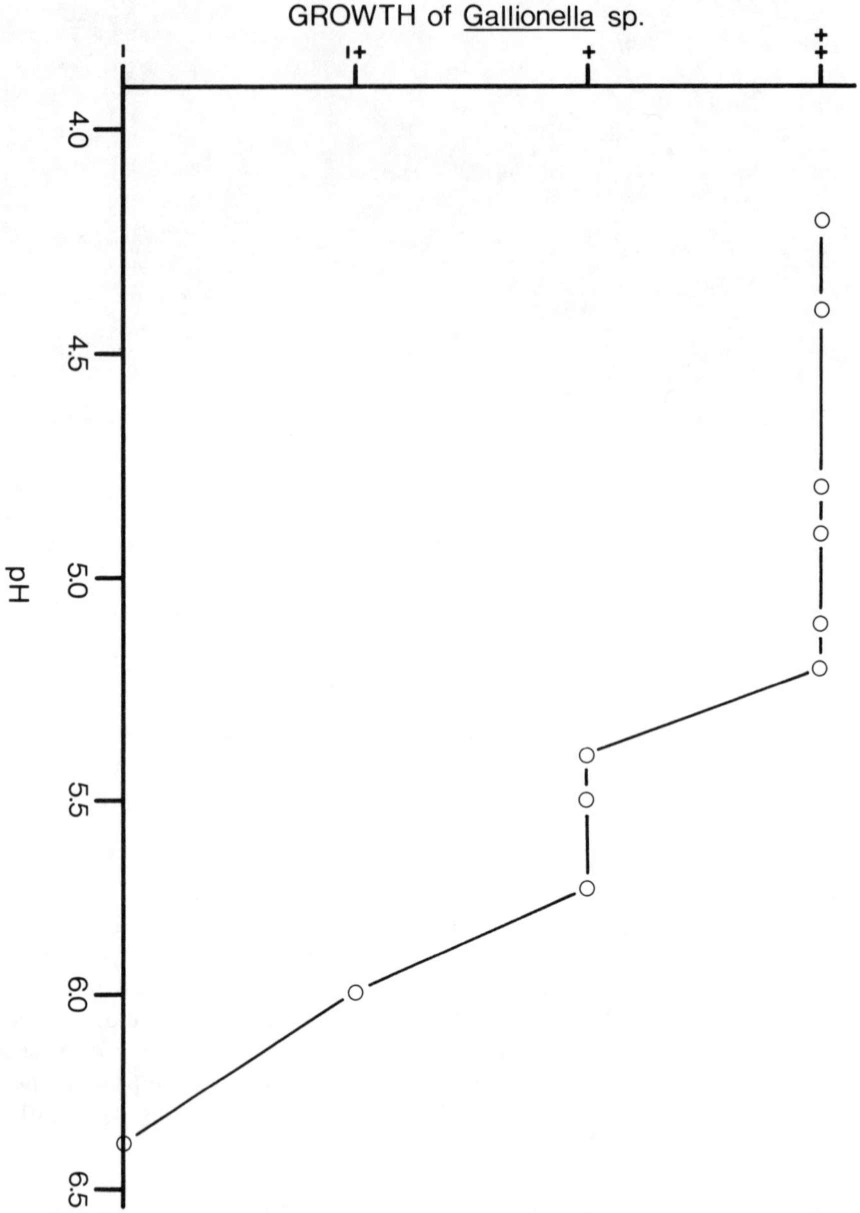

FIG. 2. The effect of pH on the growth of *Gallionella*. The results of three separate experiments are shown. The culture media contained varying amounts of carbon dioxide and the pH was adjusted with either NaOH or HCl. There was no effect on growth because of the amount of CO_2 added except as it affected pH.

KEY:

++ = heavy growth. Equivalent to Fig. 1A or 1B.
\+ = moderate growth. Less than or equal to Fig. 1C.
± = Poor growth. Equivalent to Fig. 1D.
− = No detectable colonies.

Amount of FeS added. No growth was observed in standard cultures in the absence of FeS. There was no observable difference in growth in cultures that contained between 0.05 g FeS and 0.5 g; therefore, approximately 0.3 g FeS was added to all standard test cultures.

Temperature. Four temperatures were used for testing the effect of temperature on *Gallionella* growth. The composite results of several tests are shown in Table 1. The strain of *Gallionella* used for these experiments is a psychrotroph similar to some of the strains previously used by others (Hanert 1981), whereas other strains have been reported to grow better at 25 C than at 20 C or lower (Kucera and Wolfe 1957). The incubation temperature used for all test cultures was 15 ± 2 C (tap water bath).

TABLE 1. *Effect of temperature on the growth of* Gallionella

Temperature	Incubation Conditions	7-Day Growth Observation[a]
3–5 C	Refrigerator	No growth
15–17 C	Tap water bath	200–300 Colonies per tube
22–24 C	Room temperature	100 Colonies per tube
29–30 C	Incubator	No growth

[a] No change occurs in the number of *Gallionella* colonies after 7 d.

Survival and growth of the heterotrophic members of the consortium. The microbial flora in the reddish layer of the dirt floor basement was partially characterized. At least 10 different morphological types of bacteria and 3 fungi were present. The black mud under the red layer contained at least 10^7 anaerobic sulfate reducers per cm^3. The environment from which *Gallionella* was isolated thus contained a rich heterotrophic microbial flora as well as anaerobic sulfate-reducing bacteria.

Hanert (1981) stated that he achieved pure cultures of *Gallionella* by repeated transfers of single washed colonies of *Gallionella.* For this method to be successful, some selection for growth of *Gallionella* versus attached heterotrophic bacteria would be needed; it is unlikely a single *Gallionella* cell would be obtained free of other bacteria unless there had been little or no growth of the initial heterotrophic contaminants during the week of incubation (Hanert 1981).

The results of an experiment to initially test this idea are shown in Fig. 3. Over the course of 10 wk, the concentration of heterotrophic co-contaminants present in the liquid phase of the tube (supernatant) as well as the heterotrophs that remain attached were determined. Initially (Fig. 3), there were less than 10^3 co-contaminants/ml introduced along with *Gallionella* in a diluted suspension of the primary inoculum (from the red layer described above). After the first 2 wk, there were 10^5 supernatant contaminants/ml and almost 10^6 attached co-contaminants/cm^2 of glass surface. After 2 wk, and at all subsequent time points, the resuspended attached heterotrophic population (about 10^6 cfu/ml) and approximately 10^4 *Gallionella* cfu/ml were diluted 1:100 during transfer. Even after repeated transfers and 100-fold dilutions, the supernatant and attached population grew to achieve the original population density (10^5 to 10^6 cells per ml or per cm^2).

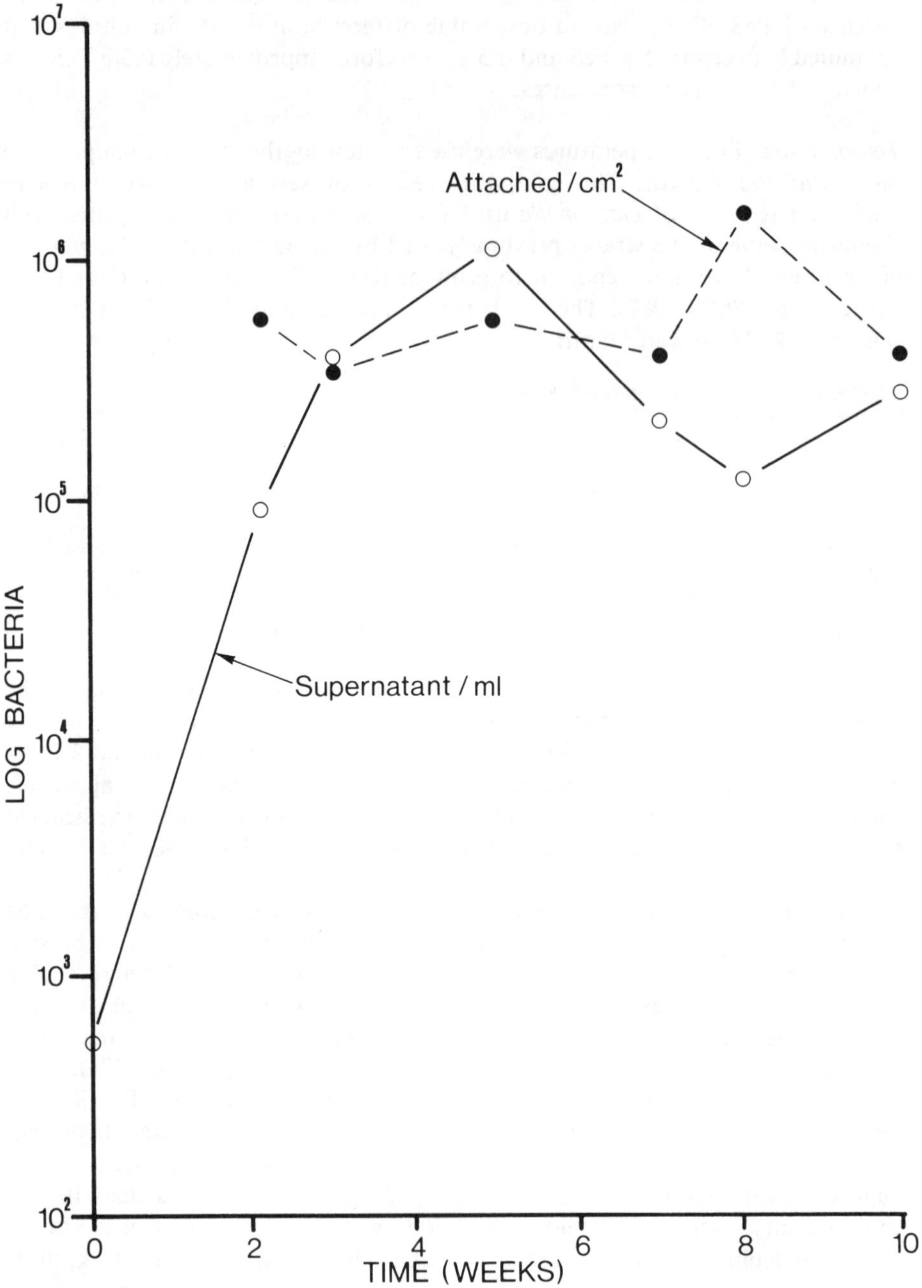

FIG. 3. Growth of heterotrophic co-contaminants during sequential transfer of *Gallionella* cultures. Supernatant counts/ml (○——○); attached counts/cm² (●——●).

Thus, this minimal medium containing only CO_2 as an added carbon source does support the growth of heterotrophs. The nature of this population changed during the repeated transfers shown in Fig. 3 until only two dominant bacteria were present and no fungi were detectable. The same two organisms that were found in the supernatant were found attached to the glass surface. Biochemical tests of these two organisms have not been successful in determining a species identification. They are oxidase-positive glucose nonfermentors and thus are pseudomonas-like but not yet identified.

Effect of biocides on Gallionella. The culture methods described above were used to evaluate two biocides against *Gallionella*. A minimal inhibitory concentration (MIC) type experiment did not seem appropriate since there are at least three parameters of *Gallionella* growth and survival that are valid to test. These three parameters are as follows:

(1) *Biocide added at time zero*. This method is equivalent to a standard MIC experiment and tests the effect of the biocide on *Gallionella* prior to attachment. An apparently effective concentration of biocide may have prevented attachment rather than been biocidal to the cells. The effect of the biocide is evaluated after 14 d of incubation.

(2) *Biocide added after 3 d of incubation*. This method tests the effect of the biocide on *Gallionella* independent of attachment. Growth is visible after about 4 d of incubation; therefore, attachment and the initiation of growth have occurred after 3 d of incubation. The effect of the biocide is evaluated after 14 d of incubation.

(3) *Biocide added after 7 d of incubation*. This method tests the ability of the biocide to kill fully developed *Gallionella* colonies. In order to evaluate this effect, biocide is added after 7 d and the culture tubes are then incubated an additional 7 d. (There is no visible effect of biocides on fully developed colonies.) The tubes are then rinsed five times with sterile water to remove the biocide, and the surface colonies are resuspended and survival tested by transfer to fresh media.

The results of such an experiment using a formaldehyde condensate biocide are shown in Fig. 4. The effectiveness of the biocide decreases as attachment and growth of *Gallionella* proceeds. The concentration of biocide needed to kill fully developed *Gallionella* colonies is two times the concentration required to kill cells prior to attachment (Fig. 4). The effect of the biocide was also evaluated against the heterotrophs present in the culture tubes. There was no effect on their survival at a biocide concentration of 50 ppm or less regardless of the time of biocide addition, but 100 ppm (not shown in Fig. 4) was significantly inhibitory.

A second biocide evaluated was a quaternary ammonium biocide. Unlike the formaldehyde condensate biocide, this biocide was effective at 1 ppm regardless of the time of addition (data not shown). Lower concentrations of this biocide (ppb) may show a similar time dependence of addition as did the formaldehyde condensate biocide. The differences observed in the effectiveness of these two biocides, however, may also suggest that a difference in test methods may be appropriate.

This report of a test method for evaluating biocides against *Gallionella* is preliminary. Much remains to be known about this complex bacterium and how best to control its growth in inappropriate environments. For example, the growth of *Gallionella* on metal surfaces should be examined. These initial studies were done using a glass surface, but corrosion and clogging due to *Gallionella* are

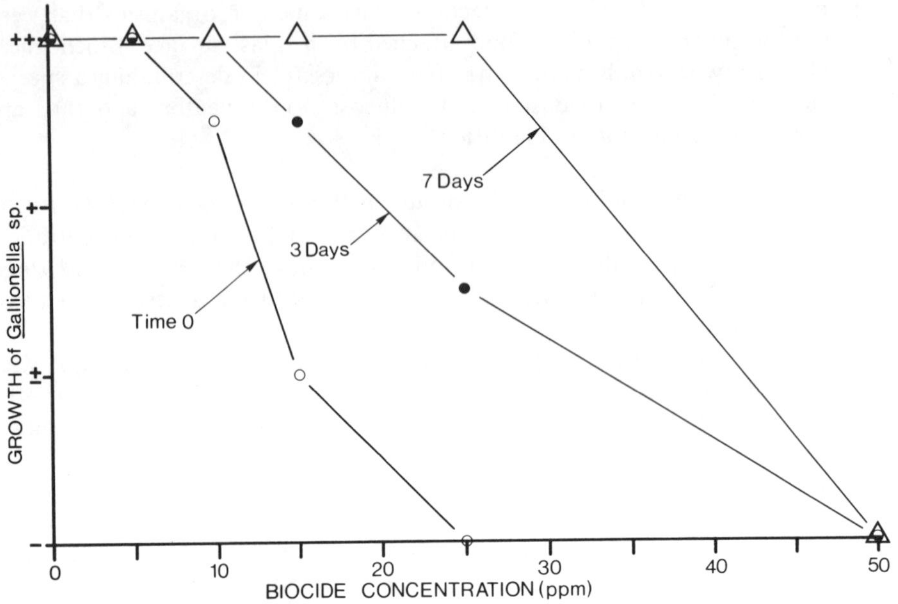

FIG. 4. Evaluation of a formaldehyde condensate biocide against *Gallionella*. The biocide was added to duplicate tubes at time zero or after 3 or 7 d of incubation. The time zero and 3-d experiments were evaluated for *Gallionella* growth after 14 d. The 7-d experiment involved 7 d of growth followed by 7 d of biocide treatment. This condition was then evaluated by determining the ability of the treated colonies to recolonize and grow in fresh media in the absence of biocide.

not likely to result from growth on glass. In addition, the parameters that are required to replace the requirement for ferrous ions provided by FeS with ferrous ions from an iron surface are not yet understood. Preliminary attempts to replace FeS with iron wire have been unsuccessful.

Gallionella is not uncommon in cold iron-rich waters, yet its growth in the laboratory is difficult to achieve. These bacteria are microaerophiles and grow best at low oxygen concentrations. The sometimes unavoidable variations in oxygen concentration were observed to result in growth as a narrow band of colonies near the bottom, middle, or top of the culture tube. This fastidious requirement for oxygen as well as the requirement for soluble ferrous iron suggests that growth of *Gallionella* in natural habitats may be enhanced by the presence of other microorganisms. For example, an iron-reducing *Pseudomonas* sp. (Obuekwe et al. 1981) might maintain a continuous supply of ferrous ions at low oxygen concentrations. Although there is no evidence for direct or indirect synergism between *Gallionella* and other microorganisms, the possibility exists that such interactions do occur and enhance the colonizing and survival capacity of *Gallionella*.

Thus, it may be appropriate for biocide test methods to include this possibility. If a biocide were effective against a synergistic heterotroph but not *Gallionella*, the laboratory test could incorrectly conclude that control of *Gallionella* during field use will be achieved. The biocides evaluated here using a mixed culture were more effective against *Gallionella* than against the pseudomonad members of the consortium. This result is not necessarily a general rule nor should it be so interpreted until a greater understanding of the growth of *Gallionella* in natural habitats is achieved.

It has not been possible during these studies to achieve a pure culture of *Gallionella* (for example, by the use of formalin; Nunley and Krieg 1968). Should biocide evaluations be done only with pure cultures of *Gallionella*? Pure culture studies are invaluable for microbiology, but they should not be used when conclusions from such studies may be misleading.

Acknowledgments

The author wishes to acknowledge the suggestion of Joe Oraveck to use commercially available granular ferrous sulfide and Leonard A. Rossmoore for his helpful advice and encouragement.

Literature Cited

Buchanan, R. E., and N. E. Gibbons, eds., 1974. *Bergey's Manual of Determinative Bacteriology,* 7th ed., pp. 160–161. Williams and Wilkins, Baltimore, MD.

Hanert, H. H. 1970. Structure and growth of *Gallionella ferruginea* during the first six hours of development. *Arch. Mikrobiol.* 75:10–24.

______. 1981. The genus *Gallionella*. Pages 509–515 *in* M. P. Starr, H. Stolp, H. G. Truper, A. Ballows, and H. G. Schlegel, eds., *The Prokaryote: A Handbook on Habits, Isolation, and Identification of Bacteria.* Springer-Verlag, Berlin, W. Germany.

Kucera, S., and R. S. Wolfe. 1957. A selective enrichment method for *Gallionella ferruginea. J. Bacteriol.* 74:344–349.

Miller, J. D. A. 1981. Metals. Pages 149–202 *in* A. H. Rose, ed., *Microbial Biodeterioration.* Academic Press, London.

Miller, J. D. A., and R. A. King. 1975. Biodeterioration of metals. Pages 83–103 *in* D. W. Lovelock and R. J. Gilbert, eds., *Microbial Aspects of the Deterioration of Materials.* Academic Press, London.

Nunley, J. W., and N. R. Krieg. 1968. Isolation of *Gallionella ferruginea* by use of formalin. *Can. J. Microbiol.* 14:385–389.

Obuekwe, C. O., D. W. S. Westlake, and F. D. Cook. 1981. Effect of nitrate on reduction of ferric iron by a bacterium isolated from crude oil. *Can. J. Microbiol.* 27:692–697.

Ridgway, H. F., E. G. Means, and G. H. Olson. 1981. Iron bacteria in drinking-water distribution systems: Elemental analysis of *Gallionella* stalks, using X-ray energy-dispensive microanalysis. *Appl. Environ. Microbiol.* 41:288–297.

CHAPTER 49

Characterization of Sulfate-Reducing Bacteria Isolated from Oilfield Waters

KATHLEEN M. ANTLOGA AND W. MICHAEL GRIFFIN

Sohio Petroleum Company, Warrensville Production Research Laboratory, Cleveland, Ohio 44128

Eight strains of sulfate-reducing bacteria, based on colony morphology, were isolated from produced water from an oilfield. The organisms were identified as two strains of *Desulfovibrio desulfuricans, D. africanus,* two strains of *Desulfotomaculum nigrificans,* and three strains of *Desulfotomaculum* sp. The organisms were identified on the basis of substrate utilization, growth temperature, presence or absence of desulfoviridin and flagella, colony, cell and spore morphologies. This work illustrates the wide variety of microorganisms loosely lumped together as the sulfate-reducing bacteria; it also illustrates that model systems (i.e., biocide tests, biodegradation studies, corrosion studies, etc.) using *Desulfovibrio desulfuricans* may underestimate the metabolic and physiological diversity of these organisms.

INTRODUCTION

Sulfate-reducing bacteria (SRB) appear to be indigenous inhabitants of oilfield waters (Davis 1967; Hamilton 1983; Moses and Springham 1982; Postgate 1979; TPC Publication No. 3 1976). SRB, although obligate anaerobes, survive in a wide variety of oxygenated environments. Their normal metabolism is based on the reduction of sulfate, which serves as an electron acceptor, to sulfide. Oilfield reservoir formation waters commonly contain sulfate at high enough levels to support SRB growth. Further information on the physiology of SRB can be obtained from a book on SRB by Postgate (1979).

The SRB are well known for their ability to cause oilfield production problems and can pose a potential health risk to operators via H_2S production. Even at very low concentrations (10 ppm) H_2S can be poisonous to humans (Hawley 1971). In addition, H_2S can react with ferrous ions present in produced water to form an iron sulfide precipitate. This precipitate, along with the bacteria themselves, may cause plugging and may stabilize oil/water emulsions making separation of the two more difficult (Davis 1967; Moses and Springham 1982; Wright 1963). Hydrogen sulfide production by SRB can ultimately sour oil and gas leading, in some cases for produced gas, to costly treatment to remove H_2S prior to transport, use and/or sale of the product (Davis 1967). In addition, high numbers of SRB in an oil production system can cause a loss of permeability within a reservoir and can cause plugging of injection faces (Doetsch 1981). Reports indicate that a combination of iron sulfide mixed with bacterial cells (e.g., SRB) and other bacterial metabolic products are one major cause of plugging (Davis 1967; Moses and Springham 1982).

The growth of SRB can lead to a very serious oilfield problem—corrosion. It has been postulated that SRB can utilize cathodic hydrogen and thus accelerate the corrosion process. Some authorities believe that the oil industry experiences considerable losses because of the anaerobic corrosion of iron and steel structures (Wakerley 1979). In 1972 it was reported that, in the United States, approximately \$500 million to \$2,000 million per year is spent for the microbial corrosion of buried pipelines (Tiller 1983). Also reported was the failure of 77% of one group of oil-producing wells because of the phenomenon of microbial corrosion (Tiller 1983). Numerous papers and review articles have been published concerning bacterial corrosion and the different mechanisms that may be involved (Farquhar 1974; Miller 1981; Minchin 1960; Sharpley 1961; Tiller 1983). Although extensive research has been carried out to define the mechanisms of microbial corrosion, a complete answer is still not available. There is hope that losses because of microbial corrosion will diminish over the next decade because of the improvement of protective systems that are being applied by pipeline users.

Most work concerning SRB in the oilfield has been conducted by oilfield service companies providing information as to the levels of SRB, location of the contamination, and possible biocide treatment programs. However, it has become apparent that a more fundamental understanding of SRB physiology and ecology is necessary to provide a rational approach to the use of biocides, facilities design, and a predictive approach to problems associated with enhanced oil recovery methods. Therefore, research designed to isolate and characterize SRB at an oil production facility was undertaken.

Materials and Methods

Cultures. Reference cultures were obtained from the American Type Culture Collection (ATCC). These cultures included *Desulfovibrio desulfuricans* (ATCC 13541) and two strains of *Desulfotomaculum nigrificans* (ATCC 19858 and 19998).

All cultures, both reference and isolated organisms, were stored at 4 C in containers completely filled with SRB medium (see Media), which was poised at −100 mV using Na_2SO_3 (1.9 g/L).

Media. The SRB medium contained: yeast extract (BBL), 1.0 g; NH_4Cl, 0.1 g; $MgSO_4 \cdot 7H_20$, 0.2 g; $FeSO_4 \cdot 7H_20$, 0.1 g; ascorbic acid, 0.1 g; K_2HPO_4, 1.0 g; sodium lactate, 4.2 ml of a 60% solution; $Fe(NH_4)_2(SO_4)_2$, 0.1 g; $Na_2SO_3,S^0{}_4)_2$1.0 g; and distilled H_2O, 1000 ml. The medium was sterilized by autoclaving at 121 C for 25 min. Both the $Fe(NH_4)_2(SO_4)_2$ and Na_2SO_3 were filter-sterilized and added after autoclaving the remaining medium components. When the SRB medium was used for liquid culture, a degreased, dry-sterilized iron nail was added: when used as a solid medium, 15 g or 30 g of Bacto Agar (Difco) was added to 1000 ml of SRB medium for mesophilic or thermophilic culture, respectively.

Facilities description. A schematic diagram of the water-handling systems of the oil production facility surveyed in this study is shown in Fig. 1. The oil-water-gas emulsion that comes out of the wells under pressure flows into the first stage separator to begin the extraction of the water and the gas from the crude oil. Water samples were obtained from the inlets and the outlets of all the major vessels in the produced water-handling facilities. This oil production facility produces substantial oil volumes at 15 to 20% water cut. The approximate composition and some characteristics of this produced water have been determined as follows: pH =7.2; specific gravity = 1.014; total dissolved solids, 25.00 g/L; sodium, 7.70 g/L; calcium, 0.74 g/L; magnesium, 0.60 g/L; iron, 0.005 g/L; potassium, 0.08 g/L; chloride, 15.00 g/L; bicarbonate, 0.40 g/L; sulfate, 0.003 g/L; and suspended solids, 0.002 g/L.

Water sampling. One liter of produced water was taken from various points throughout the oil production facility (Fig. 1). Before a sample was taken the sample port was purged until the temperature of the effluent reached the operating temperature of the vessel being sampled. The water samples were returned immediately to the laboratory for processing.

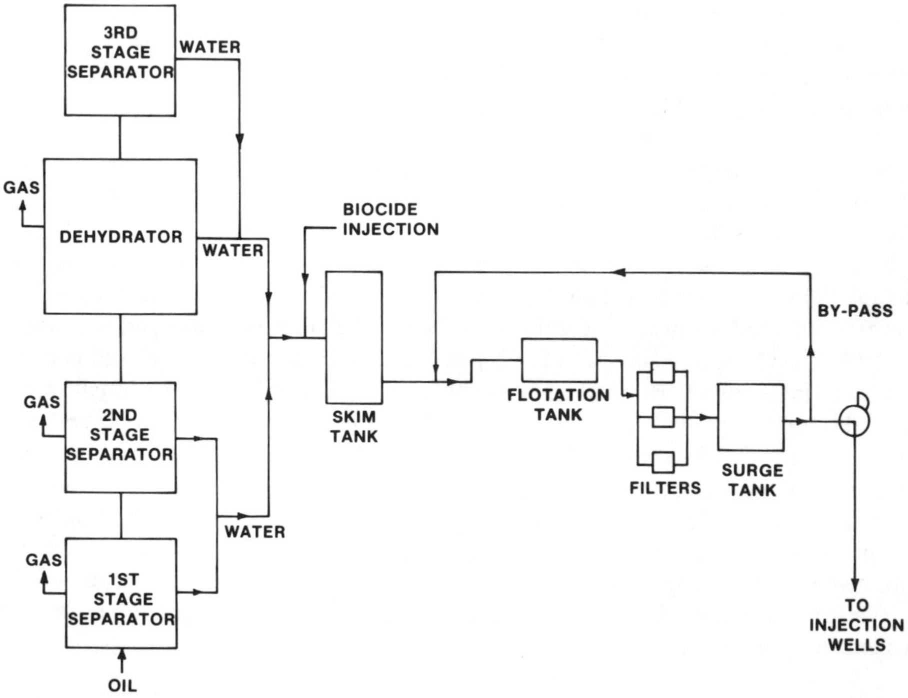

FIG. 1. Schematic diagram of produced water handling facilities.

One milliliter of produced water was transferred to a serum vial containing SRB medium: four replicate samples were performed. Two serum vials were incubated at 30 C, the others at 55 C. The samples were incubated until the medium had a black precipitate, (FeS), indicative of H_2S production due to "SRB" bacterial growth.

Isolation procedure. An aliquot from each culture positive for SRB growth was transferred to solid SRB medium. The cultures were streaked for isolation repeatedly through successive transfers. Each was incubated for at least 1 wk under anaerobic conditions (Brewer Anaerobic Systems, BBL) at the culture's original incubation temperature.

Identification method. All morphological descriptions refer to typical vegetative cells and spores after incubation for 3 to 7 d at 55 C or 30 C, respectively. Cell morphology was determined by phase contrast microscopy and the Gram stain. Flagella staining was carried out using Leifson's method (Doetsch 1981). Spore morphology was determined using the Schaeffer-Fulton method for staining spores (Murry and Robinow 1981), phase contrast microscopy, and transmission electron microscopy. Colony morphology was described according to *Bergey's Manual* (Buchanan and Gibbons 1974). Desulfoviridin presence was determined by the method of Postgate (1979).

The following substrates were tested for their ability to support each isolate's growth as sole carbon source: malate, formate, acetate, and glucose. Each substrate was substituted for lactate in the SRB medium at a concentration based on equivalent number of carbons. Pyruvate was tested as a sole carbon source without the presence of sulfate in the medium ($MgCl_2 \cdot 6H_2O$ and $FeCl_2 \cdot 4H_2O$ were substituted for $MgSO_4 \cdot 7H_2O$, $FeSO_4 \cdot 7H_2O$, $Fe(NH_4)_2(SO_4)_2$, and Na_2SO_3).

Electron Microscopy. Sporulating cells of isolates 4, 5, 6, 7, and 8 and the ATCC *Desulfotomaculum* strains were prepared for transmission electron microscopy by washing cells three times with sterile distilled water (5,000 × g, 20 min, 25 C). The specimens were prefixed for 4 h at 25 C in 5.0% (v/v) glutaraldehyde-formaldehyde buffered with 0.2 *M* sodium phosphate buffer, pH 8.0. After several buffered washes, the specimens were enrobed in sterile agar and postfixed in 1.0% (w/v) buffered osmium for 2 h at 25 C. Each specimen was rinsed in buffer two times then with distilled water to remove the osmium. Specimens were dehydrated in a graded series of acetone, after which the specimens were embedded in Spurrs resin (Spurr 1969). The entire fixation, dehydration, and embedding procedure was carried out at room temperature.

Thin sections were cut on a Sorvall MT 5000 Ultramicrotome with a DuPont diamond knife and retrieved on 300-mesh, copper EM grids. All sections were stained with 0.5% (w/v) uranyl acetate (30 min) and with 0.5% (w/v) lead citrate (10 min) and observed in a Hitachi Transmission Electron Microscope at an accelerating potential of 75 kV.

Results and Discussion

Isolation Sites

A total of 47 produced water samples were screened for sulfate-reducing bacteria (SRB). The samples were taken throughout the oil production facility at various separators, dehydrators, and in and out of all the major vessels in the water plant (Fig. 1). A list of samples yielding the final eight representative isolates are shown in Table 1.

TABLE 1. *Original isolation site for the SRB strains*

Strain No.	Vessel	Pressure (psig)	Temperature F (C)
1	Injection pump	600	140–185 (60–85)
2	Separator (2nd stage)	150	155 (68)
3	Separator (3rd stage)	90	150 (65)
4	Dehydrator	90	150 (65)
5	Skim tank	10–12	135 (57)
6	Cartridge filter	20	135 (57)
7	Skim Tank	10–12	135 (57)
8	Dehydrator	90	150 (65)

The operating temperatures of the vessels from which SRB were isolated (Table 1) suggest that any microorganism isolated from these vessels would be thermotolerant or thermophilic. However, both mesophilic and thermophilic SRB were found throughout the oil production facilities regardless of the operating temperatures of the vessels or lines. The phenomenon of isolating mesophilic microorganisms from environments having temperatures above the mesophile's maximum growth temperature may be because (1) the SRB has considerable adaptability (Miller 1981; Sudbury 1957); (2) the mesophilic microorganisms are thriving in system "cold spots" (dead leg areas); or (3) the combination of temperature and pressure may allow the microorganisms to survive. The three alternatives are not mutually exclusive and there may be a combination of reasons for the mesophilic SRB growth within the system.

The third alternative, temperature and pressure effects on microorganisms, is gaining a great deal of attention by petroleum microbiologists. Traditionally, it was thought that bacteria could "adapt" and grow at high temperatures or high pressures and that a combination of the two conditions was lethal (Marquis 1976). Zobell (1958), on the other hand, isolated SRB from cores from deep oil deposits that survived 104 C and 1000 ATM. In addition, mesophilic bacteria have been isolated from deep ocean vents that are under similar conditions as described by Zobell (1958). Thus a synergism may be postulated that may account for the isolation of the mesophiles from the oil production facilities.

The isolation of mesophilic SRB from thermophilic environments was unexpected. If these organisms are surviving in "cold spots" of the production system, then severe localized damage (corrosion) may occur. If, however, the organisms have "adapted" to high temperatures and pressures or there is a symbiotic relationship between pressure and temperature, then the possibility of widespread damage to the produced water handling systems exists if the SRB are not controlled. The level of SRB in this system is lowered by the use of biocides.

The SRB characterized in this study were isolated from areas under pressure as high as 600 psig (Table 1). No isolation of SRB was done at *in situ* pressures. These SRB appeared to have adapted a wide range of barotolerance within the produced water handling system but it is difficult to say if these barotolerant SRB function well at high pressures. Little is known about the effects of increased pressure on the structure, survival, growth, and biochemical activities of microorganisms (Zobell 1970). Zobell (1970) reported that many environmental factors, such as incubation temperature, chemical composition of the medium, and gas tension all play a part in the growth of microorganisms at high pressures.

It has been mentioned that bacterial life in deep ocean waters is more resistant to the high pressures because of a decrease in cell size. The organisms responded to starvation by decreasing their cell size to develop some type of metabolic dormancy (Marquis 1976). Sulfate-reducing baceteria in the oil production facility may be responding in this same fashion.

Classification of the Isolates as SRB

Any bacterium isolated from produced water was considered an SRB if all of the following statements were true: (1) the isolate was an obligate anaerobe; (2) the isolate grew on SRB medium designed to support SRB growth; and (3) during growth on the SRB medium, the isolate produced H_2S, which was detected by its reaction with iron in the medium to form iron sulfide. All of the isolates were found to be motile gram-negative rods.

Colony and Cell Morphology

After an organism was classified as an SRB, the isolate was characterized by colony and cell morphology. Representatives of each class of colony and cell morphology were chosen for further identification. The descriptions of the cell and colony morphology of the eight representative samples are shown in Table 2.

Postgate (1979) recommended using standard growth conditions when characterizing SRB; examination at 2–6 d for mesophiles and 1–2 d for thermophiles or the nearest practical equivalent. Standard growth conditions are important because SRB morphology can vary in response to age and environment (Postgate 1979). It was found that when culturing on SRB medium, examination of the cells and colonies was best after 7 d incubation for the mesophilic isolates and after 3 d for the thermophilic isolates.

Isolates 1–3 had the curved cell morphologies typical of *Desulfovibrio*. Isolates 4–8 had the straight rod-shaped cell morphologies consistent with those described for members of the genus *Desulfotomaculum*. All of the isolates could be distinguished by one or more features of their colony morphology.

Separation of Desulfovibrio from Desulfotomaculum

In a comprehensive survey by Skyring et al. (1977) only 25 biochemical and physiological characteristics out of 116 tested were found to be taxonomically useful. The classification of SRB has been reviewed and updated by Postgate (1979). Drawing on these reports, 10 characteristics were used to classify the SRB isolates. The characteristics included growth temperature, 30 or 55 C, presence of spores, flagella location, desulfoviridin presence, and carbon substrate utilization. Until recently the state of SRB taxonomy was unsatisfactory, mainly because of the inability to obtain pure cultures and to recognize mixtures of SRB and non-SRB. As the data base for SRB taxonomy increased, it became apparent that there were very few diagnostic characteristics, thus identification of SRB species can be a formidable task.

The information used to classify each SRB isolate as *Desulfovibrio* or *Desulfotomaculum* was based on the characteristic of the *Desulfotomaculum* genus to form spores and the *Desulfovibrio* genus to be nonsporeforming and to possess the desulfoviridin pigment. Spore formation provides an absolute distinction be-

TABLE 2. *Morphology of SRB produced water isolates*

Isolate Number	Cell Morphology[a]		Pigmentation	Form	Colony Morphology[b]		Appearance	Opacity	Size (mm)
	Shape	Size			Elevation	Margin			
1	Curved Rod	1 μm × 3–4 μm	Grey	Circular-Irregular	Convex	Undulate	Glossy	Translucent	0.5–1
2	Curved Rod	1 μm × 3–4 μm	Grey	Irregular-Filamentous	Convex	Undulate-Erose	Glossy	Opaque	0.2–0.5
3	Curved Rod	1 μm × 3–4 μm	Grey	Circular	Convex	Entire	Glossy	Translucent	0.5–1
4	Straight Rod	0.5–1 μm × 3–6 μm	Grey	Rhizoid	Umbonate	Filamentous	Dull	Translucent	1
5	Straight Rod	0.5–1 μm × 3–6 μm	Grey	Circular-Filamentous	Flat	Erose	Dull	Translucent	1–1.5
6	Straight Rod	0.5–1 μm × 3–6 μm	Grey	Rhizoid	Flat	Filamentous-Curled	Dull	Translucent	1–1.5
7	Straight Rod	0.5 μm × 3–6 μm	Grey	Irregular	Convex	Erose	Glossy	Translucent	1.5
8	Straight Rod	0.5 μm × 5–4 μm	Grey	Irregular	Umbonate	Erose	Glossy	Translucent	2

[a] Phase contrast examination with 7 d cultures of strains 1–6 or 3 d cultures of strains 7 and 8.
[b] For 7 d cultures of strains 1–6 or 3 d cultures of strains 7 and 8 grown on basic SRB agar.

tween the genera *Desulfotomaculum* and *Desulfovibrio* (Postgate 1979). The desulfoviridin test is considered a reliable test for members of the genus *Desulfovibrio* in that the presence of the pigment indicates *Desulfovibrio,* but the absence does not rule the organism out. Some authors (Miller and Saleh 1964; Rosanova and Nazina 1976) have found strains of *Desulfovibrio* that lack this characteristic.

Desulfovibrio

The results of the identification of the *Desulfovibrio* isolates are shown in Table 3. The initial classification of isolates 1–3 as *Desulfovibrio* was justified since they grew at mesophilic temperatures (30 C), did not form spores, and possessed the pigment desulfoviridin. Isolates 1 and 3 were identified as *Desulfovibrio desulfuricans* according to Postgate's (1979) working classification of SRB. These two isolates differed only slightly in colony morphology, and Isolate 1 demonstrated the ability to use glucose as a substrate while Isolate 3 did not (Table 3). Postgate (1979) reported *Desulfovibrio desulfuricans* growth with glucose plus sulfate as plus or minus (±). Both isolates were very similar to the *Desulfovibrio desulfuricans* ATCC 13541 and differed in only one trait: the isolates could use formate while the ATCC stain could not grow on the compound. The SRB classification key (Postgate 1979) also lists growth with formate plus sulfate as plus or minus (±) for *Desulfovibrio desulfuricans.* Isolate number 2 was identified as *Desulfovibrio africanus,* possessing the characteristic lophotrichous flagella and the ability to grow with lactate, malate, and formate. This isolate also grew with glucose plus sulfate, although Postgate (1979) mentioned that this species characteristic has not been reported positive or negative.

TABLE 3. *Identification of* Desulfovibrio *strain*

	SRB Isolate Number			
	1	2	3	13541[a]
Growth temperature	30 C	30 C	30 C	30 C
Spores	—	—	—	—
Flagella	Polar	Lophotricous	Polar	Polar
Desulfoviridin	+	+	+	+
Carbon Compound Utilization				
Lactate[b]	+	+	+	+
Malate[b]	+	+	+	+
Formate[b]	+	+	+	−
Acetate[b]	−	−	−	−
Glucose[b]	+	+	−	+
Pyruvate[c] w/o $SO_4^{=}$	+	−	+	+
Identification:	*Desulfovibrio desulfuricans*	*Desulfovibrio africanus*	*Desulfovibrio desulfuricans*	*Desulfovibrio desulfuricans*

[a] American Type Culture Collection (ATCC) strain number of *Desulfovibrio desulfuricans.*
[b] Growth medium contained carbon compound plus sulfate.
[c] Growth medium contained carbon minus sulfate.

Desulfotomaculum

Isolates were classified as *Desulfotomaculum* on the basis of spore formation. However, upon close investigation, two types of spore morphology were found: (1) round, produced always at the terminal end of the cell (Fig. 2a, b, c), and (2) oval, with concomitant endospores located centrally to subterminally (Fig. 3a, b).

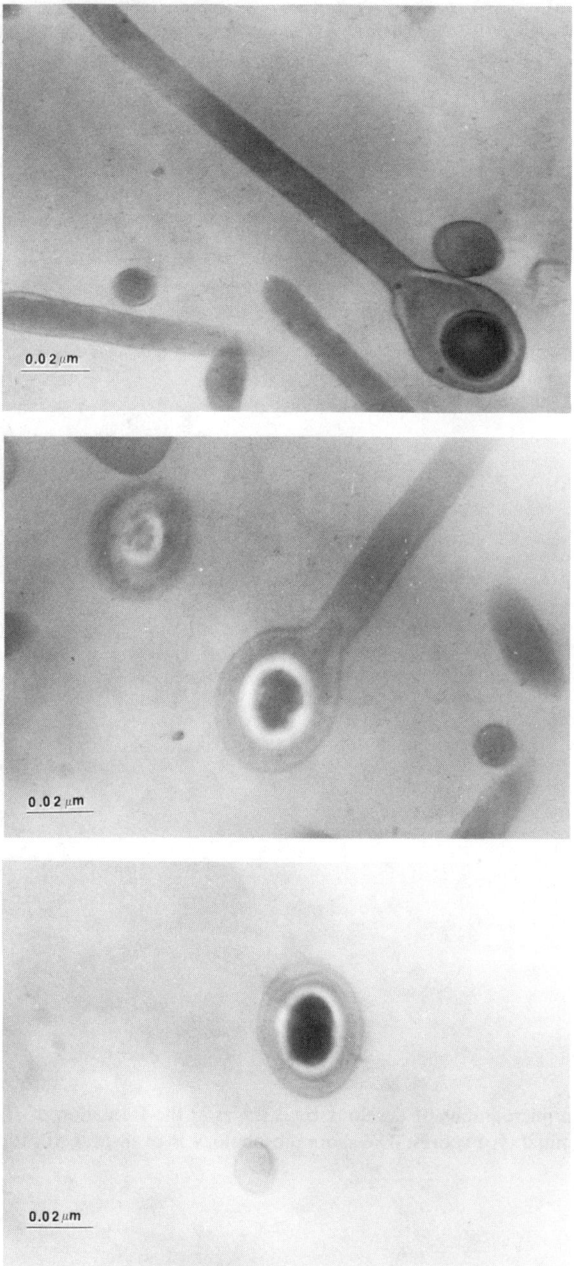

FIG. 2a, b, c. Electron micrographs of developmental stages in the formation of *Desulfotomaculum* terminal type spores. The spore morphology is round. Magnification bars represent 0.02 μm.

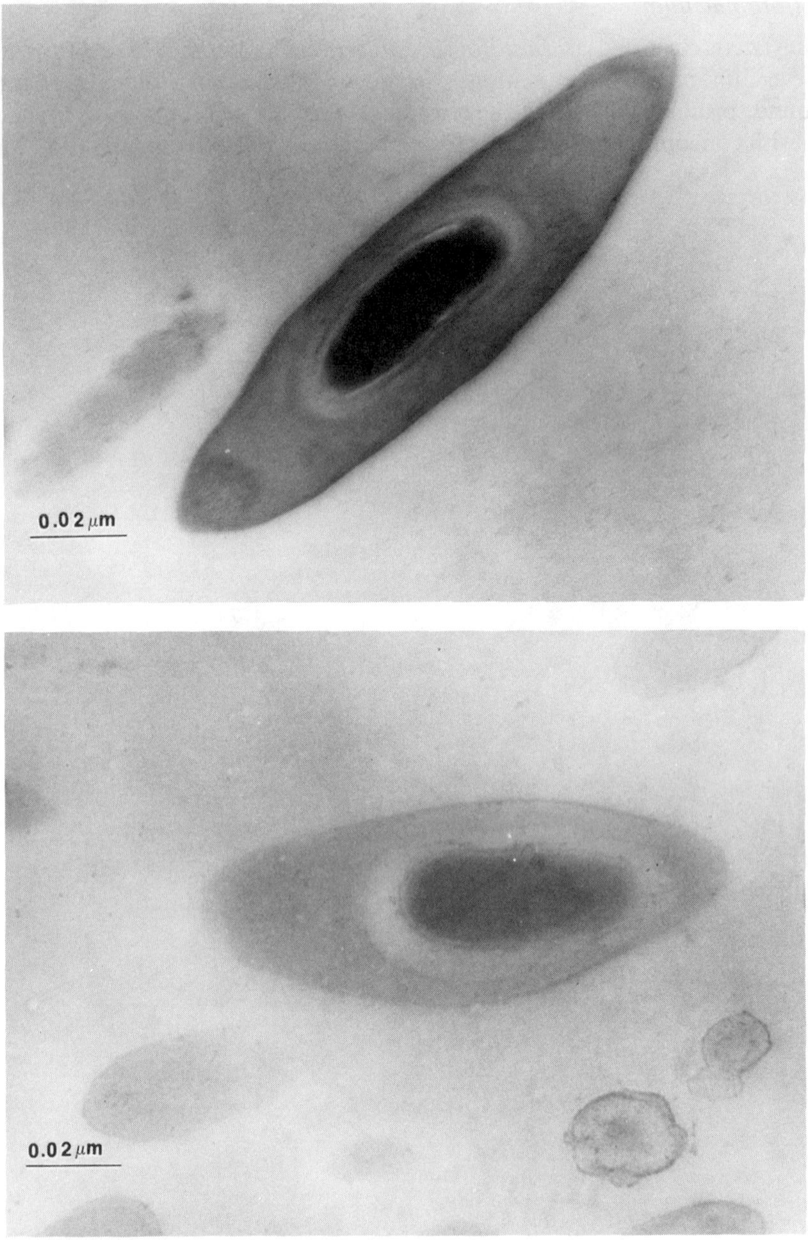

FIG. 3a, b. Electron micrographs of developmental stages in the formation of *Desulfotomaculum* central to subterminal type spores. The spore morphology is oval. Magnification bars represent 0.02 μm.

The former spore type is characteristic of isolates 4–6 and the latter is characteristic of isolates 7 and 8. The separation of the two types of *Desulfotomaculum* by spore morphology also corresponds to growth temperature, isolates 4–6 growing at 30 C, and isolates 7 and 8 growing at 55 C.

Table 4 shows the results of the classification of the mesophilic *Desulfotomaculum* strains. Isolates 4 and 6 had identical responses to all tests except that their colony morphologies differed (Table 2). Isolate 5 differed only because of the inability to use malate and its colony morphology. None of these isolates fit the species identification scheme proposed by Postgate (1979). They most closely resembled the characteristics of *Desulfotomaculum orientis*—except that they possessed polar flagella and *D. orientis* does not. It should be mentioned that flagella morphology is a good taxonomic criterion (Postgate 1979), but SRB cells do lose flagella easily if handled roughly. Both isolates 4 and 6 were able to use malate, although there was no mention (Postgate 1979) that any *Desulfotomaculum* species could use this substrate. In *Bergey's Manual* (Buchanan and Gibbons 1974) it was reported that *D. orientis* formed round spores, located centrally or paracentrally and on rare occasions terminally. Isolates 4–6 formed only terminal spores and were always round in appearance (Fig. 2a, b, c). Because of the inability to classify Isolates 4–6 into Postgate's (1979) scheme, these isolates were classified only as *Desulfotomaculum* sp. until sufficient new SRB strains have been studied for the proposal of a new classification key.

TABLE 4. *Identification of* Desulfotomaculum *strains (Terminal spore formers)*

	SRB Isolate Number		
	4	5	6
Growth temperature	30 C	30 C	30 C
Spores: Location	Terminal	Terminal	Terminal
Shape	Round	Round	Round
Flagella	Polar	Polar	Polar
Desulfoviridin	—	—	—
Carbon Compound Utilization			
Lactate	+	+	+
Malate	+	−	+
Formate	−	−	−
Acetate	−	−	−
Glucose	−	−	−
Pyruvate[a] w/o $SO_4^{=}$	−	−	−
Identification:	*Desulfotomaculum*	*Desulfotomaculum*	*Desulfotomaculum*

[a] Growth medium contained carbon compound minus sulfate.

Table 5 shows the classification of the thermophilic *Desulfotomaculum* strains. SRB Isolates 7 and 8 grew optimally at approximately 55 C with no growth occurring at mesophilic temperatures. Isolate 7 differed from the thermophilic ATCC strains in that it used glucose in addition to lactate, malate, and pyruvate (without sulfate) as substrates. Isolate 8 was not able to grow with pyruvate (without sulfate) but could use glucose, lactate, and malate. Isolates 7 and 8 also differed slightly in their colony morphologies (Table 2). Both isolates formed oval-shaped spores centrally to subterminally in the cell while causing a slight swelling of the cell (Fig. 3a, b). A comparison of the morphological features, the utilization of organic substrates, and the thermophilic nature of the two isolates suggested that these cultures were close to the species *Desulfotomaculum nigrificans* and were subsequently classified as such.

TABLE 5. *Identification of* Desulfotomaculum *strains*

	SRB Isolate Number			
	7	8	19998[a]	19858[a]
Growth temperature	55 C	55 C	55 C	55 C
Spores: Location	Cen-Subt[b]	Cen-Subt	Cen-Subt	Cen-Subt
Shape	Oval	Oval	Oval	Oval
Flagella	Perit[c]	Perit	Perit	Perit
Desulfoviridin	—	—	—	—
Carbon Compound Utilization				
Lactate	+	+	+	+
Malate	+	+	+	+
Formate	−	−	−	−
Acetate	−	−	−	−
Glucose	+	+	−	−
Pyruvate[d] w/o $SO_4^{=}$	+	−	+	+
Identification:	*D. nigrificans*	*D. nigrificans*	*D. nigrificans*	*D. nigrificans*

[a] American Type Culture Collection (ATCC) strain number of *Desulfotomaculum nigrificans.*
[b] Cen-Subt denotes central-subterminal.
[c] Perit denotes peritrichous flagella.
[d] Growth medium contained compound minus sulfate.

Conclusions

Because of the wide range of SRB isolated from the oil production facilities and the impact of SRB on oilfield operations, the origin of their growth substrates becomes important. In laboratory culture the SRB have a limited range of organic carbon substrates: lactate, pyruvate, and malate, with slower growth on a few sugars and alcohols (Moses and Springham 1982; Postgate 1979). Formation waters are unlikely to contain significant quantities of the carbon sources used in laboratory culture, but there must be minimal levels of growth substrates present to support SRB growth in oil production systems.

Al'tovskii et al. (1961) found that waters associated with oil reservoirs contained from 6.6–34.7 mg/l organic carbon. Napthenic acids were found in the water associated with oil but not subsurface formations barren of oil (Al'tovskii et al. 1962). Water close to the oil-water contact contained organic carbon at the level of 24.7 mg/l and napthenic acid concentration was as high as 900 mg/l. Formic and acetic acids have also been found in waters coproduced with petroleum (Davis 1967). In addition, aliphatic acids, amino acids, and various organic acids have been found in produced waters. It can be seen that oilfield waters could contain compounds that the SRB could use. It is also possible that the necessary carbon sources are supplied by other microorganisms growing in the oil production facilities and the reservoir.

This study shows the metabolic and physiological diversity of the SRB in just one type of system. Many model systems such as biocide tests, biodegradation studies, and corrosion studies use only a single SRB strain as a test culture. For example, a very common laboratory evaluation using SRB is the "biocide test." The RP-38 (biocide test procedure) of the American Petroleum Institute (API 1975) suggests the use of *Desulfovibrio desulfuricans* NRRL B-4304 as the test organism. It is obvious that this organism does not adequately represent the real situation. Therefore, tests based on *D. desulfuricans* ATCC 13541 may lead to poorly designed biocide treatment programs. A more appropriate system would be to use a mixed culture directly from the system to be treated.

Literature Cited

Al'tovskii, M. E., Z. I. Kuznetsova, and V. M. Shvets. 1961. *Origin of Oil and Oil Deposits.* Consultants Bureau, New York.

Al'tovskii, M. E., E. L. Bykova, Z. I. Kuznetsova, and V. M. Schvets. 1962. *Organic Substances and Microflora of Groundwaters and their Significance in Process of Oil and Gas Origin.* Gostoptekhizdat, Moscow.

API. 1975. Recommended Practice for Biological Analysis of Subsurface Injection Waters. American Petroleum Institute, Washington, DC.

Buchanan, R. E., and N. E. Gibbons. 1974. *Bergey's Manual of Determinative Bacteriology.* 8th ed. Williams & Wilkins Company, Baltimore, MD.

Davis, J. B. 1967. *Petroleum Microbiology.* Elsevier Publishing Company. Amsterdam.

Doetsch, R. N. 1981. Determinative methods of light microscopy. Pages 21–33 *in* Phillip Gerhardt, ed., *Manual of Methods for General Bacteriology.* American Society for Microbiology, Washington, DC.

Farquhar, G. B. 1974. The sulfate-reducing bacteria and oilfield bacterial corrosion "A review of the current state-of-the-art." *Corrosion* 74.

Hamilton, W. Allan. 1983. Sulphate-reducing bacteria and the offshore oil industry. *Trends in Biotechnol.* 1:36–40.

Hawley, G. G. 1971. *The Condensed Chemical Dictionary,* 8th ed. Van Nostrand Reinhold Company, New York.

Marquis, R. E. 1976. High-pressure microbial physiology. *Adv. Microb. Physiol.* 14:159.

Miller, J. D. A. 1981. Metals in microbial biodegradation. *Economic Microbiology,* Vol. 6, Academic Press, New York.

Miller, J. D. A., and A. M. Saleh. 1964. A sulphate-reducing bacterium containing cytochrome C_3 but lacking desulfoviridin. *J. Gen. Microbiol.* 37:419–423.

Minchin, L. T. 1960. Bacterial corrosion. *Pipeline Eng.* December:19–24.

Moses, V., and D. G. Springham. 1982. *Bacteria and the Enhancement of Oil Recovery.* Applied Science Publishers, London.

Murry, R. G. E., and C. F. Robinow. 1981. Specimen preparation for light microscopy. Pages 17–20 *in* Phillip Gerhardt, ed., *Manual of Methods for General Bacteriology.* American Society for Microbiology, Washington, DC.

Postgate, J. R. 1979. *The Sulphate-Reducing Bacteria.* Cambridge University, Cambridge.

Rozanova, E. P., and T. N. Nazina. 1976. Mesophilic rod-like non-sporeforming bacterium which reduces sulfates. *Mikrobiol.* 45:825–830.

Sharpley, J. M. 1961. Microbiological corrosion in waterfloods. *Corrosion* 17:846+-390+.

Skyring, G. W., H. E. Jones, and D. Goodchild. 1977. The taxonomy of some new isolates of dissimilatory sulfate-reducing bacteria. *Can. J. Microbiol.* 23:1415–1425.

Spurr, A. R. 1969. A low viscosity epoxy resin embedding medium for electron microscopy. *J. Ultrastr. Res.* 26:31–43.

Sudbury, J. D. 1957. External casing corrosion—where is it? How bad is it? *World Oil* May:210–222.

Tiller, A. K. 1983. Aspects of microbial corrosion. Pages 115–119 *in* R. N. Parkins, ed., *Corrosion Processes.* Applied Science Publishers, London.

TPC Publication No. 3. 1976. The role of bacteria in the corrosion of oilfield equipment. National Association of Corrosion Engineers.

Wakerley, Donald S. 1979. Microbial corrosion in UK industry. A preliminary survey of the problem. *Chem. Ind.* 19:656–658.

Wright, Charles C. 1963. The evaluation of bactericides in the field. *Prod. Month.* 27:2–5.

Zobell, C. E. 1958. The ecology of sulphate-reducing bacteria. *Prod. Month. Penn. Oil Prod. Assoc.* 22:12–29.

________. 1970. Pressure effects on morphology and life processes of bacteria. Pages 85–130 *in* A. M. Zimmerman, ed., *High Pressure Effects on Cellular Processes.* Academic Press, New York.

CHAPTER 50

Development of Culture Media, a Sporulation Procedure, and an Indirect Immunofluorescent Antibody Technique for Sulfate-Reducing Bacteria

W. Michael Griffin and Kathleen M. Antloga

Sohio Petroleum Company, Warrensville Production Research Laboratory, 4440 Warrensville Center Road, Cleveland, Ohio 44128

Nicholas Santoro, Alan P. Bakaletz, Melvin S. Rheins, and Olli H. Tuovinen

Department of Microbiology, The Ohio State University, Columbus, Ohio 43210

Several culture media were tested for their ability to support growth of pure cultures of *Desulfovibrio* and *Desulfotomaculum* species. Both marine and freshwater isolates were included in the study. The primary criteria in the selection of suitable growth media were (1) the choice and concentration of reducing agents, and (2) the reduction of iron sulfide precipitation to facilitate the cell harvest. In addition, a procedure was developed for preparing *Desulfotomaculum* spore suspensions. Polyvalent antisera were raised against antigens of vegetative cells and spores of sulfate-reducing bacteria (SRB). Fluorescent-labeled antibodies were then used to test the cross-reactivity among the different SRB cultures as well as among bacteria of other genera. Various bacterial isolates from oilfield-produced water were tested for reactivity with the prepared antisera. The isolates generally were agglutinated but to a low titer with the test antisera.

Introduction

Many corrosion problems of iron-based alloys have been attributed to the activities of the sulfate-reducing bacteria (SRB). While the mechanisms by which SRB catalyze localized corrosion are not completely elucidated, the possible financial impact of SRB corrosion has stimulated much research on the subject (Minchin 1960; Sharpley 1961; Farquhar 1974; Miller 1981). Though the SRB have been implicated as the cause of corrosion in many instances, only a few actual case histories have been documented (Doig and Water 1951; Koger 1956; Allred et al. 1959; Haugen and Purdy 1976). In addition to corrosion, the SRB are implicated in other operational problems in the oilfield; e.g., production of H_2S, plugging of filters and, in some cases, the oil reservoir. Hydrogen sulfide production is particularly important to the oilfield operators since it is corrosive, and a potential health hazard, and it can form iron sulfide, which has been implicated in stabilizing oil/water emulsions.

Because of the major economic impact which the SRB can have on petroleum production, the oilfield operator usually reacts to the presence of SRB by treatment with a biocide (Huddleston and Allred 1960). While chemical costs may be of minimal concern for the small field, for fields in which millions of barrels of water are injected into the reservoir daily, chemical use optimization is essential. Therefore, an accurate, rapid detection method is needed to provide relevant information for chemical optimization.

There are no rapid detection methods for SRB available for standard use in surveillance programs. Detection methods based on the growth of the organisms may be highly selective for samples of heterogeneous populations but require long incubation periods before the growth can be recorded. Similarly, incubation, although shorter than the traditional culture methods, is required for use of radiolabeling techniques based on determining the acid-volatile loss of ^{35}S. Impedance methodology (Oremland and Silverman 1979) may allow short-term incubation and automation, but the calibration and specificity of this methodology have not been determined for SRB.

Immunofluorescent detection techniques have been developed for routine monitoring purposes of many different microorganisms in environmental samples. Their primary advantage is the expediency of detection but, in many cases, the cross-reactivity of the immunofluorescence requires additional confirmation tests.

In the present paper, an indirect immunofluorescent method is reported. The method was specifically developed for several species of *Desulfovibrio* and *Desulfotomaculum*. In addition, the paper describes preparative procedures for production of vegetative cells and spore suspensions for raising antisera. The cross-reactivity of the antibody was tested toward several different bacteria.

Materials and Methods

Bacteria and Growth Conditions (SRB)

The SRB used in this study included the freshwater strains *Desulfotomaculum nigrificans* ATCC 19998 and 19858; *Desulfovibrio desulfuricans* (13541); and the marine strains, *Desulfovibrio desulfuricans* subsp. *aestuarii* (17990), *Desulfovibrio aestuarii* (14563), and *Desulfovibrio salexigens* (14822). Both *D. nigrificans* ATCC cultures were found to be heterogeneous. Two spore-forming strains isolated from these cultures, *D. nigrificans* 1998-1-T and *D nigrificans* 19859-3-C, were used for spore isolation studies. In addition, a number of *Desulfovibrio* and *Desulfotomaculum* strains were isolated from oilfield produced water. Bacteria were routinely subcultured on lactate in Medium B (Postgate 1979). Sodium chloride (2.5% w/v) was added to Medium B (for the marine strains).

Bacteria were subcultured every 2 wk. *Desulfotomaculum nigrificans* ATCC 19998 and 19858 (19858-1-T; 19858-3-C) strains were grown at 55 C. *Desulfovibrio* species ATCC 17990, 14822, and 13541 were grown at 30 C; *D. aestuarii* ATCC 14563 was grown at 37 C. Blackening of the medium resulting from FeS formation served as an indication of growth. Sterile medium could be stored at 30 C for up to 1 wk under N_2 or in head-space free test tubes. Stock cultures were checked monthly for aerobic and anaerobic contamination. Trypticase Soy Agar (BBL) or Nutrient Agar (BBL) plates were used for routine aerobic checks. Anaerobic contamination was tested for by using the following plate medium (KW) containing (g/l): sodium lactate, 4.6; NH_4Cl, 0.1; $MgSO_4 \cdot 7H_2O$, 0.2; K_2HPO_4, 1.0; ascorbic acid, 0.1; yeast extract, 1.0; agar, 30. After autoclaving the medium, filter-sterilized solutions containing 0.5 g $Fe(NH_4)_2(SO_4)_2$ and 0.1 g Na_2SO_3 were added per liter of medium. Nonblack colonies were identified as contaminants and the respective cultures were discarded.

Mass production of bacteria to be used as the antigen source were carried out in Medium C (Postgate 1979), Medium SG (Campbell et al. 1957), and the medium described by Hayward and Stadtman (1959) because they permitted appreciable growth yield and produced a minimal amount of insoluble iron sulfide. Cysteine-HC1 (0.025% w/v) or $Na_2S \cdot 9H_2O$ (0.025% w/v) was used to poise the medium E_h. Resazurin was used in all growth media at concentrations ranging from 0.1 and 0.2 mg/100 ml. All cultures were grown in 2-liter flasks equipped with a modified rubber bung that allowed purging of the medium with sterile nitrogen gas after autoclaving the solution. Cells were harvested from the growth medium by centrifugation at 10,000 × g for 20 min at 23 C. Cells were washed three times with phosphate-buffered saline (PBS; 10 m*M* potassium phosphate, 150 m*M* NaCl, pH 7.6) and resuspended in PBS containing 6% formalin. Washed cell suspensions were standardized by direct microscopic counts, using Norris-Powell diluent for dilutions.

Bacteria and Growth Conditions (Non-SRB)

The non-SRB organisms used included ATCC strains of *Pseudomonas aeruginosa* (10145), *Enterobacter aerogenes* (13048), *Bacillus subtilis* (6051), *Micrococcus luteus* (381), *Staphylococcus epidermidis* (12228), *Escherichia coli* (25922), *Acinetobacter calcoaceticus* (19606), *Lactobacillus acidophilus* (4356), *Clostridium sporogenes* (3584), *Klebsiella pneumoniae* (13883), and *Propionibacterium acnes* (11827). In addition, nine non-SRB produced water isolates were used, viz., *Propionibacterium* (three strains), *Clostridium* (two strains), *Peptococcus, Enterobacter* and *Staphylococcus* (two strains).

The ATCC cultures 10145, 13048, 6051, 381, 12228, 25922, 19606, and 13883 and the isolated produced water isolates *Enterobacter* and *Staphylococcus* (two strains) were stored on nutrient agar (Difco) slants. *Lactobacillus acidophilus* 4356 was stored on *Lactobacillus* agar (Difco) slants. The ATCC anaerobic cultures 3584 and 11827 and the *Propionibacterium, Clostridium,* and *Peptococcus* produced water isolates were maintained on thioglycollate agar (Difco) slants. All cultures were stored at 4 C.

Spore Isolation by Density Gradient Centrifugation

Desulfotomaculum was grown in the peptone-glucose SG medium described by Campbell et al. (1957). Cells were harvested after 48 h incubation and washed three times with PBS. The cells were then resuspended in 30 ml of PBS supplemented with 0.05% (w/v) sodium thioglycollate and incubated under a nitrogen atmosphere at 72 C for 48 h. Following heat treatment, the cells were centrifuged and resuspended in 5 ml of PBS and passed three times through a

French Press at 140 MPa. The suspension was washed twice before Renografin gradient centrifugation (Tamir and Gilvarg 1966; Dean and Douthit 1974). All of these manipulations were conducted at room temperature.

Renografin-76 (Squibb and Sons) was received as a 76% (v/v) solution and was diluted with distilled water to obtain solutions of desired densities. Gradients were prepared in 1- × 3-inch cellulose nitrate tubes (Beckman Instruments, Inc.). Samples 1–2 ml) were layered on top of the gradient tubes and placed in an SW27 rotor. The tubes containing the gradients were centrifuged in a Beckman model L3-40 ultracentrifuge at 20 C for 45 min at 23,000 rpm.

Isolation of Spores Using a Two-Phase Polyethylene Glycol System

Subterminal and terminal spore-forming *Desulfotomaculum* strains were cultured in modified Erlenmeyer flasks containing 2 liters of the SG medium (Campbell et al. 1957). The flasks were inoculated by washing plates containing *D. nigrificans* 19998-1-T terminal and 19858-3-C subterminal spore-forming cells. Both spore-forming strains were subcultured three times to test for contaminant anaerobes (containing 0.1 g/l $Fe(NH_4)_2(SO_4)_2$) before they were used as inoculum for liquid cultures. The culture flasks containing the SC medium were incubated at 55 C for 48 h and used as inoculum for 10 liters of SG medium in 12-liter fermentors (New Brunswick Microferm). The culture in the fermentor was maintained at 55 C. The fermentor was gassed at a rate of 2 liters N_2/min and stirred at 150 rpm. After 48 h the culture was harvested, washed, and resuspended in a solution containing 0.025% (w/v) of both ascorbic acid and sodium thioglycollate in PBS. The suspension was then transferred to a 125-ml Hypo-vial (Pierce Chemical Co.) and incubated at 85 C for 48 h under a nitrogen atmosphere.

After heat shocking, the cells and spores were centrifuged at 7,000 × *g* for 20 min at 23 C and resuspended in 5–10 ml of PBS. Vegetative cells were broken by passing the suspension three times through a French Press at 140 MPa. The suspension was centrifuged at 7,000 × *g* for 20 min and the pellet was resuspended in 30 ml PBS containing 0.5 mg/ml lysozyme. The resulting mixture was then incubated at 52 C for 2 h on a shaker at 150 rpm.

After treatment with lysozyme, the suspension was centrifuged at 7,000 × *g* for 20 min. Spores were separated from the remaining vegetative cells and cellular debris using the aqueous two-phase system W developed by Sacks (1969). System W was prepared by mixing 9 ml of 3 *M* phosphate buffer (1.76 *M* K_2PO_4) with 11 ml of polyethylene glycol (11.0 g PEG 1000 dissolved in 11 ml H_2O). After standing at room temperature for several minutes, the two phases separated. Spore suspension (5 ml) was carefully pipetted into the upper phase, making the final volume 25 ml in the tube. The spore suspension was gently mixed into the upper phase and was centrifuged at 1,500 × *g* for 2 min at room temperature. The upper phase was collected, centrifuged, and washed twice with distilled water. The spore suspension was standardized by microscopic counts under phase contrast and stored in distilled water at 4 C until it was used for immunization.

Antisera Production

White New Zealand rabbits (2–3 kg) were used for the production of the antisera. Injections of vegetative cells or spores were given in doses of 1–3 × 10^9

organisms. The initial doses of antigen (vegetative cells or spores) were emulsified with equal volumes of Freund's complete adjuvant (Difco) and given as intramuscular injections. Booster injections, beginning at 4 wk, were given intravenously as aqueous suspensions of the antigens at 2-wk intervals. On alternate weeks, 25–50 ml samples of blood were collected. Serum was recovered and stored frozen for later processing.

Serum samples from individual rabbits were pooled. The immunoglobulin fraction of the serum was concentrated by ammonium sulfate precipitation method (Good et al. 1980). The precipitate was re-dissolved in borate-buffered saline (170 m*M* borate, 120 m*M* NaCl, pH 8.4) and subjected to ultrafiltration to remove the sulfate ions.

Antibody titers of the final solutions were determined by agglutination using the Microtiter system (Cooke Laboratory Products). Doubling dilutions of antisera were prepared in 50 μl volumes of phosphate buffer saline. Fifty microliters of the antigen suspensions were then added to each well. The plates were incubated overnight at 4 C, after which the agglutination titers were determined. Antibody titers were determined for heterologous as well as homologous antigen suspensions. Antibody solutions were stored frozen at −20 C for later use in the immunofluorescence assay.

Immunofluorescence Assay

The prepared rabbit antibody was tested in an indirect fluorescent antibody assay (IFA) using goat antirabbit-globulin conjugated with fluorescein isothiocyanate (GIBCO Laboratories) as the secondary antibody. Smears of the test antigen suspensions were made on glass slides, air dried, and fixed in methanol for 30 s. The smears were then covered with the primary antibody and incubated at room temperature in a moist chamber for 30 min. The slides were rinsed gently with PBS, and then immersed in PBS for 30 min with one change of buffer. The smears were then covered with the secondary antibody, incubated, and washed as indicated. Finally, the slides were rinsed with distilled water, air dried, and covered with glycerol mounting fluid, pH 9.0 (Difco), and a coverslip. The smears were examined under a Zeiss Fluorescence microscope with epi-illumination.

Results and Discussion

Mass Production of Sulfate-Reducing Bacteria for Antisera Preparation

Desulfovibrio ATCC strains 13541, 14822, 14563, and 17990 grew best in Medium C (Postgate 1979) containing lactate. Growth without precipitation of ferrous sulfide was possible with glucose in the medium as described by Hayward and Stadtman (1959). *Desulfotomaculum nigrificans* ATCC 19998 and 19858 grew with peptone-glucose in the SG medium described by Campbell et al. (1957) without FeS precipitation. The procedure for performing microscopic counts using Norris-Powell diluent followed by dilution in 0.1 *N* HCl was modified for the marine strains because the dilution in 0.1 *N* HCl caused excessive clumping of the cells. The final dilution in HCl was omitted when counting marine strains.

Harvesting of marine strains by centrifugation resulted in pellets which resisted dispersion. Extensive clumping of cells was observed during microscopic enumeration of the marine strains. However, when glucose was replaced by lactate in the medium of Hayward and Stadtman (1959) the foregoing effects were eliminated.

Growth of Sulfate-Reducing Bacteria on Agar Media

Agar plates were examined for growth after 7 d incubation. The TSA plus salts medium (Iverson 1966) supported the growth of all *Desulfovibrio* ATCC strains, but it was not as suitable for supporting the growth of *Desulfotomaculum*. No growth differences were observed using either an N_2 or H_2 + CO_2 atmosphere (BBL Gas-Paks). The medium used for the detection of anaerobic contaminants supported the growth of both *Desulfovibrio* and *Desulfotomaculum*. Sulfate-reducing bacteria produced shiny, black colonies on this medium. These colonies were readily distinguished from the gray to white colonies produced by contaminants. When the concentration of ferrous iron in the above medium was reduced from 1.75 m*M* to 0.35 m*M*, the sulfate-reducing bacteria formed colonies that were indistinguishable from contaminants. Ferrous iron (0.35 m*M*) was included in the agar medium to grow the bacteria for inoculating liquid media. This avoided the problem of FeS contaminating the biomass.

It was found that KW medium supported the growth of both *Desulfovibrio* and *Desulfotomaculum* and was useful in routine checks for contaminating anaerobes. All the SRB tested consistently produced brown-pigmented colonies, whereas contaminant colonies appeared white to gray. The same results were observed using atmospheres of either N_2 or H_2 + CO_2 (BBL Gaspacks).

Isolation of Spores by Density Gradient Centrifugation

Attempts to purify spores using Renografin density gradient centrifugation resulted in poor cell spore separation with widespread bands containing either round or ellipsoidal spores. The density of each fraction was determined from a standard curve of refractive index versus density of Renografin at 25 C. ATCC 19998 was resolved into two major bands in density ranges of 1.334–1.273 g/ml and 1.226–1.199 g/ml; bands contained both round and ellipsoidal spores. ATCC 19858 spores were resolved into two major bands in density ranges of 1.372–1.257 g/ml and 1.236–1.212 g/ml; bands contained both round and ellipsoidal spores. Spores remained viable in Renografin for one month (tested for by growth in Postgate's Medium B after 24 h incubation). Separation of spores by density gradient centrifugation resulted in bands, which contained not only spores but also significant amounts of cell debris. Since the goal of the spore purification was to provide spores for antisera production, other methods for isolating spores were investigated.

Isolation of Spores Using Two-Phase Systems

Experiments were conducted to investigate the effect of incubation at elevated temperature on *Desulfotomaculum nigrificans* cells as a method for lysing vegetative cells and causing the release of mature spores. Cell suspensions were incubated for 24 h at 70, 75, and 85 C. Temperatures of 70 and 75 C did not result in as high a percent sporulation as did 85 C (determined microscopically). Approximately 70% sporulation was observed in suspensions that were heat treated for 48 h and only 40% in those that were incubated for 24 h. Cell suspensions were incubated for 48 h at 85 C in all following experiments.

After heat treatment and breaking the remaining intact vegetative cells with the French Press, the suspension was centrifuged in the two-phase system W described by Sacks (1969). Upon centrifugation, the cell debris and remaining vegetative cells collected at the interface of the two-phase system. The upper phase contained a clean spore suspension and was aspirated off leaving behind a portion of the upper phase, interface material, and lower phase. The upper phase was then centrifuged and the spores were washed with distilled water. The spore suspension was standardized by microscopic counts and stored at 4 C in distilled water until it was used to raise antisera.

Evaluation of Antisera Reactivity

The *Desulfotomaculum* strains were similar immunologically (Table 1) and cross-reacted extensively. Little cross-reaction was detected for these antibodies when tested against the *Desulfovibrio* species. The agglutination titers also indicated a high degree of cross-reactivity in the antisera raised against the vegetative cells and spores of *Desulfotomaculum* species. Since the vegetative cell preparations used for immunization contained spores (less than 10% as observed by phase contrast microscopy) and the spore preparations most likely had some contaminating cell wall material, the cross-reactivity of the antisera would be expected. These results differ from those of Smith (1982) who found cross-reactivity between *Desulfotomaculum* and some *Desulfovibrio* species. However, the results reported here are similar to those reported by Postgate (1979), where no cross-reactivity between the *Desulfovibrio* and *Desulfotomaculum* strains tested was noted. Also, in two earlier reports (Campbell et al. 1957; Postgate and Campbell 1963) *Desulfovibrio orientis* and Coleman's organism, which were later classified in the genus *Desulfotomaculum* (Campbell and Postgate 1965), did not cross-react with any *Desulfovibrio* strains tested.

The marine species of *Desulfovibrio* (ATCC 17990, 14563, and 14822) were similar with respect to their antigen reaction (Table 1). The cross-reactivity of the marine species was low against the freshwater species of *Desulfovibrio desulfuricans* (ATCC 13541) and the *Desulfotomaculum* strains. The antiserum to the freshwater *Desulfovibrio* had reactivity only to its homologous antigen.

The results, based on the agglutination titers, show that the SRB used in the present work to raise the antisera can be divided into three major groups. The freshwater strain *Desulfovibrio desulfuricans* showed little cross-reactivity when tested with other antisera, the marine *Desulfovibrio* strains were similar immunologically, and the two *Desulfotomaculum* strains were extremely homologous. The taxonomic significance of this cross-reactivity would require further study since only one strain of the freshwater *Desulfovibrio* was used.

Indirect Immunofluorescence Assay

The antisera were tested in the IFA against homologous and heterologous antigens. The patterns of reactivity were similar to those obtained in the agglutination assays (i.e., three groups with respect to reactivity: *Desulfotomaculum* species, freshwater *Desulfovibrio*, and marine *Desulfovibrio*). The antisera were used at

TABLE 1. *Agglutination titers against prepared antisera*

Antiserum Against ATCC Strain		Agglutination Titer[a]							
		19858 cells	19858 spores	19998 cells	19998 spores	13541	17990	14563	14822
D. nigrificans	19858								
	cells	256	160	256	80	<2	4	<2	2
	spores		160		80			20	
D. nigrificans	19998								
	cells	256	160	128	160	<2	<2	<2	2
	spores		80		80			20	
D. desulfuricans	13541	4	<10	8	<10	128	<2	<2	2
D. desulfuricans aestuarii	subsp. 17990	2	<10	8	<10	<2	64	256	32
D. aestuarii	14563	8		16		<2	64	512	64
D. salexigens	14822	2	<10	8	10	2	256	128	512

[a] Highest dilution factor of the antisera with positive reactivity.

the highest dilution that still gave bright, uniform fluorescence with the homologous organism. Because of the dilution, the minimal cross-reactions observed in the agglutination assay were eliminated in the IFA. In smears prepared from mixtures of the bacteria, the organisms homologous with the test antiserum used were readily distinguished from nonreactive organisms (Fig. 1A, B, C, D).

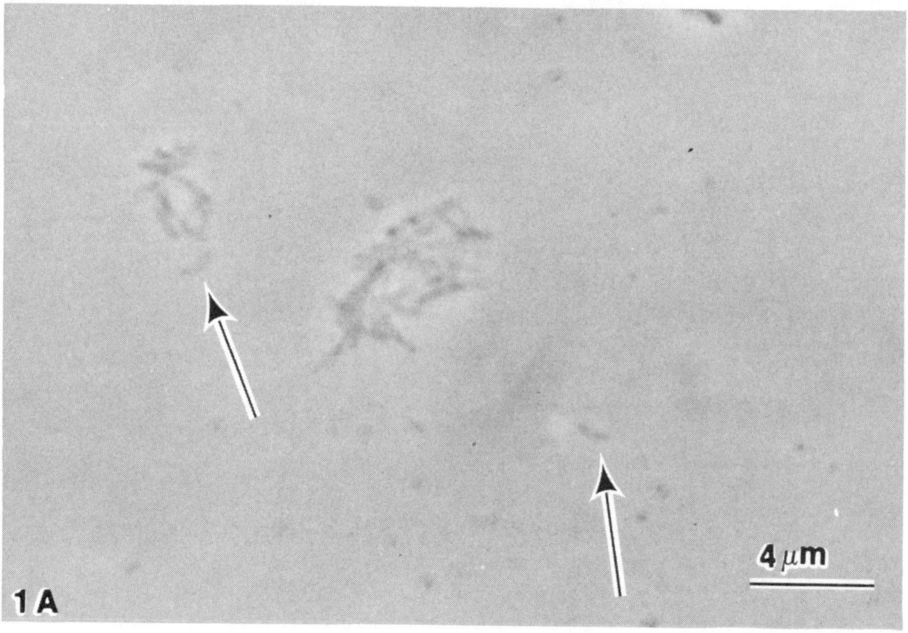

FIG. 1 A. Phase-contrast photomicrograph of an FA-stained mixed suspension containing *D. nigrificans* ATCC 19858, *D. desulfuricans* ATCC 13541, and Antisera 19858. Arrows indicate individual 13541 organisms.

Antisera Agglutination Titers Against Environmental Isolates of SRB

The reactivity of the antisera toward mesophilic strains of SRB isolated from oilfield produced water ranged from an agglutination titer of 2 to 64 (Table 2). For the mesophilic *Desulfotomaculum* isolates, a low degree of homology was repeatedly observed. Two thermophilic isolates of *Desulfotomaculum* were tested and showed homology with respect to antisera raised against the vegetative cells and spores of the *Desulfotomaculum* cultures 19858 and 19998. In both cases there was also a relatively high cross-reactivity with the *Desulfovibrio* antisera (Table 2).

Antisera Agglutination Titers Against Non-SRB

The reactivity of the SRB antisera was tested against non-SRB. If considerable cross-reactivity with non-SRB was found it would limit the usefulness of the technique. Antisera was tested against ATCC strains (Table 3) and organisms isolated from oilfield produced water (Table 4). Little cross-reactivity was ob-

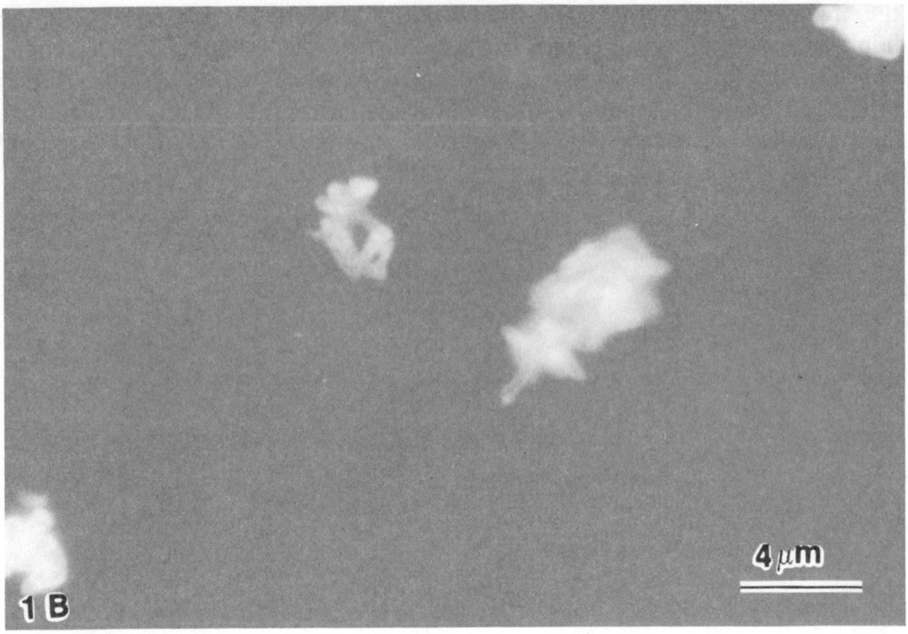

B. Same field as 1A under UV illumination.

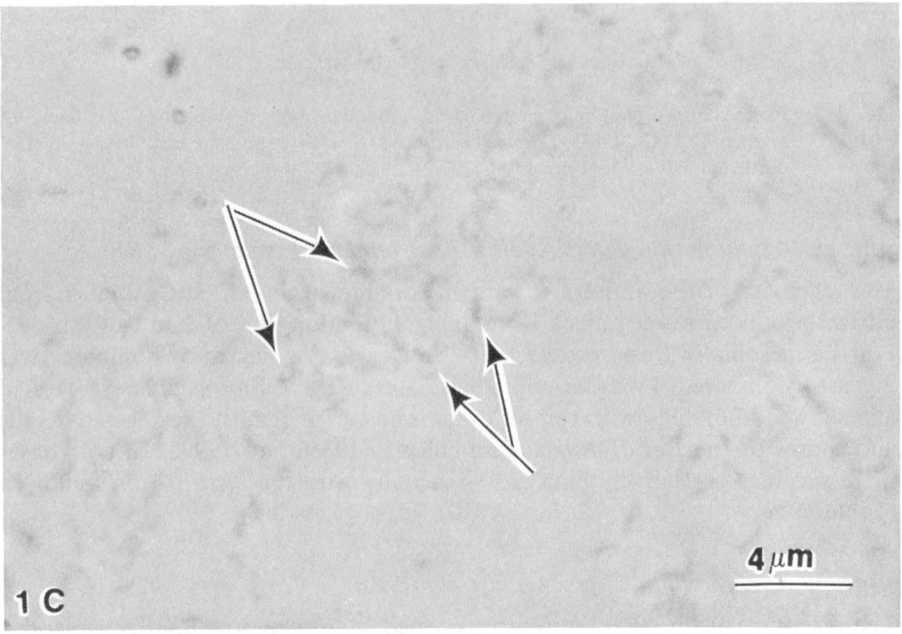

C. Phase-contrast photomicrograph of an FA-stained mixed suspension containing *D. nigrificans* ATCC 19858, *D. desulfuricans* ATCC 13541, and Antisera 13541. Arrows indicate individual 13541 organisms.

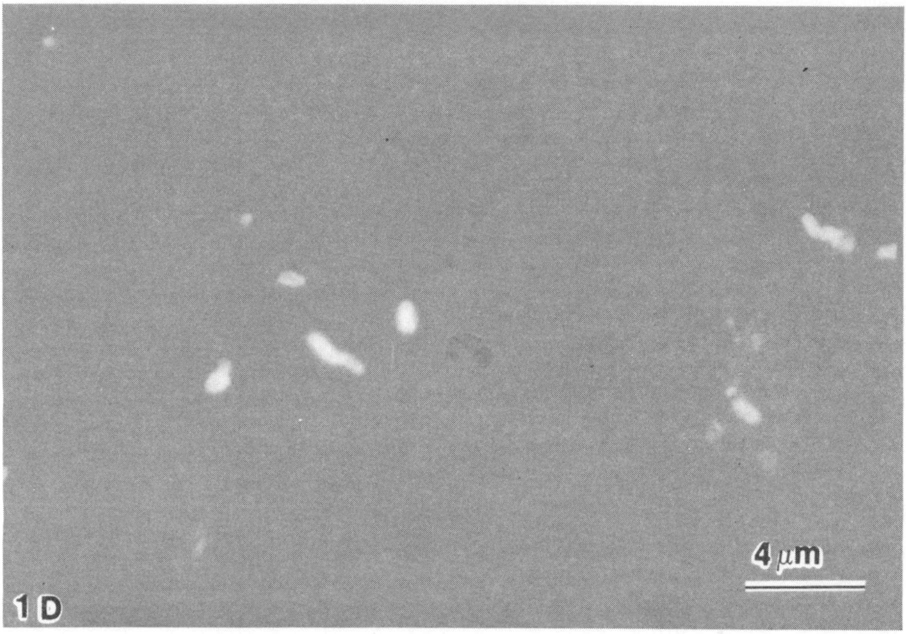

D. Same field as 1A under UV illumination.

served for most of the organisms tested. Notable exceptions can be seen for *Bacillus* (ATCC 6051) and *Clostridium* (ATCC 3584; Strain 1). Also, *Propionibacterium* ATCC 11827 cross-reacted extensively with antisera against *Desulfovibrio* ATCC 14563. All three of these organisms are commonly found in the oilfield (Carlson et al. 1961) and precautions must be taken if this technique is used in the field.

Conclusions

An IFA assay has been developed that has been shown to be useful for identifying SRB. While cross-reactivity of the antisera with *Clostridium, Bacillus,* and *Propionibacterium* was observed, it was generally of a low level. By using the highest dilution of antisera possible in the IFA assay, this cross-reactivity could be eliminated. The use of polyvalent sera is advantageous if genus and species differentiation is not required. The minimal cross-reactivity between the antisera and the wild strains of SRB was unexpected. This could be due to the presence of an extracellular coat. However, antisera produced against the wild strains of SRB would result in greater sensitivity of the IFA and make the technique more applicable to the oilfield.

Based on the results in the present work, suitable media are now available for growing large quantities of SRB for harvest without interference by the ferrous sulfide precipitate. The technique described for sporulation does not rely upon the use of liver broth (Lin and Lin 1970), or mushroom compost infusion medium (Donnelly and Busta 1980) and thereby eliminates the uncontrollable factors associated with these methods.

TABLE 2. *Reactivity of the antisera toward sulfate-reducing bacteria isolated from oilfield-produced water*

Produced Water	Antisera							
Isolates	19998	19998(s)[a]	19858	19858(s)[a]	13451	17990	14822	14563
Desulfotomaculum strains								
M[b]-22681b	2	4	4	16	4	32	16	2
M-22681d	2	8	8	8	2	8	2	4
M-22681h	2	4	2	8	2	8	2	2
T-23381	32	32	64	16	16	16	32	16
T-23681	64	64	64	64	16	16	32	16
Desulfovibrio strains								
M-30381A	2	4	2	8	2	26	2	4
M-30581A	2	8	2	16	4	32	32	2

[a] s, spore antisera.
[b] M, mesophilic; T, thermophilic.

TABLE 3. *Reactivity of antisera toward ATCC test strains*

ATCC	Antisera							
Strains	19998	19998(s)[a]	19858	19858(s)[a]	13451	17990	14822	14563
Klebsiella (13883)	2	2	2	2	2	2	2	2
Enterobacter (13048)	2	2	2	2	2	2	2	2
E. coli (25922)	2	2	2	2	2	2	16	2
Clostridium (3584)	2	64	32	128	16	128	64	64
Bacillus (6051)	4	16	2	64	16	32	32	64
Staphylococcus (12228)	2	2	2	2	2	2	2	2
Pseudomonas (10145)	2	2	2	2	2	2	2	2
Acinetobacter (19606)	2	2	2	2	2	2	2	2
Propionibacterium (11827)	2	2	2	2	2	16	4	128
Lactobacillus (4356)	2	2	2	2	2	2	2	2
Micrococcus (381)	2	2	2	2	2	2	2	2

[a] s, spore antisera

TABLE 4. *Reactivity of the antisera toward anaerobes isolated from produced water*

Produced Water Isolates	Antisera							
	19998	19998(s)[a]	19858	19858(s)[a]	13451	17990	14822	14563
Strict *Anaerobes*								
Propionibacterium Strain 1	2	2	2	2	2	16	2	2
Clostridium Strain 1	16	16	16	8	32	16	16	8
Propionibacterium Strain 2	2	4	2	2	2	2	2	2
Clostridium Strain 2	2	2	2	2	2	2	2	2
Peptococcus	2	2	2	2	2	2	2	2
Propionibacterium Strain 3	2	2	2	2	2	2	2	2
Facultative *Anaerobes*								
Enterobacter	2	2	2	2	2	2	2	2
Staphylococcus Strain 1	2	2	2	2	2	2	2	2
Staphylococcus Strain 2	2	2	2	2	2	2	2	2

[a] s, spore antisera

Acknowledgments

The authors would like to thank Dr. David Brooks for his help on the conceptual design of the project and Ms. T. C. Li for her laboratory assistance.

Literature Cited

Allred, R. C., D. C. Olson, and J. D. Sudbury. 1959. Corrosion is controlled by bactericide treatment. *Prod. Monthly* Nov:111–112.

Campbell, L. L., and J. R. Postgate. 1965. Classification of the spore-forming sulfate-reducing bacteria. *Bacteriol. Rev.* 29:359–363.

Campbell, L. L., H. A. Frank, and E. R. Hall. 1957. Studies on thermophillic sulfate-reducing bacteria, I. Identification of *Sporovibrio desulfuricans* as *Clostridium nigrificans. J. Bacteriol.* 73:516–521.

Carlson, W., E. O. Bennett, and J. A. Rowe. 1961. Microbial flora in a number of oilfield water-injection systems. *Soc. Petrol. Engrs.* 1:71–80.

Dean, D. H., and H. A. Douthit. 1974. Buoyant density heterogeneity in spores of *Bacillus subtilis:* Biochemical and physiological basis. *J. Bacteriol.* 117:601–610.

Doig, K., and A. Water. 1951. Bacterial casing corrosion in the Ventura field. *Corrosion* 7:212–224.

Donnelly, L. S., and F. F. Busta. 1980. Heat resistance of *Desulfotomaculum nigrificans* spores in soy protein infant formula preparations. *Appl. Environ. Microbiol.* 40:721–725.

Farquhar, G. B. 1974. The sulfate-reducing bacteria and oilfield bacterial corrosion "A Review of the Current State-of-the-Art." Pages 1/1–1/13 *in Corrosion* 74, NACE Paper No. 1, March 4–8, 1974, Palmer House, Chicago, IL.

Good, A. H., L. Wofsy, J. Kimura, and C. Henry. 1980. Purification of immunoglobulins and their fragments. Pages 278–286 *in* B. B. Mishell and S. M. Shiigi, eds., *Selected Methods in Cellular Immunology.* W. H. Freeman and Company, San Francisco, CA.

Haugen, A. K., and I. L. Purdy. 1976. Coating waterflood supply line stops corrosion and reduces horsepower. *Oil Gas J.* 74:148–155.

Hayward, H. R., and T. C. Stadtman. 1959. Anaerobic degradation of chlorine by an anaerobic cytochrome-producing bacterium *Vibrio cholincus* n. sp. *J. Bacteriol.* 78:557–561.

Huddleston, R. L., and R. C. Allred. 1960. Effective chemicals for microbial control in secondary recovery. *Soc. of Petrol. Eng.* Paper No. SPE-65.

Iverson, W. P. 1966. Growth of *Desulfovibrio* on the surface of agar media. *Appl. Microbiol.* 14:529–534.

Koger, W. C. 1956. Casing corrosion in Hugotan gas field. *Corrosion* 12:507–512.

Lin, C. C., and K. C. Lin. 1970. Spoilage bacteria in canned food. II. Sulfide spoilage bacteria in canned mushrooms and a versatile medium for the enumeration of *Clostridium nigrificans. Appl. Microbiol.* 19:283–286.

Miller, J. D. A. 1981. Metals. Pages 149–202 *in* A. H. Rose, ed., *Economic Microbiology,* Vol. 6. Academic Press, New York.

Minchin, L. T. 1960. Bacterial corrosion. *Pipeline Engineer.* December: D-19–D-24.

Oremland, R. S., and M. P. Silverman. 1979. Microbiological sulfate reduction measured by automated electrical impedance technique. *Geomicrobiol. J.* 1:355–372.

Postgate, J. R. 1979. *The Sulfate-Reducing Bacteria.* Cambridge University Press, Cambridge, MA.

Postgate, J. R., and L. L. Campbell. 1963. Identification of Coleman's sulfate-reducing bacterium as a mesophilic relative of *Clostridium nigrificans. J. Bacteriol.* 86:274–279.

Sacks, L. E. 1969. Modified two-phase system for partition of *Bacillus macerans* spores. *Appl. Microbiol.* 18:416–419.

Sharpley, J. M. 1961. Microbiological corrosion in waterfloods. *Corrosion* 17:836–390.

Smith, A. D. 1982. Immunofluorescence of sulphate-reducing bacteria. *Arch. Mikrobiol.* 133:118–121.

Tamir, H., and C. Gilvarg. 1966. Density gradient centrifugation for the separation of sporulating forms of bacteria. *J. Biol. Chem.* 241:1085–1090.

CHAPTER 51

Rapid Sedimentation of Microbial Suspensions

ROBERT H. DAVIS AND STEPHEN A. BIRDSELL

Department of Chemical Engineering, University of Colorado, Boulder, Colorado 80309

Yeast cells were separated from aqueous suspension by gravity sedimentation within channels that were (1) vertical, and (2) inclined from the vertical. As predicted by theory, the rate of separation of the cells from suspension increased as the angle of the channel inclination from the vertical was increased and as the spacing between the walls of the channel was decreased. Experiments were first performed in a channel with walls spaced 0.5 cm apart. As the angle of inclination of these walls from the vertical was increased from 25° to 57°, the time required to settle 75% of the yeast cells was reduced from 220 min to 100 min. With the channel vertical, over 2,000 min were required to settle 75% of the cells. The spacing between the channel walls was then reduced to 0.2 cm. When the channel was inclined to 60° from the vertical, only 32 min were necessary to settle 75% of the cells. Gravity sedimentation in inclined channels offers a very promising way to rapidly separate yeast cells from suspension.

INTRODUCTION

The separation of microorganisms from bioreactor effluent is a key step in the downstream processing of microbial suspensions. Common laboratory techniques such as centrifugation and filtration become very expensive when applied on the industrial scale. Gravity sedimentation provides an inexpensive method for collecting the cells. However, this process is slow, because the cells are small and are only slightly more dense than the culture broth. A class of devices, which facilitates this solid-liquid separation, consists of sedimentation vessels having inclined walls. These settlers are composed of either a single narrow tube or channel inclined from the vertical or a large tank containing several closely spaced, tilted plates. Clarification rates in inclined settlers are often one or two orders of magnitude larger than those observed in vertical settlers.

Enhanced sedimentation in vessels with inclined walls was first documented by Boycott (1920) and recently reviewed by Davis and Acrivos (1985). The enhancement in the sedimentation rate (the "Boycott effect") results from an increase in the surface area available for settling. When a vessel contains inclined surfaces, the particles settle not only onto the vessel bottom but also onto the upward-facing walls. These particles then form thin sediment layers that slide rapidly toward the bottom of the vessel due to gravity, as shown in Fig. 1. Simultaneously, a clarified fluid layer is produced beneath each downward-facing wall. These particle-free layers are more buoyant than the bulk suspension, and the fluid flows rapidly to the top.

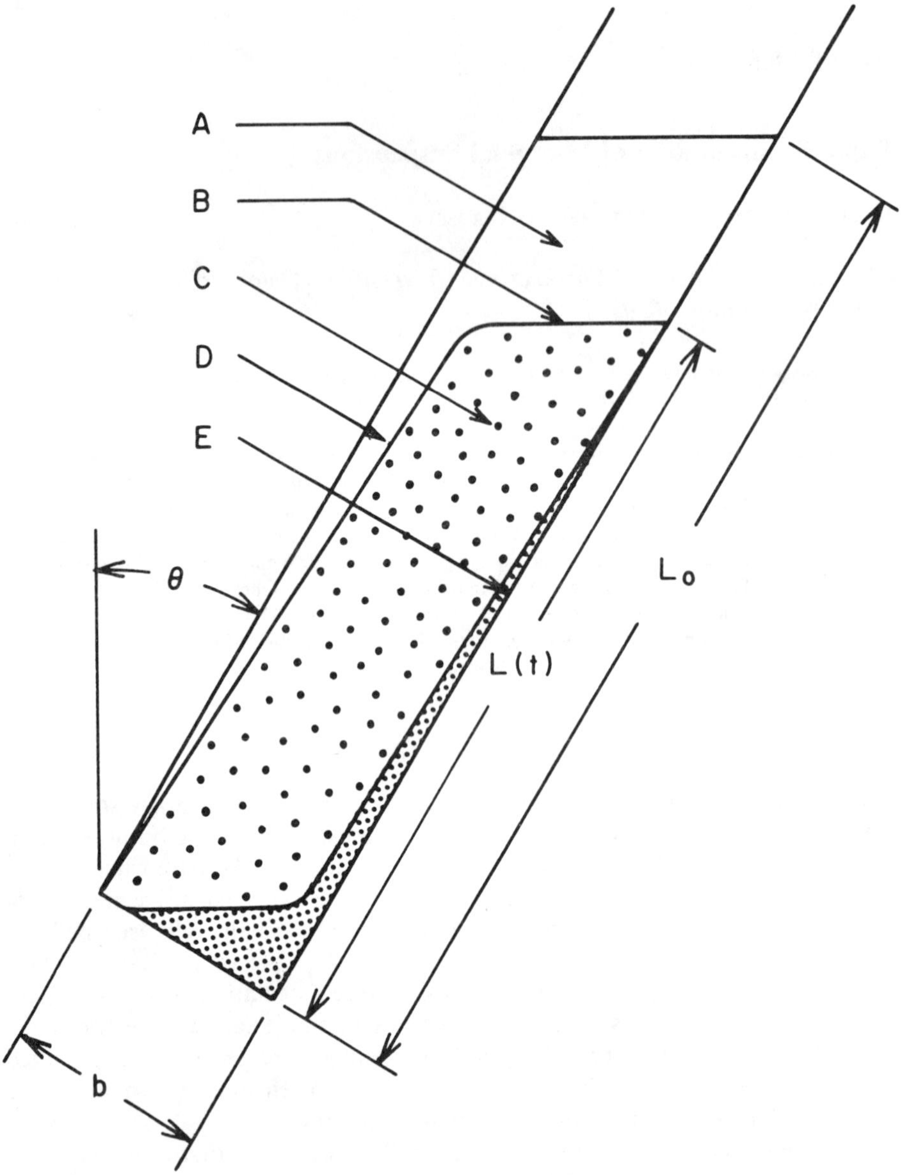

FIG. 1. The different regions that develop during sedimentation in an inclined channel: (A) region of particle-free fluid above the suspension, (B) interface between the particle-free fluid and the suspension, (C) suspension, (D) thin, particle-free fluid layer beneath the downward-facing surface, (E) concentrated sediment. L_0 is the length of the portion of the vessel filled with suspension at time t = 0, L(t) is this length at a later time, θ is the angle of inclination of the vessel walls from the vertical, and b is the spacing between these walls.

The clarification rate in inclined settlers is given by the PNK theory (Ponder 1925; Nakamura and Kuroda 1937). This theory states that the rate of production of clarified fluid is equal to the vertical settling velocity of the particles multiplied by the horizontal projection of the channel area available for settling. Thus, for the parallel-plate geometry of Fig. 1,

$$S^* = v_o\, b(\cos\theta + \frac{L}{b}\sin\theta), \tag{1}$$

where S* is the volumetric rate of production of clarified fluid per unit depth in the third dimension of the vessel, v_o is the vertical settling velocity, θ is the inclination angle of the plates from the vertical, b is the spacing between the plates, and L is the length of the portion of the vessel filled with suspension. This expression was originally proposed on purely geometric arguments; more recently, Acrivos and Herbolzheimer (1979) have used the continuity equation of fluid mechanics to verify its validity. A summary of further mathematical theories for the flow of particles and fluid within inclined settlers can be found in the review by Davis and Acrivos (1985). Equation (1) is strictly valid only if all of the particles fall with the same velocity, v_o. If there is a distribution of shapes or sizes (such as in a suspension of microorganisms), then there will be an accompanying distribution of settling velocities. The sedimentation of such polydisperse suspensions in inclined channels has been described by Davis et al. (1982).

For steady-state continuous settling, equation (1) predicts the volumetric rate at which a suspension can be clarified in a channel of length L. In batch settling, however, L changes with time. In this case, the clarified fluid accumulates at the top of the channel and causes the top of the suspension to fall at the rate

$$\frac{dL}{dt} = \frac{-S^*}{b} = -v_o\,(\cos\theta + \frac{L}{b}\sin\theta). \tag{2}$$

Integrating equation (2) yields

$$L(t) = L_o e^{-(v_o t \sin\theta)/b} - b\cot\theta(1 - e^{-(v_o t \sin\theta)/b}). \tag{3}$$

The majority of experimental studies of sedimentation in inclined tubes and channels have focused on suspensions of inert solid particles; a few experiments have demonstrated the utility of the Boycott effect for microbial suspensions (Walsh and Bungay 1979; Bungay and Millspaugh 1984). The purpose of this study was to investigate the sedimentation of *Saccharomyces cerevisiae* in channels having inclined walls. Specifically, the primary objectives were (1) to compare the results of new yeast sedimentation experiments with theoretical predictions, and (2) to provide a quantitative determination of the effect of process variables (e.g., angle of inclination, spacing between the channel walls) on sedimentation rates.

Materials and Methods

A small vessel having dimension $L_o \leq 20$ cm and b = 0.5 cm was constructed from plexiglass. A spacer insert was also made so that b could be reduced to 0.2 cm. A culture of *Saccharomyces cerevisiae* was grown on a stir-plate at 31 C in a rich glucose medium for 20 h and then allowed to stand for 24 h (or until the release of CO_2 bubbles ceased). Microscopic observation revealed approximately 5×10^7 cells/ml and no flocculation of the cells.

Batch sedimentation experiments of the yeast cells in their growth medium were performed at 20 C with the vessel vertical as well as inclined at several different angles from the vertical. As the settling proceeded, the interface separating the suspension from the clarified fluid at the top of the channel became very diffuse, presumably due to a variation in the settling velocities among the yeast cells. Since this was difficult to measure visually, a laser beam (0.063 cm diameter, 2.0 mW) was passed horizontally through the channel, and the intensity of the transmitted light was read with a photodiode and powermeter. By mounting the laser beam and photodiode to a track that ran along the channel, turbidity readings were used to determine the number density of yeast cells as a function of distance along the channel at any given time. For the experiments with $b = 0.5$ cm, measurements of the location of the diffuse interface were performed every 20 min. Each set of measurements involved taking six to eight readings of the transmitted light intensity at 1.0 cm intervals along the channel; this process took about 4 min. This method was not feasible for the experiments with $b = 0.2$ cm, because the sedimentation took place too rapidly. Instead, the laser beam was placed at a fixed location along the channel. The yeast concentration was then measured as a function of time as the suspension sedimented past the beam. Subsequently, the beam was moved to another location along the channel and the experiment was repeated.

Results and Discussion

Data from inclined settling experiments with $L_0 = 17.8$ cm and $b = 0.5$ cm are shown in Fig. 2. The data points represent the location within the diffuse interface where the number density of cells was one-half of that in the original suspension. The "error bars" are not statistical confidence intervals but rather depict the range over which the number density of cells was one-fourth to three-fourths of that in the original suspension; the spreading of the interface between the top of the suspension and the clarified fluid was significant. The solid lines in Fig. 2 are the theoretical predictions for the descension of the top of the suspension as given by equation (3) with $v_0 = 0.37$ cm/h. The experimental data and predicted curves correlate reasonably well. In particular, the rate of sedimentation increased as the angle of inclination from the vertical was increased. From Fig. 2, the time required for 75% of the yeast to settle was 220 min. for $\theta = 25°$ and only 100 min for $\theta = 57°$. In contrast, the settling rate of the cells when the vessel was vertical was 0.37 cm/h; i.e., 2,160 min were required to settle 75% of the yeast.

In Fig. 3, the rate of descent of the top of the suspension for a more narrow channel ($b = 0.2$ cm) is shown. Reducing the channel width increased the rate of sedimentation, as predicted by the theory. In fact, for $\theta = 60°$, 75% of the yeast was removed from suspension in only 32 min. The data in Fig. 3 generally lie below the prediction from equation (3) using $v_0 = 0.37$ cm/h. As a check, the vertical settling experiment was repeated. The new value of v_0 was 0.60 cm/h. A likely explanation for the observed increase in the vertical settling velocity is that the culture continued to ferment slowly, thereby lowering the density of the fluid. In addition, microscopic observation of the culture revealed that small flocs con-

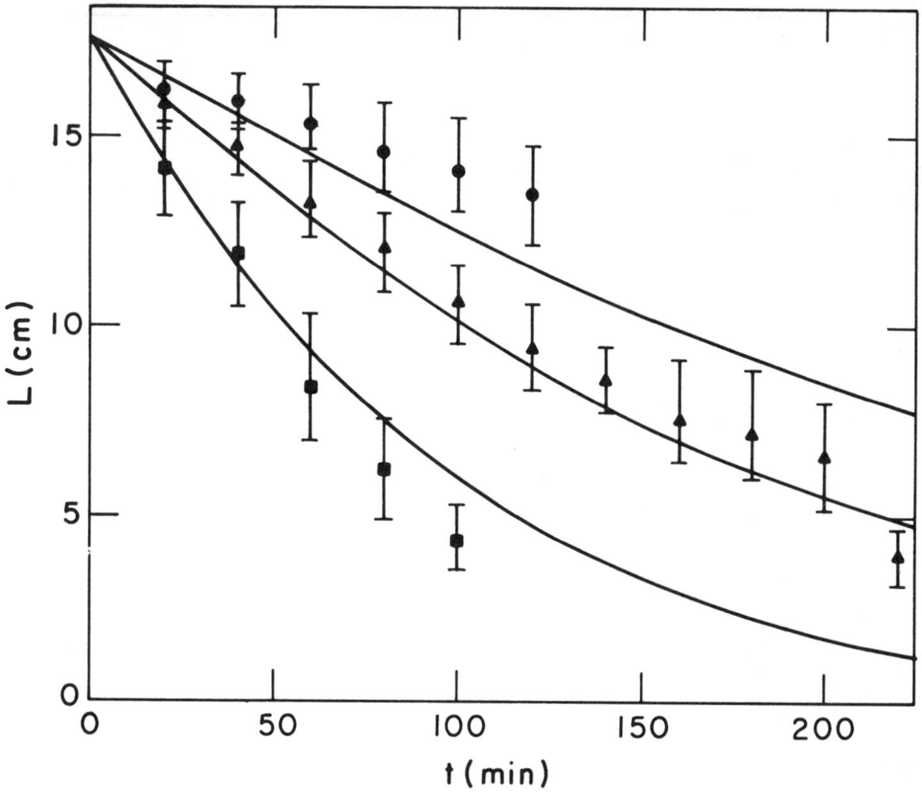

FIG. 2. The descent of the top of the suspension as a function of time during inclined sedimentation with $L_o = 17.8$ cm and $b = 0.5$ cm; ●, experimental data for $\theta = 15°$; ▲, experimental data for $\theta = 25°$; ■, experimental data for $\theta = 57°$. The solid lines are the corresponding theoretical predictions from equation (3) with $v_o = 0.37$ cm/h.

taining several cells had formed; this also would increase the sedimentation velocity. As shown in Fig. 3, the experimental data for $\theta = 60°$ are in excellent agreement with the theory using this new vertical settling velocity.

Conclusions

Sedimentation in inclined channels offers a promising and inexpensive way to separate yeast cells from suspension. The theory and supportive experiments show that the settling rate is enhanced by increasing the angle of inclination from the vertical and decreasing the spacing between the plates of an inclined channel. However, if angles greater than about 60° are used, yeast cells do not readily slide down the lower wall. In addition, for continuous-feeding inclined settlers, the spacing between the parallel plates of the channel should not be less than a few millimeters, as, in this case, the upward flow of fluid remixes the sediment layer with the suspension.

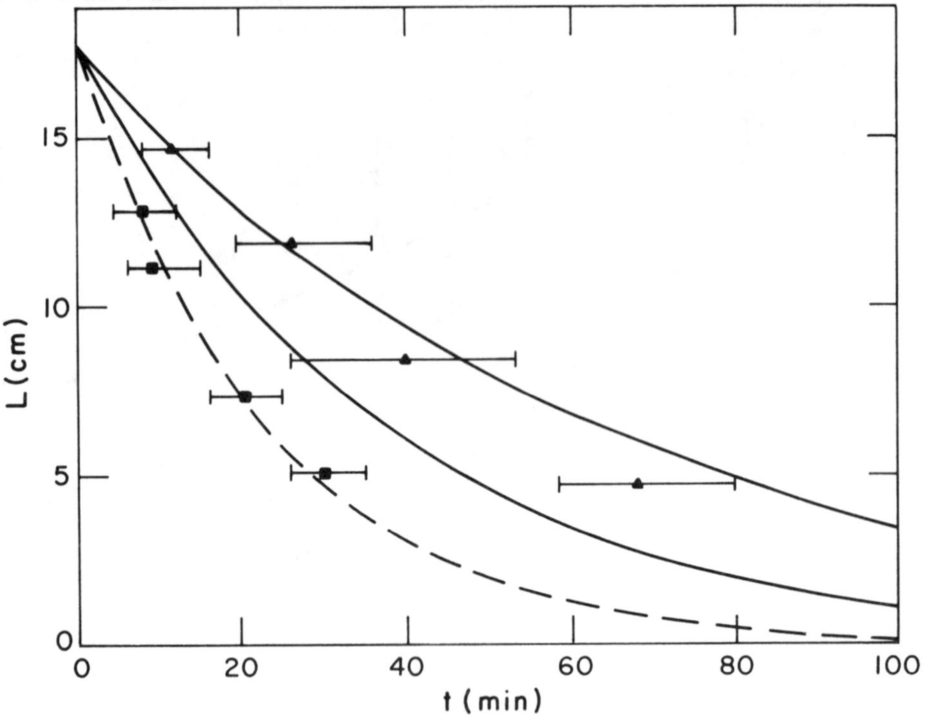

FIG. 3. The descent of the top of the suspension as a function of time during inclined sedimentation with $L_o = 17.8$ cm and $b = 0.2$ cm; ▲, experimental data for $\theta = 30°$; ■, experimental data for $\theta = 60°$. The solid lines are the corresponding theoretical predictions from equation (3) using $v_o = 0.37$ cm/h. The dashed line is the theoretical prediction for $\theta = 60°$ using $v_o = 0.60$ cm/h in equation (3).

Acknowledgments

This research was supported by a Grant-in-Aid from the University of Colorado and by the AMOCO Foundation.

Literature Cited

Acrivos, A., and E. Herbolzheimer. 1979. Enhanced sedimentation in settling tanks with inclined walls. *J. Fluid Mech.* 2:435–457.

Boycott, A. E. 1920. Sedimentation of blood corpuscles. *Nature* 104:532.

Bungay, H. R., and M. P. Millspaugh. 1984. Cross-flow lamellar settlers for microbial cells. *Biotechnol. Bioeng.* 26:640–641.

Davis, R. H., and A. Acrivos. 1985. Sedimentation of noncolloidal particles at low Reynolds numbers. *Annu. Rev. Fluid Mech.* 17:91–118.

Davis, R. H., E. Herbolzheimer, and A. Acrivos. 1982. The sedimentation of polydisperse suspensions in vessels having inclined walls. *Intl. J. Multiphase Flow* 8:571–585.

Nakamura, N., and K. Kuroda. 1937. La cause de l'acceleration de la vitesse de sedimentation des suspensions dans les recipients inclines. *Keijo J. Med.* 8:256–296.

Ponder, E. 1925. On sedimentation and rouleaux formation. *Q. J. Exp. Physiol.* 15:235–252.

Walsh, T. J., and H. R. Bungay. 1979. Shallow-depth sedimentation of yeast cells. *Biotechnol. Bioeng.* 21:1081–1084.

CHAPTER 52

Isolation and Salinity Responses of Actinomycetes from Louisiana Coastal Wetlands

P. D. Zawodny, S. P. Meyers, and R. J. Portier

Louisiana State University, Baton Rouge, Louisiana 70803-7503

Field and microcosm approaches were evaluated for selective isolation of actinomycetes from several aquatic habitats along the salinity continuum of the Louisiana Barataria Bay drainage basin. Morphology, chromogenicity, pigmentation patterns, chitin use, and halotolerance were used to characterize the isolates. Use of a membrane technique in a chitin medium proved to be the most effective isolation technique. Actinomycete biomass was estimated to be 10^4–10^5 CFU (colony forming units) per gram sediment (wet weight). Streptomycetes were predominant in sediment samples, comprising 89% of the total, with approximately 75% of the isolates chitinolytic. Those from brackish marsh and salt marsh sites were more salt tolerant than freshwater swamp isolates and sporulated at higher salinities. Organisms from microcosms exhibited less diversity than those from field sites. A shift occurred in saltwater-amended microcosms toward actinomycetes with higher salt tolerance for growth and sporulation. Ecological significance of the actinomycetes in terms of salinity effect is discussed.

Introduction

Investigations over the past 20 yr have shown salt marshes to be sites of high rates of primary production with associated complex biogeochemical processes (Teal 1962; Odum and de la Cruz 1967; Armstrong 1975). It is well recognized that microbial populations have vital roles in transfer of energy and recycling of materials (Teal 1962; Hood 1973). Studies by Hood and Meyers (1977) showed the energy-rich structural polymer, chitin, to be rapidly mineralized in the lower Mississippi River delta estuary. The activity of the chitin decomposers was postulated to comprise a significant process in which carbon and nitrogen is released from the large chitinous pool for use by heterotrophs (Roach and Silvey 1959; Grein and Meyers 1958; Okazaki and Okami 1972; Hood 1973; Walker and Colwell 1975; Cross 1981; Veiga et al. 1983; Willingham et al. 1966).

Although actinomycetes have been consistently recovered from coastal waters and sediments (Grein and Meyers 1958), little is known of their biomass activity and ecology compared with freshwater systems. Those from intertidal regions may be either dormant forms from terrestrial or marine habitats, active immigrants, or representatives of a self-propagating indigenous microflora. Recently, Aumen (1980) examined chitin material from freshwater streams and observed a population of streptomycetes as secondary colonizers that proliferated as chitin degradation progressed. A role for such organisms in degradation of chitinous material in aquatic ecosystems was suggested.

The studies of Hunter (1978) and Hunter et al. (1981) have provided the most comprehensive characterization of salt marsh actinomycete populations. Her investigations included a wide range of isolation methodologies to recover actinomycete types from water, sediments, and living and dead plant material. Marsh actinomycete isolates comprised members of the genus *Streptomyces, Micromonospora,* and *Nocardia,* especially streptomycete types. The presence of a distinct actinomycete salt marsh population was postulated.

In the present investigation, field and microcosm approaches were evaluated for recovery of actinomycetes from estuarine sediments in the Louisiana Barataria Bay drainage system. Samples collected were examined for preliminary characterization of the estuarine actinomycete population. The response of these organisms under various salinities was evaluated for expression of euryhaline types.

Materials and Methods

Sample collection. Three stations along the Barataria Bay drainage system were chosen to include a freshwater swamp, a brackish marsh, and a saline marsh. Bordered by Mississippi River tributaries on the north, east, and west, and by the Gulf of Mexico on the south, this region of low, flat wetlands follows a salinity continuum of 0 ppt in the upland areas to 24–30 ppt at the Gulf of Mexico. At the saline site, salinity fluctuates diurnally and seasonally, ranging from 8–28 ppt. Extremes in salinity occur during periods of unusual water movement. The brackish marsh site, influenced by both freshwater run-off and marine water, is dominated by the hydrology of the river-fed channel system. The salinity at this study site was 7–17 ppt during the sampling period. The freshwater sample station was situated in a relatively pristine hardwood swamp forest in a portion of the deltaic wetlands fed by river and rain water and not exposed to marine intrusion.

Sites were sampled during spring and summer months for evaluation of actinomycete isolation procedures. Collections were made throughout the year for assessment of actinomycete populations. Sediment was collected with a soil core sampler or spatula from the top 10 cm of the soil profile and placed in sterile plastic bags. Samples were kept on ice during transport to the laboratory and were processed within 24 h.

Microcosm studies. The microcosm system used is that developed and described by Portier (1982) and Portier and Meyers (1982). Each unit consisted of a 2-liter glass vessel and teflon lid with appropriate probe and sampling ports. Natural site water was aged in the microcosm units for 72 h prior to introduction of sediment to allow volatilization and mixing to reach equilibrium. Thirty grams of sediment were added to 1.8 l of water. Two series of microcosm studies were run. In Study 1, natural fresh water and swamp sediment were used as inoculum. Sterile aged sea water was used to sequentially increase salinity in the microcosm chambers to 4, 9, 15, 19, and 24 ppt. Samples were taken at days 1, 3, 5, 13, 18, and 24. Sediment and water mixture were allowed to equilibrate for at least 48 h after each addition before assay. Study 2 differed in that 1.0 g crab chitin was added to one member of each pair of microcosms.

Sterile air from a compressor was continuously passed through the frittered glass bottom for aeration and mixing. Sterile site water was added via a peristaltic pump at the rate of 0.21 ml/min. Water was simultaneously withdrawn to maintain constant volume. Particulates were retained by a sediment trap. Temperature was controlled at 28 C and pH was monitored by a pH controller (Horizon Ecology Co. Model 5997-20). A top sampling port allowed removal of sediment and water for microbiological assay.

Isolation techniques. Microcosm samples and topmost sediment from field sites were assayed for actinomycetes by three methods, i.e., membrane filter, desiccation, and pour plate. Three isolation media, chitin, SCN, and Jensen's (modified), were used. Chitin and starch-casein-nitrate (SCN) media were used with the membrane filter isolation method, while the chitin medium was employed with the desiccation isolation method. Jensen's agar was used with the pour plate technique. The salinity of all isolation media was similar to that of the original sample. Duplicate samples were plated, in triplicate, from each dilution and for all isolation methods. These were incubated at 28 C, and examined for actinomycete growth after 2-3 wk to allow development of slow-growing organisms.

Evaluation of isolation methodologies. Sediment from the top 10 cm of marsh and swamp sites was subjected to a variety of physical, chemical, and enrichment pretreatment procedures (Fig. 1). These isolation methodologies were compared

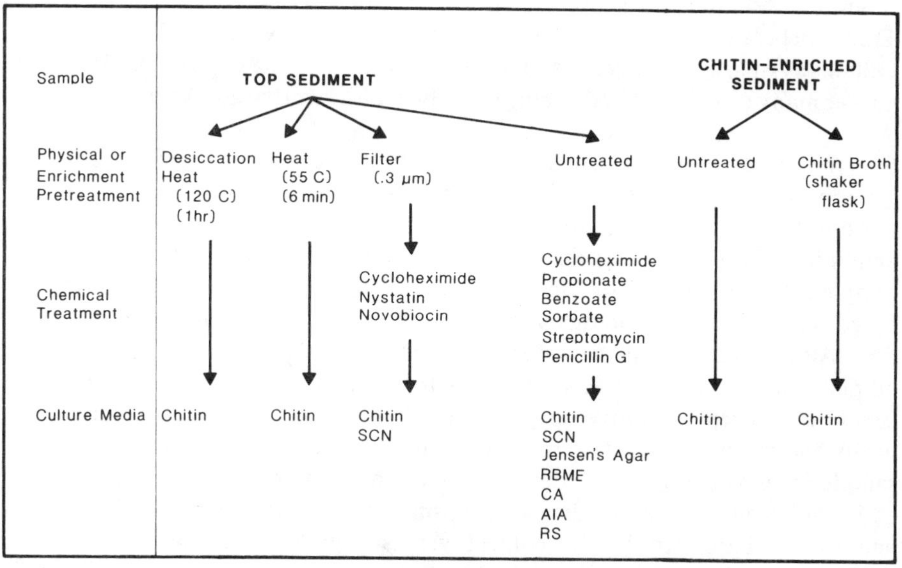

FIG. 1. Isolation methodologies tested for actinomycete recovery. Starch casein nitrate (SCN); Cellulose asparagine (CA); actinomycete isolation agar (AIA); Rose bengal malt extract (RBME); Rao and Subrahmanyan's Agar (RS).

for their ability to select for aerobic actinomycetes in a variety of coastal sediments. Recently, Cross (1982) and Goodfellow and Williams (1983) have reviewed isolation techniques for recovery of actinomycetes from a variety of materials.

Enrichment pretreatments were carried out *in situ* and in shaker flasks. At the salt marsh study site, ground crab chitin (500–250 μm) was placed on the surface of the sediment at low tide. A plastic (37 cm diam) bucket, with the bottom removed and side holes to permit flushing, was used to contain the enriched site. Chitin-enriched and nonenriched sediment was sampled for three consecutive days. Samples were added to 50 ml of chitin broth in 250-ml flasks, and subsamples removed after 3, 6, and 24 h of shaking.

Desiccation (Makkar and Cross 1982), heat (Rowbotham and Cross 1977), and membrane filter (Hirsch and Christensen 1983) procedures were used to reduce the extant bacterial population. The membrane filter method involved placing a membrane filter between the material to be assayed and the surface of the agar media for a brief incubation period. Sediment samples were desiccated at room temperature or at 35 C in a convection oven for 48 h. Several dry sediment samples were subsequently heated to 120 C for 1 h. For the moist heat treatment, diluted sediment was heated for 6 min in a diluent maintained in a 55 C water bath.

Chemical treatments included a variety of antifungal and antibacterial agents incorporated into the isolation media. The substance employed and quantities per plate were: sodium propionate, 60 mg; sodium benzoate, 0.5 mg; sorbate, 0.5 mg; penicillin G, 0.05 mg; streptomycin, 0.5 mg; cycloheximide, 1.0 mg; cycloheximide plus nystatin, 1.5 mg each; and novobiocin, 0.5 mg.

Several isolation media were compared using distilled or natural sea water. These included chitin agar, starch-casein-nitrate (SCN) agar, Jensen's Agar, Cellulose-Asparagine agar, actinomycete isolation agar (AIA), Rao and Subrahmanyan's Agar (R&S), and Rose Bengal-Malt Extract Agar.

Selection of actinomycete isolates. Every effort was taken to treat each set of isolation plates identically so that the actinomycetes isolates accurately reflected the number of distinctive colonies observed. Equal attention was given to examination of the surface of each set of plates and to selection of isolates to avoid sample bias. For each set of triplicate plates, different actinomycete colony types, based on colonial morphology and spore mass color, were transferred for isolation. Although streptomycetes were frequently observed upon casual inspection, all plates were scanned at 40X and 100X for other actinomycete forms to ensure isolation of a representative range of microbial types. Isolates, differentiated by genus and pigmentation patterns, were separated according to the origin of the sample from which they were derived. Field samples were grouped by station and by isolation method and media. Microcosm isolates were grouped by experiment and then subgrouped by (1) control versus experimental pairs, (2) the day isolated, and (3) isolation method and media. Gross morphological differences on maintenance media allowed comparison and identification of similar types recovered from different samples or by dissimilar isolation methodology. Axenic cultures were obtained by transferring individual colonies to agar media of the same composition and salinity as that of the isolation media.

Growth and sporulation of isolates on saline media. Twenty-nine actinomycetes from swamp or marsh sediments, and 51 from microcosms, were tested for salt tolerance. SCN agar, adjusted to pH 7.4, was prepared with artificial Instant Ocean® sea salt and distilled water to provide salinities of 0, 10, 20, 40, 70, 100, and 150 ppt. Plates were streaked with washed cells or a spore suspension (Shirling and Gottlieb 1966), incubated at 28 C, and examined after 10 d for growth and sporulation.

Results and Discussion

Recovery of actinomycetes from sediments. A total of 1,100 isolates were obtained from estuarine sediments, 75% of which were chitinolytic. The freshwater swamp isolates exhibited a slightly higher percentage of chitinolytic streptomycetes than did combined brackish and saline marsh isolates. A total of 80% of the isolates from freshwater samples, field and microcosm, were chitin users, compared with 65% of those from sediments exposed to sea water. The organisms from saline sediments were tested in fresh and saline media. No difference in chitinoclastic activities was observed. Of the 782 organisms from untreated sediments, 467 were from the freshwater swamp sampling station, 185 from the brackish marsh site, and 130 from the salt marsh station. Those collected during the microcosm studies accounted for an additional 318 isolates.

The diversity of actinomycetes obtained is a reflection of isolation methodologies used as well as the composition of the microflora of the samples. Similar isolation methods and media were used at each site to facilitate recovery of morphologically comparable organisms from the diverse sediments. However, the diluents and media were adjusted with respect to salinity to approximate *in situ* conditions. A variety of cultures were observed on media of contrasting salinity to minimize overlooking atypical colony formation on saltwater media.

The higher number of actinomycetes from freshwater samples reflects an apparent decrease in the diversity of actinomycetes between freshwater and saline habitats. This may be because of an actual decrease in diversity as marine factors and time/spatial transition in environmental factors become more pronounced. Macrophyte diversity in estuaries follows this pattern. A second factor may be that of the isolation media that did not uniformly satisfy the growth requirements of the viable populations. It is also possible that in brackish and salt marsh sediments, there exists a higher proportion of "stressed" organisms or those present in low concentrations that failed to develop on the isolation plates.

A representative group of 269 isolates was selected on the basis of differential colonial and pigment formation on BA media for preliminary taxonomic studies. This group included 230 *Streptomyces,* 3 *Streptoverticillium,* 13 *nocardioforms, 3 Micromonospora,* 1 *Micropolyspora,* and 10 unidentified cultures.

All samples examined yielded primarily streptomycetes. Those from freshwater and saline sediments were predominately gray-spored. In contrast, the majority of brackish isolates were green or yellow-green-spored. The red-, yellow-, blue-, and brown-spored cultures comprised fewer than 30% of the streptomycetes. Chromogenicity (melanin formation) was relatively rare among field isolates. Only two of the 12 saline isolates tested were melanin producers. However, morphologically similar types, which formed dark brown/black pigments on complex media, were frequently recovered from saline samples.

Specific relationships between spore mass color or chromogenicity and survival or growth in estuaries are unknown. However, Tresner et al. (1968) have observed that halotolerance was statistically associated with various taxonomic features. White- and yellow-spored streptomycetes exhibited higher salt tolerance than other groups. Nonchromogenic streptomycetes were more salt tolerant than were melanin producers. Broad halotolerance may give organisms an adaptive advantage in saline regions.

In vitro exposure of actinomycete populations to sea water. Freshwater sediment maintained in microcosms was exposed to sea water and periodically assayed for actinomycetes. Isolates recovered were not as diverse as those from salt marsh sediments. Those from freshwater swamp sediment adjusted to a salinity of 24 ppt were similar to isolates from the same sediment sampled at lower salinity levels as well as from control microcosms. However, over half of the isolate types from control microcosms were not recovered from sediments exposed to sea water. Of the organisms collected during the microcosm study, 70% of those from controls and 90% from experimental units were gray- or yellow-green-spored.

The number of isolates obtained in Microcosm Studies 1 and 2 are shown in Figs. 2 and 3, respectively. The salinity of experimental units is noted for each sampling period. The two studies demonstrate a peak in the number of isolates after 5 d, decreasing in 3 to 5 d. Similar trends were observed in both control and experimental microcosms, although population peaks appear to occur later in ex-

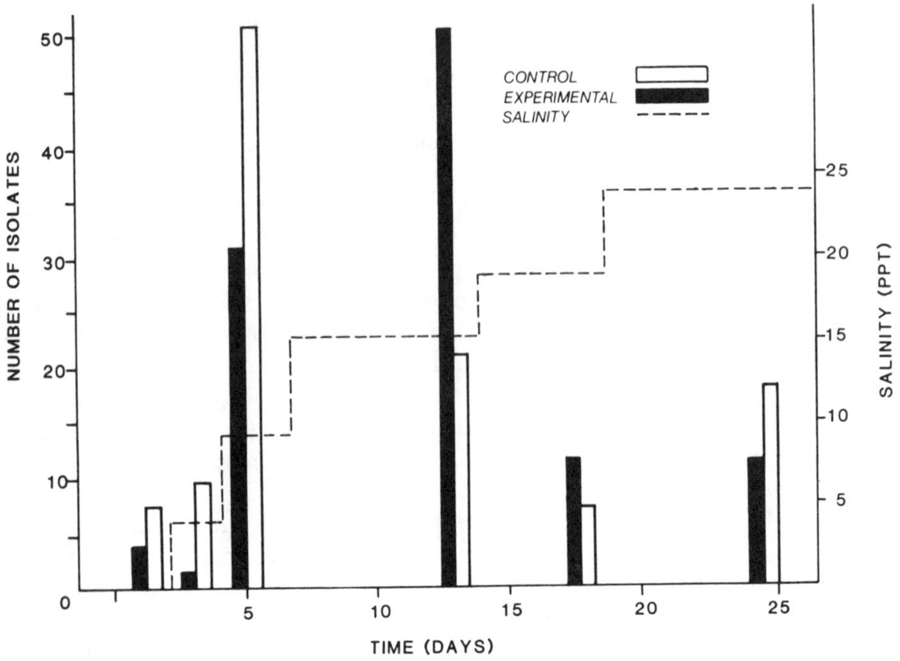

FIG. 2. Histogram of number of isolates collected—Microcosm study 1.

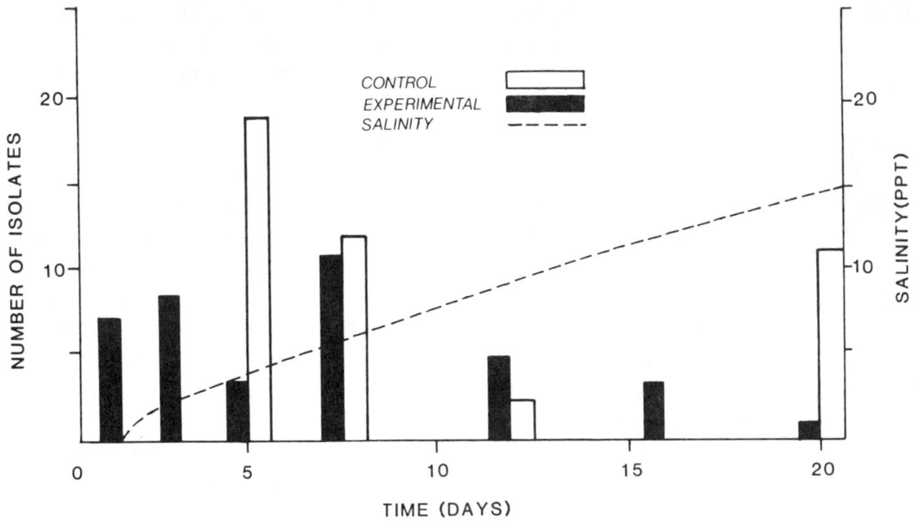

FIG. 3. Histogram of number of isolates collected—Microcosm study 2.

perimental units. The increase in isolates at 5–10 d may be because of a variety of factors, including containerization, changes in competing populations, or delayed responses to changes in oxygen and nutrient level, or disruption of the sediment matrix with dispersal of actinomycete fragments or spores. These two studies showed similar responses by actinomycete populations to salt water addition. Yet, in the first study salinity levels were raised discretely by increments of 4 to 6 ppt and in the second, salt water was added continuously. Fewer isolates were obtained during the latter study. This may be because of development of a population of filamentous fungi that rapidly overgrew isolation plates. Plates inoculated with material from the control microcosms on days 1, 3, and 16 were covered by such fungi with no actinomycete colonies observed.

Results from these studies showed a shift in populations toward actinomycetes with higher salt tolerance limits. Clearly, the microcosm could maintain actinomycete populations that respond to environmental stresses. This indicates its potential use as an enrichment step to select for organisms with desired capacities (or for those not recovered under standard isolation conditions).

Comparison of isolation methodologies. Of the isolation methodologies tested, the desiccation and membrane filter techniques with chitin media, plus antibiotics, proved to be most selective for actinomycetes. Both techniques were used in the assessment of field and microcosm sediment samples for viable aerobic actinomycetes. SCN medium, plus novobiocin, also was used in combination with the membrane filter techniques. Novobiocin reduced the number of streptomycetes and allowed expression of micromonosporae possibly present at lower concentrations. The desiccation technique reduced vegetation growth and was developed for recovery of actinoplanetes and other actinomycetes whose spores resist inactivation under dry conditions. Jensen's agar pour-plates were used with microcosm samples. After 2 wk, bacteria decreased in number and slowly formed restricted colonies, which did not interfere with actinomycete development.

Cycloheximide, 35 mg/ml, adequately retarded filamentous fungal growth on both freshwater and saline Jensen's agar. These methods were tested with sediment inoculated with axenic cultures to demonstrate recovery of known species of *Streptomyces, Nocardia,* and *Actinoplanes.*

The membrane filter isolation technique was developed recently by Hirsch and Christensen (1983) in which isolation of *Streptomycetaceae, Micromonosporaceae, Nocardiaceae,* and *Actinoplanaceae* was reported. Penetration of branching hyphae allowed establishment of actinomycete colonies on both surfaces of the filter. The present study supports the observations of Hirsch and Christensen (1983) on the effectiveness of the filter membrane technique in isolation of actinomycete populations. Results suggest that use of the membrane filter procedure, together with an appropriate medium or sample treatment, may provide a valuable tool for selective isolation of organisms of special interest from a variety of natural materials.

The total number of isolates recovered by selected isolation techniques and media is presented in Table 1. Over half of the actinomycetes collected were from membrane filter/chitin agar isolation plates. Distinctive differences were observed between numbers and types of actinomycetes on chitin and SCN plates inoculated by the membrane filter method. A greater number of isolates were obtained from chitin plates. Although the desiccation isolation method provided only 8% of the actinomycetes identified to genus, this group included 27% of the nonstreptomyces. All methods and media were selective for actinomycetes over bacteria and filamentous fungi. A wide variety of streptomycetes was observed on all isolation plates except on starch-casein-nitrate (plus 25 mg/ml novobiocin) media. In this instance, streptomycete growth was restricted with poor sporulation. Small, fragmenting, nonsporulating white and yellow colonies were frequently encountered that exhibited typical streptomycete morphology upon transfer.

TABLE 1. *Isolation methods and actinomycete types recovered*

Method	Media	Number of Isolates	No. of Isolates Identified	% Non-Streptomyces
membrane filter	chitin	636	191	19
membrane filter	SCN[a]	179	13	1
desiccation	chitin	177	44	8
pour plate	chitin	108	21	2
		1100	269	

[a] SCN = starch casein nitrate.

All pour-plate and spread-plate isolations, with the above exceptions, were overgrown rapidly by filamentous fungi and/or bacteria. Although Actinomycete Isolation Agar, Rao and Subrahmanyan's Agar, and cellulose-asparagine agar exhibited actinomycete colonies, they were not as selective for actinomycetes as was chitin and SCN media. Most of the antimicrobial agents tested were ineffective against interfering microbial populations with exception of cycloheximide, nystatin, and novobiocin. The rose bengal-malt extract agar was completely overgrown by bacteria with no actinomycetes observed. The membrane filter procedure and the desiccation technique, in combination with selective media and antibiotics, were the most effective isolation procedures tested.

Few nonstreptomycetes were recovered. Such organisms may have been present at levels below the 10^4–10^5 range examined, or the methods may not have provided necessary growth conditions for their effective isolation. These results indicate a need for future development of selective media and treatments to improve techniques to recover nonstreptomycetes.

Salt tolerance. A group of 49 microcosm isolates were tested for salt tolerance. Each isolate was examined on a series of media of increasing salinity for the maximal concentration of salt allowing expression of the morphological characteristics of growth and sporulation. Isolates were grouped according to source of the sample material. Those from microcosms were divided into control and experimental sets. The former group included organisms from the control microcosms throughout the study period. Isolates from the experimental microcosms were further subdivided by salinity level. The percentage of isolates in each group capable of growth/sporulation at the specified sea salt concentration is reported in Table 2. Over 90% of the control microcosm isolates did not grow above 40 ppt ASW. All experimental microcosm isolates from salinity levels of 15 and 24 ppt grew at 40 ppt ASW or higher. These data suggest that organisms recovered had become acclimated to salt water; or they suggest that those which predominated in freshwater conditions were inhibited by sea water, allowing more euryhaline organisms to dominate. Okami and Okazaki (1974, 1975) have shown the ability of actinomycetes to attain increased salt tolerance, suggesting that the data obtained in the present study are attributed to acclimation mechanisms of the actinomycetes.

None of the cultures tested grew in 150 ppt ASW. In contrast, all grew on freshwater media. Each isolate exhibited colony morphology that was not uniform on all media. Heaviest growth of substrate and aerial mycelium generally was observed on freshwater media with development becoming more sparse with increasing salt concentrations. Aerial hyphae were observed at salt concentrations above levels that limited sporulation. At the upper range of salt tolerance for specific isolates, aerial hyphae disappeared, and growth consisted of a thin, diffuse network of substrate mycelium. The media provided growth requirements for all organisms tested. However, the presence of sea salts influenced colonial development. Increased salt concentration decreased the ability of all isolates to produce spores and vegetative mycelia although salt tolerance limits varied between test organisms. Isolates from brackish marsh and salt marsh sites were more salt tolerant than freshwater swamp forms and were capable of sporulation at higher salinities.

The effect of salinity on expression of actinomycete populations may allow organisms to survive and grow in regions that do not allow reproduction by spore development. Viable actinomycetes may be limited to sites or microzones of active metabolism. This would lead to disruption of the mycelium and dispersal of the less resistant vegetative form. This would reduce survival and distribution of those actinomycetes with sporulation mechanisms sensitive to *in situ* salinity levels. Lechevalier (1981) has suggested that some actinomycetes have developed specialized mechanisms for survival under harsh conditions.

TABLE 2. *Percentage of microcosm isolates exhibiting growth or sporulation at specific sea salt concentrations*

Number of Isolates	Microcosm	Response	Nonsporulating	ppt ASW[a]						
				0	10	20	40	70	100	150
13	control	growth		100%	85%	85%	70%	8%	0%	0%
		sporulation	15%	85%	70%	47%	8%	0%	0%	0%
18	experimental (9 ppt)	growth		100%	89%	89%	89%	33%	0%	0%
		sporulation	27%	73%	62%	34%	17%	0%	0%	0%
12	experimental (15 ppt)	growth		100%	100%	100%	100%	50%	0%	0%
		sporulation	25%	75%	67%	42%	34%	8%	0%	0%
6	experimental (14 ppt)	growth		100%	100%	100%	100%	17%	0%	0%
		sporulation	0%	100%	83%	83%	17%	0%	0%	0%

[a] ASW = Artificial Sea Water (ppt ASW is equivalent to 0/00 natural sea water).

TABLE 3. *Percentage of field isolates growth or sporulation at specific sea salt concentrations*

Number of Isolates	Microcosm	Response	Nonsporulating	ppt ASW[a]						
				0	10	20	40	70	100	150
9	fresh	growth		100%	100%	100%	100%	22%	0%	0%
		sporulation	44%	56%	56%	45%	11%	0%	0%	0%
9	brackish	growth		100%	100%	100%	100%	78%	44%	0%
		sporulation	33%	67%	67%	67%	45%	11%	0%	0%
7	saline	growth		100%	100%	100%	100%	100%	29%	0%
		sporulation	4%	86%	86%	86%	43%	14%	0%	0%

[a] ASW = Artificial Sea Water (ppt ASW is equivalent to 0/00 natural sea water).

Freshwater, brackish, and salt marsh isolates were examined for salt tolerance (Table 3). All isolates tested grew in 40 ppt ASW, and nearly 80% of the freshwater isolates grew in 70 or 100 ppt ASW. Sporulation was inhibited more by lower salt concentrations than by growth, but tolerance patterns followed those observed for growth responses. Earlier, Grein and Meyers (1958) observed that species of actinomycetes of terrestrial origin are characterized by halotolerants.

Conclusions

The present investigation indicates that actinomycete populations may vary in composition throughout an estuary. This may be an important source of organisms with biochemical activities expressed over a wide range of environmental conditions. Workers in Japan (Okazaki et al. 1975; Okami et al. 1976) have demonstrated the recovery of actinomycetes from coastal regions. Under marine cultural conditions, these organisms exhibited biochemical activities significantly different from those observed on standard media. Emphasis on further development of selective media and use of innovative sample treatment should allow more definitive analyses of the occurrence and ecological role of actinomycetes from saline ecosystems.

Acknowledgments

The active support of the Louisiana Sea Grant Program and Merck, Sharpe, and Dohme (Rahway, NJ) in the development of these investigations is readily acknowledged.

Literature Cited

Armstrong, W. 1975. Waterlogged soils. Pages 181–218 *in* J. R. Etherington, ed. *Environmental and Plant Ecology*. John Wiley and Sons, London.

Aumen, N. G. 1980. Microbial succession on a chitinous substrate in a woodland stream. *Microb. Ecol.* 6:317–327.

Cross, T. 1981. Aquatic actinomycetes: A critical survey of the occurrence, growth and role of actinomycetes in aquatic habitats. *J. Appl. Bacteriol.* 50:397–423.

______. 1982. Actinomycetes: A continuing source of new metabolites. *Dev. Ind. Microbiol.* 23:1–18.

Goodfellow, M., and S. T. Williams. 1983. Ecology of actinomycetes. *Annu. Rev. Microbiol.* 37:189–216.

Grein, A., and S. P. Meyers. 1958. Growth characteristics and antibiotic production of actinomycetes isolated from littoral sediments and materials suspended in sea water. *J. Bacteriol.* 76:457–463.

Hirsch, C. E., and D. L. Christensen. 1983. A novel method for the selective isolation of actinomycetes. *Appl. Environ. Microbiol.* 46:925–929.

Hood, M. A. 1973. Chitin degradation in the salt marsh environment. Ph.D. thesis. Louisiana State University, Baton Rouge, LA, 156 p.

Hood, M. A. and S. P. Meyers. 1977. Rates of chitin degradation in an estuarine environment. *J. Oceanogr. Soc., Jap.* 33:328–334.

Hunter, J. C. 1978. Actinomycetes of a salt marsh. Ph.D. thesis. Rutgers University, New Brunswick, NJ, 271 p.

Hunter, J. C., D. E. Eveleigh, and G. Cassella. 1981. Actinomycetes of a salt marsh. Pages 195–200 *in* K. P. Schaal and G. Pulverer, eds., *Actinomycetes.* Proceedings of the Fourth International Symposium on Actinomycete Biology. Gustav Fischer Verlag, New York.

Lechevalier, M. P. 1981. Ecological associations involving actinomycetes. Pages 159–166 *in* K. P. Schaal and G. Pulverer, eds., *Actinomycetes.* Proceedings of the Fourth International Symposium on Actinomycete Biology. Gustav Fischer Verlag, New York.

Makkar, N. S., and T. Cross. 1982. Actinoplanetes in soil and on plant litter from freshwater habitats. *J. Appl. Bacteriol.* 52:209–218.

Odum, E. P., and A. A. de la Cruz. 1967. Particulate organic detritus in a Georgia salt marsh-estuarine ecosystem. Pages 383–388 *in* G. H. Lauff, ed., *Estuaries.* Am. Assoc. Adv. Sci., Washington, DC.

Okami, Y., and T. Okazaki. 1974. Studies on marine microorganisms. III. Transport of spores of actinomycete into shallow sea mud and the effect of salt and temperature on their survival. *J. Antibiot.* 27:240–247.

Okami, Y., T. Okazaki, T. Kitahara, and H. Umezawa. 1976. Studies on marine microorganisms. V. A new antibiotic, aplasmomycin, produced by a streptomycete isolated from shallow sea mud. *J. Antibiot.* 29:1019–1025.

Okazaki, T. T., Kitachara, and Y. Okami. 1975. Studies on marine microorganisms. IV. A new antibiotic SS-228 Y produced by *Chainia* isolated from shallow sea mud. *J. Antibiot.* 28:176–184.

Okazaki, T., and Y. Okami. 1972. Studies on marine microorganisms. II. Actinomycetes in Sagami Bay and their antibiotic substances. *J. Antibiot.* 25:461–466.

______. 1975. Actinomycetes tolerant to increased NaCl concentration and their metabolites. *J. Ferment. Technol.* 53:833–840.

Portier, R. J. 1982. Correlative field and laboratory microcosm approaches in ascertaining xenobiotic effect and fate in diverse aquatic microenvironments. Ph.D. thesis. Louisiana State University, Baton Rouge, LA, 212 p.

Portier, R. J., and S. P. Meyers. 1982. Use of microcosms for analyses of stress-related factors in estuarine-wetland ecosystems. Pages 375–387 *in* B. Gopal, R. E. Turner, R. G. Wetzel and D. F. Whigham, eds., *Wetlands: Ecology and Management.* Proceedings of the First International Wetlands Conference. National Institute of Ecology and International Scientific Publications. Jaipur, India.

Roach, A. W., and J. K. G. Silvey. 1959. The occurrence of marine actinomycetes in Texas Gulf Coast substrates. *Am. Midl. Nat.* 62:482–499.

Rowbotham, T. J., and T. Cross. 1977. Ecology of *Rhodococcus coprophilus* and associated actinomycetes in freshwater and agricultural habitats. *J. Gen. Microbiol.* 100:231–240.

Shirling, E. G., and D. Gottlieb. 1966. Methods for characterization of *Streptomyces* species. *Int. J. Syst. Bacteriol.* 16:313–340.

Teal, J. M. 1962. Energy flows in the salt marsh ecosystem of Georgia. *Ecology* 43:614–624.

Tresner, H. D., J. A. Hayes, and E. J. Backus. 1968. Differential tolerance of streptomyces to sodium chloride as a taxonomic aid. *Appl. Microbiol.* 16:1134–1136.

Veiga, M., A. Esparis, and J. Fabregas. 1983. Isolation of cellulytic actinomycetes from marine sediments. *Appl. Environ. Microbiol.* 46:286–287.

Walker, J. D., and R. R. Colwell. 1975. Factors affecting enumeration and isolation of actinomycetes from Chesapeake Bay and southwestern Atlantic Ocean sediments. *Mar. Biol.* 30:192–201.

Willingham, C. A., A. W. Roach, and J. K. G. Silvey. 1966. Comparative studies of substrate degradation by marine-occurring actinomycetes. *Am. Midl. Nat.* 75:232–241.

CHAPTER **53**

Characterization of Bacterial Populations in an Industrial Cooling System

CYNTHIA A. LIEBERT AND MARY A. HOOD

Department of Biology, University of West Florida, Pensacola, Florida 32514

A study was undertaken to examine the bacterial populations in water cooling units of an industrial textile plant. Water samples from the units were taken during 60-d operational cycles and plated onto Plate Count agar (Difco). In addition, stainless steel sampling plates were placed into the cooling units, removed at selected intervals, and cultured. One hundred and twenty-three strains were isolated and characterized. A series of 43 tests were used to differentiate the strains, and cellular fatty acid content of representative genera was determined. The culturable organisms comprised six predominant genera: *Sphaerotilus, Gallionella, Caulobacter, Pseudomonas, Flavobacterium, Corynebacterium,* and *Cytophaga.* Colonization of the stainless steel sampling plates was observed after 1, 7, 21, and 59 d using scanning electron microscopy. The initial colonizers were iron bacteria, while the secondary colonizers were elongate, flexible rods. Subsequent colonization was by *Caulobacter*-like organisms and by rods approximately 3 μm in length enmeshed in exopolymer material.

INTRODUCTION

Bacterial contamination of air-cooling systems has been associated with a variety of respiratory disorders (Banaszak et al. 1970; Cordes et al. 1980; Flaherty et al. 1984a,b). The presence of bacteria is also responsible for biofilm formation that results in fouling and corrosion problems leading to inefficient operation of the cooling units (Birchall 1979; Friend and Whitekettle 1980). Bacteria, often the pioneer colonizers of surfaces in many aquatic systems (Zobell 1943), become attached to substrates, resulting in the formation of a slime matrix that facilitates subsequent microbial colonization by enhancing adhesion and producing a protective microenvironment (Alexander 1971).

The object of this study was to complete the description of the biofouling process begun in a previous study (Liebert et al. 1983), which enumerated the bacteria present in four air-cooling/air-washing units at a nylon-producing textile facility. In this phase of the study, populations sampled at discrete time intervals were characterized by determining morphological and biochemical traits.

MATERIALS AND METHODS

Sample collection. Representative samples of circulating, chilled water (17 C, pH 6.4) were collected in sterile glass containers at selected intervals, i.e., 1, 21, and 59 d. Stainless steel sampling plates, 2.5 × 2.5 cm, were submerged in cooling-unit water and removed at the same time intervals. The plates were then scraped, suspending the accumulated biofilm in 10.0 ml of phosphate-buffered distilled water (0.3 m*M* KH_2PO_4, pH 7.4). Stainless steel discs, 0.8 cm^2, were also submerged with the sampling plates and removed after 1, 7, 21, and 59 d. All samples were immediately transported to the laboratory and processed within 1 h of collection.

Isolation of microorganisms. Water and biofilm samples were diluted in phosphate-buffered distilled water and triplicate samples of appropriate dilutions were spread onto Plate Count agar (PCA) (Difco) and Nutrient agar (NA) (Difco). Culture plates of each medium were inoculated and incubated at 20 C for 14 d. After incubation, colonies were randomly selected and streaked onto PCA to isolate individual colonies.

Cultural and morphological characteristics. Colony morphology and pigmentation were observed when grown on PCA for 48 h at 20 C. Isolated colonies were picked and restreaked on PCA to check for purity. Gram stains were performed on 24 h cultures from PCA plates. Dimensions, motility, and arrangement of cells were observed with light and transmission electron microscopy. Motility was determined by use of wet-mount preparations of 24 h cultures grown in 1% peptone water (PW) at 20 C. Gliding motility was determined by the method of Perry (1973). Cellular morphology was also observed after 5 and 10 d growth in 1% PW. Transmission electron microscopy was used to observe whole mounts of isolates with a Philips EM 201 transmission electron microscope operated at 40 kV. Fluorescent pigmentation was evaluated after 72 h growth on medium B (Smibert and Krieg 1981) under ultraviolet light (254 nm). Temperature optimum was determined by incubating isolates in 1% PW at selected temperatures, i.e., 10, 20, 30, and 40 C, and observing for visible growth. Growth on MacConkey's agar and SS agar (Difco) was also evaluated. One hundred and twenty-three strains were selected to represent similar strains for biochemical assessment.

Biochemical characteristics. All biochemical tests were performed at room temperature (ca. 23 C). Carbohydrate metabolism was determined by the production of acid in media incubated aerobically and anaerobically (sealed with petroleum jelly) using Hugh-Leifson medium (Difco) amended with 1% glucose. Acid production from growth on Board and Holding (1960) medium plus 0.5% selected carbohydrates, i.e., arabinose, cellobiose, fructose, galactose, lactose, maltose, rhamnose, sucrose, and xylose, was also determined. Cytochrome oxidase was determined by testing for oxidation of tetramethyl-*p*-phenylene diamine dihydrochloride. Nitrate reduction was evaluated after 1, 3, and 7 d growth in nitrate broth (Difco) (Smibert and Krieg 1981). To determine amylase activity, NA plates amended with 0.2% soluble starch were developed after 72 h of growth with Gram's iodine solution. To detect gelatin hydrolysis, culture plates containing NA amended with 0.4% gelatin were inoculated, incubated for 72 h, and flooded with Frazier's solution. NA plates plus 10% skim milk were inoculated and observed after 72 h for zones of clearing to evaluate caseinase activity. Urea Agar

Base (Difco) slants were inoculated and observed daily up to 7 d to detect urease production. Lecithinase and lipase activity were evaluated on Egg Yolk agar (Smibert and Krieg 1981) after a 7-d incubation period. The methods of Lewin and Lounsbery (1969) were employed for the determination of alginate, cellulose, carboxymethylcellulose, and agar hydrolysis. Chitinase activity on chitin agar (Bennett and Hood 1980) was also determined after a 14-d incubation period. Twenty additional tests were performed on selected representative cultures by employing API-20E strips (Analytab Products, Inc., Plainview, NY).

Lipid analysis. The fatty acid content of selected representative isolates was compared to those of three reference strains (Liebert et al. 1984). Cells were harvested from a 48 h one-half strength NB and lyophilized. Cellular fatty acid profiles (Fautz et al. 1979; Moss 1975; Moss et al. 1974) were prepared by saponification of 4.0 mg dried bacterial cells with 5% (w/v) NaOH in 50% aqueous methanol. The liberated fatty acids were neutralized with 1.0 *N* HCl and extracted with chloroform:hexane (1:5, v/v). After drying, the fatty acids were converted to methyl esters with 10% BF_3-methanol at 70 C for 15 min. The methyl esters were then removed using three extractions of chloroform:hexane (1:5, v/v) and concentrated to 100 μl by nitrogen evaporation at room temperature. Aliquots of 5.0 μl were used for GC injection. Gas liquid chromatograph analyses were performed with a Varian Model 2100 gas chromatograph equipped with flame ionization detectors and dual differential electrometers. For lipid analysis, glass columns (12 ft × 2 mm ID, packed with 3% SP-2100 DOH) were used. After sample injection at 150 C, the instrument was programmed to 250 C (4 C/min), using a sensitivity of 4×10^{-10} AFS. N_2 at a flow rate of 20 ml/min was employed as the carrier gas.

Scanning electron microscopy. Stainless steel plates plus attached biofilm were first chemically fixed (2% glutaraldehyde), then critically dried (serially dehydrated 95% EtOH, then exposed to 100% Freon), and vacuum coated with gold particles (Hyatt and Zirkin 1973). The surfaces were viewed with a Jeol JSM-15 scanning electron microscope operated at 15 kV.

Results and Discussion

Three genera, *Pseudomonas, Flavobacterium,* and *Caulobacter*, appeared to be the most abundant in both waters and biofilms during the early stages of cooling-unit operation (1 d) (Table 1). These same genera were also abundant during the entire 60-d cycle.

Species of *Caulobacter* were observed in the waters, and by 59 d they were the major genus comprising the population of colonizers (Table 1). These gram-negative bacteria formed rosettes (Fig. 2C) and were observed to adhere to the stainless steel plates by the formation of holdfasts (Fig. 3D). In pure culture, the isolates were typical of the described genus *Caulobacter* (Poindexter 1974) in that they were oxidative, oxidase negative (Table 2), and divided by transverse asymmetrical fission (of stalked cells) giving rise to daughter cells, each having a single polar flagellum. The strains examined in this study produced small (<0.1 cm).

yellow colonies on PCA (Table 2). Furthermore, lipid analysis of one representative strain (WF-749) revealed a similar fatty acid profile (Table 3) to *C. crescentus* (Chow and Schmidt 1974).

TABLE 1. *Bacterial composition of water and biofilm accumulated on stainless steel plates in cooling units during a 60-d operational cycle*

	% isolates per sample					
	Day 1		Day 21		Day 59	
Genus	water	biofilm	water	biofilm	water	biofilm
Acinetobacter	2	–	1	–	1	1
Alcalagenes	1	–	–	–	–	2
Bacillus	–	–	–	–	3	1
Caulobacter	13	34	18	40	26	51
Corynebacterium	–	–	3	–	4	1
Cytophaga	–	–	6	21	1	–
Flavobacterium	21	18	19	4	16	3
Moraxella	–	–	5	–	3	1
Pseudomonas	55	31	42	19	38	22
Sphaerotilus	–	7	–	11	–	13
unclassified	8	10	6	5	8	5

– = no isolates observed.

The most abundant genus in waters, and a primary colonizer on the surface of stainless steel plates, was *Pseudomonas* (Table 1). Many isolates were typical species of *Pseudomonas* (Oberhofer et al. 1977; Palleroni 1984; Stanier et al. 1966), although an atypical *Pseudomonas* species (WF-816) was the most predominant one in both waters and biofilms. This gram-negative bacterium produced a small (<0.1 cm) red colony, was strictly respiratory, was motile, demonstrated a weak oxidase reaction and reduced nitrate (Table 2). The lipid profile of this organism (Table 3) demonstrated a fatty acid profile similar to the reference strain, *Flavobacterium capsulatum*, ATCC 14666. However, *F. capsulatum*, with a GC value of 63.0% (McMeekin and Shewan 1978) and a fatty acid profile similar to other *Pseudomonas* strains (Dees and Moss 1975) is now considered to belong to the genus *Pseudomonas*. The major fatty acid present in both WF-816 and *F. capsulatum* was methyl oleate. Interestingly, this fatty acid was also the major component detected in the *Caulobacter* species, WF-749 (Table 3).

The genus, *Flavobacterium,* was also a primary colonizer on stainless steel plates but declined in relative abundance with time (Table 1). As described by McMeekin and Shewan (1978) and Weeks et al. (1984), these isolates were gram-negative, nonmotile rods; were oxidase positive; produced a yellow pigment on solid media; and produced caseinase, gelatinase, and lipase (Table 2). The representative isolate, WF-629, was similar to the reference strain, *F. meningosepticum,* ATCC 13253, with the exception of one biochemical reaction, urease, and the percent fatty acid Ci-17:1 (Table 3).

Cytophaga species were relatively rare in the waters but appeared as secondary colonizers on the surface of the stainless steel plates. They comprised 21% of the

TABLE 2. *Characteristics of selected isolates from cooling waters and accumulated biofilms*

Characteristics	Isolates									
	Acinetobacter sp. WF-913	*Alcaligenes* sp. WF-710	*Bacillus* sp. WF-983	*Caulobacter* sp. WF-749	*Corynebacterium* sp. WF-940	*Cytophaga* sp. WF-361	*Flavobacterium* sp. WF-629	*Moraxella* sp. WF-931	*Pseudomonas* sp. WF-816	*Sphaerotilus* sp. WF-892
Cell diameter, μm	1.0	1.0	2.0	0.7	1.0	0.6	0.7	1.0	1.0	1.3
Cell length, μm	1.0–1.2	1.5–3.0	5.0–10.0	2.0–8.0	4.0–6.0	4.0–6.0	4.0–10.0	1.0–1.5	2.0–4.0	3.0–5.0
Sheath	–	–	–	–	–	–	–	–	–	+
Stalk/Rosette formation	–	–	–	+	–	–	–	–	–	+
Gram reaction	–	–	+	–	+	–	–	–	–	–
Colony morphology										
pigment	–	–	–	+(Y)	+(O)	+(Y)	+(Y)	–	+(R)	–
form	I	C	I	C	C	I	C	C	C	F
elevation	R	Co	Fl	Co	Co	Co	Co	Co	Co	Co
margin	I	E	I	E	E	U	E	U	E	F
Growth range, °C	10–40	10–30	20–40	20–30	10–30	10–30	20–30	20–30	20–30	10–30
Growth on MAC agar	+	+	–	–	–	–	–	+	–	–
Growth on SS agar	–	–	–	–	–	–	–	–	–	–
Motility	–	+	+	+	–	–(gl)	–	–	+	+
Oxidase	–	+	+	–	–	+	+	+	+w	+
O/F reaction	OX	Alk	NS	OX	NS	F	OX	Alk	Alk	NS
Nitrite production	–	–	+	+	+	+	–	–	+	+
Denitrification	–	–	–	–	–	+	–	–	–	–
DNase	–	–	–	+	–	+	+	–	–	–
Starch hydrolysis	–	–	+	–	–	+	+	–	+w	+
Caseinase	–	–	+	–	+	+	+	–	+w	+
Gelatinase	–	–	+	+	+	+	+	–	–	+
Lipase	–	–	+	+	–	–	+	–	–	–
Lecithinase	–	–	–	–	–	–	+	–	–	–

TABLE 2. *Characteristics of selected isolates from cooling waters and accumulated biofilms*

Characteristics	Isolates									
	Acinetobacter sp. WF-913	*Alcaligenes* sp. WF-710	*Bacillus* sp. WF-983	*Caulobacter* sp. WF-749	*Corynebacterium* sp. WF-940	*Cytophaga* sp. WF-361	*Flavobacterium* sp. WF-629	*Moraxella* sp. WF-931	*Pseudomonas* sp. WF-816	*Sphaerotilus* sp. WF-892
Arginine dihydrolase	–	–	NT	–	NT	+	–	–	–	–
Lysine	–	–	NT	–	NT	–	–	–	–	–
Ornithine	–	–	NT	–	NT	–	–	–	–	–
ONPG	–	–	NT	+	NT	+	+	–	–	–
Citrate utilization	–	–	NT	–	NT	–	NT	–	–	–
H_2S	–	–	NT	–	NT	–	NT	–	–	–
Urease	–	–	NT	–	NT	–	NT	+	–	–
Indole	–	–	NT	–	NT	–	NT	+	–	–
VP test	–	–	NT	–	NT	–	NT	–	–	–
Hydrolysis of										
Alginate	–	–	–	–	–	+	NT	–	–	–
Cellulose	–	–	–	–	–	–	NT	–	–	–
CMC	–	–	+	–	–	–	NT	–	–	–
Agar	–	–	–	–	–	–	NT	–	–	–
Chitin	–	–	+	–	–	+	NT	–	–	–
Acid Reaction										
Arabinose	+	–	–	+	–	–	NT	–	+	–
Cellobiose	–	–	–	+	–	–	NT	–	–	–
Fructose	–	–	–	+	–	–	NT	–	–	–
Galactose	+	–	–	+	–	–	NT	–	+	–
Maltose	–	–	+	+	–	+	NT	–	–	–
Rhamnose	–	–	–	+	–	–	NT	–	–	–
Sucrose	–	–	–	+	–	–	NT	–	–	–
Xylose	+	–	–	+	–	–	NT	–	–	–

TABLE 2. *Characteristics of selected isolates from cooling waters and accumulated biofilms*

+ = positive	(O) = orange	R = raised	OX = oxidative reaction
− = negative	(R) = red	Co = convex	Alk = alkaline reaction
w = weak	I = irregular	Fl = flat	NS = no reaction
NT = not done	C = circular	E = erose	F = fermentation reaction
(Y) = yellow	F = filamentous	U = undulate	

TABLE 3. *Lipid analysis of reference strains and selected isolates from cooling waters*[a]

Fatty acids	*Caulobacter* sp. WF-749	*Cytophaga johnsonae* ATCC 17061	*Cytophaga* sp. WF-361	*Flavobacterium meningosepticum* ATCC 13253	*Flavobacterium* sp. WF-629	*Flavobacterium (Pseudomonas) capsulatum* ATCC 14666	*Pseudomonas* sp. WF-816
12:0	−	−	−	T	−	−	−
14:0	−	T	T	T	−	T	T
i-15:1	−	3.5	5.7	−	−	−	−
i-15:0	−	45.2	44.5	65.6	56.0	T	T
15:1	−	T	T	−	−	−	−
15:0	−	4.4	11.5	−	−	−	−
2-OH 14:0	−	−	−	−	−	−	−
3-OH 14:0	−	T	T	−	−	−	T
16:1	2.1	22.4	18.3	8.8	T	7.0	5.1
16:0	22.0	14.9	8.7	9.4	T	14.0	9.1
i-17:1	−	3.5	5.7	5.3	36.6	−	−
17:0	−	−	−	−	−	T	T
2-OH 16:0	−	−	−	−	−	−	−
18:1	63.4	T	T	T	−	72.0	75.4
18:0	6.7	T	T	T	−	3.0	5.8
20.0	−	−	−	−	−	−	−

[a] Values in percent of total lipids; T = less than 2.0%; − = not detected.

isolates in the biofilm collected at day 21 (Table 1). The *Cytophaga* species were gram-negative, nonmotile, and oxidase positive; demonstrated gliding motility; produced a yellow pigment on solid media; and were capable of degrading polysaccharides including carboxymethylcellulose and chitin (Table 2). Such characteristics are typical of *Cytophaga* (Christensen 1977; Lewin and Lounsbery 1969; Mitchell et al. 1969; Reichenbach and Dworkin 1981). The lipid profile of one *Cytophaga* isolate, WF-361, was similar to the reference strain, *Cytophaga johnsonae,* ATCC 17061.

Colonization of the stainless steel plates was observed directly using SEM (Figs. 1, 2, and 3). The initial colonizers appeared to be iron bacteria (Figs. 2A and B). *Sphaerotilus* and *Gallionella* were observed and cultured from biofilms at both 1 and 7 d, although low levels, especially in the waters, may be a function of the medium. Secondary colonizers consisting of elongate, flexible rods similar to the *Cytophaga* species were cultured from the day 21 samples (Fig. 2C). Although *Caulobacter* species were detected in all samples, they were most apparent by SEM in the day 59 sample (Fig. 2D).

The scanning electron micrographs, as well as the characterization of the at-

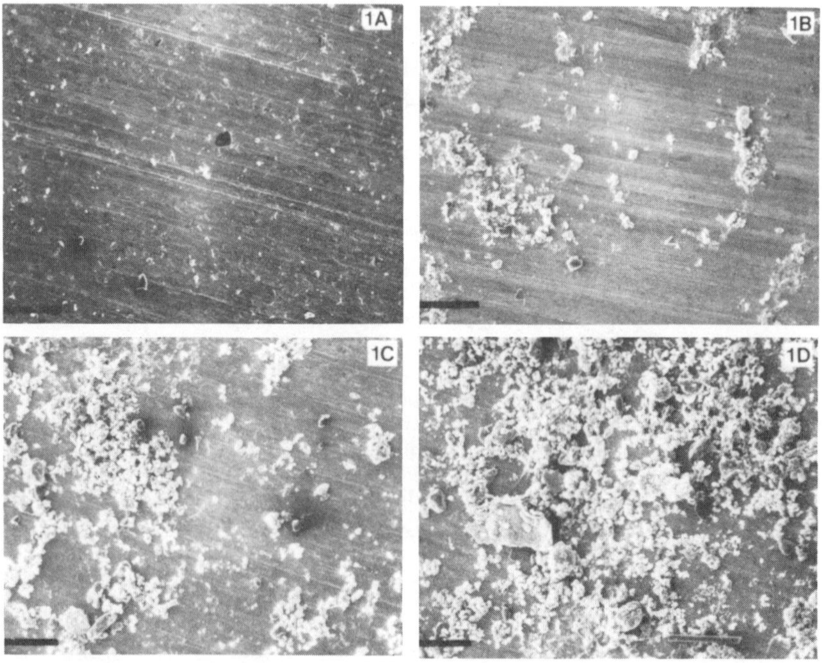

FIG. 1. Scanning electron micrographs of stainless steel discs at low magnification. (A) 1 d of exposure, (B) 7 d of exposure, (C) 21 d of exposure, and (D) 59 d of exposure. Bar = 100 μm.

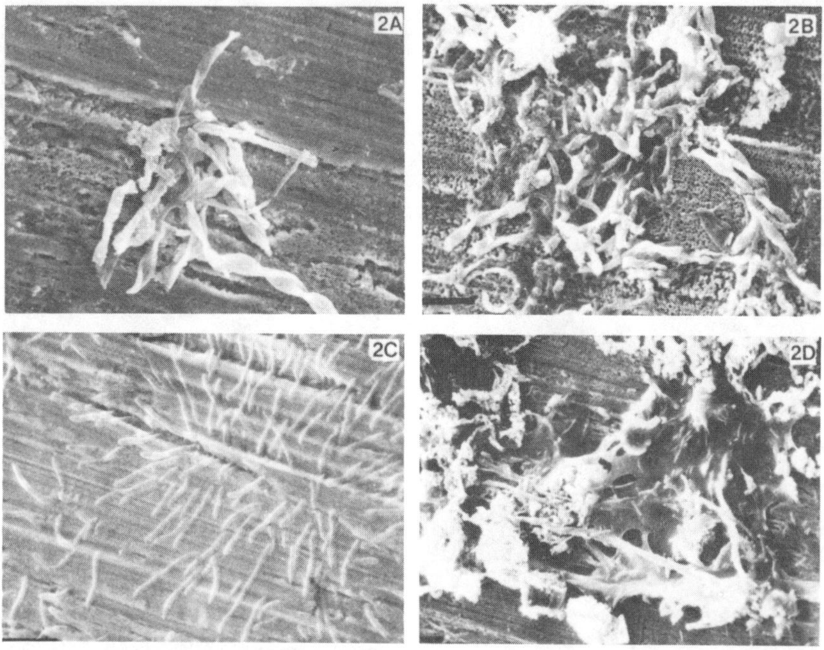

FIG. 2. Scanning electron micrographs showing predominant organisms attached to stainless steel plates at different sample periods. (A) and (B) Iron bacteria at 1 d and 7 d of exposure, (C) *Cytophaga/Flavobacterium*-like bacteria at 21 d of exposure, and (D) *Pseudomonas*-like bacteria and *Caulobacter,* along with iron bacteria, at 59 d of exposure. Bar = 5 μm.

tached culturable bacteria, revealed a pattern of increased diversity in the biofilm. Five identifiable genera were detected after 1 d of exposure, whereas nine genera were detected after 59 d of exposure. The pattern of colonization consisted initially of typical rods (*Pseudomonas* and *Flavobacterium*) and iron bacteria, (*Sphaerotilus* and *Gallionella*) (Figs. 2A and B). The stalked *Caulobacter* were also primary colonizers. *Cytophaga* and additional *Flavobacterium* species appeared as secondary colonizers on the surface of the stainless steel plates. By the final day, bacterial diversity was very high and nearly all genera were represented.

Acknowledgments

This research was supported by a grant from Monsanto Fibers and Intermediates Co. We thank the following for technical assistance: Fred Deck for the lipid analyses, Rex Davidson for the scanning electron micrographs, and Karen Bishop for sample collection. We also thank Dr. Dennis K. Flaherty for valuable suggestions and coordination efforts.

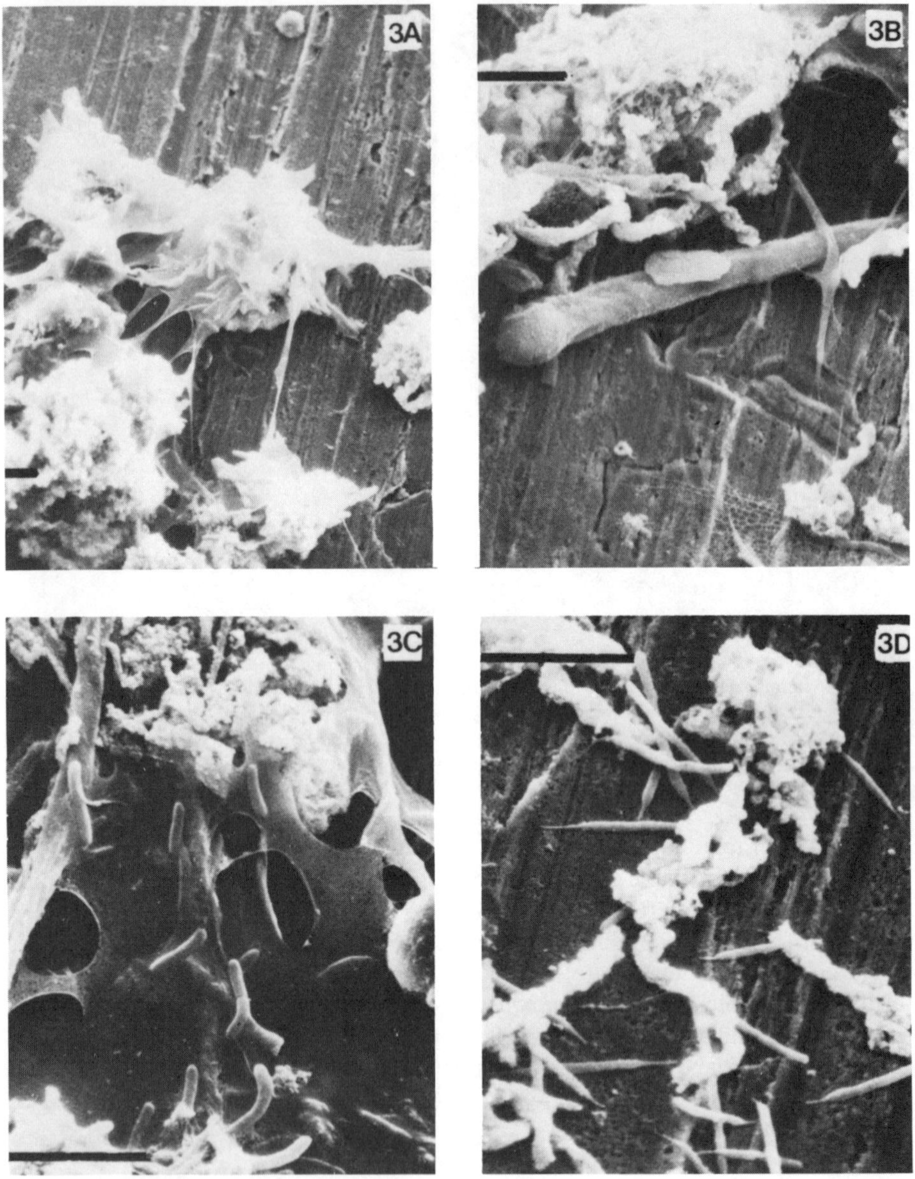

FIG. 3. Scanning electron micrographs showing increased diversity in the 59 d exposure sample. (A) and (C) *Pseudomonas*-like bacteria enmeshed in exopolymer, (B) Iron bacteria, (D) Stalked *Caulobacter* species. Bar = 5 μm.

Literature Cited

Alexander, M. 1981. *Microbial Ecology*. John Wiley and Sons, Inc., New York, 511 p.

Banaszak, E. F., W. H. Thiede, and N. Jordan. 1970. Hypersensitivity pneumonitis due to contamination of an air conditioner. *N. Eng. J. Med.* 283:271–276.

Bennett, C. B., and M. A. Hood. 1980. Effects of cultural conditions on the production of chitinase by a strain of *Bacillus megaterium*. *Dev. Ind. Microbiol.* 21:357–364.

Birchall, G. A. 1979. Control of fouling within water cooling systems. *Effluent Water Treat. J.* 19:571–578.

Board, R. G., and A. J. Holding. 1960. The utilization of glucose by aerobic gram-negative bacteria. *J. Appl. Bacteriol.* 23:xi–xii.

Chow, T. C., and J. M. Schmidt. 1974. Fatty acid composition of *Caulobacter crescentus*. *J. Gen. Microbiol.* 83:369–373.

Christensen, P. J. 1977. The history, biology, and taxonomy of the *Cytophaga* group. *Can. J. Microbiol.* 23:1599–1653.

Cordes, L. G., D. W. Fraser, P. Skaliy, C. A. Perlino, W. R. Elsea, G. F. Mallison, and P. S. Hayer. 1980. Legionnaires' disease outbreak at the Atlanta, Georgia, country club: Evidence for spread from an evaporative condenser. *Am. J. Epidemiol.* 111:425–431.

Dees, S. B., and C. W. Moss. 1975. Cellular fatty acids of *Alcaligenes* and *Pseudomonas* species isolated from clinical specimens. *J. Clin. Microbiol.* 1:414–419.

Fautz, E., G. Rosenfelder, and L. Grotjahn. 1979. Isobranched 2- and 3-hydroxy fatty acids as characteristic lipid constituents of some gliding bacteria. *J. Bacteriol.* 140:852–858.

Flaherty, D. K., F. H. Deck, J. Cooper, K. Bishop, P. A. Winzenburger, L. R. Smith, L. Bynum, and W. B. Witmer. 1984a. Bacterial endotoxin isolated from a water spray air humidification system as a putative agent of occupation-related lung disease. *Infect. Immunol.* 43:206–212.

Flaherty, D. K., F. H. Deck, M. A. Hood, C. Liebert, F. Singleton, P. Winzenburger, K. Bishop, L. R. Smith, L. M. Bynum, and W. B. Witmer. 1984b. A *Cytophaga* species endotoxin as a putative agent of occupation-related lung disease. *Infect. Immunol.* 43:213–216.

Friend, P. L., and W. K. Whitekettle. 1980. Biocides and water cooling towers. *Dev. Ind. Microbiol.* 21:123–131.

Hyatt, M. A., and B. R. Zirkin. 1973. Critical point-drying method. Pages 297–313 *in* M. A. Hayat, ed., *Principles and Techniques of Electron Microscopy: Biological Applications,* Vol. 3. Van Nostrand Reinhold Co., New York.

Lewin, R. A., and D. M. Lounsbery. 1969. Isolation, cultivation and characterization of *Flexibacteria*. *J. Gen. Microbiol.* 58:145–170.

Liebert, C. A., M. A. Hood, F. H. Deck, K. Bishop, and D. K. Flaherty. 1984. Isolation and characterization of a new *Cytophaga* species implicated in an outbreak of work-related lung disease. *Appl. Environ. Microbiol.* 48:936–943.

Liebert, C. A., M. A. Hood, P. A. Winter, F. L. Singleton. 1983. Observations on biofilm formation in industrial air-cooling units. *Dev. Ind. Microbiol.* 24:509–517.

McMeekin, T. A., and J. M. Shewan. 1978. A review of taxonomic strategies for *Flavobacterium* and related genera. *J. Appl. Bacteriol.* 45:321–332.

Mitchell, T. G., M. S. Hendrie, and J. M. Shewan. 1969. The taxonomy, differentiation and identification of *Cytophaga* species. *J. Appl. Bacteriol.* 32:40–50.

Moss, C. W. 1975. Gas-liquid chromatography as an analytical tool in microbiology. *Public Health Lab.* 33:81–83.

Moss, C. W., M. A. Lambert, and W. H. Merwin. 1974. Comparison of rapid methods for analysis of bacterial fatty acids. *Appl. Microbiol.* 28:80–85.

Oberhofer, T. R., J. W. Rowen, and G. F. Cunningham. 1977. Characterization and identification of gram-negative, nonfermentative bacteria. *J. Clin. Microbiol.* 5:208–220.

Palleroni, N. J. 1984. Genus I. *Pseudomonas* Migula 1894. Pages 144–218 *in* N. R. Krieg and J. G. Holt, eds., *Bergey's Manual of Determinative Bacteriology,* 9th ed. The Williams and Wilkins Co., Baltimore, MD.

Perry, L. B. 1973. Gliding motility in some non-spreading *Flexibacteria. J. Appl. Bacteriol.* 36:227–232.

Poindexter, J. S. 1974. Genus *Caulobacter* Henrici and Johnson 1935. Pages 153–155 *in* R. E. Buchanan and N. E. Gibbons, eds., *Bergey's Manual of Determinative Bacteriology,* 8th ed. The Williams and Wilkins Co., Baltimore, MD.

Reichenbach, H., and M. Dworkin. 1981. Introduction of the gliding bacteria. Pages 315–379 *in* M. P. Starr, H. Stolp, H. G. Triiper, A. Balows, and H. G. Schelgel, eds., *The Prokaryotes.* Springer-Verlag, New York.

Smibert, R. M., and N. R. Krieg. 1981. General characterization. Pages 409–443 *in* P. Gerhardt, ed., *Manual of methods for General Bacteriology.* American Society for Microbiology, Washington, DC.

Stanier, R. Y., N. J. Palleroni, and M. Doudoroff. 1966. The aerobic Pseudomonads: A taxonomic study. *J. Gen. Microbiol.* 43:159–271.

Weeks, O. B., R. J. Owen, and T. A. McMeekin. 1984. Genus *Flavobacterium* Bergey et al. Pages 357–364 *in* N. R. Krieg and J. G. Holt, eds., *Bergey's Manual of Determinative Bacteriology,* 9th ed. The Williams and Wilkins Co., Baltimore, MD.

Zobell, C. E. 1943. The effect of solid surfaces upon bacterial activity. *J. Bacteriol.* 46:39–56.

CHAPTER **54**

Nitrous Oxide Production and Denitrification Potential of Oceanic Waters

STEVEN J. SCHROPP,* JOHN R. SCHWARZ, AND LAUREL A. LOEBLICH

Texas A&M University at Galveston
Galveston, Texas 77553

The oceans appear to be a source of atmospheric N_2O, the N_2O being generated by one or more biological processes, including denitrification, nitrification, or assimilatory nitrate reduction. We have measured N_2O production from several oceans to determine whether marine denitrification can be a source of N_2O. Water samples were taken at the depths of suspended particle maxima in the Caribbean Sea, the Mediterranean Sea, and the western North Atlantic. Suspended particles in the samples were size fractionated, by filtration, and incubated anaerobically, under argon, in sealed serum vials with combinations of added nitrate (0.9 m*M*) and acetylene (0.01 atm); N_2O accumulation was measured by gas chromatography. N_2O production was detected in samples from all stations, with up to 91% of the denitrification product being N_2O. N_2O was not detected when samples were filtered through 0.2 μm Nuclepore filters. N_2O production rates in nitrate-supplemented, acetylene-blocked samples (an estimate of denitrification potential) ranged from <1.0 to 1.4×10^5 nL N_2O liter^{-1} day^{-1}. The production of N_2O in <3-μm size fractions suggests that facultative, denitrifying bacteria are free in the water or that they are easily dislodged from particulate microenvironments. The ubiquitous nature of facultative, denitrifying bacteria in the oceans suggests that the denitrifying respiratory pathway is a useful function in these microorganisms, perhaps being used when they encounter anaerobic microhabitats, such as within particulate matter.

INTRODUCTION

The presence of nitrous oxide (N_2O) in the atmosphere has been of considerable interest since Crutzen (1970) showed that nitrogen oxides have a controlling effect on the concentration of ozone in the stratosphere. The N_2O in the atmosphere is generated by microbial activity at the earth's surface, where the most important sources are denitrification and nitrification in soil and water (McElroy et al. 1977; Hahn and Junge 1977). The oceans play an important role in the global N_2O budget. The amount of dissolved N_2O in the upper water column often exceeds that which would be expected based on its being in equilibrium with atmospheric N_2O (Junge et al. 1971; Hahn 1974; Hahn 1975; Yoshinari 1976; Cohen 1977; Elkins et al. 1978; Singh et al. 1979; Elkins 1980; Pierotti and Rasmussen 1980). Frequently, dissolved N_2O maxima are found at the depth of the thermocline or pycnocline (Cohen and Gordon 1978; Hahn 1975; Pierotti and Rasmussen 1980).

*Present address: Department of Oceanography, Texas A&M University, College Station, Texas 77843

Since the dissolved N_2O concentrations are too large to be the result of physical processes such as bubble injection or mixing (Bieri et al. 1968; Seiler and Schmidt 1974; Hahn 1974), the excess N_2O in seawater must be the result of biological activity. The three most likely biological sources of N_2O in the ocean are: denitrification, nitrification, and assimilatory nitrate reduction (Hahn 1975). Current work in our laboratory is focused on the role of anaerobic processes in oceanic environments; therefore, we are interested in the anaerobic production of N_2O by facultative denitrifiers. The occurrence of denitrification in oxygenated water of the open ocean would suggest the presence of anaerobic microhabitats.

Denitrification is a respiratory process in which nitrogen oxides, instead of oxygen, are used as terminal electron acceptors and are reduced to gaseous end products (Knowles 1982). Nitrogen is the major end product of denitrification but nitrous oxide is a freely diffusible intermediate that can either be released or further reduced to nitrogen (Firestone et al. 1980). Acetylene blocks the reduction of N_2O to N_2 (Balderston et al. 1976) and acetylene blockage can be used to measure denitrification potential (Haines et al. 1981).

Denitrification occurs in oxygen deficient areas of the eastern tropical Pacific Ocean, Arabian Sea, and Cariaco Trench (Brandhorst 1959; Codispoti and Richards 1976; Deuser et al. 1978; Hashimoto et al. 1983). Denitrification may also take place in anaerobic microhabitats, such as particulate matter, within oxygenated water (Knowles 1978; Schropp and Schwarz 1983). Estimates of oceanic denitrification rates have been based on indirect methods such as electron transport system activity (Codispoti and Richards 1976; Codispoti and Packard 1980) or calculation from nitrate anomalies (Deuser et al. 1978). ^{15}N methods of measuring denitrification rates have also been used to a limited extent (Goering and Dugdale 1966; Goering 1968). No one, however, has conducted an extensive investigation of the distribution of denitrifying activity or N_2O production in the open ocean.

We have shown that N_2O can be released by oceanic bacteria during denitrification and that denitrifiers are present in both oxygen-deficient and oxygenated water (Schropp and Schwarz 1983). Our hypothesis is that particulate matter in the ocean provides anaerobic microenvironments where denitrifying bacteria can function. We now present the results of work that extends the geographical range of our observations and allows us to make estimates of potential denitrification rates from a wide range of oceanic areas. The goals of this research were to demonstrate the widespread occurrence of N_2O-producing, denitrifying microorganisms in the open ocean; to determine whether such microorganisms are associated with particulate matter; and to make an estimate of the denitrification potential of different regions.

Materials and Methods

The data presented in this report were collected on four separate cruises: U.S.N.S. *Bartlett* cruise 1309-80, August 25–September 14, 1980, in the Mediterranean Sea; R. V. *Gyre* cruise 80-G-12, December 2–16, 1980, in the Caribbean Sea; U.S.N.S. *Lynch* cruise 710-82, June 4–25, 1982, in the Sargasso Sea, and

R. V. *Gyre* cruise 83-G-7, June 25-July 7, 1983, in the western North Atlantic. Station locations are listed in Table 1.

TABLE 1. *Station locations and sample collection methods*

Cruise	Station	Location: Latitude	Location: Longitude	Depth (m)	Sample Collection: Fractions (μm) < 3[a]	3-35	> 3	> 35	Method[b]
Bartlett 1309-80	2	37° 00.9′ N	18° 30.9′ E	15				+	BN
	4	32° 58.4′ N	18° 01.9′ E	22				+	BN
	6	39° 13.5′ N	11° 44.8′ E	23				+	BN
	9	38° 18.9′ N	4° 57.9′ E	38				+	T
	10	36° 53.4′ N	0° 01.5′ E	26				+	T
	11	36° 05.5′ N	4° 14.7′ W	25				+	T
	12	34° 12.3′ N	7° 49.7′ W	38				+	T
Gyre 80-G-12	1	15° 25.0′ N	63° 51.8′ W	38				+	T
	2	15° 00.5′ N	69° 59.3′ W	65				+	T
	3	15° 27.4′ N	79° 59.3′ W	50				+	T
	5	18° 23.5′ N	81° 28.0′ W	55				+	T
Lynch 710-82	9	26° 00.0′ N	60° 00.0′ W	20			+		BRF
	11	25° 01.0′ N	65° 00.0′ W	26	+		+		BRF
	15	25° 59.5′ N	75° 11.7′ W	36			+		BRF
Gyre 83-G-7	2	40° 03.0′ N	69° 06.0′ W	29	+	+		+	BRF
	4	35° 54.5′ N	69° 26.5′ W	43	+		+		BRF
	5	37° 06.1′ N	70° 24.8′ W	33	+		+		BRF

[a] The < 3-μm fraction was the effluent from a reverse filter equipped with a 3-μm pore diameter membrane filter.

[b] BN = bottle collection, net filtration; T = net tow; BRF = bottle collection, reverse filtration.

Sample collection. Since we expected the denitrifying bacteria to be associated with particulate matter, we collected samples from the depths of nearsurface particle maxima as determined by nephelometry or transmissometry. When this information was not available we took samples at the depth of the thermocline or pycnocline because particle maxima are associated with these physical features (Pak et al. 1980; Bishop et al. 1977). Three collection and concentration methods were employed during the course of this study in an effort to determine the most effective way to sample for denitrifiers. In the Mediterranean Sea, the samples at Stations 2, 4, and 6 were collected in 30-liter Niskin bottles. The bottles were mounted on a rosette that was equipped with a nephelometer or transmissometer; with this equipment, the location of the sample in relation to particle maxima could be determined. Once on board ship, the water was drained through a 35-μm mesh plankton net; in this way, the particulates in up to 700 liters of seawater were collected and concentrated in a final volume of about 2 liters. Since there was a relatively small amount of particulate matter visible in the concentrate, we decided to modify the technique to collect a greater number of particles. At the remainder of the stations in the Mediterranean Sea and at all of the stations in the Caribbean sea, we collected samples by towing the same plankton net at the depth of the particle maxima. The amount of water filtered was determined using a net-mounted flowmeter or calculated from the ship's speed and duration of the tow.

Although we observed a large amount of particulate matter (including phytoplankton, zooplankton, and aggregates) and a greater absolute N_2O production in these samples, the results indicated that the increase in N_2O production was not proportional to the increase in the volume of seawater passing through the net. Therefore, on the cruises in the Sargasso Sea and western North Atlantic, we used a reverse flow filtration system (Schropp and Schwarz 1983) to gently concentrate particles. The samples were collected in 30-liter Niskin bottles, and then the water was passed through the reverse flow filters. The filters were equipped with 3-μm Nuclepore filters or 35-μm mesh nylon netting. Two or more filter units could be connected in series for size fractionation of particles in the samples. The particle sizes of the fractions collected at each station are indicated in Table 1.

Sample processing. After collecting and concentrating each seawater sample, 25-ml aliquots of the concentrated material were pipetted into each of 12 sterile 50-ml serum vials. The vials were sealed with butyl rubber stoppers and aluminum caps. In addition, 12 control vials were prepared by filtering the concentrated material through 0.2-μm Nuclepore filters and using the filtrate in the same manner as the experimental material.

The anaerobic incubation technique used in this study was derived from that of Miller and Wolin (1974). The sample-containing vials were flushed a minimum of 15 min with oxygen-free argon that was delivered to the vials through a manifold system that incorporated a column of heated copper turnings for oxygen removal (Hungate 1969). The manifold was constructed of 750 mm $\times$ 50 mm stainless steel pipe drilled and tapped to accept 13 brass valves (Nupro Co.). Attached to each valve with plastic tubing was a 1-ml plastic syringe barrel from which the plunger and handle had been removed. The syringe barrel was filled with sterile cotton. Sterile 50-mm, 22-ga stainless steel needles on the syringe barrels were inserted into the vials through the rubber stoppers and gas was vented through disposable 26-ga needles.

After gassing, the 12 experimental vials were divided into four groups, which were injected with combinations of KNO_3 solution (0.9 m*M* final concentration) and acetylene (0.01 atm final concentration). Acetylene at 0.01 atm blocks the reduction of N_2O to N_2 (Balderston et al. 1976). The combinations (KNO_3/acetylene, + or −) for each group were as follows: Group 1 (−/−); Group 2 (−/+); Group 3 (+/−); Group 4 (+/+). The 12 control vials were treated in an identical fashion. Each experimental and control condition was thus prepared in triplicate. The preparation of both acetylene-blocked and non-blocked vials allows us to estimate the proportion of the denitrification product that might normally appear as N_2O (Firestone et al. 1980). The vials were then incubated in the dark at room temperature and their headspace gas was analyzed periodically for N_2O over periods of from 6 to 60 d.

Gas analysis. N_2O was measured on a Hewlett-Packard 5840 gas chromatograph equipped with an electron capture detector (ECD). The analytical column was 2 m $\times$ 3 mm stainless steel packed with Porapak Q (Waters Associates, Inc.). A carrier gas of 95% argon/5% methane was used at a flow rate of 30 ml min^{-1}. The ECD was operated at 250 C; the oven and injection port at 80 C. Headspace gases were removed for analyses with a gas-tight syringe (Hamilton). The 1-ml syringe was flushed and filled with argon, inserted into a vial, and pumped several times to ensure adequate mixing of the headspace gases. A 1.0-ml gas sample was withdrawn, and 0.5 ml of the sample was injected into the gas chromatograph. The amount of N_2O in the samples was determined by comparison with a gas standard that contained 500 ppm N_2O in argon carrier (Linde Division, Union Carbide Corp.)

N_2O production amounts are presented here as the amount of N_2O produced per liter of seawater (nL N_2O $liter^{-1}$) before concentration, calculated using the equation: N_2O production per liter = [total N_2O produced (nL)/volume of sample in vial (liters)] × [final volume of the concentrated material (liters)/volume of seawater filtered (liters)]. The total amount of N_2O produced in each vial (nL N_2O) was determined by calculating the amount of dissolved N_2O using the solubility coefficients of Weiss and Price (1980) and correcting for N_2O removed during previous analyses. Each point in the figures represents net N_2O production, i.e., the mean amount of N_2O in three experimental vials minus the mean amount of N_2O produced, if any, in the corresponding control vials.

We calculated maximum denitrification rates in the nitrate-supplemented, acetylene-blocked vials, based on the amount of N_2O produced from the start of incubation to the time of gas analysis. Our results are presented in terms of nL N_2O $liter^{-1}$ day^{-1} and represent maximum potential denitrification rates. Haines et al. (1981) used a similar technique to measure denitrification rates and denitrification potential in marine sediment.

Results

N_2O production was detected in samples from all stations. Nitrate-supplemented, acetylene-blocked samples consistently produced the greatest amount of N_2O. N_2O was usually not detected in the control vials but, when present, it was $<3\%$ of the amount in the corresponding experimental vials. N_2O production in samples from the four cruises is shown in Figs. 1–4. The maximum potential denitrification rates varied considerably, apparently depending on the collection and concentration method as well as on regional differences in denitrification activity. Maximum potential denitrification rates are listed in Table 2.

In the Mediterranean Sea (Fig. 1) there were differences, apparently related to the method of sample collection, between Stations 2, 4, and 6 (bottle collections) and Stations 9, 10, 11, and 12 (net tow collections). Stations 2, 4, and 6 had less total N_2O produced in the vials but had a greater N_2O production per liter of original seawater. N_2O was detected in as early as 2 d in samples collected by the net tow method, but N_2O production was usually not apparent until after four or more days in samples collected in Niskin bottles. At five of the seven stations, some N_2O was produced in the nitrate-supplemented, nonblocked samples, but no N_2O production could be detected in any of the samples containing only the *in situ* nitrate concentrations (nonsupplemented vials). The amount of N_2O produced in nitrate-supplemented, nonblocked vials was from 0 to 23% of that in nitrate-supplemented, acetylene-blocked vials. N_2O was not consumed in four of the five nitrate-supplemented, nonblocked vials in which it was produced.

TABLE 2. *Maximum potential denitrification rates*[a] *(nL N_2O L^{-1} d^{-1})*

Cruise	Station	Time (days)	Denitrification Rates Size Fractions (µm)		
			<3	>3	>35
Bartlett 1309–80	2	62			28
	4	7.6[b]			28
	6	3.4			54
	9	3.6			5
	10	2.2			38
	11	40[b]			1
	12	40[b]			2
Gyre 80–G–12	1	3.5			48
	2	7			6
	3	6.3			2
	5	3.3			1
Lynch 710–82	9	12.3		3517	
	11	10	1355	1264	
	15	4.8		59	
Gyre 83–G–7	2	1.9	141543	436	551
		5.7	48782	1508	4779
	4	5.1	9786	1348	
	5	3.6	18821	0	

[a] Measured in nitrate-supplemented, acetylene-blocked vials from the start of incubation to the time indicated.

[b] This was the first time these samples were analyzed

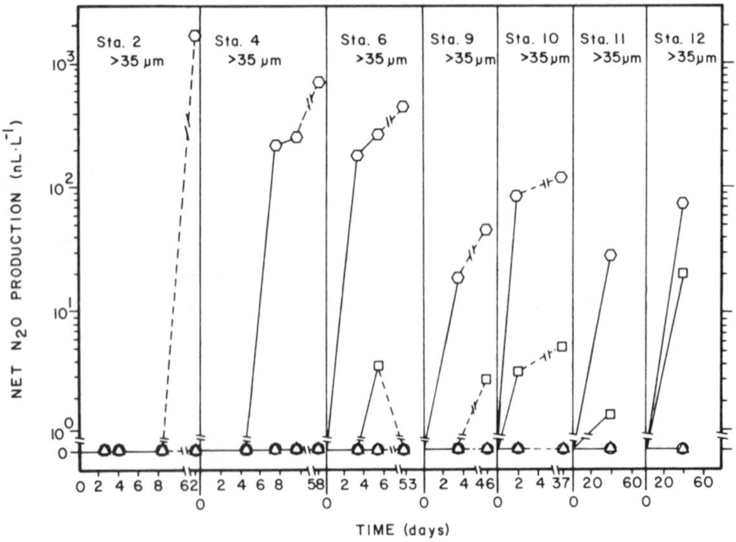

FIG. 1. Net N_2O production in the Mediterranean Sea, U.S.N.S. *Bartlett* cruise 1309–80. Station number and size fraction are indicated above each section. Nitrate-supplemented (0.9 m*M*), acetylene-blocked (0.01 atm), ⬡; nitrate supplemented, non-acetylene-blocked, □; non-nitrate-supplemented, acetylene-blocked, △; non-nitrate-supplemented, non-acetylene-blocked, ○.

Denitrification rates appeared to be uniform throughout the Mediterranean, except at Station 9, which had an unusually low rate. The low rates at Stations 11 and 12 do not represent true maximum rates because the first N_2O analyses of these samples could not be performed aboard ship and were not made until 40 d after taking the samples.

In the Caribbean Sea (Fig. 2) net N_2O production per liter was in the same range as for the Mediterranean Sea Stations 9, 10, 11, and 12, where the same collection technique was used. At three of four stations, N_2O was produced in the nitrate-supplemented, nonblocked vials, in amounts ranging from <1.0 to 91% of that in nitrate-supplemented, acetylene-blocked vials. N_2O produced in the nitrate-supplemented, nonblocked vials was not consumed at two of these stations. The maximum potential denitrification rate was greatest at Station 1 and decreased from east to west in the Caribbean Sea.

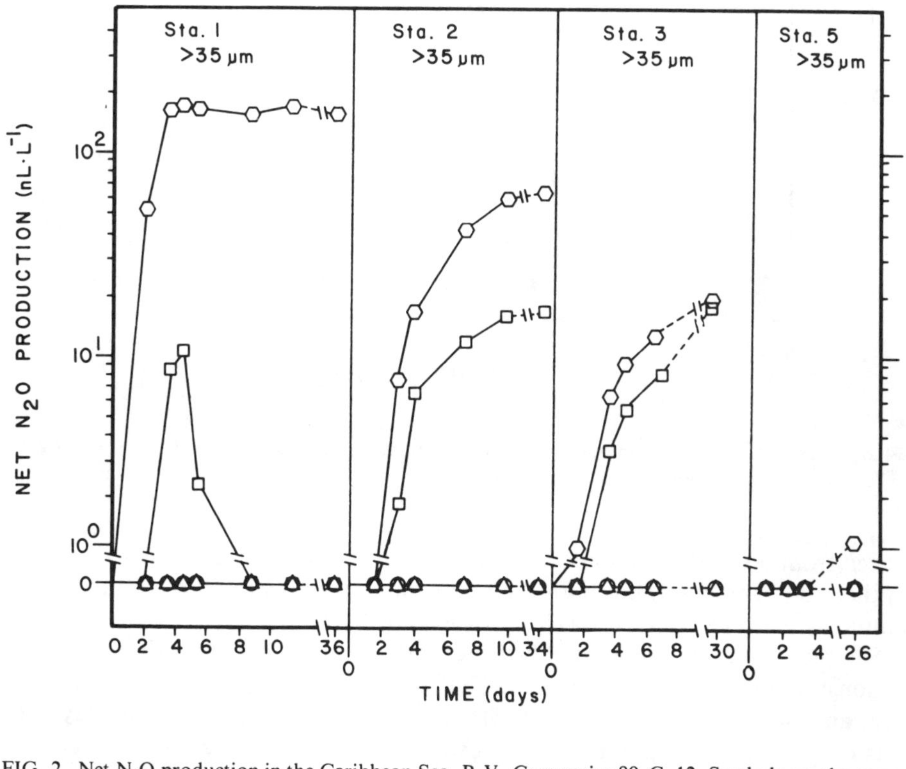

FIG. 2. Net N_2O production in the Caribbean Sea, R.V. *Gyre* cruise 80–G–12. Symbols are the same as in Fig. 1.

In the Sargasso Sea N_2O production (Fig. 3) and maximum potential denitrification rates were greater than in the Mediterranean Sea or Caribbean Sea, but this was probably due to the more gentle concentration of the reverse flow filtration technique used on this cruise. At Station 11 we prepared both <3-μm and >3-μm fractions, and the results for each were similar; N_2O was pro-

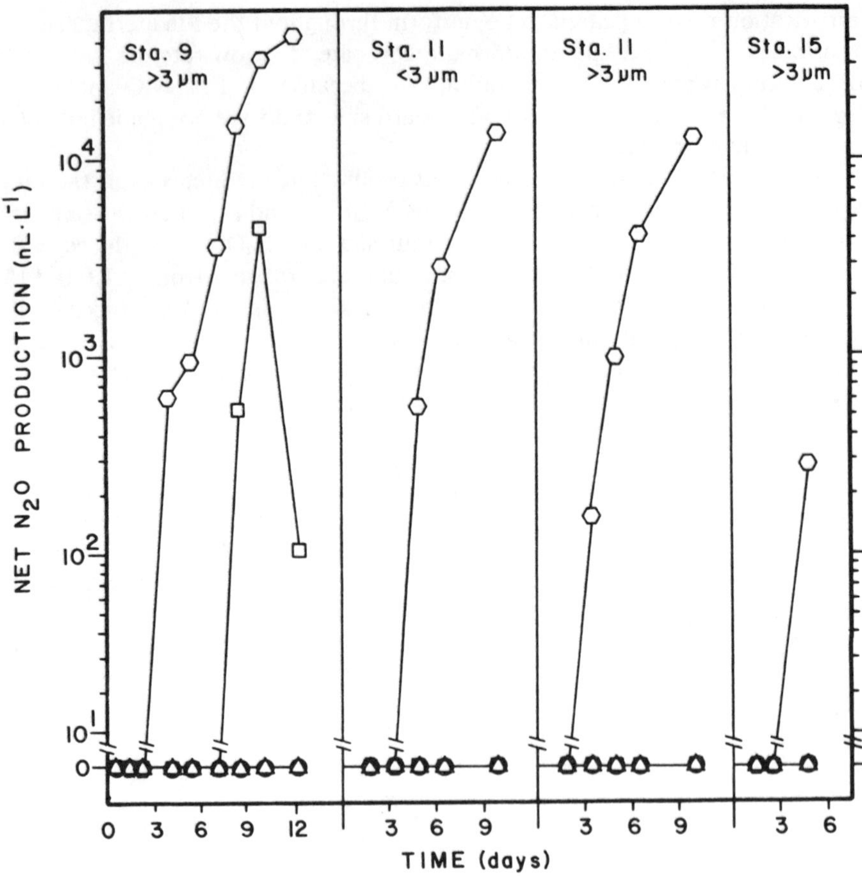

FIG. 3. Net N_2O production in the western North Atlantic Ocean (Sargasso Sea), U.S.N.S. *Lynch* cruise 710–82. Symbols are the same as in Fig. 1.

duced from both size fractions. N_2O production was detectable about 40 h earlier in the >3-μm fraction, but both the final amount of N_2O per liter and the maximum N_2O production rate were nearly identical.

Samples from the western North Atlantic were taken from three distinct regions; Station 2 was inshore of the Gulf Stream, Station 4 was in the Sargasso Sea, and Station 5 was in the Gulf Stream. N_2O production began quickly and was detectable within 2 d at all stations (Fig. 4). At Stations 2 and 4, N_2O was present in all types of experimental vials. In all but one instance, N_2O was subsequently consumed in all vials except those of the nitrate-supplemented, acetylene-blocked group. N_2O production occurred in the <3-μm samples at all three stations. The inshore station had the greatest N_2O production per liter and the greatest potential denitrification rate. The denitrification rate in the Sargasso Sea >3-μm fraction was the same order of magnitude as denitrification rates measured from the *Lynch* cruise samples taken from the Sargasso Sea the previous year.

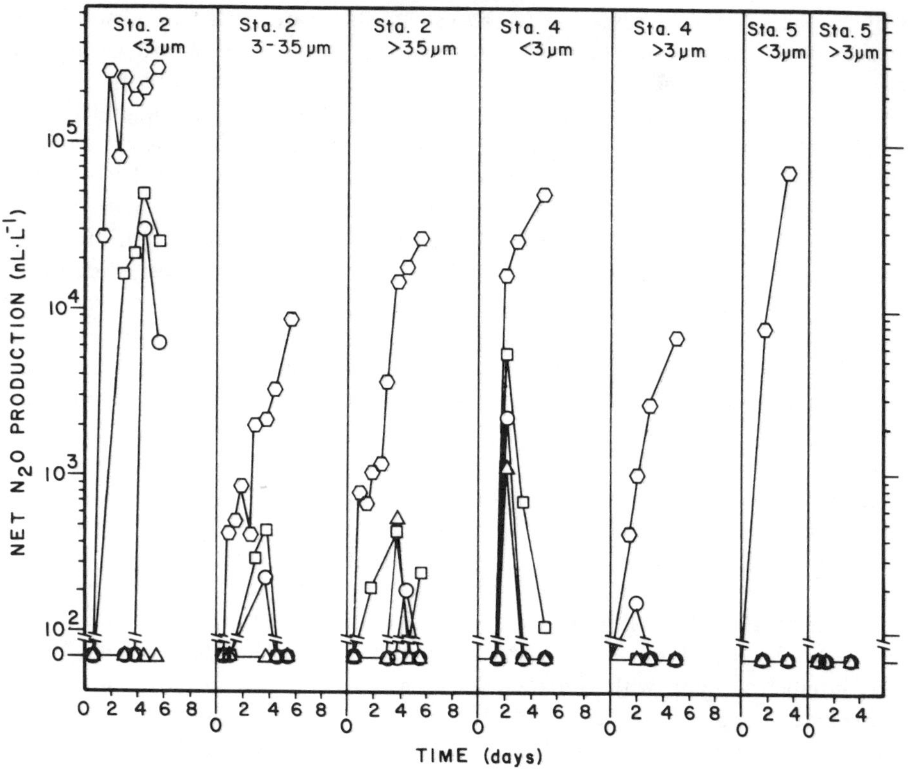

FIG. 4. Net N_2O production in the western North Atlantic Ocean, R.V. *Gyre* cruise 83–G–7. Symbols are the same as in Fig. 1.

Discussion

N_2O production in samples from all stations indicates that denitrifying microorganisms are ubiquitous in oceanic waters. We purified nearly 100 isolates of N_2O-producing bacteria from these cruises; all were gram-negative rods. The biologic source for the gas in our samples was further confirmed by the lack of N_2O production in water filtered through 0.2 μm Nuclepore membrane filters. Denitrification was the only possible biologic source of the N_2O production in our samples because dark anaerobic incubation would have inhibited the other two potential N_2O-producing processes, i.e., nitrification and assimilatory nitrate uptake by phytoplankton. At all stations, addition of nitrate to the samples stimulated N_2O production, as would be expected if denitrification was the source of the N_2O.

The acetylene inhibition technique has proven to be a simple, reliable method for measuring denitrification rates in oceanic samples. The drawback is that we measure potential and not *in situ,* denitrification rates. Other methods of measuring denitrification rates, however, also have their problems. Electron transport system-based denitrification rates can only be obtained for oxygen deficient water masses because, in order to use the technique, one must assume that denitrification is the dominant respiratory process in the water mass (Codispoti and Richards 1976). ^{15}N methods can provide an accurate estimate of *in situ* denitrification rates in some cases, but are cumbersome and require the use of a shore-based mass spectrophotometer (Goering 1968; Goering and Dugdale 1966).

We had originally hypothesized that most of the denitrifying microorganisms in the ocean would be attached to particulate matter and, therefore, tried several sample collection methods that increased the concentration of particulate matter. It became evident, however, that an increase in the particle concentration factor did not result in a proportional increase in N_2O production. The best evidence of this was from samples taken in the Mediterranean Sea. At the four stations where the net tow collection was used, the amount of visible particulate matter in the samples and the total N_2O produced was much greater than at the stations where bottle collections were used, but when the results were converted to N_2O production per liter, N_2O production was less in the tow collections than at the bottle collection stations. In view of this, as well as the observation that the <3-μm size fractions consistently produced N_2O, we concluded that denitrifying microorganisms were free in the water column or, if they were attached to particles, they were easily dislodged during the collection and concentration procedure.

Measurements of denitrification rates in oxygen-deficient regions of the ocean made using the ^{15}N technique have yielded values ranging from 5.5×10^3 to 2×10^5 nL N_2O liter^{-1} day^{-1} (Goering and Dugdale 1966; Goering 1968). Measurements made with the electron transport system method have yielded lower rates, ranging from 30 to 500 nL N_2O liter^{-1} day^{-1} (Codispoti and Packard 1980; Codispoti and Richards 1976). The only information available about potential denitrification rates in an oxygenated region of the ocean can be calculated from the data of Schropp and Schwarz (1983); in the Caribbean Sea, they obtained maximum potential denitrification rates of 20 to 260 nL N_2O liter^{-1} day^{-1}. The potential denitrification rates reported in this paper, ranging from 1 to 1.4×10^5 nL N_2O liter day^{-1}, fall within the range of those obtained using the methods mentioned above. The significance of our results is that they were all obtained from well-oxygenated regions, indicating that the facultative denitrifiers are present in such regions. Given a suitable oxygen-deficient microhabitat, i.e., particulate matter, the denitrifiers could be active and could produce N_2O in otherwise oxygenated seawater.

Is denitrification a source of N_2O in the open ocean? Denitrification undoubtedly occurs in oxygen-deficient regions such as the eastern tropical Pacific, Arabian Sea, and Cariaco Trench (Brandhorst 1959; Codispoti and Richards 1976; Deuser et al. 1978; Hashimoto et al. 1983). Cohen and Gordon (1978) claimed that N_2O is consumed during oceanic denitrification because dissolved N_2O minima exist in the core of the oxygen-deficient region of the eastern tropical Pacific. A more likely possibility, however, is that N_2O is produced and released during denitrification and then may ultimately be consumed if the oxygen concentration drops to extremely low levels, i.e., <0.1 mL liter^{-1} (Pierotti and Rasmussen 1980). Our data support the latter possibility. In our nitrate-supplemented, nonblocked samples, N_2O was produced and, in most instances, it was not subsequently consumed.

Oxygen-deficient oceanic areas, although large, comprise only a small percentage of the total ocean area. N_2O is supersaturated in areas of the ocean that are well oxygenated (Hahn 1975; Yoshinari 1976; Elkins et al. 1978; Singh et al. 1979). Can denitrification be a source of this N_2O? We have shown that denitrifiers are present in oxygenated water throughout the oceans. The widespread occurrence of denitrifiers, which are facultative microorganisms, suggests that the denitrifying respiratory pathway is a useful metabolic process. Given the presence of denitrifiers, we can speculate about the conditions under which denitrification and N_2O production might take place in an oxygenated water column. Knowles (1978) suggested that denitrification could take place in an anaerobic microenvironment, such as a particle, that is located in a larger, oxygenated environment. In the ocean, potential anaerobic microenvironments exist in the form of particulate matter called marine snow.

Marine snow is a term for fragile, amorphous, aggregate particles up to several millimeters in width. It is difficult to collect marine snow because the particles disintegrate when conventional sampling devices are used (Trent et al. 1978; Silver et al. 1978; Jannasch 1973). Nevertheless, the particles can be collected by SCUBA divers and studies of marine snow particles show them to be sites of nutrient enrichment (Shanks and Trent 1979) and of enriched populations of bacteria, phytoplankton, and protozoa (Silver et al. 1978; Silver et al. 1984; Trent et al. 1978; Prezelin and Alldredge 1983). Marine snow sinks at a relatively slow rate, less than 100 m per day, as determined by calculation or direct measurement. Since the density of the particles is close to that of seawater, their *in situ* sinking rate may be significantly retarded due to turbulence or density changes and the particles may accumulate at density gradients (Alldredge 1979; Shanks and Trent 1980).

When we consider all of these facts, the ubiquitous distribution of denitrifiers, the nature of marine snow, and the existence of dissolved N_2O maxima at the thermocline or pycnocline (Cohen and Gordon 1978; Hahn 1975; Pierotti and Rasmussen 1980), we conclude that denitrification, and thus N_2O production, is indeed possible in oxygenated water. Our results show that N_2O is produced by marine denitrifiers and that it can be a substantial portion of the total denitrification product. The denitrifying microorganisms are free in the water column but they could colonize particulate matter and then, when microbial respiration depleted the oxygen within a particle, make use of the denitrifying respiratory pathway. In the ocean, N_2O that diffused from an anaerobic microsite would stand little chance of being consumed in the aerobic water surrounding the microsite.

Acknowledgments

This work was supported by contract N00014–80–C–00113 from the Oceanic Biology section of the Office of Naval Research. We thank the officers and crews of the research vessels *Lynch, Bartlett,* and *Gyre,* without whose cooperation this work would not have been possible. We also thank Dr. J. M. Brooks (Texas A&M University, College Station, TX) and Dr. D. M. Reid (NORDA, NSTL Station, MS) for providing the rosette sampling gear and ancillary chemical and physical data.

Literature Cited

Alldredge, A. L. 1979. The chemical composition of macroscopic aggregates in two neritic seas. *Limnol. Oceanogr.* 24:855–866.

Balderston, W. L., B. Sherr, and W. J. Payne. 1976. Blockage by acetylene of nitrous oxide reduction in *Pseudomonas perfectomarinus. Appl. Environ. Microbiol.* 31:504–508.

Bieri, R. H., M. Koide, and E. D. Goldberg. 1968. Noble gas contents of marine waters. *Earth Planet. Sci. Lett.* 4:329–340.

Bishop, J. K. B., J. M. Edmond, and D. R. Ketten. 1977. The chemistry, biology and vertical flux of particulate matter from the upper 400 m of the equatorial Atlantic Ocean. *Deep-Sea Res.* 24:511–548.

Brandhorst, J. W. 1959. Nitrification and denitrification in the eastern tropical North Pacific. *J. Cons., Cons. Int. Explor. Mer.* 25:3–20.

Codispoti, L. A., and T. T. Packard. 1980. Denitrification rates in the eastern tropical South Pacific. *J. Mar. Res.* 38:453–477.

Codispoti, L. A., and F. A. Richards. 1976. An analysis of the horizontal regime of denitrification in the eastern tropical North Pacific. *Limnol. Oceanogr.* 21:379–388.

Cohen, Y. 1977. Shipboard measurement of dissolved nitrous oxide in seawater by electron capture gas chromatography. *Anal. Chem.* 49:1238–1240.

Cohen, Y., and L. I. Gordon, 1978. Nitrous oxide in the oxygen minimum of the eastern tropical North Pacific: Evidence for its consumption during denitrification and possible mechanisms for its production. *Deep-Sea Res.* 25:509–524.

Crutzen, P. J. 1970. The influence of nitrogen oxides on the atmospheric ozone content. *Quart. J. R. Met. Soc.* 96:320–325.

Deuser, W. G., E. H. Ross, and Z. J. Mlodzinska. 1978. Evidence for and rate of denitrification in the Arabian Sea. *Deep-Sea Res.* 25:435–445.

Elkins, J. W. 1980. Determination of dissolved nitrous oxide in aquatic systems by gas chromatography using electron-capture detection and multiple phase equilibration. *Anal. Chem.* 52:263–267.

Elkins, J. W., S. C. Wofsy, M. B. McElroy, C. E. Kolb, and W. A. Kaplan. 1978. Aquatic sources and sinks for nitrous oxide. *Nature* 275:602–606.

Firestone, M. K., R. B. Firestone, and J. M. Tiedje. 1980. Nitrous oxide from soil denitrification: Factors controlling its biological production. *Science* 208:749–751.

Goering, J. J. 1968. Denitrification in the oxygen minimum layer of the eastern tropical Pacific Ocean. *Deep-Sea Res.* 15:157–164.

Goering, J. J., and R. C. Dugdale. 1966. Denitrification rates in an island bay in the equatorial Pacific Ocean. *Science* 154:505–506.

Hahn, J. 1974. The North Atlantic Ocean as a source of atmospheric N_2O. *Tellus* 26:160–168.

Hahn, J. 1975. N_2O measurements in the northeast Atlantic Ocean. *Meteor Forsch.-Ergebnisse* A 16:1–14.

Hahn, J., and C. Junge. 1977. Atmospheric nitrous oxide: A critical review. *Zeitschrift fur Naturforschung* 32:190–214.

Haines, J. R., R. M. Atlas, R. P. Griffiths, and R. Y. Morita. 1981. Denitrification and nitrogen fixation in Alaskan continental shelf sediments. *Appl. Environ. Microbiol.* 41:412–421.

Hashimoto, L. K., W. A. Kaplan, S. C. Wofsy, and M. B. McElroy. 1983. Transformations of fixed nitrogen and N_2O in the Cariaco Trench. *Deep-Sea Res.* 30:575–590.

Hungate, R. E. 1969. A roll tube method for cultivation of strict anaerobes. Pages 117–132 *in* J. R. Norris, and D. W. Ribbons, eds., *Methods in Microbiology,* Vol. 3B. Academic Press, Inc., London.

Jannasch, H. W. 1973. Bacterial content of particulate matter in offshore surface waters. *Limnol. Oceanogr.* 18:340–342.

Junge, C., B. Bockholt, K. Schutz, and R. Beck. 1971. N_2O measurements in air and seawater over the Atlantic. *Meteor Forsch.-Ergebnisse* B 6:1–11

Knowles, R. 1978. Common intermediates of nitrification and denitrification, and the metabolism of nitrous oxide. Pages 367–371 *in* D. Schlessinger, ed., *Microbiology 1978.* American Society for Microbiology, Washington, D. C.

Knowles, R. 1982. Denitrification. *Microbiol. Rev.* 46:43–70.

McElroy, M. B., S. C. Wofsy, and Y. L. Yung. 1977. The nitrogen cycle: perturbations due to man and their impact on atmospheric N_2O and O_3. *Philos. Trans. R. Soc. London* B 277:159–181.

Miller, T. L., and M. J. Wolin. 1974. A serum bottle modification of the Hungate technique for cultivating obligate anaerobes. *Appl. Environ. Microbiol.* 27:985–987.

Pak, H., L. A. Codispoti, and R. V. Zaneveld. 1980. On the intermediate particle maxima associated with oxygen-poor water off western South America. *Deep-Sea Res.* 27:783–797.

Pierotti, D., and R. A. Rasmussen. 1980. Nitrous oxide measurements in the eastern tropical Pacific Ocean. *Tellus* 32:56–72.

Prezelin, B. B., and A. L. Alldredge. 1983. Primary production of marine snow during and after an upwelling event. *Limnol. Oceanogr.* 28:1156–1167.

Schropp, S. J., and J. R. Schwarz. 1983. Nitrous oxide production by denitrifying microorganisms from the eastern tropical North Pacific and Caribbean Sea. *Geomicrobiol. J.* 3:17–31.

Seiler, W., and U. Schmidt. 1974. Dissolved nonconservative gases in seawater. Pages 219–243 *in* E. D. Goldberg, ed., *The Sea,* Vol. 5. John Wiley & Sons, Inc., New York.

Shanks, A. L., and J. D. Trent. 1979. Marine snow: Microscale nutrient patches. *Limnol. Oceanogr.* 24:850–854.

Shanks, A. L., and J. D. Trent. 1980. Marine snow: Sinking rates and potential role in vertical flux. *Deep-Sea Res.* 27:137–143.

Silver, M. W., A. L. Shanks, and J. D. Trent. 1978. Marine snow: Microplankton habitat and source of small-scale patchiness in pelagic populations. *Science* 201:371–373.

Silver, M. W., M. Gowing, D. C. Brownlee, and J. O. Corliss. 1984. Ciliated protozoa associated with oceanic sinking detritus. *Nature* 309:246–248.

Singh, H. B., L. J. Salas, and H. Shigeishi. 1979. The distribution of nitrous oxide (N_2O) in the global atmosphere and the Pacific Ocean. *Tellus* 31:313–320.

Trent, J. D., A. L. Shanks, and M. W. Silver. 1978. *In situ* and laboratory

measurements on macroscopic aggregates in Monterey Bay, California. *Limnol. Oceanogr.* 23:626–635.
Weiss, R. R., and B. A. Price. 1980. Nitrous oxide solubility in water and seawater. *Mar. Chem.* 8:347–359.
Yoshinari, T. 1976. Nitrous oxide in the sea. *Mar. Chem.* 4:189–202.

CHAPTER **55**

Histological and Metabolite Analysis of Toxin-Sensitive Quail and Survivors from Injection of Fertile Eggs with Aspergillus-Derived Mycotoxin

W. V. DASHEK +*[1], E. T. SHANKS, JR.*, W. R. STATKIEWICZ*, M. J. GIANOPOLUS*, C. E. O'REAR**, AND G. C. LLEWELLYN***

+ *Department of Biology, Atlanta University, Atlanta, Georgia 30314*

* *Department of Biology, West Virginia University, Morgantown, West Virginia 26506*

** *Department of Forensic Sciences, George Washington University, Washington, D.C. 20052*

*** *Department of Biology, Virginia Commonwealth University, Richmond, Virginia 23284 and Bureau of Toxic Substances Information, Virginia Department of Health, Richmond, Virginia 23219*

Aflatoxins (AFTs), secreted by *Aspergillus parasiticus* and *A. flavus*, are carcinogens, mutagens, and teratogens. Combined oral AFTs dosing and genetic selection result in toxin-resistant quail. Here, we describe attempts to both generate such quail by an egg assay and evaluate the survivors for resistance by both histopathological and metabolite analysis following rechallenging them with AFTs. Although survivors were generated, a dose-response curve was not. Hematoxylin and erythrosin-stained sections of livers from dosed sensitive and *in situ* survivors revealed fat accumulation, cellular necrosis, bile duct proliferation, vacuolization, mild hepatitis, and pleiomorphic hepatocyte mitochondria when compared to unexposed sensitive *in situ* survivor quail. Putative aflatoxin B_1 (AFB_1) was $CHCl_3$-extracted from both stomach and excrement homogenates and pepsin-digests of both stomach and liver. Digests' AFB_1 comigrated with AFB_1 upon thin-layer chromatography. Marked differences in AFB_1 tissue distribution between sensitive and survivor quail were not apparent. These results suggest that the *in situ* egg assay may be useful for generating AFTs-resistant quail provided that the assay is either modified and/or combined with a genetic selection program.

INTRODUCTION

Aflatoxins (AFTs) are secondary metabolites of *Aspergillus flavus* and *A. parasiticus*, two fungi that can grow upon agroeconomic crops (Dashek and Llewellyn 1982) thereby contaminating animal feeds (Scott 1978). These toxins, which are carcinogenic, mutagenic, and teratogenic (Stoloff 1977), can adversely affect the poultry industry since Aves are very susceptible to AFTs (Cavalherio 1981).

Japanese quail can suffer from aflatoxicosis (Dashek et al. 1983), but recently Marks and Wyatt (1979) obtained aflatoxin B_1 (AFB_1)-resistant quail through

[1]Address correspondence to Dr. W. V. Dashek, Biology Department, 223 Chestnut Street, S.W., Atlanta University, Atlanta, GA 30314.

combined oral dosing and subsequent genetic selection. An *in situ*, chick embryo assay involving AFB_1 dosing of the embryo via a puncture into the shell has been reported also (Horwitz et al. 1975).

The present paper (1) reports our preliminary attempt to adapt the chick embryo assay to quail for generation of resistance and (2) compares both liver histopathology and AFB_1 tissue distribution for both toxin-sensitive and *in situ* egg survivors. The liver was selected because it appears to be the primary target of AFTs (Hsieh et al. 1977).

Materials and Methods

Rearing and Maintenance of Quail

Coturnix coturnix japonica (colony of M. W. Schein) were mated, eggs were incubated, and hatchlings were raised to maturity as previously described by Dashek et al. (1983). Quail were fed Southern States (Morgantown, WV) starting and growing mash (medicated A+) *ad libitum* for 6 wk when they were either mated or intubated (shown later).

Generation of in situ *AFB_1 Survivors*

Each egg was swabbed with 95% ethanol prior to both locating the air-cell's center and drilling a hole into the shell. Sterile forceps served both to remove shell fragments and to make a hole into the shell's membrane for the entrance of a Finpipette's tip. Two and a half, 5, 10, and 15 μl EtOH (carrier either containing or lacking AFB_1 (Grade A, Calbiochem., LaJolla, CA) were injected into the egg. The AFB_1 concentrations were 0.025, 0.050, 0.100, and 0.150 μg egg^{-1}. Following injection, the holes were sealed with two layers of sterile cellophane tape. Unopened controls also received tape over the shell approximating the air cell's position. Eggs were incubated for 16 d at 30 C, 80% relative humidity, and ambient light with automatic turning every 13 h. At 16 d, eggs were transferred from incubating racks to hatching trays for 3 to 4 d when hatched quail were removed.

Properties and Verification of Aflatoxin Stock Concentration Light Microscopic Analyses

An AFTs "working solution" was prepared by concentrating 100 ml of "stock" (Llewellyn et al. 1973) containing 806, 28, 3, and 27 μg/ml AFB_1, AFB_2, AFG_1, and AFG_2, respectively, in an Erlenmeyer flask with the subsequent addition of 57 ml H_2O to yield 3.0 mg/ml. To verify this concentration, 10, 50, and 100 μl of solution were spotted together with 10 μg each AFB_1, AFB_2, AFG_1, and AFG_2 (Applied Sciences Lab. Inc., State College, PA) onto Adsorbosil + 1 hard plates. Toxins were separated by thin-layer chromatography (TLC) within sealed, equilibrated, but unlined chambers containing either 200 ml chloroform plus 6 ml, 95% methanol or chloroform-acetone-H_2O (88.0 + 12.0 + 1.5). The latter solvent separates AFB_1, AFR, AFQ_1, AFM_1, and AFB_{2a} (Roebuck and Wogan 1977). Fluorescent spots visualized with a Mineralight (Ultraviolet Products Inc., South Pasadena, CA) were scraped off the plate and the presumed toxins eluted from the absorbent with methanol. To quantify the toxins, their maximal absorbances from absorption spectra were compared to those for authentic AFB_1,

AFB_2, AFG_1, and AFG_2 by employing both standard curve construction and linear regression analysis. The stock's concentration was 0.3 mg/ml.

Electron microscopic and metabolite analyses. Aflatoxin B_1 (Southern Regional Research Laboratory, New Orleans, LA), whose purity was assessed by TLC as it both comigrated with authentic AFB_1 and fluoresced blue under ultraviolet (UV) light. The toxin was dissolved within either peanut oil or propylene glycol by stirring, and the "stocks" concentration (0.33 to 1.22 mg AFB_1/ml) was quantified through both absorbances and standard curve construction as before.

When $[^3H]$-AFB_1 was used, the "stock" solution (containing 0.75 mg/ml "cold" AFB_1) was such that each quail received 5.25×10^7 cpm/kg body wt.

Administration and recovery of aflatoxin light microscopy. Quail were intubated through the crop at 0.3 mg/kg body wt with a carrier (H_2O) containing AFTs. Control quail were intubated with carrier only. To ensure that the intubation was performed correctly, quail were intubated with safranin and the feces examined for voided dye the following day.

Electron microscopy and AFB_1 tissue distribution. Quail were intubated with either the AFB_1 "stock" at 2.75 mg/kg body wt (containing or lacking $[^3H]$-AFB_1 (spc. act. 5 Ci/mmol, Moraveck, City of Industry, CA) or a carrier (peanut oil or propylene glycol) alone and manually decapitated 6 h later. The blood flow was collected in heparinized glassware, hemolyzed by distilled H_2O, and placed onto ice. For microscopy, the top of the liver's left lobe was excised and immersion fixed in 2.5% glutaraldehyde buffered with pH 7.4, 0.2 *M* sodium cacodylate (Sabatini et al. 1963) and processed as below. The liver's remainder, the stomach, intestines, and the kidneys were excised, minced into distilled H_2O on ice, and homogenized in an "ice-cold" Virtis Model 23 homogenizer.

Blood, organs, and excrement homogenates were either stored at −10 C, lyophilized, and chloroform-extracted (1:1, V:V) within separatory funnels or extracted immediately. In either case, the extracts were reduced in volume under an air current before spotting onto heat-activated (110 C, 30 min) Adsorbosil+1 TLC plates. Plates were also spotted with authentic AFB_1, developed, removed from the chamber, air-dried, and examined with an UV light as before. If $[^3H]$-AFB_1 was not used, blue fluorescent compounds that comigrated with authentic AFB_1 were scraped from the plate and dissolved in 3 ml methanol in test tubes; the toxins were quantified by their absorbances at 362 nm.

When $[^3H]$-AFB_1 was employed, the plate's area at the R_f for AFB_1 for each sample was scraped from the plate and dissolved in 10-ml scintillation fluid (0.2 g POPOP, 4 g PPO toluene^{-1}). In some cases, the entire plate was divided into rows each consisting of 19 cm segments that were added to fluid. Counts per minute from a Beckman LS-230 counter were corrected for background following counting of samples for either 20 min or 0.2% preset error and examination of external standard ratio to ensure the absence of quench.

Each sample's aqueous phase was mixed with 0.3% pepsin (1:1) and the pH adjusted to 1.6 prior to incubation for 16 h at 37 C. Following incubation, the samples were chloroform-extracted within separatory funnels and then processed in the same manner as the homogenate chloroform extracts.

Pathology

Light microscopy. Segments (0.50–0.70 × 1.25–1.50 cm) of freshly excised livers were fixed in either chloroform-acetic and acid-ethanol, 30:60:10 or 5% formaldehyde saturated with $CaCO_3$ for 1.0–4.5 h at 25 ± 2 C. Following fixation, the livers were washed twice with absolute ethanol, cleared in two changes of xylene (20 min each change), and then infiltrated into 3:1, 1:1, and 1:3 xylene:paraplast (1 h each), with final embedment in paraplast.

Sections of livers from nondosed quail were stained with both hematoxylin and erythrosin B (Galigher and Kosloff 1964) within the same dishes and times as those from dosed quail. Prior to staining, sections were deparaffinized by passage through two changes of absolute toluene and one of alcohol toluene (1:1) for 2 min each change. Deparaffinized sections were hydrated by transferring them through the changes (100, 95, 70, and 30% ethanol, 2 min each) followed by 1–3 distilled H_2O washes.

Transmission electron microscopy. Freshly excised liver segments were fixed 4.5 h at 4 C in 2.5% glutaraldehyde buffered with 0.2 *M* sodium cacodylate, pH 7.4. Additional manipulations were at 25 C unless otherwise specified. Following fixation, the tissue was rinsed twice in buffer (1 min each rinse), postfixed, and stained for 2 h with 1% aqueous uranyl acetate. After rinsing twice with buffer (1 min each rinse), the segments were passed through 30, 50, 70, 90, and 95% ethanol (1 min in each) followed by three 1-min rinsings with absolute ethanol. The tissue was then either (a) stored in L. R. White resin overnight followed by 5 min in fresh resin and embedment in another change of resin with polymerization at 60 C for 20–24 h or (b) left in propylene oxide:Epon-Araldite 5 (1:1) overnight and embedment for 24 h in fresh resin with polymerization at 60 C.

Thin sections were cut with a Porter-Blum ultramicrotome, collected upon formvar-coated grids, counterstained with either 5% uranyl acetate in absolute ethanol or saturated uranyl acetate solution (aqueous) followed by lead citrate (Reynolds 1963). Sections were both viewed and photographed with an RCA-EMU3G microscope at either 50 or 100 KV.

Nine livers were examined, five AFB_1-treated and four controls. Three tissue blocks from each liver were used, and approximately 10 micrographs were taken of sections from each. Thus, interpretations were from about 250 micrographs.

Results

Generation of in situ *Egg Survivors*

Table 1 presents percent embryo mortality as a function of injected AFB_1 concentration. While survivors were obtained in both experiments, a dose-response curve was not.

Histopathology of Sensitive and in situ *Survivors*

Light microscopy. Figs. 1A-D compare liver histopathology of toxin-dosed sensitive and *in situ* survivors with undosed sensitive and *in situ* survivors. The histopathology is based upon hematoxylin and erythrosin staining. The livers of

TABLE 1. *Percent embryo mortality following injection of EtOH[a] with and without AFB_1 into opened eggs*

		% Survivor		
Condition of Egg	Experiment 1	EtOH plus AFB_1 minus EtOH minus AFB_1	Experiment 2	EtOH plus AFB_1 minus EtOH minus AFB_1
Unopened	42.9		46.2	
Opened but not injected	30.8		NP	
Opened injected with:				
5.0 µl EtOH minus AFB_1	25.0		45.0	
10.0 µl EtOH minus AFB_1	40.0		60.0	
15.0 µl EtOH minus AFB_1	NP		55.0	
2.5 µl EtOH plus AFB_1	30.0		NP	
5.0 µl EtOH plus AFB_1	40.0	15.0	45.0	0.0
10.0 µl EtOH plus AFB_1	10.0	0.0	25.0	35.0
15.0 µl EtOH plus AFB_1	NP[b]		50.0	50.0

[a] EtOH concentration was 100%; AFB_1 concentrations were 0.15 µg/15.0 µl EtOH.
[b] NP = not performed.

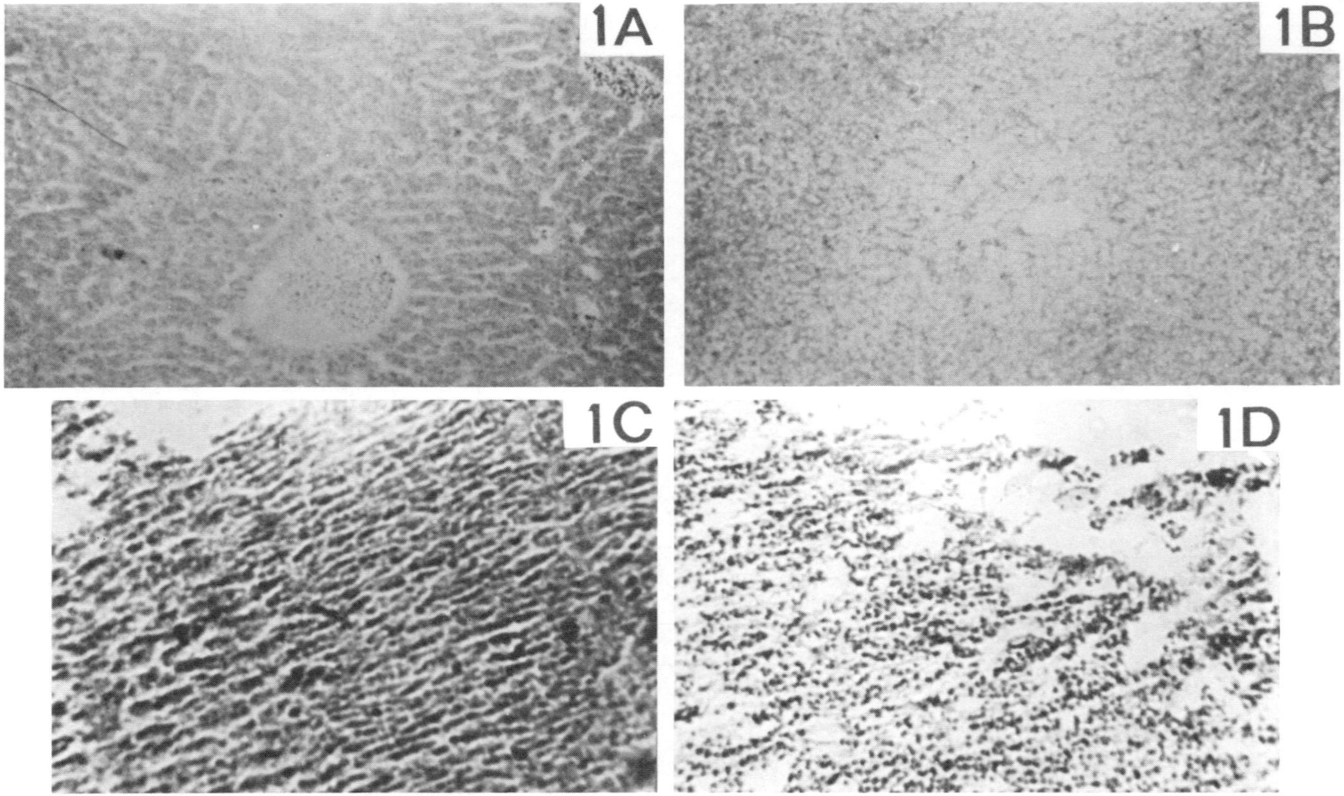

FIG. 1. Hematoxylin and erythrosin-stained sections of livers derived from dosed (B,D) and nondosed (A,C) quail, X 80. *In situ* egg survivors (C,D) and sensitive quail (A,B). Variants of Fig. 1A,B appear in Dashek et al. 1983).

both dosed sensitive (Fig. 1B) and *in situ* survivor (Fig. 1D) quail exhibited bile duct proliferation, cellular necrosis, and vacuolization. The sinusoids between rows of parenchymal cells were collapsed. Instead, the thickenings about the hepatocytes gave the appearance of a reticulum ramifying throughout each liver lobule. Other observations not readily apparent included an increased amount of fat and mild hepatitis for dosed quail. These abnormalities were not present in either nondosed sensitive (Fig. 1A) and *in situ* survivors (Fig. 1C).

Electron microscopy. The most striking effect of AFB_1 treatment (2.75 mg AFB_1/kg body wt, 6 h) for both sensitive and survivors was a change in mitochondrial configuration wherein mitochondria appeared to encircle areas of cytoplasm (Fig. 2B). In addition, mitochondria formed both horseshoe and forked configurations that were observed only rarely within hepatocytes of nondosed quail (Fig. 2A). Serial sections of liver tissue from dosed quail seemed to suggest that horseshoe-shaped mitochondria "rolled up" to form encirclements when the same mitochondrion was examined in different sectioning planes. Nuclear segregation, often reported as an ultrastructural effect of short-term AFB_1 treatment of hepatic parenchymal cells, was not observed.

AFB_1 Distribution in Tissues

Sensitive quail. The AFTs distribution in blood, excrement, and liver is displayed in Table 2. Both the liver and excrement contained a fluorescent compound with an R_f similar to that of authentic AFB_1. Figs. 3A-B shows that the absorption spectra of the liver and excrement compounds were similar with a symmetrical peak not being obtained for either compound. However, the compounds did absorb at 362 nm.

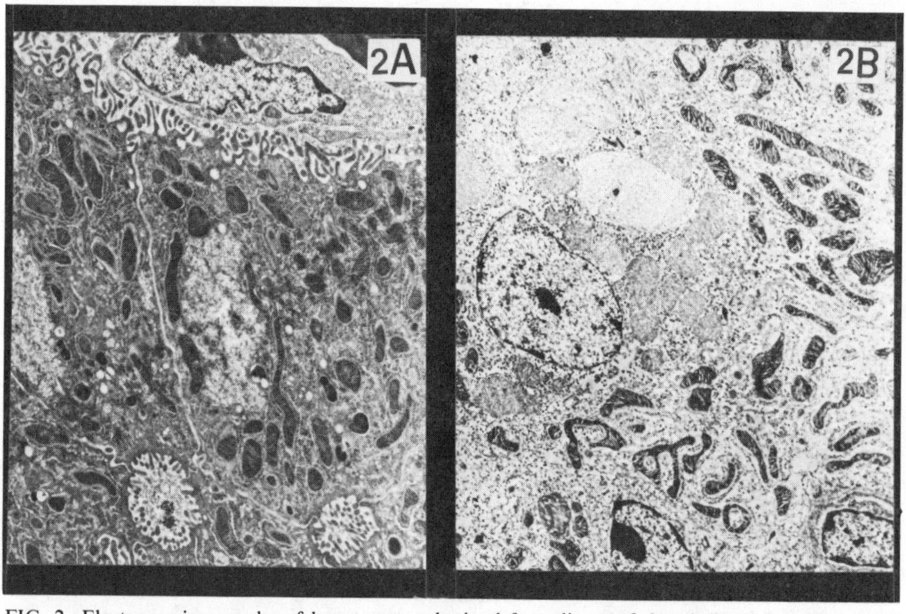

FIG. 2. Electron micrographs of hepatocytes obtained from livers of dosed (B) and nondosed (A) *in situ* egg survivors. X 6,480 (A) and 5,400 (B). Pleiomorphic mitochondria were also seen in dosed-sensitive quail (data not shown). Variants of Fig. 2A-C appear in Shanks et al. (submitted).

TABLE 2. *R_{fa} of fluorescent compounds originating from pepsin treatment of aqueous fractions of blood, excrement, and liver from quail intubated with aflatoxins*

Time	Sample	R_{fa}	Aflatoxin standards	R_{fa}
(h)				
6	blood	0.65	AFB_1	0.85
		0.59	Aflatoxicol	0.68
	excrement	0.56	AFO_1	0.42
		0.84	AFP_1	0.32
	liver	0.84, 0.22, 0.16, 0.08	AFM_1	0.19
24	blood	none	AFB_{2a}	
	excrement	none		
	liver	none		
48	blood	none		
	excrement	none		
	liver	none		
120	blood	none		
	excrement	none		
	liver	0.95, 0.16		

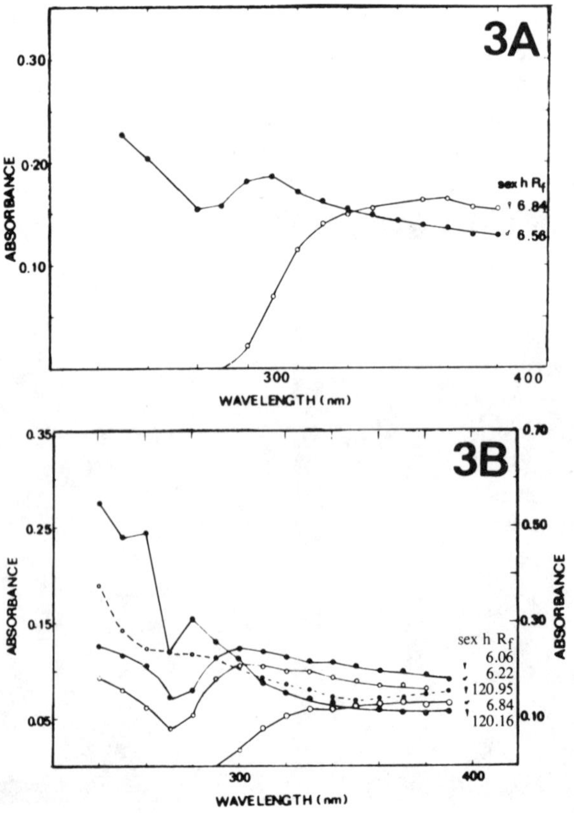

FIG. 3. A-B Ultraviolet absorption spectra of fluorescent compounds released by pepsin-treatment of aqueous phases of feces (A) and liver (B) from sensitive-quail.

In situ *survivor quail.* Unlabeled toxin: ultraviolet fluorescent compounds were extracted by chloroform from both the liver and stomach homogenates of a female (bird 1) quail that had been intubated with AFB_1 in peanut oil (Table 3). These compounds possessed somewhat lower R_fs and absorption maxima as well as substantially "flatter" UV-absorption spectra than those for authentic AFB_1 (Fig. 4A).

TABLE 3. *Distribution of AFB_1 in selected quail organs and excrement 6 h after dosing*[a]

Sample	R_f Sample/R_f AFB_1	Absorption maxima[b]	μg
Female			
Bird 1			
excrement ($CHCl_3$-extractable)	0.62/0.70	265,355	1.43
stomach (organ and contents)			
$CHCl_3$-extractable	0.45/0.70	265,355	1.72
$CHCl_3$-extractable after pepsin digestion	0.58/0.59	355	1.65
liver			
$CHCl_3$-extractable after pepsin digestion	0.55/0.59	265,360	2.35
Bird 2			
stomach contents $CHCl_3$-extractable	0.61/0.54	265,355	
Male			
Bird 3			
stomach contents $CHCl_3$-extractable	0.62/0.54	265,355	

[a] Peanut Oil as AFB_1 carrier.
[b] Absorption maxima of AFB_1 are 265,360 nm.

Putative AFB_1 that was chloroform-extracted from pepsin-digests of both the stomach and liver of the quail comigrated with authentic AFB_1 upon TLC. Although the UV-spectrum of the liver compound exhibited the same maxima as authentic AFB_1, small quantities also exhibited a "flat" spectrum.

Quantification of the presumed AFB_1 that was recovered from this bird revealed that of the samples tested, the greatest amount (2.35 μg) was in the liver and could be liberated for chloroform extraction by pepsin digestion. This digestion released 1.65 μg of putative AFB_1 from the stomach. Chloroform extraction of both excrement and stomach homogenates yielded 1.43 and 1.72 μg, respectively. No "aflatoxin-like" compounds were found in either blood or intestine samples.

In two subsequent experiments (Table 3), presumed AFB_1 was chloroform-extractable only from the stomach contents of one male and one female. In both cases, the blue fluorescent compound exhibited both an R_f and absorption maxima similar to those for authentic AFB_1 (data not shown).

Toxin. An assay of the distribution of tritium from [^3H]-AFB_1 upon TLC revealed that approximately 46% of the plate's radioactivity was at R_fs other than

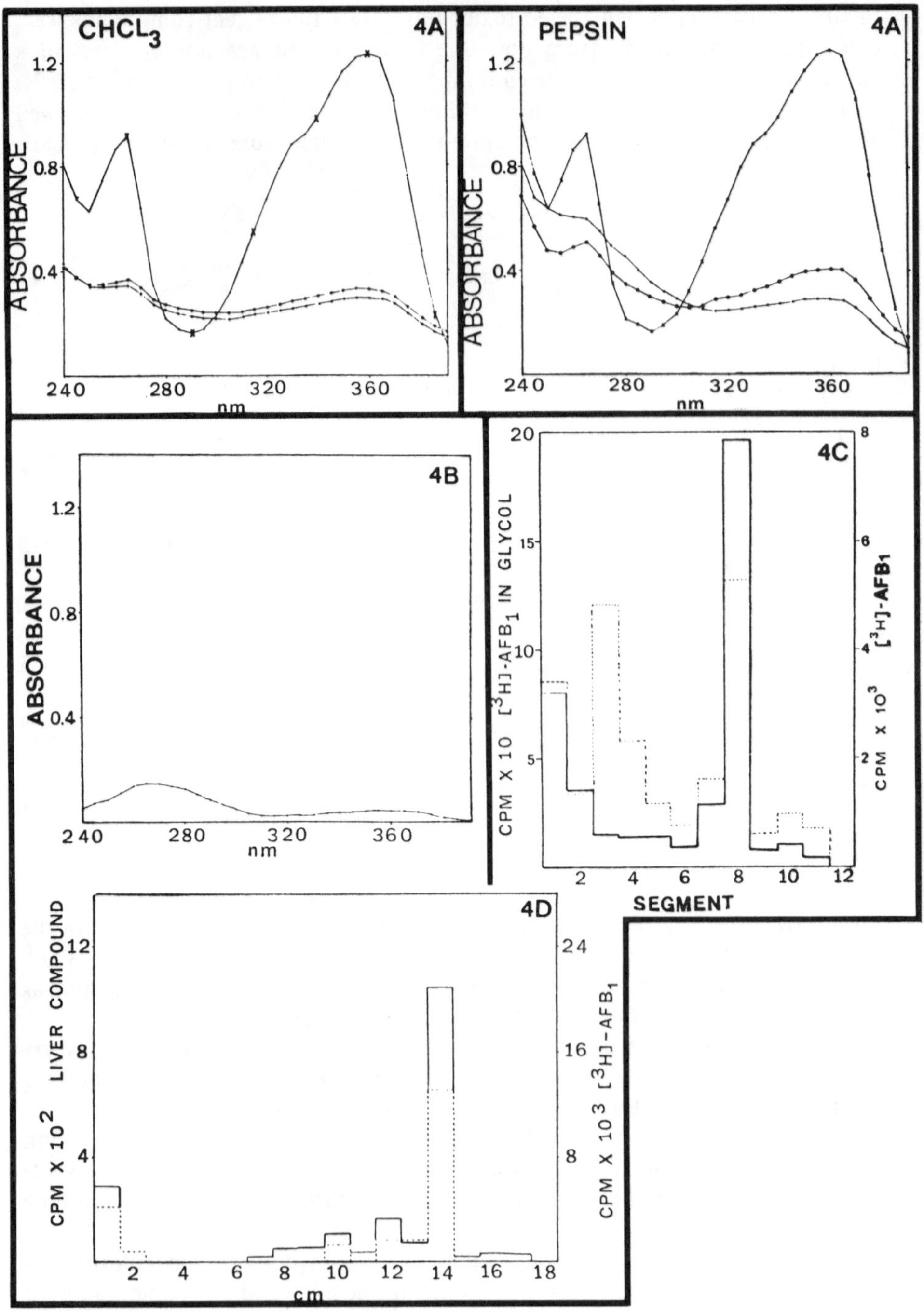

FIG. 4. (A) Absorption spectra of authentic AFB_1 and chloroform extracts of quail liver and stomach pepsin-digests. X--X = authentic AFB_1, o--o = liver, •---• = stomach. (B) Ultraviolet absorption spectra of authentic AFB_1. (C) Distribution of [^{3}H] following TLC of [^{3}H]-AFB_1 dissolved in either methanol or propylene glycol — [^{3}H]-AFB_1 in methanol,--[^{3}H]-AFB_1 in propylene glycol. (D) TLC distribution of radioactivity from the liver of a quail intubated with [^{3}H]-AFB_1.

that for AFB_1. Furthermore, this was exacerbated by mixing of [^{3}H]-AFB_1 with nonradioactive AFB_1 in propylene glycol as a stock solution with 70% of the radioactivity occurring as R_fs other than that for AFB_1 (Fig. 4C). Subsequently, radioactivity was quantified only for those quail-derived compounds that comigrated with authentic AFB_1.

A compound chloroform-extracted from the liver homogenate of a female quail comigrated with authentic AFB_1. Following TLC, 652 cpm were detected at the R_f for AFB_1. The distribution of radioactivity among various R_fs within this sample upon the TLC plate was quite similar to that for authentic [^{3}H]-AFB_1 (Fig. 4D). No radioactivity was detected at the R_f for AFB_1 in chloroform extracts of either the homogenates or pepsin-digests of stomach, intestine, or excrement, nor in the chloroform-extract of the pepsin-digest of the liver from quail no. 2.

Discussion

Was resistance conferred upon mature quail by exposing them to AFTs while they were embryos? Although AFB_1 survivors were obtained from an *in situ* egg assay (Table 1), comparisons of hematoxylin and erythrosin-stained sections of livers derived from either dosed survivors (Fig. 1D) or dosed sensitive (Fig. 1B) quail revealed aflatoxicosis and thus contradict our suggestion that injection of fertile eggs with AFB_1 might confer resistance. This was confirmed by electron microscopy, which showed that hepatocytes from both quail types contained pleiomorphic mitochondria, thereby supporting the findings of Kelly and Mora (1976) and Obidoa and Siddiqui (1978) who reported aflatoxin-induced mitochondrial aberrations for fowl. Because toxin-resistant quail were obtained by Marks and Wyatt (1979) through a combined oral dosing and genetic selection program, it may be possible to derive such quail via a combined *in situ* egg assay and selection program. Perhaps fewer generations would be required to produce toxin-resistant quail by the egg method.

What could be the mechanism by which quail could attain resistance? Some possibilities that may explain the development of resistance include the following: (1) failure of the toxin to reach the liver due to either an inability to be transported to it or a failure of the toxin to transverse the membrane, (2) complexing of intracellular AFB_1 with glutathione to render the toxin less toxic (Novi 1981), and/or (3) loss of the capacity to form AFB_1-epoxide (purported carcinogenic (Swenson et al. 1974) form of AFB_1) in resistant quail. Table 1 and Fig. 3, as well as Table 2 and Fig. 4, suggest that AFB_1 reaches the liver in both sensitive and resistant quail. Therefore, the second and third alternatives should be examined. Although the formation of a glutathione complex could be detected (Novi 1981), the formation of the epoxide would be difficult to quantify because of its liability. However, its occurrence, or lack thereof, could be inferred through detection of aflatoxin dihydrodiols (Wogan and Busby 1980).

Acknowledgments

This work was aided by Grants IN-27B from the American Cancer Society (WVD), The Sigma Xi (ETS), and Virginia Commonwealth University (GCL) as well as a donation for cancer research from the Ladies Auxiliary, West Virginia Veterans of Foreign Wars (WVD). We are grateful to Dr. H. Voelz for his technical assistance with various aspects of the electron microscopy. Thanks are extended to Ms. Pamela Caldwell and Mrs. Barbara Spain for the manuscript typing, to Mr. Ron Gamber for technical assistance, and to Mr. James J. Testaguzza for photography. We thank Mr. S. Barker for histochemical advice.

Literature Cited

Cavalherio, A. C. L. 1981. Aflatoxins and aflatoxicosis: A review. *World's Poult. Sci. J.* 37:34–38.

Dashek, W. V., and G. C. Llewellyn. 1982. Aflatoxins and plants. *Postepy Microbiologii.* 21:65–64.

Dashek, W. V., S. M. Barker, W. R. Statkiewicz, E. T. Shanks, and G. C. Llewellyn. 1983. A histochemical analysis of liver cells from short term, aflatoxin-dosed and nondosed *Coturnix coturnix japonica.* Aflatoxin-sensitive quail. *Poult. Sci.* 62:2347–2359.

Galigher, A. E., and E. N. Kosloff. 1964. *Essentials of Practical Microtechnique.* Lea and Febiger, Philadelphia, PA.

Hsieh, D. P. H., J. J. Wong, Z. A. Wong, C. Michas, and B. H. Ruebner. 1977. Hepatic transformations of aflatoxin and its carcinogenicity. Pages 697–707 *in Cold Spring Harbor Meeting on Origins of Human Cancer.* Cold Spring Laboratory.

Horwitz, W., A. Senzel, and H. Reynolds. 1975. Natural poisons. *In Official Methods of Analysis.* Assoc. Offic. Anal. Chemists Soc., Washington, DC.

Kelly, V. C., and E. C. Mora. 1976. Ultrastructural changes induced by chronic aflatoxicosis in chickens. *Poult. Sci.* 55:317–324.

Lillie, R. D., and H. M. Fullmer. 1976. *Histopathologic Technic and Practical Histochemistry.* McGraw-Hill, New York.

Llewellyn, G. C., W. W. Carlton, J. E. Robbers, and W. G. Hansen. 1973. A rapid and simplified method for the preparation of pure aflatoxin B_1. *Dev. Ind. Microbiol.* 14:325–335.

Marks, H. L., and R. D. Wyatt. 1979. Genetic resistance to aflatoxin in Japanese quail. *Science* 206:1329–1330

Novi, A. M. 1981. Regression of aflatoxin B_1-induced hepatocellular carcinomas by reduced glutathione. *Science* 212:541–542.

Obidoa, O., and H. T. Siddiqui. 1978. Aflatoxin inhibition of avian hepatic mitochondria. *Biochem. Pharmacol.* 27:547–550.

Reynolds, E. S. 1963. The use of lead citrate at high pH as an electron-opaque stain in electron microscopy. *J. Cell Biol.* 17:208–212.

Roebuck, B. D., and G. N. Wogan. 1977. Species comparison of *in situ* metabolism of aflatoxin B_1. *Cancer Res.* 37:1649–1686.

Sabatini, D., K. Bensch, and R. J. Barnett. 1963. Cytochemistry and electron microscopy. *J. Cell Biol.* 17:19–58.

Scott, P. M. 1978. Mycotoxins in feeds and ingredients and their origin. *J. Food Prot.* 41:385–398.

Shanks, E. T., W. R. Statkiewicz, G. C. Llewellyn, and W. V. Dashek. Tissue distribution and hepatic ultrastructural effects of aflatoxin B_1 in Japanese quail. *Postepy Microbiologii* (submitted).

Sheehan, D. C., and B. B. Hrapchak. 1973. *Theory and Practice of Histotechnology*. The C. V. Mosby Co., St. Louis, MO.

Stoloff, L. 1977. Aflatoxins—an overview. *In* J. V. Rodericks, C. W. Hesseltine, and M. S. Mehlman, eds., *Mycotoxins in Human and Animal Health,* Pathotox Publishers, Inc., Park Forest South, IL.

Swenson, D. H., E. C. Miller, and J. A. Miller. 1974. Aflatoxin B^1 2,3-oxide: Evidence for its formation in rat liver *in situ* and by human liver microsomes *in situ. Biochem. Biophys. Res. Commun.* 60:1036–1043.

Wogan, G. N., and W. F. Busby, Jr. 1980. Naturally occurring carcinogens. Pages 329–369 *in* I. E. Liener, ed., *Toxic Constituents of Plant Foodstuffs.* Academic Press, New York.

CHAPTER 56

Mycotoxic-Induced Hormonal Responses in Plant Cells Treated with Aflatoxin B_1, Sterigmatocystin, Patulin, and Shikimate

L. B. WEEKLEY, T. D. KIMBROUGH*, J. D. REYNOLDS*, C. E. O'REAR** AND G. C. LLEWELLYN* [1]

*University of Wyoming, Laramie, Wyoming; *Virginia Commonwealth University, Richmond, Virginia; and **Department of Forensic Sciences, The George Washington University, Washington, D.C.*

The influences of toxic secretions from several fungi were evaluated on tryptophan metabolism of plant cells. Single cell suspension cultures of *Sedum morganianum* were incubated with aflatoxin B_1 (AFB_1), sterigmatocystin (STR), patulin (PAT), or shikimic acid (SA). Tissue levels of 5-hydroxytryptamine (5-HT), and 5-hydroxyindole-3-acetic acid (5-HIAA) were quantified. The activities of tryptophan-5-hydroxylase and IAA-5-hydroxylase were also measured. Incubation of cultures with SA decreased cellular 5-HT but increased 5-HIAA significantly without affecting enzymatic activity. Incubation of cells with 10 μg PAT/ml reduced both cellular 5-HT and 5-HIAA levels, decreased IAA hydroxylase activity, and increased tryptophan hydroxylase activity. Addition of SA did not markedly alter these trends except to suppress tryptophan hydroxylase activity. Sterigmatocystin lowered 5-HT while 5-HIAA, tryptophan hydroxylase and IAA hydroxylase activity were enhanced. Supplementation with SA increased both 5-HT and 5-HIAA levels. Aflatoxin B_1 reduced the activities of tryptophan-5-hydroxylase, but elevated levels of 5-HT and 5-HIAA and IAA-5-hydroxylase. In contrast, SA plus AFB_1 raised enzyme activities while 5-HIAA levels were diminished. These results suggest that mold metabolites affect enzymatic activities and formation of serotonin-related products in this plant cell system.

INTRODUCTION

Sterigmatocystin (STR) is a metabolic product of several fungal species including *Aspergillus versicolor, A. nidulans,* and *Bipolaris* spp. (Holzapfel et al. 1966). It is related chemically to the aflatoxins (AFTs), which are natural secretions of some *A. flavus* and *A. parasiticus* isolates. Aflatoxins are known to be carcinogenic, mutagenic, and teratogenic; they occur in low levels within many nut and grain products (Diener and Davis 1969; Dashek and Llewellyn 1982). Sterigmatocystin has been shown to be carcinogenic when administered subcutaneously to rats where it was estimated to be 1/250 as carcinogenic as AFTs (Dickens et al. 1966).

Patulin (PAT) is another mycotoxin produced by a variety of *Aspergillus* and *Penicillium* spp. Historically PAT also has been used at times as an antibiotic,

[1]Correspondence to G. C. Llewellyn, Bureau of Toxic Substances Information, Virginia, Department of Health, 109 Governor Street, Richmond, VA 23219

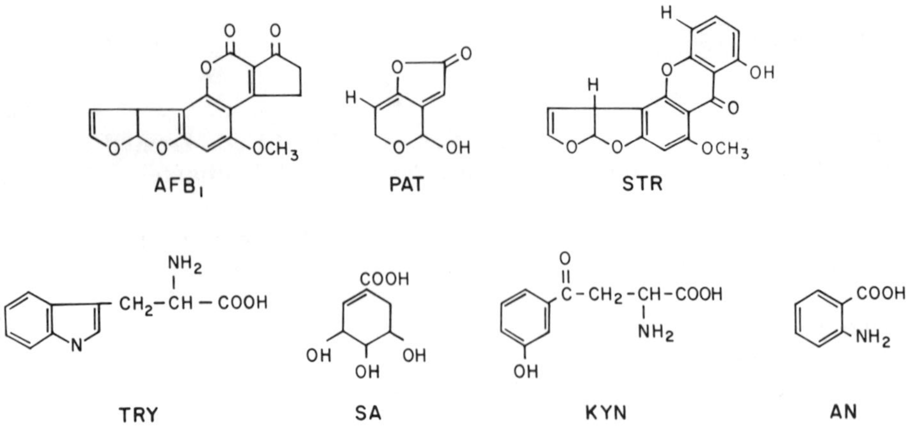

FIG. 1. Mycotoxin and indole structures showing aflatoxin B_1 (AFB_1), patulin (PAT), sterigmatocystin (STR), L-tryptophan (TRY), shikimic acid (SA), kynurenine (KYN), and anthranilic acid (AN).

thus it is not thought to be a highly toxic mycotoxin (Carlton and Szczech 1978).

There is a paucity of information regarding the effects of various mycotoxins on plant systems. However, several studies (Truelove et al. 1970; Dashek and Llewellyn 1974) have demonstrated that AFT alters amino acid uptake in plants, which could have secondary effects on plant growth and metabolism. Indeed, Reiss (1970) found that aflatoxin B_1 (AFB_1) accelerated the action of indole-3-acetic acid (IAA) in a *Pisum sativum* bioassay and suggested that AFB_1 behaved as an IAA synergist. Furthermore, Reiss (1977) demonstrated that AFB_1 inhibited the *Kalanchoe diagremontiana* root elongation while the germination of *Vigna sinensis* seeds was completely inhibited by 50 μg AFB_1/ml. This can be reversed by treatment with IAA (Adekunle and Bassir 1973).

The specific pathway of tryptophan (TRP) synthesis in higher plants has not been established. However, in *E. coli* TRP synthesis commences with anthranilate, which is formed in a sequence of reactions leading to other aromatic amino acids (Bender 1975). In other systems, and particularly those using mammalian models, TRP is degraded primarily by two pathways, one involves oxidation of TRP to 5-hydroxytryptophan (5-HTP) followed by decarboxylation to 5-hydroxytryptamine (5-HT, serotonin, SER). The other pathway involves oxidation to kynurenine, which is converted to a series of intermediates and byproducts. One of these is anthranilic acid, which is also a byproduct of shikimic acid (SA) metabolism (Gibson and Pittard 1968).

The available evidence supports the view that AFB_1 affects IAA action in plant systems (Dashek and Llewellyn 1983). Therefore, several studies were conducted to determine the effects of AFB_1, STR, and PAT on IAA metabolism in cell suspension cultures of *Sedum morganianum,* a small succulent plant lending itself readily to such investigations.

Materials and Methods

Young leaves from adult *Sedum morganianum,* purchased commercially, were maintained in constant light at 18–23 C for at least 2 wk prior to their use. A single-cell suspension culture was obtained by slicing up the leaves into small pieces; they were later incubated in White's medium containing 5,000 units pectinase (Sigma Chemical Co., St. Louis, MO) for 60 min at 25 ± 1 C in a metabolic shaker. Then the suspension was centrifuged (4,000 × *g*, 5 min) to pack the intact cells into a pellet while cellular fragments remained in solution. To remove residual pectinase, the pellet was washed four times with 5 ml of White's medium (White 1963). Cells were finally suspended in White's medium and an aliquot counted with a hemacytometer. The cellular concentration was adjusted to 2×10^6 cells/ml by use of Pearson's Square.

To study the effect of various mycotoxins and the L-tryptophan precursor, shikimic acid (shikimate), on cellular metabolism, either the shikimate or AFT plus vehicle were added to 1 ml of cellular suspension cultures and incubated at 25 C for 1 h. To terminate the experiment, the cell suspension cultures were centrifuged and the pellet was washed three times with control White's medium. Cells were then disrupted by sonication and the supernatant was assayed for either indole content or tryptophan-5-hydroxylase and IAA-5-hydroxylase activities.

To assay indole content, the pellet was weighed and extracted into acidified-*n*-butanol. The extract was centrifuged and a measured aliquot of the supernatant was returned to a second tube containing *n*-heptane and 0.1 *N* HCl. After shaking and centrifugation, the 5-HT was extracted into the aqueous layer, then assayed colorimetrically by the *o*-phthaldehyde (OPT) method (Tachiki and Aprison 1975). The upper organic layer was shaken with 0.033 *M* $NaHCO_3$ to extract 5-HIAA, which was assayed fluorometrically (Tachiki and Aprison 1975). Results are expressed as ng/mg wet wt ± S.D.

Tryptophan-5-hydroxylase and IAA-5-hydroxylase were assayed as described previously (Reynolds et al. 1984). Cellular protein content was quantified by the method of Lowry et al. (1951). The activities are expressed as ng product/mg protein/min.

Results

Incubation of *Sedum morganianum* cells in the presence of 100 μg/ml SA caused a diminution in cellular 5-HT content while cellular 5-HIAA level was increased dramatically. In contrast, tryptophan hydroxylase and IAA-hydroxylase activities were not significantly altered by the addition of SA (Tables 1 and 2).

Incubation of *Sedum* cell suspension cultures with 10 μg/ml PAT induced reductions in both cellular 5-HT and 5-HIAA levels and IAA hydroxylase activity; this effect was not altered by the presence of SA. If cells were incubated with PAT, tryptophan hydroxylase activity was enhanced, while incubation with both PAT and SA decreased tryptophan hydroxylase activity (Tables 1 and 2).

Incubation of *Sedum* cell suspension culture with 0.5 μg/ml of STR lowered cellular 5-HT levels, while 5-HIAA quantities as well as the activities of both

TABLE 1. *Percent change in metabolite levels or enzymatic activity of* Sedum morganianum *cells following incubation with shikimate, patulin, sterigmatocystin or aflatoxin*

Treatments of Plant Cells[a]	Tryptophan-5-hydroxylase[b] (TRP-5-hydroxylase)	5-hydroxytryptamine[c] (5-HT)	Indole-3-acetic acid-5-hydroxylase[d] (IAA-5-hydroxylase)	5-hydroxyindole-3-acetic acid[e] (5-HIAA)
Control	100.0[a]	100.0[a]	100.00[a]	100.0[a]
Control + Shikimic Acid (SA) (100 μg/ml)	12.7 ± 3.4[f]	−43.7 ± 3.6[f]	15.0 ± 3.7	262.5 ± 9.3[g]
Patulin (PAT) (10 μg/ml)	28.9 ± 1.8[f]	−58.4 ± 2.1[f]	−27.2 ± 1.2[f]	−51.9 ± 2.6[f]
PAT + SA	−57.5 ± 3.5[g]	−30.4 ± 3.4[f]	53.3 ± 5.3[g]	−11.8 ± 1.8
Sterigmatocystin (STR) (0.5 μg/ml)	136.5 ± 6.4[g]	−29.4 ± 5.6[f]	444.1 ± 4.2[g]	72.9 ± 6.4[g]
STR + SA	136.4 ± 7.1[g]	48.9 ± 4.3[g]	322.2 ± 6.3[g]	101.9 ± 5.9[g]
Aflatoxin B_1 (10 μg/ml)	−54.0 ± 5.2[f]	43.9 ± 2.9[f]	22.7 ± 1.3[f]	55.8 ± 3.7[f]
AFB_1 + SA	150.0 ± 3.9[g]	−15.8 ± 1.7	23.3 ± 2.7[f]	−36.0 ± 4.2[f]

[a] Cells were adjusted to a concentration of 2×10^6/ml and incubated for 1 h at 27 C. Each value is expressed as the % change from control values and presented as the mean and standard deviation (n = 6)
[b] Results are expressed as ng 5-HT/mg protein/10 min at 27 C
[c] Results are expressed as μg 5-HT/g wet weight
[d] Results are expressed as ng 5-HIAA/mg protein/10 at 27 C
[e] Results are expressed as μg 5-HIAA/g wet weight
[f] Significantly different from controls at $p > 0.01$
[g] Significantly different from controls at $p > 0.001$

TABLE 2. *Summary of effects of shikimate, patulin, sterigmatocystin or aflatoxin B_1 on tryptophan-5-hydroxylase activity, 5-hydroxytryptamine level, indole-3-acetic acid-5-hydroxylase activity, and 5-hydroxyindole-3-acetic level*

Treatments of Plant Cells	Tryptophan-5-hydroxylase (TRP-5-hydroxylase)	5-hydroxytryptamine (5-HT)	Indole-3-acetic acid-5-hydroxylase (IAA-5-hydroxylase)	5-hydroxyindole-3-acetic acid (5-HIAA)
Shikimic Acid (SA)	No Effect[a]	Decrease	No Effect[a]	Increase
Patulin (PAT) (10 μg/ml)	Increase	Decrease	Decrease	Decrease
PAT + SA	Decrease	Decrease	Decrease	No Effect[a]
Sterigmatocystin (STR) (0.5 μg/ml)	Increase	Decrease	Increase	Increase
STR + SA	Increase	Increase	Increase	Increase
Aflatoxin B_1 (AFB_1) (10 μg/ml)	Decrease	Increase	Increase	Increase
AFB_1 + SA	Increase	No Effect[a]	Increase	Decrease

[a] No effect equals not significant at $p > 0.01$.

tryptophan hydroxylase and IAA-hydroxylase were raised. Shikimic acid plus STR promoted elevations in both 5-HT and 5-HIAA levels and in both tryptophan hydroxylase and IAA hydroxylase activities (Tables 1 and 2).

When a *Sedum* cell suspension culture was incubated with 10 μg/ml AFB_1, tryptophan hydroxylase activity was inhibited while IAA-hydroxylase activity as well as 5-HT and 5-HIAA levels were increased. On the other hand, incubation of a cell suspension with AFB_1 and SA reduced cellular 5-HIAA levels, while tryptophan hydroxylase and IAA hydroxylase activities were enhanced (Tables 1 and 2).

Discussion

The results of these experiments, are summarized in Table 2, complement and extend the observations of Reiss (1970; 1977) and Adekunle and Bassir (1973), who found that AFB_1 may act either as an IAA agonist or antagonist in a dose-dependent manner.

Significantly, AFB_1 increased cellular levels of 5-HT and 5-HIAA, both of which are metabolites closely related to IAA and which may exhibit hormonal activities in their own right (Reynolds et al. 1984). On the other hand, the decline in tryptophan hydroxylase activity following AFB_1 treatment coupled with enhanced product levels implies that AFB_1 may also alter substrate availability. However, the addition of SA, the precursor to many amino acids in plants, plus AFB_1, caused a diminution in 5-HIAA levels but increased the activities of both of the rate-limiting synthesis enzymes.

The fact that STR treatment enhances both IAA-hydroxylase and 5-HIAA levels suggests an enzyme-mediated rise in metabolites. A depression of 5-HT amounts, together with the elevated tryptophan hydroxylase activity, indicates changes in substrate availability. Indeed, given the fact that some AFT's (i.e., AFB_1) alter amino acid uptake (Truelove 1970; Young et al. 1978), it is not unreasonable to propose that STR may act in this manner. However, a direct demonstration of STR-altered amino acid uptake is required to establish this point. On the other hand, STR plus SA enhanced the activity of both synthesizing enzymes, raising the possibility that SA may indirectly stimulate protein synthesis via increased amino acid formation. Indeed, preliminary experiments (data not shown) have indicated that incubation of a *Sedum* cell suspension culture with cycloheximide reduced both tryptophan hydroxylase and IAA-hydroxylase activities. However, further work is needed to clearly delineate the mechanism.

Patulin decreased 5-HT levels and IAA-hydroxylase activity in *Sedum* cell suspension cultures but increased tryptophan hydroxylase activity thereby raising the possibility that PAT may differentially affect enzyme activities and substrate synthesis or availability. Also, it was noted that PAT plus SA decreased the activity of both synthesizing enzymes, which may contribute to the reduction in cellular metabolite levels.

In summary, these experiments demonstrate that mycotoxins often differentially affect enzymes involved in the synthesis of plant indoles and the indole levels found in tissues. However, whether these differences are due to either altered hormone synthesis or degradation are not clear. Furthermore, the fact that the addition of SA alters differentially some of these responses supports the view that substrate availability may be important.

Acknowledgments

We are grateful to Dr. W. V. Dashek for editing the manuscript and for his many helpful suggestions.

Literature Cited

Adekunle, A. A., and O. Bassir. 1973. The effect of aflatoxin B_1 and palmotoxin B_0 and G_0 on the germination and leaf colour of the cowpea *(Vigna sinensis). Mycopathol. Mycol. Appl.* 51:299–305.

Bender, D. A. 1975. *Amino Acid Metabolism.* Pages 280–429. Wiley-Interscience, New York.

Carlton, W. W., and G. M. Szczech. 1978. Mycotoxicoses in laboratory animals. Pages 333–462 *in* T. D. Wyllie and L. G. Morehouse, eds. *Mycotoxic Fungi, Mycotoxins, Mycotoxicoses: An Encyclopedic Handbook,* Vol. 1, Marcel Dekker, Inc., New York.

Dashek, W. V., and G. C. Llewellyn. 1974. The influence of the carcinogen aflatoxin B_1 on the metabolism of germinating lily pollen. Pages 351–360 *in* H. F. Linskens, ed. *Fertilization in Higher Plants.* North Holland Pub. Co., Amsterdam.

________. 1982. Aflatoxins and plants. *Microbiologii* 2:65–84.

________. 1983. Mode of action of the hepatocarcinogens, aflatoxins in plant systems: A review. *Mycopathologia* 81:83–94.

Dickens, F., H. E. H. Jones, and H. B. Wanyford. 1966. Oral, subcutaneous and intratracheal administration of carcinogenic lactones and related substances: The intratracheal administration of cigarette tar in the rat. *Fr. J. Cancer* 20:137–140.

Diener, U. L., and N. D. Davis. 1969. Aflatoxin formation by *Aspergillus flavus.* Pages 307–327 *in* L. A. Goldblatt, ed., *Aflatoxin-Scientific Background, Control and Implications.* Academic Press, New York.

Gibson, F., and J. Pittard. 1968. Pathways of biosynthesis of aromatic amino acids and vitamins and their control in microorganisms. *Bacteriol. Rev.* 32:465–492.

Holzapfel, C. E., I. F. H. Purchase, P. S. Steyn, and L. Gouws. 1966. The toxicity and chemical assay of sterigmatocystin, a carcinogenic mycotoxin and its isolation from two new fungal sources. *S. Afr. J. Med. J.* 40:1100–1101.

Lowry, O. H., N. J. Rosebrough, A. L. Farr, and R. J. Randall. 1951. Protein measurement with the Folin phenol reagent. *J. Biol. Chem.* 193:265–273.

Reiss, J. 1970. Forderung der aktivitat von B. indolylessig saive durch aflatoxin B_1. *Z. fur Pflanzenphysiol.* 64:260–262.

________. 1977. Effects of mycotoxins on the development of epiphyllous buds of *Kalanchoe daigremontiana. Z. fur Pflanzenphysiol.* 82:446–449.

Reynolds, J. D., T. D. Kimbrough, and L. B. Weekley. 1984. Evidence for enzymatic 5-hydroxylation of indole-3-acetic acid *in vitro* by extracts of *Sedum morganianum*. *Z. fur Pflanzenphysiol.* 112:465–470.

Tachiki, K. H., and M. H. Aprison. 1975. Fluorometric assay for 5-hydroxytryptophan with sensitivity in the picomole range. *Anal. Chem.* 47:7–13.

Truelove, B., D. E. Davis, and O. C. Thompson. 1970. The effects of aflatoxin B_1 on protein synthesis by cucumber cotyledon discs. *Can. J. Bot.* 48:485–591.

White, P. R. 1963. The nutrients. Pages 59–61 *in The Cultivation of Animal and Plant Cells.* Ronald Press Co., New York, NY.

Young, J. W., W. V. Dashek, and C. C. Llewellyn. 1978. Aflatoxin B_1 influence on excised soya-bean root growth, ^{14}C-leucine uptake and incorporation. *Mycopathologia* 66:91–97.

CHAPTER 57

Effects of Temperature on the Potency of Ethanol as an Inhibitor of Growth and Membrane Function in *Zymomonas mobilis*

K. M. Dombek, A. S. Benschoter, and L. O. Ingram*

Department of Microbiology and Cell Science, University of Florida, Gainesville, Florida 32611

Previous studies with *Zymomonas mobilis* have shown that cell growth and ethanol production are reduced at elevated temperatures. The rate of ethanol production during fermentation is directly related to cell mass and to cellular activity per mass unit. Thus it is likely that the reduction in growth at elevated temperature is in part responsible for the decreased rate of fermentation. In this study, we have investigated the effects of elevated temperature and ethanol on the growth of *Z. mobilis*. The minimal concentration of ethanol required to inhibit the growth of this organism at 40 C was found to be 4% while 8% was required to inhibit growth at 30 C. Conversely, the addition of ethanol to the growth medium dramatically reduced the maximal growth temperature of *Z. mobilis*. Both ethanol and increased temperature caused an increase in membrane fluidity and in the leakage of intracellular components. The increased sensitivity of growth to ethanol at 40 C was correlated with an increase in the potency of ethanol as a membrane fluidizing agent and as a permeabilizing agent, facilitating the leakage of magnesium ions, nucleotides, and proteins. The retention of these components within the cell is essential for homeostasis, growth and metabolism. Thus our results are consistent with the hypothesis that the decrease in the minimum inhibitory concentration of ethanol for growth at 40 C and the reduction of the maximal growth temperature in the presence of ethanol result from an enhancement of membrane leakage. Since the retention of metabolites, cofactors, and enzymes is also essential for fermentation, it is tempting to speculate that some of the adverse effects of elevated temperature on fermentation also result from similar damage to the cell membrane.

Introduction

Zymomonas mobilis has been reported to have many advantages over yeasts for the fermentative production of ethanol, including higher rates of substrate conversion to alcohol and increased thermal tolerance (Rogers et al. 1982). Thermal stress has been reported to adversely affect ethanolic fermentations by both organisms. Fermentation is an exergonic process and the removal of excess heat is essential during large-scale fermentations by mesophilic microorganisms. With *Z. mobilis*, fermentation temperatures above 30 C to 34 C decrease the amount of cell growth, the rate of ethanol production, the efficiency of conversion, and the maximal level of ethanol that can be achieved (Charley et al. 1983; King and Hossain 1982; Laudrin and Goma 1982; Lee et al. 1981; Ohta et al. 1981). Typically, the final ethanol concentration achieved at 40 C is half that which can be ob-

*Address correspondence to Dr. L. O. Ingram.

tained during fermentation at 30 C in the presence of excess glucose. These results suggest that the sensitivities of cell growth and fermentation to inhibition by ethanol are increased at elevated temperatures.

In our study with *Z. mobilis*, we have determined the effects of incubation temperature on the minimum concentration of ethanol required to inhibit growth and the effects of ethanol on the maximal growth temperature. Increasing temperature and the addition of ethanol have been shown to decrease the degree of order within biological membranes and alter membrane function (Eaton et al. 1982; Dombek and Ingram 1984). Thus the effects of ethanol and temperature on membrane fluidity and membrane leakage (magnesium ions, nucleotides, and proteins) have been investigated as a possible basis for growth inhibition.

Materials and Methods

Organism and growth conditions. Z. mobilis strain CP4 was generously supplied by Dr. Arie Ben-Bassat of Cetus Corporation (Berkeley, CA). This organism was grown and maintained at 30 C in the complex medium described by Skotnicki et al. (1981) with 100 g/liter of glucose, as described previously (Carey et al. 1983). Growth was monitored by measuring optical density at 550 nm using a Bausch and Lomb Spectronic 70 spectrophotometer. Alcohol was added as indicated to cooled medium after autoclaving and concentrations were expressed as a percentage of volume.

To determine the effects of incubation temperature on ethanol tolerance, an overnight culture of *Z. mobilis* (grown at 30 C) was diluted 1:100 into culture tubes containing the appropriate concentrations of ethanol. These were incubated at 20 C, 30 C, or 40 C, and growth was measured as optical density (550 nm) after 24, 48, and 72 h. Values reported are averages of at least two experiments after 48 h and did not change appreciably during subsequent incubation.

To determine the effect of ethanol on the maximal growth temperature, an overnight culture of *Z. mobilis* (grown at 30 C) was diluted 1:100 into fresh medium containing 0%, 2%, 5%, and 7% ethanol. These cultures were distributed into culture tubes and incubated at different temperatures. Growth was measured after 24, 48, and 72 h. Values reported are averages of three experiments after 48 h and did not change appreciably during subsequent incubation.

Membrane fluidity. A cell membrane fraction was prepared from an exponentially growing culture (30 C; optical density 0.6 to 0.8). Cells were harvested by centrifugation (5,000 × *g*, 4 C, 10 min), washed once in ¼ volume of 0.01 *M* Tris-HC1 (pH 7.5), and resuspended in $^1/_{10}$ volume of buffer containing 10% Ficoll. Lysozyme (1 mg/ml) was added and cells were incubated for 30 min at room temperature. The cell suspension was disrupted using a precolled French pressure cell (20,000 psi, three passages). Unbroken cells and large debris were removed by centrifugation at 3,000 × *g* for 5 min and membranes harvested by centrifugation at 100,000 × *g* for 1 h.

Fluorescence measurements were performed using an SLM series 4000 polarization fluorimeter as described previously (Dombek and Ingram 1984;

Esko et al. 1977). Under the conditions of these experiments, the fluorescence polarization of the membrane probe, diphenylhexatriene, is inversely related to membrane fluidity although other factors such as polarity and fluorescence lifetime also affect the behavior of probe fluorescence. Cuvette temperature was controlled using a Neslab refrigerated circulator and was measured within the cuvette using a thermistor. Ethanol was added directly to the cuvette containing the membrane suspension and was allowed to equilibrate for 5 min prior to measurements. The values of polarization (P) that are reported represent averages from three experiments in which 30 determinations were made on each sample. The standard deviation for these measurements was less than 0.003 P.

Cell leakage. Cells were grown at 30 C in 250-ml spinner bottles with constant agitation (150 rpm) and harvested (optical density of 0.8) by centrifugation in a warm centrifuge. The cell pellet was washed once with 0.05 *M* sodium phosphate buffer (pH 6.5) and resuspended in 40 ml of buffer. At zero time, this suspension was diluted into an equal volume of buffer containing the appropriate ethanol concentration and incubated at either 20 C, 30 C, or 40 C. Samples (1.5 ml) were taken at various times, and cells were removed using an Eppendorf microcentrifuge (10,000 × *g*; 10 min). Nucleotides present in the supernatant were measured as absorbance at 260 nm. The spectrum of these supernatants was similar to that for adenine nucleotides (not shown). Magnesium was measured using a spectrophotometric assay system supplied by the American Monitor Corporation (Indianapolis, IN). Initially, the accuracy of this procedure for our samples was verified by atomic absorption analyses. Protein was determined using the dye-binding assay described by Spector (1978).

Total releasable magnesium and protein were determined in supernatants in which the cell suspensions had been shaken vigorously with 0.2 ml of chloroform to disrupt the membrane. After this treatment, 100% of the cellular magnesium and 70% of the cellular protein were recovered in the supernatant. Values reported for proteins and for magnesium are expressed as a percentage of the total released by this chloroform treatment. Since it is likely that chloroform treatment also caused the release of some nucleic acids, measured absorbance for nucleotides was not converted to a percentage of total. All leakage results represent averages of three experiments.

RESULTS

Growth. Figure 1A shows the effects of incubation temperature on the minimum inhibitory concentration of ethanol for the growth of *Z. mobilis* strain CP4. The minimum inhibitory concentration for growth decreased with increasing incubation temperature. Growth was prevented by 10% ethanol at 20 C, by 8% ethanol at 30 C and by 4% ethanol at 40 C.

Figure 1B shows the converse experiment, the effects of ethanol on the maximum growth temperature. In the absence of added ethanol, limited growth occurred at 42.5 C and growth was prevented by incubation at temperatures of 43.5 C and above. The addition of ethanol to the growth medium caused a reduction in maximal growth temperature. The highest temperature at which growth oc-

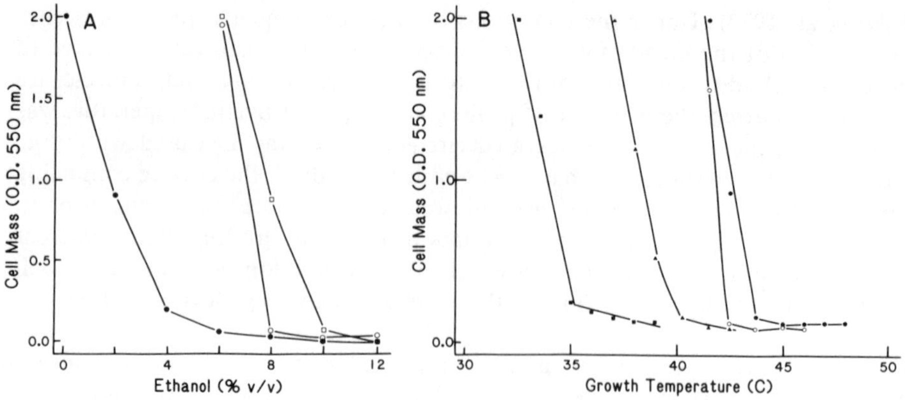

FIG. 1. Effects of ethanol and incubation temperature on the growth of *Z. mobilis* strain CP4. A. Effect of incubation temperature on the minimal inhibitory concentration of ethanol. Symbols: □, 20 C; ○, 30 C; ●, 40 C. B. Effects of ethanol on maximum growth temperature. Symbols: ●, no added ethanol; ○, 2% 6 ethanol; ▲, 5% ethanol; ■, 7% ethanol.

curred was 41.5 C with 2% ethanol, 39 C with 5% ethanol and 34 C with 7% ethanol. Poor growth was obtained at 30 C in the presence of higher ethanol concentrations.

Membrane fluidity. Figure 2A shows the effects of ethanol and of assay temperature on the fluidity of membranes from cells grown at 30 C. Membrane fluidity is inversely related to the reported values of polarization (P). The decrease in polarization observed with increasing assay temperatures is indicative of an increase in membrane fluidity. A similar, less pronounced decrease in polarization (increase in fluidity) was observed in response to added ethanol. The measured change in membrane fluidity caused by the addition of 5% ethanol to

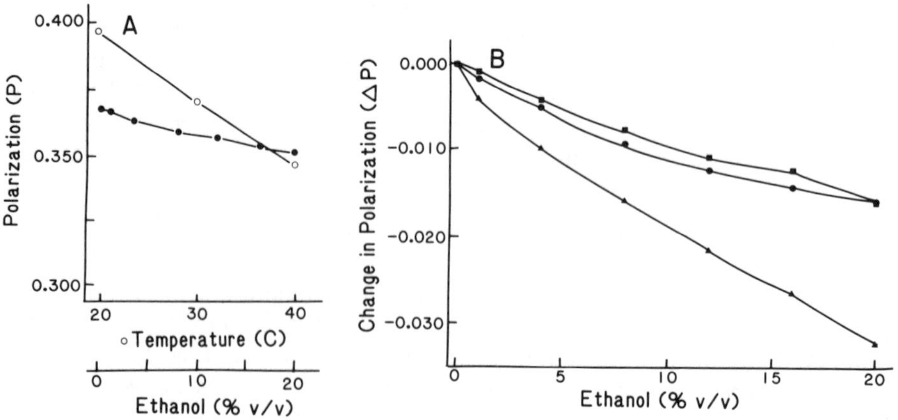

FIG. 2. Effects of ethanol and assay temperature on the fluidity of membranes isolated from cells grown at 30 C. A. Effects of temperature and of ethanol on polarization. Symbols: ●, assayed at 30 C with added ethanol; ○, assayed in the absence of ethanol at different temperatures. B. Effects of assay temperature on the ethanol-induced change in polarization. The initial polarization value of membranes in the absence of ethanol was subtracted from that measured after ethanol addition. Symbols: ■, assayed at 20 C; ●, assayed at 30 C; ▲, assayed at 40 C.

membranes assayed at 30 C is roughly equivalent to the change caused by a one-degree increase in incubation temperature.

Figure 2B shows the effects of assay temperature on the ethanol-induced change in polarization in membranes isolated from cells grown at 30 C. This was computed by subtracting the initial polarization in the absence of ethanol from that measured in the presence of added ethanol. The negative sign of these changes is indicative of an increase in fluidity. Although the measured changes in polarization at 20 C were consistently less than observed at 30 C, the standard deviations for these changes overlapped and were not significant at the 95% confidence level. However, the addition of ethanol to membranes being held at 40 C resulted in more than twice the magnitude of change observed at 20 C and 30 C. Thus the potency of ethanol as a membrane-fluidizing agent at 40 C was roughly double that observed at the two lower temperatures.

A similar decrease in polarization (0.07–0.08 P) was caused by 10% ethanol at 20 C, by 8% ethanol at 30 C, and by 4% ethanol at 40 C. These conditions correspond to the minimum inhibitory concentrations of ethanol for cell growth during incubation at the respective temperatures (Fig. 1A).

Leakage of cellular constituents. We have examined the effects of ethanol and incubation temperature on the leakage of magnesium ions, nucleotides, and proteins. Cells from exponential cultures grown at 30 C were resuspended in 0.05 *M* sodium phosphate buffer (pH 6.5) and incubated at either 20 C, 30 C, or 40 C, with and without ethanol. Samples were removed hourly for 3 h to determine the extent of leakage of cellular constituents. The extent of leakage increased with incubation time in all cases, and results after 3 h of incubation are summarized in Fig. 3. The leakage of magnesium ions was greater than observed for proteins in

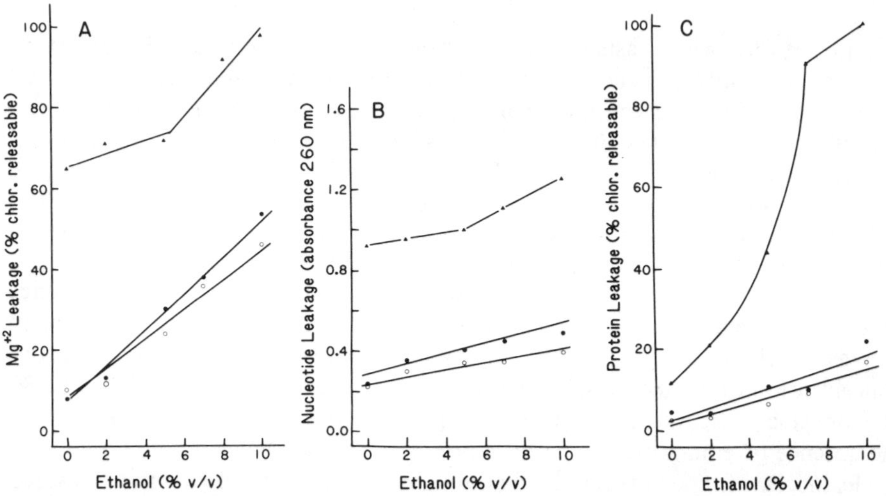

FIG. 3. Effects of ethanol and assay temperature on the leakage of cells grown at 30 C. A. Magnesium. B. Nucleotides (absorbance at 260 nm). C. Protein. Symbols for A, B, and C: ○, 20 C; ●, 30 C; ▲, 40 C.

all but one case (10% ethanol at 40 C) where both were completely released into the supernatant. The addition of ethanol to buffer-containing cells and incubation at higher temperatures increased the rates of cellular leakage.

The leakage of cells incubated at 20 C was slightly lower than that observed at 30 C in most cases. Leakage at these two temperatures was much lower than that observed following incubation at 40 C. In the absence of ethanol, little leakage was observed at 20 C or 30 C. However, over 65% of the intracellular magnesium ions and substantial amounts of nucleotides and proteins were released into buffer at 40 C even in the absence of ethanol.

The addition of increasing concentrations of ethanol caused a dose-dependent increase in the release of magnesium ions, nucleotides, and proteins during incubation at 20 C and 30 C (Fig. 3). At 40 C, the addition of 2% and 5% ethanol had little effect on the already substantial leakage of magnesium ions and nucleotides (Figs. 3A and 3B). The addition of 7% and 10% ethanol to cells incubated at 40 C caused the release of the remaining cellular magnesium and further increases in the release of nucleotides. Protein leakage at 40 C increased in a dose-dependent fashion with ethanol addition. Over 90% of the chloroform-releasable protein was obtained in the supernatant at this temperature following incubation in the presence of 7% ethanol, indicating a loss of membrane integrity.

The effects of ethanol and assay temperature can be roughly correlated with the growth studies. Cell leakage was much more extensive in the presence of ethanol at 40 C than at either 20 C or 30 C, consistent with the decrease in the alcohol concentration required to inhibit growth at 40 C (Fig. 1A). In general, leakage in the presence of ethanol at 20 C was slightly less than that observed at 30 C, consistent with the small increase in the alcohol tolerance of growth at 20 C as compared to 30 C (Fig. 1A).

Discussion

Previous studies with yeasts by Dr. Van Uden and coworkers have demonstrated that the ethanol tolerance of growth is dramatically reduced at elevated temperatures and that the maximal growth temperature of these organisms is depressed by the presence of ethanol (Van Uden 1984). We have obtained similar results using *Z. mobilis* strain CP4. Both the presence of ethanol and incubation above the optimal growth temperature are types of environmental stress. The combined effects of these stress factors appear to be more than additive for the inhibition of growth in both yeast (Van Uden 1984) and *Z. mobilis.*

Our results with *Z. mobilis* indicates that the decrease in the alcohol tolerance of growth, which is observed with increasing growth temperature, results from an increase in the potency of ethanol at this temperature. The minimal inhibitory concentration of ethanol at 40 C was 4%, half the level tolerated at 20 C and 30 C. Conversely, the maximum growth temperature was reduced from over 40 C in the absence of ethanol to 35 C in the presence of 7% ethanol.

The increase in the potency of ethanol as an inhibitor of cell growth at elevated growth temperatures may be due in part to increases in the effective membrane concentration of ethanol. Ethanol is an amphipathic molecule and partitions between hydrophobic environments such as the cell membrane and the aqueous milieu. In model systems, increasing temperatures are known to shift this partitioning toward higher concentrations in the hydrophobic environment (Janoff and Miller 1982). Consistent with this shift, the potency of ethanol as a fluidizing agent in isolated membranes was highest at 40 C. At this temperature, equivalent amounts of ethanol caused twice the change in polarization observed at 30 C. However, further reduction of assay temperature to 20 C had little effect on ethanol-induced fluidization of membranes. The minimum inhibitory concentrations of ethanol for growth at 20 C, 30 C, and 40 C caused equivalent changes in the fluidity (decreased in P) in membranes (isolated from cells grown at 30 C) assayed at these respective temperatures.

Pang et al. (1979) demonstrated an excellent correlation between ethanol-induced increase in membrane fluidity and increased cation leakage. At 20 C and 30 C, zero to 10% ethanol caused a dose-dependent increase in the leakage of magnesium ions, nucleotides, and proteins from cells of *Z. mobilis*. At 40 C, however, concentrations of ethanol above 5% caused a loss of membrane integrity as indicated by the release of the bulk of the soluble intracellular proteins and cell death. These results are consistent with the hypothesis that at 40 C, membrane integrity serves to limit ethanol tolerance for growth. Although complete membrane disruption does not occur even with added ethanol at the lower temperatures examined, the extent of leakage observed in the presence of similar concentrations of ethanol at 20 C was generally less than that observed at 30 C, consistent with leakage as a determinant of ethanol tolerance for growth at these temperatures also. The lack of strict correspondence between the extent of leakage and minimum inhibitory concentrations of ethanol at different temperatures may be due in part to the artificial conditions of these assays (buffer), which undoubtedly influence the observed leakage.

One of the most important functions of the plasma membrane is to serve as a barrier to prevent the loss of essential ions and metabolites. Previous studies by Enequist et al. (1981) have demonstrated that ethanol uncouples various cellular processes from the proton-motive force generated across the membrane of *E. coli* by increasing ion leakage. An analogous electrochemical gradient exists across the membrane of *Z. mobilis* (Barrow et al. 1984), which is probably involved in the active transport of essential ions and metabolites. Subsequent studies with *E. coli* in our laboratory (Ingram and Buttke 1984; Eaton et al. 1982) have demonstrated that ethanol increased the leakage of nucleotides and implicated cell leakage as the basis of ethanol killing. It is likely that the ethanol-induced increase in the leakage of magnesium ions and nucleotides in *Z. mobilis* is indicative of a general increase in the leakage of all intracellular ions and metabolites, including the hydrogen and potassium ions typically involved in transport systems.

Laudrin and Goma (1982) and King and Hossain (1982), among others, have demonstrated that maintaining fermentations at temperatures above 30 C to 34 C results in a decrease in the rate of ethanol production and a reduction in the final ethanol concentration that can be achieved. The rate of production of ethanol by microorganisms is directly related to both microbial mass and the metabolic activity per mass unit (Cooney 1983). During batch fermentations, the mass of fermentative cells is determined by the size of the inoculum and by the extent to which growth continues as fermentation proceeds and products of fermentation accumulate. At elevated temperatures, the enhanced sensitivity of growth to inhibition results in a reduction in cell mass (Laudrin and Goma 1982; King and Hossain 1982), which may be partly responsible for the decreased rate of substrate conversion to ethanol.

The intracellular retention of nucleotides and magnesium ions as cofactors is essential for glycolysis and ethanol production. Fluorescence studies demonstrated that the addition of ethanol disturbed the physical organization of membranes from *Z. mobilis*, increasing membrane fluidity. The addition of ethanol to cells of *Z. mobilis* also increased the rate of leakage of these essential components at all temperatures examined, indicating that ethanol decreased the effectiveness of the cell membrane as a hydrophobic barrier. Thus it is tempting to speculate that the inhibition of fermentation by ethanol in *Z. mobilis* may also result in part from the ethanol-induced increase in cellular leakage.

ACKNOWLEDGMENTS

This investigation was supported by grant PCM-8204928 from the National Science Foundation, by the Weston Research Centre, and by the Florida Agricultural Experiment Station (publication number 5130).

LITERATURE CITED

Barrow, K. D., J. G. Collins, R. S. Norton, P. L. Rogers, and G. M. Smith. 1984. ^{31}P Nuclear magnetic resonance studies of the fermentation of glucose to ethanol by *Zymomonas mobilis. J. Biol. Chem.* 259:5711–5716.

Carey, V. C., S. K. Walia, and L. O. Ingram. 1983. Expression of the lactose transposon (Tn951) in *Zymomonas mobilis. J. Bacteriol.* 46:1163–1168.

Charley, R. G., J. E. Fein, B. H. Lavers, H. G. Lawford, and G. R. Lawford. 1983. Optimization of process design for continuous ethanol production by *Zymomonas mobilis* ATCC 29191. *Biotechnol. Lett.* 5:169–174.

Cooney, C. L. 1983. Bioreactors: Design and operation. *Science* 219:728–733.

Dombek, K. M., and L. O. Ingram. 1984. Effects of ethanol on the *Escherichia coli* plasma membrane. *J. Bacteriol.* 157:233–239.

Eaton, L. C., T. F. Tedder, and L. O. Ingram. 1982. Effects of fatty acid composition on the sensitivity of membrane functions to ethanol in *Escherichia coli. Substance and Alcohol Actions/Misuse* 3:77–87.

Enequist, H.G., T. R. Hirst, S. Harayama, S. J. S. Hardy, and L. L. Randall. 1981. Energy is required for the maturation of exported proteins in *Escherichia coli. Eur. J. Biochem.* 116:227–233.

Esko, J. D., J. R. Gilmore, and M. Glaser. 1977. Use of a fluorescent probe to determine the viscosity of LM cell membranes with altered phospholipid compositions. *Biochemistry* 16:1881–1890.

Janoff, A. S., and K. W. Miller. 1982. A critical assessment of the lipid theories of general anaesthetics action. Pages 417–476 *in* D. Chapman, ed., *Biological Membranes.* Academic Press, New York, N.Y.

King, F. G., and M. A. Hossain. 1982. The effect of temperature, pH, and initial glucose concentration on the kinetics of ethanol production by *Zymomonas mobilis* in batch fermentation. *Biotechnol. Lett.* 4:531–536.

Ingram, L. O., and T. M. Buttke. 1984. Effects of alcohols on microorganisms. *Adv. Microbiol. Physiol.* In Press.

Laudrin, I., and G. Goma. 1982. Ethanol production by *Zymomonas mobilis:*

Effect of temperature on cell growth, ethanol production and intracellular ethanol accumulation. *Biotechnol. Lett.* 4:537–542.

Lee, K. J., M. L. Skotnicki, D. E. Tribe, and P. L. Rogers. 1981. The effect of temperature on the kinetics of ethanol production by strains of *Zymomonas mobilis. Biotechnol. Lett.* 3:291–296.

Ohta, K., K. Supanwong, and S. Hayashida. 1981. Environmental effects on ethanol tolerance of *Zymomonas mobilis. J. Ferment. Technol.* 59:435–439.

Pang, K. -Y., T. L. Chang, and K. W. Miller. 1979. On the coupling between anesthetic induced membrane fluidization and cation permeability in lipid vesicles. *Mol. Pharmacol.* 15:729–738.

Rogers, P. L., K. J. Lee, M. L. Skotnicki, and D. E. Tribe. 1982. Ethanol production by *Zymomonas mobilis. Adv. Biochem. Eng.* 12:37–84.

Skotnicki, M. L., K. J. Lee, D. E. Tribe, and P. L. Rogers. 1981. Comparison of ethanol production in different *Zymomonas* strains. *Appl. Environ. Microbiol.* 41:889–893.

Spector, T. 1978. Refinement of the coomassie blue method of protein quantitation. *Anal. Biochem.* 86:142–146.

Van Uden, N. 1984. Temperature profiles of yeasts. *Adv. Microbiol. Physiol.* In Press.

CHAPTER **58**

Purification of Large Plasmids by Ion Exchange Chromatography

Douglas E. Dennis* and Suzanne Esterline

Genetic Engineering/Biotechnology Program, Cedar Crest College, Allentown, Pennsylvania 18104

A method has been developed whereby large plasmids may be purified utilizing ion exchange chromatography. The method represents an inexpensive alternative to cesium chloride (CsCl) gradient purification. The method employs an accepted procedure for the crude extraction of *Pseudomonas* plasmids, which is followed by additional steps which remove RNA, chromosomal DNA, and contaminating material that migrates in the range of the plasmids on agarose gels. The extract from this preparation is then placed onto a commercially available ion exchange resin and remaining RNA, chromosomal DNA, and plasmid DNA are differentially eluted by the application of a salt gradient. This procedure requires approximately the same length of time as CsCl gradient purification. Plasmids obtained by this procedure are more pure than those isolated by CsCl gradient centrifugation, but are isolated in reduced yields.

Introduction

Due to the discovery in recent years that many *Pseudomonas* and related species carry large biodegradative plasmids (Chakrabarty 1976; Don and Pemberton 1981; Franklin et al. 1981; Pierce et al. 1981; Vandenbergh et al. 1981; Furukawa and Chakrabarty 1982; Serdar et al. 1982), the study of these plasmids has taken on new meaning. Because these plasmids are so large, 20 Mdal to greater than 300 Mdal (Chakrabarty 1976), they have necessitated the development of new methods of isolation (Sharp et al. 1972; Humphreys et al. 1975; Currier and Nester 1976; Palchaudhuri and Chakrabarty 1976; Fennewald et al. 1978; Hansen and Olsen 1978). Previous procedures developed for *Escherichia coli* plasmids rely on a differential precipitation step in which high molecular weight chromosomal DNA is precipitated and the relatively small plasmids (1–10 Mdal) remain in the supernate (Birnboim and Doly 1979). This type of procedure, however, will not suffice for large plasmids, because the large plasmid DNA is coprecipitated with the chromosomal DNA (Hansen and Olsen 1978).

Over the past few years the most commonly used method to isolate large plasmids is one developed by Hansen and Olsen (1978). This technique takes advantage of the fact that chromosomal DNA is bound to bacterial membranes, and can be removed from lysates by precipitation of the membrane-chromosomal complex. Plasmid DNA remains in the supernate and is then precipitated by the

* Address correspondence to D. E. Dennis, Department of Biology, James Madison University, Harrisonburg, Virginia.

addition of polyethylene glycol. Plasmid DNA obtained in this manner is not pure, but is "enriched" (Hansen and Olsen 1978). This preparation contains large amounts of protein, RNA, and some chromosomal DNA. In order to perform genetic manipulations, such as restriction, ligation, and transformation, the plasmid DNA must be further purified by CsCl gradient centrifugation, requiring the use of an ultracentrifuge. Many laboratories which have an interest in large biodegradative plasmids cannot afford nor have reasonable access to ultracentrifuges. This prompted us to search for an alternative procedure for isolating large plasmids in a pure enough form to allow genetic manipulation. Our quest centered not on the development of a radically new technique, but rather on the integration of recently defined isolation methods for small plasmids into the framework of the already proven Hansen and Olsen technique (1978). The techniques of particular interest included an acid-phenol extraction technique developed by M. Zasloff et al. (1978) and ion exchange chromatography techniques developed by Bethesda Research Laboratories (Best et al. 1981; Thompson et al. 1983; NACS Applications Manual). In the acid phenol technique, protein and RNA are removed from a lysate of bacterial cells and the lysates are then subjected to several phenol extractions at a pH of 4 and a NaCl concentration of 75 n*M*. Covalently closed circular DNA molecules remain in the aqueous phase and linear chromosomal DNA molecules are sequestered into the phenol phase. Zasloff et al. (1978) has successfully utilized this technique on covalently closed circular plasmids to a size of 25 Mdal, but the procedure has not been applied to larger plasmids. The second technique which we incorporated into the protocol also relies on the isolation of covalently closed circular DNA molecules. In this procedure a protein and RNA-free preparation from a bacterial lysate is eluted through a commercially available ion exchange resin with 0.5 *M* NaCl. RNA passes through the column, while chromosomal DNA and plasmid DNA are bound. Supercoiled molecules bind to the column less tightly, so that plasmids may be differentially eluted by the passage of a salt gradient through the column. Plasmid DNA elutes slightly ahead of chromosomal DNA and linearized plasmid DNA. This paper describes a protocol that incorporates these two procedures and the Hansen and Olsen procedure. The technique allows one to obtain plasmids which are more pure than those purified by CsCl gradient centrifugation.

Materials and Methods

Plasmid purification procedures. We strongly urge anyone who might attempt this extraction to first refer to the original papers from which this collation is drawn (Hansen and Olsen 1978; Zasloff et al. 1978; NACS Applications Manual). A thorough understanding of the original experimentation is necessary for successfully performing this extraction. Fig. 1 shows a brief outline of each step. These are described in detail below.

Growth of cells. Cells are prepared by inoculating 50 ml of Luria broth plus 0.4% glucose contained in a 125-ml flask. Bacteria from a recent plate culture are used to inoculate the medium and the culture is incubated overnight with vigorous shaking. Next day, the entire contents of the flask are added to 250 ml of Luria

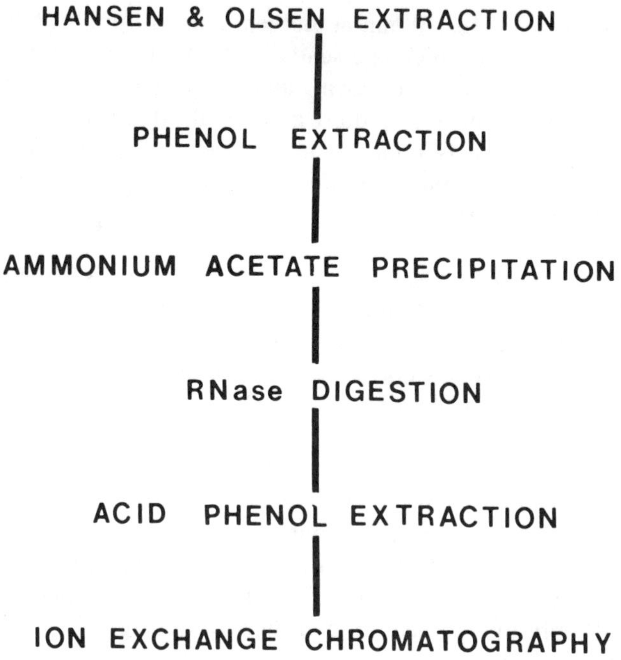

FIG. 1. Outline of large plasmid extraction procedure.

broth plus 0.4% glucose contained in a one-liter flask and the cells are again incubated overnight at 30 C with vigorous shaking. Cell densities at the time of harvest are generally 2×10^7 to 2×10^8 cells per ml.

Hansen and Olsen extraction. A crude preparation of plasmid enriched DNA was prepared according to Hansen and Olsen (1978). All volumes were scaled up 7.5-fold to compensate for the 300-ml culture we used as compared to the 40 ml cultures used by Hansen and Olsen. These volumes conveniently allowed one 300-ml culture to be extracted in an Oak Ridge tube, at a final voloume of approximately 45 ml. The final PEG pellet was resuspended in 15 ml of TNE buffer (10 m*M* Tris-HC1 (pH 8.0), 0.1 *M* NaCl, 1 m*M* EDTA). All extraction procedures performed were done as gently and quickly as possible to minimize plasmid breakage. Precautions taken included using large bore plastic pipets, pipetting solutions slowly to reduce convection, maintenance of the sample on ice whenever possible, and gentle mixing during the phenol extractions. If a cold room is available, all steps after the Hansen and Olsen procedure should be done at 4 C, unless otherwise specified.

Phenol extraction. The crude plasmid preparation was extracted three times with equal volumes of 49:59:2 mixture of phenol:chloroform:isoamyl alcohol (phenol saturated with TNE buffer). Mixing was done by placing the screw-cap tube horizontally on a reciprocating shaker at a speed of 1 cycle per second, and the sample extracted for 5 min, 7.5 min and 10 min, respectively. Separation of the phases was accomplished by centrifugation of $12{,}000 \times g$ in a small angle-head rotor (50 ml tube capacity) for 10 min at ambient temperature. The final extraction results in an interphase which is essentially clean. If clean preparations are not obtained, this indicates that the starting amount of bacterial cells might have been too large, and the sample may require additional extractions. After the final extraction, the upper phase is removed and placed in another tube and two volumes of ice-cold 95% ethanol are added. The nucleic acids are allowed to precipitate for 20 min on ice.

Ammonium acetate precipitation. The ethanol precipitate is pelleted by centrifugation at 12,000 × *g* for 20 min in the small angle head rotor at 4 C. The resulting pellet is placed in a speed vac (or other vacuum desiccator) and allowed to dry completely. If the pellet is not completely dry, or if a large amount of protein is left in the sample, the pellet will be difficult to resuspend. This will result in breakage of the plasmid DNA and will subsequently result in a poor yield. The dried pellet is resuspended in 7 ml of TE buffer (10 m*M* Tris-HC1 (pH 7.2), 1 m*M* EDTA) and 1.39 g of ammonium acetate are added and dissolved by gentle mixing. The volume of the solution is then brought to a total of 10 ml with TE buffer and placed on ice for 20 min. The preparation is centrifuged at 12,000 × *g* for 20 min at 4 C in the small angle-head rotor. This supernate is poured off, the pellets are dried for 5–10 min under vacuum, and then are resuspended in 5 ml of TE buffer containing 0.05 *M* NaCl.

RNase digestion. A small amount of the sample is diluted and measured for its O.D. 260 and 1 unit of RNase T1 is added for each A260 unit in the original sample. The sample is incubated for 15 min at 37 C.

Phenol extraction. The concentration of the solution is adjusted to 0.1 *M* with respect to NaCl by the addition of 0.01 volume of 5 *M* NaCl and extracted with phenol-chloroform-isoamyl alcohol (49:49:2, phenol saturated with TNE buffer) until the interphase is completely free of debris. This usually can be accomplished in two to three extractions. Extractions are performed by placing the tube horizontally on a reciprocating shaker at a speed of one cycle per second for 10 min. The phases are separated by centrifugation at 12,000 × *g* for 10 min at 4 C. The upper phase is removed and placed in another tube. Two volumes of ice-cold 95% ethanol are added to the tube and the sample is then placed at −20 C for 30 min. The sample is centrifuged for 30 min at 12,000 × *g* at 4 C. The supernate is poured off, and the pellet is washed with 80% ice-cold ethanol, recentrifuged, and dried under vacuum for 5 min.

Acid-phenol extraction. The pellet is resuspended in 5 ml of low molarity TE buffer (1 m*M* Tris-HC1 (pH 8), 1 m*M* EDTA), 0.05 volume (0.25 ml) of 1 *M* sodium acetate (pH 4.0) and 0.05 volume of 1.5 *M* NaCl (0.25 ml). The solution is extracted twice with phenol that has been equilibrated with 50 m*M* sodium acetate (pH 4). It is essential that only redistilled phenol be used for this extraction. Separation of the phases is accomplished by centrifugation at 6000 × *g* in a small angle-head rotor for 5 min at 4 C. After the extraction, the pH of the solution is raised by the addition of 0.05 volume of 1 *M* Tris-HC1 (pH 8.6). One-tenth volume of 2.5 *M* sodium acetate (pH 5.2) is added and the nucleic acids are precipitated by the addition of two volumes of ice-cold 95% ethanol. The tubes are placed at −20 C for 30 min. The nucleic acids are pelleted by centrifugation at 12,000 × *g* for 30 min at 4 C and the resulting pellet is dried under a vacuum for 5 min.

Ion exchange chromatography. The pellet is resuspended in 10 ml of TE buffer containing 0.5 *M* NaCl (TE is 10 m*M* Tris (pH 7.2), 1 m*M* EDTA) and applied at a flow rate of 0.5 ml/min to a NACS-37 (Bethesda Research Laboratories) column (12 cm length × 1 cm i.d.) equilibrated with 0.5 *M* NaCl in TE buffer. The eluate is monitored in line at an optical density of 254 nm (O.D. 254). Once the sample has been loaded, the column is flushed with 0.5 *M* NaCl in TE buffer until the RNA peak has been eluted and the O.D. has returned to baseline. At this time, a 300 ml gradient of 0.5 *M* NaCl in TE to 0.8 *M* NaCl in TE is begun at a flow rate of approximately 0.5 ml/min. Plasmids usually elute at a molarity of approximately 0.6 *M* NaCl (one-third of the way through the gradient), and chromosomal DNA elutes very closely behind it. The appropriate fractions are collected, pooled and diluted to an approximate molarity of 0.2 *M* NaCl in TE buffer.

Concentration of plasmid DNA. Plasmid DNA is concentrated using Schleicher and Schuell elutips. Briefly, the DNA is passed through a previously equilibrated elutip in 0.2 *M* NaCl in TE buffer using a syringe. The DNA is then eluted in a volume of 0.4 ml by passing 1 *M* NaCl in TE buffer through the elutip. The sample is diluted to a final molarity of 0.1 *M* NaCl and two volumes of ice-cold 95% ethanol are added. The sample is placed at −20 C for at least 4 h. The DNA is pelleted by centrifugation at 12,000 × *g* for 30 min at 4 C. The pellet is dried and resuspended in an appropriate amount of TE.

Purification of plasmids by CsCl gradient centrifugation. A crude preparation of plasmid DNA was prepared as per Hansen and Olsen (1978). This preparation was then purified by CsCl gradient centrifugation as per Maniatis et al. (1982), pages 93–94. Each 12 ml centrifuge tube contained 0.5 liters of material from the original culture.

Gas electrophoresis. Agarose gel electrophoresis was done using the method of Meyers et al. (1976) on a horizontal gel apparatus.

Restriction digests. Restriction endonuclease reactions were performed following the procedure of Maniatis et al. (1982), pages 103–104.

DNA ligation. DNA ligation was performed on phenol-extracted digests using a 1:3 ratio of vector to insert. Reactions were done in 20 μl overnight at 12 C. The ligation buffer used was that of Maniatis et al. (1982) page 246.

DNA transformation. Transformation of pUC-13 DNA was done utilizing the procedure of Hanahan (1983).

Reagents and bacteria. Pseudomonas diminuta and *Flavobacterium* (ATCC #27551) were supplied by L. Johnson of Cytox, *Pseudomonas aeruginosa* PA 025 by S. Krawiec of Lehigh University, *Pseudomonas putida* (TOL-120 Mdal) by D. K. Chaterjee of the University of Illinois, and *Pseudomonas putida* (TOL-78 Mdal) by the American Type Culture Collection (#33015). Redistilled phenol, restricted enzymes, NACS-37, DNA ligase, and Hind III-cut lambda DNA were obtained from Bethesda Research Laboratories. Elutips were obtained from Schleicher & Schuell, pUC-13 plasmid DNA was obtained from PL-Biochemicals.

Results and Discussion

Analysis of Individual Steps of the Procedure

To determine the effect of each of the major steps of this procedure, RP1, a 40-Mdal plasmid of *Pseudomonas aeruginosa* was extracted using the prescribed protocol. After each major step, aliquots amounting to one-thirtieth of the starting material were saved and subsequently analyzed by electrophoresis on a 0.7% agarose gel (Fig. 2). After the Hansen and Olsen procedure (lane 2), the sample

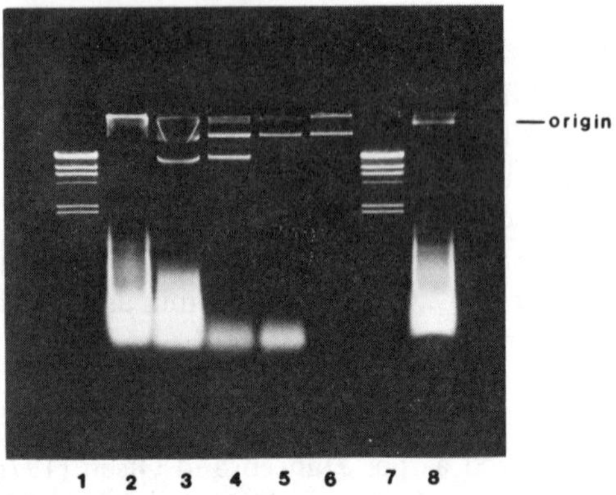

FIG. 2. Electrophoretic analysis of RP1 DNA from individual steps in the extraction procedure. Lanes 1 and 7, Hind III digest of lambda DNA (fragment sizes in kilobases are 23, 9.4, 6.7, 4.4, 2.3, and 2.0); lane 2, RP1 after Hansen and Olsen extraction; lane 3, RP1 after phenol extraction and ammonium acetate precipitation; lane 4, RP1 after RNase digestion; lane 5, RP1 after acid phenol extraction; lane 6, RP1 after ion exchange chromatography; lane 8, ammonium acetate pellet.

consists of large amounts of low and high molecular weight RNA, plasmid and chromosomal DNA (barely discernible), and protein (not discernible. The phenol extractions and the ammonium acetate precipitation eliminate most of the protein, much of the high molecular weight RNA, and has the added effect of removing impurities tightly bound to the plasmid. This results in cleaner DNA bands (lane 3). To ensure that part or all of the sample was not being lost in the ammonium acetate precipitate, the pellet was resuspended and one-thirtieth of the pellet was run in lane 8. Neither chromosomal DNA nor plasmid DNA can be seen in this lane. The RNase digestion eliminates most of the RNA from the sample and results in an even clearer definition of the chromosomal DNA and plasmid DNA bands (lane 4). The acid-phenol step effectively eliminates the chromosomal DNA while leaving only plasmid DNA and low molecular weight RNA (lane 5). Finally, the RNA was removed by passage of the sample over the NACS-37 column. This resulted in a highly purified plasmid DNA preparation (lane 6).

Comparison of RP1 Purified by CsCl Centrifugation and RP1 Purified by Extraction

The next question we asked was: How pure are these extracted plasmids in relation to CsCl gradient purified plasmids? To answer this, a one-liter culture of *Pseudomonas aeruginosa* PA025, containing RP1, was divided in half; one of the halves was purified by CsCl gradient centrifugation and the other half was purified by our procedure. The plasmid DNA obtained from these procedures was analyzed by electrophoresis in 0.7% agarose gels and by passage through the NACS-37 ion exchange column (Fig. 3). Gel electrophoresis reveals that RP1

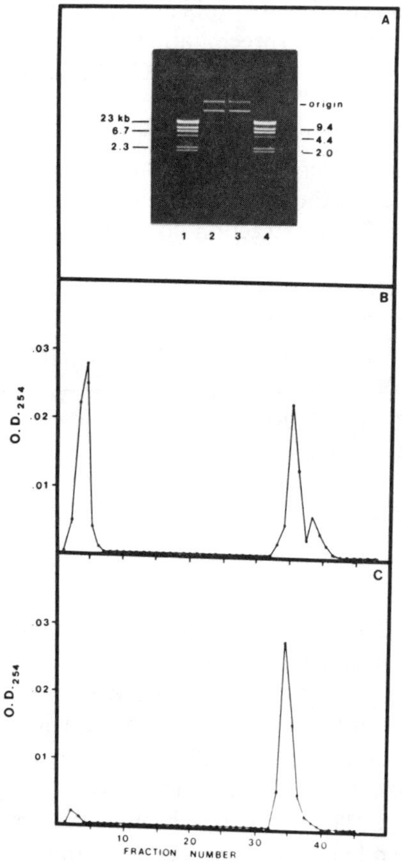

FIG. 3. Comparison of CsCl purified RP1 to extracted RP1. A—Lanes 1 and 4, Hind III digest of lambda DNA; lane 2, RP1 purified by CsCl gradient centrifugation; lane 3, RP1 purified by the extraction procedure in this paper. B—NACS-37 ion exchange chromatography of RP1 purified by CsCl gradient centrifugation. C—NACS-37 ion exchange chromatography of RP1 purified by the extraction procedure in this paper.

purified by CsCl gradient centrifugation contains a small amount of contaminating cellular DNA, while the extracted RP1 is essentially free of chromosomal DNA (Fig. 3A). Moreover, when the gradient-purified RP1 is analyzed by NACS-37 ion exchange chromatography, not only is a chromosomal DNA peak evident, but a large peak that elutes in the 0.5 *M* NaCl-TE wash before the gradient is started (Fig. 3B, gradient started at fraction 8). This peak is composed of mostly protein and a small amount of RNA (which could be visualized on a gel when concentrated). RP1 that had been purified by the extraction procedure prescribed in this paper essentially showed only one peak because of plasmid DNA (Fig. 3C). The protein and RNA in the CsCl-purified RP1 is likely to have come from two sources. First, their presence may be because of entrapment by the plasmid band, which presents an actual physical barrier within the gradient. Proteins that are moving upward and RNA that is moving downward may be hindered by this barrier and become entrapped. Second, this contamination may be due to actual binding of the proteins or RNA to the plasmid DNA. The extraction procedure has circumvented these problems because there is no such physical barrier, and the ammonium acetate precipitation step conveniently removes many impurities that are tightly bound to DNA (NACS Applications Manual).

Yield

The yields of the plasmids were analyzed by averaging the area under the ion exchange peak and calculating the expected result from one liter of culture (Table 1). Yields varied greatly from extraction to extraction and from plasmid to

TABLE 1. *Estimated yield of five different large plasmid species.*

Plasmid	Host	Size (Mdal)	Estimated yield/liter
PCS1	*Pseudomonas diminuta*	44	12 μg
RP1	*Pseudomonas aeruginosa*	40	25 μg
TOL	*Pseudomonas putida*	78	15 μg
TOL	*Pseudomonas putida*	120	15 μg
unnamed	*Flavobacterium sp.* approx.	40	60 μg

plasmid, but at no time did the yield surpass the yield from CsCl gradient centrifugation, and in most instances this yield was 50 to 70% of that obtained by CsCl gradient centrifugation. Yields for larger plasmids tended to be lower. The reason for this may reside in the means by which the acid-phenol and the ion exchange resin select out the plasmids. Both techniques rely on the presence of covalently closed circular plasmid DNA to allow it to be segregated from chromosomal DNA. Large plasmids tend to be most susceptible to breakage, particularly during an extended procedure such as this, which requires much manipulation. Therefore, plasmids that are very large (greater than 100 Mdal) may not be as amenable to this procedure as plasmids in the size range of 20–100 Mdal. We have attempted to limit the number and severity of the manipulations, but this has yielded mixed results. Currently, various experimental modifications are being implemented in hopes of obtaining more consistent yields.

Applicability to Other Plasmids

To determine whether or not this procedure could be applied to other plasmids, four bacterial species containing large plasmids were grown and plasmid DNA was extracted according to the prescribed protocol (Fig. 4). The extraction procedure resulted in the purification of plasmids from all four bacterial strains. However, the extraction was much more effective for plasmids in the size range of 20–100 Mdal than for larger plasmids. The upper size limit of this procedure has not been determined, but in certain extractions we have received good yields of a multimer of a *Flavobacterium* plasmid that is approximately 144 Mdal in size (data not shown).

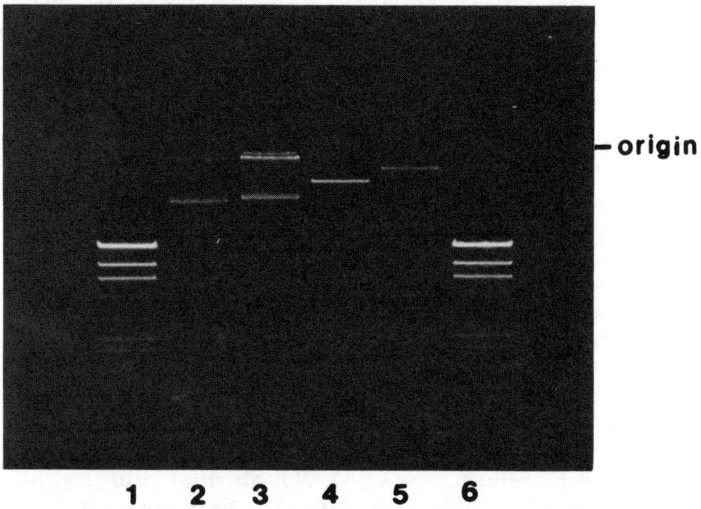

FIG. 4. Electrophoretic analysis of four different plasmids isolated by the ion exchange chromatography procedure. Lanes 1 and 6, Hind III digest of lambda DNA; lane 2, *Flavobacterium* plasmid; lane 3, pCS1; lane 4, TOL (78 Mdal); lane 5, TOL (120 Mdal).

Functionality of DNA

The final result of any plasmid purification procedure must be plasmid DNA that is fully functional in its ability to be restricted, ligated, and free of defects that would disallow expression of its constituent genes. To determine whether or not this was the case for the extraction procedure used in this paper, *Flavobacterium* plasmid DNA that had been purified by this procedure was restricted with several different endonucleases and each separate digest was ligated into an appropriately digested pUC-13 plasmid. A 20 μl ligation reaction in which there was 1 μg of pUC-13 and 3.5 μg of digested *Flavobacterium* plasmid yielded more than 1,000 colonies that contained inserts, upon transformation (unpublished data). Furthermore, when these recombinants were analyzed for the production of parathion hydrolase (a gene that we suspected might reside on the plasmid, based on literature reports of a plasmid-encoded parathion hydrolase in *Pseudomonas diminuta* (Serdar et al. 1982)) more than 50 clones were isolated that produced parathion hydrolase, thereby indicating that the cloned DNA was functionally expressed (paper to be submitted).

Conclusions

A procedure has been developed that results in plasmids that are significantly purer than plasmids purified by CsCl gradient centrifugation. This procedure does not rely on the utilization of an ultracentrifuge and can be performed in a time period comparable to that of CsCl gradient purification. However, it is not expected that this procedure will supplant CsCl gradient purification because it has limited utility for the very large plasmids (> 100 Mdal). It is, however, a very useful procedure for laboratories that would like to conduct research in the area of large plasmid biology but that cannot bear the expense of an ultracentrifuge or do not have reasonable access to such an instrument.

Acknowledgments

The authors would like to thank Ken Pidcock for useful discussions and Layne Johnson for his critical review of this paper.

Literature Cited

Best, A. N., D. P. Allison, and G. D. Novelli. 1981. Purification of supercoiled DNA of plasmid Col E1 by RPC-5 Chromatography. *Anal. Biochem.* 114:235.

Birnboim, H. C., and J. Doly. 1979. A rapid alkaline extraction procedure for screening recombinant plasmid DNA. *Nucleic Acids Res.* 7:513.

Chakrabarty, A. M. 1976. Plasmids in *Pseudomonas. Annu. Rev. Genet.* 10:7–30.

Currier, T. C., and E. W. Nester. 1976. Isolation of covalently closed circular DNA of high molecular weight from bacteria. *Anal. Biochem.* 76:431–441.

Don, R. H., and J. M. Pemberton. 1981. Properties of six pesticide degradation plasmids isolated from *Alcaligenes paradoxus* and *Alcaligenes eutrophus. J. Bacteriol.* 145:681–686.

Fennewald, M., W. Prevatt, R. Meyer, and J. Shapiro. 1978. Isolation of Inc P-2 plasmid DNA from *Pseudomonas aeruginosa. Plasmid* 1:164–173.

Franklin, F. C. H., M. Bagdasarian, and K. N. Timmis. 1981. Manipulation of degradative genes of soil bacteria. Pages 109–130 *in* T. Leisinger, A. M., Cook, R. Hutter, and J. Nuesch, eds., *Microbial Degradation of Xenobiotics and Recalcitrant compounds.* Academic Press, London.

Furukawa, K., and A. M. Chakrabarty. 1982. Involvement of plasmids in total degradation of chlorinated biphenyls. *Appl. Environ. Microbiol.* 44:619–626.

Hanahan, D. 1983. Studies on transformation of *Escherichia coli* with plasmids. *J. Mol. Biol.* 166:557–580.

Hansen, J. B., and R. H. Olsen. 1978. Isolation of large bacterial plasmids and characterization of the P2 incompatibility group plasmids pMG1 and pMG5. *J. Bacteriol.* 135:227–238.

Humphreys, G. O., G. A. Willshaw, and E. S. Anderson. 1975. A simple method for the preparation of large quantities of pure plasmid DNA. *Biochim. Biophys. Acta* 383:457–463.

Maniatis, T., E. F. Fritsch, and J. Sambrook. 1982. *Molecular cloning: A labor-*

atory manual. Pages 93–94. Cold Spring Harbor Publications, Cold Spring Harbor, N. Y.

Meyers, J. A., D. Sanchez, L. P. Elwell, and S. Falkow. 1976. Simple agarose electrophoretic method for the identification and characterization of plasmid deoxyribonucleic acid. *J. Bacteriol.* 127:1529–1537. (Erratum, *J. Bacteriol.* 129:1171, 1977.)

NACS Application Manual, Bethesda Research Laboratories, Rockville, Maryland.

Palchaudhuri, S., and A. Chakrabarty. 1976. Isolation of plasmid deoxyribonucleic acid from *Pseudomonas putida. J. Bacteriol.* 126:410–416.

Pierce, G. E., T. J. Facklam, and J. M. Rice. 1981. Isolation and characterization of plasmids from environmental strains of bacteria capable of degrading the herbicide 2,9-D. *Dev. Ind. Microbiol.* 22:401–408.

Serdar, C. M., D. T. Gibson, D. M. Munnecke, and J. H. Lancaster. 1982. Plasmid involvement in parathion hydrolysis by *Pseudomonas diminuta. Appl. Environ. Microbiol.* 44:246–249.

Sharp, P. D., M. -T. Hsu, E. Ohtsuba, and N. Davidson. 1972. Electron microscope heteroduplex studies of sequence relations among plasmids of *Escherichia coli.* I. Structure of G-prime factors. *J. Mol. Biol.* 71:471–397.

Thompson, J. A., R. W. Blakesley, K. Doran, C. J. Hough, and R. D. Wells. 1983. Purification of nucleic acids by RPC-5 ANALOG chromatography: Peristaltic and gravity flow applications. *Methods in Enzymology.* Recombinant DNA, Part B. Academic Press, London.

Vandenbergh, P. A., R. H. Olsen, and J. F. Colaruotolo. 1981. Isolation and genetic characterization of bacteria that degrade chloroaromatic compounds. *Appl. Environ. Microbiol.* 42:737–739.

Zasloff, M., G. D. Ginder, and G. Felsenfeld. 1978. A new method for the purification of identification of covalently closed circular DNA molecules. *Nucleic Acids Res.* 5:1139–52.

CHAPTER 59

Immobilized Fungal Whole Cells as a Beta-Glucosidase Source

KWANG-PIN TSAI AND D. F. DAY*

Audubon Sugar Institute, Louisiana State University, Baton Rouge, Louisiana 70803-7305

Whole cells of *Aspergillus terreus* (ATCC 20514) were immobilized in agar beads and used as a source of beta-glucosidase. Optimal activity was observed at pH 4.6 and 50 C. Temperature stability was not changed significantly by immobilization. Km's (app.) of 4.55 m*M* and 1.00 m*M* were found for beta-glucosidase in immobilized cells when cellobiose and *p*-nitrophenyl beta-D-glucoside, respectively, were used as substrates. An activation energy of 13,100 cal/mol was found for free beta-glucosidase, and it increased to 20,100 cal/mol for beta-glucosidase in the immobilized cells. Approximately 90% of the activity was retained after 30 d operation at 35 C. A higher productivity could be reached by operating at 50 C but with a much shorter half-life, 20 d. The beta-glucosidase in the immobilized cells had a higher operating stability compared to other immobilized beta-glucosidase systems and both its optimal pH and temperature were similar to that of the *Trichoderma* cellulase system. This system could be feasible for industrial use with an increase in specific activity.

INTRODUCTION

Cellulose can be decomposed by cellulase, which is composed of three enzymes, 1,4-beta-D-glucan glucanohydrolase, 1,4-beta-D-glucan cellobiohydrolase, and beta-glucosidase. Cellobiose, which is an intermediate of cellulose hydrolysis, inhibits the activity of 1,4-beta-D-glucan glucanohydrolase and 1,4-beta-glucosidase, and is hydrolyzed by beta-glucosidase to glucose (Ladisch et al. 1983). *Trichoderma reesei*, one of the highest yielding of the cellulase-producing fungi has a major drawback for use in commercial saccharification of cellulase in that the amount of beta-glucosidase it produces is rate limiting (Mandels and Andreotti 1978). Supplementation with a beta-glucosidase from another source helps reduce this inhibition and enhances the saccharification capability of the *Trichoderma* cellulase system (Joglekar et al. 1983). The beta-glucosidase produced from *Aspergillus terreus* was found to have a low Michaelis-Menten constant, a high specific activity, and a temperature and pH optimum that were similar to the *Trichoderma* cellulase system. The enzyme has been reported to have potential as supplement for *Trichoderma* cellulase in industrial saccharification systems (Workman and Day 1982).

This investigation evaluates the possibility of using the beta-glucosidase in *Aspergillus terreus* in the form of immobilized whole cells for the production of glucose from cellobiose.

*Author to whom correspondence should be addressed.

MATERIALS AND METHODS

Microorganism. Stock cultures of *Aspergillus terreus* (ATCC 20514) were maintained at room temperature on potato dextrose agar (Difco) plates.

Medium. The medium for batch culture was a modified preparation of Czapek according to Dox (Raper and Fennell 1965). The medium contained 3.5 × 10^{-2} *M* sodium nitrate, 6.7 × 10^{-3} *M* potassium chloride, 5.7 × 10^{-3} *M* dipotassium phosphate, 2.0 × 10^{-3} *M* magnesium sulfate, 4.0 × 10^{-5} *M* ferrous sulfate, and 0.056 *M* cellobiose in distilled water. The same medium containing 3% agar was used for cell immobilization.

Immobilization. Sterilized distilled water (20 ml) was added to a 7-d-old sporulated culture of *Aspergillus terreus* on a potato dextrose agar plate. A spore suspension was collected and inoculated into 180 ml of 3% agar sol at 50 C, which contained sufficient nutrients to obtain the previously listed concentrations, post-inoculation. This mixture was stirred and then dispensed into a 200-ml jacketed vessel. The spores and nutrient-containing agar sol were then pumped dropwise through a 0.01-inch nozzle into cold vegetable oil. The beads obtained had an average size of 3.30 ± 0.26 mm. These beads were collected in a funnel and sterilized water was used to wash to remove any residual oil. The beads were then spread out in a sterilized container and were incubated in a humidified atmosphere at 37 C for 4 d.

Enzyme Assay. The beta-glucosidase activity was determined using *p*-nitrophenyl-beta-D-glucoside (PNPG) and cellobiose as substrates. Assay was by the methods of Hagerdal et al. (1979) and Workman and Day (1982). For assay of enzymatic activity in immobilized fungal beads 1 g of prewashed beads was added to 50 ml of substrate whose concentration was 10 m*M* in a 0.05 *M* sodium citrate buffer (pH 4.8) that had been preheated to 50 C, in a 500-ml Erlenmeyer flask. After the addition of the beads, the flask was put into a shaking water bath at 50 C shaking at 100 rpm. A sample, 0.5 ml, was withdrawn every 30 min and assayed for product. One enzyme unit (EU) is defined as the production of one μmol of glucose per minute from the substrate.

Michaelis-Menten constant. The apparent Michaelis-Menten constant was determined by the method of Lineweaver and Burk (1934). Six 1-g portions of beads were reacted with six different concentrations of substrate. Each portion of beads was prewashed three times with an equal volume of test substrate for a total time of 1 h. After this pretreatment, the beads were reacted with desired concentrations of the substrate. Both PNPG and cellobiose were used as substrates. The apparent Km of the enzyme in the non-entrapped cells was determined without pretreatment. All of the reactions were carried out at 50 C, pH 4.8, and 100 rpm agitation.

Arrhenius plot. An Arrhenius plot was constructed from assays using PNPG as substrate. The tested temperatures covered a range from 25 C to 50 C in intervals of 5 C.

Operational stability. Beads, 160 g, were washed with 500 ml of 0.05 *M* sodium citrate buffer solution. These beads were then packed into a jacketed column 2.6 cm in diameter and 40 cm in length. This produced a column with a reaction volume of 43.2 ml. A sterilized substrate solution containing 8 m*M* cellobiose and 0.05 *M* sodium citrate buffer was fed into the bottom of the column and the effluent was collected from the top. The reactor was allowed to run from 20 to 50 d for each tested temperature.

RESULTS

The enzyme in the immobilized cells and in free cells was found to have its highest activity at pH 4.6. The purified enzyme has a pH optimum of pH 4.8 (Workman and Day 1982).

The highest enzyme activity in the immobilized cells was found at 50 C. At temperatures higher than 50 C, the enzyme activity decreased sharply. This is the same value that was reported for the purified beta-glucosidase from *Aspergillus terreus* (Workman and Day 1982). Temperature stability of the immobilized enzyme was determined by incubation at test temperatures for 5 h, and then assayed for residual activity. It was found that 96%, 49.9%, and 13.3% of activity remained after incubation at 50 C, 55 C, and 62 C, respectively, for 5 h.

A comparison of Km values is shown in Table 1. The Km values tended to increase for enzyme in the immobilized cells. Activation energies of 13,100 cal/mol and 13,300 cal/mol were found for free enzyme and enzyme in free cells; the activation energy, between the range of 25 C and 40 C, for the enzyme in the immobilized cells was found to be 20,100 cal/mol (Fig. 1).

TABLE 1. *Comparison of Km*

	Free Enzyme		Enzyme in Immob. Cells		Enzyme in Free Cells
Substrate	PNPG	Cellobiose	PNPG	Cellobiose	PNPG
km	078	0.40	1.00	4.55	1.00

A column was constructed as described previously. At a flow rate of 0.4 mg/ml, 1.8 mg/ml, and 1.2 mg/ml, and productivity was found to be 1.33 g/h/l, 1.00 g/h/liter and 0.67 g/h/liter for temperatures of 50 C, 35 C, and 25 C, respectively. The highest productivity for this reactor was 3.61 g/h/liter at a flow rate of 2.2 ml/min at 50 C.

The column had a half-life of 480 h at 50 C. The same half-life was found when it was operated at 42 C; however, the conversion rate did not decrease significantly after 40 d operation at 25 C. At 35 C, a 90% conversion rate was maintained after 30 d operation. This gives an apparent half-life of 150 d with a productivity of 1.00 g/h/liter when operating at 35 C.

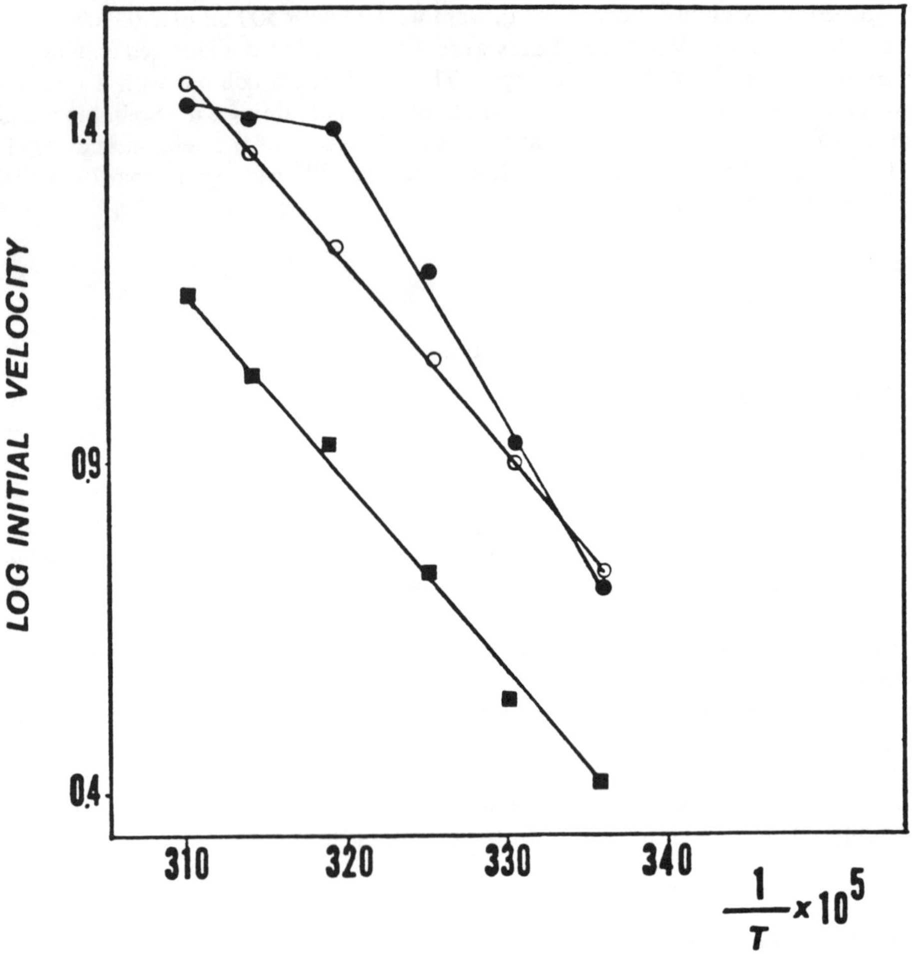

FIG. 1. An Arrhenius plot for the determination of the activation energy for free enzyme (open circle), for enzyme in immobilized cells (closed circle), and for enzyme in cells (closed square).

Discussion

Aspergillus terreus grew significantly faster in beads than in batch culture (Fig. 2). This was probably because of the higher specific surface area of the beads minimizing oxygen limitation. Enzyme production followed a similar pattern in both batch culture and beads. The amount of enzyme produced per unit cell decreased initially and then increased as growth approached the stationary phase. As the beta-glucosidase is an inducible enzyme in *Aspergillus terreus*, this increase in specific enzyme production near the stationary phase may be a result by catabolite repression.

Enzyme production per unit cell increased more slowly in the immobilized system than in the batch culture. This may be because three times more cell mass

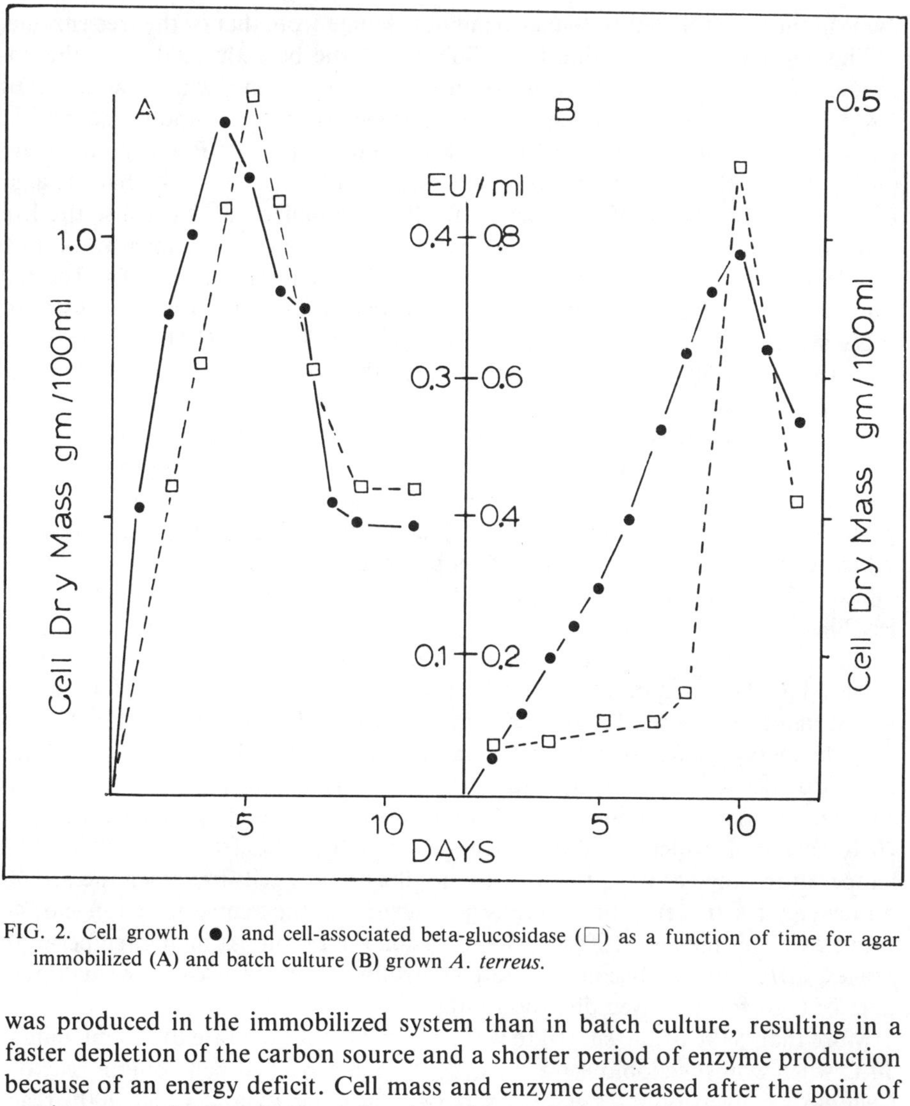

FIG. 2. Cell growth (●) and cell-associated beta-glucosidase (□) as a function of time for agar immobilized (A) and batch culture (B) grown *A. terreus*.

was produced in the immobilized system than in batch culture, resulting in a faster depletion of the carbon source and a shorter period of enzyme production because of an energy deficit. Cell mass and enzyme decreased after the point of maximum production for both batch and immobilized cells. The enzyme levels dropped for 3 d and then held stable in immobilized cells; however, the enzyme activity was lost after 6 d in batch culture. This difference may be due to a confinement effect to autolytic enzymes by agar resulting in a lower rate of degradation of beta-glucosidase.

The optimum pH of beta-glucosidase in both the immobilized cells and the free cells was 4.6, which is shifted slightly to the acidic side, compared to the optimum pH of the free enzyme, 4.8. A pH shift in the immobilized cell system is not unusual and may be explained by a different microenvironment for the enzyme in the cell than in the bulk solution. This small shift is not very significant, since at pH 4.8, the enzyme in the immobilized cells had 95% of their maximum activity.

The temperature optimum for the enzyme in the immobilized cell was 50 C, which was the same as that of the free enzyme. Temperature stability of the enzyme in the immobilized system also did not change from that of the free enzyme.

The approximate Km value for cellobiose of the beta-glucosidase in the immobilized cell was 4.55 m*M*. Compared to the free enzyme, which had a Km of 0.4 m*M* for cellobiose, the Km of beta-glucosidase in the immobilized cell increased 11.4-fold. The approximate Km value using PNPG for the beta-glucosidase in the immobilized cell increased slightly. It is unlikely that the agar affected the Km value of the enzyme in the immobilized cells because the Km value for PNPG of the beta-glucosidase in free cells was also shown to be 1.00 m*M*, the same as that of the beta-glucosidase in the immobilized cells. The cell wall probably plays an important role in this decreasing rate of mass transfer of substrate to enzyme. It was found that the apparent Km for cellobiose was larger than that for PNPG. However, the Km of the free enzyme for cellobiose is smaller than that of the free enzyme for PNPG. This may be a function of the products produced. Decomposition of one molecule of cellobiose produces two molecules of glucose. Decomposition of one molecule of PNPG produces only one molecule of glucose. The microenvironment in the location of the beta-glucosidase would accumulate more glucose when cellobiose was used as substrate because of the cell wall diffusion barrier. As glucose is a competitive inhibitor, the apparent Km value for cellobiose would increase more than that for PNPG.

The activation energy found for the free enzyme was 13,100 cal/mol, which corresponded to a temperature coefficient, Q_{10}, of 2.05. A Q_{10} value of about 2.0 with a range of between 1.7 and 2.5 is generally found for most enzymes. The activation energy found for the enzyme in the free cells was 13,300 cal/mol. Compared to the activation energy for the free enzyme, there was no significant change. The activation energy for beta-glucosidase in the immobilized cell was 20,100 cal/mol, which corresponded to a temperature coefficient of 3.03 in the temperature range of 25 C to 40 C. Immobilization of cells increased the activation energy for this enzyme. There is nonlinearity in the temperature range of 40 C to 50 C. This may result from pore diffusion limitation. At higher temperature ranges, diffusion may become the rate-controlling step because of the small pore sizes of the agar increased diffusion rates.

More than 83% of the substrate was converted at a flow rate of 0.4 ml/min at 50 C when 8 m*M* cellobiose was passed through a packed bed column reactor. Cellobiose conversion rates at 50 C were twice as much as at 25 C, when the reactor was operated at a flow rate of 0.4 ml/min. Although a higher conversion rate could be attained, a shorter half-life was shown at 50 C, 20 d, compared to 25 C. The advantage of running the reactor at 25 C was its operational stability. The conversion rate did not decrease significantly after 50 d operation at 25 C; however, 35 C may be a better temperature for operation of this reactor. A conversion rate, which was 75% of that obtained at 50 C and a retention of 90% of its activity after 30 d operation, was found for 35 C. Appropriate economic analysis would be required to define the optimum operating conditions for any scaled-up process.

There are advantages and disadvantages for using this beta-glucosidase in a saccharification process. The advantages are (1) The process of immobilization is easy and enzyme extraction and purification are not necessary; (2) cell mass was immobilized directly after germination of spores; (3) mycelia developed much faster in agar beads than in batch culture; (4) the enzyme in the immobilized cells was more resistant to autolysis; (5) more than 90% of the enzyme produced was immobilized in the beads; (6) the optimum temperature did not change, and the optimum pH changed minimally from the optimum pH for *Trichoderma* cellulase; and the most important (7) the immobilized enzyme was reusable.

Some disadvantages were also apparent: (1) *Aspergillus terreus* tended to produce more cell mass and less enzyme in beads than in batch culture; (2) the specific activity in beads was low and was about 0.5 EU/g beads (this may be overcome by increasing the carbon source and other nutrients in agar before spores are immobilized in agar beads); and (3) the Km value with cellobiose as substrate was increased about 10-fold—this was also found for other immobilized beta-glucosidase systems.

The use of agar beads in a saccharification reactor is probably not feasible because of low strength of agar beads. However, a model reactor was suggested by Ghose and Kostick (1970), and was reconsidered by Ladisch et al. (1983), which could be used with this system. This model suggested that saccharification be carried out with high concentrations of cellulase with the continuous removal of sugar syrups by ultrafiltration. However, the sugar syrup produced by this reactor would contain a high concentration of cellobiose. This could be reduced by passing this filtrate through a reactor containing these immobilized *Aspergillus terreus* cells.

Generally speaking, use of this beta-glucosidase as immobilized cells in agar beads is feasible. However, in order to be economical, the specific activity of beta-glucosidase in agar beads should probably be increased.

Literature Cited

Ghose, T. K., and J. A. Kostick. 1970. A model for continuous enzymatic saccharification of cellulose with simultaneous removal of glucose syrup. *Biotechnol. Bioeng.* 12:921–946.

Hagerdal, B., H. Harris, and E. K. Pye. 1979. Association of beta-glucosidase with intact cells of *Thermoactinomyces. Biotechnol. Bioeng.* 21:345–355.

Joglekar, A. V., N. G. Karanth, and M. C. Srinivasan. 1983. Significance of beta-D-glucosidase in the measurement of exo-beta-D-glucosidase activity of cellulolytic fungi. *Enzymol. Microbiol. Technol.* 5:25–29.

Ladisch, M. R., K. W. Lin, M. Voloch, and G. T. Tsao. 1983. Process considerations in the enzymatic hydrolysis of biomass. *Enzymol. Microbiol. Technol.* 5:82–102.

Lineweaver, H., and D. Burk. 1934. The determination of enzyme dissociation constants. *J. Am. Chem. Soc.* 56:658–666.

Mandels, M., and R. E. Andreotti. 1978. Problems and challenges in the cellulose to cellulase fermentation. *Process Biochem.* 13:6–13.

Raper, K. B., and D. I. Fennell. 1965. Cultivation. Pages 35–55 *in* K. B. Raper and D. I. Fennell, eds., *The Genus Aspergillus.* William and Wilkins Co., Baltimore, MD.

Workman, W. E., and D. F. Day. 1982. A β-glucosidase from *Aspergillus terreus*: Purification and properties. *Appl. Environ. Microbiol.* 44:1289–1295.

CHAPTER 60

Alginate Lyase-Secreting Bacteria Associated with the Algal Genus *Sargassum*

JAMES F. PRESTON III, TONY ROMEO, JOHN C. BROMLEY, RALPH W. ROBINSON, AND HENRY C. ALDRICH

Department of Microbiology and Cell Science, University of Florida, Gainesville, Florida 32611

Alginase-secreting bacteria associated with actively growing tissues of the pelagic Phaeophyta species *Sargassum fluitans* and *Sargassum natans* have been enumerated and shown to represent a major component of the aerotolerant heterotrophic bacterial flora associated with these species. Seven alginase secreting bacteria, six from actively growing tissues of *S. fluitans* and *S. natans* and one from beached decaying *Sargassum* tissue, have been isolated and characterized to determine their metabolic potential and the levels and specificities of both the intra- and extracellular alginate lyase activities. All are gram-negative polarly flagellated rods, which are either oxidative and oxidase positive or fermentative and oxidase negative. Buoyant density determination of isolated DNA has established GC mol% values ranging from 44.3 to 47.4. The properties of the oxidative organisms are consistent with those expected for the genus *Alteromonas* while the fermentative organisms contain at least one property that exempts them from inclusion in an established genus.

INTRODUCTION

Alginate-degrading enzymes have been identified and partially characterized from a number of marine invertebrates where they appear to function in the digestion of the marine brown algae (Favorov and Vaskovsky 1971; Nakada and Sweeny 1967; Favorov et al. 1979; Elyakova and Favorov 1974; Muramatsu et al. 1977; Muramatsu and Katayose 1979; Muramatsu and Egawa 1982; Muramatsu 1984). These enzymes are generally, if not exclusively, alginate lyase activities that generate an unsaturated bond between carbon atoms 4 and 5 of either D-mannuronate or L-guluronate residues upon cleavage of the 1–4 glycosidic linkages of hetero and/or homopolymers derived from alginate. Analogous activities have been identified in several bacterial species (Preiss and Ashwell 1962; Kashiwabara et al. 1969; Davidson et al. 1976; Davidson et al. 1977; Quatrano and Caldwell 1978; Doubet and Quatrano 1982; Doubet and Quatrano 1984; Hansen et al. 1984; Sutherland and Keen 1981; Boyd and Turvey 1977), which have defined specificities with respect to substrate composition and sites of polymer cleavage. Some of these enzymes have been used to dissect the matrices and cell walls of Phaeophyta species (Quatrano et al. 1980) to define the structural role of alginates.

Alginase-secreting bacteria have been isolated from soil and decaying tissues of brown algae (Thjotta and Kass 1945; Meland 1962) as well as actively growing tissues of *Fucus* (Quantrano and Caldwell 1978; Doubet and Quatrano 1982). The earlier designations of the genera *Alginomonas, Alginovibrio,* and *Alginobacter* (Thjotta and Kass 1945; Meland 1962), which were based upon the use of alginate for growth, have not led to their general acceptance as *bona fide* genera, and they have been included as species *incertae sedis* of the genus *Pseudomonas* in the 8th edition of *Bergey's Manual* (Doudoroff and Palleroni 1974). Species assigned to the genera *Azotobacter, Bacillus, Benekia,* and *Klebsiella* are capable of secreting alginase (Davidson et al. 1977; Hansen et al. 1984; Riechelt and Baumann 1973; Boyd and Turvey 1977), and an isolate with properties similar to members of the genus *Alteromonas* has been characterized with respect to the substrate specificities of extracellular and intracellular alginate lyase activities produced by this organism (Quatrano and Caldwell 1978).

Here we show that most of the aerobic and/or aerotolerant bacteria associated with actively growing tissues of the pelagic species of *Sargassum, S. fluitans,* and *S. natans* are gram-negative polar flagellates that secrete alginate lyase. These flagellates may be divided into two groups, one of which includes oxidative organisms with properties consistent with those expected for the genus *Alteromonas,* and a second, which includes facultative organisms with metabolic properties that exclude them from the genus *Benekia* and a DNA base composition that excludes them from the genus *Photobacterium.*

Materials and Methods

Chemicals and reagents. The isolation of DNA used redistilled phenol (Molecular Biology grade, International Biotechnologies Inc.), Analar CsCl (Gallard Schlesinger), electrophoretic grade acrylamide (Pierce Chemical), N,N′-methylene-bisacrylamide (Sigma Chemical, recrystallized from acetone), lysozyme (Sigma, 40,000 units/mg), and Ribonuclease A (Sigma, 80 units/mg). Alginate was obtained from Fisher Scientific Company as a purified grade originally isolated from species of *Macrocystis.* All of the other chemicals were analytical reagent grade or high purity biochemicals. DNA isolation buffer SSC (Kohne 1968) was prepared as a 10X stock solution containing 1.5 *M* NaCl, 0.15 *M* Na_3citrate, adjusted to pH 7.0 with HC1.

Enumeration and isolation of bacteria. Actively growing tissues of *Sargassum fluitans* and *Sargassum natans* were collected within 400 yd of the shore approximately 1 mile south of Delray Beach, FL. Specimens with a relatively large number of growing tips and a low level of associated epiphytic hydroids and bryozoa were selected and transported to Gainesville, FL. The specimens were maintained at room temperature in artificial sea water with aeration until they were analyzed, which was within 3 d of their collection.

To remove bacteria from the algal tissue prior to their enumeration, specimens of the healthiest appearing tissues (1 g wet wt) were suspended in 5 ml of sterile instant ocean (Provasoli's enriched sea water supplemented with 0.27 mg/liter KI, Provasoli 1968; Polne-Fuller et al. in press) and subjected to mild sonication in a sonic water bath (Branson model B-22-4, 125 watt) for 60 s at room temperature.

Serial dilutions of the sonicate were prepared and spread with a glass rod on 2% agar plates containing either rich medium (1% D-glucose, 1% yeast extract, 0.8% nutrient broth, in PESI (Provasoli's enriched sea water supplemented with 0.27 mg/liter KI, Provasoli 1968; Polne-Fuller et al. in press)) or alginate medium (1% sodium alginate in PESI). Cultures were incubated for 48 h at room temperature and detectable colonies were scored and examined for the secretion of alginase as evidenced by the appearance of clearing zones on the alginate plates.

A number of defined colonies were transferred to liquid cultures containing alginate medium and grown to provide turbid cultures (optical densities of 0.15 to 0.42, 600 nm, 1.00 cm light path) and then streaked to assure the isolation of pure cultures. One isolate was obtained from decaying *Sargassum* tissue.

Electron microscopy. Scanning electron microscopy was carried out on 48 h cultures of the various isolates grown on alginate medium containing 2% agar. A colony sample was fixed for 1 h in 2% glutaraldehyde, rinsed, and then fixed in 2% OsO_4 for 1 h. After dehydration in ethanol, they were critical point dried and examined and photographed with an Hitachi S450 scanning electron microscope.

For determination of flagellar organization, cultures were grown for 24 h in liquid media containing 1% D-glucose, 1% yeast extract in PESI at room temperature with shaking (100 rpm). Cells were transferred from a drop of culture to a formvar coated grid, stained with 1% uranyl acetate, and examined and photographed with a JEOL 100-CX transmission electron microscope.

Metabolic properties of alginase-secreting bacteria. Oxidative versus fermentative capability was determined in agar stabs (Hugh and Leifson 1953) containing either glucose, fructose, mannitol, or alginate as carbon sources. Oxidase was determined by three procedures, including those described by Gordon and McLeod (1928), Kovacs (1956), and Stanier et al. (1966). In using the latter procedure, cells derived from 2-d cultures on 1% alginate, 2% agar medium were evaluated after treatment with toluene (Baumann et al. 1972).

DNA base composition. Procedures for the isolation of DNA were adapted from those described by Marmur (1961) and Kohne (1968), with the use of hot phenol (Massie and Zimm 1965). Specifically, cells that had been isolated from exponential phase cultures grown on alginate medium (0.1% alginate, PESI, with Ca and Mg concentrations reduced 10-fold) were harvested by centrifugation, suspended in 30 volumes of distilled water, centrifuged, and stored at −70 C. Cells (the following indicated quantities are based upon 1 g wet wt) were thawed and suspended in 10 ml of 0.1 *M* EDTA, 0.15 *M* NaCl, pH 8.0, upon passing through an 18-gauge syringe needle several times. Lysozyme (5 mg) was added and the mixture was incubated at 37 C for 1 h with gyratory shaking at 50 rpm. The digested cell suspension was brought to 2% sodium dodecylsulfate (SDS) with the addition of 25% SDS, and incubated for 10 min at 60 C with occasional swirling. After cooling to 55 C, an equal volume of phenol saturated with SSC at 55 C was added dropwise with swirling. After incubating for 10 min at 55 C, the mixture was cooled to 0 C and centrifuged in the cold in a swinging bucket rotor at 12,000 × *g* for 10 min to provide phase separation. Most (about 80%) of the upper aqueous phase was removed and reextracted with phenol as described above. The aqueous solution was made 1.0 *M* with respect to $NaClO_4$ with the dropwise addition of 5.0 *M* $NaClO_4$ and mixed with an equal volume of 1% 2-octanol in chloroform at 180 rpm in a gyratory shaker for 15 min at room temperature. The phases were separated upon centrifugation at 2,000 × *g*, 10 min, 25 C. The aqueous phase was removed, mixed with two volumes of cold (−20 C) 95% ethanol, and the DNA spooled on a glass rod. Infrequently only a small amount of DNA was recovered by spooling, in which case the alcoholic solution was allowed to stand overnight at −20 C, and the precipitated nucleic acids were collected by centrifugation at 2,000 × *g* for 10 min at 0 C.

The insoluble nucleic acid fraction was dissolved in 1 or 2 ml of 0.1 × SSC, and 0.1 volume of Ribonuclease A solution (0.5 mg/ml in water, heated at 85 C for 10 min to inactivate DNAse activity) were added. The solution was incubated at 37 C with gyratory shaking, 150 rpm for 2 h. The solution was made to 1 × SSC with the addition of 10 × SSC and extracted with an equal volume of phenol as described above. The resulting phenol phase was back extracted with an equal volume of 1 × SSC and the DNA precipitated with the addition of two volumes of cold ethanol to the combined aqueous phases. The DNA was collected by spooling and/or centrifugation, dissolved in 1 ml of 0.1 × SSC, and extracted briefly three times with three volumes of diethyl ether: 95% ethanol (3:1, v/v). The DNA concentration was estimated by measuring absorbance at 260 nm and using the approximating relationship of 0.045 mg/ml for an absorbance of 1.0 (1.00 cm light path, Hotchkiss 1957).

The buoyant densities of the DNA's were determined in CsCl density gradients, which were generated by centrifugation and stabilized as acrylamide gels as previously described (Preston and Boone 1973). After polymerization the gels were washed for 1 h with a continuous stream of distilled water. Riboflavin fluorescence was removed upon incubation in 15 ml 0.01 *M* NaOH, 50% formamide under 12,000 lux of cool white fluorescent light at 25 C with gyratory shaking, 50 rpm. After washing again with water, the gels were incubated in ethidium bromide staining solution (1 μg/ml ethidium bromide, 10 m*M* Tris-HC1 pH 8.0, 20 m*M* NaCl, 5 m*M* EDTA pH 7.7) in the dark at room temperature for several hours. Stained DNA bands were visualized with a Photodyne Model 3-4400 310 nm transilluminator (Berlin, WI) with excitation by four no. 3-4408 bulbs. Photographs were taken with Polaroid high speed type 57 Land film. The following DNA standards were used to calibrate the gradient: *Bdellovibrio stolpii*, 1.7007 g/cm^3; *Escherichia coli*, 1.710 g/cm^3; *Pseudomonas aeruginosa* 385, 1.727 g/cm^3; bacteriophage ϕ25, 1.742 g/cm^3. With the exception of *E. coli* DNA, the standards were kindly provided by Dr. Manley Mandel, M. D., Anderson Medical Center, Houston, TX. The distance of each band from the top of the gel divided by the distance of the ϕ25 band was plotted against the known density of each to provide a linear relationship from which the density of an unknown DNA could be accurately interpolated. An unknown DNA was run with as many internal standard DNA's as could be completely resolved. The GC mol% was estimated using the formula described by Shildkraut et al. (1962).

Assays for alginase activity. Alginase-secreting bacteria were identified by their ability to generate a clear zone around a growing colony on a Petri dish containing 1% alginate in PESI 2% agar medium. The contrast of the clearing zone could be accentuated by treating the plates with dilute HCl, allowing an estimation of the relative levels of activity secreted by each organism.

Alginate lyase was quantified by the spectrophotometric determination at 548 nm of chromogen formed upon reaction of thiobarbituric acid (TBA) with periodate-treated products (Preiss and Ashwell 1962). Endolytic activities were detected by capillary viscometry (McKie and Brandts 1982). Unless otherwise noted, reactions containing enzyme fraction, 0.1% sodium alginate, 0.05 *M* KCl, and 0.03 *M* sodium phosphate pH 8.0 were incubated at 19.2 C. Viscometric determinations were made every 10–30 min; 0.1 ml aliquots were removed periodically and assayed for lyase-generated products. For the viscometric assays, native alginate was centrifuged at 100,000 × *g*. Homopolymers of G and M were prepared by partial HCl hydrolysis of alginate and separated by differential acid precipitation (Haug et al. 1966). Fractions of defined size ranges were prepared by gel filtration on Sephadex G-50. ^{13}C NMR analysis indicated that poly M blocks contained less than 15% as guluronate moieties while the poly G blocks contained less than 5% as mannuronate.

For the partial purification of alginate lyase activities, bacteria were grown at 22 C in 1 liter of liquid alginate medium with rapid gyratory shaking. Cells were grown to late exponential phase, harvested by centrifugation, frozen in liquid nitrogen, and stored at −70 C as a source of intracellular alginase. The spent media were concentrated by tangential flow filtration using a Millipore Pellicon cassette system with a polysulfone membrane that allowed retention of molecules with molecular weights greater than 10 kd. The concentrated material was dialyzed against distilled, deionized water at 0 C to obtain the extracellular alginate lyase activities.

For intracellular enzymes, cells were thawed, suspended in four volumes of 0.10 *M* sodium phosphate buffer, pH 7.5, and disrupted with a French pressure cell at 16,000 lb/in^2. The brei was centrifuged at 10,000 × *g*, 0 C, 10 min, and the supernate was brought to 2% streptomycin sulfate and mixed at 0 C for 10 min. The precipitated nucleic acids were removed by centrifugation and protein fraction containing alginate lyase was precipitated with ammonium sulfate (to 65% saturation). The protein was collected by centrifugation at 10,000 × *g*, 0 C, 10 min; dissolved in pH 8 phosphate buffer; and dialyzed in the cold against this same buffer for analysis of substrate specificities. Extracellular fractions were dialyzed against the same buffer for comparative assays. Activities were measured at room temperature.

Results

Alginase-secreting bacteria associated with pelagic Sargassum *species.* As shown in Table 1, the bacteria that are associated with healthy tissues of the two pelagic species of *Sargassum,* i.e., *S. fluitans* and *S. natans*, are predominantly those capable of growing on alginate as the primary carbon source. More than 90% of the colonies growing on the alginate plates showed clearing zones indicative of the secretion of alginase activities. The relatively mild sonication used to obtain the sample in which the bacteria were enumerated removes most of the bacteria observed by scanning electron microscopy to be associated with the *Sargassum* tissues (data not shown).

TABLE 1. *Comparison of total and alginate-metabolizing bacteria associated with pelagic* Sargassum *species*[a]

	Young Tissue[b]		Old Tissue[c]	
Species	R Medium[d]	A Medium[e]	R Medium	A Medium
S. fluitans	2.9×10^7	1.8×10^7	5.5×10^7	5.2×10^7
S. natans	2.6×10^7	6.1×10^7	7.6×10^7	1.3×10^8

[a] Viable bacteria were determined after spreading on 15 ml of appropriate medium 0.1 ml of an aqueous wash obtained from mild sonication of tissue in sterile sea water. Incubation at room temperature for 48 h was followed by enumeration of colonies. Values are given as viable bacteria/g wet tissue.

[b] Young tissue is defined as the most rapidly growing tips of plants, including stems and leaves.

[c] Old tissue is defined as internal stems and leaves.

[d] R medium is rich medium containing 1% yeast extract, 1% glucose, 0.8% nutrient broth, and 2% agar in PES.

[e] A medium is alginate-based medium containing 1% sodium alginate and 2% agar in PESI.

The abilities of different isolates to secrete alginase activities is shown in Fig. 1. With the exception of the FM isolate, all were obtained from healthy tissues of recently collected *S. fluitans* or *S. natans*. The size of the cleared zone surrounding each colony allows an estimation of the secreting activity of each isolate, since

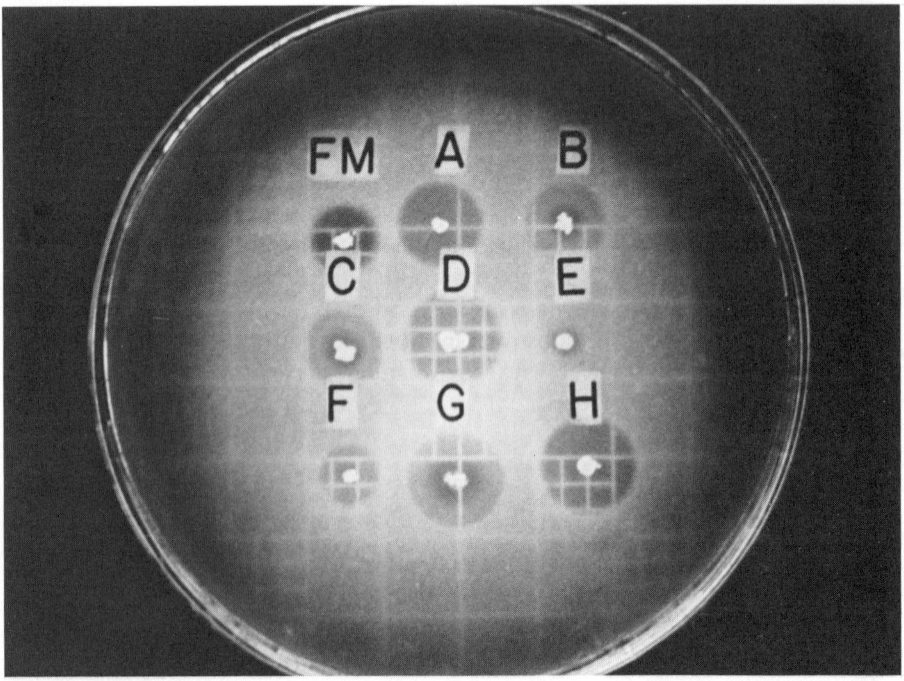

FIG. 1. Alginase secreting bacterial isolates from *S. fluitans* and *S. natans*. Exponential phase cultures growing in liquid medium containing 1% alginate, PESI, were spot inoculated on 1% alginate, PESI, 2% agar plates that were incubated for 48 h at room temperature and flooded with 0.1 *N* HCl.

each colony represents growth from a spotted inoculum obtained from an exponential phase liquid culture.

Negatively stained cells obtained from exponential cultures grown on a rich liquid medium were examined by transmission electron microscopy and established that all of the alginase-secreting isolates shown in Fig. 1 were polarly flagellated rods. They were also shown to be gram-negative, and thin sections indicated a cell wall structure characteristic of gram-negative bacteria (data not shown). Scanning electron microscopy of 1 to 2 d cultures grown on alginate agar allowed a morphological comparison as shown in Fig. 2. Two of the oxidative isolates, B and F, are thinner and/or longer than the fermentative isolates, A and G. However, the fermentative isolate D has dimensions nearly identical to F, and the oxidative isolate C is similar to fermentative isolate A.

The isopycnic banding of DNA isolated from each species is shown in Fig. 3. A plot of the relative positions of each band, calculated as $R_{\phi 25}$ values, versus the known densities of the standards provided a strictly linear relationship that allowed an accurate determination of the buoyant densities of the unknown DNA's. The following DNA densities (g/cm^3) were determined to an estimated accuracy of $\pm$ 0.0005 g/cm^3: A, 1.7045; B, 1.7045; C, 1.7040; D, 1.7060; F,

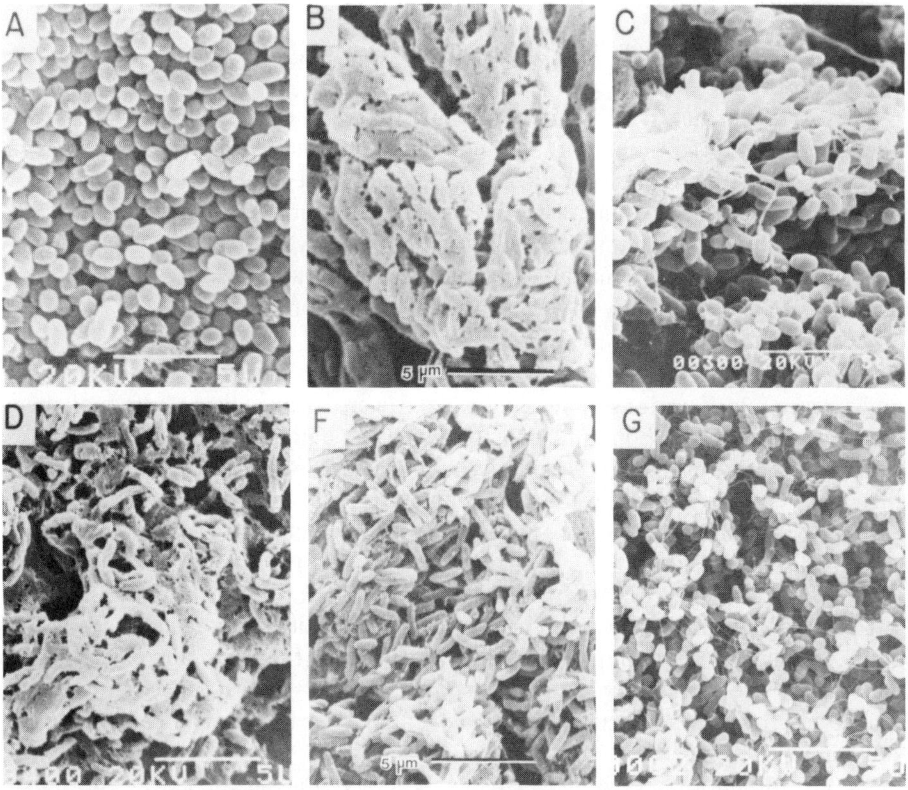

FIG. 2. Scanning electron microscopy of isolates obtained from *S. fluitans* and *S. natans*. Cultures were grown on 1% alginate, PESI, 2% agar for 48 h at room temperature, fixed, dried, and examined.

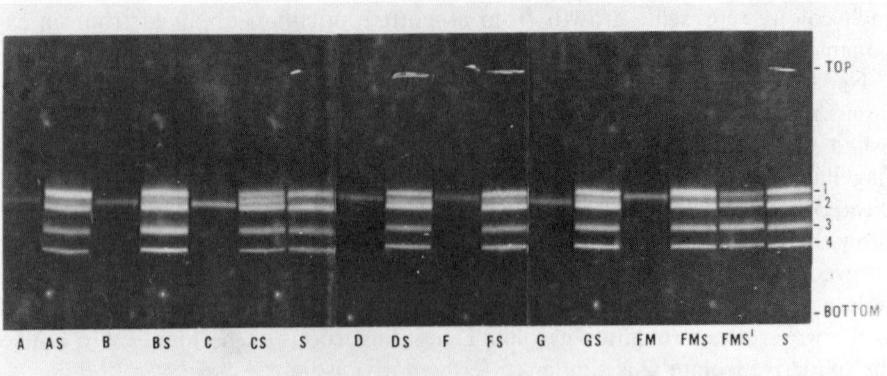

FIG. 3. Isopycnic centrifugation of DNA from alginase-secreting bacterial isolates. DNA bands were detected with ethidium bromide after immobilization in CsCl-acrylamide gelled gradients. Unknown DNA samples from isolates A, B, C, D, F, G, and FM were run with and without a mixture of internal standard DNA's. S refers to a mixture of standard DNA's that gave the following numbered bands: (1) *Bdellovibrio stolpii,* 1.7007 g/cm^3, 0.93 μg; (2) *E. coli,* 1.710 g/cm^3, 1.45 μg; (3) *Pseudomonas aeruginosa,* 1.727 g/cm^3, 1.95 μg; and (4) ϕ_{25}, 12.742 g/cm^3, 0.45 μg. S′ refers to the same mixture lacking the *B. stolpii* DNA. AS, BS, etc. indicate runs including the unknown and standard DNA's.

1.7055; G, 1.7065; FM, 1.7035. These values were used to determine the GC content of the DNA.

The morphological and metabolic properties of the seven isolates described above are compared along with GC contents and secreted alginase activities in Table 2. All of the fermentative isolates are oxidase negative, even when assayed after treatment with toluene. The GC mol% values are very close for all of the isolates, both the fermentative as well as the oxidative organisms. The fermentative isolates secrete as much if not more alginase activity than the oxidative organisms, as estimated by the sizes of the clearing zones.

Comparisons of intracellular and extracellular alginate lyase activities. The intra- and extracellular fractions of selected isolates were assayed using homopolymers of L-guluronate and D-mannuronate as well as native alginate as substrates (Table 3). With this assay the oxidative organisms (FM, B, and C) show most of their poly G lyase activity in the extracellular fraction, while the opposite is true for the fermentative isolate A. The levels of intracellular activity (units/g wet wt of cells) are significantly greater for FM and A than for B and C. The levels of extracellular activity toward poly M are comparable for all the organisms tested.

Upon normalizing the activities to the lyase rate with native alginate, the oxidative organisms show similar substrate specificities; the intra- and extracellular fractions of the fermentative A isolate show different substrate specificities, with the latter fraction essentially devoid of activity with poly L-guluronate as substrate.

As an estimate of the endo- versus exolytic activities in the extra- and intracellular fractions, the increase with time of incubation in the specific fluidity, ϕ_{sp}, and the formation of periodate-generated TBA reactive products were plotted against each other (Fig. 4). The ratio of endo- to exolytic activity is proportional

TABLE 2. *Morphological and biochemical properties of alginase-secreting bacteria associated with* Sargassum *species*

Isolate	Morphology[a]	Oxidative[b]/ Fermentative	Clearing[c] Zone Diameter (mm)	Oxidase[d] Reaction	% GC
FM	short rod	oxidative	8.4	+	44.3
A	0.9–1.1 × 1.6–2.2	fermentative	10.8	−	45.4
B	0.5–0.6 × 2.1–2.6	oxidative	9.6	+	45.4
C	0.7–0.9 × 1.6–2.0	oxidative	9.6	+	44.9
D	0.4–0.6 × 1.4–2.1	fermentative	12.6	−	46.9
F	0.46 × 1.2–2.0	oxidative	7.2	+	46.4
G	0.5 × 1.1–1.3	fermentative	12.0	±	47.4

[a] Morphologies and dimensions determined by measurements from scanning electron micrographs with the exception of isolate FM, which was analyzed with the optical microscope. All isolates were gram-negative.
[b] Oxidative/fermentative evaluation was based upon standard liquid media.
[c] Clearing zones indicative of alginate lyase secretion were measured on 2% agar plates containing 1.0% sodium alginate in Provasoli's enriched medium after incubation for 4 d at room temperature.
[d] Exponential phase cultures were assayed in the three oxidase tests described in the methods section.

TABLE 3. *Substrate specificities of bacterial lyases at pH 8.0*

	Intracellular[a]			Extracellular[b]		
Isolate[c]	Poly G	Poly M	Alginate	Poly G	Poly M	Alginate
FM	0.190	0.527	0.235	0.300	0.569	0.647
A	0.689	1.140	0.764	0.000	0.370	0.225
B	0.026	0.091	0.032	0.276	0.287	0.367
C	0.015	0.042	0.018	0.423	0.373	0.648
Isolate[d]						
FM	0.80	2.24	1.00	0.46	0.89	1.00
A	0.90	1.46	1.00	0.00	1.64	1.00
B	0.81	2.84	1.00	0.75	0.78	1.00
C	0.83	2.33	1.00	0.65	0.58	1.00

[a] Intracellular activities were obtained after disrupting bacteria in a French pressure cell followed by partial purification to remove anionic polymers. This fraction may include activities bound to the cell surface as well as those that are truly intracellular.
[b] Extracellular activities were measured in the medium after concentration with dialysis but without further purification.
[c] Values in the upper panel are μmols of unsaturated product formed per min per g (wet wt) of cells. All activities presented were calculated after subtracting activities observed in the absence of added substrate.
[d] Values in the lower panel are calculated as ratios of lyase activity dependent upon added substrate divided by the activity dependent upon native alginate.

to the slope. For both organisms this ratio is greatest for the extracellular enzymes. The oxidative FM organism shows a particularly striking partitioning of endolytic and exolytic activities in the extracellular and intracellular fractions, respectively.

Discussion

The identification of both fermentative as well as oxidative organisms associated with *Sargassum fluitans* was at first surprising, since the bacteria that are associated with *Fucus* have been designated as only oxidative (Quatrano and Caldwell 1978). On the other hand, both fermentative and oxidative species of gram-negative polar flagellates have been isolated from seawater as well as algal tissues, and have been designated as *Alginovibrio aquatilis* and *Alginovibrio norvegicus,* respectively (Meland 1962). *Alginovibrio* has not received general acceptance as a genus, and the marine genera that are characterized as gram-negative polarly flagellated rods and include alginase-producing isolates are *Alteromonas* and *Pseudomonas* for the oxidative and *Benekia* and *Photobacterium* for the fermentative organisms (Reichelt and Baumann 1973; Baumann et al. 1972; Sutherland and Keen 1981). Of these, only members of the genus *Benekia,* i.e., *B. harveyi,* (Reichelt and Baumann 1973) have been shown to secrete alginase; an oxidative alginase-secreting organism with morphological properties consistent with *Alteromonas* and *Pseudomonas* had a GC content of 38 mol%, thus precluding its inclusion with these established genera (Quatrano and Caldwell 1978). Since few of the marine organisms that have been used to provide a source of extensively purified alginate lyase activities have been characterized to an extent that would allow their designation as something more than a marine pseudomonad (Preiss and Ashwell 1962; Kashiwabara et al. 1969; Davidson et al. 1976), with the exception noted above (Quatrano and Caldwell 1978), it is difficult to assess the genetic potential for the synthesis of alginase activities.

The properties of the oxidative species FM, B, C, and F have allowed these to be included in the genus *Alteromonas,* which includes a range of GC content from 43.2 to 48.0 mol% (Baumann et al. 1972). The morphological and metabolic properties of the fermentative isolates A, D, and G are consistent with those of the genus *Photobacterium*. However, the GC content of these organisms is 45.4, 46.9, and 47.4 mol%, respectively, which compares with a range of 39.8 to 42.9 mol% for previously characterized *Photobacterium* species (Reichelt and Baumann 1973). It is, therefore, premature to assign these isolates to a designated genus.

The substrate specificities of the intracellular lyase activities for four of the isolates examined show a preference for poly M over both poly G and native alginate, and a preference for poly M over poly G was observed for the "bound enzyme" fraction for all of the isolates obtained by Doubet and Quatrano (1982) from *Fucus.* The extracellular activities of the oxidative isolates showed significant lyase activity toward poly G, although in contrast to some of the *Fucus* isolates, these activities were generally less than those found with poly M as substrate. The extracellular fraction from the fermentative isolate A is unique; its difference from all other fractions examined here is the complete absence of activity with poly G and yet strong activity with poly M and native alginate. This activity may be similar to those isolated as extracellular enzymes from different soil bacteria (Kooiman 1954; Davidson et al. 1977; Hansen et al. 1984) and represents a ready source of specific activity for the characterization and/or modification of alginate.

The tendency of the extracellular fractions to be richer in endolytic activity than the intracellular activities from the same organism (Fig. 4) is not surprising, since it would be expected that smaller oligomers generated through the endolytic

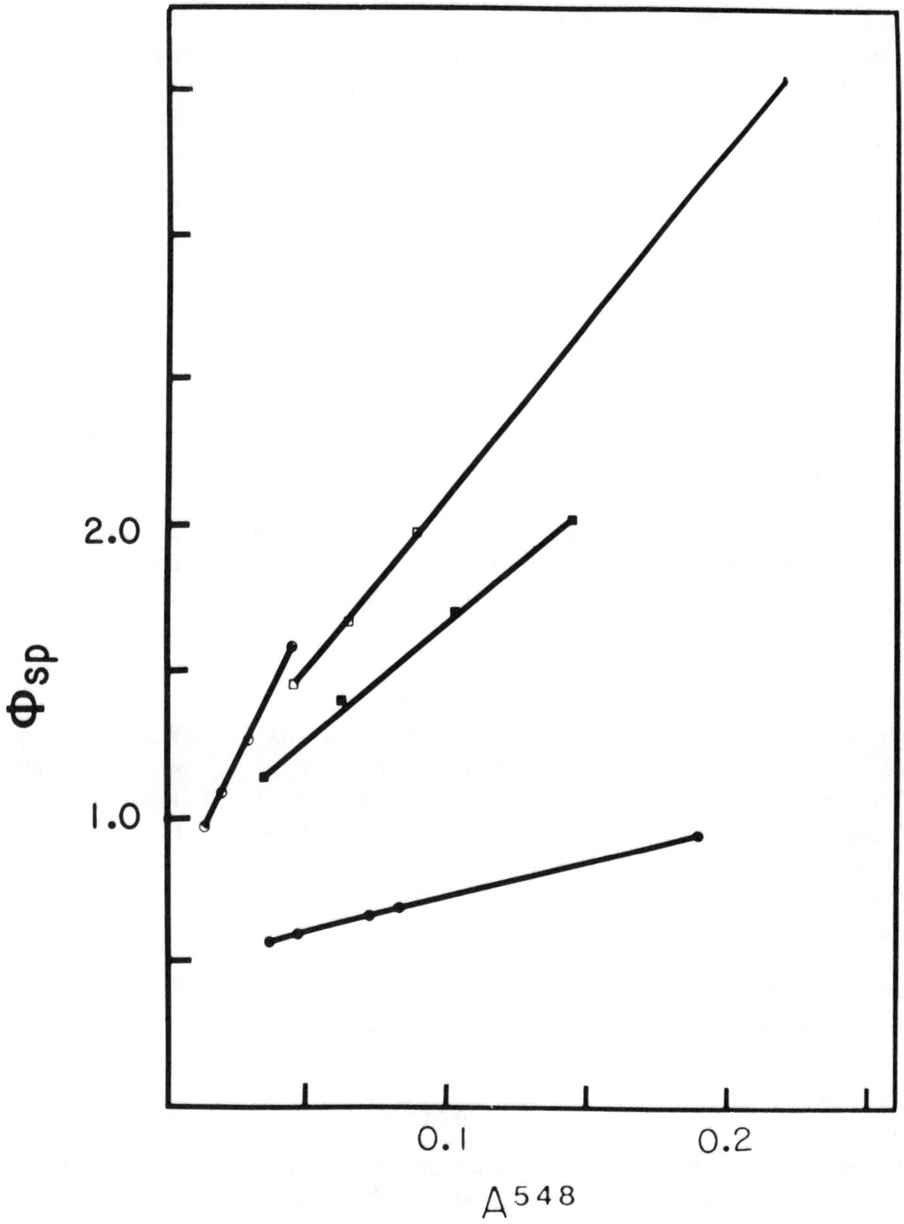

FIG. 4. Comparison of extracellular and intracellular fractions with respect to endolytic and exolytic activities. Specific fluidity (ϕ_{sp}) is plotted against lyase activity as determined by A^{548} detection of TBA reactive compounds generated upon periodate oxidation of enzyme-treated alginate. Open symbols refer to extracellular fractions; closed symbols refer to intracellular fractions. Circles refer to FM isolate; squares refer to A isolate.

cleavage of native alginate would be taken up more readily than the intact alginate. The extracellular fraction from isolate A, in containing an endolytic activity toward native alginate and a specificity for D-mannuronate moieties, is similar to that observed for one of two alginase activities purified from the hepatopancreas of the Pacific abalones (Nakada and Sweeney 1967) as well as the extracellular enzymes characterized from phage-infected *Azotobacter vinelandii* (Davidson et al. 1977) and *Bacillus circulans* (Hansen et al. 1984).

The actively growing tissues of pelagic *Sargassum* species are thus a rich source of bacteria with a varied genetic potential for the synthesis of alginase enzymes. As in the case of those isolated from *Fucus,* these bacteria may be expected to provide a battery of specific enzymes to dissect the cell wall and the intercellular matrices of different species of the Phaeophyta and are now being exploited for the generation of viable protoplasts of *Sargassum* species to permit approaches to genetic manipulation through somatic cell hybridization.

The large number of alginase-secreting bacteria associated with actively growing tissue of *Sargassum fluitans* and *Sargassum natans* suggests these may play an important ecological role in the turnover of the pelagic *Sargassum* species. Since *Sargassum* species are not grazed to a significant degree, at least not relative to other species of marine algae, including other Phaeophyta species (Fine 1970; Ryland 1974), it is likely that these bacteria represent an important agent responsible for limiting the proliferation of pelagic *Sargassum* species in the ocean. The fermentative isolates that secrete alginase are now being evaluated for their potential in enhancing the bioconversion of Phaeophyta as a source of biomass to methane.

Acknowledgments

We are grateful for the generous gift of DNA standards from Dr. Manley Mandel, M. D., Anderson Medical Center, and for the ^{13}C NMR analysis of poly-L-guluronate and poly D-mannuronate preparations by Dr. John E. Gander, Department of Microbiology and Cell Science, University of Florida. H. J. Jiminez provided the FM organism that he isolated from decaying *Sargassum* tissues. We thank Ms. Tonie Henry for assisting in the preparation of this manuscript. This work was supported by the Gas Research Institute and the Institute of Food and Agricultural Sciences, University of Florida, as CRIS project MCS-02183-BI and represents Journal Series No. 5967, University of Florida Institute of Food and Agricultural Sciences Experiment Station.

Literature Cited

Baumann, L., P. Baumann, M. Mandel, and R. D. Allen. 1972. Taxonomy of aerobic marine eubacteria. *J. Bacteriol.* 110:402–429.

Boyd, J., and J. R. Turvey. 1977. Isolation of a poly-α-L-guluronate lyase from *Klebsiella aerogenes. Carbohydr. Res.* 57:163–171.

Davidson, I. W., I. W. Sutherland, and C. J. Lawson. 1976. Purification of an alginate lyase from a marine bacterium. *Biochem. J.* 159:707–713.

Davidson, I. W., C. J. Lawson, and I. W. Sutherland. 1977. An alginate lyase from *Azotobacter vinelandii* phage. *J. Gen. Microbiol.* 98:223–229.

Doubet, R. S., and R. S. Quatrano. 1982. Isolation of marine bacteria capable of producing specific lyases for alginate degradation. *Appl. Environ. Microbiol.* 44:754–756.

Doubet, R.S., and R. S. Quatrano. 1984. Properties of alginate lyases from marine bacteria. *Appl. Environ. Microbiol.* 47:699–703.

Doudoroff, M., and N. J. Palleroni. 1974. *Pseudomonas.* Pages 217–243 *in* R. E. Buchanan and M. E. Gibbons, eds., *Bergey's Manual of Determinative Bacteriology, 8th ed.* Williams & Wilkins Co., Baltimore, MD.

Elyakova, L. A., and V. V. Favorov. 1974. Isolation and certain properties of alginate lyase VI from the mollusk *Littorina* sp. *Biochim. Biophys. Acta* 358:341–354.

Favorov, V. V., and V. E. Vaskovsky. 1971. Alginases of marine invertebrates. *Comp. Biochem. Physiol.* 38:689–696.

Favorov, V. V., E. I. Vozhova, V. A. Denisenko, and L. A. Elyakova. 1979. A study of the reaction catalyzed by alginate lyase VI from the sea mollusk, *Littorina* sp. *Biochim. Biophys. Acta* 569:259–266.

Fine, M. L. 1970. Faunal variation on pelagic *Sargassum. Mar. Biol.* 7:112–122.

Gordon, J., and J. W. McLeod. 1928. The practical application of the direct oxidase reaction in bacteriology. *J. Pathol. Bacteriol.* 31:185–190.

Haug, A., B. Larsen, and O. Smidsrod. 1966. A study of the constitution of alginic acid by partial acid hydrolysis. *Acta Chem. Scand.* 20:183–190.

Hansen, J. B., R. S. Doubet, and J. Ram. 1984. Alginase enzyme production by *Bacillus circulans. Appl. Environ. Microbiol.* 47:704–709.

Hotchkiss, R. D. 1957. Methods for characterization of nucleic acid. Pages 708–715 *in* S. P. Colowick and N. O. Kaplan, eds., *Methods in Enzymology.* Academic Press, New York.

Hugh, R., and E. Leifson. 1953. The taxonomic significance of fermentative versus oxidative metabolism of carbohydrates by various gram negative bacteria. *J. Bacteriol.* 66:24–26.

Kashiwabara, Y., H. Suzuki, and K. Nisizawa. 1969. Alginate lyases of Pseudomonads. *J. Biochem.* 66:503–512.

Kohne, D. E. 1968. Isolation and characterization of bacterial ribosomal RNA cistrons. *Biophys. J.* 8:1104–1118.

Kooiman, P. 1954. Enzymic hydrolysis of alginic acid. *Biochim. Biophys. Acta* 13:338–340.

Kovacs, N. 1956. Identification of *Pseudomonas pyocyanea* by the oxidase reaction. *Nature* 178:703.

Marmur, J. 1961. A procedure for the isolation of deoxyribonucleic acid from micro-organisms. *J. Mol. Biol.* 3:208–218.

Massie, H. R., and B. H. Zimm. 1965. The use of hot phenol in preparing DNA. *Proc. Natl. Acad. Sci. U.S.A.* 54:1641–1643.

Meland, S. M. 1962. Marine alginate-decomposing bacteria from North Norway. *Nytt Mag. Bot.* 10:53–80.

McKie, J. E., and J. F. Brandts. 1982. High precision capillary viscometry. *Meth. Enzymol.* 26:257–288.

Muramatsu, T., S. Hirose, and M. Katayose. 1977. Isolation and properties of alginate lyase from the mid-gut gland of wreath shell *Turbo cornutus. Agric. Biol. Chem.* 41:1939–1946.

Muramatsu, T., and M. Katayose. 1979. A mode of action of alginate lyase from *Turbo cornutus* on sodium alginate. *Agric. Biol. Chem.* 43:287–291.

Muramatsu, T., and K. Egawa. 1982. Chemical composition of alginate lyases from the mid-gut gland of *Turbo cornutus* and chemical modification of cys, trp, and lys residues. *Agric. Biol. Chem.* 46:883–889.

Muramatsu, T. 1984. Additional evidence for substrate specificities of alginate lyase isozymes from *Turbo cornutus. Agric. Biol. Chem.* 48:811–813.

Nakada, H. I., and P. C. Sweeny. 1967. Alginic acid degradation by eliminases from Abalone hepatopancreas. *J. Biol. Chem.* 242:845–851.

Polne-Fuller, M., N. Saga, and A. Gibor. 1984. Algal cell, callus, and tissue cultures and selection of algal strains. Proceedings of the Algal Biomass Symposium, Golden, CO, 1984 (in press).

Preiss, J., and G. Ashwell. 1962. Alginic acid metabolism in bacteria. I. Enzymatic formation of unsaturated oligosaccharides and 4-deoxy-L-erythro-5-hexoseulose uronic acid. *J. Biol. Chem.* 237:309–316.

Preston, J. F., and D. R. Boone. 1973. Analytical determination of the buoyant density of DNA in acrylamide gels after preparative CsCl gradient centrifugation. *FEBS Lett.* 37:321–324.

Provasoli, L. 1968. Media and prospects for cultivation of marine algae. Pages 63–75 *in* A. Watanabe and A. Hattori, eds., *Cultures and Collections of Algae.* Japan Society Plant Physiology, Tokyo.

Quatrano, R. S., and B. A. Caldwell. 1978. Isolation of a unique marine bacterium capable of growth on a wide variety of polysaccharides from macroalgae. *Appl. Environ. Microbiol.* 36:979–981.

Quantrano, R. S., J. J. Peterman, and R. S. Doubet. 1980. Deposition of polysaccharides in cell walls of early embryos of *Fucus.* Pages 136–143 *in* I. A. Abbott, M. S. Foster, and L. F. Eklund, eds., *Pacific Seaweed Aquaculture: Proceedings of a Symposium on Useful Algae.* California Sea Grant College Program, Institute of Marine Resources, University of California, La Jolla.

Reichelt, J. L., and P. Baumann. 1973. Taxonomy of the marine, luminous bacteria. *Arch. Mikrobiol.* 94:283–330.

Ryland, J. S. 1974. Observations on some epibionts of gulf-weed, *Sargassum natans* (L.) Meyen. *J. Exp. Mar. Biol. Ecol.* 14:17–25.

Schildkraut, C. L., J. Marmur, and P. Doty. 1962. Determination of the base composition of deoxyribonucleic acid from its buoyant density in CsCl. *J. Mol. Biol.* 4:430–443.

Stanier, R. Y., N. J. Palleroni, and M. Doudoroff. 1966. The aerobic Pseudomonads: A taxonomic study. *J. Gen. Microbiol.* 43:159–271.

Sutherland, I. W., and G. A. Keen. 1981. Alginases from *Benekia pelagia* and *Pseudomonas* spp. *J. Appl. Biochem.* 3:48–57.

Thjotta, Th., and E. Kass. 1945. A study of alginic acid destroying bacteria. Videnskaps-Akademi I Oslo 1. Mat. Naturv, Klasse 5:3–20.

CHAPTER **61**

Production of Gentisate Intermediates Using Immobilized *Salmonella typhimurium*

FREDERICK E. GOETZ* AND LIN SUN-CHIANG**

Biology Department, Mankato State University, Mankato, Minnesota 56001

Salmonella typhimurium metabolizes *m*-hydroxybenzoate (MHBA) and gentisate (GA) via the Gentisic Acid pathway. Cells were grown in batch culture with MHBA as the sole carbon source and were immobilized in carrageenan. Cells immobilized in 2.5% carrageenan containing 4 m*M* calcium ion at a ratio of bacterial protein to carrageenan of 1.25/1.0 (w/w) showed the best retention of enzyme activity. Optimal conditions for operation of a bioreactor using these cells were pH 7.4, 35 C, at an MHBA concentration of 5 m*M* and a flow rate of 0.25 ml/min. A mutant with a mutation in the gentisate dioxygenase gene was used to demonstrate the feasibility of producing GA from MHBA using immobilized mutant cells. In 60 h 16% of the MHBA was converted to GA.

INTRODUCTION

Bacteria have an almost unlimited capacity to metabolize a wide variety of organic compounds (Dagley 1971). The use of this metabolic diversity is one of the goals of biotechnology. Of particular interest are the oxygenase enzymes that are used by microorganisms to metabolize aromatic and aliphatic compounds. These enzymes and the metabolic pathways of which they are a part have considerable potential for use in detoxifying pollutants (Munnecke et al. 1982) and in the manufacture of a variety of chemicals (Demain 1977). Problems associated with the preparation and use of purified enzymes for these processes can often be bypassed by using intact immobilized bacteria. Applications for and the advantages of using immobilized microorganisms have been reviewed by Fukui and Tanaka (1982).

Unfortunately, the genetics and biochemistry of many of the organisms, e.g., Bacillaceae and Pseudomonadaceae, that are best suited for these applications are poorly developed. In contrast, the molecular biology of the Enterobacteriaceae is well developed, but the organisms have a limited capacity to catabolize aromatic compounds. However, we have recently demonstrated (manuscript in preparation) that *Salmonella typhimurium* uses the Gentisate pathway, Fig. 1, to catabolize *m*-hydroxybenzoate (MHBA) and gentisate (GA).

We have used this organism to investigate the effects of immobilization in carrageenan on the activity of the mono- and dioxygenases in intact cells. A bioreactor

*Corresponding author.

**Current address: Department of Biochemistry and Biophysics, University of Houston, Houston, Texas 77004.

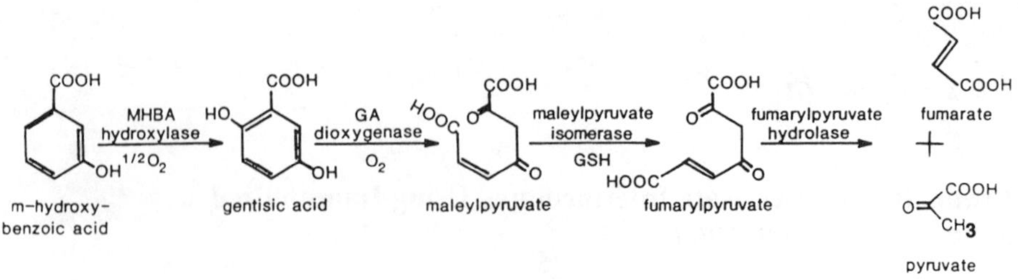

FIG. 1. Reaction sequence for the catabolism of MHBA and GA via the *gentisate pathway*.

using these immobilized cells was then used to determine the conditions necessary to obtain the maximum utilization of MHBA. A mutant blocked at the dioxygenase was immobilized and used to investigate the feasibility of producing GA from MHBA using the bioreactor.

Materials and Methods

Organisms, growth, and harvesting. Salmonella typhimurium strain TA 1530 (Ames et al. 1975) was used to optimize the immobilization conditions and bioreactor operation. Mutants unable to grow on *m*-hydroxybenzoic acid (MHBA) or gentisic acid (GA) were selected on MHBA/TTC plates (Bouchner and Savageau 1977) following mutagenesis of TA 1530 with ethylmethane sulfonate (Miller 1972). Mutants with a mutation in the gentisic acid dioxygenase were isolated by growing the mutants in minimal media with succinate and MHBA. These cells were then screened for monooxygenase and dioxygenase activity using an oxygen electrode. One such mutant EMS-11 has an active MHBA oxygenase and an inactive GA dioxygenase. Bacteria were grown in minimal media that contained, per liter: 3.9 g KH_2PO_4, 6.5 g K_2HPO_4, 3.5 g $NaNH_4HPO_4 \cdot H_2O$, 0.2 g $MgSO_4 \cdot 7H_2O$, and 40 mg histidine and 1 mg biotin. MHBA was used at a final concentration of 0.05%. EMS-11 was grown in the same medium with 0.05% succinate as the carbon source. Cells were grown on a rotary shaker at 37 C in one liter batch cultures for 24 h. High levels of enzymes were obtained by adding MHBA to a final concentration of 0.05% to the parent and EMS-11 2 h prior to harvesting. The cells were harvested by centrifugation and washed twice in Tris buffer (0.05 *M*, pH 7.4, 5 m*M* $MgSO_4$) and resuspended in the same buffer.

Immobilization. Cells were mixed with carrageenan (Gelcarin CIC, FMC Corporation) dissolved in Tris buffer containing $CaCl_2$ (see Results) and maintained at 40 C. This suspension was then added dropwise using a 1-ml tuberculin syringe (needle removed) to a beaker of ice-cold 0.3 *M* KCl in Tris buffer. The solution formed beads ca.3 mm in diameter on contact with the cold KCl solution. After 2–3 h the beads were harvested by filtration and stored at 5 C in Tris buffer.

Analytical methods. Protein was determined by the Lowry method (Lowry et al. 1951). Monooxygenase and dioxygenase activities were measured with an oxygen electrode (Yellow Springs Instrument Company YSI Model 53) following the addition of MHBA or GA to the probe chamber to a final concentration of 0.03% (w/v). Immobilized cells were incubated for 12–24 h with aeration in Tris buffer to respire endogenous pools of MHBA and/or GA prior to measuring oxygenase activity. MHBA and GA concentrations were measured using HPLC. The stationary phase was Partisil scx 10 μ and the mobile phase 0.075 M KH_2PO_4 pH 3.5. The flow rate was 1 ml/min and MHBA and GA were detected with a UV monitor set at 256 nm.

Bioreactor. The reactor was similar to that described by Margaritis and Rowe (1983). However, the feed solution was continuously aerated with sterile air so that the concentration of oxygen would not limit the activity of the oxygenases. In addition, no screen was used in the column and no gas-liquid separator was used. The column measured 20 cm × 1 cm and effluent could be recycled through the column. Substrate was dissolved in Tris buffer. When EMS-11 was used the buffer also contained succinate, histidine, and biotin.

RESULTS

Optimization of immobilization conditions. The effects of varying the percent carrageenan, the ratio of protein to carrageenan and the concentration of calcium (as calcium chloride) on the retention of oxygenase activity by the immobilized cells are shown in Fig. 2. These experiments indicated that the maximum amount of both monooxygenase and dioxygenase activity was retained in the cells immobilized in 2.5% carrageenan containing 4 mM Ca^{+2} at a ratio of bacteria (total protein) to carrageenan of 1.25/1.00. Beads were also surface hardened by treatment with either 1% 1,6-hexanediamine or 1% glutaraldehyde. These treatments reduced the monooxygenase activity 30–70% and 100%, respectively. However, without surface hardening the beads did not maintain their integrity in the bioreactor. To overcome this and extend the lifetime of the beads the operating buffer was made up with 0.3M KCl.

Optimization of bioreactor operating conditions. Figs. 3 and 4 show the effects of varying the MHBA concentration, flow rate, temperature, and pH on the operation of the reactor. These experiments were done using cells immobilized under the optimal conditions determined in the previous section. The results indicated that the optimal conditions for operation of the column were 35 C, pH 7.2, a flow rate of 0.25 ml/min, and a MHBA concentration of 5 mM. When operated under these conditions the monooxygenase had a half-life of 7 d as opposed to a half-life of 2.5 d in free cells.

Synthesis of GA. EMS-11 was immobilized as described previously and the bioreactor operated using the conditions determined in the previous section. MHBA was recycled through the column. The concentration of MHBA and GA were determined at regular intervals using HPLC. The results are shown in Table 1. After 60 h 16% of the MHBA had been converted to GA.

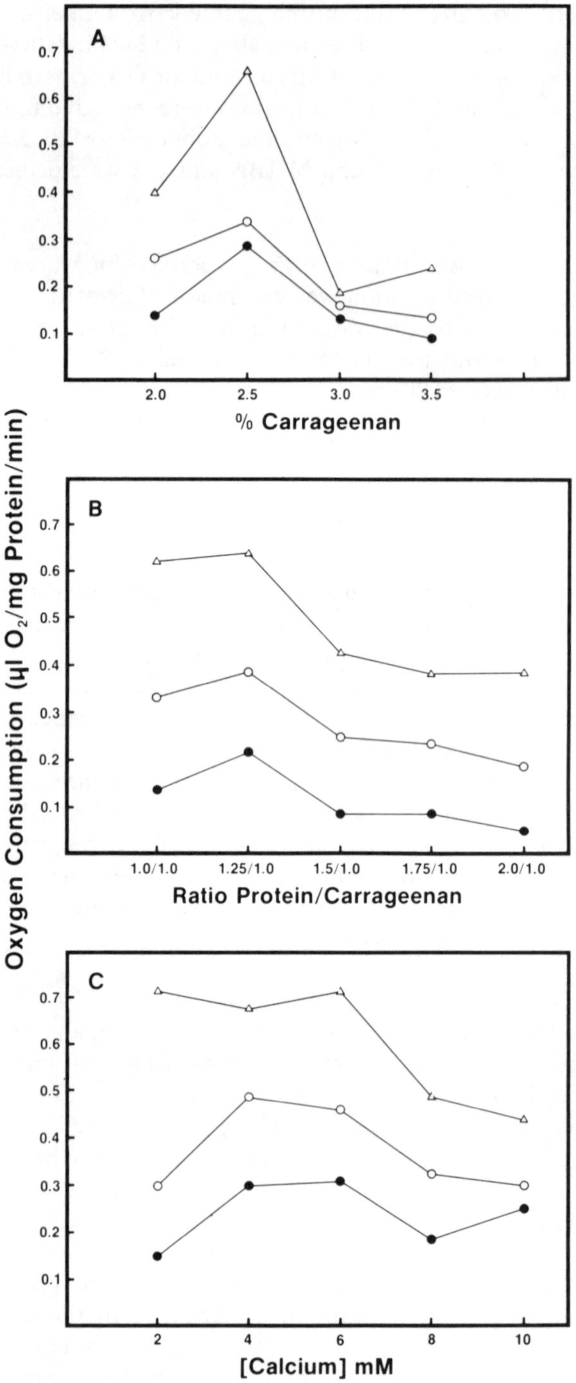

FIG. 2. Oxygen consumption of immobilized *S. typhimurium* with △ gentisic acid, ○ *m*-hydroxybenzoic acid, and ● no substrate as a function of (A) carrageenan, (B) ratio of total bacterial protein to carrageenan, and (C) calcium (as $CaCl_2$).

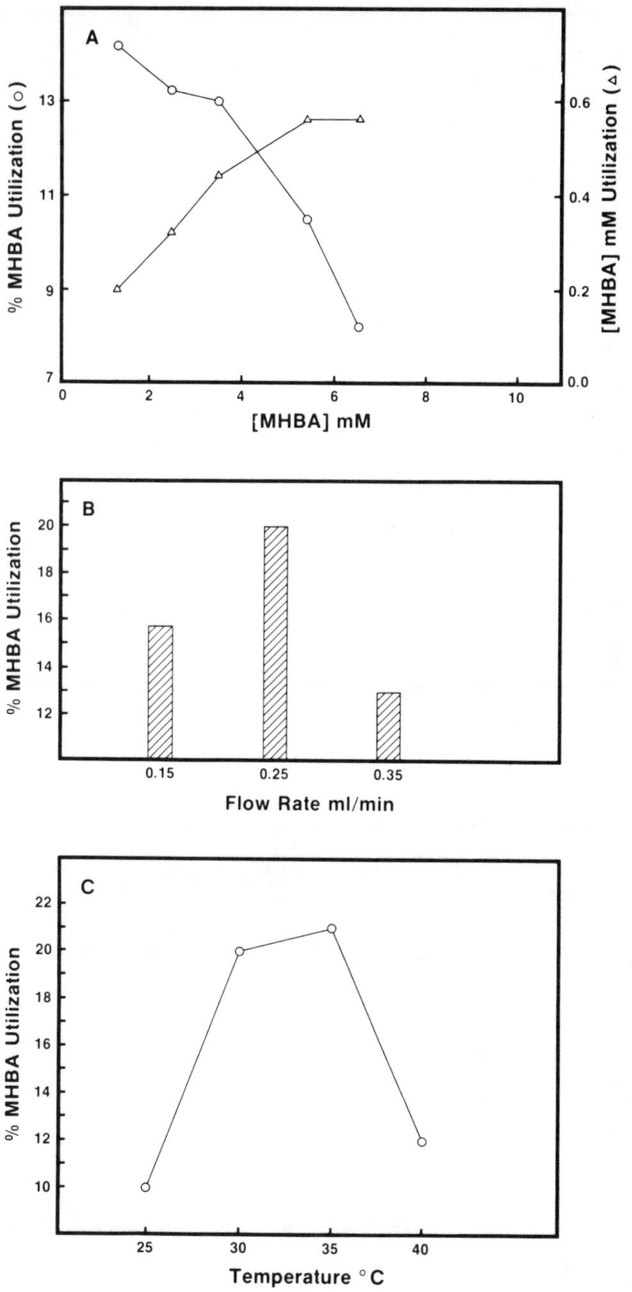

FIG. 3. *m*-Hydroxybenzoic acid utilization as a function of (A) MHBA concentration, (B) flow rate, and (C) temperature.

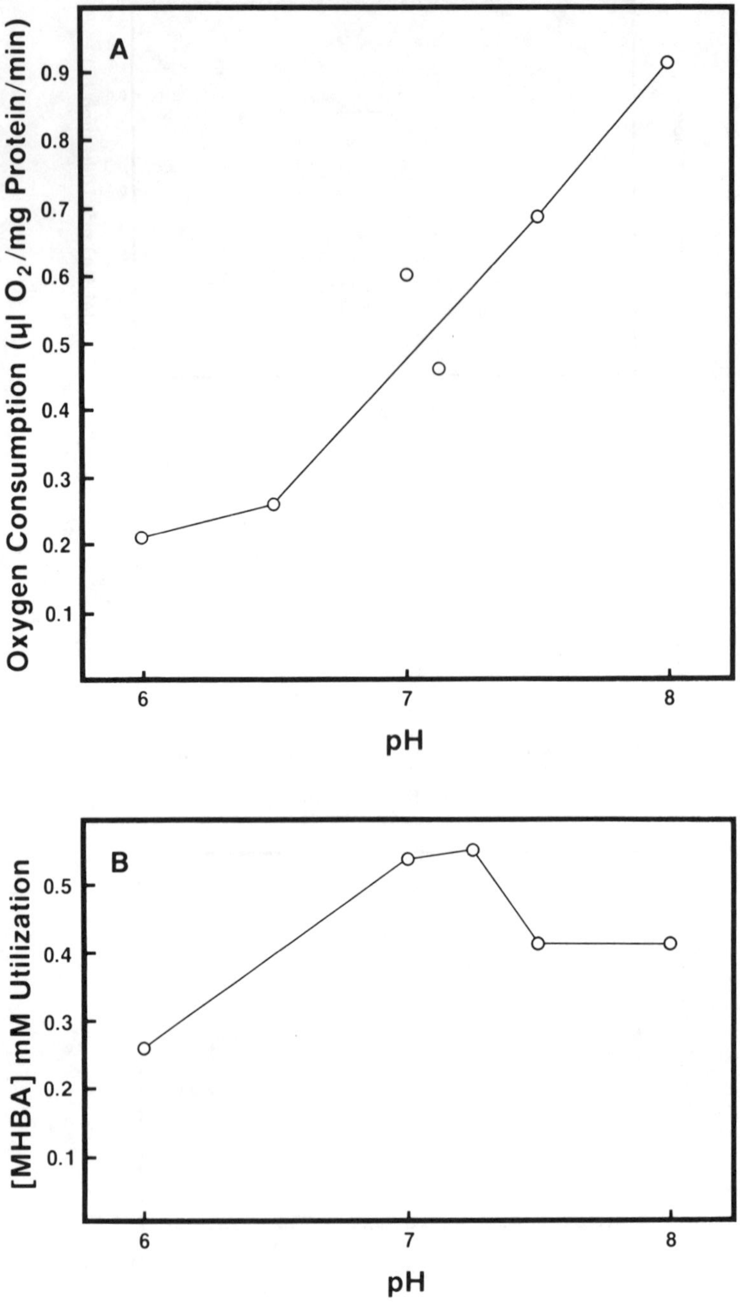

FIG. 4. Effect of pH on the utilization of MHBA in (A) free cells measured by oxygen consumption and (B) immobilized cells determined by using HPLC to measure the concentration of MHBA.

Discussion

The results presented here demonstrate that the immobilization conditions had a significant effect on the activity of the enzymes of the Gentisate pathway. In particular these enzymes were inactivated by cross-linking agents commonly used to harden carrageenans. However, by adding potassium chloride to the running buffer the integrity of the beads was retained without inactivating the enzymes. The results (Fig. 2C) also showed that 2–4 m*M* $CaCl_2$ enhanced the activity of the monooxygenase. If this enzyme is membrane bound this effect might be due to perturbations of the membrane by Ca^{+2} that made the enzyme more accessible to the substrate. Increased loading of the gel with either carrageenan or bacteria resulted in less activity (Figs. 2A and 2B). Since increased loading probably results in decreased porosity of the matrix, this would make cells in the inner part of the beads less accessible to the substrate. The results shown in Fig. 1A also suggest that carrageenan up to a concentration of 2.5% enhances the activity of the oxygenases.

The operation of the bioreactor was also sensitive to a number of parameters (Figs. 3 and 4). All of the parameters investigated exhibited optimum values. However, temperatures greater than 40 C and a pH less than 6 or greater than 8 resulted in the irreversible inactivation of the monooxygenase. The decreased utilization of MHBA at a low flow rate (0.15 ml/min) was unexpected. At high flow rates this would be expected because of shorter residence time of the substrate in the column. However, the immobilized cells continued to use oxygen even in the absence of substrate (Fig. 2). Thus, because of this intrinsic oxygen consumption, oxygen may become the limiting substrate at low flow rates. These results demonstrated that immobilized intact cells can be used to completely oxidize either MHBA or GA. This would be very difficult and expensive (if not impossible) with purified enzymes. In addition we have shown (Table 1) that a mutant can be used to synthesize GA from MHBA. Methods to enhance this process are being investigated.

TABLE 1. *EMS-11 catalyzed synthesis of GA from MHBA*

Time (hours)	Concentration[a] (m*M*)	
	MHBA	GA
0	5.19	0
12	4.99	0.20
24	4.88	0.31
36	4.65	0.54
48	4.53	0.66
60	4.41	0.78

[a] Concentrations determined using HPLC.

Acknowledgments

We would like to thank J. T. Whitt of FMC Corporation, Marine Colloids Division, for providing us with a sample of Gelcarin CIC carrageenan. This research was supported by a grant to FEG and a graduate assistantship to LSC from Mankato State University.

Literature Cited

Ames, B. N., J. McCann, and E. Yamasaki. 1975. Methods for detecting carcinogens and mutagens with the Salmonella/mammalian-microsome mutagenicity test. *Mutat. Res.* 31:347–364.

Bouchner, B. R., and M. A. Savageau. 1977. Generalized indicator plate for genetic, metabolic and taxonomic studies with microorganisms. *Appl. Environ. Microbiol.* 33:434–444.

Dagley, S. 1971. Catabolism of aromatic compounds by microorganisms. *Adv. Microb. Physiol.* 6:1–46.

Demain, A. L. 1977. The health of the fermentation industry: A prescription for the future. *Dev. Ind. Microbiol.* 18:72–77.

Fukui, S., and A. Tanaka. 1982. Immobilized microbial cells. *Annu. Rev. Microbiol.* 36:145–172.

Lowry, O. H., N. J. Rosenbrough, A. L. Farr, and R. J. Randall. 1951. Protein measurement with the Folin phenol reagent. *J. Biol. Chem.* 193:265–275.

Margaritis, A., and G. E. Rowe. 1983. Ethanol production using *Zymomonas mobilis* immobilized in different carrageenan gels. *Dev. Ind. Microbiol.* 24:329–336.

Miller, J. H. 1972. Experiments in Molecular Genetics. Cold Spring Harbor Laboratory. Cold Spring Harbor, NY, p. 138.

Munnecke, D. M., L. M. Johnson, H. W. Talbot, and S. Barik. 1982. Microbial metabolism and enzymology of selected pesticides. Pages 1–32 *in* A. M. Chakrabarty, ed., *Biodegradation and Detoxification of Environmental Pollutants.* CRC Press, Inc. Boca Raton, FL.

CHAPTER **62**

A Scanning Electron Microscopy Study of Crystalline Structures on Commercial Cheese

CLINTON J. WASHAM*[1], THOMAS J. KERR[2], VERNON J. HURST[3], AND WILLIAM E. RIGSBY[4]

[1]*Department of Dairy Science,* [2]*Department of Microbiology,* [3]*Department of Geology, and* [4]*Electron Microscope Laboratory, University of Georgia, Athens, Georgia 30602*

Surface crystalline bloom on the surfaces of various retail packaged natural and processed cheeses were studied using Scanning Electron Microscopy and X-ray diffraction analysis. Crystals were studied intact on cheese surfaces, on the surfaces of packaging film, and after removal from cheese surfaces and recrystallization. Crystals on Cheddar, marble Cheddar, and sliced Cheddar were identified as calcium lactate. Those on smoked processed Swiss were identified after recrystallization as sodium orthophosphate dihydrate. Factors contributing to crystal formation are discussed.

INTRODUCTION

Although crystals and granules have been observed in and on cheese since the early history of cheese-making, considerable disagreement with regard to the composition and origin of these structures still exists.

Babcock et al. (1903) reported the occurrence of white specks in Cheddar cheese. The incidence of these specks was reportedly increased by low concentration of rennet and salt, use of skimmed milk or low-fat milk, manufacture of sweet-curd cheese, and curing at 4.4 C as compared to 15.5 C. No adverse effect on the texture or flavor of the cheese was noted. Van Slyke and Publow (1910) suggested that the white spots they found in cheese were a calcium soap resulting from the reaction between calcium and the free fatty acids liberated by bacteria. A year later white specks in Roquefort cheese were reported by Dox (1911). After isolation and subsequent recrystallization, they were identified as tyrosine plus a lesser amount of leucine.

X-ray diffraction analysis of white specks from Cheddar (Tuckey et al. 1938a,b) revealed crystal structure spacings nearly identical to those of calcium lactate. The following year New Zealand researchers (McDowall and McDowell 1939) also reported the white specks in Cheddar to be calcium lactate. However, white particles similarly isolated from Cheddar (Dorn and Dahlberg 1942) gave a negative test for the lactate radical. Subsequent chemical analyses proved the particles to be primarily tyrosine. The confusion over the identity of the white specks in Cheddar was deepened by the chemical and X-ray diffraction results of Shock et al. (1948), who concluded that the specks contained both tyrosine and calcium

*Present address: Carlin Foods Corporation, Dairy Division, St. Louis, MO 63146.

lactate, as well as cystine, other amino acids, and magnesium. They also stated that the X-ray diffraction patterns for specks from the surface and from the interior of the cheese were identical.

The divergence in findings by these earlier investigators led Harper et al. (1953) to study the specks on different lots of Cheddar to determine the heterogeneity of their chemical composition. A study of the white particles collected separately from six different lots of Cheddar revealed the presence of calcium lactate and tyrosine in five samples, cystine in four samples, and leucine and isoleucine in one sample. Further evidence for the heterogeneity of the crystals was provided by Dolezalek (1956), who proposed that the grains in Roquefort consist of a tyrosine core surrounded by a layer composed of a variety of amino acids, including leucine, but not cystine.

Swiatek and Jaworski (1959) examined four different types of cheese, including Roquefort, using histochemical techniques and microscopy. Their observations suggested that the "kernals" were deposits of calcium phosphate.

Scientists in Australia (Conochie et al. 1960) used a variety of analytical techniques to show that crystals from the surface of a mature (15 month) rindless Cheddar were composed almost entirely of tyrosine, while crystals from a less mature cheese were identified as calcium lactate. Farrer and Hollberg (1960) supported this finding by identifying the white deposits on the surface of cheese wrapped in flexible packaging material as calcium lactate. They also suggested that tyrosine crystals would tend to form only on cheese of advanced age. Doubt was cast on these conclusions by Italian workers studying Grana Padano (Bianchi et al. 1974; Beretta et al. 1976). Two different physical forms of specks were observed in hard grating cheese. The first form was termed "granules" and was reported to occur in different crystalline forms in the surface of cheese slices during early stages of curing. Tyrosine was determined to comprise 47.8 to 82.5% of these "granules." The second form of specks was termed "spots." Only after curing for 16 to 18 months were "spots" easily detected. They contained several amino acids, but the most predominant were leucine and isoleucine. No explanation was proposed for the origin of "spots," but granules were attributed, at least in part, to the action of the high population of bacteria in and around them.

Light and electron microscope studies of Cheddar cheese (Brooker et al. 1975) revealed the occurrence of microscopic as well as macroscopic crystals. Both types of crystals were found in one-month-old Cheddar with the microscopic type being far more abundant. The macroscopic crystals were identified as calcium lactate and the microscopic aggregates as crystalline calcium phosphate.

Other studies (Bolcato et al. 1969) have implicated rennin in the formation of both lipoprotein granules and calcium phosphate granules. A more recent paper (Kalab 1980) has proposed that decayed lactic acid bacteria may serve as the nuclei for crystal formation, at least in Cheddar.

This project was undertaken because of the frequency at which the crystal bloom defect is encountered in retail market cheese and because there is a need to better define the problem to manufacturers, retailers, and consumers.

Materials and Methods

Samples. Consumer cuts of various brands of Cheddar, marble Cheddar, sliced

Cheddar, and smoked processed Swiss cheese, which already exhibited surface crystalline bloom defect, were purchased from supermarkets in Georgia, Missouri, Iowa, and Wisconsin. These samples were kept refrigerated prior to preparation for SEM and X-ray diffraction studies.

Sample preparation. For electron microscopy, small rectangular blocks, approximately 3 × 2 × 1 mm, were removed from areas of cheese exhibiting surface crystallization. The blocks were fixed in cacodylic acid-buffered glutaraldehyde (2%) and dehydrated by the procedure of Eino et al. (1976), except acetone was replaced by ethanol. After critical point drying in a Polaron apparatus, the blocks were secured to SEM studs with silver paint. The samples were coated with gold:palladium in a Hummer sputter coater (Technics, Inc., Alexandria, VA).

Recrystallization. Crystals were picked off the cheese with a stainless steel spatula and deposited in an acid-cleaned, 50-ml glass beaker. A minimal amount of a 1:1 mixture of distilled-deionized water and chloroform was added to the crystals and the mixture was stirred for one min. The chloroform layer was removed and the water solution pipetted onto a watch glass, which was placed in a desiccator under negative pressure for 24 h. Crystals were removed from the watch glass and analyzed by X-ray diffraction and examined by SEM. Samples examined by SEM were placed on SEM studs with double-stick scotch tape and coated as described above.

SEM observations and EDAX analysis. A Cambridge Mark IIA Stereoscan electron microscope operating at 10 kV with supporting energy-dispersive X-ray equipment was used to view and study the crystals. Micrographs were taken using a Polaroid camera with type 665 land film. An EDAX 707B analyzer and EDAX X-ray detector were used to obtain elemental composition data from energy-dispersive analysis of the X-rays emitted upon excitation of the sample by the primary electron beam.

X-ray diffraction analysis. For X-ray diffraction analysis, samples of crystals and granules were either picked from the cheese surfaces or recrystallized as described above and made into cell mounts. All samples were run on a Philips X-ray Diffractometer with Pulse Height Analyzer, using Ni-filtered Cuk_α radiation and a scan rate of 2° 2θ/min. The 2θ values of peaks on the X-ray diffraction patterns were converted to interplanar spacings, d-spacings, which were compared with lists of d-spacings of known materials in the ASTM Powder Diffraction File (1969) to identify the X-rayed material.

Results and Discussion

Fairly typical visual appearance of crystalline bloom on commercial cheese is presented in Fig. 1. Its similarity to mold growth and the potential negative impact on sales are evident. A low magnification micrograph of crystals on the surface of sliced, mild Cheddar is presented in Fig. 2A. Greater magnification (2B) revealed long, thread-like structures with interspersed granular material. This heterogeneous appearance might provide some insight into the lack of agreement on the

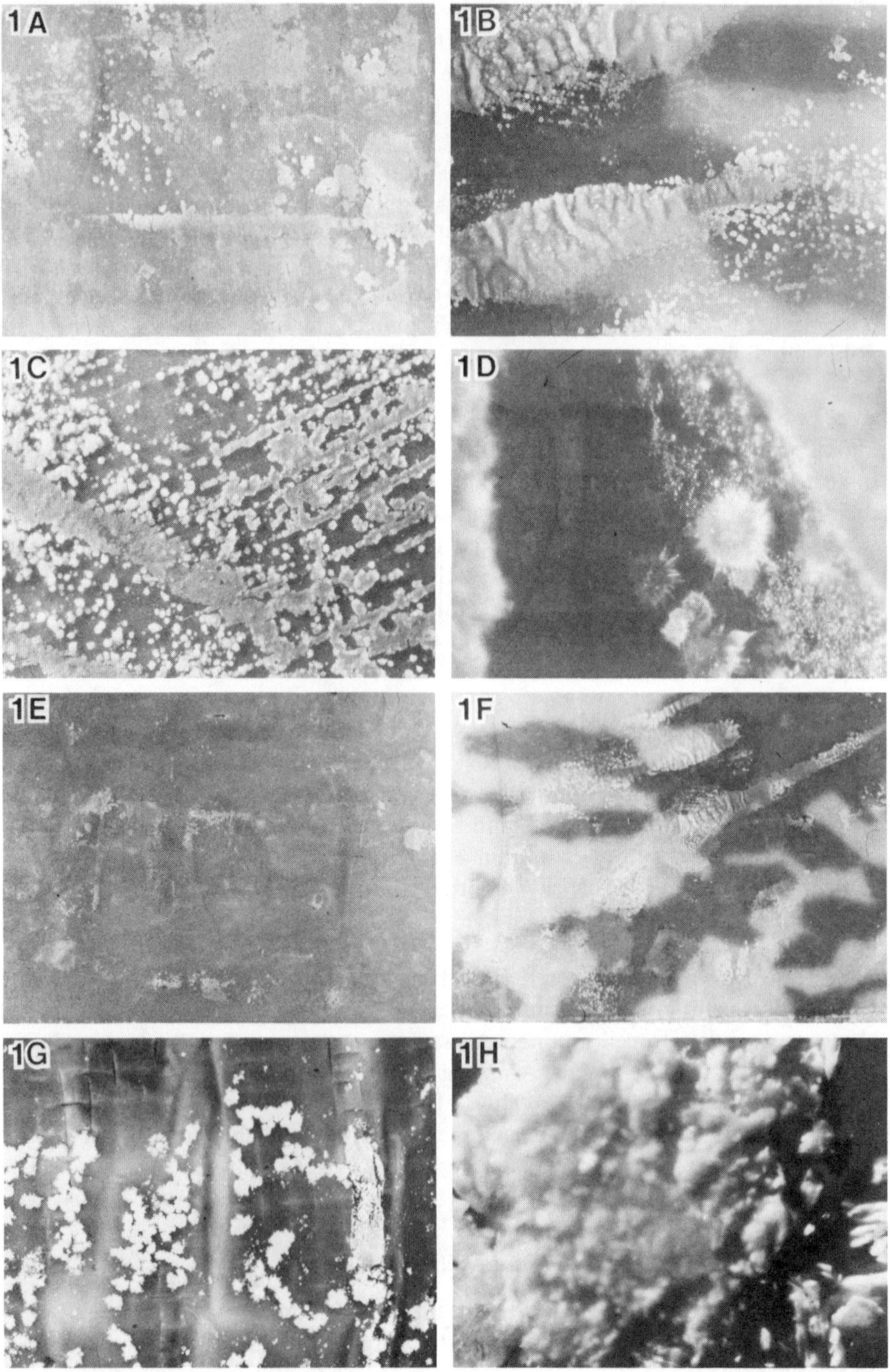

FIG. 1. Macroscopic appearance of crystals on commercially obtained mild Cheddar (A,B), sliced Cheddar (C,D), marble Cheddar (E,F), and smoked processed Swiss (G,H). Pictures B, D, F, and H are magnified to provide greater detail.

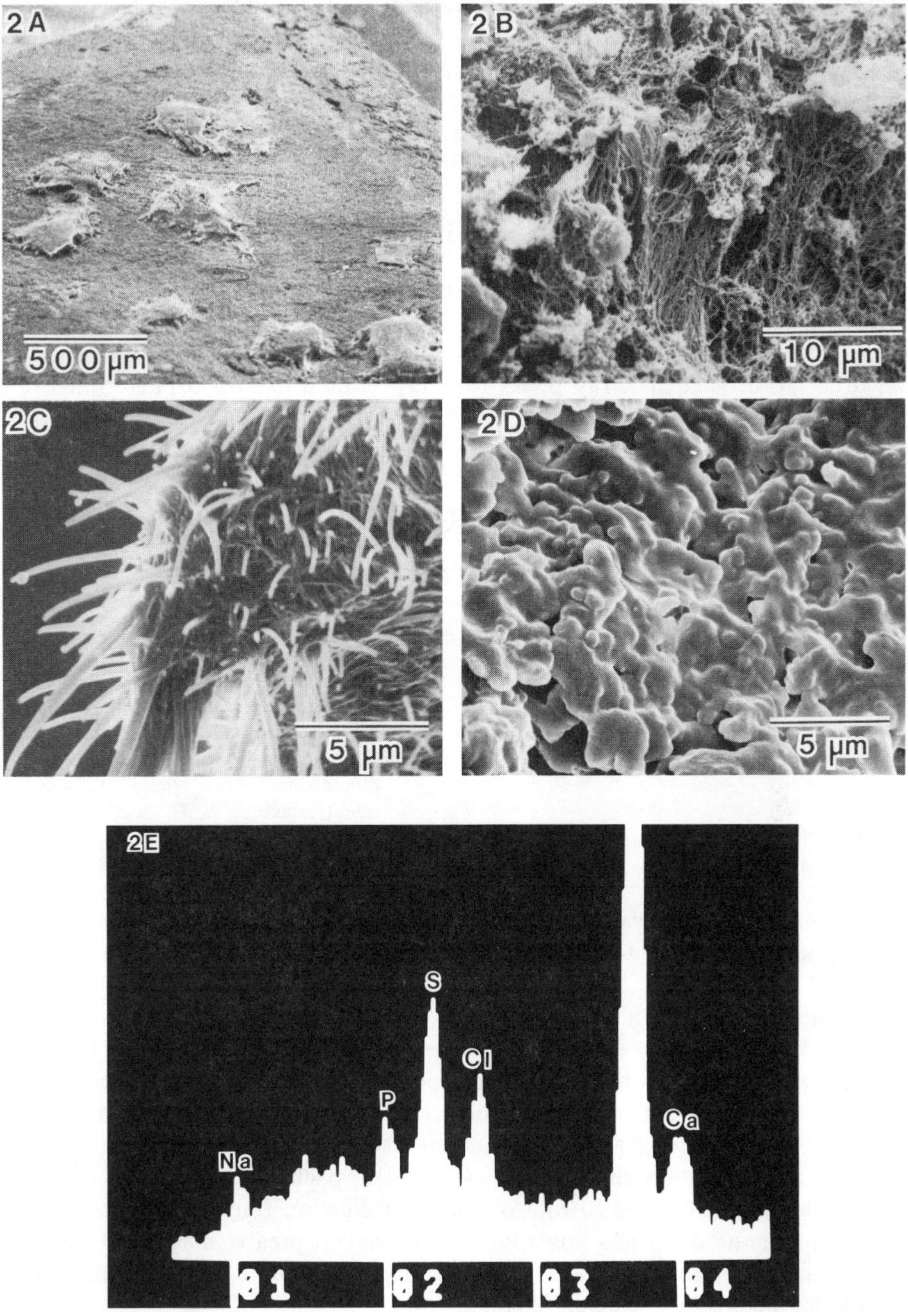

FIG. 2. Micrographs of crystals on the surface of sliced cheddar (A) at 50 X and (B) at 2,600 X. C is the topside of a calcium lactate crystal found on retail Cheddar cheese (5,000 X). D is a micrograph of the bottom of the same crystal at 5,000 X showing the adhering cheese. 2E is the EDAX spectrum of the crystals found on the surface of sliced Cheddar cheese, which suggests calcium lactate contaminated with calcium phosphate and other crystals.

identity of crystal bloom by various investigators. X-ray spectrum (2E) and X-ray diffraction data presented in Table 1 verify the presence of calcium lactate. EDAX data indicate the presence of phosphorus, probably in the form of very fine calcium phosphate crystallites. The height of X-ray diffraction peaks decreases rapidly with decreasing crystallite size; therefore, crystallites finer than 0.08 μm might yield no X-ray diffraction peaks.

Recrystallization of crystalline bloom from cheese may yield better crystallized materials, but it does not always yield the same materials, i.e., recrystallization may result in the formation of crystals different from those initially present. For example, granules from smoked processed Swiss cheese were shown by X-ray diffraction to consist of potassium tartrate, epsomite, and minor amounts of dipotassium tartrate hemihydrate plus oxalate, while the same granules after recrystallization yielded X-ray peaks for only sodium orthophosphate dihydrate (Table 1).

TABLE 1. *Identification of secondary crystallites composing granules or occurring as scattered crystals on the surface of various cheeses*

Source of crystals	Components identified by X-ray diffraction
Cheddar	calcium lactate
Recrystallized from Cheddar	calcium lactate
Marble	calcium lactate and trace of NaCl(?)
Recrystallized from marble	calcium lactate
Sliced Cheddar	calcium lactate and trace of NaCl(?)
Recrystallized from sliced Cheddar	calcium lactate
Smoked processed Swiss cheese	potassium tartrate, epsomite, minor dipotassium tartrate hemihydrate plus minor oxalate
Recrystallized from smoked processed Swiss cheese	sodium orthophosphate dihydrate

Earlier studies have examined crystals picked from the surface of cheese. This technique was evaluated in this study by examining both the top and bottom surfaces of crystals from Cheddar with the SEM. The top surface (2C) consisted of long, needle-like crystals typical of calcium lactate (Table 1 and Fig. 5), as reported by Washam et al. (1982). The appearance of the undersurface (2D) was typical of amorphous Cheddar. What appeared to be a single mass of crystalline material picked carefully from the surface of cheese is evidently heterogeneous, containing some cheese attached to its undersurface.

Tabular crystals on the interior of a plastic packaging film were carefully picked and studied using SEM. At low magnification (Fig. 3A), a flat, tightly knit, unstructured surface was apparent. However, greater magnification (Fig. 3B) along the edges and cracks revealed long, needle-like crystals, which were again characteristic of calcium lactate.

Figs. 3C and 3D present two magnifications of a typical crystal on the surface of marble Cheddar. The microscopic appearance is not typically that seen previously on sliced or consumer-cut Cheddar. Although long, thread-like crystals were not found, the crystals were still acicular, and the EDAX spectrum (Fig. 3E) again supported their identification as calcium lactate. The modified crystal habit might arise from the manufacturing procedure of making separate

5

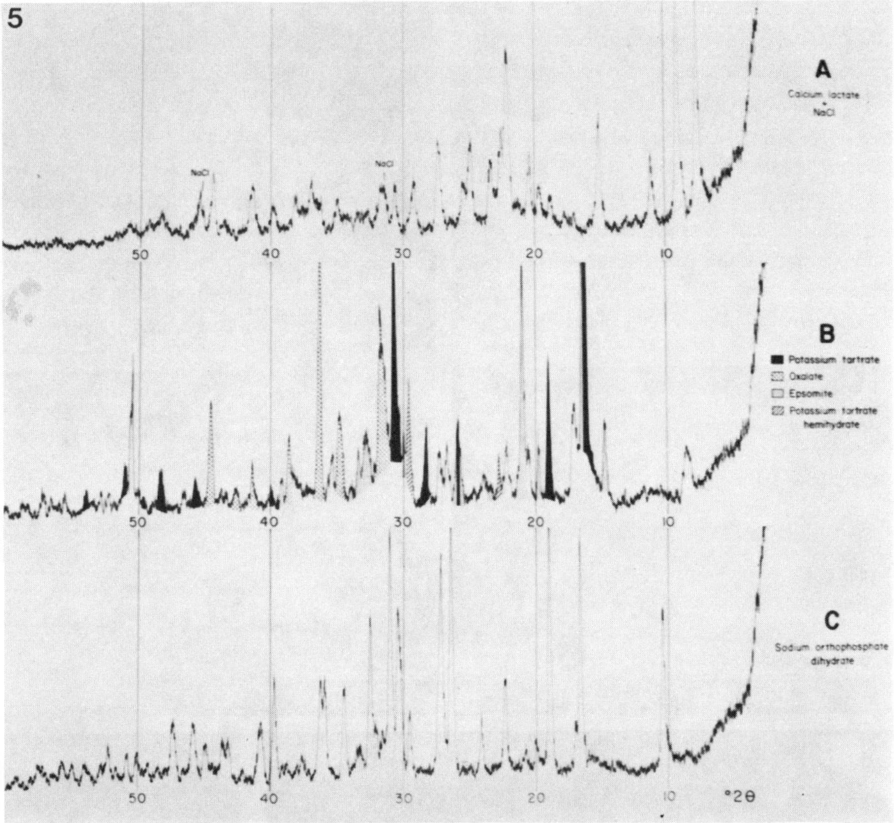

FIG. 5. X-ray diffraction patterns of crystallites from sliced Cheddar cheese (5A), smoked processed Swiss cheese (5B), and recrystallized crystallites from smoked processed Swiss Cheddar cheese (5C).

vats of yellow and white Cheddar and mixing the two prior to hooping. This means the salt, mineral, lactose, protein, fat, and moisture contents, as well as the pH, are different between the two types of curd comprising the finished cheese. These differences might provide fundamentally sound reasons for variations in the morphology and rate of crystal growth.

Processed cheese has come under much less consumer criticism for crystal bloom than has natural cheese. This probably results from the careful standardization of moisture content and the generous use of phosphate emulsifying salts. The chelation of calcium by the phosphates (Heertje et al. 1981) should greatly diminish the chance for calcium lactate crystallization. If the moisture content and temperature are controlled, calcium phosphate should not crystallize out of processed cheese. However, there are some specialty or gourmet processed cheeses that occasionally exhibit crystal bloom. One such cheese is a smoked processed Swiss. The smoking process dehydrates the surface and reduces the water content below the critical level. As a result, crystals of potassium tartrate, epsomite, and minor dipotassium tartrate hemihydrate plus minor oxalate grow on the surface (Table 1). Although unsightly, the crystal bloom does not connote

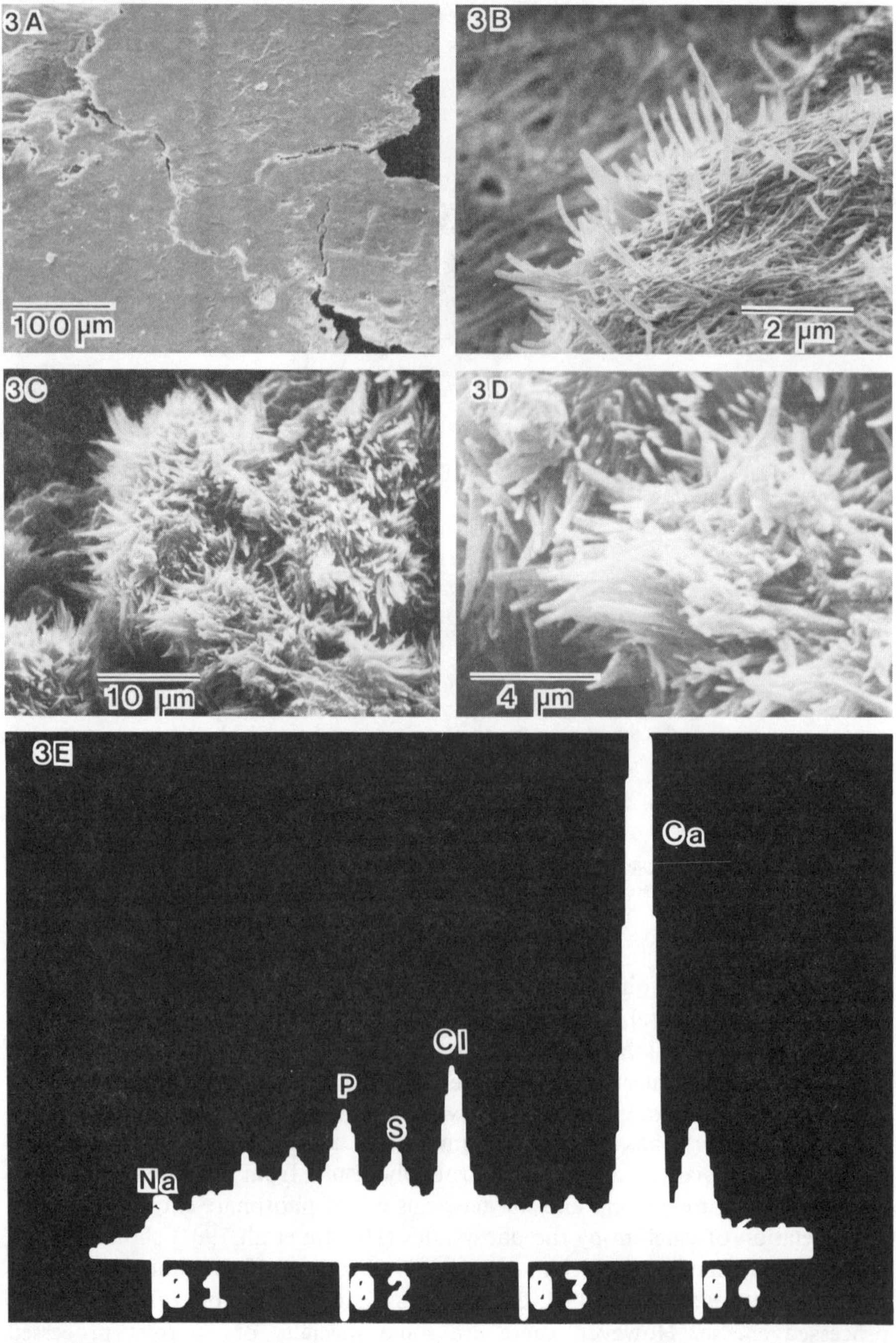

FIG. 3. Micrographs 3A and 3B are of a crystal found on the packaging film surface next to Cheddar cheese. 3A appears as a dense flat particle with no apparent substructure (180 X) while 3B shows needle-like crystalline substructure of calcium lactate at 10,000 X. Micrographs 3C (2,300 X) and 3D (5,750 X) are of crystals found on retail marble cheese. 3E is an EDAX spectrum of the crystals on marble cheese, which shows high concentrations of calcium suggesting calcium lactate. The significant chloride peak also suggests the presence of sodium chloride.

a chemical difference from cheese not exhibiting this bloom. This defect could be avoided by using liquid smoke instead of natural hickory smoke, but much of its consumer appeal might be lost. When the crystals from smoked Swiss cheese were recrystallized, a completely different crystal was formed with a different composition (Fig. 4A), EDAX spectrum (Fig. 4B), and X-ray diffraction pattern (Fig. 5 and Table 1).

An EDAX spectrum (Fig. 4C) of the surface of Cheddar cheese free of bloom indicates a predominance of sodium, phosphorus, sulfur, chlorine, and calcium. An EDAX spectrum (Fig. 4D) of the surface of Cheddar cheese below a crystal bloom shows depletion of both sodium and chlorine. Under the experimental conditions, it was not possible to quantitate the phosphorus, sulfur, and calcium concentrations.

The X-ray diffraction patterns of crystals and granules picked from cheese surfaces show many narrow, intense peaks, indicating that the material is well crystallized and the crystallites are larger than 0.5 μm. The identification by X-ray diffraction of a crystal is virtually positive when: (a) more than four diffraction peaks correspond exactly with those of a reference mineral; (b) the relative intensities correspond exactly with those of the reference mineral; and (c) there are no major peaks of the reference mineral missing from the X-ray diffraction pattern of the crystal.

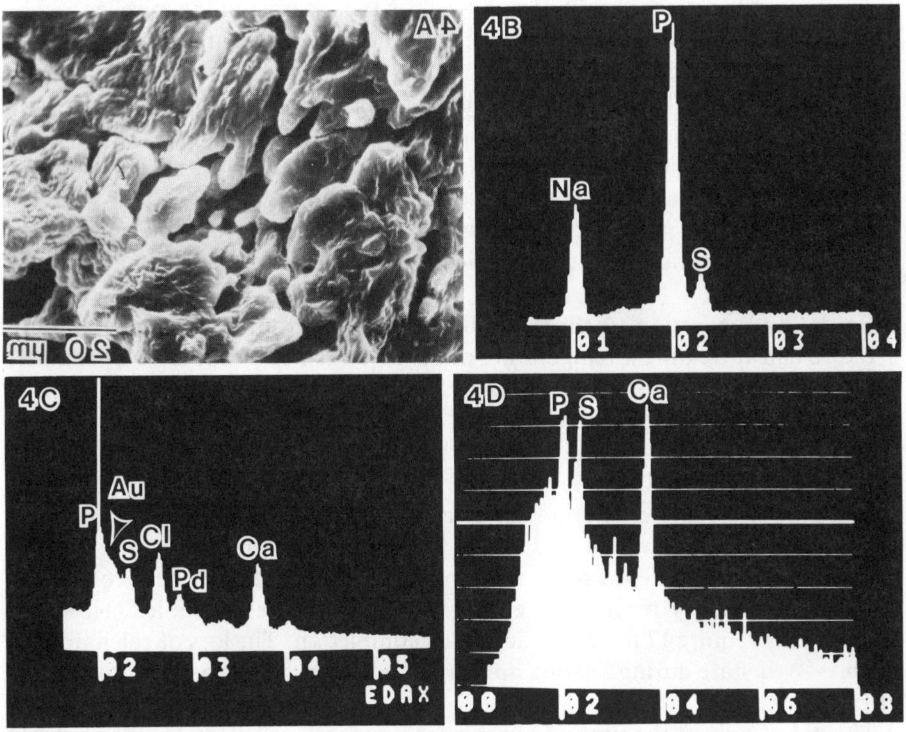

FIG. 4. Micrograph 4A shows the appearance of recrystallized crystals from smoked processed Swiss cheese (1,000 X). 4B is the EDAX spectrum of this crystal indicating the crystals are sodium orthophosphate dihydrate. 4C is a typical EDAX spectrum from Cheddar cheese. 4D is an EDAX spectrum of Cheddar cheese beneath a crystal bloom.

Table 1 provides a definite identification for most of the crystals on the cheese used in this study. The exact matching of two peaks, as NaCl in Pattern A of Fig. 5, is a strong indication but cannot be said to be definitive. Whenever this is the case, the identifications in Table 1 are followed by a question mark (?).

Any attempt to describe the mechanism of crystal formation in or on cheese must be concerned with the water content, pH, temperature, mineral concentrations and their solubilities, manufacturing procedures, etc.

Cheese is manufactured from milk and, in the early stages of ripening, contains the primary components of milk: fat, protein, lactose, minerals, and water. In the simplest terms, cheese can be thought of as partially dehydrated milk although most of the lactose and whey proteins have been removed. This concept, however simplified, will benefit the initial consideration of the fundamental principles involved.

The primary protein in cheese is casein, which exists in milk as spherical units called micelles. Approximately 6% of the dry weight of the micelles is composed of mineral matter, primarily calcium phosphate, which is referred to as colloidal calcium phosphate (CCP). Fox and Mulvihill (1982) reported that there were 70,600 calcium atoms, 30,100 inorganic phosphate groups, and 25,000 organic phosphate groups per micelle. There are about 10^{14} micelles per ml of milk. Schmidt (1980) stated that the submicelles are held together by CCP.

The caseinate system in milk is stabilized by the k-casein in the micelles and by colloidal calcium phosphate. It has a high negative zeta-potential of -18 to -20 mV and a very high degree of hydration. One pound of protein will contain about 0.5 to 4.0 lb of water. Only about 0.5 lb of water is actually bound by the casein micelles; the remainder is occluded within the micelles (Fox 1982). This water serves as the dissolving medium for the minerals, amino acids, etc.

Although the papers of Czulak et al. (1969) and Fox (1980, 1982) were not directed toward the phenomenon of crystal formation, they provide an excellent background presentation of the chemical events that occur prior to and during crystallization.

Before its conversion to cheese, milk is increased from a silo temperature of 3–6 C to a heat treatment temperature of 60–80 C (generally about 72 C) and held for varying periods of time. Because the solubility of calcium phosphate decreases as the temperature increases, it may undergo micro-precipitation on the surface of the micelles. The micelles become somewhat less labile to calcium ions.

In general, skim milk contains about 32 m*M* calcium and 30 m*M* phosphate anion. At pH 6.5–6.7 only 3 m*M* of calcium and 5 m*M* of phosphate are in the free ionic state (Bloomfield and Morr 1983). The concentrations of dissolved calcium and phosphate decrease when milk is heated. This transfer to the colloidal state is slowly reversed when the milk is subsequently cooled. However, reversal may require 12 to 48 h to approach completion. The loss of calcium from the dissolved state during heating appears to be at least twice as great as the loss of phosphate.

The pH of milk at the time of heating has a pronounced effect on the observed shifts in minerals between the colloidal and dissolved states. When heat-treated milk is cooled for cheese-making or for interim storage, some of the colloidal calcium and phosphate begin to move into the dissolved state. As the pH decreases because of the metabolism of lactose to lactic acid, colloidal calcium phosphate becomes more soluble and is removed with the whey at dipping. Most of the lactose, being soluble, is found in the whey and is likewise lost at dipping. There seems to be a critical time while the curd is being stirred in the whey that may determine, to a large extent, whether crystals will later show up on the cheese.

The whey has a higher concentration of lactose, a lower pH, and a lower concentration of lactic acid than does the curd. A prolonged stir-out tends to promote diffusion of the lactose into the curd (Czulak et al. 1969). Lactose in the curd serves as a substrate for lactic acid production and further depression of the curd pH. Lower pH, in turn, causes greater solubilization of the calcium phosphate and, on some occasions, detrimental effects on the cheese body.

The water phase inside the curd is primarily a continuous system associated with the aggregated network of casein micelles. Microbial fermentation raises the concentration of lactic acid in this phase and solubilizes additional calcium phosphate. Calcium lactate is subsequently formed, especially in the liquid-filled intercurd spaces, where more activity by lactic acid bacteria occurs. The continued production of lactic acid during pressing and ripening causes additional solubilization of calcium phosphate in the water phase.

As the cheese is pressed and aged, the water phase tends to equilibrate throughout the cheese. However, the diffusion of water from pockets of high moisture tends to leave behind areas of high calcium lactate and calcium phosphate. Since calcium lactate is more soluble in hot whey than in cold whey, it may be the first compound to crystallize as the curd cools during ripening. Distribution of sodium chloride on the surface of curd prior to hooping causes more calcium phosphate to migrate to the surface of the curd particles since it is more soluble in the presence of salt. In a contrasting effect, salting of the curd in the whey to slow the rate of fermentation, as is sometimes done, may result in a greater loss of calcium phosphate in the whey, leaving less in the curd.

During further curing of cheese, the pH frequently increases as additional chemical and metabolic activity occurs. Simultaneously, the protein undergoes degradation and becomes less able to bind calcium phosphate. This may explain why some cheese consumers actually consider crystalline bloom to indicate a cheese will be more flavorful and have a less-rubbery body. These factors, coupled with the loss of moisture from high calcium phosphate areas through equilibrium diffusion and evaporation from exposed surfaces, tend to promote crystal growth.

An additional contributing factor in some manufacturing plants might be the application of steam in the vat jacket during Cheddaring and milling. The surfaces of the curd in contact with the vat experience significant temperature increase and, presumably, accompanying decrease in calcium phosphate solubility.

Crystal formation and growth are difficult to duplicate consistently in pilot plant trials. However, some of the contributing factors have been discussed in this paper with identification of crystals that were found on consumer cuts in retail outlets. Careful consideration of these factors and the information provided by Czulak et al. (1969) and Olson (1984) should help to minimize their occurrence.

Literature Cited

American Society for Testing and Materials. 1969. *Powder Diffraction File* (Indexes). Philadelphia, PA.

Babcock, S. M., H. L. Russell, A. Vivian, and U. S. Baer. 1903. Condition affecting the development of white specks in cold-curd cheese. Pages 180–183 *in Wis. Agric. Expt. Sta. 19th Annu. Rep.*

Beretta, G., G. Caserio, M. A. Bianchi, and M. Gennari. 1976. Amminoacidi presentinei granuli e nei noduli del formaggio grana padano ottenuto da latte addizionato di formaldeide e da latte trattao con raggi ultravioletti. *Il Latte* 1:19–22.

Bianchi, A., G. Beretta, G. Caserio, and G. Giolitti. 1974. Amino acid composition of granules and spots in Grana padano cheeses. *J. Dairy Sci.* 57:1504–1508.

Bloomfield, V. A., and C. V. Morr. 1973. Structure of casein micelles: Physical methods. *Neth. Milk Dairy J.* 27:103–120.

Bolcato, V., P. Spettoli, and G. Grinzato. 1969. Rennin, the enzyme responsible for the formation of granules lipoproteins in cheeses. *Milchwissenschaft* 24:71–73.

Brooker, B. E., D. G. Hobbs, and A. Turvey. 1975. Observations on the microscopic crystalline inclusion in Cheddar cheese. *J. Dairy Res.* 42:341–348.

Conochie, J., J. Dzulak, A. J. Lawrence, and W. F. Cole. 1960. Tyrosine and calcium lactate crystals on rindless cheese. *Aust. J. Dairy Technol.* 15:120.

Czulak, J., J. Conochie, B. J. Sutherland, and H. J. M. Van Leeuwen. 1969. Lactose, lactic acid, and mineral equilibria in Cheddar cheese manufacture. *J. Dairy Res.* 36:93–102.

Dolezalek, D. J. 1956. A contribution to the determination of the chemical composition of grains in Roquefort cheese. *Proc. XIV Intern. Dairy Congr.* 2:175–180.

Dorn, F. L., and A. C. Dahlberg. 1942. Identification of the white particles found on ripened Cheddar cheese. *J. Dairy Sci.* 25:31–36.

Dox, A. W. 1911. The occurrence of tyrosine crystals in Roquefort cheese. *J. Am. Chem. Soc.* 33:423–435.

Eino, M. F., D. A. Biggs, D. M. Irvine, and D. W. Stanley. 1976. A comparison of microstructures of Cheddar cheese curd manufactured with calf rennet, bovine pepsin, and porcine pepsin. *J. Dairy Res.* 43:113–115.

Farrer, K. T. H., and W. C. J. Hollberg. 1960. Calcium lactate on rindless cheese. *Aust. J. Dairy Technol.* 15:151–152.

Fox, P. F. 1980. Heat-induced changes in milk preceding coagulation. *J. Dairy Sci.* 64:2127–2137.

______. 1982. *Developments in Dairy Chemistry-1.* Pages 189–223. Applied Science Publishers, New York.

Fox, P. F., and D. M. Mulvihill. 1982. Milk protein: Molecular, colloidal, and functional properties. *J. Dairy Res.* 49:679–693.

Harper, W. J., A. M. Swanson, and H. H. Sommer. 1953. Observations on the chemical composition of white particles in several lots of Cheddar cheese. *J. Dairy Sci.* 36:368–372.

Heertje, I., M. J. Boskamp, F. van Kleef, and F. H. Gortemaker. 1981. The microstructure of processed cheese. *Neth. Milk Dairy J.* 35:177–179.

Kalab, M. 1980. Decayed lactic bacteria—A possible source of crystallization nuclei in cheese. *J. Dairy Sci.* 63:301–304.

McDowall, F. H., and A. K. R. McDowell. 1939. The white particles in mature Cheddar cheese. *J. Dairy Res.* 10:118–119.

Olson, N. F. 1983. Minimizing salt crystal formation. *Dairy Field,* February.

Schmidt, D. G. 1980. Colloidal aspects of casein. *Neth. Milk Dairy J.* 34:42.

Shock, A. A., W. J. Harper, A. M. Swanson, and H. H. Sommer. 1948. What's in those "white specks" on Cheddar? Pages 31–32 *in Wis. Agric. Expt. Sta. Bull.* 474.

Swiatek, A., and J. Jaworski. 1959. Histochemische untersuchungen uber die verteilung von mineralsalzen im kase. *XV Int. Dairy Congr.* 1509–1518.

Tuckey, S. L., H. A. Ruehe, and G. L. Clark. 1938a. X-ray diffraction analysis of white specks in Cheddar cheese. *J. Dairy Sci.* 21:161.

______. 1938b. An x-ray diffraction analysis of Cheddar cheese. *J. Dairy Sci.* 21:777–789.

Van Slyke, L. L., and C. A. Publow. 1910. *The Science and Practice of Cheese-Making.* The Orange Judd Co., New York.

Washam, C. J., T. J. Kerr, and V. J. Hurst. 1982. Microstructure of various chemical compounds crystallized on Cheddar Cheese. *J. Food Prot.* 45:594–596.

CHAPTER 63

Fermentation Studies with Haemophilus influenzae

CHRISTINE E. CARTY*, RALPH MANCINELLI**, ARPI HAGOPIAN, FRANCIS X. KOVACH, EVELYN RODRIGUEZ, PAMELA BURKE, NANCY R. DUNN, WILLIAM J. MCALEER, ROBERT Z. MAIGETTER, AND PETER J. KNISKERN

Department of Virus and Cell Biology, Merck Sharp and Dohme Research Laboratories, West Point, Pennsylvania 19486

The capsular polysaccharide of (PRP) *H. influenzae* type b has been used as an effective vaccine. There were two fermentation problems associated with PRP production: low yields (e.g., 70–80 μg/ml) and a culture medium that contained blood-group reactive substances. Media screening studies were conducted in 2-liter shake flasks. PRP concentrations of cell-free broths were determined by rocket immunoelectrophoresis and by immunochemical rate nephelometry. Highest yields were achieved by soy peptone medium (258 μg/ml); lower amounts occurred when casamino acids or heart infusion broth were used (197 μg/ml and 91 μg/ml, respectively). The PRP from the soy medium had a higher molecular weight than that from heart infusion broth. The Ross and Eagan strains were compared and gave similar yields. Scale-up from 2-liter flasks to fermenters was tested; both gave similar yields. The higher molecular weight PRP isolated from these fermentation broths has been tested alone and in combination with a meningococcal protein carrier isolated in our laboratories. Clinical evaluation of these products is under way.

INTRODUCTION

The major cause of endemic bacterial meningitis is infection by *Haemophilus influenzae* type b (Robbins 1978). Most cases occur in children between the ages of 2 months and 4 yr. Antibiotic therapy has reduced the mortality rate from 90% to 5%; but at least 30% of the survivors suffer permanent neurological sequelae including blindness, deafness, and mental retardation (Sell et al. 1972). The principal virulence factor of *H. influenzae* type b is capsular polysaccharide—polyribosylribitolphosphate (PRP) (Crisel et al. 1975). Vaccines consisting of PRP alone or with a carrier protein have been effective in the prevention of disease in adults and children (Schneerson et al. 1980; Smith et al. 1975).

Our objective was to develop a process to produce a high yield of PRP containing minimal levels of blood-group reactive components. A 2-liter shake flask system was used to evaluate different culture media. Since PRP is released into the culture medium after its synthesis (Anderson et al. 1976), we isolated PRP from the clarified culture fluid.

*Person to whom reprint requests should be sent.

**Department of Analytical Research, Merck Sharp and Dohme Research Laboratories, Rahway, N.J. 07065.

MATERIALS AND METHODS

Strains. The Ross and Eagan strains of *H. influenzae* type b were obtained from Dr. G. Schiffman, State University of New York. Both strains were preserved at −50 C or less. Culture identities were verified by agglutination with specific antisera and by Gram stain.

Culture media. Four different media were tested. Heart infusion broth (HIB; Difco Laboratories, Detroit, MI) supplemented with hemin chloride and nicotinamide adenine dinucleotide (NAD) was used for initial studies. Modified heart infusion broth (MHIB) contained soy peptone powder (Sheffield Products, Memphis, TN) in place of dehydrated heart infusion broth. Casamino acids medium was the same as that described by Anderson and Smith (1977). A medium, designated MP, contained the following per liter of distilled water: soy peptone, 10 g; diafiltered yeast extract, 10 ml; NaCl, 5 g; K_2HPO_4, 2.5 g; Na_2HPO_4, 13.1 g; NaH_2PO_4, 3.3 g; dextrose, 5 g; hemin chloride, 10 mg; and NAD, 10 mg.

Cultivation. A frozen vial of culture was transferred to chocolate agar plates supplemented with IsoVitalex (BBL, Cockeysville, MD) and incubated for approximately 18 h at 37 C in the presence of 6–8% CO_2. Cells were transferred to 2-liter Erlenmeyer flasks containing 800 ml of culture medium and incubated at 36 C in a shaker incubator. The pH was checked every 2 h and was adjusted to pH 6.5 with 2 *N* NaOH. After 6–8 h of incubation, the culture was transferred to a second 2-liter shake flask or to a fermenter. Two types of fermenters were employed. The first was a 70-liter, air-overlay, nonbaffled fermenter (Bilthoven Unit System, Contact Ridderkerk, The Netherlands) containing 40 liters of culture medium and operated at 36 C, 750 rpm, and 8 liters of air per min. Subsequent studies employed a 16-liter, air-sparged, baffled fermenter (New Brunswick Scientific Co., Edison, NJ) containing 10 liters of culture medium and operated at 36 C, 200 rpm, and 2 liters of air per min.

Sampling and harvest. Samples were tested for optical density (O.D.) at 660 nm, glucose (Beckman Glucose Analyzer II) and pH. After cell growth was completed, the culture was inactivated with thimerosal (1:10000, w/v) for at least 2 h. The cells were centrifuged at 10,000 × *g* and the supernatant used for isolation of PRP.

Polysaccharide concentration and molecular size. Polysaccharide levels of cell-free fermentation broths were determined by rocket immunoelectrophoresis (Maigetter et al. 1979) or by immunochemical rate nephelometry. The antiserum used was boro-anti-*H. influenzae* type b. Molecular sizing was done with high performance size exclusion chromatography with Sepharose® 4B.

Polysaccharide purification. PRP was isolated from thimerosal inactivated, cell-free culture supernatants as described elsewhere (Kniskern et al. 1984).

Detection of blood group reactive substances. Blood group reactive components were measured by hemagglutination assay.

Results and Discussion

Both the Ross and Eagan strains of *H. influenzae* type b have been used in vaccine-related studies (Robbins, personal communication). As shown in Table 1, both strains grew similarly in heart infusion broth, reaching an $O.D._{660}$ greater than 1.0 in 7 h. Since the Ross strain produced approximately 15% more PRP than the Eagan strain, it was used for subsequent studies.

TABLE 1. *PRP production by two strains of* H. influenzae *type b grown in heart infusion broth*

Strain	$O.D._{660}$	Polysaccharide (μg/ml broth)
Eagan	1.2	91
Ross	1.3	109

Although heart infusion broth is an excellent culture medium, it contains blood-group reactive substances. These substances are generally found in animal products. Since these substances may be harmful and can be copurified with polysaccharides, soy peptone (a plant product) was substituted for heart infusion broth. Soy-based medium supported growth and polysaccharide production equivalent to levels in the casamino acids medium used by Anderson (165 vs. 197 μg/ml). These polysaccharide levels were approximately twice as high as those for heart infusion broth (Table 2). Even higher yields of polysaccharide were achieved when soy medium was supplemented with diafiltered yeast extract (MP medium). When this formulation was used, a concentration of 258 μg/ml was obtained.

TABLE 2, *Shake flask studies of* H. influenzae *type b (Ross strain) in various culture media*

Culture Medium	$O.D._{660}$	Polysaccharide (μg/ml broth)
Heart Infusion Broth	1.1	91
Modified Heart Infusion Broth	1.3	165
Casamino Acids-Yeast Extract	1.3	197
MP Medium	1.8	258

Since good yields of polysaccharide had been obtained in 2-liter flasks, scale-up fermentation studies were initiated in two types of fermenters. One was an air-sparged unit containing baffles, the other had an air-overlay system without baffles. These two systems were compared to determine if air delivery and fermenter configuration had an effect on polysaccharide production (Table 3). Although cell growth was higher in the air-sparged system ($O.D._{660}$ 1.5 vs. 0.6), more polysaccharide was produced in the air-overlay fermenter (198 μg/ml vs. 149 μg/ml). A possible explanation is that the reduced shear associated with air-overlay systems enhances polysaccharide synthesis or minimizes its breakdown.

TABLE 3. *PRP production by* H. influenzae *type b grown under various conditions in MP medium*

Size Configuration	$O.D._{660}$	Polysaccharide (μg/ml broth)
2L/shake flask	1.8	258
16L/air-sparge	1.5	149
70L/air-overlay	0.6	198

PRP isolated from MP medium differed from that derived from heart infusion broth in two ways. First, there was a definite and reproducible shift to the production of high molecular weight (Kd = 0.0 – 0.10 vs. Kd = 0.3 – 0.4) PRP (Table 4). Secondly, preparations of PRP from MP medium contained no detectable blood group substances (data not shown). The effect(s) of the increased molecular size on the immunogenicity of the PRP is being evaluated in clinical trials.

TABLE 4. *The molecular weight of PRP isolated from heart infusion broth and MP medium*

Culture Medium	Batch No.	Kd on Sepharose® 4B
Heart Infusion Broth	1	0.34
	2	0.38
	3	0.42
MP Medium	4	0.09
	5	0.08
	6	0.09

Conclusion

Replacement of animal-derived products in the culture medium with those isolated from plants resulted in a fermentation process that produced large amounts of high molecular weight, blood group-free PRP.

Acknowledgment

We thank Ms. A. Wolfe for assistance with blood group assays.

Literature Cited

Anderson, P., J. Pitt, and D. H. Smith. 1976. The synthesis and release of polyribophosphate by *Haemophilus influenzae* type b *in vitro. Infect. Immunol.* 13:581–589.

Anderson, P., and D. H. Smith. 1977. Isolation of the capsular polysaccharide from culture supernatant of *Haemophilus influenzae* type b. *Infect. Immunol.* 15:472–477

Crisel, R. M., R. S. Baker, and D. E. Dorman. 1975. Capsular polymer of *Haemophilus influenzae* type b. *J. Biol. Chem.* 250:4926–4930.

Kniskern, P. J., S. Marburg, and R. L. Tolman. 1984. U.S. Pat. No. 608738.

Maigetter, R. Z., P. J. Kniskern, R. J. Mancinelli, and D. J. Carlo. 1980. Abstr. of the 80th Annu. Meet. of the Amer. Soc. for Microbiol. Miami Beach, FL.

Robbins, J. B. 1978. Vaccines for the prevention of encapsulated bacterial diseases: current status, problems and prospects for the future. *Immunochem.* 15:839–854.

Schneerson, R., O. Barrera, A. Sutton, and J. B. Robbins. 1980. Preparation, characterization, and immunogenicity of *Haemophilus influenzae* type b polysaccharide-protein conjugates. *J. Exp. Med.* 152:361–376.

Sell, S. H. W., R. E. Merrill, E. O. Doyne, and E. P. Zimsky. 1972. Long-term sequelae of *Haemophilus influenzae* meningitis. *Pediatrics* 49:206–211.

Smith, D. H., G. Peter, D. L. Ingram, and P. Anderson. 1975. Responses of children immunized with the capsular polysaccharide of *Haemophilus influenzae* type b. *Pediatrics* 52:637.

CHAPTER 64

Production of Vitamin B_{12} by Fermentation of an Industrial Waste-Product of the Mexican Lime (*Citrus aurantifolia* Swingle)

I. Fermentation Kinetics of *Propionibacterium shermanii* ATCC 13673

L. SANTANA-CASTILLO, J. L. PEREZ-MENDOZA*, AND
F. GARCIA-HERNANDEZ

Departamento de Biotecnología
Instituto de Investigaciones Biomédicas, UNAM,
Apdo. Postal 70228, 04510 México, D.F., México

In the industrialization of the Mexican lime (*Citrus aurantifolia* Swingle), specifically during the process of obtaining distilled essential oil, a distillation residue (D.R.) is generated, the volume of which reaches, nationally, approximately 80 million liters annually. Since an industrial use of the D.R. would not only exploit the nutrients it contains but would also help to alleviate the problem of its disposal, our research group has used this waste-product to produce vitamin B_{12} by fermentation. In the present work, our principal objective was the study of the kinetics of fermentation by *P. shermanii* ATCC 13673 in a culture medium based on this agroindustrial waste-product. Experiments were carried out in a 5-liter fermentor under three distinct aeration conditions: anaerobic, aerobic, and anaerobic-aerobic. The volumetric and specific rates and yields were calculated from the concentration vs. time curves for each fermentation. The results showed that, regardless of the aeration conditions, *P. shermanii* ATCC 13673 presented similar specific growth rates ($\mu \cong 0.04\ h^{-1}$) during the first 48 h of fermentation. For vitamin B_{12} production, the aerobic fermentation was more efficient than was either of the fermentations carried out under the other aeration conditions and gave a production of 2.6 mg/liter, a productivity of 21.9 μg/liter, and a yield of 802 μg/g cells.

INTRODUCTION

In the use and industrialization of agriculture products, enormous volumes of residues are generated that are either wasted or only partially exploited and that present serious risks of environmental contamination. Such residues originating from renewable sources have a great potential for use in biotechnology to obtain diverse products that are important to the food, chemical, pharmaceutical, and cattle feed industries.

An example of this are the subproducts produced in the industrialization of the Mexican lime (*Citrus aurantifolia* Swingle) during the process of obtaining distilled essential oil, in which a distillation residue (D.R.) is generated, the volume of which reaches, nationally, approximately 870 million liters annually. To date, the disposal of this residue is a serious problem, the composition of the D.R., and the increasing need to reduce importation of vitamin B_{12} gave rise to a line of

*Departamento de Tecnologías Básicas Agroindustriales Subdirección de Investigación y Docencia. CONAFRUT, SARH, México, D.F.

investigation in which this residue from the lime processing industry is used for the production of B_{12} by fermentation.

The residue was subjected to different treatments in order to condition it to the fermentative process that was carried out using strains of *Propionibacterium shermanii* ATCC 13673 and *Propionibacterium freudenreichii* ATCC 6207. From the results of these tests, treatment with NaOH and *P. shermanii* were selected for use in the development of the fermentation conditions for obtaining vitamin B_{12}. A fermentation medium was established and the effects of anaerobiosis and anaerobiosis-aerobiosis as well as the effects of the addition of the precursor 5,6-dimethyl benzimidazol on the production of vitamin B_{12} were tested (Pérez-Mendoza and García-Hernández 1983).

These experiments together with those reported (Pérez-Mendoza 1981) show that *P. shermanii* presents different growth rates (μ) possibly due to being supplied with different carbon sources, such as sugars and lactic and citric acids, in the culture medium. It was observed that the growth of the microorganism was slowed during the anaerobic phase but increased again during the aerobic phase, suggesting that *P. shermanii* consumes the sugars and lactic acid under anaerobic conditions and possibly the citrates under aerobic conditions.

Based on the foregoing findings, the object of the present work was to study the kinetics of *P. shermanii* ATCC 13673 in a culture medium containing D.R. as carbon source and complemented with corn steep liquor as nitrogen source, under different aeration conditions.

Materials and Methods

Microorganism. Propionibacterium shermanii ATCC 13673, a strain known as a good vitamin B_{12} producer (Florent and Ninet 1979), was used. In order to use the microorganism in the fermentation medium based on distilled residue, this strain was adapted in the following manner: The strain was grown on Krebs' medium (sodium lactate-yeast extract (Lapage et al. 1970)) containing, per liter of Krebs', 0.5 g cysteine and 4 ml of a 0.025% solution of resazurin (Lennette et al. 1974) and was stored in screw-top tubes containing 10 ml of the same medium.

From one of the stock tubes, 1 ml of the bacterial suspension was transferred to 10 ml of the same medium and under the same conditions described above. After incubation for 48 h, 10 ml of the bacterial suspension was transferred to a 250-ml Erlenmeyer flask containing 190 ml of the culture medium based on the D.R. (see below) and was incubated for 96 h at 29 C in a rotary shaker (1.5 cm radius; 125 rpm). After this incubation, three consecutive transfers were made by inoculating the same culture medium, under the same conditions, with the strain to give a concentration of 10%. From the last culture, 20-ml aliquots of the bacterial suspension were poured into screw-top tubes and stored at 4 C. These aliquots were used to prepare the inocula for the fermentations.

Treatment of the distillation residue. The D.R. was filtered through glass wool to eliminate pulp and seeds and then through filter paper (Whatman 41) until a clear liquid was obtained. The pH of the clarified D.R. was adjusted to 7.0 with NaOH.

Preparation of the fermentation medium. The fermentation medium was obtained by mixing the D.R. (500 ml), corn steep liquor (60 ml), $CoCl_2$ (5 mg), and distilled water to make 1 liter of medium. This medium was autoclaved at 121 C for 20 min. During sterilization, some solids were formed but they were eliminated by aseptic decantation. The final pH of the fermentation medium was 7.0.

Preparation of the inoculum. An aliquot (20 ml) of the microorganisms, which had been previously adapted to the fermentation medium and then stored at 4 C (see above), was inoculated into a 250-ml Erlenmeyer flask containing 180 ml of fermentation medium and was incubated at 29 C in a rotary shaker (1.5 cm radius; 125 rpm) for 60 to 72 h.

Fermentations. For the fermentations, a 5-liter glass jar (3.5-liter operating volume) was used in a fermentor (New Brunswick Scientific Co., model M-19) equipped with automatic pH and temperature controls. An incubation temperature of 29 C and agitation speed of 250 rpm were employed. The inoculum was adjusted to give an initial optical density of 1.0 at 540 nm (Spectronic 20, Bausch and Lomb, path length 1 cm). The fermentations were run as follows: (a) 160 h, under anaerobic conditions provided by an aeration rate of 0.01 VVM (vol of sterile air/vol of medium/minute); (b) 168 h, under aerobic conditions provided by an aeration rate of 0.1 VVM; and (c) 72 h, under anaerobic conditions (0.01 VVM) and then 96 h, under aerobic conditions (0.1 VVM). Samples were taken periodically, and the pH of the fermentation medium was automatically maintained at 7.0 with either 20% NaOH or 20% HCl.

Analytical determinations. Microbial growth was measured by optical density at 540 nm and the dry weight was calculated from a calibration curve (Wang et al. 1979). Total reducing sugars were determined by colorimetry (Ting 1956). Total and nondialyzable nitrogen were analyzed by the Kjeldahl method (AOAC 1975). Samples were dialyzed against distilled water for 24 h. Dialyzable nitrogen was calculated by difference between pre- and postdialysis values. Citric, succinic, and lactic acids were measured in a gas-liquid chromatograph (Varian 3700-FID; column: 3% OV-17 on 80/100 Chromosorb WHP) (Pérez-Mendoza and García-Hernández, unpublished data). Vitamin B_{12} was determined in cells by spectrophotometry (Fisher 1954). All experiments were carried out at least in duplicate. Assays carried out with vitamin B_{12}, or citric, succinic, and lactic acid, were done in quadruplicate; all other assays were done at least in duplicate. The results are expressed as the means of the replicate values.

Definitions and bases of calculations. According to the terminology and definitions proposed by Gaden (Luedeking 1967; Wang et al. 1979), fermentation rates may be defined on two different bases, volumetric and specific rates. Typical units for volumetric rates are grams of product produced per liter per hour, grams of substrate used per liter per hour, or grams of cells produced per liter per hour. The volumetric rate at any instant in a batch fermentation may be determined by measuring the slope of the appropriate concentration-time curve. A specific rate may be defined as the rate of exchange of concentration per unit of cellular material. To determine a specific rate, one divides the volumetric rate at any instant by the bacterial density. The volumetric terms disappear and typical units are grams of product produced per hour per gram of cells, grams of substrate used per hour per gram of cells, or grams of cells produced per hour per gram of cells. In this manner, volumetric and specific rates were calculated for 12 h intervals from the concentration vs. time curves of our fermentations.

The expression and calculation of the yields presented in this work were obtained according to Stouthamer (1969). For the calculation of the yields ($^{Y}x/mol$) for the consumption of citric acid, the molar yields obtained in the fermentations in which the D.R. had been substituted by glucose (under the same aeration conditions) were used (Santana-Castillo 1983). The calculated values for the anaerobic fermentation were:

$^{Y}x/mol_G$ = 38.16 g of cells/mol of glucose consumed;
$^{Y}x/mol_L$ = 4.77 g of cells/mol of lactic acid consumed.

The value of 4.77 (one-eighth of 38.16), corresponding to the molar yield of lactic acid, was calculated taking into account that, for *P. shermanii*, the ratio of the molar yields for glucose and lactate was 8:1 (Lee et al. 1974; DeVries et al. 1973). The value of 38.16 falls within the range of molar yields obtained under anaerobic conditions for the propionic acid bacteria (Bauchop and Elsden 1961).

The values for the aerobic fermentation were:

$^{Y}x/mol_G$ = 69.59 g of cells/mol of glucose consumed;
$^{Y}x/mol_L$ = 8.70 g of cells/mol of lactic acid consumed.

From these yields, the amount of cells (in g) was calculated from both consumption of the reducing sugars (using glucose as reference) and lactic acid in each of the fermentations. For each fermentation, the value obtained by subtracting the sum of these respective values from the total was considered as the quantity of cells resulting from the consumption of citric acid. Using this calculated quantity of cells and the citric acid consumption, both the molar and gram cellular yields for citric acid were calculated for each experiment.

Results and Discussion

Anaerobic fermentation. The results are shown in Fig. 1. Under anaerobic conditions, the microorganisms grew exponentially during the first 48 h after the start of fermentation (2.52 g/liter). There was no marked stationary phase since the bacteria continued to grow, although more slowly, until the end of the fermentation (3.75 g/liter). The specific rates of growth (μ) decreased between 120 and 144 h (Fig. 4). At 48 h, the microorganisms consumed all the lactic acid (11.63 g/liter) and part of the sugars that were present (5.0 g/liter). The residual sugars (approximately 12 g/liter) in the D.R. are not metabolizable (Santana-Castillo 1983). During the fermentation, the citric acid and the nitrogen were consumed slowly, 5.0 and 0.35 g/liter, respectively. The production of vitamin B_{12} (2.0 mg/liter) followed a pattern similar to that of the growth curve. In addition to synthesizing vitamin B_{12}, the microorganism produced 1.8 g/liter of succinic acid in 72 h; more succinic acid was not produced during the remainder of the fermentation.

These data were analyzed as a function of their specific rates (q), which showed that μ and the consumption of substrates (q_s) were zero before the fermentation

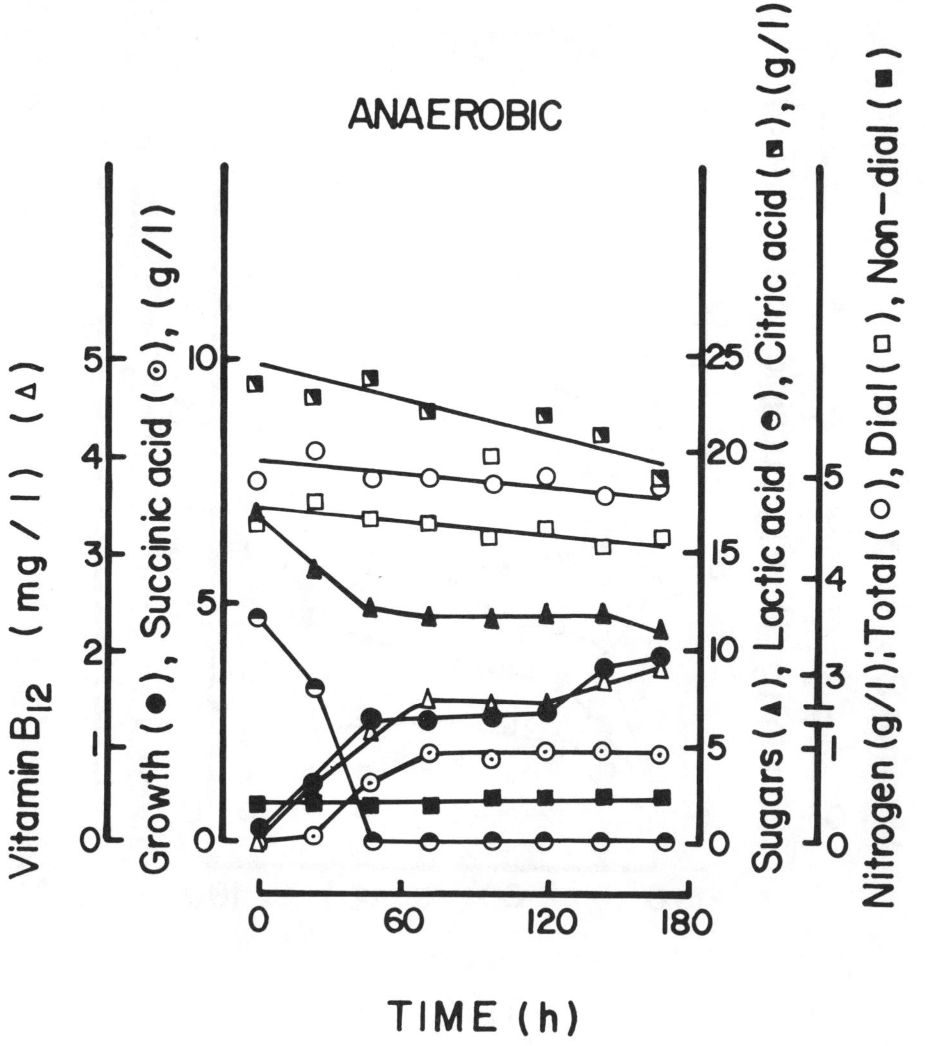

FIG. 1. Concentration vs. time curves of the anaerobic fermentation of a medium based on distillation residue by *Propionibacterium shermanii* ATCC 13673.

ended, with the exception that the q_s of citric acid was diminished after 48 h, after which time it remained constant. This implies that citric acid is the substrate that supported the metabolic activities after 48 h. The specific rates of production (q_p) for B_{12} and succinic acid were associated and semiassociated, respectively, with the growth of the microorganism and were classified as Type I and Type II fermentations, respectively, following Gaden (Luedeking 1967).

Aerobic fermentation. The results are shown in Fig. 2. In this fermentation, the growth (4.25 g/liter), the consumption of lactic acid (11.63 g/liter), sugars (4.65 g/liter), and nitrogen (0.35 g/liter) had a pattern similar to that of the anaerobic fermentation; whereas, the consumption of citric acid (17.28 g/liter) was

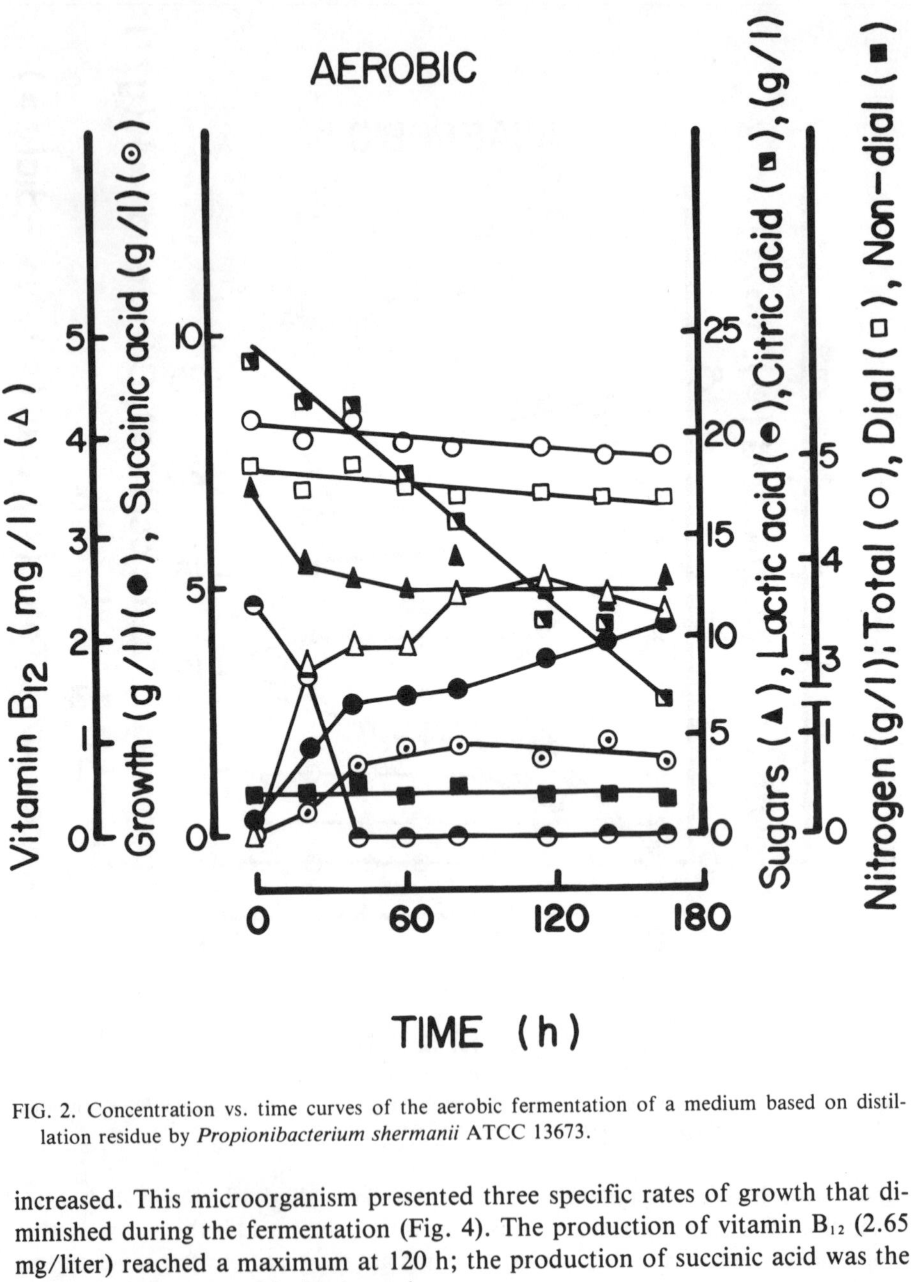

FIG. 2. Concentration vs. time curves of the aerobic fermentation of a medium based on distillation residue by *Propionibacterium shermanii* ATCC 13673.

increased. This microorganism presented three specific rates of growth that diminished during the fermentation (Fig. 4). The production of vitamin B_{12} (2.65 mg/liter) reached a maximum at 120 h; the production of succinic acid was the same as in the anaerobic fermentation.

Analyzing the results as a function of the specific rates, the q_s and μ showed a comportment similar to that encountered under anaerobic conditions, with the exception that μ showed a slight increase between 96 and 132 h of fermentation and the q_s of citric acid did not remain constant after 48 h but diminished slightly until the end of the fermentation. On the other hand, the specific rate of production (q_p) for vitamin B_{12} showed a kinetics associated with growth (Type I fermentation) at 0 to 48 h and a nonassociated kinetics (Type III fermentation) at 48 to

96 h with a maximum q_s at 72 h, the precise moment when μ approached zero. Thus, it appeared that a low specific growth rate promoted the synthesis of vitamin B_{12}. The specific rate of production of succinic acid showed a maximum at 24 h, corresponding to a Type II fermentation.

Anaerobic-aerobic fermentation. The results are presented in Fig. 3. The results of the fermentation carried out both anaerobically (72 h) and aerobically (96 h) demonstrated that all the parameters of the first stage of the fermentation were similar to those encountered under anaerobic conditions. In the second stage of

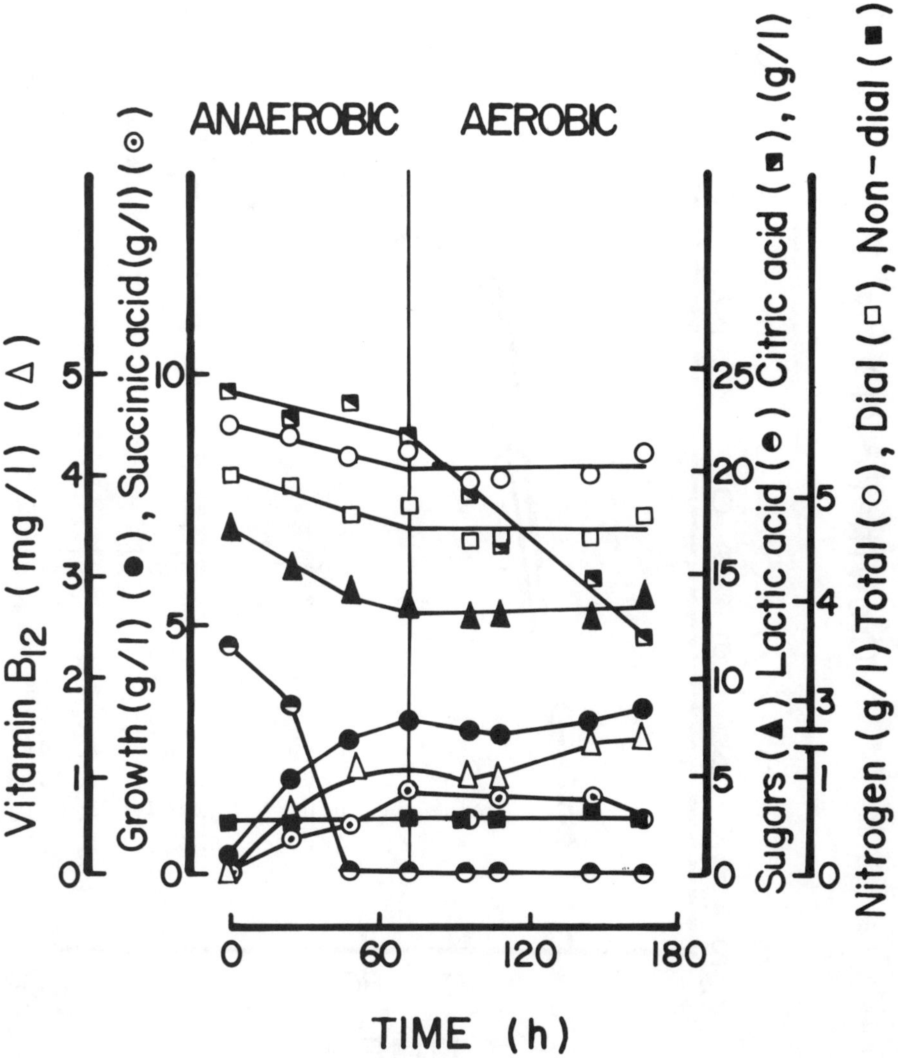

FIG. 3. Concentration vs. time curves of the anaerobic-aerobic fermentation of a medium based on distillation residue by *Propionibacterium shermanii* ATCC 13673.

fermentation, very little of the nutrients, with the exception of citric acid, were consumed, which apparently gave rise to a slight increase in growth and production of vitamin B_{12}. Under these conditions, the specific growth rates were diminished in a manner similar to that of the other two fermentations (Fig. 4).

The analysis of the results of this experiment as a function of the specific rates showed that growth and the consumption of substrates were similar to those of the anaerobic fermentation and that the kinetics of production of vitamin B_{12} and of succinic acid were associated and semiassociated, respectively, to the growth, thus corresponding to Type I and Type II fermentations. In the second stage of the fermentation, the specific rates of growth and of consumption of substrates were practically zero. The exception was citric acid, which was slowly consumed,

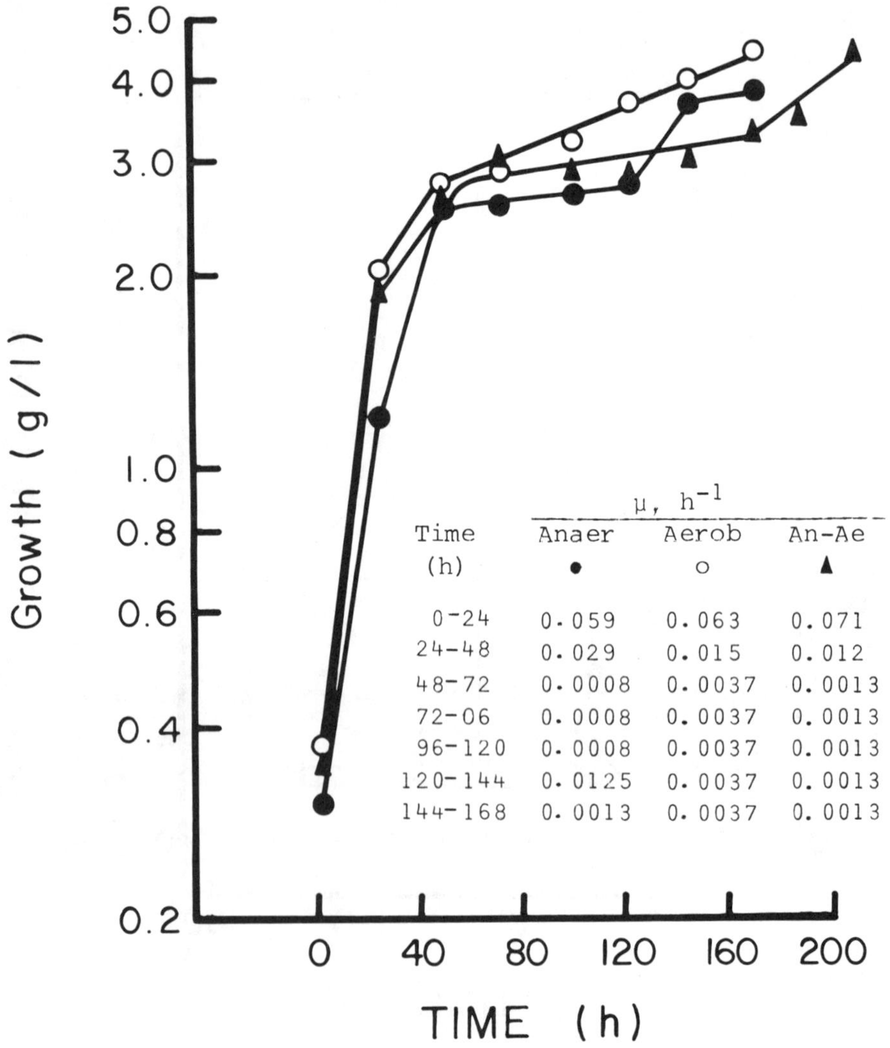

Time (h)	μ, h^{-1} Anaer ●	Aerob ○	An-Ae ▲
0-24	0.059	0.063	0.071
24-48	0.029	0.015	0.012
48-72	0.0008	0.0037	0.0013
72-06	0.0008	0.0037	0.0013
96-120	0.0008	0.0037	0.0013
120-144	0.0125	0.0037	0.0013
144-168	0.0013	0.0037	0.0013

FIG. 4. Specific growth rates (μ) of *Propionibacterium shermanii* ATCC 13673 in a medium based on distillation residue under different aeration conditions.

implying that it was the substrate that supported the metabolic activities in the second stage of the fermentation. There was a slight increase in the specific rate of production of vitamin B_{12} between 96 and 160 h of fermentation. Therefore, in the second stage, the production of vitamin B_{12} was not associated with the growth.

Yields and volumetric rates. These values were calculated as described in Materials and Methods. The results are shown in Tables 1 and 2.

TABLE 1. *Yields of* P. shermanii *ATCC 13673 under different aeration conditions in a medium based on a distillation residue from the processing of the Mexican lime* (Citrus aurantifolia *Swingle)*

Yields	Units	Anaerobiosis	Aerobiosis	Anaerob.-Aerob.
$Y_{x/s}$	g cells/g Glucose[a]	0.21	0.39	0.21
	g cells/g Lactic Acid	0.05	0.10	0.05
	g cells/g Citric Acid	0.30	0.05	0.12
$Y_{x/mol}$	g cells/mol Glucose	38.16	69.59	38.16
	g cells/mol Lactic Acid	4.77	8.70	4.77
	g cells/mol Citric Acid	62.92	10.05	24.47
$Y_{x/ATP}$	g cells/mol $ATP_{(G)}$	9.54	17.40	9.54
	g cells/mol $ATP_{(L)}$	9.54	17.40	9.54
	g cells/mol $ATP_{(C)}$	9.54	17.40	9.54
$Y_{x/N}$	g cells/g Nitrogen	9.48	12.13	9.51
$Y_{p/x}$	$\mu g B_{12}$/g cells	557.07	802.00	449.70
$Y_{p/s}$	$\mu g B_{12}$/mol substrate[b]	9763.29	11938.40	6673.04
$Y_{p/N}$	$\mu g B_{12}$/g Nitrogen	5252.37	6560.00	4279.45
$Y_{Suc/x}$	g Succinic acid/g cells	0.83	0.71	0.61
$Y_{Suc/s}$	g Succinic acid/mol substrate	10.90	9.23	6.71

[a] Reducing sugars; glucose used as reference.
[b] Mol total substrate = mol glucose + mol lactic acid + mol citric acid.

TABLE 2. *Volumetric Rates of* P. shermanii *ATCC 13673 under different aeration conditions in a medium based on a distillation residue from the processing of the Mexican lime* (Citrus aurantifolia *Swingle)*

Average Volumetric Rates[a]	Anaerobiosis	Aerobiosis	Anaerob.-Aerob.
$\mu g_{B_{12}}$/l/h	10.940	21.900	8.000
g succinic acid/l/h	0.020	0.020	0.020
g cells/l/h	0.020	0.020	0.020
g sugar/l/h	0.070	0.080	0.060
g lactic acid/l/h	0.240	0.300	0.240
g citric acid/l/h	0.020	0.110	0.060
g nitrogen/l/h	0.002	0.002	0.015

[a] These rates were calculated from the maximum consumption of substrate or from the maximum production of metabolites.

The production yields (Table 1) showed that, under aerobic conditions, the quantity of cells produced per gram citric acid ($^{Y}x/s = 0.05$) was less than those produced under the other two aeration conditions (0.30 and 0.12), implying that the aerobic fermentation was less efficient in producing cellular mass from citric acid. Also, in the aerobic fermentation, the quantity of ATP produced from the citric acid was low (0.6 mol of ATP/mol of citric acid). However, the highest yield of vitamin B_{12} ($^{Y}p/s$) produced in this series of fermentation was obtained under aerobic conditions.

Under aerobic conditions, the volumetric rates for vitamin B_{12} and for the consumption of sugars, of lactic acid, and of citric acid were higher than those obtained in the other two fermentations in which different aeration conditions were used (Table 2). All other volumetric rates were similar, regardless of aeration conditions.

The highest yield of vitamin B_{12} ($^{Y}p/x = 802.0$ μg/g cells) obtained in this work is approximately double that reported by Perlman (1977), 385.0 μg/g cells. However, from the point of view of production and productivity (2.6 mg/liter and 21.9 μg/liter/h, respectively), these results were less than those reported by Speedie and Hull (1960), 23 mg/liter and 136 μg/liter/h. Our continued work in these studies will be directed to increasing the production of vitamin B_{12}, possibly by exploiting the succinic acid that is produced during the fermentation since it is one of the precursors in the biosynthetic pathway of vitamin B_{12} (Florent and Ninet 1979).

Conclusions

Independent of aeration conditions, the microorganism presented similar specific growth rates (μ) during the first 48 h of fermentation. In all the fermentations, the microorganism consumed the three carbon sources present in the medium at the same time. Succinic acid was produced in all fermentations. The highest quantity of vitamin B_{12} was obtained in the aerobic fermentation. In general, *Propionibacterium* produced the highest yield of vitamin B_{12} when the fermentation was carried out in two stages (anaerobic-aerobic); however, our results demonstrated that aerobic fermentation was more efficient. The kinetics of production of vitamin B_{12} for each fermentation was as follows: anaerobic, associated with growth; aerobic, a combination of associated and nonassociated with growth; and anaerobic-aerobic, associated with growth during the first stage and nonassociated with growth during the second stage of the fermentation.

Acknowledgments

We thank P. Pérez-Gavilán for use of the chromatograph, N. H. Vázquez-Díaz and J. Pérez-Gavilán for their help in running the chromatograph, and M. A. López-Alpizar for typing the manuscript; all are from the Departamento de Biotecnología, Instituto de Investigaciones Biomédicas, UNAM, México. We also thank FIDEFRUT, México, for the distillation residue and Véronica Yakoleff of the Departamento de Inmunología, Instituto de Investigaciones Biomédicas, UNAM, Mexico, for the translation of the manuscript.

This work was presented at the 1984 SIM meeting held at Fort Collins, CO, with financial support from the Consejo Nacional de Ciencia y Tecnología (CONACyT), México.

Literature Cited

AOAC. 1975. *Official Methods of Analysis.* 12th ed. Association of Official Analytical Chemists, Washington, DC.

Bauchop, T., and S. R. Eldsden. 1961. Microbial growth in relation of their energy supply. *J. Gen. Microbiol.* 23:457–469.

DeVries, W., W. M. C. Van Wijck-Kapteijn, and A. H. Stouthamer. 1973. Generation of ATP during cytochrome-linked anaerobic electron transport in propionic acid bacteria. *J. Gen. Microbiol.* 76:31–34.

Fisher, R. A. 1954. Rapid spectrophotometric determination of vitamin B_{12} in microbial material. *Agric. Food Chem.* 11:951–953.

Florent, J., and L. Ninet. 1979. Vitamin B_{12}. Pages 457–519 *in* H. J. Peppler and D. Perlman, eds., *Microbial Technology.* Vol. 1, 2nd ed. Academic Press, Inc., New York.

Lapage, S. P., J. E. Shelton, and T. G. Mitchell. 1970. Media for the maintenance and preservation of bacteria. Pages 109–126 *in* J. R. Norris and D. W. Ribbons, eds., *Methods in Microbiology.* Academic Press, Inc., London.

Lee, In Hee, A. G. Fredrickson, and H. M. Tsuchiya. 1974. Growth of *Propionibacterium shermanii. Appl. Microbiol.* 28:831–835.

Lennette, E. H., E. H. Spaulding, and J. P. Truant. 1974. *Manual of Clinical Microbiology.* 2nd ed. Chaps. 37 and 38, Pages 363–375. American Society for Microbiology.

Luedeking, R. 1967. Fermentation process kinetics. Pages 202–210 *in* N. Blakebrough, ed., *Biochemical and Biological Engineering Science.* Vol. 1. Academic Press, Inc., New York.

Pérez-Mendoza, J. L. 1981. M. Sc. Thesis. Utilización de Subproductos Agroindustriales de Limón Mexicano por Vía Fermentativa para la obtención de Vitamina B_{12}. *Comisión Nacional de Fruticultura,* SARH, México.

Pérez-Mendoza, J. L., and F. García-Hernández. 1983. Fermentation of a waste-product from the industrial processing of the lime (*Citrus aurantifolia* Swingle) for vitamin B_{12} production. *Biotechnol. Lett.* 5:259–264.

Perlman, D. 1977. Microbial production of therapeutic compounds. Pages 283–287 *in* H. J. Peppler, ed., *Microbial Technology.* R. E. Krieger Publishing Co., Inc., New York.

Santana-Castillo, L. 1983. B. Sc. Thesis. Cinética de Crecimiento de *Propionibacterium shermanii* en un Subproducto de la Industria del Limón Mexicano. Universidad Nacional Autonoma de México, México.

Speedie, J. D., and G. W. Hull. 1960. Cobalamin producing fermentation process. U.S. Pat. 2,951,017, August.

Stouthamer, A. H. 1969. Determination and significance of molar growth yields. Pages 629–663 *in* J. R. Norris, and D. W. Ribbons, eds., *Methods in Microbiology.* Vol. 1. Academic Press, Inc., New York.

Ting, S. V. 1956. Rapid colorimetric methods for simultaneous determination of total reducing sugar and fructose in citrus juices. *Agric. Food Chem.* 4:363–366.

Wang, D. I. C., C. L. Cooney, A. L. Demain, P. Dunnill, A. E. Humphrey, and M. D. Lilly. 1979. *Fermentation and Enzyme Technology.* John Wiley & Sons, New York.

CHAPTER 65

Growth of Selected Yeasts on Enzyme-Hydrolyzed Potato Starch

E. Kombila-M., B. H. Lee, and R. E. Simard

Departement de sciences et technologie des aliments and Centre de recherche en nutrition, Universite Laval, Sainte-Foy (Quebec) G1K 7P4 and Agriculture Canada Research Station, Saint-Jean (Quebec) J3B 6Z8

A comparative study was carried out on the growth and yield of *Saccharomycopsis fibuligera, Saccharomyces diastaticus, Schwanniomyces alluvius,* and *Candida utilis* in different concentrations (2 to 20%) of α-amylase hydrolyzed potato starch with and without control of pH. The pH controlled fermentation did not significantly ($p < 0.05$) increase cell yields but the best results in batch culture were obtained with substrate levels between 4 and 8%, regardless of strain. The highest biomass (15 g/liter) and yield (31%) were obtained at 4 and 6% with *S. Diastaticus,* but the yields were generally lower than 26% with the other yeasts tested. The insufficient fermentation was not because of the unavailability of nitrogen source but because of repression of glucoamylase by metabolites in the medium. The fed-batch culture did not improve the biomass yields of *S. diastaticus.*

Introduction

Among the substrates that could be utilized as raw materials for the production of single cell protein (SCP), starches are especially attractive since they are easily metabolized by many yeasts. There is also a growing interest in the use of potato wastes because of the increase in costs of high-protein feeds.

A few processes have been described for the direct fermentation of potato wastes by *S. fibuligera* and *C. utilis* (Stogman 1976) and of cassava wastes by *C. lipolytica* (Revuz and Voisin 1980). However, the slow process by direct starch fermentation by yeasts is the prime rate-limiting factor. This inefficient process may be improved by simultaneous enzymatic hydrolysis and fermentation by an important food yeast.

In a previous work (Lee and Simard 1984), we reported the growth response of *S. fibuligera* on α-amylase hydrolyzed potato starch in batch and continuous cultures. The maximum productivities of 0.155 g/liter/h in batch and 0.290 g/liter/h in continuous culture with *S. fibuligera* were obtained at 30 g/liter of substrate.

In another attempt to select an appropriate yeast that exhibits a good yield on high concentrations of starch hydrolysates, several strains were compared for their growth response in different concentrations of substrates.

Materials and Methods

Yeast strains and maintenance. *Saccharomycopsis fibuligera* (ATCC-9947), *Schwanniomyces alluvius* (NRC-2509), *Saccharomyces diastaticus* (Labatt culture collection, Ontario), and *Candida utilis* (Universite Laval, Quebec) were checked for purity and maintained at 4 C on potato starch agar (PSA).

Growth medium. Yeast strains were grown in a basal medium (Haggstrom and Dostalek 1981) containing 0.5% yeast extract, 0.13% KH_2PO_4, 0.13% K_2HPO_4, 0.05% $MgSO_4 \cdot 7H_2O$, 0.5% $(NH_4)_2SO_4$, and 1% starch hydrolysate (pH 5.0).

Starch liquefaction. A 40% (w/v) suspension of soluble potato starch (Sigma) was prepared by agitation in 0.02 *M* phosphate buffer (pH 6.9) in the presence of 0.05% $CaCl_2$. When the temperature reached 50 C, 0.2% (v/v) of Thermamyl (α-amylase, Novo) was added and the temperature was raised to 90 C with continuous stirring for 1 h.

Batch culture conditions. A 1-ml portion of culture (approx. 10^8 cells/ml) activated in basal medium was transferred to four 500-ml baffled shake-flasks containing 100 ml of the same medium. Culture conditions, unless otherwise specified, were maintained at 21 C, pH 5.0, 300 rpm for 18 h.

To study the effects of substrate concentration (from 20 to 200 g/liter) without pH control, 1 ml of inoculum was added to 500-ml baffled shake-flasks each containing 100 ml of Haggstrom and Dostalek's (1981) medium. Cultures were grown at 21 C for 48 h in the same medium. To study the effects of substrate concentration with pH control, the inoculum was added at 1% (v/v) level in a 1-liter fermentor (Bio-Flo model C30, New Brunswick) with a working volume of 300 ml. Culture conditions were maintained at 30 C, pH 5.0, 300 rpm with aeration at 0.9 vvm for 48 h. Samples of 10 ml were withdrawn for the analyses.

In the experiments on the effects of carbon/nitrogen ratio, a basal medium was made up with 6% starch hydrolysate and supplemented with varying quantities (0.5 to 4.0% w/v) of ammonium sulfate without pH control.

Fed-batch culture. To the 1-liter Bio-Flow fermentor was added 297.5 ml of basal medium without substrate followed by 3.5 ml of inoculum containing 4.6×10^8 cells per ml. Hydrolyzed starch (40%) was added at a rate of 2.19 ml/h for 24 h. The conditions were the same as described for batch culture, but the pH was kept at 5.0 by the addition of 0.2 *N* NH_4OH. Samples were taken at the start and after 24 and 48 h.

Analytical methods. Growth was measured by cell counts using a Levy Chamber hemacytometer. The dry weight of the cells was determined by sampling 10 ml of culture, centrifuging, incubating the pellet in 5 ml of glucoamylase (Novo, final 1.0 unit) for 1 h at 55 C, washing the pellet and drying for 18 h at 105 C. The presence of starch was qualitatively tested with iodine solution. The concentration of reducing sugars was determined by the method of dinitrosalicylic acid (Bernfeld 1955) at 520 nm. The results were converted to starch equivalents (glucose $\times$ 0.9 = starch). The overall yield was calculated by dividing the resultant biomass (X) by the amount of starch consumed (Y = X − XoSo − S) where Xo and So are the cell and substrate concentration at t = 0.

Results and Discussion

Batch culture. The batch growth of four yeasts on starch hydrolysate ranging from 20 to 200 g/liter without pH control shows that, with the exception of *S. fibuligera,* the cell concentrations increased linearly up to 8% hydrolysates and then decreased with increasing substrate concentration. The maximum cell concentrations obtained were at 6 and 8% of substrate for *S. diastaticus, S. alluvius,* and *C. utilis,* whereas the growth of *S. fibuligera* was generally poor on this substrate. A comparison of biomass and yield obtained in different concentrations showed that the maximum biomass (10 g/liter) and yield (28.6%) were produced at 40 g/liter of substrate with *S. diastaticus.* On the other hand, *C. utilis,* which produced the highest cell concentration, yielded the lower biomass (4–7 g/liter) at this substrate level. A possible explanation for this low yield could be starvation of the cells due to the insufficient fermentable sugar available in the starch hydrolysate for optimum growth of the nonamylolytic, *C. utilis.* The cell concentration of *S. alluvius* (6–9 × 10^8 cells/ml) was higher than that of *S. fibuligera* (4.6 × 10^8 cells/ml) at 4, 6, and 8% of substrates but yield of *S. alluvius* was less than than of *S. fibuligera.*

As the yield of biomass cannot be related to substrate assimilation in experiments where pH of the medium is not constant, further experiments were conducted in 6 and 8% of substrate with controlled pH (Table 1). The results showed that the regulation of pH and temperature (from 21 to 30 C) did not significantly affect the final utilization of starch hydrolysate and the yield. Nevertheless, the maximum biomass (15.3 g/liter) and yield (31%) were obtained with *S. diastaticus,* but this value is much lower than the theoretical value. Operating costs for SCP production would rise steeply unless the process was operated at above 50% μmax (Abbott and Clamon 1973). As the rate of starch fermentation would be limited in a batch culture, this probably represents a reasonable estimate of the maximum growth rate obtained in high concentrations of

TABLE 1. *pH controlled fermentation of potato starch hydrolysate by yeasts*

Yeasts	Cells ($N \times 10^8$/ml)	S_0 (g/liter)	S (g/liter)	$S_0 - S$ (g/liter)	$X - X_0$ (g/liter)	Y (%)
S. fibuligera	5.8	60	30.42	29.58	7.71 ± 0.27	26.10
	3.0	80	31.70	48.30	5.94 ± 0.45	12.30
C. utilis	17.0	60	30.46	29.54	7.49 ± 0.19	25.36
	16.0	80	28.56	51.44	10.16 ± 1.60	19.75
S. diastaticus	8.0	60	10.60	49.40	15.29 ± 2.20	30.95
	4.2	80	12.42	67.58	7.74 ± 0.57	11.45
S. alluvius	6.3	60	13.36	46.64	12.43 ± 1.30	26.65
	6.5	80	12.12	67.88	10.73 ± 3.20	15.81

S_0 = original starch hydrolysate; S = residual starch; $S_0 - S$ = assimilated starch; $X - X_0$ = biomass; Y = yield ($X - X_0/S_0 - S$)

substrate. The relatively high residual of substrate in these experiments suggests that glucoamylase activity may have been limited by catabolite repression, as a negative glucose effect on glucoamylase has been described before (Searle and Tubb 1981).

Since another possible explanation for these low yields is nitrogen limitation (C/N ratio), the effect of ammonium sulfate on growth was examined. Increasing the concentrations of ammonium sulfate without pH control (Table 2) showed that insufficient fermentation by *S. diastaticus* was not because of nitrogen limitation. Since the medium components containing ammonium sulfate and yeast extracts were autoclaved, it was also important to establish whether or not browning reactions during autoclaving had a significant effect on the availability of nitrogen. Comparison of experiments with solutions of ammonium sulfate and yeast extract filter-sterilized and added separately to the autoclaved medium showed that nonenzymatic browning did not exert its influence on the biomass yields (data not shown).

TABLE 2. *Effect of ammonium sulfate on growth and yield of* S. diastaticus *without pH controlled fermentation (48 h)*

$(NH_4)_2SO_4$ (%)	Cells ($N \times 10^8$/ml)	$X - X_0$ (g/liter)	S (g/liter)	$S_0 - S$ (g/liter)	Y (%)
0.5	1.73	12.7	35.2	24.8	51.0
1.0	1.40	11.1	27.5	32.5	34.2
2.0	1.25	10.1	32.4	27.6	36.6
3.0	1.21	9.3	29.7	30.3	30.5
4.0	1.19	9.2	16.4	43.6	21.1

$X - X_0$ = biomass; S = residual starch; $S_0 - S$ = assimilated starch; Y = yield ($X - X_0/S_0 - S$)

During the fermentation of high concentrations of starch and dextrin, ethanol may be produced in sufficient amounts to inhibit yeast growth in addition to the inhibitory effect of various metabolites including glucose and dextrin on α-amylase and glucoamylase.

Besides α-amylase and glucoamylase, a debranching enzyme such as pullulanase is essential for the complete hydrolyses of starch (Sills and Stewart 1982) and this enzyme activity may have been limited. *S. diastaticus* is a typical brewing yeast leaving the larger dextrins unfermented, and three genes of glucoamylase have been found to control the fermentation of dextrin (Errat and Stewart 1981). Although Ghoul and Engasser (1982) increased the yield of *C. utilis* by fed-batch culture (50% biomass conversion yield after 20 h) in α-amyloase and glucoamylase-hydrolyzed cassava starch, the fed-batch experiment with *S. diastaticus* did not increase the yield significantly (Table 3). Other experiments to improve yields are underway.

TABLE 3. *Yield and residual substrate for fed-batch culture*

Fermentation time	X (g/liter)	S (g/liter)	$S_0 - S$ (g/liter)	Y (%)
0	0.78	—	—	—
24	9.82	39.86	20.14	44.86
48	14.24	10.66	49.34	27.28

X = biomass; S = residual starch; $S_0 - S$ = assimilated starch; Y = yield ($X/S_0 - S$)

Acknowledgments

Contribution No. 17 of centre de recherche alimentaire de St-Hyacinthe, Agriculture Canada.

Literature Cited

Abott, B. J., and A. Clamon. 1973. The relationship of substrate, growth rate and maintenance coefficient to single cell protein production. *Biotechnol. Bioeng.* 15:117–127.

Bernfeld, P. 1955. Amylase α and β. *Meth. Enzymol.* 1:149–158.

Errat, J. A., and G. G. Stewart. 1981. Fermentation studies using *Saccharomyces diastaticus* yeast strains. *Dev. Ind. Microbiol.* 22:557–587.

Ghoul, M., and J. R. Engasser. 1983. Nouveau procede d'enrichissement proteique de manioc par hydrolyse enzymatique et culture de *Candida utilis. Microbiol. Aliment. Nutr.* (France) 1:277–283.

Haggstrom, M. H., and M. Dostalek. 1981. Regulation of a mixed culture of *Streptococcus lactis* and *Saccharomycopsis fibuligera. Eur. J. Appl. Microbiol. Biotechnol.* 12:216–219.

Lee, B. H., and R. E. Simard. 1984. Growth of *Saccharomyces fibuligera* on enzyme hydrolyzed potato starch. *Dev. Ind. Microbiol.* 25:459–466.

Revuz, B., and D. Voisin. 1980. Le manioc proteine. *Ind. Aliment. Agric.* 97:1079–1084.

Searle, B. A., and R. S. Tubb. 1981. Regulation of amyloglucosidase production by *Saccharomyces diastaticus. J. Inst. Brew.* 87:244–247.

Sills, A. M., and G. G. Stewart. 1982. Production of amylolytic enzymes by several yeast species. *J. Inst. Brew.* 88:313–316.

Stogman, H. 1976. Production of symba-yeast from potato wastes. Pages 167–179 *in* G. C. Birch, K. J. Parker and J. T. Worgam, eds., *Food from Waste.* Applied Science Publ., London.

CHAPTER 66

Preservation of Antibiotic Production by Representative Bacteria and Fungi

R. L. MONAGHAN AND S. A. CURRIE

Merck and Company, P.O. Box 2000, Rahway, New Jersey 07065

A 2-year study was performed to discover the effect of preservation methods on recovery of antibiotic production and viability of eight microorganisms. We tested *Bacillus subtilis, Pseudomonas fluorescens, Streptomyces flavogriseus, Streptomyces cattleya, Micromonospora echinospora, Nocardia lactamdurans, Verticillium lamellicola,* and a *Cephalosporium* species. These organisms produce bacillin, oxamycin, epithienamycin, thienamycin, gentamycin, cephamycin C, the antifungal MSD A43F, and cephalosporin C plus penicillin N. The preservation techniques used were freezing on slants, freezing vegetative cells, lyophilization, drying on soil, drying on paper discs, and drying on silica gel. Overall frozen slants and frozen vegetative cells resulted in the best recovery of antibiotic production.

INTRODUCTION

The preservation of microorganisms is essential so that the discoveries made by microbiologists can be shared by investigators over reasonable periods of time and distance. The first requirement of a good preservation method is that the cells being preserved remain viable. In a recent review (Onions 1983), we are reminded that there are a number of preservation methods available to the microbiologist that are suitable for the retention of viable cultures. As members of a research organization that has screened for and discovered antibiotics we were particularly interested in the question of what preservation methods were suitable for retaining both viability and antibiotic production. Since our screening operation included *Streptomyces* species, other actinomycetes, other bacteria, and fungi, for this study we chose two representatives of each group that made antibiotics. We also selected for this study some organisms that had exhibited problems when preserved as lyophilized cultures.

MATERIALS AND METHODS

Microorganisms. The microorganisms used in this study are listed in Table 1. The source of all cultures was lyophilization tubes maintained at Merck and Co.

Preservation. A single slant of each culture was used to inoculate liquid and solid growth media from which all subsequent preserved culture samples were derived. Lyophilization tubes were prepared from slants of the above organisms. The

TABLE 1. *Cultures preserved*

Culture Name	Antibiotic(s) Produced	Reference
Streptomyces cattleya	thienamycin	Kahan et al. 1979
Streptomyces flavogriseus	epithienamycin	Stapley et al. 1981
Bacillus subtilis	bacillin	Foster and Woodruff 1946
Pseudomonas fluorescens	oxamycin	Stapley et al. 1968
Verticillium lamellicola	MSD A43F	Onishi et al. 1980
Cephalosporium sp.	cephalosporin C + penicillin N	S. B. Zimmerman (personal communication)
Micromonospora echinospora	gentamycin	Weinstein et al. 1963
Nocardia lactamdurans[a]	cephamycin c	Stapley et al. 1972

[a] Reclassified based upon cell wall analysis.

growth was scraped into a 15% skim milk solution. Glass ampules were filled with 0.15-ml aliquots, frozen in a dry ice solvent bath, and dried under vacuum overnight. Ampules were sealed under vacuum and stored at 4 C.

Frozen vegetative cells were prepared from cultures grown in a liquid growth medium used to start each fermentation. After growth, glycerol was added as a cryoprotective agent to a final concentration of 10%. Screw-capped borosilicate glass vials (containing 2 ml) were frozen and stored in the vapor phase of a liquid nitrogen freezer.

Frozen slants were prepared using cultures grown on agar (15 ml per 22 × 175-mm tube) slanted to provide maximum surface area. Slants were frozen and stored in a mechanical freezer at −70 C.

Soil tubes were prepared using a mixture of 3.5 g of an African violet soil mixture (Clinton Nursery Products, Clinton, CT) plus 0.5 g $CaCO_3$. Cells were scraped from slants and mixed with the autoclaved soil. The soil was dried with daily turning at 28 C. After drying (14 d) soil tubes were stored at 4 C.

Paper discs were prepared by saturating 0.5-inch discs of Whatman #1 paper with 0.15 ml of the liquid culture used to start each fermentation. Discs were air dried and then were stored at 4 C.

Silica gel tubes were prepared by adding 1 ml of the liquid culture used to start each fermentation to a 2-cm deep layer of silica gel in a 16 × 125-mm test tube immersed in an ice bath. Silica gel tubes were then stored at 4 C.

Fermentation. All cultures of liquid growth medium were inoculated by using the entire contents of a preserved source. The bacterial growth medium was composed of: 1 g dextrose, 10 g soluble starch, 3 g beef extract, 5 g Ardamine pH, 5 g NZAmine E, 0.05 g $MgSO_4 \cdot 7H_2O$, 0.182 g KH_2PO_4, 0.190 g Na_2HPO_4, per 1000 ml distilled water, adjusted to pH 7–7.2 with NaOH then 0.5 g $CaCO_3$. The fungal growth medium was composed of: 10 ml of a trace elements solution (which contains 1 g $FeSO_4 \cdot 7H_2O$, 1 g $MnSO_4O \cdot 4H_2O$, 0.025 g $CuCl_2 \cdot 2H_2O$, 0.1 g $CaCl_2$, 0.056 g H_3BO_3, 0.019 g $(NH_4)_6MoO_2 \cdot 4H_2O$, and 0.2 g $ZnSO_4 \cdot 7H_2O$ per 1000 ml distilled water), 5 g cornsteep liquor, 40 g tomato paste, 10 g oat flour, and 10 g glucose per 1000 ml distilled water, pH 6.8.

The growth medium was incubated with agitation at 28 C for 1 d for *Bacillus subtilis* and *Pseudomonas fluorescens*. Two days of incubation were used for *Streptomyces cattleya, Streptomyces flavogriseus, Nocardia lactamdurans,* and

Verticillium lamellicola. Three days of incubation were used for *Micromonospora echinospora* and the *Cephalosporium* sp. After growth, a 5% inoculum was used to inoculate antibiotic production media. The antibiotic production medium for *B. subtilis* contained 20 g tomato paste, 10 g primary yeast, 20 g dextrin, and 0.005 g $CoCl_2 \cdot 6H_2O$ per 1000 ml distilled water, pH 7.2–7.4. Antibiotic titer was estimated after 2 d incubation with agitation at 28 C.

The antibiotic production medium for *P. fluorescens* was the same as the fungal growth medium described above. Antibiotic titer was estimated after 2 d incubation with agitation at 28 C.

The antibiotic production medium for *S. cattleya* contained 10 g glycerol, 15 g cornsteep liquor, 10 g cottonseed meal, 15 g distillers solubles, 0.01 g $CoCl_2 \cdot 6H_2O$, and 2.5 ml Polyglycol P2000 per 1000 ml distilled water, pH 7.3, then 3 g/L $CaCO_3$. Antibiotic titer was estimated after 4 d incubation with agitation at 28 C.

The antibiotic production medium for *S. flavogriseus* was the same as that used for *B. subtilis* described above. Antibiotic titer was estimated after 4 d incubation with agitation at 28 C.

The antibiotic production medium for *N. lactamdurans* contained 4.8% cornstarch, 0.5% distillers solubles, 0.1% soybean meal, 0.8% glycerol, 0.5% NZAmine E, 0.01% $FeSO_4 \cdot 7H_2O$, 0.05% Na_2SO_4, 0.05% DL lysine, and 1000 ml tap water, pH 7.0. Antibiotic titer was estimated after 4 d incubation with agitation at 28 C.

The antibiotic production medium for *M. echinospora* contained 30 g soybean meal, 40 g cerelose, 1 g $CaCO_3$ and 0.008 g $CoCl_2 \cdot 6H_2O$ per 1000 ml tap water, pH 7.0. Antibiotic titer was estimated after 4 d incubation with agitation at 28 C.

The antibiotic production medium for *V. lamellicola* contained 60 g sucrose, 10 g Bacto peptone, 2 g Ardamine pH per 1000 ml distilled water, pH 6.0. Antibiotic titer was estimated after 4 d incubation with agitation at 25 C.

The antibiotic production medium for *Cephalosporium* sp. contained 10 g cornsteep liquor, 20 g Edamine, 40 g glucose, and 2.5 ml mineral oil per 1000 ml distilled water, pH 6.8, then 10 g/L $CaCO_3$. Antibiotic titer was estimated after 3 d incubation with agitation at 28 C.

Antibiotic assays. Antibiotic titer was estimated by measuring the zone of inhibition caused by the microbial broth in a disc diffusion assay against indicator organisms known to be sensitive to the antibiotic produced by each preserved organism. Broth samples originating from all six preservation methods were tested on each plate to minimize the effect of plate-to-plate assay variations.

Viability. Viability was estimated by inoculating the growth medium with the entire contents of a particular preservation method. After shaking to disperse the inoculum a one-ml sample was immediately removed and used to inoculate solid growth medium at appropriate dilutions. (The remainder was used after growth to inoculate antibiotic production media.)

At each time period (just after preservation, and 2 yr later), at least two representatives of each preservation method were opened. Each was plated at a minimum of three dilutions with three plates at each dilution to estimate viable counts. Each preservation sample was tested in at least duplicate flasks for antibiotic production. Antibiotic production for each production flask was then estimated from a minimum of two assay plates.

Results

The results for all eight microorganisms were combined by averaging the number of viable cells and the antibiotic activity produced as measured by zones of inhibition for each preservation method. Frozen slants as prepared in this study had the highest counts just after preservation and after 2 yr (Table 2).

TABLE 2. *Preservation of antibiotic production and viability of eight microorganisms*

Preservation Method	Antibiotic Activity (Average Zone mm)		Viability (Average Count × 10^7)	
	Initial	Two Years[a]	Initial	Two Years[a]
Lyophilization	22.8	10.8	24.8	14.8
Frozen Cells	22.8	18.7	340.8	52.5
Frozen Slants	23.5	19.8	358.3	118.6
Dried Soil	19.0	16.1	10.4	11.9
Dried Discs	22.6	7.6	2.0	0.3
Silica Gel	18.4	11.8	290.5	0.6
Original Lyophilized Culture Sources	23.9	ND	ND	ND

[a] *Pseudomonas* test 1.5 yr.

Frozen slants were not the preferred method for general long-term viability retention. Drying on soil and lyophilization retained viability better than all other methods. Only dried soil tubes, however, retained viable counts $>10^2$ organisms per tube for all eight organisms. The larger size of these tubes may be justified especially with cultures that lose viability over time.

Frozen slants produced the best levels of antibiotic production just after preservation and after 2 yr of storage (Table 2). Frozen vegetative mycelia had the next best retention of antibiotic production. All eight microorganisms produced antibiotic activity when preserved as frozen slants or as frozen vials under the rigid requirement that activity be produced after standard growth and production conditions. The other four methods had at least two failures to produce antibiotic activity out of the 16 tests conducted.

There were marked differences in the general response of each microorganism to preservation of antibiotic synthesis (Table 3). *N. lactamdurans, M. echinospora,* and the *Cephalosporium* sp. lost over time the ability to produce significant levels of antibiotic activity under standard fermentation conditions. Only an initial loss in synthetic activity was seen in *P. fluorescens*. Loss of antibiotic production was minimal in the *Streptomyces cultures.*

Some organisms' specific preservation results were exceptions to the general trends. For the two *Streptomyces* cultures, lyophilization was good for viability retention but was one of the worst methods for retention of antibiotic synthesis. For retention of antibiotic production, frozen vegetative cells were the method of choice for *S. cattleya* and *S. flavogriseus.* For *B. subtilis* and *P. fluorescens,*

TABLE 3. *Average antibiotic zone—All preservation methods*

Organism	Control[a]	Initial	Two Years[b]
Bacillus subtilis	30.0	31.0	23.4
Pseudomonas fluorescens	19.5	10.4	10.8
Cephalosporium sp.	32.5	22.8	12.6
Verticillium lamellicola	23.0	23.8	16.9
Streptomyces flavogriseus	24.5	24.8	22.2
Streptomyces cattleya	21.5	20.1	16.7
Nocardia lactamdurans	20.0	19.2	8.4
Micromonospora echinospora	20.5	19.5	2.0

[a] Control = Original lyophilization tube.
[b] *Pseudomonas* test 1.5 years.

lyophilization resulted in the optimum retention of viability and antibiotic synthesis. Lyophilization of the two fungi resulted in both poor initial counts and low retention of antibiotic synthesis. Frozen slants, frozen vegetative cells, and cells dried on soil preserved the fungi's ability to produce antibiotic activity in this study. For *N. lactamdurans* and *M. echinospora,* frozen vegetative cells and soil tubes were the best methods of preservation. These methods retained levels of antibiotic activity comparable to the original control fermentation for *N. lactamdurans* but not for *M. echinospora.* A culture selection step or a culture propagation step may be required to recover original antibiotic levels from *M. echinospora.*

Conclusions

Frozen slants and frozen vegetative cells appear to offer the best features of viability retention and retention of antibiotic production for general microorganism preservation during the short term. The mechanical breakdown that ended this study at 2 yr strongly argued for a backup to mechanical freezers. Dried soil preservation should be considered when ultracold storage is not practical. Maximum preservation of viability is not necessarily the same as preservation of maximum antibiotic production.

Acknowledgments

We wish to thank Dr. Judith A. Bland (Merck and Co.) for providing us with plates of assay organisms for this study.

Literature Cited

Foster, J. W., and H. B. Woodruff. 1946. Bacillin, a new antibiotic substance from a soil isolate of *Bacillus subtilis. J. Bacteriol.* 51:363–369.

Kahan, J. S., F. M. Kahan, R. Goegelman, S. A. Currie, M. Jackson, E. O. Stapley, T. W. Miller, A. K. Miller, D. Hendlin, S. Mochales, S. Hernandez, H. B. Woodruff, and J. Birnbaum. 1979. Thienamycin, a new β-lactam antibiotic. I. Discovery, taxonomy, isolation and physical properties. *J. Antibiot.* 32:1–12.

Onions, A. H. S. 1983. Preservation of Fungi. Pages 373–390 *in* J.E. Smith, D. R. Berry, and B. Kristiansen, eds., *The Filamentous Fungi, Vol. 4, Fungal Technology*. E. Arnold, London.

Onishi, J. C., G. L. Rowin, and J. E. Miller, Jr. 1980. Antibiotic A43F. U.S. Pat. 4,201,771.

Stapley, E. O., T. W. Miller, and M. Jackson. 1968. Production of the Antibiotic D-4-Amino-3 Isoxazolidone by bacteria. *Antimicrob. Agents Chemother.* 1968:268–273.

Stapley, E. O., M. Jackson, S. Hernandez, S. B. Zimmerman, S. A. Currie, S. Mochales, J. M. Mata, H. B. Woodruff, and D. Hendlin. 1972. Cephamycins, a new family of β-lactam antibiotics. I. Production by actinomycetes, including *Streptomyces lactamdurans,* sp n. *Antimicrob. Agents Chemother.* 2:122–131.

Stapley, E. O., P. J. Cassidy, J. Tunac, R. L. Monaghan, M. Jackson, S. Hernandez, S. B. Zimmerman, J. M. Mata, S. A. Currie, D. Daoust, and D. Hendlin. 1981. Epithienamycins-novel β-lactams related to Thienamycin. I. Production and antibacterial activity. *J. Antibiot.* 34:628–636.

Weinstein, M. J., G. M. Leudemann, E. M. Oden, and G. H. Wagman. 1963. Gentamycin, a new broad-spectrum antibiotic complex. *Antimicrob. Agents Chemother.* 1963:1–7.

CHAPTER 67

Improved Cloning and Transfer of *Pseudomonas* Plasmid DNA

G. E. Pierce,* J. B. Robinson,* G. E. Garrett,*
D. K. Terman,* and S. A. Sojka**

**Battelle Columbus Laboratories, Columbus, Ohio 43201
and **Occidental Chemical Corporation, Grand Island, New York 14072*

An improved procedure for cloning and banking of plasmids of *Pseudomonas* spp is described. DNA fragments of these plasmids, generated by *Hind* III endonuclease, were cloned into the cloning vector pVK100. This vector has single restriction sites allowing insertional inactivation of tetracycline and kanamycin markers and contains the RK2 replication functions allowing it to be maintained in a wide range of gram-negative bacteria. The hybrid plasmids were banked in *Escherichia coli* K-12 strain AC80. The hybrid plasmids were transferred into *Pseudomonas putida* strain KT2440 by conjugal cotransfer with the helper plasmid pRK2013. This system provides a technique for establishing a clone bank of *Pseudomonas* DNA in *E. coli* for ease of maintenance, while expression can be monitored in *Pseudomonas* hosts.

Introduction

The transformation of *P. putida* (Chakrabarty et al. 1975; Nagahari and Sakaguchi 1978) and *P. aeruginosa* (Mercer and Loutit 1979) with plasmid and bacteriophage DNA is well documented. Bagdasarian et al. (1981) have also reported the transformation of both *P. putida* and *P. aeruginosa* strains with broad range, high copy number, RSF1010 derived vectors. However, the frequency of transformation of environmental pseudomonad plasmid DNA, even when RSF1010 derived vectors are used, can be quite low. It is also evident, for some DNA sequences, that insertion into a high copy number plasmid may not be advantageous and may result in decreased stability. Plasmid vectors containing the RK2 replication functions are low copy number plasmids but are maintainable in a wide range of gram-negative bacteria (Kahn et al. 1979). The cloning vectors pRK290 (Ditta et al. 1980) and pVK100 (Knauf and Nester 1982) both contain the RK2 replication functions and thus can be maintained in a broad range of gram-negative bacteria; however, they are no longer self-transmissible and, therefore, require the use of a helper plasmid (PRK2013) to be mobilized (Ditta et al. 1981). Because the helper plasmid pRK2013 will only replicate in *E. coli*, the use of the binary vector system provides the researcher the ability to first transform DNA into *E. coli* at high frequency and then to mobilize the plasmid into any number of gram-negative hosts, again at high frequency.

For the cloning of pseudomonad DNA we employed the vector pVK100 because it contains two selectable markers (tetracycline and kanamycin) whereas pRK290 only contains one selectable marker. The above binary cloning system has been used in our laboratory to clone sequences of DNA from plasmids that encode for the degradation of chlorinated hydrocarbons. This technique will facilitate our research into the mechanisms and regulation of plasmid encoded degradation and will enable us to construct improved and expanded degradative pathways.

Materials and Methods

Bacterial strains and plasmids. Wild type bacterial strain *Pseudomonas* sp. strain H-5, active against chlorotoluene compounds, was originally isolated from soil samples from western New York (Vandenbergh et al. 1981). The cloning vector pVK100 (Tc^R, Km^R), a derivative of the RK2 plasmid (Knauf and Nester 1982), supplied by M. Lidstrom (University of Washington, Seattle, WA), was used to generate a clone bank of *Pseudomonas* sp. strain H-5 plasmid DNA. *E. coli* K12 strain AC80 (Tc^S, Km^s), supplied by L. Bopp (General Electric, Niskayana, NY), was employed as the host for all transformations. The helper plasmid pRK2013 (Km^R), in *E. coli* CSR 603, used to mobilize pVK100 hybrid plasmids in all conjugations, was supplied by M. Lidstrom. *P. putida* strain 2440 (Tc^S, Km^S), used as the host in all conjugations, was supplied by M. Bagdasarian.

Media and cultural conditions. Pseudomonas sp. strain H-5 was maintained at 25 C on TN agar (Vandenbergh et al. 1981) supplemented with 150 ppm of 3,4-dichlorotoluene (Pierce et al. 1983). The plasmid pVK100, in *E. coli* HB101, was maintained at 37 C on L-agar (Bolivar and Backman 1979) supplemented with 25 μg of tetracycline per ml. *E. coli* CSR 603 (pRK2013) and *E. coli* K12 strain AC80 were maintained at 37 C on L-agar. *P. putida* strain 2440 was maintained at 30 C on M9e agar (Bolivar and Backman 1979).

Transformants of *E. coli*, containing pVK100 with inserted H-5 plasmid DNA, were maintained at 30 C on L-agar supplemented with 25 μg of tetracycline per ml. Transconjugants of *P. putida* 2440, containing pVK100 with inserted H-5 plasmid DNA, were maintained at 30 C on M9e agar supplemented with 50 μg of tetracycline per ml.

Selection of presumptive transconjugants was conducted at 30 C using King's *Pseudomonas* Selective Base plus CN supplement (Oxoid, St. Louis, MO) to which 100 μg of tetracycline per ml had also been added.

Isolation and purification of plasmid DNA. E. coli and *Pseudomonas* spp. plasmid DNA was isolated according to the Ish-Horowitz modification of the Birnboim and Doly procedure (Maniatis et al. 1982). Plasmid DNA from *Pseudomonas* spp. was further purified by cesium chloride isopycnic ultracentrifugation prior to restriction and ligation procedures.

Gel electrophoresis. For gel electrophoresis, horizontal gels (11 × 14 cm, H1 Gel Unit, BRL, Bethesda, MD) of 0.5–0.7% agarose were employed. Gels were run submerged at 5 V/cm in TEA buffer; 40 m*M* Tris and 2 m*M* EDTA adjusted to pH 8.0 with glacial acetic acid. Gels were either stained by incorporating ethidium bromide in the gel, at 0.5 μg/ml, or by staining after electrophoresis with ethidium bromide, 1.0 μg/ml, in TEA buffer. Gels were photographed under transillumination (wavelength, 310nm).

Restriction and ligation conditions. Restriction endonuclease *Hind* III, T_4-DNA ligase, and all other restriction enzymes were obtained from New England Biolabs. Reaction conditions used were those described by the supplier. Restrictions were conducted at 37 C for 60–120 min then heated at 70 C for 10 min to inactivate the restriction enzyme. Ligation reactions were incubated overnight at 16 C. The total volume of the ligation reaction was 10–20 μl, with a 2-molar ratio excess of insert to vector DNA. Ligation was terminated by heating in a 70 C water bath for 10 min.

Transformation and conjugation. E. coli strain AC80 was transformed according to the procedure of Kushner (1978) as modified by Pierce et al. (1984). Presumptive transformants were selected from L-agar supplemented with 10 μg of tetracycline per ml. Presumptive transformants were then picked onto L-agar supplemented with tetracycline and kanamycin to verify insertional inactivation of the kanamycin resistance gene of pVK100. Three-way conjugal matings were performed according to the procedure of Ditta et al. (1980) by mixing 10^9 cells each of donor, helper, and host and then filtering the suspension onto 0.45 μm filters (Millipore). The filters were then incubated at 30 C on nonselective medium (L-agar) for 3–6 h before plating on selective medium (described above).

Results

The plasmid profiles and *Hind* III generated plasmid digests of plasmid DNA from presumptive transformants of *E. coli* AC80 (pVK100::*Hind* III-H-5-plasmid) are shown, respectively, in Figs. 1 and 2. From Figs. 1 and 2 it is apparent that the presumptive transformants, TF254, TF357, and TF669 contained insert DNA. Using the above techniques, a total of eight distinct inserts of *Hind* III generated H-5 plasmid DNA have been inserted into pVK100 and cloned into *E. coli* AC80.

Cloned H-5 plasmid DNA was transferred to *P. putida* 2440 using the three-way conjugal mating described above. Depending on the size of the *Hind* III generated H-5 plasmid DNA inserted, the frequency of transconjugants that were tetracycline resistant was 2×10^{-2} to 5×10^{-3} per recipient. Whole plasmid and *Hind* III plasmid digests of the presumptive transconjugants generated from transformants TF254 and TF357 are shown in Fig. 3. The comparison between the transconjugant A4, generated from transformant TF254 and TF254 is more easily made when the gel, shown in Fig. 4, is examined. The two transconjugants A4 and A22 were generated from the transformant TF254, while the transconjugant B2 was generated from the transformant TF357. Fig. 3 shows that the plasmid profiles of A4 and A22 are not identical. The apparent difference seen between the plasmids of A4 and A22 is not unusual even though they were generated from the same transformant. Nakazawa (in press) reports that such differences are common when *Pseudomonas* DNA is cloned into *E. coli* and then

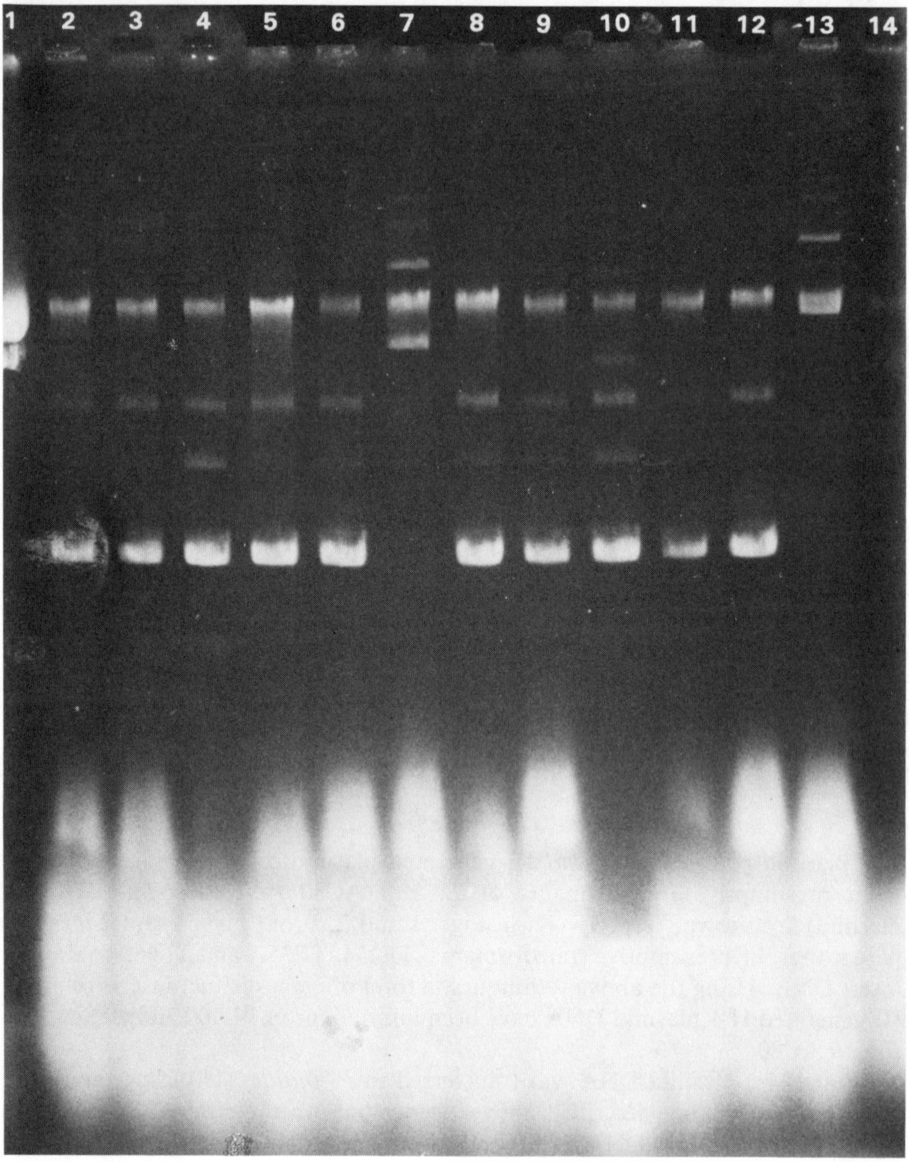

FIG. 1. Agarose gel electrophoresis of plasmid DNA from *E. coli* transformants containing pVK100::*Hind* III H-5 plasmid DNA. The gel 0.5% agarose was run for 5 h at 75 V and then stained with ethidium bromide. Tracks from left to right are (1) pVK100 (2) Transformant (TF) 17 (3) TF53 (4) TF84 (5) TF109 (6) TF142 (7) TF254 (8)TF286 (9) TF322 (10) TF357 (11) TF359 (12) TF664 (13) TF669 (14) H-5.

transferred back into a pseudomonad. Preliminary evidence suggests that the difference is caused by deletions (up to 200 bases) but that these deletions have been reported to occur only in the vector and not in the inserted pseudomonad DNA (Nakazawa, in press; Bagdasarian, in press).

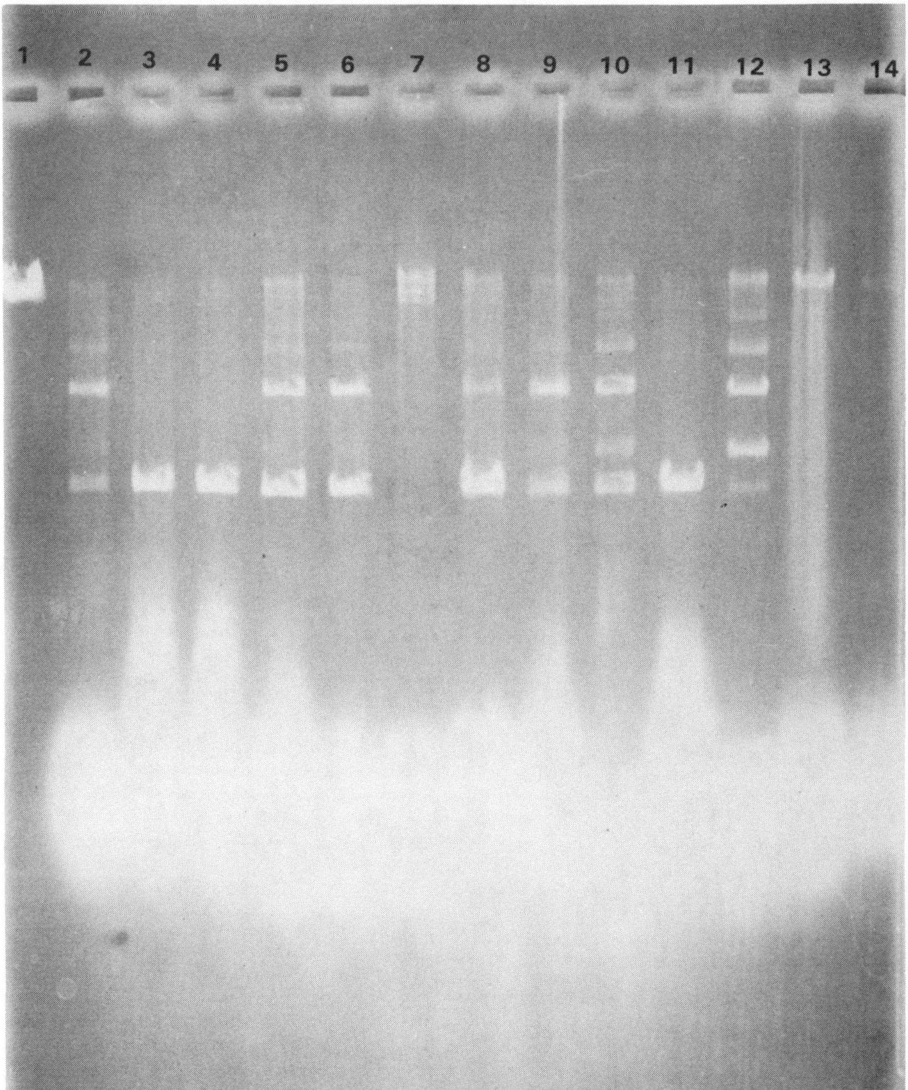

FIG. 2. Agarose gel electrophoresis of *Hind* III restricted plasmid DNA from transformants. The gel 0.5% agarose was run for 3.5 h at 75 V and then stained with ethidium bromide. Tracks from left to right are (1) pVK100 (2) TF17 (3) TF53 (4) TF84 (5) TF109 (6) TF142 (7) TF254 (8) TF286 (9) TF322 (10) TF357 (11) TF359 (12) TF664 (13) TF669 (14) H-5.

Discussion

The ability to clone pseudomonad plasmid DNA in *E. coli* and then transfer that cloned DNA back into a pseudomonad or some other gram-negative bacterium provides a powerful mechanism for the manipulation of DNA. The ability to

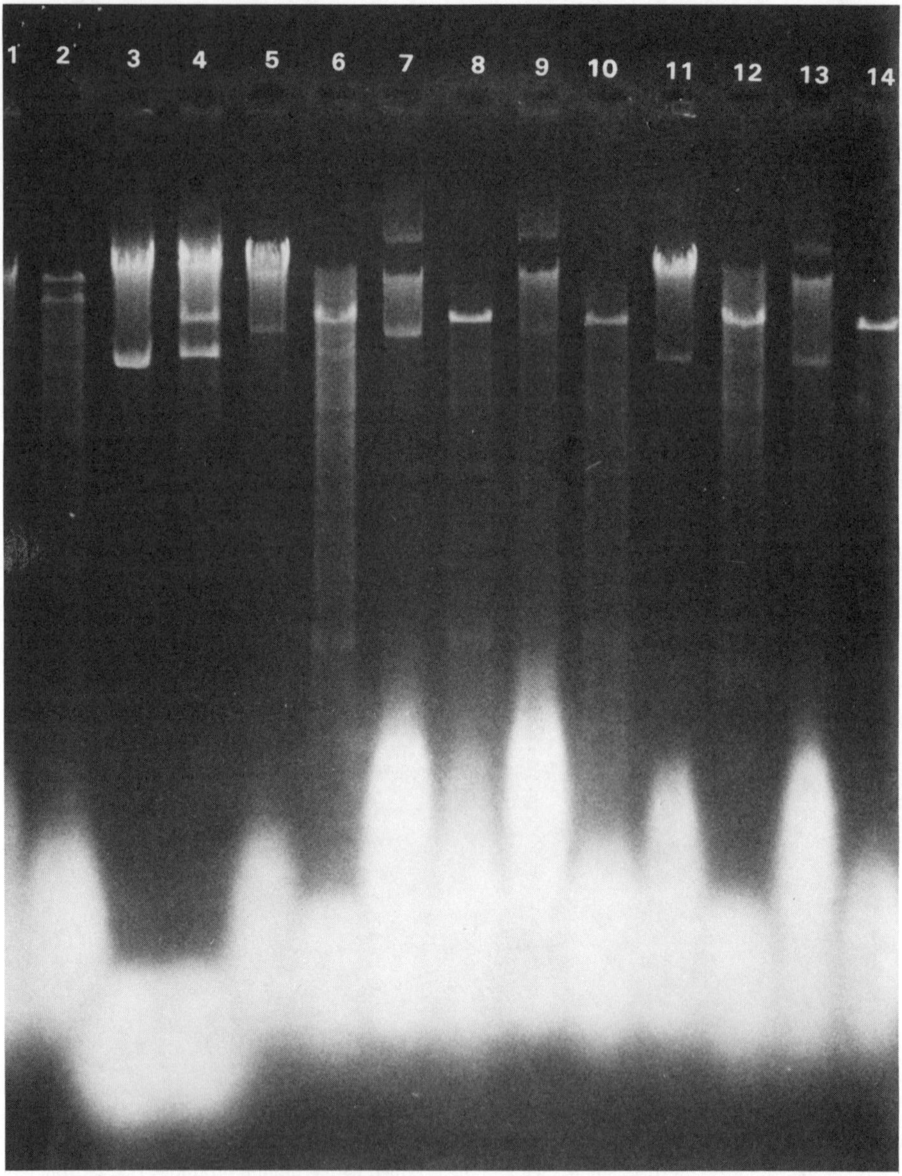

FIG. 3. Agarose gel of whole and *Hind* III restricted plasmid DNA from transformants (TF) and transconjugants (TC). The gel 0.5% agarose was run for 16 h at 20 V and then stained with ethidium bromide. Tracks from left to right are (1) H-5, (2) H-5/*Hind* III, (3) pVK100, (4) pVK100/*Hind* III, (5) TF254, (6) TF254/*Hind* III, (7) TC A4, (8) TC A4/*Hind* III, (9) TC A22, (10) TC A22/*Hind* III, (11) TF357, (12) TF357/*Hind* III, (13) TC B2, (14) TC B2/*Hind* III.

transform selected species of *Pseudomonas* together with the potential of conjugal transfer, when mobilizable broad host range vectors are used should result in an expansion of our knowledge of *Pseudomonas* genetics, physiology, and biochemistry.

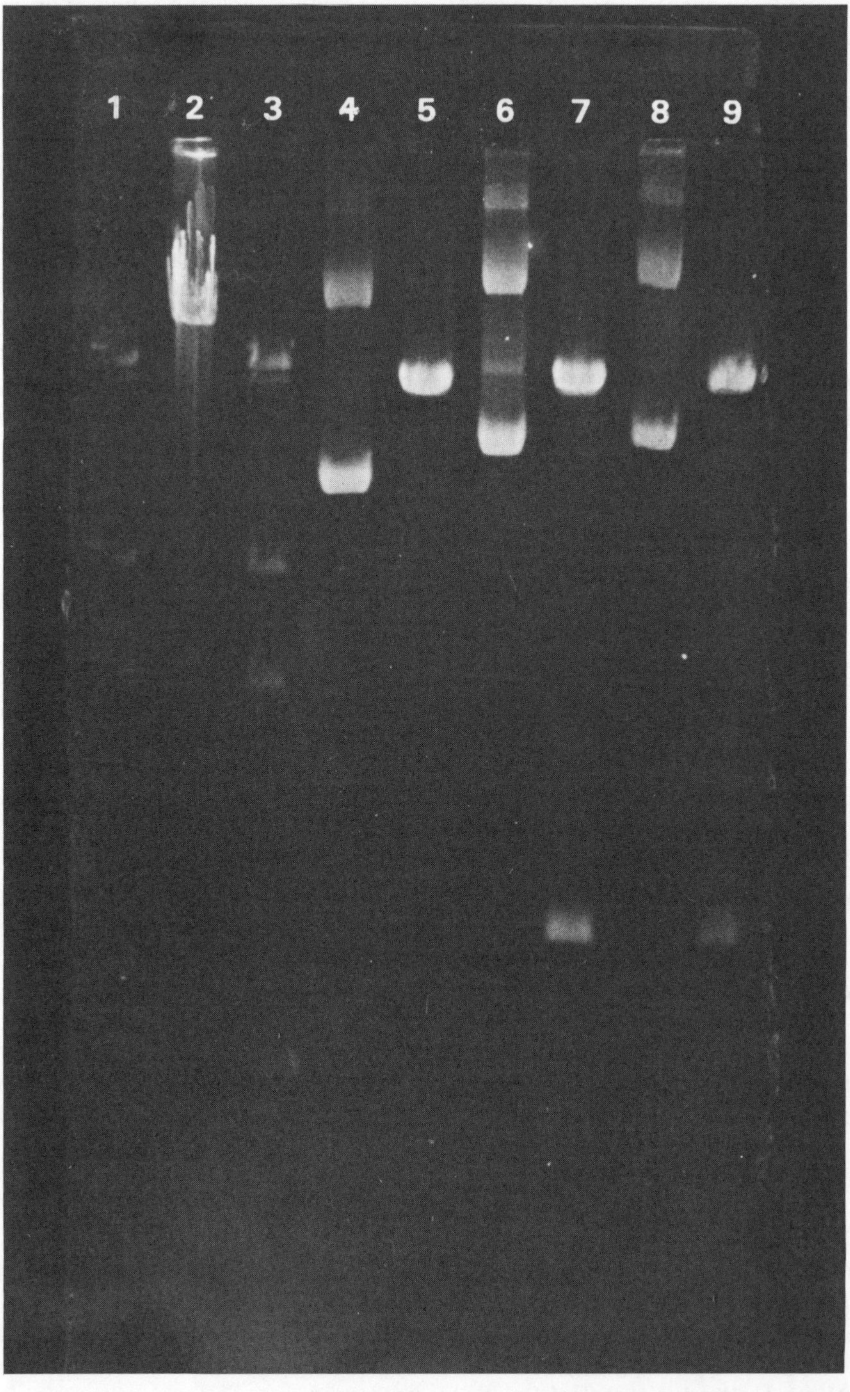

FIG. 4. Agarose gel of whole and *Hind* III restricted plasmid DNA from transformant TF254 and one of its transconjugants, A4. The gel 0.5% agarose was run for 18 h at 22 V and then stained with ethidium bromide. Tracks from left to right are (1) λDNA (2) λ/*Hind* III, (3) pVK100, (4) pVK100/*Hind* III, (5) TF254, (6) TF254/*Hind* III, (7) TC A4, (8) TC A4/*Hind* III.

Using the binary vector system and first cloning in *E. coli* and followed by conjugation into a pseudomonad host, we have achieved overall frequencies higher than when we had directly cloned into a pseudomonad host using transformation alone.

As yet, it is unknown if the deletions seen when pseudomonad DNA is cloned in *E. coli* and then transferred via conjugation are unique to the pseudomonads or what the incidence of deletion is or if deletion is avoidable. Even with deletion, the ability to conjugally transfer pseudomonad DNA provides the physiologist the opportunity to express cloned pseudomonad DNA in a variety of hosts under varying conditions thus facilitating the study of complex plasmid encoded traits.

Experiments are also underway to assess the degradative potential of *P. putida* 2440 strains that now contain cloned segments of H-5 plasmid DNA. These experiments should provide useful information regarding the potential of constructing improved degradative strains and also provide a mechanism to investigate the individual enzymes of a degradative pathway.

Literature Cited

Bagdasarian, M. *Pseudomonas* plasmid workshop. *In Plasmids in Bacteria: A Research Conference.* May 14–18. University of Illinois. Urbana, IL (In press).

Bagdasarian, M., R. Lurz, B. Ruckert, F. C. H. Franklin, M. M. Bagdasarin, J. Frey, and K. N. Timmis. 1981. Specific-purpose plasmid cloning vectors II. Broad host range, high copy number, RSF1010 derived vectors, and a host-vector system for genc cloning in *Pseudomonas. Gene* 16:237–247.

Bolivar, F., and K. Backman. 1979. Plasmids of *Escherichia coli* as cloning vectors. Pages 262–263 *in* R. Wu, ed., *Methods in Enzymology.* Academic Press, New York.

Chakrabarty, A. M., D. E. Mylroie, D. A. Friello, and J. G. Vacca. 1975. Transformation of *Pseudomonas putida* and *Escherichia coli* with plasmid-linked drug resistance factor DNA. *Proc. Nat. Acad. Sci.* 72:3647–3651.

Ditta, G., S. Stanfield, D. Corbin, and D. R. Helsinki. 1980. Broad host range DNA cloning system for gram-negative bacteria: Construction of a gene bank of *Rhizobium melioti. Proc. Nat. Acad. Sci.* 77:7347–7351.

Ditta, G., S. Stanfield, D. Corbin, and D. R. Helsinki. 1981. Cloning DNA from *Rhizobium melioti* using a new broad host range binary vehicle system. *In* J. M. Lyons, ed., *Basic Life Sciences.* 17:31–40.

Kahn, M., R. Kolter, C. Thomas, D. Figurski, R. Meyer, E. Remaut, and D. R. Helsinki. 1979. Plasmid cloning vehicles derived from plasmids ColE1, F, R6K and RK2. *In* R. Wu, ed., *Methods in Enzymology.* 68:268–280.

Knauf, V. C., and E. K. Nester. 1982. Wide host range cloning vectors: A cosmid clone bank of an *Agrobacterium* Ti plasmid. *Plasmid.* 8:45–54.

Kushner, S. R. 1978. An improved method for transformation of *Escherichia coli* with ColE1 derived plasmids. *In Proc. of Int. Symp. on Genet. Eng.* Elsevier/North-Holland Biomedical Press, Amsterdam.

Manaitis, T., E. F. Fritsch, and J. Sambrook. 1982. *Molecular Cloning: A Laboratory Manual.* Cold Spring Harbor Laboratory. Cold Spring Harbor, NY, pp. 88–91.

Mercer, A., and J. S. Loutit. 1979. Transformation and transfection of *Pseudomonas aeruginosa:* Effects of metal ions. *J. Bacteriol.* 140:37–42.

Nagahari, K., and K. Sakaguchi. 1978. RSF1010 plasmid as a potentially useful vector in *Pseudomonas* species. *J. Bacteriol.* 134:1527–1529.

Nakazawa, T. *Pseudomonas* plasmid workshop. *In Plasmids in Bacteria: A Research Conference.* May 14–18. University of Illinois. Urbana, IL (In press).

Pierce, G. E., J. B. Robinson, and J. R. Colaruotolo. 1983. Substrate diversity of *Pseudomonas* spp. containing chlorotoluene degradative plasmids. *Dev. Ind. Microbiol.* 24:499–507.

Pierce, G. E., J. B. Robinson, G. E. Garrett, and S. A. Sojka. 1984. Cloning of the chlorotoluene gene. *Dev. Ind. Microbiol.* 25:597–602.

Vandenbergh, P. A., R. H. Olsen, and J. R. Colaruotolo. 1981. Isolation and characterization of bacteria that degrade chloroaromatic compounds. *Appl. Environ. Microbiol.* 42:737–739.

Author Index

Subject Index

NOTES

NOTES

NOTES

NOTES

NOTES

NOTES

NOTES